W9-CXX-312

CRC HANDBOOK of THERMOPHYSICAL and THERMOCHEMICAL DATA

David R. Lide
Former Director, Standard Reference Data
National Institute of Standards and Technology

Henry V. Kehiaian
Research Director
University of Paris 7-CNRS
Institute of Topology and System Dynamics

CRC Press
Boca Raton Ann Arbor London Tokyo

Library of Congress Cataloging-in-Publication Data

Lide, David R., 1928-
CRC handbook of thermophysical and thermochemical data / authors, David R. Lide, Henry V. Kehiaian.
p. cm.
Includes bibliographical references and index.
ISBN 0-8493-0197-1
1. Matter--Thermal properties--Handbooks, manuals, etc. 2. Thermodynamics--Handbooks, manuals, etc. 3. Thermochemistry--Handbooks, manuals, etc. I. Kehiaian, H. V. (Henry V.) II. Title.
QC173.397.L53 1994
541.3′6′0212- -dc20 93-36909
CIP

International Standard Book Number 0-8493-0197-1
Library of Congress Card Number 93-36909
Printed in the United States of America 2 3 4 5 6 7 8 9 0
Printed on acid-free paper

TABLE OF CONTENTS

INTRODUCTION

Thermodynamics is one of the few sciences that possesses a highly developed and elegant theoretical framework and also has widespread and immensely important industrial applications. While the theory was well established a century ago, the practical applications of thermodynamics have continued to grow. These applications touch on virtually every industry and have a strong influence on our efforts to understand and control environmental pollution. At the research level, thermodynamics plays a prominent role in fields ranging from astrophysics to biochemistry.

The practical applications of thermodynamics generally require data on properties of real physical systems. In most cases these data rest on laboratory measurements of various thermodynamic properties. While it sometimes is possible to calculate thermodynamic properties from microscopic parameters (e.g., ideal gas heat capacity may be calculated quite accurately for atoms and simple molecules from molecular structure parameters), the bulk of the data in use comes from laboratory measurements of calorimetric or volumetric parameters. Over the years, a very significant effort has been devoted to compiling such data from the original literature, reanalyzing and evaluating the data, and preparing tables and graphs for both scientific and industrial use. Hundreds of such compilations have been published, and electronic versions of thermodynamic databases have come into use in recent years.

The objective of this book is to give an overview of the most important thermodynamic properties (as well as the related transport properties) for a range of pure substances and mixtures. It is obviously impossible in a single volume to cover more than a small sample of the substances and mixtures for which thermodynamic and transport property data exist. We have attempted to select systems that have high industrial and laboratory importance and that are representative of different classes of chemical compounds. Thus a user may be able to find data on a chemically related substance even if the compound of direct interest is not listed. Furthermore, extensive references are given in each table to larger compilations and databases, in the hope that this book can serve as an entry point to the sizable literature on thermophysical and thermochemical properties.

One of the practical difficulties in presenting thermodynamic data is that the properties are generally dependent on temperature, pressure, and in the case of mixtures, composition. Thus comprehensive tables of thermodynamic properties require a very large number of pages. We have dealt with this problem by fitting the property values to equations representing the functional dependence on these parameters and tabulating the coefficients of the equations. Furthermore, a disk accompanies the book on which the equations have been programmed and the coefficients for each system stored. The user can calculate property values from this disk and can generate tables and graphs which cover any range of temperature, pressure, or composition at any desired interval (within the valid range of the data). A large selection of units is also offered. Therefore, the combination of book and disk provides data which would require many volumes to present in conventional printed form.

Other features of the book should be mentioned. While it is assumed that the user is familiar with the laws, concepts, and basic equations of thermodynamics, care has been taken to define each quantity for which data are given in an unambiguous way. Terminology and symbols recommended by the International Union of Pure and Applied Chemistry (IUPAC) have been used throughout. All data in the book (and the default values from programs on the disk) are given in SI units. The introduction to each table gives a brief discussion of the quantities tabulated and points out the variation in the data with environmental parameters and, when useful generalizations can be made, with chemical structure. The IUPAC name, symbol, and SI unit are listed for each quantity at the beginning of each table.

Most of the data in the book have been drawn from evaluated sources, i.e., from compilations in which the authors have correlated data extracted from the literature, resolved discrepancies, and selected recommended values. The major sources used are listed with each table. When it was necessary to go to the original literature, we have used our own judgement in the selection of values. While it has not been feasible to give an uncertainty for every value in the book, we have adjusted the significant figures in the printed tables to give a rough indication of the accuracy.

Finally, many of the tables in the book are related to each other, and the same thermodynamic quantity may appear in more than one table. In such cases we have attempted to maintain internal consistency to the extent possible. However, there are circumstances where slightly different values of a property may be found in different tables. For example, a normal boiling temperature may appear in one table as a selected "best value" while a non-identical number appears elsewhere as a value derived from a best fit of vapor pressure data over a range of temperatures. All discrepancies of this type are believed to be within the uncertainty of the data.

Since usage of chemical names and synonyms varies so widely, most of the tables are arranged by molecular formula following the Hill convention. In this convention the molecular formula is written with C first (if present), H second, and then all other elements in alphabetical order of their chemical symbols. The sequence of entries in the table follows alphabetical order of the symbols in the formula and the number of atoms of each element, in ascending order. We have deviated from the strict Hill convention as used, for example, by Chemical Abstracts Service, by first listing all substances that do not contain carbon and then continuing with the carbon-containing substances. In this way the list of inorganic substances, which is generally shorter, is not broken up by the long list of organic compounds.

As an aid in identifying chemical compounds, an extensive Substance List appears at the end of the book, which includes the systematic names and common synonyms for the compounds covered in the various tables. The Substance List also gives molecular weights and Chemical Abstracts Service Registry Numbers.

In a compilation of this scope it is impossible to avoid errors, even after careful checking. The authors will be grateful to users who call our attention to mistakes and make other suggestions for improvements.

The authors are grateful to a number of people who made important contributions to this work. Dr. Christine Kehiaian assisted in the data selection and correlation; Mr. John Miller was responsible for the layout of the Substance List; and Mr. Jean-Claude Fontaine developed the database management system. Ms. Lauren Bascom and Ms. Joanne Chen assisted in data entry and editing. Finally, Mr. Paul Gottehrer of CRC Press did a masterful job of typesetting from the computer files supplied by the authors.

INSTRUCTIONS ON USING THE DISKETTE

The CRCTHERM Program on the diskette can be used only in conjuction with the book. A property is selected from a table in the book, and the user refers to the corresponding *Property Code*, consisting of four capital letters, e.g., VLEG. He locates the particular chemical system of interest (a pure substance or mixture) in the table and notes the *System Number* (SN) given there. He then opens the *Property Code Menu* and selects the Property Code and then opens the *System Number Window* and enters the SN at the indicated prompt. The values displayed by the program are calculated from the smoothing equations given in the book. The results are viewed on the screen in tabular and/or graphical form and can be printed and/or saved to disk. The detailed procedure is described below.

Any property is characterized by following *Physical Quantities*:

- Y(i) — One or several interrelated quantities, functions of the variable X and the fitted *Coefficients* A(i) of the Smoothing Equation (i = 1 to 6).
- X — A single independent variable.

Certain properties are characterized additionally by:

- P(1) — A quantity having a constant value in the representation (*Parameter*).
- V(i) — *Auxiliary Values* in the Smoothing Equation (i = 1 or 2).

Each physical quantity has a specific *Symbol* and may be expressed - if not zero-dimensional - either in *SI Units* or in some user-selected *Current Units*. The calculations can be performed for one or several selected variables Y(i), within a given X-Range of variation of X (X-min, X-max), with a given step (X-step), or for any discrete value of X (X-value).

SYSTEM REQUIREMENTS

- Hardware: IBM 286 or higher or 100% compatible microcomputer; DOS 3.0 or higher.
- Memory: 640 K RAM.
- Hard disk space: 1 MB.
- Drive: One 5 1/4″ double density floppy diskette drive.
- Graphics: VGA. Color adapter is recommended but not necessary.

INSTALLING THE PROGRAM

If there is a READ.ME file on the installation diskette, please read it for information not included in these Instructions.

The diskette contains an installation program. Put the diskette in the proper floppy diskette drive, change to that drive, type INSTALL, and then follow the procedure indicated on the screen. You may install the system on the hard disk partition of your choice, provided the required amount of free space is available (this is verified by the installation program).

OPENING THE PROGRAM

After installing the CRCTHERM Program, you can calculate properties for chemical systems that you specify.

- Type CT and press <Enter>. (See below for options which may be specified.)

After the title screen, the Program displays the *Property Window* with the Physical Quantities and Units, and - eventually - the *Chemical System Window*, with the X-value and the X-Range that were selected the last time the Program was run, and the X-step. You can directly perform calculations for this selection. Otherwise, you open the Property Code Menu and/or the Units Menu and/or the System Number Window and select a new property, unit, or system.

You can move the cursor in any of the fields of the Menus by means of the Arrow, PageUp, PageDn, Home, and End Keys. You can select your options, either by scrolling or by typing text, and pressing <Enter>.

DEFINING THE OBJECT

You need to define first the Property, the Physical Quantities and their Units, and the Chemical System. For this purpose, use the following commands:

- Press P to open the Property Code Menu and change the Property.
- Press S to open the System Number Window and enter or change the System Number.

These commands can be used from any place in the program, at any time, except when the cursor is in a field for typing text outside the Property Window. In that case, first:

- Press Esc to return to the Property Window.

Selecting Property

After pressing P, scroll with the cursor to the desired Property Code or, simpler, type the initial letters of the Property Code. Each typed letter moves the cursor to the corresponding Code. A few letters may suffice to find the desired Property. An audio bip signal indicates that the letter does not correspond to an existing Property Code. Press <Enter> to select the Property and open automatically the System Number Window. Press Esc if you wish to select specific physical quantities and/or change the units.

An alphabetical list of property codes with page references and brief descriptions is given at the end of these instructions.

Selecting Physical Quantity

If you wish to tabulate or graph only certain variables, Y(i), place the cursor on the line of the corresponding quantity in the Num column. Each time you press <Enter> you select or unselect that property. This is highlighted by, respectively, the appearance or disappearance of the askerisk (*) sign in the column.

Changing Units of Physical Quantities

If you wish to change the unit of a physical quantity, place the cursor on the line of the corresponding quantity in the Current Unit column and press <Enter>. You display the appropriate Unit Menu. Select the desired unit by scrolling and pressing <Enter>. You access the SI Unit directly by pressing the Home Key. For some complex physical quantities you may be prompted to select units for some more basic quantities; e.g., in the case of density, you have to specify the units of mass and volume. A quick way to restore the SI Unit of a physical quantity is to place the cursor on the line of the corresponding quantity in the Current Unit column and press <Delete>, instead of <Enter>.

Selecting System

After pressing S, type the System Number and press <Enter>. A bip signal indicates that the number does not correspond to an existing System Number for the property that you have specified (note that a given system has different SN's in different tables). The Insert Key permits you to toggle Insert/Overwrite mode.

The System Number may be obtained from the appropriate table in the book.

CALCULATING RESULTS

You next need to define the X-Range and carry out the calculation. You can then display the data both in a Numerical Output File and as a Graph.

Defining the X-Value or X-Range of Calculation

The Default line displays the lower (X-min) and upper (X-max) range of calculation, an X-value within this range, and a minimum step (X-step) in the selected unit of X. The Selected line displays round values for X-min and X-max, the corresponding X-value, and an X-step generating approximately 10 X-values within the X-range. You can change X-min, X-max, X-step, or X-value within the permitted range by placing the cursor in the appropriate column of the Selected line and typing the desired numbers. The numerical values need not be written in exponential form. Instead of typing the entire new number, you may wish to change the old one. Just press <Enter> to pass in the Overwrite edit mode. The Insert Key permits you to toggle Insert/Overwrite mode. A bip signal indicates that the typed number is outside the allowed range. Any typed number is rounded to the accuracy of the Default X-step. You can change quickly X-min, X-max, or X-value to the corresponding Default values by placing the cursor on that value and pressing <Delete>.

- Press C to calculate data in the X-Range and the X-step of the variable X given in the Selected line and display on the Numerical Output File.
- Press G to calculate data in the given X-Range and display as a Graph.
- Press Esc to stop the calculation before completion.
- To calculate results for a discrete Selected X-value place the cursor in the appropriate column and press <Enter>; you pass in the

Overwrite edit mode (see above), eventually change the value, and press <Enter> again. Another way is to place the cursor in the appropriate column, type the desired X-value, and press <Enter>. The calculated values are stored in the Numerical Output File. You can enter other discrete X-values and each time you press <Enter> the calculated results are appended to the Numerical Output File.

Numerical Output File

The Numerical Output File with the headings and all the calculated values is displayed on the screen after the first calculation. You may delimit a block from this file (or from the zoomed file described below), e.g., with all or part of the numerical values, by pressing the <Shift> key and moving the cursor by means of the Arrow, PageUp, PageDn, Home or End Keys. The delimited block is highlighted on the screen.

- Press Z to zoom the Numerical Output File. (To return, press Z again.)
- Press R to print the Numerical Output File or the delimited block. You are then prompted to specify the type of printer (see below).
- Press W to write the Numerical Output File or the delimited block in an external file. You are then prompted to accept or modify the default path, filename, and file format.
    - Press F to toggle the file format from an extended ASCII to a tab delimited format, for use by spreadsheet programs.
    - Press P to change the path.
    - Press N to change the filename.
    - Press W again to accept the default or the modified values.
- Press E to clear the Numerical Output File.

Changing the Property, the Units, or the System Number clears automatically the Numerical Output File. To keep a record of it, you should print or write it to a file in advance.

Graph

If you press G from any place of the Program, except when the cursor is in a field for typing text outside the Property Window, you calculate results within the selected X-Range and the results are presented graphically. Each independent variable Y(i) is graphed separately as a function of X. Press PageUp or PageDn to display all the selected Y(i)'s.

- Press G to display/remove the grid.
- Press L to change the lower limit of Y(i) by typing another value.
- Press U to change the upper limit of Y(i) by typing another value.
- Press D or <Delete> to restore the lower and upper limits of Y(i).
- Press R to print the Graph. You are then prompted to specify the type of printer (see below).
- Press Esc to return to the Property Window.

QUITTING THE PROGRAM

Press Q (followed by Y) to quit the Program.

PRINTER SPECIFICATION

Although CRCTHERM supports most of the commonly used printers, it may be necessary to experiment. When the command to print an output file or graph is given, you are prompted to specify the type of printer language required by your printer. The choices are:

- FX80 — The language recognized by Epson and many other dot matrix printers.
- HPGL — HP Graphics Language, used with many curve plotters.
- PCL — Printer Command Language, used with most laser printers.
- DJ500 — HP DeskJet 500 printer.

When you have specified one of these, it remains your default until you make a change. You may wish to try each of these options to determine the most safisfactory print result. In case of problems, consult your printer manual.

OPTIONS

When starting the program from the DOS prompt, several options may be invoked by attaching switches to the CT command. The options can be listed by entering CT /?. Note that a space must always be typed after CT. These options are:

- CT /M — instructs a color monitor to display in black-and-white mode.
- CT /S — turns off the sound.
- CT /Wnnn — where nnn is a number in the range 78 to 120, specifies the number of columns to be printed. The default is 100 columns.
- CT /Pnnn — where nnn is a number in the range 20 to 200, specifies the number of lines per printed page. The default value is 55.
- CT /C — cancels the condensed print mode.
- CT /Y — checks for possible corruption in the data files. If a problem is reported, the program should be reinstalled from the original diskette (or from a back-up diskette).
- CT /D — restores all the initial defaults.
- CT /? — displays this information.

Several switches may be used at the same time, such as CT /S /W80, as long as a space is included before each /. Once an option has been invoked, it remains in effect when the program is restarted. To change options, first return to the initial default by using CT /D, then restart with the new set of switches.

PROPERTY CODES

Code	Table	Page	Program calculates:
CPEX	3.3.2	393	Heat capacity of binary liquid mixtures as function of composition
CPGT	2.3.1	89	Gas heat capacity as function of temperature
CPLT	2.3.2	93	Liquid heat capacity as function of temperature
CPST	2.3.3	97	Solid heat capacity as function of temperature
HETX	3.3.1	375	Enthalpy of mixing of binary liquid mixtures as function of composition
LLEX	3.1.4	351	Liquid-liquid equilibrium temperature of binary mixtures as function of composition
PCXX	3.1.3	339	Critical pressure of binary liquid mixtures as function of composition
PVTL	2.2.2	81	Liquid density and molar volume as function of temperature
SLTP	2.1.5	63	Melting point as function of pressure
STLT	2.5.1	203	Surface tension as function of temperature
TCGT	4.2.1	415	Gas thermal conductivity as function of temperature
TCLT	4.2.2	417	Liquid thermal conductivity as function of temperature
TCST	4.2.3	419	Solid thermal conductivity as function of temperature
TCXX	3.1.3	339	Critical temperature of binary liquid mixtures as function of composition
VETX	3.2.1	363	Density and molar volume of binary liquid mixtures as function of composition
VIBT	2.2.1	69	Second virial coefficients as function of temperature
VIGT	4.1.1	407	Gas viscosity as function of temperature
VILT	4.1.2	409	Liquid viscosity as function of temperature
VLEG	3.1.1	211	Vapor pressure and related VLE properties of binary mixtures as function of composition
VLPT	2.1.3	49	Vapor pressure and enthalpy of vaporization as function of temperature
VLTP	2.1.3	49	Boiling point and enthalpy of vaporization as function of pressure
VSPT	2.1.4	61	Sublimation pressure as function of temperature

Section 1
Symbols, Units, and Terminology

1.1. NAMES AND SYMBOLS FOR THERMODYNAMIC AND TRANSPORT PROPERTIES

The recommended names and symbols for physical quantities encountered in the treatment of thermophysical and thermochemical data are listed in Table 1.1.1. These have been extracted from a more comprehensive list published by the International Union of Pure and Applied Chemistry; they are reprinted with permission of IUPAC. With a few minor exceptions, symbols recommended by the International Union of Pure and Applied Physics (IUPAP) and the International Organization for Standardization (ISO) agree with these.

General rules for expressing physical quantities are given first, followed by the table of names and symbols for general quantities, thermodynamic properties, and transport properties. The expression in the *Definition* column of this table is given as an aid in identifying the quantity but is not necessarily the complete or unique definition. The *SI unit* column gives one (not necessarily unique) expression for the coherent SI unit for the quantitiy. Other equivalent unit expressions, including those which involve SI prefixes, may be used.

REFERENCES

1. Mills, I., Ed., *Quantities, Units, and Symbols in Physical Chemistry*, IUPAC, Blackwell Scientific Publications, Oxford, 1988.
2. Cohen, E. R., and Giacomo, P., *Symbols, Units, Nomenclature, and Fundamental Constants in Physics*, Document IUPAP-25, 1987; also published in *Physica*, 146A, 1—68, 1987.
3. *ISO Standards Handbook 2: Units of Measurement*, International Organization for Standardization, Geneva, 1982.

GENERAL RULES

The value of a physical quantity is expressed as the product of a numerical value and a unit, e.g.:

$$T = 300 \text{ K}$$
$$V = 26.2 \text{ cm}^3$$
$$C_p = 45.3 \text{ J mol}^{-1} \text{ K}^{-1}$$

The symbol for a physical quantity is always given in italic (sloping) type, while symbols for units are given in Roman type. Column headings in tables and axis labels on graphs may conveniently be written as the physical quantity symbol divided by the unit symbol, e.g.:

$$T/K$$
$$V/\text{cm}^3$$
$$C_p/\text{J mol}^{-1} \text{ K}^{-1}$$

The values in the table column or graph axis are then pure numbers.

Subscripts to symbols for physical quantities should be italic if the subscript refers to another physical quantity or to a number, e.g.:

C_p — heat capacity at constant pressure
B_n — nth virial coefficient

Subscripts that have other meanings should be in Roman type:

m_p — mass of the proton
E_k — kinetic energy

TABLE 1.1.1
RECOMMENDED NAMES AND SYMBOLS

Name	*Symbol*	*Definition*	*SI unit*
GENERAL QUANTITIES			
number of entities (e.g. molecules, atoms, ions, formula units)	N		1
amount (of substance)	n	$n_B = N_B/L$	mol
Avogadro constant	L, N_A		mol^{-1}
mass of atom, atomic mass	m_a, m		kg
mass of entity (molecule, or formula unit)	m_f, m		kg
atomic mass constant	m_u	$m_u = m_a(^{12}C)/12$	kg
molar mass	M	$M_B = m/n_B$	$kg\ mol^{-1}$
relative molecular mass (relative molar mass, molecular weight)	M_r	$M_{r,B} = m_B/m_u$	1
molar volume	V_m	$V_{m,B} = V/n_B$	$m^3\ mol^{-1}$
mass fraction	w	$w_B = m_B/\Sigma m_i$	1
volume fraction	ϕ	$\phi_B = V_B/\Sigma V_i$	1
mole fraction, amount fraction, number fraction	x, y	$x_B = n_B/\Sigma n_i$	1
(total) pressure	p, P		Pa
partial pressure	p_B	$p_B = y_B p$	Pa
mass concentration (mass density)	γ, ρ	$\gamma_B = m_B/V$	$kg\ m^{-3}$
number concentration, number density of entities	C, n	$C_B = N_B/V$	m^{-3}
amount concentration, concentration	c	$c_B = n_B/V$	$mol\ m^{-3}$
solubility	s	$s_B = c_B$ (saturated solution)	$mol\ m^{-3}$
molality (of a solute)	$m, (b)$	$m_B = n_B/m_A$	$mol\ kg^{-1}$
surface concentration	Γ	$\Gamma_B = n_B/A$	$mol\ m^{-2}$
stoichiometric number	ν		1
extent of reaction, advancement	ξ	$\Delta\xi = \Delta n_B/\nu_B$	mol
degree of dissociation	α		1
THERMODYNAMIC PROPERTIES			
heat	q, Q		J
work	w, W		J
internal energy	U	$\Delta U = q + w$	J
enthalpy	H	$H = U + pV$	J
thermodynamic temperature	T		K
Celsius temperature	θ, t	$\theta/°C = T/K - 273.15$	°C
entropy	S	$dS \geqslant dq/T$	$J\ K^{-1}$
Helmholtz energy, (Helmholtz function)	A	$A = U - TS$	J

TABLE 1.1.1
RECOMMENDED NAMES AND SYMBOLS (continued)

Name	*Symbol*	*Definition*	*SI unit*
Gibbs energy, (Gibbs function)	G	$G=H-TS$	J
Massieu function	J	$J=-A/T$	$\mathrm{J\,K^{-1}}$
Planck function	Y	$Y=-G/T$	$\mathrm{J\,K^{-1}}$
surface tension	γ, σ	$\gamma=(\partial G/\partial A_s)_{T,p}$	$\mathrm{J\,m^{-2}}$, $\mathrm{N\,m^{-1}}$
molar quantity X	X_m	$X_m=X/n$	(varies)
specific quantity X	x	$x=X/m$	(varies)
pressure coefficient	β	$\beta=(\partial p/\partial T)_V$	$\mathrm{Pa\,K^{-1}}$
relative pressure coefficient	α_p	$\alpha_p=(1/p)(\partial p/\partial T)_V$	$\mathrm{K^{-1}}$
compressibility,			
isothermal	κ_T	$\kappa_T=-(1/V)(\partial V/\partial p)_T$	$\mathrm{Pa^{-1}}$
isentropic	κ_S	$\kappa_S=-(1/V)(\partial V/\partial p)_S$	$\mathrm{Pa^{-1}}$
linear expansion coefficient	α_l	$\alpha_l=(1/l)(\partial l/\partial T)$	$\mathrm{K^{-1}}$
cubic expansion coefficient	α, α_V, γ	$\alpha=(1/V)(\partial V/\partial T)_p$	$\mathrm{K^{-1}}$
heat capacity,			
at constant pressure	C_p	$C_p=(\partial H/\partial T)_p$	$\mathrm{J\,K^{-1}}$
at constant volume	C_V	$C_V=(\partial U/\partial T)_V$	$\mathrm{J\,K^{-1}}$
ratio of heat capacities	γ, (κ)	$\gamma=C_p/C_V$	1
Joule–Thomson coefficient	μ, μ_{JT}	$\mu=(\partial T/\partial p)_H$	$\mathrm{K\,Pa^{-1}}$
second virial coefficient	B	$pV_m=RT(1+B/V_m+\cdots)$	$\mathrm{m^3\,mol^{-1}}$
compression factor (compressibility factor)	Z	$Z=pV_m/RT$	1
partial molar quantity X	X_B, (X'_B)	$X_B=(\partial X/\partial n_B)_{T,p,n_{j\neq B}}$	(varies)
chemical potential (partial molar Gibbs energy)	μ	$\mu_B=(\partial G/\partial n_B)_{T,p,n_{j\neq B}}$	$\mathrm{J\,mol^{-1}}$
absolute activity	λ	$\lambda_B=\exp(\mu_B/RT)$	1
standard chemical potential	μ°, μ^{o}		$\mathrm{J\,mol^{-1}}$
standard partial molar enthalpy	H_B°	$H_B^\circ=\mu_B^\circ+TS_B^\circ$	$\mathrm{J\,mol^{-1}}$
standard partial molar entropy	S_B°	$S_B^\circ=-(\partial\mu_B^\circ/\partial T)_p$	$\mathrm{J\,mol^{-1}\,K^{-1}}$
standard reaction Gibbs energy (function)	$\Delta_r G^\circ$	$\Delta_r G^\circ=\sum_B \nu_B\mu_B^\circ$	$\mathrm{J\,mol^{-1}}$
affinity of reaction	A, ($\mathcal{A}$)	$A=-(\partial G/\partial\xi)_{p,T}=-\sum_B \nu_B\mu_B$	$\mathrm{J\,mol^{-1}}$
standard reaction enthalpy	$\Delta_r H^\circ$	$\Delta_r H^\circ=\sum_B \nu_B H_B^\circ$	$\mathrm{J\,mol^{-1}}$
standard reaction entropy	$\Delta_r S^\circ$	$\Delta_r S^\circ=\sum_B \nu_B S_B^\circ$	$\mathrm{J\,mol^{-1}\,K^{-1}}$
equilibrium constant	K°, K	$K^\circ=\exp(-\Delta_r G^\circ/RT)$	1
equilibrium constant,			
pressure basis	K_p	$K_p=\prod_B p_B^{\nu_B}$	$\mathrm{Pa}^{\Sigma\nu}$

TABLE 1.1.1
RECOMMENDED NAMES AND SYMBOLS (continued)

Name	*Symbol*	*Definition*	*SI unit*
concentration basis	K_c	$K_c=\prod_B c_B{}^{\nu_B}$	$(\text{mol m}^{-3})^{\Sigma\nu}$
molality basis	K_m	$K_m=\prod_B m_B{}^{\nu_B}$	$(\text{mol kg}^{-1})^{\Sigma\nu}$
fugacity	$f, \tilde{p}$	$f_B=\lambda_B \lim_{p\to 0} (p_B/\lambda_B)_T$	Pa
fugacity coefficient	ϕ	$\phi_B=f_B/p_B$	1
activity and activity coefficient referenced to Raoult's law,			
(relative) activity	a	$a_B=\exp\left[\frac{\mu_B-\mu_B{}^*}{RT}\right]$	1
activity coefficient	f	$f_B=a_B/x_B$	1
activities and activity coefficients referenced to Henry's law, (relative) activity,			
molality basis	a_m	$a_{m,B}=\exp\left[\frac{\mu_B-\mu_B{}^\circ}{RT}\right]$	1
concentration basis	a_c	$a_{c,B}=\exp\left[\frac{\mu_B-\mu_B{}^\circ}{RT}\right]$	1
mole fraction basis	a_x	$a_{x,B}=\exp\left[\frac{\mu_B-\mu_B{}^\circ}{RT}\right]$	1
activity coefficient,			
molality basis	γ_m	$a_{m,B}=\gamma_{m,B}m_B/m^\circ$	1
concentration basis	γ_c	$a_{c,B}=\gamma_{c,B}c_B/c^\circ$	1
mole fraction basis	γ_x	$a_{x,B}=\gamma_{x,B}x_B$	1
ionic strength,			
molality basis	I_m, I	$I_m=\frac{1}{2}\Sigma m_B z_B{}^2$	mol kg^{-1}
concentration basis	I_c, I	$I_c=\frac{1}{2}\Sigma c_B z_B{}^2$	mol m^{-3}
osmotic coefficient,			
molality basis	ϕ_m	$\phi_m=(\mu_A{}^*-\mu_A)/(RTM_A\Sigma m_B)$	1
mole fraction basis	ϕ_x	$\phi_x=(\mu_A-\mu_A{}^*)/(RT\ln x_A)$	1
osmotic pressure	Π	$\Pi=c_BRT$ (ideal dilute solution)	Pa

(i) *Symbols used as subscripts to denote a chemical process or reaction*

These symbols should be printed in roman (upright) type, without a full stop (period).

vaporization, evaporation (liquid→gas)	vap
sublimation (solid→gas)	sub
melting, fusion (solid→liquid)	fus
transition (between two phases)	trs
mixing of fluids	mix
solution (of solute in solvent)	sol
dilution (of a solution)	dil
adsorption	ads
displacement	dpl
immersion	imm
reaction in general	r
atomization	at
combustion reaction	c
formation reaction	f

TABLE 1.1.1
RECOMMENDED NAMES AND SYMBOLS (continued)

Name	*Symbol*	*Definition*	*SI unit*
(ii) *Recommended superscripts*			
standard		$\ominus$, o	
pure substance		*	
infinite dilution		∞	
ideal		id	
activated complex, transition state		‡	
excess quantity		E	
TRANSPORT PROPERTIES			
flux (of a quantity X)	J_X, J	$J_X = A^{-1}\,\mathrm{d}X/\mathrm{d}t$	(varies)
volume flow rate	q_V, $\dot{V}$	$q_v = \mathrm{d}V/\mathrm{d}t$	$\mathrm{m^3\,s^{-1}}$
mass flow rate	q_m, $\dot{m}$	$q_m = \mathrm{d}m/\mathrm{d}t$	$\mathrm{kg\,s^{-1}}$
mass transfer coefficient	k_d		$\mathrm{m\,s^{-1}}$
heat flow rate	ϕ	$\phi = \mathrm{d}q/\mathrm{d}t$	W
heat flux	J_q	$J_q = \phi/A$	$\mathrm{W\,m^{-2}}$
thermal conductance	G	$G = \phi/\Delta T$	$\mathrm{W\,K^{-1}}$
thermal resistance	R	$R = 1/G$	$\mathrm{K\,W^{-1}}$
thermal conductivity	λ, k	$\lambda = J_q/(\mathrm{d}T/\mathrm{d}l)$	$\mathrm{W\,m^{-1}\,K^{-1}}$
coefficient of heat transfer	h, (k, K, α)	$h = J_q/\Delta T$	$\mathrm{W\,m^{-2}\,K^{-1}}$
thermal diffusivity	a	$a = \lambda/\rho c_p$	$\mathrm{m^2\,s^{-1}}$
diffusion coefficient	D	$D = J_n/(\mathrm{d}c/\mathrm{d}l)$	$\mathrm{m^2\,s^{-1}}$

1.2. SI UNITS AND CONVERSION FACTORS

The International System of units (SI) was adopted by the 11th General Conference on Weights and Measures (CGPM) in 1960. It is a coherent system of units built from seven *SI base units*, one for each of the seven dimensionally independent base quantities: they are the meter, kilogram, second, ampere, kelvin, mole, and candela, for the dimensions length, mass, time, electric current, thermodynamic temperature, amount of substance, and luminous intensity, respectively, The definitions of the SI base units are given below. The *SI derived units* are expressed as products of powers of the base units, analogous to the corresponding relations between physical quantities but with numerical factors equal to unity.

In the International System there is only one SI unit for each physical quanitity. This is either the appropriate SI base unit itself or the appropriate SI derived unit. However, any of the approved decimal prefixes, called *SI prefixes*, may be used to construct decimal multiples or submultiples of SI units.

It is recommended that only SI units be used in science and technology (with SI prefixes where appropriate). Where there are special reasons for making an exception to this rule, it is recommended always to define the units used in terms of SI units. This section was reprinted with the permission of IUPAC.

Definitions of SI Base Units

Meter—The meter is the length of path travelled by light in vacuum during a time interval of 1/299 792 458 of a second (17th CGPM, 1983).

Kilogram — The kilogram is the unit of mass; it is equal to the mass of the international prototype of the kilogram (3rd CGPM, 1901).

Second—The second is the duration of 9 192 631 770 periods of the radiation corresponding to the transition between the two hyperfine levels of the ground state of the cesium-133 atom (13th CGPM, 1967).

Ampere— The ampere is that constant current which, if maintained in two straight parallel conductors of infinite length, of negligible circular cross-section, and placed 1 meter apart in vacuum, would produce between these conductors a force equal to 2×10^{-7} newton per meter of length (9th CGPM, 1948).

Kelvin—The kelvin, unit of thermodynamic temperature, is the fraction 1/273.16 of the thermodynamic temperature of the triple point of water (13th CGPM, 1967)

Mole—The mole is the amount of substance of a system which contains as many elementary entities as there are atoms in 0.012 kilogram of carbon-12. When the mole is used, the elementary entities must be specified and may be atoms, molecules, ions, electrons, other particles, or specified groups of such particles (14th CGPM, 1971).

Examples of the use of the mole:

1 mol of H_2 contains about 6.022×10^{23} H_2 molecules, or 12.044×10^{23} H atoms
1 mol of HgCl has a mass of 236.04 g
1 mol of Hg_2Cl_2 has a mass of 472.08 g
1 mol of Hg_2^{2+} has a mass of 401.18 g and a charge of 192.97 kC
1 mol of $Fe_{0.91}S$ has a mass of 82.88 g
1 mol of e^- has a mass of 548.60 μg and a charge of –96.49 kC
1 mol of photons whose frequency is 10^{14} Hz has an energy of about 39.90 kJ

Candela—The candela is the luminous intensity, in a given direction, of a source that emits monochromatic radiation of frequency 540×10^{12} hertz and that has a radiant intensity in that direction of (1/683) watt per steradian (16th CGPM, 1979).

TABLE 1.2.1
NAMES AND SYMBOLS FOR THE SI BASE UNITS

Physical quantity	Name of SI unit	Symbol for SI unit
length	meter	m
mass	kilogram	kg
time	second	s
electric current	ampere	A
thermodynamic temperature	kelvin	K
amount of substance	mole	mol
luminous intensity	candela	cd

TABLE 1.2.2
SI DERIVED UNITS WITH SPECIAL NAMES AND SYMBOLS

Physical quantity	Name of SI unit	Symbol for SI unit	Expression in terms of SI base units
frequency[1]	hertz	Hz	s^{-1}
force	newton	N	$m\ kg\ s^{-2}$
pressure, stress	pascal	Pa	$N\ m^{-2} = m^{-1}\ kg\ s^{-2}$
energy, work, heat	joule	J	$N\ m = m^2\ kg\ s^{-2}$
power, radiant flux	watt	W	$J\ s^{-1} = m^2\ kg\ s^{-3}$
electric charge	coulomb	C	A s
electric potential, electromotive force	volt	V	$J\ C^{-1} = m^2\ kg\ s^{-3}\ A^{-1}$
electric resistance	ohm	Ω	$V\ A^{-1} = m^2\ kg\ s^{-3}\ A^{-2}$
electric conductance	siemens	S	$\Omega^{-1} = m^{-2}\ kg^{-1}\ s^3\ A^2$
electric capacitance	farad	F	$C\ V^{-1} = m^{-2}\ kg^{-1}\ s^4\ A^2$
magnetic flux density	tesla	T	$V\ s\ m^{-2} = kg\ s^{-2}\ A^{-1}$
magnetic flux	weber	Wb	$V\ s = m^2\ kg\ s^{-2}\ A^{-1}$
inductance	henry	H	$V\ A^{-1}\ s = m^2\ kg\ s^{-2}\ A^{-2}$
Celsius temperature[2]	degree Celsius	°C	K
luminous flux	lumen	lm	cd sr
illuminance	lux	lx	$cd\ sr\ m^{-2}$
activity[3] (radioactive)	becquerel	Bq	s^{-1}
absorbed dose[3] (of radiation)	gray	Gy	$J\ kg^{-1} = m^2\ s^{-2}$
dose equivalent[3] (dose equivalent index)	sievert	Sv	$J\ kg^{-1} = m^2\ s^{-2}$
plane angle[4]	radian	rad	$1 = m\ m^{-1}$
solid angle[4]	steradian	sr	$1 = m^2\ m^{-2}$

(1) For radial (circular) frequency and for angular velocity the unit rad s^{-1}, or simply s^{-1}, should be used, and this may not be simplified to Hz. The unit Hz should be used only for frequency in the sense of cycles per second.
(2) The Celsius temperature θ is defined by the equation:

$$\theta/°C = T/K - 273.15$$

The SI unit of Celsius temperature interval is the degree Celsius, °C, which is equal to the kelvin, K. °C should be treated as a single symbol, with no space between the ° sign and the letter C. (The symbol °K, and the symbol °, should no longer be used.)
(3) The units becquerel, gray, and sievert are admitted for reasons of safeguarding human health.
(4) The units radian and steradian are described as 'SI supplementary units'. However, in chemistry, as well as in physics,

TABLE 1.2.2
SI DERIVED UNITS WITH SPECIAL NAMES AND SYMBOLS (continued)

they are usually treated as dimensionless derived units, and this was recognized by CIPM in 1980. Since they are then of dimension 1, this leaves open the possibility of including them or omitting them in expressions of SI derived units, In practice this means that rad and sr may be used when appropriate and may be omitted if clarity is not lost thereby.

TABLE 1.2.3
SI PREFIXES

To signify decimal multiples and submultiples of SI units the following prefixes may be used.

Submultiple	Prefix	Symbol	Multiple	Prefix	Symbol
10^{-1}	deci	d	10	deca	da
10^{-2}	centi	c	10^{2}	hecto	h
10^{-3}	milli	m	10^{3}	kilo	k
10^{-6}	micro	μ	10^{6}	mega	M
10^{-9}	nano	n	10^{9}	giga	G
10^{-12}	pico	p	10^{12}	tera	T
10^{-15}	femto	f	10^{15}	peta	P
10^{-18}	atto	a	10^{18}	exa	E

Prefix symbols should be printed in roman (upright) type with no space between the prefix and the unit symbol.

Example kilometer, km

When a prefix is used with a unit symbol, the combination is taken as a new symbol that can be raised to any power without the use of parentheses.

Examples
$1\ \text{cm}^3 = (0.01\ \text{m})^3 = 10^{-6}\ \text{m}^3$
$1\ \mu\text{s}^{-1} = (10^{-6}\ \text{s})^{-1} = 10^{6}\ \text{s}^{-1}$
$1\ \text{V/cm} = 100\ \text{V/m}$
$1\ \text{mmol/dm}^3 = 1\ \text{mol m}^{-3}$

A prefix should never be used on its own, and prefixes are not to be combined into compound prefixes.

Example pm, not μμm

The names and symbols of decimal multiples and sub-multiples of the SI base unit of mass, the kg, which already contains a prefix, are constructed by adding the appropriate prefix to the word gram and symbol g.

Examples mg, not μkg; Mg, not kkg

The SI prefixes are not to be used with °C.

TABLE 1.2.4
CONVERSION FACTORS

To convert a quantity expressed in a non-SI unit to a value in SI units, multiply the value in the non-SI unit by the factor k in the middle column. For the inverse process, multiply the value in the SI unit by the factor k^{-1} in the last column to obtain the value in the non-SI unit.

TABLE 1.2.4
CONVERSION FACTORS (continued)

Non-SI unit	k 1 (non-SI unit) = k (SI unit)	k^{-1} 1 (SI unit) = k^{-1} (non-SI unit)
Length	**SI unit, m**	
Å (angstrom)	0.1×10^{-9}*	0.1×10^{11}*
cm (centimeter)	0.1×10^{-1}*	0.1×10^{3}*
in (inch)	0.254×10^{-1}*	0.3937008×10^{2}
ft (foot)	0.3048*	0.3280840×10
Area	**SI unit, m^2**	
cm^2 (square centimeter)	0.1×10^{-3}*	0.1×10^{5}*
in^2 (square inch)	0.64516×10^{-3}*	0.1550003×10^{4}
ft^2 (square foot)	0.9290304×10^{-1}*	0.1076391×10^{2}
Volume	**SI unit, m^3**	
cm^3 (cubic centimeter)	0.1×10^{-5}*	0.1×10^{7}*
dm^3 (cubic decimeter)	0.1×10^{-2}*	0.1×10^{4}*
in^3 (cubic inch)	$0.16387064 \times 10^{-4}$*	0.6102374×10^{5}
ft^3 (cubic foot)	0.2831685×10^{-1}	0.3531467×10^{2}
L (liter)	0.1×10^{-2}*	0.1×10^{4}*
mL (milliliter)	0.1×10^{-5}*	0.1×10^{7}*
UKgal (UK gallon)	0.45461×10^{-2}	0.21997×10^{3}
USgal (US gallon)	0.37854×10^{-2}	0.26417×10^{3}
Mass	**SI unit, kg**	
g (gram)	0.1×10^{-2}*	0.1×10^{4}*
mg (milligram)	0.1×10^{-5}*	0.1×10^{7}*
t (tonne)	0.1×10^{4}*	0.1×10^{-2}
lb (pound)	0.45359237*	0.2204623×10
Density	**SI unit, kg m^{-3}**	
g cm^{-3} (gram per cubic centimeter)	0.1×10^{4}*	0.1×10^{-2}*
g L^{-1} (gram per liter)	0.1×10*	0.1×10*
lb in^{-3} (pound per cubic inch)	0.2767991×10^{5}	0.3612728×10^{-4}
lb ft^{-3} (pound per cubic foot)	0.1601847×10^{2}	0.6242795×10^{-1}
lb $UKgal^{-1}$ (pound per UK gallon)	0.99776×10^{2}	0.100224×10^{-1}
lb $USgal^{-1}$ (pound per US gallon)	0.1198264×10^{3}	0.8345406×10^{-2}
Time	**SI unit, s**	
min (minute)	0.6×10^{2}*	0.1666667×10^{-1}
h (hour)	0.36×10^{4}*	0.2777778×10^{-3}
d (day)	0.864×10^{5}*	0.1157407×10^{-4}

TABLE 1.2.4
CONVERSION FACTORS (continued)

Non-SI unit	k 1 (non-SI unit) = k (SI unit)	k^{-1} 1 (SI unit) = k^{-1} (non-SI unit)
Force	**SI unit, N (newton, kg m s^{-2})**	
dyn (dyne)	0.1×10^{-4}*	0.1×10^{6}*
kgf (kilogram-force)	0.980665×10*	0.1019716
lbf (pound-force)	0.44482×10	0.22481
Pressure	**SI unit, Pa (pascal, kg m^{-1} s^{-2})**	
bar	0.1×10^{6}*	0.1×10^{-4}*
atm (atmosphere)	0.101325×10^{6}*	0.9869233×10^{-5}
dyn cm^{-2} (dyne per square centimeter)	0.1*	0.1×10^{2}*
kgf cm^{-2} (kilogram-force per square centimeter)	0.980665×10^{5}*	0.1019716×10^{-4}
lbf in^{-2} (p.s.i., pound-force per square inch)	0.6894757×10^{4}	0.1450377×10^{-3}
lbf ft^{-2} (pound-force per square foot)	0.4788026×10^{2}	0.2088543×10^{-1}
inHg (inch of mercury)	0.3386388×10^{4}	0.2952999×10^{-3}
mmHg (millimeter of mercury, torr)	0.1333224×10^{3}	0.7500617×10^{-2}
Energy	**SI unit, J (joule, kg m^2 s^{-2})**	
erg	0.1×10^{-6}*	0.1×10^{8}*
cal_{IT} (I.T. calorie)	0.41868×10*	0.2388459
cal_{th} (thermochemical calorie)	0.4184×10*	0.2390057
kW h (kilowatt hour)	0.36×10^{7}*	0.2777778×10^{-6}
L atm (liter atmosphere)	0.101325×10^{3}*	0.9869233×10^{-2}
ft lbf (foot pound-force)	0.1355818×10	0.7375622
hp h (horse power hour)	0.2684519×10^{7}	0.3725062×10^{-6}
Btu_{IT} (British thermal unit)	0.1055056×10^{4}	0.9478172×10^{-3}
Viscosity (absolute)	**SI unit, Pa s (pascal second, kg m^{-1} s^{-1})**	
P (poise)	0.1*	0.1×10^{2}*
cP (centipoise)	0.1×10^{-2}*	0.1×10^{4}*
lbf s in^{-2} (pound-force second per square inch)	0.6894757×10^{4}	0.1450377×10^{-3}
lb ft^{-1} s^{-1} (pound per foot second)	0.1488164×10	0.6719689
Viscosity (kinematic)	**SI unit, m^2 s^{-1}**	
St (stokes)	0.1×10^{-3}*	0.1×10^{5}*
cSt (centistokes)	0.1×10^{-5}*	0.1×10^{7}*
ft^2 s^{-1} (square foot per second)	0.9290304×10^{-1}*	0.1076391×10^{2}

TABLE 1.2.4
CONVERSION FACTORS (continued)

Non-SI unit	k 1 (non-SI unit) = k (SI unit)	k^{-1} 1 (SI unit) = k^{-1} (non-SI unit)
Thermal conductivity	**SI unit, W m^{-1} K^{-1} (watt per meter kelvin, kg m K^{-1} s^{-3})**	
W cm^{-1} K^{-1} (watt per centimeter kelvin)	0.1×10^{3}*	0.1×10^{-1}*
cal$_{th}$ s^{-1} cm^{-1} °C^{-1} (calorie per second centimeter degree Celsius)	0.4184×10^{3}*	0.2390057×10^{-2}
kcal$_{th}$ h^{-1} m^{-1} °C^{-1} (kilocalorie per hour meter degree Celsius)	0.116222×10*	0.8604223
Btu$_{IT}$ h^{-1} ft^{-1} °F^{-1} (British thermal unit per hour foot degree Fahrenheit)	0.173073×10	0.5777909
Diffusion coefficient	**SI unit, m^2 s^{-1}**	
cm^2 s^{-1} (square centimeter per second)	0.1×10^{-3}*	0.1×10^{5}*
in^2 s^{-1} (square inch per second)	0.64516×10^{-3}*	0.1550003×10^{4}
ft^2 h^{-1} (square foot per hour)	0.258064×10^{-4}*	0.3875008×10^{5}
Surface tension	**SI unit, N m^{-1} (newton per meter, kg s^{-2})**	
mN m^{-1} (millinewton per meter)	0.1×10^{-2}*	0.1×10^{4}*
dyn cm^{-1} (dyne per centimeter)	0.1×10^{-2}*	0.1×10^{4}*

* These factors are exact.

1.3. INTERNATIONAL TEMPERATURE SCALE OF 1990

A new temperature scale, the International Temperature Scale of 1990 (ITS-90), was officially adopted by the Comité International des Poids et Mesures (CIPM), meeting 26—28 September 1989 at the Bureau International des Poids et Mesures (BIPM). The ITS-90 was recommended to the CIPM for its adoption following the completion of the final details of the new scale by the Comité Consultatif de Thermométrie (CCT), meeting 12—14 September 1989 at the BIPM in its 17th Session. The ITS-90 became the official international temperature scale on 1 January 1990. The ITS-90 supersedes the present scales, the International Practical Temperature Scale of 1968 (IPTS-68) and the 1976 Provisional 0.5 to 30 K Temperature Scale (EPT-76).

The ITS-90 extends upward from 0.65 K, and temperatures on this scale are in much better agreement with thermodynamic values that are those on the IPTS-68 and the EPT-76. The new scale has subranges and alternative definitions in certain ranges that greatly facilitate its use. Furthermore, its continuity, precision, and reproducibility throughout its ranges are much improved over that of the present scales. The replacement of the thermocouple with the platinum resistance thermometer at temperatures below 961.78°C resulted in the biggest improvement in reproducibility.

The ITS-90 is divided into four primary ranges:

1. Between 0.65 and 3.2 K, the ITS-90 is defined by the vapor pressure-temperature relation of ^3He, and between 1.25 and 2.1768 K (the λ point) and between 2.1768 and 5.0 K by the vapor pressure-temperature relations of ^4He. T_{90} is defined by the vapor pressure equations of the form:

$$T_{90} / \mathrm{K} = A_0 + \sum_{i=1}^{9} A_i \left[\left(\ln(p / \mathrm{Pa}) - B \right) / C \right]^i$$

   The values of the coefficients A_i, and of the constants A_o, B, and C of the equations, are given below.
2. Between 3.0 and 24.5561 K, the ITS-90 is defined in terms of a ^3He or ^4He constant volume gas thermometer (CVGT). The thermometer is calibrated at three temperatures: at the triple point of neon (24.5561 K), at the triple point of equilibrium hydrogen (13.8033 K), and at a temperature between 3.0 and 5.0 K, the value of which is determined by using either ^3He or ^4He vapor pressure thermometry.
3. Between 13.8033 K (–259.3467°C) and 1234.93 K (961.78°C), the ITS-90 is defined in terms of the specified fixed points given below, by resistance ratios of platinum resistance thermometers obtained by calibration at specified sets of the fixed points, and by reference functions and deviation functions of resistance ratios which relate to T_{90} between the fixed points.
4. Above 1234.93 K, the ITS-90 is defined in terms of Planck's radiation law, using the freezing-point temperature of either silver, gold, or copper as the reference temperature.

Full details of the calibration procedures and reference functions for various subranges are given in Reference 1.

REFERENCES

1. Preston-Thomas, H., *Metrologia*, 27, 3—10, 1990 (Official release on ITS-90); errata in *Metrologia*, 27, 107, 1990.
2. McGlashan, M. L., *J. Chem. Thermodynamics*, 22, 653—663, 1990 (General description of ITS-90 and its relation to previous scales).
3. Rusby, R. L., *J. Chem. Thermodynamics*, 23, 1153—1161, 1991 (Conversion from previous scales to ITS-90).
4. Lide, D. R., Editor, *CRC Handbook of Chemistry and Physics, 74th Edition*, CRC Press, Boca Raton, FL, 1993 (Conversion from previous scales; ITS-90 thermocouple tables).

TABLE 1.3.1
DEFINING FIXED POINTS OF THE ITS-90

Material[a]	Equilibrium state[b]	Temperature	
		T_{90}/K	t_{90}/°C
He	VP	3 to 5	–270.15 to –268.15
e-H_2	TP	13.8033	–259.3467
e-H_2 (or He)	VP (or CVGT)	≈17	≈ –256.15
e-H_2 (or He)	VP (or CVGT)	≈20.3	≈ –252.85
Ne[c]	TP	24.5561	–248.5939
O_2	TP	54.3584	–218.7916
Ar	TP	83.8058	–189.3442
Hg[c]	TP	234.3156	–38.8344
H_2O	TP	273.16	0.01
Ga[c]	MP	302.9146	29.7646
In[c]	FP	429.7485	156.5985
Sn	FP	505.078	231.928
Zn	FP	692.677	419.527
Al[c]	FP	933.473	660.323
Ag	FP	1234.93	961.78
Au	FP	1337.33	1064.18
Cu[c]	FP	1357.77	1084.62

[a] e-H_2 indicates equilibrium hydrogen, that is, hydrogen with the equilibrium distribution of its ortho and para states. Normal hydrogen at room temperature contains 25% para hydrogen and 75% ortho hydrogen.

[b] VP indicates vapor pressure point; CVGT indicates constant volume gas thermometer point; TP indicates triple point (equilibrium temperature at which the solid, liquid, and vapor phases coexist); FP indicates freezing point, and MP indicates melting point (the equilibrium temperatures at which the solid and liquid phases coexist under a pressure of 101 325 Pa, one standard atmosphere). The isotopic composition is that naturally occurring.

[c] Previously, these were secondary fixed points.

TABLE 1.3.2
VALUES OF COEFFICIENTS IN THE VAPOR PRESSURE EQUATIONS FOR HELIUM

Coef.or constant	^{3}He 0.65—3.2 K	^{4}He 1.25—2.1768 K	^{4}He 2.1768—5.0 K
A_0	1.053 447	1.392 408	3.146 631
A_1	0.980 106	0.527 153	1.357 655
A_2	0.676 380	0.166 756	0.413 923
A_3	0.372 692	0.050 988	0.091 159
A_4	0.151 656	0.026 514	0.016 349
A_5	–0.002 263	0.001 975	0.001 826
A_6	0.006 596	–0.017 976	–0.004 325
A_7	0.088 966	0.005 409	–0.004 973
A_8	–0.004 770	0.013 259	0
A_9	–0.054 943	0	0
B	7.3	5.6	10.3
C	4.3	2.9	1.9

Section 2
Thermodynamic Properties of Pure Substances

2.1. PHYSICAL CONSTANTS AND PHASE BEHAVIOR

Most pure substances at moderate pressures exhibit a phase diagram of the type shown schematically in Figure 1a. The horizontal dashed line in this figure represents normal atmospheric pressure. The solid, liquid, and gas phases are in equilibrium at the triple point (TP), and the liquid-gas phase boundary ends at the critical point (CP). The overall phase behavior is conveniently described by the physical constants indicated in the figure: the normal melting point temperature T_m (where "normal" signifies a pressure of 101 325 Pa), normal boiling point temperature T_b, critical temperature T_c, and critical pressure P_c. Other solid phases usually exist at higher pressures; for some substances, especially metals, other solid phases are significant even at ambient pressure.

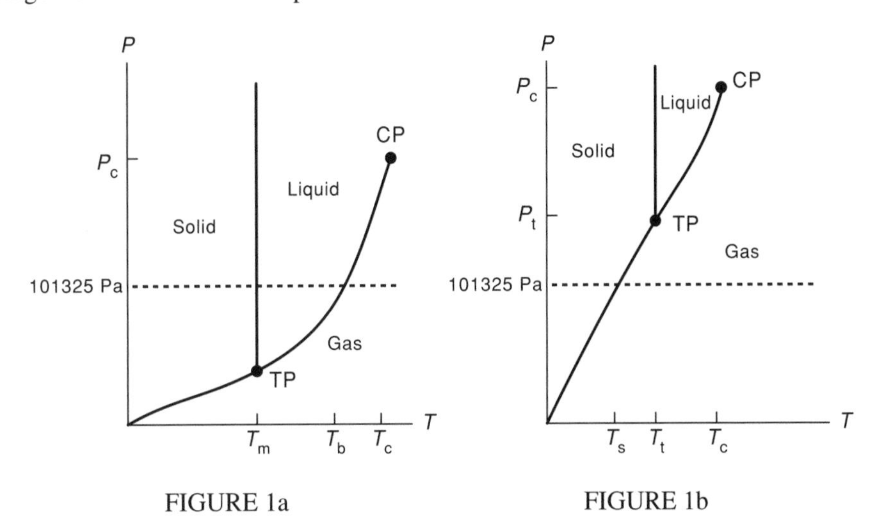

FIGURE 1a FIGURE 1b

Some substances, including such common compounds as carbon dioxide, show a somewhat different phase behavior as illustrated in Figure 1b. In such cases the normal melting and boiling points are undefined because of the steepness of the solid-gas phase boundary. Here the most useful physical constants are the triple point temperature and pressure, T_t and P_t, and the normal sublimation temperature T_s, which is the temperature at which the sublimation pressure of the solid reaches atmospheric pressure.

The tables in this section summarize the physical constants of some common substances and give details of the liquid-gas phase boundary (vapor pressure), the solid-gas boundary (sublimation pressure), and the solid-liquid boundary (variation of melting temperature with pressure).

TABLE 2.1.1
SUMMARY OF PHYSICAL CONSTANTS OF IMPORTANT SUBSTANCES

This table gives the normal melting and boiling temperatures, the liquid-gas critical constants, and the density at 298.15 K (25°C) for a selected list of about 1000 important substances. The properties are defined as follows:

Normal Melting Point Temperature: Temperature at which the solid and liquid phases are in equilibrium at a pressure of 101325 Pa. For most substances, this differs by less than 0.1 K from the triple point temperature, T_t, at which the solid, liquid, and gas phases are in equilibrium. However, for a few highly volatile solids such as carbon dioxide the sublimation pressure of the solid reaches 101325 Pa at a temperature below T_t, so that the normal melting point is undefined. In such cases the value in this table is followed by "t" and represents the triple point temperature (see Table 2.1.2. for triple point temperatures and pressures of other substances).
Normal Boiling Point Temperature: Temperature at which the liquid and gas phases are in equilibrium at a pressure of 101325 Pa. A value followed by "s" is a sublimation temperature, where the solid and gas are in equilibrium at 101325 Pa. See Table 2.1.3 for boiling point temperatures as a function of pressure.
Critical Temperature: Temperature above which the liquid phase does not exist.
Critical Pressure: Pressure corresponding to the critical temperature on the liquid-gas phase boundary.
Critical Molar Volume: Volume of one mole of the substance at the critical temperature and pressure.
Density: Ratio of mass to volume. The density value in the table refers to the solid phase if the normal melting point $T_m > 298.15$ K and to the liquid phase if $T_m < 298.15$ K. The pressure is taken to be nominal atmospheric pressure (about 100 kPa) except for those density values preceded by an * , where the applicable pressure is the saturation pressure of the liquid (greater than atmospheric) at 298.15 K. See Table 2.2.2 for liquid densities as a function of temperature.
Note on temperature scales: The values in this table come from many sources and were published over a period of many years. Thus it has not been possible to correct all values to the ITS-90 temperature scale (see Section 1.3). Only in the case of the elements are the values given consistently on the ITS-90 scale. For temperatures up to 1300 K, the difference between the previous scale, IPTS-68, and ITS-90 is less than 0.25 K.

Substances are arranged in the Hill order, with substances that do not contain carbon preceding those that do contain carbon.

Physical quantity	Symbol	SI unit
Normal melting point temperature	T_m	K
Normal boiling point temperature	T_b	K
Critical temperature	T_c	K
Critical pressure	P_c	Pa
Critical molar volume	V_c	$m^3 mol^{-1}$
Density at 298.15 K	ρ_o	$kg m^{-3}$

REFERENCES

1. *DIPPR Data Compilation of Pure Compound Properties*, Design Institute for Physical Property Data, American Institute of Chemical Engineers, 1987.
2. Budavari, S., Editor, *The Merck Index, Eleventh Edition*, Merck & Co., Rahway, NJ, 1989.
3. Stevenson, R. M.; Malanowski, S., *Handbook of the Thermodynamics of Organic Compounds*, Elsevier, New York, 1987.
4. Riddick, J. A.; Bunger, W. B.; Sakano, T. K., *Organic Solvents, Fourth Edition*, John Wiley & Sons, New York, 1986.
5. Lide, D. R., *CRC Handbook of Chemistry and Physics*, 74th Edition, CRC Press, Boca Raton, FL, 1993.
6. Weast, R. C.; Grasselli, J. G., *Handbook of Data on Organic Compounds*, CRC Press, Boca Raton, FL, 1989.
7. Donnay, J. D. H.; Ondik, H. M., *Crystal Data Determinative Tables, Third Edition*, Vol. 1, 1972; Vol. 2, 1973; Vol. 3, 1978; Vol. 4, 1978, Joint Committee on Powder Diffraction Standards, Philadelphia.
8. Chase, M. W., et al., *JANAF Thermochemical Tables, Third Edition, J. Phys. Chem. Ref. Data 14, Suppl. 1*, 1985.
9. Dinsdale, A. T., "SGTE Data for Pure Elements", *CALPHAD*, 15, 317-425, 1991.

TABLE 2.1.1
SUMMARY OF PHYSICAL CONSTANTS OF IMPORTANT SUBSTANCES (continued)

Molecular formula	Name	T_m/ K	T_b/ K	T_c/ K	P_c/ MPa	V_c/ cm³ mol⁻¹	ρ_o/ g cm⁻³
Ac	Actinium	1324	3471				10
Ag	Silver	1234.93	2435				10.5
AgBr	Silver bromide	705	1775				6.47
AgCl	Silver chloride	728	1820				5.56
AgI	Silver iodide	831	1779				5.68
Al	Aluminum	933.47	2792				2.70
AlB_3H_{12}	Aluminum trihydride-tris(borane)	208.6	317.6				
$AlBr_3$	Aluminum tribromide	370.6	528	763	2.89	310	3.21
$AlCl_3$	Aluminum trichloride	463		620	2.63	257	2.48
AlI_3	Aluminum triiodide	464	655	983		408	3.98
Al_2O_3	Dialuminum trioxide	2327					3.97
Ar	Argon	83.80	87.30	150.87	4.898	75	
As	Arsenic	1090 t	887 s	1673		35	5.78
$AsBr_3$	Arsenic tribromide	304.2	494				3.40
$AsCl_3$	Arsenic trichloride	257	403	654		252	2.150
AsF_3	Arsenic trifluoride	267.2	330.9				2.7
AsF_5	Arsenic pentafluoride	193.3	219.9				
AsH_3	Arsane	157	210.6	373.1			
AsI_3	Arsenic triiodide	414.0	697				4.73
Au	Gold	1337.33	3129				19.3
B	Boron	2348	4273				2.34
BBr_3	Boron tribromide	228	364	581		272	2.6
BCl_3	Boron trichloride	166	285.80	455	3.87	239	
BF_3	Boron trifluoride	146.3	172	260.8	4.98	115	
BI_3	Boron triiodide	316	483	773		356	3.96
B_2Cl_4	Tetrachlorodiborane(4)	180.5	338				
B_2F_4	Tetrafluorodiborane(4)		239.3				
B_2H_6	Diborane(6)	107.6	180.77	289.8	4.05		
B_2O_3	Diboron trioxide	723					2.46
$B_3H_6N_3$	Cyclotriborazane	215	326				0.80
B_4H_{10}	Tetraborane(10)	153	291				
B_5H_9	Pentaborane(9)	226.5	333				0.60
B_5H_{11}	Pentaborane(11)	150	336				
B_6H_{10}	Hexaborane(10)	210.8	381				0.67
Ba	Barium	1000	2170				3.62
$BaCl_2$	Barium dichloride	1235					3.9
BaF_2	Barium difluoride	1641					4.893
BaO_4S	Barium sulfate	1623					4.49
Be	Beryllium	1560	2744				1.85
$BeCl_2$	Beryllium dichloride	688	820				1.90
BeF_2	Beryllium difluoride	825	1432				2.1
BeI_2	Beryllium diiodide	753	760				4.32
BeO	Beryllium oxide	2780					3.01
Bi	Bismuth	544.55	1837				9.79
$BiBr_3$	Bismuth tribromide	491	726	1220		301	5.72
$BiCl_3$	Bismuth trichloride	503	712.5	1179	12.0	261	4.75
BrCs	Cesium bromide	909					4.43
BrF_3	Bromine trifluoride	281.92	398.9				2.803
BrF_5	Bromine pentafluoride	212.6	313.91				2.460
BrH	Hydrogen bromide	186.34	206.77	363.2	8.55		
BrI	Iodine bromide	313	389	719		139	4.3
BrIn	Indium bromide	563	929				4.96
BrK	Potassium bromide	1007					2.74
BrLi	Lithium bromide	825					3.464
BrNa	Sodium bromide	1014					3.200
BrRb	Rubidium bromide	955	1613				3.35
BrTl	Thallium bromide	733	1092				7.5
Br_2	Bromine	265.9	331.9	588	10.34	127	3.1028
Br_2Ca	Calcium dibromide	1015					3.38
Br_2Cd	Cadmium dibromide	841	1117				5.19
Br_2Hg	Mercury dibromide	509	595	1012			6.05
Br_2Pb	Lead dibromide	644	1165				6.69
Br_2Sn	Tin dibromide	488	912				5.12

TABLE 2.1.1
SUMMARY OF PHYSICAL CONSTANTS OF IMPORTANT SUBSTANCES (continued)

Molecular formula	Name	T_m/ K	T_b/ K	T_c/ K	P_c/ MPa	V_c/ cm³ mol⁻¹	ρ_o/ g cm⁻³
Br_2Zn	Zinc dibromide	667	970				4.5
Br_3Ga	Gallium tribromide	394.6	552	806.7		303	
Br_3HSi	Tribromosilane	200	382	610.0		305	2.7
Br_3P	Phosphorus tribromide	233	446.10	711		300	2.8
Br_3Sb	Antimony tribromide	369.7	560.26	904		300	4.35
Br_4Ge	Germanium tetrabromide	299.2	459.50	718		392	
Br_4Si	Silicon tetrabromide	278.3	427	663		382	2.8
Br_4Sn	Tin tetrabromide	304	478	744		417	3.7
Br_4Ti	Titanium tetrabromide	312	503	795.7		391	3.37
Br_4Zr	Zirconium tetrabromide	723		805		424	3.98
Br_5Ta	Tantalum pentabromide	538	622	974		461	4.99
Ca	Calcium	1115	1757				1.54
$CaCl_2$	Calcium dichloride	1045	2208.6				2.22
CaF_2	Calcium difluoride	1691	2806.5				3.18
CaI_2	Calcium diiodide	1052					3.96
CaO	Calcium oxide	3200					3.34
CaO_4S	Calcium sulfate	1723					2.96
Cd	Cadmium	594.22	1040				8.69
$CdCl_2$	Cadmium dichloride	837	1233				4.08
CdF_2	Cadmium difluoride	1383	2021				6.33
CdI_2	Cadmium diiodide	660	1015				5.64
Ce	Cerium	1072	3697				8.16
Ce_2O_3	Dicerium trioxide	2503					6.2
ClCs	Cesium chloride	918	1570				3.988
ClF	Chlorine fluoride	117.5	172.0				
$ClFO_3$	Perchloryl fluoride	126	226.40	368.4	5.37	161	
ClF_3	Chlorine trifluoride	196.81	284.90				
ClF_3Si	Chlorotrifluorosilane		203.10	307.7	3.46		
ClF_5	Chlorine pentafluoride		260	416	5.27	233	
ClH	Hydrogen chloride	158.97	188.15	324.7	8.31	81	*0.804
ClH_3Si	Chlorosilane		242.7				
ClH_4N	Ammonium chloride	793		1155	163.5		1.519
ClK	Potassium chloride	1043					1.988
ClLi	Lithium chloride	878	1656				2.07
ClNO	Nitrosyl chloride	211.6	267.60	440			
ClNa	Sodium chloride	1073.5	1738				2.17
$ClNaO_3$	Sodium chlorate	521					2.5
ClO_2	Chlorine dioxide	214	284				
ClRb	Rubidium chloride	990	1663				2.76
ClTl	Thallium chloride	703	1080				7.0
Cl_2	Chlorine	171.6	239.18	416.9	7.991	124	*1.394
Cl_2Cr	Chromium dichloride	1087	1573				2.88
Cl_2CrO_2	Chromyl dichloride	176.6	390				1.91
Cl_2Cu	Copper dichloride	703					3.4
Cl_2H_2Si	Dichlorosilane	151	281.4				
Cl_2Hg	Mercury dichloride	549	577	973		174	5.6
Cl_2Hg_2	Dimercury dichloride	816					6.97
Cl_2O	Dichlorine oxide	152.5	275.3				
Cl_2OS	Sulfinyl dichloride	172	348.7				1.631
Cl_2OSe	Selenosyl dichloride	281.6	453	730	7.09	235	2.430
Cl_2O_2S	Sulfonyl dichloride	222	342.5				1.680
Cl_2Pb	Lead dichloride	774	1224				5.98
Cl_2Sn	Tin dichloride	520	896				3.90
Cl_2Zn	Zinc dichloride	563	1005				2.907
Cl_3FSi	Trichlorofluorosilane		285.40	438.6	3.58		
Cl_3Ga	Gallium trichloride	351.0	474	694		263	2.47
Cl_3HSi	Trichlorosilane	144.9	306	479		268	1.3313
Cl_3OP	Phosphoryl trichloride	274	378.6				1.645
Cl_3OV	Vanadyl trichloride	194	400				
Cl_3P	Phosphorus trichloride	161	349.10	563		264	1.574
Cl_3PS	Thiophosphoryl chloride	236.9	398				
Cl_3Sb	Antimony trichloride	346.5	493.4	794		272	3.14
Cl_4Ge	Tetrachlorogerman	223.6	359.70	553.2	3.861	330	1.88

TABLE 2.1.1
SUMMARY OF PHYSICAL CONSTANTS OF IMPORTANT SUBSTANCES (continued)

Molecular formula	Name	T_m/ K	T_b/ K	T_c/ K	P_c/ MPa	V_c/ cm³ mol⁻¹	ρ_o/ g cm⁻³
Cl_4Hf	Hafnium tetrachloride			725.7	5.42	314	
Cl_4ORe	Rhenosyl tetrachloride	302.4	496	781		362	
Cl_4OW	Tungstenosyl tetrachloride	484	500.70	782		338	
Cl_4Si	Silicon tetrachloride	205	330.80	508.1	3.593	326	1.5
Cl_4Sn	Tin tetrachloride	240	387.30	591.9	3.75	351	2.3
Cl_4Te	Tellurium tetrachloride	497	660	1002	8.56	310	3.0
Cl_4Th	Thorium tetrachloride	1043	1194				4.59
Cl_4Ti	Titanium tetrachloride	248	409.60	638	4.66	339	1.73
Cl_4V	Vanadium tetrachloride	247.4	425				
Cl_4Zr	Zirconium tetrachloride	710		778	5.77	319	2.80
Cl_5Mo	Molybdenum pentachloride	467	541	850		369	2.93
Cl_5Nb	Niobium pentachloride	477.8	527.20	803.5	4.88	397	2.73
Cl_5P	Phosphorus pentachloride		433	646			
Cl_5Ta	Tantalum pentachloride	489	512.50	767		402	3.68
Cl_6W	Tungsten hexachloride	548	619.90	923		422	3.52
Co	Cobalt	1768	3200				8.86
CoO	Cobalt oxide	2103					6.5
Cr	Chromium	2180	2944				7.15
Cr_2O_3	Dichromium trioxide	2603					5.22
Cs	Cesium	302.9	944				1.93
Cu	Copper	1357.77	2835				8.96
CuO	Copper oxide	1719					6.5
CuO_4S	Copper sulfate						3.6
Dy	Dysprosium	1684	2834				8.55
Dy_2O_3	Didysprosium trioxide	2681					8.0
Er	Erbium	1802	3135				9.07
Er_2O_3	Dierbium trioxide	2691					8.6
Eu	Europium	1095	1869				5.24
FH	Hydrogen fluoride	189.79	292.65	461	6.48	69	*0.9580
FH_3Si	Fluorosilane		174.5				
FK	Potassium fluoride	1131	1775				2.48
FLi	Lithium fluoride	1121.3	1946				2.640
FNO	Nitrosyl fluoride	140.6	213.2				
FNO_2	Nitryl fluoride	107	200.7	349.5			
FNa	Sodium fluoride	1265	1977				2.78
FRb	Rubidium fluoride	1106					3.2
FTl	Thallium fluoride	595	928				8.36
F_2	Fluorine	53.49	84.95	144.13	5.172	66	
F_2HN	Difluoroazane		250	403			
F_2N_2	*cis*-Difluorodiazene		167.40	272	7.09		
F_2N_2	*trans*-Difluorodiazene		161.70	260	5.57		
F_2O	Oxygen difluoride	49.3	128.40	215			
F_2OS	Sulfinyl difluoride	143.6	229.3				
F_2Pb	Lead difluoride	1103	1566				8.44
F_2Zn	Zinc difluoride	1145	1773				4.9
F_2Xe	Xenon difluoride		387.50	631	9.32	148	
F_3HSi	Trifluorosilane		178				
F_3N	Nitrogen trifluoride	66.36	144.40	234.0	4.46	126	
F_3NO	Trifluoroazane oxide	112	185.60	303	6.43	147	
F_3P	Phosphorus trifluoride	121.6	171.6	271.2		4.33	
F_3PS	Thiophosphoryl trifluoride	124.3	220.90	346.0	3.82		
F_3Sb	Antimony trifluoride	565	649				4.38
F_4N_2	Tetrafluorohydrazine	108.6	199	309	3.75		
F_4S	Sulfur tetrafluoride	148	232.70	364			
F_4Si	Silicon tetrafluoride	182.9	187	259.0	3.72		
F_4Th	Thorium tetrafluoride	1383	1953				6.1
F_4U	Uranium tetrafluoride	1309	1690				6.7
F_4Xe	Xenon tetrafluoride	387	388.90	612	7.04	188	
F_5I	Iodine pentafluoride	282.58	373.6				3.19
F_5Nb	Niobium pentafluoride	353	502	737	6.28	155	2.7
F_5P	Phosphorus pentafluoride		188.5				
F_5Ta	Tantalum pentafluoride	368.2	502.3				5.0
F_5V	Vanadium pentafluoride	292.6	321.4				2.50

TABLE 2.1.1
SUMMARY OF PHYSICAL CONSTANTS OF IMPORTANT SUBSTANCES (continued)

Molecular formula	Name	T_m/ K	T_b/ K	T_c/ K	P_c/ MPa	V_c/ cm³ mol⁻¹	ρ_o/ g cm⁻³
F_6Ir	Iridium hexafluoride	317	326				4.8
F_6Mo	Molybdenum hexafluoride	290.6	307	473	4.75	226	2.54
F_6S	Sulfur hexafluoride	222.4 t	209.3 s	318.69	3.77	199	
F_6Se	Selenium hexafluoride		227	345.5			
F_6Te	Tellurium hexafluoride		234	356			
F_6U	Uranium hexafluoride	337.20 t	329.6 s	505.8	4.66	250	5.09
F_6W	Tungsten hexafluoride	275.4	290	444	4.34	233	
Fe	Iron	1811	3134				7.87
FeO	Iron oxide	1650					6.0
FeS	Iron sulfide	1463					4.3
Fe_3O_4	Triiron tetraoxide	1870					5.2
Ga	Gallium	303.0	2477				5.91
GaI_3	Gallium triiodide	485	613	951		395	4.5
Gd	Gadolinium	1587	3537				7.90
Ge	Germanium	1211.40	3106				5.3234
GeH_4	Germane	108	183	312.2	4.95	147	
GeI_4	Germanium tetraiodide	417	650	973		500	4.4
Ge_2H_6	Digermane		303.9				
HI	Hydrogen iodide	222.38	237.60	424.0	8.31		
HKO	Potassium hydroxide	679	1600				
HLiO	Lithium hydroxide	744.3	1899				1.5
HNO_3	Nitric acid	231.5	356				1.55
HN_3	Hydrogen azide	193	308.8				
HNaO	Sodium hydroxide	596	1661				2.13
H_2	Hydrogen	13.81	20.38	32.97	1.293	65	
H_2O	Water	273.15	373.15	647.14	22.06	56	0.9970
H_2O_2	Dihydrogen peroxide	272.72	423.3				1.4
H_2O_4S	Sulfuric acid	283.46	610				1.8
H_2S	Dihydrogen sulfide	187.6	212.84	373.2	8.94	99	*0.769
H_2Se	Dihydrogen selenide	207.42	231.90	411	8.92		
H_2Te	Dihydrogen telluride	224	271				
H_3N	Ammonia	195.41	239.72	405.5	11.35	72	*0.602
H_3NO	Hydroxylamine	306.2	331				1.21
H_3O_2P	Phosphonous acid	299.6	403				1.49
H_3O_3P	Phosphorous acid	347.5	473				1.65
H_3O_4P	Phosphoric acid	315.5	680				
H_3P	Phosphine	140	185.40	324.5	6.54		
H_3Sb	Stibine	185	256				
H_4N_2	Hydrazine	274.5	386.70	653	14.7		1.0036
$H_4N_2O_3$	Ammonium nitrate	442.7					1.72
H_4Si	Silane	88	161				
H_4Sn	Stannane		221.3				
H_6Si_2	Disilane	140.8	258.8				
H_8Si_3	Trisilane	155.7	326				0.7
He	Helium		4.22	5.19	0.227	57	
Hf	Hafnium	2506	4876				13.3
HfI_4	Hafnium tetraiodide	722		916		528	5.6
Hg	Mercury	234.32	629.88	1750	172.00	43	13.5336
HgI_2	Mercury diiodide	532	627	1072			6.3
Ho	Holmium	1745	2967				8.80
IIn	Indium iodide	624	985				5.3
IK	Potassium iodide	954	1596				3.12
ILi	Lithium iodide	742	1444				4.06
INa	Sodium iodide	928	1577				3.67
IRb	Rubidium iodide	915	1573				3.55
ITl	Thallium iodide	713	1097				7.1
I_2	Iodine	386.8	457.5	819		155	4.933
I_2Pb	Lead diiodide	683	1145				6.16
I_2Sr	Strontium diiodide	811	2046				4.4
I_2Zn	Zinc diiodide	719	898				4.74
I_3Sb	Antimony triiodide	441	674	1102			4.92
I_4Si	Silicon tetraiodide	393.6	560.50	944		558	4.1
I_4Sn	Tin tetraiodide	416	637.50	968		531	4.46

TABLE 2.1.1
SUMMARY OF PHYSICAL CONSTANTS OF IMPORTANT SUBSTANCES (continued)

Molecular formula	Name	T_m/ K	T_b/ K	T_c/ K	P_c/ MPa	V_c/ cm^3 mol^{-1}	ρ_o/ g cm^{-3}
I_4Ti	Titanium tetraiodide	423	650	1040		505	
In	Indium	429.75	2345				7.31
Ir	Iridium	2719	4701				22.5
K	Potassium	336.53	1032				0.89
KNO_3	Potassium nitrate	610					2.11
Kr	Krypton	115.79	119.93	209.41	5.50	91	
La	Lanthanum	1193	3728				6.15
Li	Lithium	453.6	1615				0.534
$LiNO_3$	Lithium nitrate	526					2.38
Lu	Lutetium	1936	3666				9.84
Mg	Magnesium	923	1363				1.74
MgO	Magnesium oxide	3099					3.6
Mn	Manganese	1519	2334				7.3
Mo	Molybdenum	2896	4912				10.2
MoO_3	Molybdenum trioxide	1074	1428				4.70
$NNaO_3$	Sodium nitrate	580					2.26
NO	Nitrogen oxide	109.5	121.41	180	6.48	58	*0.505
NO_2	Nitrogen dioxide (see N_2O_4)						
N_2	Nitrogen	63.15	77.35	126.21	3.39	90	
N_2O	Dinitrogen oxide	182.3	184.67	309.57	7.255	97	
N_2O_4	Dinitrogen tetraoxide	263.8	294.30	431	10.1	167	
N_2O_5	Dinitrogen pentaoxide	303	320				2.0
Na	Sodium	370.87	1156				0.97
Na_2O_4S	Disodium sulfate	1157					2.7
Nb	Niobium	2750	5017				8.57
Nd	Neodymium	1289	3339				7.01
Nd_2O_3	Dineodymium trioxide	2593					7.2
Ne	Neon	24.56	27.07	44.4	2.76	42	
Ni	Nickel	1728.3	3186				8.90
OTl_2	Dithallium oxide	573	773				
OZn	Zinc oxide	2248					5.6
O_2	Oxygen	54.36	90.17	154.59	5.043	73	
O_2S	Sulfur dioxide	197.6	263.13	430.8	7.884	122	
O_2Se	Selenium dioxide	613					3.9
O_2Si	Silicon dioxide	1883	2503				2.6481
O_3	Ozone	80	161.80	261.1	5.57	89	
O_3S	Sulfur trioxide	289.9	318	491.0	8.2	127	1.92
O_3Y_2	Diyttrium trioxide	2712					5.03
O_3Yb_2	Diytterbium trioxide	2708					9.2
O_4Os	Osmium tetraoxide	314	408	678			5.0
O_5P_2	Diphosphorus pentaoxide	693					2.30
O_5V_2	Divanadium pentaoxide	943	2073				3.35
O_7Re_2	Dirhenium heptaoxide	570	723	942		334	6.103
Os	Osmium	3306	5285				22.5
P	Phosphorus	317.30	550	994			2.69
Pb	Lead	600.61	2022				11.3
PbS	Lead sulfide	1386.5					7.60
Pd	Palladium	1828	3236				12.0
Pm	Promethium	1315	3273				7.26
Pr	Praseodymium	1204	3783				6.77
Pt	Platinum	2041.5	4098				21.5
Pu	Plutonium	913	3501				19.7
Rb	Rubidium	312.46	961				1.53
Rh	Rhodium	2237	3968				12.4
Rn	Radon	202	211.4	377	6.28		
Ru	Ruthenium	2607	4423				12.1
S	Sulfur	388.36	717.75	1314	20.7		2.07
STl_2	Dithallium sulfide	721	1640				8.39
Sb	Antimony	903.78	1860				6.68
Sc	Scandium	1814	3103				2.99
Se	Selenium	495	958	1766	27.2		4.79
Si	Silicon	1687	3538				2.3290
Sm	Samarium	1345	2063				7.52

TABLE 2.1.1
SUMMARY OF PHYSICAL CONSTANTS OF IMPORTANT SUBSTANCES (continued)

Molecular formula	Name	T_m/ K	T_b/ K	T_c/ K	P_c/ MPa	V_c/ cm³ mol⁻¹	ρ_o/ g cm⁻³
Sn	Tin	505.08	2875				7.28
Sr	Strontium	1050	1655				2.64
Ta	Tantalum	3290	5731				16.4
Tb	Terbium	1632	3494				8.23
Tc	Technetium	2430	4538				11
Te	Tellurium	722.66	1261				6.24
Th	Thorium	2023	5061				11.7
Ti	Titanium	1941	3560				4.5
Tl	Thallium	577	1746				11.8
Tm	Thulium	1818	2219				9.32
U	Uranium	1408	4404				19.1
V	Vanadium	2183	3680				6.0
W	Tungsten	3695	5828				19.3
Xe	Xenon	161.40	165.11	289.73	5.84	118	
Y	Yttrium	1799	3609				4.47
Yb	Ytterbium	1097	1467				6.90
Zn	Zinc	692.68	1180				7.14
Zr	Zirconium	2127.85	4682				6.52
C	Carbon	4247 t	4098 s				3.51
$CBrClF_2$	Bromochlorodifluoromethane	113.6	269.14	426.88	4.254	246	
$CBrCl_3$	Bromotrichloromethane	267.4	378				2.012
$CBrF_3$	Bromotrifluoromethane	101	215.26	340.2	3.97	196	
$CBrN$	Cyanogen bromide	325	334.6				2.005
CBr_2F_2	Dibromodifluoromethane	163.0	295.9	471.3			
CBr_3Cl	Tribromochloromethane	328	431.6				
CBr_4	Tetrabromomethane	363.2	462.6				3.42
$CCaO_3$	Calcium carbonate	1612.2					2.711
$CClF_3$	Chlorotrifluoromethane	92	191.7	302	3.870	180	*0.830
$CClN$	Cyanogen chloride	266.60	286				
CCl_2F_2	Dichlorodifluoromethane	117	243.3	384.95	4.136	217	*1.311
CCl_2O	Carbonyl dichloride	145.2	281	455	5.67	190	*1.3719
CCl_3F	Trichlorofluoromethane	162.04	296.78	471.2	4.41	248	*1.4760
CCl_4	Tetrachloromethane	250	349.8	556.6	4.516	276	1.5844
CF_2O	Carbonyl difluoride	161.89	188.58				
CF_4	Tetrafluoromethane	89.56	145.07	227.6	3.74	140	
$CHBrCl_2$	Bromodichloromethane	216	363				1.970
$CHBr_3$	Tribromomethane	281.20	422.36				2.8761
$CHClF_2$	Chlorodifluoromethane	115.73	232.40	369.3	4.99	169	1.3113
$CHCl_2F$	Dichlorofluoromethane	138	282.05	451.58	5.18	196	
$CHCl_3$	Trichloromethane	209.5	334.33	536.4	5.47	239	1.480
CHF_3	Trifluoromethane	117.97	191.0	299.3	4.858	133	*0.673
CHI_3	Triiodomethane	392	491				4.008
CHN	Hydrogen cyanide	259.7	298.81	456.7	5.39	139	0.684
CH_2BrCl	Bromochloromethane	185.20	341.20				1.925
CH_2Br_2	Dibromomethane	220.60	370				2.48
CH_2Cl_2	Dichloromethane	176.5	312.79	510	6.10		1.3168
CH_2F_2	Difluoromethane	137	221.45	351.6	5.830	121	*0.960
CH_2I_2	Diiodomethane	279.2	455				3.3079
CH_2N_2	Diazomethane	128	250				
CH_2O	Methanal	181	254.05				
CH_2O_2	Methanoic acid	281.4	373.71	588			1.214
CH_3Br	Bromomethane	179.4	276.7				
CH_3Cl	Chloromethane	175.4	248.95	416.25	6.679	139	*0.911
CH_3Cl_3Si	Trichloromethylsilane	183	339.27	517	3.28	348	1.267
CH_3F	Fluoromethane	131.3	194.7	317.8	5.88	113	*0.566
CH_3I	Iodomethane	206.70	315.57	528			2.2650
CH_3NO	Methanamide	275.70	493				1.1291
CH_3NO_2	Nitromethane	244.60	374.35	588	5.87	173	1.1313
CH_4	Methane	90.69	111.65	190.53	4.604	99	
CH_4N_2O	Urea	405.8					1.32
CH_4O	Methanol	175.47	337.7	512.64	8.092	118	0.7866
CH_4S	Methanethiol	150	279.10	470.0	7.23	145	
CH_5N	Methylamine	179.71	266.82	430.7	7.614		*0.656

TABLE 2.1.1
SUMMARY OF PHYSICAL CONSTANTS OF IMPORTANT SUBSTANCES (continued)

Molecular formula	Name	T_m/ K	T_b/ K	T_c/ K	P_c/ MPa	V_c/ cm^3 mol^{-1}	ρ_o/ g cm^{-3}
CH_6N_2	Methylhydrazine	220.7	364.1	567	8.24	271	
CH_6Si	Methylsilane	116.6	215.60	352.5			
CI_4	Tetraiodomethane	444					4.3
CN_4O_8	Tetranitromethane	286.9	399.2				1.6229
CNa_2O_3	Disodium carbonate	1131.2					2.54
CO	Carbon oxide	68	81.6	132.91	3.499	93	
COS	Carbonyl sulfide	134.3	223	375	5.88	137	
CO_2	Carbon dioxide	216.58 t	194.7 s	304.14	7.375	94	*0.720
CS_2	Carbon disulfide	161.6	319.37	552	7.90	173	1.2555
$C_2Br_2ClF_3$	1,2-Dibromo-1-chloro-1,2,2-trifluoroethane		366	560.7	3.61	368	
$C_2Br_2F_4$	1,2-Dibromotetrafluoroethane	162.7	320.50	487.8	3.393	341	2.163
C_2ClF_3	Chlorotrifluoroethene	115	245.0	379	4.05	212	
C_2ClF_5	Chloropentafluoroethane	173.71	234.0	353.2	3.229	252	*1.288
$C_2Cl_2F_4$	1,1-Dichlorotetrafluoroethane	216.5	277	418.6	3.30	294	*1.455
$C_2Cl_2F_4$	1,2-Dichlorotetrafluoroethane	179	276.9	418.78	3.252	297	*1.455
$C_2Cl_3F_3$	1,1,1-Trichlorotrifluoroethane	287.3	318.95				1.571
$C_2Cl_3F_3$	1,1,2-Trichlorotrifluoroethane	238	320.8	487.3	3.42	325	1.5642
C_2Cl_4	Tetrachloroethene	250.80	394.2	620.2			1.6130
$C_2Cl_4F_2$	Tetrachloro-1,2-difluoroethane	299	366	551			1.6447
C_2Cl_6	Hexachloroethane	460	460				2.080
C_2F_4	Tetrafluoroethene	130.6	197.50	306.5	3.94	172	
C_2F_6	Hexafluoroethane	172.4	194.9	293		222	
C_2HCl	Chloroethyne	147	243				
C_2HClF_2	1-Chloro-2,2-difluoroethene	134.6	254.60	400.6	4.46	197	
C_2HClF_4	1-Chloro-1,1,2,2-tetrafluoroethane	156	263	399.9	3.72	244	
C_2HCl_3	Trichloroethene	188.40	360.34	544.2	5.02		1.4578
C_2HCl_3O	Trichloroethanal	215.6	370.9				1.505
$C_2HCl_3O_2$	Trichloroethanoic acid	330.6	470.7				
C_2HCl_5	Pentachloroethane	244	433.03				1.6749
$C_2HF_3O_2$	Trifluoroethanoic acid	257.90	344.9	491.3	3.258	204	1.485
C_2H_2	Ethyne	192.40 t	188.43 s	308.33	6.139	113	*0.377
$C_2H_2Br_4$	1,1,2,2-Tetrabromoethane	273	516.70				2.9529
$C_2H_2Cl_2$	1,1-Dichloroethene	150.6	304.7				1.18
$C_2H_2Cl_2$	*cis*-1,2-Dichloroethene	193	333.78	544.2			1.2649
$C_2H_2Cl_2$	*trans*-1,2-Dichloroethene	223.3	320.82	516.5	5.51		1.2444
$C_2H_2Cl_2O$	Chloroacetyl chloride	251	379				1.413
$C_2H_2Cl_4$	1,1,1,2-Tetrachloroethane	202.94	403.3				1.5346
$C_2H_2Cl_4$	1,1,2,2-Tetrachloroethane	229.3	418.3	661.15			1.5872
$C_2H_2F_2$	1,1-Difluoroethene	129	187.4	302.9	4.46	154	
C_2H_2O	Ketene	122	223.34				
$C_2H_2O_2$	Ethanedial	288	323.5				1.134
C_2H_3Br	Bromoethene	135.3	288.9				
C_2H_3BrO	Acetyl bromide	177	349				1.65
C_2H_3Cl	Chloroethene	119.36	259.34				
$C_2H_3ClF_2$	1-Chloro-1,1-difluoroethane	142.3	263.40	409.6	4.332	231	*1.107
C_2H_3ClO	Acetyl chloride	160.30	323.90				1.100
C_2H_3ClO	Chloroethanal	256.8	358.6				
$C_2H_3ClO_2$	Chloroethanoic acid	334.4	462.50				
$C_2H_3Cl_3$	1,1,1-Trichloroethane	242.7	347.24	545	4.30		1.3303
$C_2H_3Cl_3$	1,1,2-Trichloroethane	236.5	387.0				1.4346
C_2H_3F	Fluoroethene	112.6	201	327.9	5.24	144	
C_2H_3FO	Acetyl fluoride	189	293.9				
$C_2H_3F_3$	1,1,1-Trifluoroethane	161.8	225.60	346.3	3.76	194	*0.963
C_2H_3N	Ethanenitrile	229.32	354.80	545.5	4.85	173	0.7765
C_2H_3NO	Methyl isocyanate	228	312.2				
$C_2H_3NaO_2$	Sodium ethanoate	597					1.528
C_2H_4	Ethene	103	169.38	282.34	5.041	131	
$C_2H_4Br_2$	1,1-Dibromoethane	210	382				2.045
$C_2H_4Br_2$	1,2-Dibromoethane	283.08	404.5	583.0	7.2		2.1687
$C_2H_4Cl_2$	1,1-Dichloroethane	176.19	330.5	523	5.07	236	1.1680
$C_2H_4Cl_2$	1,2-Dichloroethane	237.6	356.6	561	5.4	225	1.2457
$C_2H_4Cl_2O$	Bis(chloromethyl) ether	231.6	379				1.31

TABLE 2.1.1
SUMMARY OF PHYSICAL CONSTANTS OF IMPORTANT SUBSTANCES (continued)

Molecular formula	Name	T_m/ K	T_b/ K	T_c/ K	P_c/ MPa	V_c/ cm^3 mol^{-1}	ρ_o/ g cm^{-3}
$C_2H_4F_2$	1,1-Difluoroethane	156	248.20	386.7	4.50	181	*0.896
C_2H_4O	Ethanal	150	293.6	466		154	
C_2H_4O	Oxirane	161.4	283.6	469	7.19	140	*0.8717
$C_2H_4O_2$	Ethanoic acid	289.7	391.0	592.71	5.786	171	1.0439
$C_2H_4O_2$	Methyl methanoate	174	304.9	487.2	5.998	172	0.9664
$C_2H_4O_3$	Peroxyethanoic acid	272.9	383				
$C_2H_4O_3$	Hydroxyethanoic acid	352.6	373				
C_2H_5Br	Bromoethane	154.5	311.6	503.9	6.23	215	1.4505
C_2H_5Cl	Chloroethane	134.4	285.4	460.4	5.3		*0.8889
C_2H_5ClO	2-Chloroethanol	205.6	401.7				1.1965
C_2H_5F	Fluoroethane	129.9	235.50	375.31	5.028		
C_2H_5I	Iodoethane	162.0	345.6				1.9244
C_2H_5N	Aziridine	195.20	329				0.832
C_2H_5NO	Ethanamide	354	495.16				
C_2H_5NO	*N*-Methylmethanamide	269.3	472.66				1.00
$C_2H_5NO_2$	Nitroethane	183.63	387.22				1.0427
C_2H_6	Ethane	90.3	184.5	305.4	4.884	148	*0.315
$C_2H_6Cl_2Si$	Dichlorodimethylsilane	257	343.4	520.4	3.49	350	1.064
C_2H_6O	Dimethyl ether	131.6	248.3	400.0	5.37	190	*0.661
C_2H_6O	Ethanol	159.0	351.44	513.92	6.132	167	0.7849
C_2H_6OS	Dimethyl sulfoxide	291.67	462.2				1.0955
$C_2H_6O_2$	1,2-Ethanediol	260	470.49	718			1.1101
$C_2H_6O_2S$	Dimethyl sulfone	382	511				
C_2H_6S	2-Thiapropane	174.8	310.48	503.0	5.53	201	0.8423
C_2H_6S	Ethanethiol	125.26	308.2	499	5.49	207	0.8315
$C_2H_6S_2$	2,3-Dithiabutane	188	382.9				1.057
C_2H_7N	Dimethylamine	180.9	280.03	437.22	5.340		*0.6501
C_2H_7N	Ethylamine	192.62	289.80	456	5.62	182	*0.677
C_2H_7NO	2-Aminoethanol	283.6	444.1				1.0136
$C_2H_8N_2$	1,1-Dimethylhydrazine	215	337.0				0.785
$C_2H_8N_2$	1,2-Ethanediamine	284.2	390				0.8931
C_2N_2	Ethanedinitrile	245.2	252.0	400	5.98		
C_3F_6	Hexafluoropropene	116.6	243.5				
C_3F_6O	Hexafluoro-2-propanone	148	245.70	357.14	2.84	329	*1.313
C_3F_8	Octafluoropropane	125.46	236.40	345.1	2.680	299	
C_3HN	Propynenitrile	278	315.6				
C_3H_3N	Propenenitrile	189.6	350.57				0.8002
C_3H_3NO	Oxazole		342.6				
C_3H_3NO	Isoxazole		368	552.0			1.073
C_3H_4	Propyne	170.4	250.16	402.38	5.628	164	*0.607
C_3H_4	Propadiene	136.87	238.70	393			*0.584
$C_3H_4Cl_2$	2,3-Dichloropropene	283	367				1.205
$C_3H_4N_2$	Imidazole	363.6	530				
$C_3H_4N_2$	Pyrazole	341	460				
C_3H_4O	Propenal	185.4	325.26				0.837
C_3H_4O	2-Propyn-1-ol	221.3	386.7				0.9450
$C_3H_4O_2$	Propenoic acid	285.4	414.2				1.046
$C_3H_4O_2$	2-Oxetanone	239.7	435				1.1420
C_3H_5Br	3-Bromopropene	154	343.2				1.391
C_3H_5Cl	*cis*-1-Chloropropene	138.3	305.9				0.930
C_3H_5Cl	*trans*-1-Chloropropene	174	310.5				0.930
C_3H_5Cl	2-Chloropropene	135.7	295.80				
C_3H_5Cl	3-Chloropropene	138.6	318.30	514			0.933
C_3H_5ClO	(Chloromethyl)oxirane	215.9	389.26				1.1746
$C_3H_5ClO_2$	Methyl chloroethanoate	241.03	402.6				1.228
$C_3H_5Cl_3$	1,2,3-Trichloropropane	258.4	430				1.382
C_3H_5N	Propanenitrile	180.26	370.6	561.3	4.26	229	0.7768
C_3H_5NO	3-Hydroxypropanenitrile	227	494				1.0404
C_3H_5NO	Propenamide	357.6	465.7				1.12
C_3H_6	Propene	87.90	225.46	364.85	4.601	181	*0.505
C_3H_6	Cyclopropane	145.7	240.34	398.3	5.579	162	*0.617
$C_3H_6Br_2$	1,2-Dibromopropane	217.9	415.0				1.9234

TABLE 2.1.1
SUMMARY OF PHYSICAL CONSTANTS OF IMPORTANT SUBSTANCES (continued)

Molecular formula	Name	T_m/ K	T_b/ K	T_c/ K	P_c/ MPa	V_c/ cm³ mol⁻¹	ρ_o/ g cm⁻³
$C_3H_6Cl_2$	1,2-Dichloropropane	172.6	369.52				1.1496
$C_3H_6Cl_2$	1,3-Dichloropropane	173.6	394.0				1.182
C_3H_6O	2-Propanone	178.3	329.20	508.1	4.700	209	0.7844
C_3H_6O	2-Propen-1-ol	144	370.23				0.850
C_3H_6O	Propanal	193	321.2	504.4	5.27	204	0.797
C_3H_6O	Methyloxirane	161.22	308	482.2	4.92	186	0.8209
C_3H_6O	Oxetane		320.7				0.8930
C_3H_6O	Methyl vinyl ether	151	278.2				
$C_3H_6O_2$	Ethyl methanoate	193.5	327.5	508.5	4.74	229	0.9153
$C_3H_6O_2$	Methyl ethanoate	175	330.02	506.55	4.75	228	0.9279
$C_3H_6O_2$	Propanoic acid	252.4	414.30	604	4.53	222	0.9881
$C_3H_6O_2$	1,3-Dioxolane	178	348.8				1.055
$C_3H_6O_3$	1,3,5-Trioxane	333.3	387.18				
C_3H_6S	Thiacyclobutane	199.9	368				1.015
C_3H_7Br	1-Bromopropane	163	344.1				1.3452
C_3H_7Br	2-Bromopropane	184	332.6				1.3060
C_3H_7Cl	1-Chloropropane	150.3	319.67	503	4.58		0.8830
C_3H_7Cl	2-Chloropropane	155.9	308.89				0.8563
C_3H_7F	1-Fluoropropane	114	275.6				
C_3H_7I	1-Iodopropane	171.8	375.7				1.740
C_3H_7I	2-Iodopropane	183	362.6				1.6946
C_3H_7N	Allylamine	184.9	326.4				0.755
C_3H_7N	Cyclopropylamine	237.7	323.6				0.820
C_3H_7NO	*N,N*-Dimethylmethanamide	212.72	426	649.6		262	0.9447
$C_3H_7NO_2$	1-Nitropropane	165	404.33				0.9961
$C_3H_7NO_2$	2-Nitropropane	181.83	393.40				0.9835
C_3H_8	Propane	85.46	231.08	369.82	4.250	203	*0.493
C_3H_8O	1-Propanol	147.0	370.3	536.78	5.168	219	0.7996
C_3H_8O	2-Propanol	183.6	355.4	508.3	4.762	220	0.7813
C_3H_8O	Ethyl methyl ether	160	280.35	437.8	4.40	221	*0.6922
$C_3H_8O_2$	1,2-Propanediol	213	460.7				1.0327
$C_3H_8O_2$	1,3-Propanediol	246.4	487.5				1.050
$C_3H_8O_2$	2-Methoxyethanol	188.0	397.8				0.9598
$C_3H_8O_2$	2,4-Dioxapentane	168.3	315				0.8538
$C_3H_8O_3$	1,2,3-Propanetriol	291.3	563.2				1.2567
C_3H_8S	2-Thiabutane	167.2	339.8	533	4.26		0.838
C_3H_8S	1-Propanethiol	159.8	340.9	536.6		286	0.836
C_3H_8S	2-Propanethiol	142.6	325.7				0.810
$C_3H_8S_2$	1,3-Propanedithiol	194	446.0				1.072
C_3H_9ClSi	Chlorotrimethylsilane	233	333	497.8	3.20	366	0.856
C_3H_9N	Propylamine	190	320.37	497.0	4.72		0.7121
C_3H_9N	Isopropylamine	178.01	304.91	471.8	4.54	221	0.6821
C_3H_9N	Trimethylamine	156.0	276.02	432.79	4.087	254	*0.627
C_3H_9NO	3-Amino-1-propanol	284	460.6				
C_3H_9NO	*DL*-1-Amino-2-propanol	274.89	432.61				
C_4F_8	Octafluorocyclobutane	232.96	267.16	388.46	2.784	324	*1.500
C_4F_{10}	Decafluorobutane	144.9	270.96	386.4	2.323	378	
C_4F_{10}	Perfluoroisobutane		273	395.4			
$C_4H_2O_3$	1-Oxa-3-cyclopenten-2,5-dione	325.9	475				1.5
$C_4H_4N_2$	Pyrazine	328	388				1.27
$C_4H_4N_2$	Pyridazine	265	481				1.102
$C_4H_4N_2$	Pyrimidine	295	396.9				
$C_4H_4N_2$	Butanedinitrile	327.6	539				1.023
C_4H_4O	Furan	187.5	304.5	490.2	5.50	218	0.9348
C_4H_4S	Thiophene	233.7	357.1	579.4	5.69	219	1.0588
C_4H_5N	Pyrrole	249.73	402.94	639.7	6.34	200	0.9656
C_4H_6	1,2-Butadiene	136.9	284.0				
C_4H_6	1,3-Butadiene	164.2	268.74	425	4.33	221	*0.6149
C_4H_6	1-Butyne	147.43	281.23	463.7			
C_4H_6	2-Butyne	240.79	300.12	488.7			0.688
C_4H_6O	Cyclobutanone	222.2	372				0.93
$C_4H_6O_2$	Oxolan-2-one	229.78	477				1.40

TABLE 2.1.1
SUMMARY OF PHYSICAL CONSTANTS OF IMPORTANT SUBSTANCES (continued)

Molecular formula	Name	T_m/ K	T_b/ K	T_c/ K	P_c/ MPa	V_c/ cm³ mol⁻¹	ρ_o/ g cm⁻³
$C_4H_6O_2$	*cis*-2-Butenoic acid	288	442				1.022
$C_4H_6O_2$	*trans*-2-Butenoic acid	345	457.8				
$C_4H_6O_2$	Vinyl ethanoate	179.9	345.7				0.927
$C_4H_6O_3$	Ethanoic anhydride	200	412.20	606	4.0		1.077
$C_4H_7ClO_2$	Ethyl chloroethanoate	252	417.4				1.151
C_4H_7N	Butanenitrile	161.2	390.77	585.4	3.88		0.7865
C_4H_8	1-Butene	87.80	267.0	419.57	4.023	240	*0.588
C_4H_8	*cis*-2-Butene	134.2	276.86	435.58	4.197	234	*0.616
C_4H_8	*trans*-2-Butene	167.62	274.03	428.63	3.985	238	*0.599
C_4H_8	2-Methylpropene	132.7	266.05	417.9	4.000	239	*0.589
C_4H_8	Cyclobutane	182.48	285.7	460.0	4.98	210	0.689
$C_4H_8Cl_2O$	Bis(2-chloroethyl) ether	221.2	451.6				1.2130
C_4H_8O	Butanal	174	348.0	537.2	4.32	258	0.798
C_4H_8O	2-Butanone	186.48	352.74	536.78	4.207	267	0.7994
C_4H_8O	Oxolane	164.76	339.1	540.1	5.19	224	0.8800
C_4H_8O	Ethyl vinyl ether	157.3	308.70	475	4.07		0.755
$C_4H_8O_2$	1,3-Dioxane	228	379.2				1.029
$C_4H_8O_2$	1,4-Dioxane	284.9	374.47	587	5.21	238	1.0286
$C_4H_8O_2$	Ethyl ethanoate	189.5	350.26	523.3	3.882	286	0.8945
$C_4H_8O_2$	Methyl propanoate	185.6	352.5	530.6	4.004	282	0.9090
$C_4H_8O_2$	Propyl methanoate	180.2	354.0	538.0	4.06	285	0.8996
$C_4H_8O_2$	Butanoic acid	267.4	436.90	624	4.03	290	0.9529
$C_4H_8O_2$	2-Methylpropanoic acid	227	427.85	605	3.7	292	0.9431
$C_4H_8O_2S$	Thiolane 1,1-dioxide	300.7	558				1.2660
C_4H_8S	Thiacyclopentane	176.99	394.1	632.0			0.9938
C_4H_9Br	1-Bromobutane	160.7	374.7				1.2686
C_4H_9Br	2-Bromobutane	160.4	364.5				1.2530
C_4H_9Cl	1-Chlorobutane	150.0	351.58				0.8810
C_4H_9Cl	2-Chlorobutane	141.8	341.4				0.866
C_4H_9I	1-Iodobutane	170	403.7				1.6072
C_4H_9N	Pyrrolidine	215.31	359.71	568.2	5.59	238	0.8538
C_4H_9NO	*N,N*-Dimethylethanamide	253	439.3				0.9365
C_4H_9NO	Morpholine	268.2	402.1				0.9959
C_4H_{10}	Butane	134.86	272.6	425.14	3.784	255	*0.573
C_4H_{10}	2-Methylpropane	134.8	261.42	407.85	3.630	257	*0.551
$C_4H_{10}O$	1-Butanol	183.3	390.88	563.05	4.423	275	0.8058
$C_4H_{10}O$	2-Butanol	158.4	372.66	536.05	4.179	269	0.8026
$C_4H_{10}O$	2-Methyl-2-propanol	298.5	355.5	506.21	3.973	275	0.7812
$C_4H_{10}O$	2-Methyl-1-propanol	165	381.04	547.78	4.300	273	0.7978
$C_4H_{10}O$	Diethyl ether	156.8	307.6	466.74	3.638	280	0.7078
$C_4H_{10}O$	Methyl propyl ether		312.2	476.25	3.801		0.727
$C_4H_{10}O$	Isopropyl methyl ether		303.92	464.48	3.762		
$C_4H_{10}O_2$	1,4-Butanediol	293.2	508				1.015
$C_4H_{10}O_2$	2-Ethoxyethanol	203	408.8				0.9247
$C_4H_{10}O_2$	2,5-Dioxahexane	215	358	536	3.87	271	0.859
$C_4H_{10}O_3$	3-Oxa-1,5-pentanediol	262.70	518.84				1.1150
$C_4H_{10}S$	1-Butanethiol	157.4	371.6	570.1		324	0.8367
$C_4H_{10}S$	3-Thiapentane	169.20	365.2	557	3.96	318	0.8312
$C_4H_{11}N$	Butylamine	224.05	350.15	531.9	4.25	277	0.7369
$C_4H_{11}N$	(2-Methylpropyl)amine	186.4	340.90	519	4.07	278	0.7297
$C_4H_{11}N$	2-Aminobutane	168.6	335.88	514.3	4.20	278	0.7200
$C_4H_{11}N$	(1,1-Dimethylethyl)amine	206.20	317.19	483.9	3.84	292	0.6901
$C_4H_{11}N$	Diethylamine	223.3	328.7	499.99	3.758		0.7016
$C_4H_{11}NO_2$	Bis(2-hydroxyethyl)amine	301	541.9				1.0899
$C_4H_{12}Si$	Tetramethylsilane	174.11	299.80	448.64	2.821	362	0.643
$C_4H_{12}Sn$	Tetramethylstannane	218.3	351	521.8	2.981		1.314
$C_4H_{13}N_3$	Bis(2-aminoethyl)amine	234	480				0.952
C_5F_{12}	Dodecafluoropentane	263	302.40	420.59	2.045	473	
$C_5H_2F_6O_2$	1,1,1,5,5,5-Hexafluoroacetylacetone		327.30	485.1	2.767		1.478
$C_5H_4O_2$	2-Furaldehyde	236.6	435.00	670	5.89		1.1554
C_5H_5N	Pyridine	231.49	388.38	620.0	5.67	243	0.9786
C_5H_6	1,3-Cyclopentadiene	188	314				0.7966

TABLE 2.1.1
SUMMARY OF PHYSICAL CONSTANTS OF IMPORTANT SUBSTANCES (continued)

Molecular formula	Name	T_m/ K	T_b/ K	T_c/ K	P_c/ MPa	V_c/ cm³ mol⁻¹	ρ_o/ g cm⁻³
$C_5H_6N_2$	2-Methylpyrazine	244	410	634.3	5.01	283	1.025
C_5H_6O	2-Methylfuran		338	527	4.72	247	0.909
$C_5H_6O_2$	2-Furylmethanol	242	444				1.1308
C_5H_8	*cis*-1,3-Pentadiene	132.3	317.2				0.688
C_5H_8	*trans*-1,3-Pentadiene	186	315.1				0.671
C_5H_8	1,4-Pentadiene	124.3	299				0.658
C_5H_8	Cyclopentene	138.0	317.39	507.0	4.802	245	0.768
C_5H_8	Spiropentane	138.5	312				0.723
C_5H_8	1-Pentyne	183	313.33	493.5			0.687
C_5H_8O	Cyclopentanone	221.8	403.9	624.5	4.60		0.944
C_5H_8O	Cyclopropyl methyl ketone	204.8	384.4				0.894
C_5H_8O	3,4-Dihydro-2-H-pyran		359	561.7	4.56	268	
$C_5H_8O_2$	Ethyl propenoate	201.9	372.5				0.917
$C_5H_8O_2$	Methyl 2-methylpropenoate	225	373.5				0.939
C_5H_9N	Pentanenitrile	176.9	414.5	610.3	3.58		0.797
C_5H_9NO	1-Methyl-2-pyrrolidinone	249	475	721.8		311	1.0251
C_5H_{10}	1-Pentene	107.9	303.11	464.78	3.527	293	0.6353
C_5H_{10}	*cis*-2-Pentene	121.7	310.08	475	3.69		0.6508
C_5H_{10}	2-Methyl-1-butene	135.58	304.3	470	3.8		0.6451
C_5H_{10}	2-Methyl-2-butene	139.39	311.72	481	3.91		0.6570
C_5H_{10}	Cyclopentane	179.3	322.4	511.7	4.508	260	0.7405
$C_5H_{10}O$	Cyclopentanol	254	413.57	619.5	4.90		0.943
$C_5H_{10}O$	2-Pentanone	196.2	375.41	561.08	3.694	301	0.8020
$C_5H_{10}O$	3-Pentanone	234	375.11	561.46	3.729	336	0.811
$C_5H_{10}O$	3-Methyl-2-butanone	181	367.48	553.4	3.85	310	0.801
$C_5H_{10}O$	Oxane	228	361	572.2	4.77	263	0.8772
$C_5H_{10}O$	2-Methyloxolane		351	537	3.76	267	0.851
$C_5H_{10}O$	Pentanal	181.6	376	566.1	3.97	313	0.805
$C_5H_{10}O$	Allyl ethyl ether		340.80	518			0.761
$C_5H_{10}O_2$	Butyl methanoate	181.6	379.3				0.884
$C_5H_{10}O_2$	Isobutyl methanoate	177.3	371.40	551	3.88	352	0.8732
$C_5H_{10}O_2$	Propyl ethanoate	180	374.69	549.7	3.36	345	0.8830
$C_5H_{10}O_2$	Isopropyl ethanoate	199.7	361.7	531			0.8711
$C_5H_{10}O_2$	Ethyl propanoate	199.2	372.2	546.0	3.362	345	0.8840
$C_5H_{10}O_2$	Methyl butanoate	187.3	375.9	554.4	3.47	340	0.8926
$C_5H_{10}O_2$	Methyl 2-methylpropanoate	188.4	365.6	540.8	3.43	339	0.8854
$C_5H_{10}O_2$	Pentanoic acid	239	458.70	643	3.58	340	0.9345
$C_5H_{10}O_2$	3-Methylbutanoic acid	243.8	449.70	629	3.40		0.926
$C_5H_{10}O_3$	2-Methoxyethyl ethanoate	203	416				1.0049
$C_5H_{11}Br$	1-Bromopentane	178	402.9				1.212
$C_5H_{11}Cl$	1-Chloropentane	174	380.9				0.877
$C_5H_{11}I$	1-Iodopentane	187.5	428				1.509
$C_5H_{11}N$	Piperidine	262.12	379.37	594.1	4.94	288	0.8578
C_5H_{12}	Pentane	143.4	309.21	469.69	3.364	311	0.6214
C_5H_{12}	2-Methylbutane	113.2	301.03	460.43	3.381	306	0.6142
C_5H_{12}	2,2-Dimethylpropane	256.5	282.63	433.8	3.197	307	*0.5852
$C_5H_{12}O$	Butyl methyl ether	157.6	343.31	512.78	3.371	329	0.741
$C_5H_{12}O$	Isobutyl methyl ether		331.7				0.727
$C_5H_{12}O$	*tert*-Butyl methyl ether	164.5	328.3	497.1	3.430		0.737
$C_5H_{12}O$	Ethyl propyl ether	145.6	336.36	500.23	3.370	339	0.735
$C_5H_{12}O$	1-Pentanol	194.2	411.13	588.15	3.909	326	0.8108
$C_5H_{12}O$	2-Pentanol	200	392.2	560.4			0.8054
$C_5H_{12}O$	3-Pentanol	204	389.40	559.6			0.8160
$C_5H_{12}O$	2-Methyl-2-butanol	264.3	375.3	545			0.8050
$C_5H_{12}O$	3-Methyl-1-butanol	155.9	404.2	579.4			0.8071
$C_5H_{12}O$	2,2-Dimethyl-1-propanol	325.6	386.6				0.808
$C_5H_{12}O_2$	3,5-Dioxaheptane	206.6	361				0.828
$C_5H_{12}O_4$	Pentaerythritol	533					0.886
$C_5H_{12}S$	1-Pentanethiol	197.4	399.8				0.846
$C_5H_{12}S$	3-Methyl-1-butanethiol		393	604			0.831
$C_5H_{13}N$	Pentylamine	218	377.4				0.750
C_6BrF_5	Bromopentafluorobenzene	242	410	601	3.0		1.948
C_6ClF_5	Chloropentafluorobenzene		391.11	570.81	3.238	376	1.567

TABLE 2.1.1
SUMMARY OF PHYSICAL CONSTANTS OF IMPORTANT SUBSTANCES (continued)

Molecular formula	Name	T_m/ K	T_b/ K	T_c/ K	P_c/ MPa	V_c/ cm^3 mol^{-1}	ρ_o/ g cm^{-3}
$C_6Cl_3F_3$	1,3,5-Trichloro-2,4,6-trifluorobenzene		471.52	684.8	3.27	448	
C_6Cl_6	Hexachlorobenzene	504.9	598				2.040
C_6F_6	Hexafluorobenzene	278.50	353.41	516.73	3.273	335	1.6144
C_6F_{10}	Decafluorocyclohexene		325.20	461.8			
C_6F_{12}	Dodecafluorocyclohexane	321.6	323.76	457.2	2.43		
C_6F_{14}	Tetradecafluorohexane	186.0	329.80	448.77	1.868	606	1.680
C_6HF_5	Pentafluorobenzene	225.8	358.89	530.97	3.531	324	1.514
C_6HF_5O	Pentafluorophenol	305.9	418.79	609	4.0	348	
C_6HF_{11}	Undecafluorocyclohexane		335.20	477.7			
$C_6H_2F_4$	1,2,3,4-Tetrafluorobenzene		367.51	550.83	3.791	313	1.423
$C_6H_2F_4$	1,2,3,5-Tetrafluorobenzene	225	357.61	535.25	3.747		1.319
$C_6H_2F_4$	1,2,4,5-Tetrafluorobenzene	277.6	363.41	543.35	3.801		1.424
$C_6H_3Cl_3$	1,2,4-Trichlorobenzene	290	486.6				1.459
$C_6H_3Cl_3$	1,3,5-Trichlorobenzene	336.6	481				1.66
$C_6H_3Cl_3O$	2,4,6-Trichlorophenol	342	519				1.490
$C_6H_4ClNO_2$	1-Chloro-2-nitrobenzene	305.6	518.6				
$C_6H_4ClNO_2$	1-Chloro-3-nitrobenzene	317.5	508.6				1.53
$C_6H_4ClNO_2$	1-Chloro-4-nitrobenzene	356.6	515				1.520
$C_6H_4Cl_2$	1,2-Dichlorobenzene	256.4	453.6				1.3003
$C_6H_4Cl_2$	1,3-Dichlorobenzene	248.3	446.14				1.2828
$C_6H_4Cl_2$	1,4-Dichlorobenzene	325.8	447.27				1.211
$C_6H_4FNO_2$	1-Fluoro-4-nitrobenzene	294	478				1.323
$C_6H_4F_2$	1,2-Difluorobenzene	239	367				1.150
$C_6H_4F_2$	1,3-Difluorobenzene	204.06	355.7				1.151
$C_6H_4F_2$	1,4-Difluorobenzene	260	362	556	4.40		1.163
C_6H_5Br	Bromobenzene	242.5	429.06	670	4.52	324	1.4882
C_6H_5Cl	Chlorobenzene	227.9	404.87	632.4	4.52	308	1.1009
C_6H_5ClO	2-Chlorophenol	282.9	448.0				1.257
C_6H_5ClO	3-Chlorophenol	305.7	487				1.218
C_6H_5ClO	4-Chlorophenol	315.8	493				1.40
C_6H_5F	Fluorobenzene	230.94	357.88	560.09	4.551	269	1.019
C_6H_5I	Iodobenzene	241.8	461.48	721	4.52	351	1.8229
$C_6H_5NO_2$	Nitrobenzene	278.8	483.9				1.1985
$C_6H_5NO_3$	2-Nitrophenol	317.9	489				
C_6H_6	Benzene	278.68	353.24	562.16	4.898	259	0.8736
C_6H_6ClN	2-Chloroaniline	259	481.99				1.206
C_6H_6ClN	3-Chloroaniline	262.7	503.6				1.210
C_6H_6ClN	4-Chloroaniline	345.6	505				1.41
$C_6H_6N_2O_2$	2-Nitroaniline	344.3	557				0.9015
$C_6H_6N_2O_2$	3-Nitroaniline	387					0.9011
$C_6H_6N_2O_2$	4-Nitroaniline	420	605				1.42
C_6H_6O	Phenol	314.0	455.02	694.2	6.13		1.132
$C_6H_6O_2$	1,4-Benzenediol	445.4	559				1.33
C_6H_6S	Benzenethiol	258.2	442.2				1.0727
C_6H_7N	Aniline	267.13	457.14	699	4.89	287	1.0175
C_6H_7N	2-Methylpyridine	206.47	402.53	621.0	4.60	292	0.9398
C_6H_7N	3-Methylpyridine	255.01	417.29	645.0	4.48	288	0.952
C_6H_7N	4-Methylpyridine	276.81	418.51	645.7	4.70	292	0.9502
$C_6H_8N_2$	Hexanedinitrile	274	568				0.9599
$C_6H_8N_2$	Phenylhydrazine	292.7	516.70				1.095
$C_6H_8N_2$	1,2-Phenylenediamine	375.6	530				
$C_6H_8N_2$	1,3-Phenylenediamine	336.6	558				1.23
$C_6H_8N_2$	1,4-Phenylenediamine	419	540				1.2
C_6H_{10}	Cyclohexene	169.6	356.13	560.48			0.806
C_6H_{10}	1,5-Hexadiene	132.4	332.60	507			0.688
C_6H_{10}	1-Hexyne	141.2	344.4				0.712
$C_6H_{10}O$	Cyclohexanone	242	428.80	653.0	4.0		0.9425
$C_6H_{10}O_2$	2-Oxepanone	255	488				1.064
$C_6H_{10}O_3$	Ethyl acetylethanoate	228	453.9				1.022
$C_6H_{10}O_4$	Hexanedioic acid	426.3	610.6				1.360
$C_6H_{10}S$	4-Thia-1,6-heptadiene	188	411.7	653			0.89
$C_6H_{11}Cl$	Chlorocyclohexane	229	415				0.995

TABLE 2.1.1
SUMMARY OF PHYSICAL CONSTANTS OF IMPORTANT SUBSTANCES (continued)

Molecular formula	Name	T_m/ K	T_b/ K	T_c/ K	P_c/ MPa	V_c/ cm^3 mol^{-1}	ρ_o/ g cm^{-3}
$C_6H_{11}F$	Fluorocyclohexane	286	374				0.923
$C_6H_{11}N$	Hexanenitrile	192.8	436.61	633.8	3.30		0.801
$C_6H_{11}NO$	Hexahydro-2-azepinone	342.4	543				
C_6H_{12}	Cyclohexane	279.7	353.88	553.5	4.07	308	0.7739
C_6H_{12}	Methylcyclopentane	130.6	344.96	532.73	3.784	319	0.745
C_6H_{12}	1-Hexene	133.39	336.63	504.1	3.206	348	0.6686
C_6H_{12}	*cis*-2-Hexene	132.0	341.9				0.6824
C_6H_{12}	2,3-Dimethyl-2-butene	198.5	346.4				0.7037
$C_6H_{12}O$	Cyclohexanol	298.61	434.25	650.0	4.26		0.9604
$C_6H_{12}O$	2-Hexanone	217.6	400.7	587.0	3.32		0.8070
$C_6H_{12}O$	3-Hexanone	217.6	396.6	582.82	3.320		0.808
$C_6H_{12}O$	4-Methyl-2-pentanone	189	389.6	571	3.27		0.7962
$C_6H_{12}O$	Butyl vinyl ether	181	367				0.785
$C_6H_{12}O$	Hexanal	217	404	591	3.46		0.829
$C_6H_{12}O_2$	Pentyl methanoate	199.6	403.60	576	3.46		0.881
$C_6H_{12}O_2$	3-Methylbutyl methanoate	179.6	396.70	578			0.873
$C_6H_{12}O_2$	Butyl ethanoate	195	399.2	579			0.8761
$C_6H_{12}O_2$	Isobutyl ethanoate	174.30	389.70	561	3.16		0.8695
$C_6H_{12}O_2$	Propyl propanoate	197.2	395.6	578			0.878
$C_6H_{12}O_2$	Ethyl butanoate	175	394.6	566	3.06	421	0.874
$C_6H_{12}O_2$	Ethyl 2-methylpropanoate	184.9	383.2	553	3.07	421	0.864
$C_6H_{12}O_2$	Methyl pentanoate		400.5	567	3.19		0.890
$C_6H_{12}O_2$	Hexanoic acid	270	478.17	662	3.20		0.923
$C_6H_{12}O_2$	4-Hydroxy-4-methyl-2-pentanone	229	441.0				0.9342
$C_6H_{12}O_3$	2,4,6-Trimethyl-1,3,5-trioxane	285.7	397.50	563			0.991
$C_6H_{12}O_3$	3-Oxapentyl ethanoate	211.4	429.5	607.3	3.166	443	0.9730
$C_6H_{13}Br$	1-Bromohexane	188.4	428.4				1.169
$C_6H_{13}Cl$	1-Chlorohexane	179	408				0.874
$C_6H_{13}I$	1-Iodohexane	198.9	454				1.433
$C_6H_{13}N$	Cyclohexylamine	255.4	407.11				0.8627
C_6H_{14}	Hexane	177.8	341.88	507.7	3.010	370	0.6548
C_6H_{14}	2-Methylpentane	119.4	333.41	497.7	3.031	367	0.650
C_6H_{14}	3-Methylpentane	110.2	336.42	504.5	3.126	367	0.6598
C_6H_{14}	2,2-Dimethylbutane	174	322.88	488.8	3.090	359	0.645
C_6H_{14}	2,3-Dimethylbutane	144.3	331.13	500.0	3.131	358	0.6570
$C_6H_{14}O$	Methyl pentyl ether		372	546.53	3.042	391	0.757
$C_6H_{14}O$	Dipropyl ether	147.0	363.23	530.6	3.028		0.7419
$C_6H_{14}O$	Diisopropyl ether	186.3	341.66	500.32	2.832	386	0.7207
$C_6H_{14}O$	1-Hexanol	228.5	430.7	610.7	3.47	381	0.8153
$C_6H_{14}O$	2-Hexanol		413.0	586.2			0.8105
$C_6H_{14}O$	4-Methyl-1-pentanol		425.0	603.5			0.8095
$C_6H_{14}O$	2-Methyl-2-pentanol	170	394.2	559.5			0.8095
$C_6H_{14}O$	4-Methyl-2-pentanol	183	404.80	574.4			0.8033
$C_6H_{14}O_2$	1,1-Diethoxyethane	173	375.40	527			0.822
$C_6H_{14}O_2$	3,6-Dioxaoctane		392.5				0.844
$C_6H_{14}O_2$	2-Butoxyethanol	198.3	441.5	633.9		424	0.8964
$C_6H_{14}O_2$	2-Methyl-2,4-pentanediol	223	470.2				0.9182
$C_6H_{14}O_3$	2,5,8-Trioxanonane	209	435				0.939
$C_6H_{14}O_4$	3,6-Dioxa-1,8-octanediol	266	558				1.12
$C_6H_{15}N$	Hexylamine	250.2	405.9				0.762
$C_6H_{15}N$	Triethylamine	158.4	362	535.6	3.032	389	0.7230
$C_6H_{15}N$	Dipropylamine	210	382.4	555.8	3.63		0.7329
$C_6H_{15}N$	Diisopropylamine	212	357.0	523.1	3.02		0.7100
$C_6H_{15}NO_3$	Tris(2-hydroxyethyl)amine	293.6	608.5				1.1205
$C_6H_{18}OSi_2$	Hexamethyldisiloxane	207	372				0.760
C_7F_8	Octafluorotoluene	207.5	377.73	534.47	2.705	428	
C_7F_{14}	Tetradecafluoromethylcyclohexane	228.4	349.50	485.91	2.019	570	1.788
C_7F_{16}	Hexadecafluoroheptane	195	355.66	474.8	1.62	664	1.725
C_7HF_{15}	1-Hydropentadecafluoroheptane		369.20	495.8			1.725
$C_7H_3F_5$	2,3,4,5,6-Pentafluorotoluene	243.3	390.6	566.52	3.126	384	1.439
C_7H_5ClO	Benzoyl chloride	272.1	470.3				1.2070
$C_7H_5Cl_3$	(Trichloromethyl)benzene	268	494				1.365

TABLE 2.1.1
SUMMARY OF PHYSICAL CONSTANTS OF IMPORTANT SUBSTANCES (continued)

Molecular formula	Name	T_m/ K	T_b/ K	T_c/ K	P_c/ MPa	V_c/ cm³ mol⁻¹	ρ_o/ g cm⁻³
C_7H_5N	Benzonitrile	260.40	464.30	699.4	4.21		1.00
$C_7H_6Cl_2$	2,4-Dichlorotoluene	259.6	474				1.241
C_7H_6O	Benzaldehyde	247	451.9	695	4.65		1.044
$C_7H_6O_2$	Benzoic acid	395.5	522.3				1.322
$C_7H_6O_2$	2-Hydroxybenzaldehyde	266	470				1.162
C_7H_7Br	4-Bromotoluene	301.6	457.4				
C_7H_7Cl	2-Chlorotoluene	237.5	432.3				1.077
C_7H_7Cl	3-Chlorotoluene	225.3	434.9				1.070
C_7H_7Cl	4-Chlorotoluene	280.6	435.14				1.064
C_7H_7Cl	Benzyl chloride	228	452				1.095
C_7H_7F	4-Fluorotoluene	217	389.7				0.992
C_7H_7NO	Benzamide	402.2	563				1.28
$C_7H_7NO_2$	2-Nitrotoluene	263	495				1.154
$C_7H_7NO_2$	3-Nitrotoluene	288.6	505				1.152
$C_7H_7NO_2$	4-Nitrotoluene	324.7	511.4				1.286
C_7H_8	Toluene	178.16	383.78	591.79	4.104	316	0.8622
C_7H_8O	2-Methylphenol	302.9	464.19	697.6	5.01		1.135
C_7H_8O	3-Methylphenol	284.9	475.42	705.8	4.56	309	1.0302
C_7H_8O	4-Methylphenol	307.8	475.13	704.6	5.15		1.154
C_7H_8O	Phenylmethanol	257.9	478.46	715	4.3		1.044
C_7H_8O	Anisole	235.6	426.8	645.6	4.25		0.9893
C_7H_9N	*N*-Methylaniline	216	469.40	701	5.20		0.9822
C_7H_9N	2-Methylaniline	256.80	473.49	707	4.37		0.9947
C_7H_9N	3-Methylaniline	241.90	476.52	707	4.28		0.9850
C_7H_9N	4-Methylaniline	316.90	473.7	706	4.58		0.957
C_7H_9N	2,3-Dimethylpyridine		434.41	655.4			0.932
C_7H_9N	2,4-Dimethylpyridine		431.6	647			0.926
C_7H_9N	2,5-Dimethylpyridine	257	430.16	644.2			0.925
C_7H_9N	2,6-Dimethylpyridine	267.0	417.2	623.8			0.9181
C_7H_9N	3,4-Dimethylpyridine		452.28	683.8			0.923
C_7H_9N	3,5-Dimethylpyridine	266.5	445.06	667.2			0.938
C_7H_9N	Benzylamine		458				0.977
$C_7H_{10}N_2$	2,4-Diaminotoluene	372	565				
$C_7H_{12}O$	4-Methylcyclohexanone	232.5	443				0.909
C_7H_{14}	Cycloheptane	265.12	391.95	604.2	3.81	353	0.8066
C_7H_{14}	Methylcyclohexane	146.5	374.08	572.2	3.471	368	0.7651
C_7H_{14}	Ethylcyclopentane	134.71	376.6	569.5	3.397	375	0.763
$C_7H_{14}O$	2-Heptanone	238	424.20	611.5	3.436		0.8111
$C_7H_{14}O$	Heptanal	229.8	425.9				0.8132
$C_7H_{14}O_2$	Pentyl ethanoate	202.3	422.40				0.8721
$C_7H_{14}O_2$	3-Methylbutyl ethanoate	194.6	415.70	599			0.8666
$C_7H_{14}O_2$	Methyl hexanoate	202	422.6				0.880
$C_7H_{14}O_2$	Ethyl pentanoate	181.9	419.2	570			0.873
$C_7H_{14}O_2$	Propyl butanoate	177.9	416.20	600			0.869
$C_7H_{14}O_2$	Propyl 2-methylpropanoate		408.60	589			
$C_7H_{14}O_2$	Isobutyl propanoate	201.7	410	592			0.87
$C_7H_{14}O_2$	Ethyl 3-methylbutanoate	173.8	408.20	588			0.861
$C_7H_{14}O_2$	Heptanoic acid	265.6	495.40	679	2.90		0.9140
$C_7H_{15}Br$	1-Bromoheptane	215	452				1.134
$C_7H_{15}Cl$	1-Chloroheptane	203.6	432				0.871
C_7H_{16}	Heptane	182.5	371.6	540.3	2.756	428	0.6795
C_7H_{16}	2-Methylhexane	154.9	363.19	530.4	2.734	421	0.6744
C_7H_{16}	3-Methylhexane	153.7	364.99	535.3	2.814	404	0.6829
C_7H_{16}	2,2-Dimethylpentane	149.3	352.3	520.5	2.773	416	0.6695
C_7H_{16}	2,3-Dimethylpentane		362.93	537.4	2.908	393	0.6909
C_7H_{16}	2,4-Dimethylpentane	153.2	353.64	519.8	2.737	418	0.6683
C_7H_{16}	3,3-Dimethylpentane	138.2	359.21	536.4	2.946	414	0.6891
C_7H_{16}	3-Ethylpentane	154.5	366.6	540.7	2.891	416	0.6940
C_7H_{16}	2,2,3-Trimethylbutane	248	354.01	531.2	2.954	398	0.6859
$C_7H_{16}O$	1-Heptanol	239	449.85	632.5	3.135	435	0.8187
$C_7H_{16}O$	(±)-2-Heptanol	243	432	611.4			0.8139
$C_7H_{16}O$	(±)-3-Heptanol	203	430	605.4			0.8170

TABLE 2.1.1
SUMMARY OF PHYSICAL CONSTANTS OF IMPORTANT SUBSTANCES (continued)

Molecular formula	Name	T_m/ K	T_b/ K	T_c/ K	P_c/ MPa	V_c/ cm³ mol⁻¹	ρ_o/ g cm⁻³
$C_7H_{16}O$	4-Heptanol	231.9	429	602.6			0.8149
$C_7H_{17}N$	Heptylamine	255	429				0.772
C_8F_{18}	Octadecafluorooctane		379.0	502	1.66		1.770
$C_8H_4O_3$	1,2-Benzenedicarboxylic anhydride	403.9	568				1.494
C_8H_7N	2-Tolunitrile	259.6	478				0.991
C_8H_7N	3-Tolunitrile	250	486				1.026
C_8H_7N	4-Tolunitrile	302.6	490.20	723			
C_8H_8	Vinylbenzene	242	418.29				0.9001
C_8H_8O	Methyl phenyl ketone	293	475	709.5		386	1.023
$C_8H_8O_2$	Methyl benzoate	258	472				1.0838
$C_8H_8O_3$	Methyl 2-hydroxybenzoate	265	496.10	709			1.179
C_8H_{10}	Ethylbenzene	178.20	409.34	617.2	3.600	374	0.8625
C_8H_{10}	1,2-Dimethylbenzene	247.9	417.6	630.3	3.730	369	0.8759
C_8H_{10}	1,3-Dimethylbenzene	225.3	412.27	617.05	3.535	376	0.8601
C_8H_{10}	1,4-Dimethylbenzene	286.3	411.52	616.2	3.511	379	0.8566
$C_8H_{10}O$	2-Ethylphenol	291	477.67	703.0			1.0146
$C_8H_{10}O$	3-Ethylphenol	269	491.57	718.8			1.023
$C_8H_{10}O$	4-Ethylphenol	318.23	491.13	716.4			0.991
$C_8H_{10}O$	2,3-Dimethylphenol	345.9	490.07	722.8			0.942
$C_8H_{10}O$	2,4-Dimethylphenol	297.68	484.13	707.6			0.980
$C_8H_{10}O$	2,5-Dimethylphenol	347.9	484.33	706.9			0.975
$C_8H_{10}O$	2,6-Dimethylphenol	318.8	474.22	701.0			0.891
$C_8H_{10}O$	3,4-Dimethylphenol	333.9	500	729.8			1.012
$C_8H_{10}O$	3,5-Dimethylphenol	336.7	494.89	715.6			0.984
$C_8H_{10}O$	Ethyl phenyl ether	243.63	442.96	647	3.42		0.9605
$C_8H_{11}N$	*N,N*-Dimethylaniline	275.60	467.20	687	3.63		0.9523
$C_8H_{11}N$	*N*-Ethylaniline	209.6	476.20	698			0.961
C_8H_{14}	1-Octyne	193.8	399.4				0.742
$C_8H_{14}O_4$	Ethyl succinate	252	490.90	663			1.035
$C_8H_{15}N$	Octanenitrile	227.5	478.40	674.4	2.85		0.810
C_8H_{16}	Cyclooctane	287.98	424.3	647.2	3.56	410	0.8320
C_8H_{16}	Ethylcyclohexane	161.8	405.0				0.784
C_8H_{16}	1,1-Dimethylcyclohexane	239.8	392.7				0.777
C_8H_{16}	*cis*-1,2-Dimethylcyclohexane	223.2	402.9				0.792
C_8H_{16}	*trans*-1,2-Dimethylcyclohexane	183	396.6				0.772
C_8H_{16}	*cis*-1,3-Dimethylcyclohexane	197.5	393.2				0.762
C_8H_{16}	*trans*-1,3-Dimethylcyclohexane	183.0	397.6				
C_8H_{16}	*cis*-1,4-Dimethylcyclohexane	185.7	397.5				0.779
C_8H_{16}	*trans*-1,4-Dimethylcyclohexane	236.2	392.5	587.7			
C_8H_{16}	1-Octene	171.4	394.44	566.7	2.675	464	0.711
$C_8H_{16}O_2$	Octanoic acid	289.4	513.1	695	2.64		0.9066
$C_8H_{16}O_2$	Propyl 3-methylbutanoate		429.10	609			0.857
$C_8H_{16}O_2$	Isobutyl butanoate		430.10	611			0.866
$C_8H_{16}O_2$	Isobutyl 2-methylpropanoate	192.4	421.80	602			0.850
$C_8H_{16}O_2$	3-Methylbutyl propanoate		433.40	611			0.865
$C_8H_{17}Br$	1-Bromooctane	218	473				1.108
$C_8H_{17}Cl$	1-Chlorooctane	215.3	454.6				0.869
C_8H_{18}	Octane	216.3	398.82	568.9	2.493	492	0.6986
C_8H_{18}	2-Methylheptane	164.16	390.81	559.7	2.484	488	0.6939
C_8H_{18}	3-Methylheptane	152.6	392.09	563.7	2.546	464	0.7018
C_8H_{18}	4-Methylheptane	152	390.87	561.8	2.542	476	0.7006
C_8H_{18}	2,2-Dimethylhexane	151.97	380.01	549.9	2.529	478	0.6911
C_8H_{18}	2,3-Dimethylhexane		388.77	563.5	2.628	468	0.7081
C_8H_{18}	2,4-Dimethylhexane		382.6	553.6	2.556	472	0.6962
C_8H_{18}	2,5-Dimethylhexane	182	382.27	550.1	2.487	482	0.6893
C_8H_{18}	3,3-Dimethylhexane	147.0	385.12	562.1	2.654	443	0.7060
C_8H_{18}	3,4-Dimethylhexane		390.88	568.9	2.692	466	0.715
C_8H_{18}	3-Ethylhexane		391.7	565.5	2.608	455	0.7095
C_8H_{18}	3-Ethyl-2-methylpentane	158.20	388.81	567.1	2.700	443	0.7152
C_8H_{18}	3-Ethyl-3-methylpentane	182.2	391.42	576.6	2.808	455	0.7235
C_8H_{18}	2,2,3-Trimethylpentane	160.89	383	563.5	2.730	436	0.7121
C_8H_{18}	2,2,4-Trimethylpentane	165.8	372.37	544.0	2.568	468	0.6878

TABLE 2.1.1
SUMMARY OF PHYSICAL CONSTANTS OF IMPORTANT SUBSTANCES (continued)

Molecular formula	Name	T_m/ K	T_b/ K	T_c/ K	P_c/ MPa	V_c/ $cm^3\ mol^{-1}$	ρ_o/ $g\ cm^{-3}$
C_8H_{18}	2,3,3-Trimethylpentane	172.22	387.9	573.6	2.820	455	0.7223
C_8H_{18}	2,3,4-Trimethylpentane	163.9	386.6	566.5	2.730	461	0.7150
C_8H_{18}	2,2,3,3-Tetramethylbutane	373.8	379.60	567.8	2.87	461	0.820
$C_8H_{18}O$	1-Octanol	257.6	468.31	652.5	2.86	490	0.8223
$C_8H_{18}O$	(±)-2-Octanol	241	453.03	638	2.9		0.8171
$C_8H_{18}O$	4-Methyl-3-heptanol	150	443	623.5			0.7940
$C_8H_{18}O$	5-Methyl-3-heptanol	181.9	445	621.2			0.8143
$C_8H_{18}O$	2-Ethyl-1-hexanol	203	457.77	641	2.8		0.8290
$C_8H_{18}O$	Dibutyl ether	177.9	413.43	584.1	3.01		0.7641
$C_8H_{18}O$	Di-*tert*-butyl ether		380.38	550			0.762
$C_8H_{18}S$	5-Thianonane	198.1	462				0.834
$C_8H_{19}N$	Dibutylamine	211	432.7	607.5	3.11		0.7571
$C_8H_{19}N$	Bis(2-methylpropyl)amine	199.6	412.84	584.4	3.20		
$C_8H_{20}Si$	Tetraethylsilane		427.90	603.7	2.602		0.762
$C_9H_6N_2O_2$	2,4-Diisocyanatotoluene	293.6	524				1.221
C_9H_7N	Quinoline	258.37	510.25	782	4.86	437	1.090
C_9H_7N	Isoquinoline	299.62	516.37	803	5.10	374	
C_9H_8	Indene	271.3	455				0.996
C_9H_{10}	Indan	221.7	451.12	684.9	3.95		0.959
C_9H_{12}	Isopropylbenzene	177.14	425.56	631.1	3.209		0.8574
C_9H_{12}	2-Ethyltoluene	192.3	438.3				0.876
C_9H_{12}	3-Ethyltoluene	177.6	434.4				0.860
C_9H_{12}	4-Ethyltoluene	210.8	435				0.857
C_9H_{12}	Propylbenzene	173.59	432.39	638.32	3.200	440	0.8579
C_9H_{12}	1,2,3-Trimethylbenzene	247.7	449.27	664.47	3.454		0.890
C_9H_{12}	1,2,4-Trimethylbenzene	229.3	442.53	649.17	3.232		0.8723
C_9H_{12}	1,3,5-Trimethylbenzene	228.4	437.89	637.25	3.127		0.8614
$C_9H_{13}N$	*N,N*-Dimethyl-2-toluidine	213	467.30	668	3.12		0.924
$C_9H_{14}O$	3,5,5-Trimethyl-2-cyclohexen-1-one	265.0	488.3				0.9196
$C_9H_{14}O_6$	1,2,3-Propanetriol tris(ethanoate)	195	532				1.154
C_9H_{18}	*trans*-1,3,5-Trimethylcyclohexane	165.7	413.70	602.2			0.776
$C_9H_{18}O$	5-Nonanone	267.2	461.60	640			0.818
$C_9H_{18}O$	Nonanal		464				0.824
$C_9H_{18}O$	2,6-Dimethyl-4-heptanone	231.6	442.5				0.802
$C_9H_{18}O_2$	Nonanoic acid	285.50	527.70	711	2.40		0.9013
$C_9H_{18}O_2$	3-Methylbutyl butanoate		452	619			0.860
$C_9H_{18}O_2$	Isobutyl 3-methylbutanoate		441.70	621			0.849
C_9H_{20}	Nonane	219.6	423.97	594.9	2.288		0.7138
C_9H_{20}	2-Methyloctane	192.78	416.43	587.0	2.310		0.705
C_9H_{20}	2,2-Dimethylheptane	160	405.8	576.8	2.350		0.707
C_9H_{20}	2,2,5-Trimethylhexane	167.39	397.24	568			0.7032
C_9H_{20}	2,2,3,3-Tetramethylpentane	263.3	413.44	607.7	2.741		0.753
C_9H_{20}	2,2,3,4-Tetramethylpentane	152.06	406.18	592.7	2.602		0.735
C_9H_{20}	3,3-Diethylpentane	240.0	419.3				0.750
C_9H_{20}	2,2,4,4-Tetramethylpentane	206.61	395.44	574.7	2.485		0.716
C_9H_{20}	2,3,3,4-Tetramethylpentane	171.0	414.72	607.7	2.716		0.751
$C_9H_{20}O$	1-Nonanol	268	486.65	671.5	2.63		0.8247
$C_{10}F_8$	Perfluoronaphthalene	360.6	482	673.1			
$C_{10}F_{18}$	Perfluorodecalin	263	415	566	1.52		
$C_{10}F_{22}$	Perfluorodecane		417.40	542	1.45		
$C_{10}H_7Cl$	1-Chloronaphthalene	270.6	532.4				1.189
$C_{10}H_8$	Azulene	372					1.175
$C_{10}H_8$	Naphthalene	353.3	491.14	748.4	4.051	413	1.1536
$C_{10}H_8O$	1-Naphthol	369	561				1.292
$C_{10}H_8O$	2-Naphthol	396	558				1.252
$C_{10}H_{10}O_4$	Dimethyl 1,2-benzenedicarboxylate	278.6	556.8				1.193
$C_{10}H_{12}$	1,2,3,4-Tetrahydronaphthalene	237.40	480.77	719.9		408	0.9671
$C_{10}H_{14}$	Butylbenzene	185.2	456.46	660.5	2.887	497	0.856
$C_{10}H_{14}$	Isobutylbenzene	221.70	445.94	650	3.05		0.8491
$C_{10}H_{14}$	2-Isopropyltoluene	201.6	451.2				0.8726
$C_{10}H_{14}$	3-Isopropyltoluene	209.4	448.2				0.857
$C_{10}H_{14}$	4-Isopropyltoluene	204.2	450.28	651	2.73		0.8525

TABLE 2.1.1
SUMMARY OF PHYSICAL CONSTANTS OF IMPORTANT SUBSTANCES (continued)

Molecular formula	Name	T_m/ K	T_b/ K	T_c/ K	P_c/ MPa	V_c/ cm³ mol⁻¹	ρ_o/ g cm⁻³
$C_{10}H_{14}$	1,2-Diethylbenzene	241.9	457				0.876
$C_{10}H_{14}$	1,3-Diethylbenzene	189.2	454.2				0.856
$C_{10}H_{14}$	1,4-Diethylbenzene	230.32	456.94	657.88	2.803		0.858
$C_{10}H_{14}$	1,2,4,5-Tetramethylbenzene	352.4	469.99	675	2.9		1.03
$C_{10}H_{14}O$	2-Isopropyl-5-methylphenol	324.6	505.70	698			0.970
$C_{10}H_{16}$	2,7,7-Trimethylbicyclo[3.1.1]hept-2-ene	209	429				0.855
$C_{10}H_{16}O$	1,7,7-Trimethylbicyclo [2.2.1]hepten-2-one	453	480				0.992
$C_{10}H_{18}$	*cis*-Bicyclo[4.4.0]decane	230.20	468.96	702.3	3.20		0.893
$C_{10}H_{18}$	*trans*-Bicyclo[4.4.0]decane	242.79	460.46	687.1			0.866
$C_{10}H_{18}$	1,3-Decadiene		442	615			
$C_{10}H_{20}$	1-Decene	206.8	443.7	616.4	2.218	584	0.737
$C_{10}H_{20}O$	(±)-2-Isopropyl-5-methylcyclo-hexanol	315	489.50	694			0.89
$C_{10}H_{20}O$	Decanal	268	481.6	674.2			
$C_{10}H_{20}O_2$	Decanoic acid	305.14	541.90	726	2.23		
$C_{10}H_{20}O_2$	Ethyl octanoate	230.0	481.70	659			0.860
$C_{10}H_{22}$	Decane	243.4	447.30	617.65	2.104		0.7264
$C_{10}H_{22}$	3,3,5-Trimethylheptane		428.8	609.7	2.317		0.721
$C_{10}H_{22}$	2,2,3,3-Tetramethylhexane	219	433.48	623.2	2.510		0.7609
$C_{10}H_{22}$	2,2,5,5-Tetramethylhexane	260.5	410.63	581.6	2.186		0.7148
$C_{10}H_{22}O$	1-Decanol	280.0	504.27	689	2.41		0.8263
$C_{10}H_{22}S$	Bis(3-methylbutyl) sulfide		484	664			0.828
$C_{11}H_{10}$	1-Methylnaphthalene	242.67	517.83	772			1.015
$C_{11}H_{10}$	2-Methylnaphthalene	307.5	514.2	761			1.001
$C_{11}H_{16}$	4-*tert*-Butyltoluene	221	463				0.857
$C_{11}H_{22}O_2$	Ethyl nonanoate	236.4	500.20	674			0.861
$C_{11}H_{24}$	Undecane	247.5	469.08	638.85	1.955		0.7365
$C_{12}H_8$	Acenaphthylene	365.6	553				0.899
$C_{12}H_9N$	Carbazole	519.3	627.85	901.8	3.93	502	1.1
$C_{12}H_{10}$	Acenaphthene	366.5	552				1.19
$C_{12}H_{10}$	Biphenyl	342	528.4	789	3.85	502	1.04
$C_{12}H_{10}N_2$	Azobenzene	340.2	566				1.2
$C_{12}H_{10}N_2O$	Azoxybenzene	309					1.159
$C_{12}H_{10}O$	Diphenyl ether	300.02	531.20	766.8			
$C_{12}H_{11}N$	Diphenylamine	326.13	575				1.16
$C_{12}H_{12}N_2$	4,4′-Biphenyldiamine	393					
$C_{12}H_{14}O_4$	Diethyl 1,2-benzenedicarboxylate	232.6	568				1.22
$C_{12}H_{18}$	Hexamethylbenzene	439.6	536.60	758			1.0630
$C_{12}H_{24}$	1-Dodecene	237.9	486.9	657.6	1.930		0.755
$C_{12}H_{24}O_2$	Dodecanoic acid	316.37	364.5				1.032
$C_{12}H_{26}$	Dodecane	263.5	489.47	658.65	1.830		0.7452
$C_{12}H_{26}O$	1-Dodecanol	297	537.7	720	2.08		0.8308
$C_{12}H_{27}N$	Tributylamine	203	489.6				0.7748
$C_{13}H_{10}$	Fluorene	387.9	568				1.202
$C_{13}H_{10}O$	Diphenyl ketone	321.03	578.5				1.103
$C_{13}H_{12}$	Diphenylmethane	298.39	537	770	2.86		1.001
$C_{13}H_{26}O_2$	Methyl dodecanoate	278.3	540	712			0.866
$C_{13}H_{28}$	Tridecane	267.76	508.62	676	1.71		0.7527
$C_{14}H_{10}$	Anthracene	488.1	614	869.3		554	1.28
$C_{14}H_{10}$	Phenanthrene	372.39	612.5	873		554	1.179
$C_{14}H_{10}O_2$	Benzil	368.01	620				1.22
$C_{14}H_{12}$	*cis*-1,2-Diphenylethene	268					
$C_{14}H_{12}$	*trans*-1,2-Diphenylethene	396	580				1.046
$C_{14}H_{14}$	1,1-Diphenylethane	255.2	545.7				0.995
$C_{14}H_{30}$	Tetradecane	279.01	526.66	693	1.61		0.7592
$C_{15}H_{32}$	Pentadecane	283.0	543.76	708	1.515		0.765
$C_{16}H_{10}$	Pyrene	424.3	677				1.268
$C_{16}H_{22}O_4$	Dibutyl 1,2-benzenedicarboxylate	238	613				1.0426
$C_{16}H_{34}$	Hexadecane	291.34	560.01	722	1.435		0.769
$C_{17}H_{36}$	Heptadecane	295	575.1	735	1.37		0.7746

TABLE 2.1.1
SUMMARY OF PHYSICAL CONSTANTS OF IMPORTANT SUBSTANCES (continued)

Molecular formula	Name	T_m / K	T_b / K	T_c / K	P_c / MPa	V_c / $cm^3\ mol^{-1}$	ρ_o / $g\ cm^{-3}$
$C_{17}H_{36}O$	1-Heptadecanol	326.9	606	780	1.50		
$C_{18}H_{12}$	Chrysene	531.3	721				1.268
$C_{18}H_{14}$	1,2-Diphenylbenzene	329.3	605	891.0	3.90	753	1.16
$C_{18}H_{14}$	1,3-Diphenylbenzene	360	636	924.9	3.51	768	1.195
$C_{18}H_{14}$	1,4-Diphenylbenzene	483.2	649	926.0	3.32	763	1.213
$C_{18}H_{34}O_2$	*cis*-9-Octadecenoic acid	286.5					0.898
$C_{18}H_{38}$	Octadecane	301.3	589.50	746	1.30		0.7791
$C_{18}H_{38}O$	1-Octadecanol	332.6		790	1.44		
$C_{19}H_{40}$	Nonadecane	305.2	603.0	758	1.23		0.776
$C_{20}H_{12}$	Perylene	546.6					1.35
$C_{20}H_{42}$	Icosane	309.9	617	769	1.16		0.7823
$C_{20}H_{42}O$	1-Icosanol	339.2		809	1.30		
$C_{24}H_{12}$	Coronene	710.4	798				1.38

TABLE 2.1.2
SOLID-LIQUID-GAS TRIPLE POINT OF PURE SUBSTANCES

A pure substance (single-component system) may appear in a single gas phase, usually in a single liquid phase (helium is an exception), and in one or several solid crystalline polymorphic phases. According to Gibbs' phase rule, a single-component system has two degrees of freedon. This means that a maximum of three phases may coexist in a state with fixed temperature T, pressure P, and molar volume V, a so-called *triple point* (solid-liquid-gas, solid-solid-liquid, solid-solid-gas, or solid-solid-solid).

This table gives the solid-liquid-gas triple point temperatures, T_t, and pressures, P_t (when available), of 91 substances.

The substances are arranged in a modified Hill order, with compounds not containing carbon preceding those that contain carbon.

Physical quantity	Symbol	SI unit
Triple point temperature	T_t	K
Triple point pressure	P_t	Pa

REFERENCES

1. Lide, D. R.,Editor, *CRC Handbook of Chemistry and Physics*, 74th Edition, CRC Press, Boca Raton, FL, 1993.
2. Riddick, J. A.; Bunger, W. B.; Sakano, T. K., *Organic Solvents, in: Techniques of Chemistry, Vol. II*, 4th Edition, John Wiley & Sons, New York, 1986.
3. *TRC Thermodynamic Tables - Hydrocarbons*, Thermodynamic Research Center: The Texas A&M University System, College Station, TX (Loose-leaf data sheets).
4. *TRC Thermodynamic Tables - Non-hydrocarbons*, Thermodynamic Research Center: The Texas A&M University System, College Station, TX (Loose-leaf data sheets).

Molecular formula	Name	T_t/K	P_t/Pa
Ar	Argon	83.806	68950
Br_2	Bromine	280.4	5879
Cl_2	Chlorine	172.17	1392
ClH	Hydrogen chloride	158.8	
H_2	Hydrogen	13.8	7042
H_2O	Water	273.16	611.73
^2H_2O	($^2H_2$)Dihydrogen oxide	276.97	661
H_2S	Dihydrogen sulfide	187.67	23180
H_3N	Ammonia	195.41	6077
Hg	Mercury	234.31	
Kr	Krypton	115.8	72920
Ne	Neon	24.556	50000
NO	Nitrogen oxide	109.54	21916
N_2	Nitrogen	63.15	12463
N_2O	Dinitrogen oxide	182.32	87852
O_2	Oxygen	54.36	146.33
O_2S	Sulfur dioxide	197.68	1674
O_3S	Sulfur trioxide	289.94	21130
S	Sulfur	388.33	
Xe	Xenon	161.4	81590
CH_3Cl	Chloromethane	175.43	
CH_3Cl_3Si	Trichloromethylsilane	197.37	
CH_4	Methane	90.694	11696
CH_4O	Methanol	175.59	
CO	Carbon oxide	68.15	15420
CO_2	Carbon dioxide	216.58	518500
C_2ClF_5	Chloropentafluoroethane	173.71	
$C_2Cl_4F_2$	Tetrachloro-1,2-difluoroethane	297.91	
$C_2H_2Cl_2F_2$	1,1-Dichloro-1,2-difluoroethane	162.99	

TABLE 2.1.2
SOLID-LIQUID-GAS TRIPLE POINT OF PURE SUBSTANCES (continued)

Molecular formula	Name	T_t/K	P_t/Pa
$C_2H_3Cl_3$	1,1,1-Trichloroethane	243.13	
$C_2H_4Br_2$	1,2-Dibromoethane	282.90	
$C_2H_4O_2$	Ethanoic acid	289.69	1277
C_2H_6OS	Dimethyl sulfoxide	291.67	
$C_2H_8N_2$	1,2-Ethanediamine	284.29	
C_3H_3N	Propenenitrile	189.63	
$C_3H_4O_3$	1,3-Dioxolan-2-one	309.49	
C_3H_6	Propene	87.89	0.000950
$C_3H_6O_2$	Propanoic acid	252.65	
C_3H_8	Propane	85.52	
C_3H_8O	1-Propanol	148.75	
C_3H_9N	Propylamine	188.36	
C_4F_8	Octafluorocyclobutane	232.95	
C_4H_5N	Pyrrole	249.74	
$C_4H_8O_2$	Butanoic acid	268.03	
C_4H_9N	Pyrrolidine	215.30	
C_4H_{10}	Butane	134.86	
C_4H_{10}	2-Methylpropane	113.55	
$C_4H_{10}O$	1-Butanol	184.51	
$C_4H_{10}O$	Diethyl ether	156.92	
$C_4H_{10}O$	2-Methyl-1-propanol	171.18	
$C_4H_{10}S$	1-Butanethiol	157.46	
$C_4H_{10}S$	3-Thiapentane	169.20	
$C_4H_{11}N$	*tert*-Butylamine	206.19	
C_5H_5N	Pyridine	231.48	
C_5H_6S	2-Methylthiophene	209.78	
C_5H_6S	3-Methylthiophene	204.18	
C_5H_{12}	2,2-Dimethylpropane	256.60	
C_5H_{12}	2-Methylbutane	113.25	
C_5H_{12}	Pentane	143.43	
$C_5H_{12}O$	1-Pentanol	195.56	
C_6F_6	Hexafluorobenzene	278.30	
$C_6H_4Cl_2$	1,4-Dichlorobenzene	326.16	
$C_6H_4F_2$	1,3-Difluorobenzene	204.03	
C_6H_5Br	Bromobenzene	242.40	
C_6H_6S	Benzenethiol	258.27	
C_6H_7N	2-Methylpyridine	206.45	
C_6H_7N	3-Methylpyridine	255.01	
C_6H_{12}	Cyclohexane	279.83	5400
$C_6H_{12}O_2$	Hexanoic acid	269.71	
C_6H_{14}	Hexane	177.83	
C_6H_{14}	3-Methylpentane	110.25	
$C_6H_{14}O$	Diisopropyl ether	187.77	
$C_6H_{14}O$	Dipropyl ether	158.36	
$C_7H_5F_3$	(Trifluoromethyl)benzene	244.14	
C_7H_5NO	Benzoxazole	302.51	
C_7H_5NS	Benzothiazole	275.65	
$C_7H_{14}O_2$	Heptanoic acid	265.83	
C_7H_{16}	Heptane	182.55	
C_8H_6S	2,3-Benzothiophene	304.48	
C_8H_{18}	Octane	216.37	
$C_9H_{18}O_2$	Nonanoic acid	285.53	
C_9H_{20}	Nonane	219.65	
$C_{10}H_8$	Naphthalene	353.42	999.6
$C_{10}H_{16}O$	(1*R*)-(+)-Camphor	453.30	51440
$C_{10}H_{22}$	Decane	243.50	
$C_{11}H_{10}$	1-Methylnaphthalene	242.70	

TABLE 2.1.2
SOLID-LIQUID-GAS TRIPLE POINT OF PURE SUBSTANCES (continued)

Molecular formula	Name	T_t/K	P_t/Pa
$C_{11}H_{22}O_2$	Undecanoic acid	301.63	
$C_{12}H_8O$	Dibenzofuran	355.31	
$C_{12}H_8S$	Dibenzothiophene	371.82	
$C_{12}H_{11}N$	2-Aminobiphenyl	322.28	
$C_{12}H_{24}$	1-Dodecene	237.93	

TABLE 2.1.3 (VLPT and VLTP)
VAPOR PRESSURE AND BOILING TEMPERATURE OF LIQUIDS

According to Gibbs' phase rule, a single-component two-phase system has a single degree of freedom. This means that the two phases may coexist along a phase-boundary curve (see Figure 1 in Section 2.1). The liquid-gas, solid-liquid, and solid-gas phase boundary curves in pressure-temperature (*P-T*) coordinates of a substance emerge from the solid-liquid-gas triple point (see Table 2.1.2) and are named, respectively, *vapor pressure curve*, *melting curve* (see Table 2.1.5), and *sublimation curve* (see Table 2.1.4). The vapor pressure curve ends at the gas-liquid critical point.

The vapor pressure P of a liquid is a monotonically increasing function of the temperature T. The inverse relationship represents the boiling temperature T of the substance as a function of the pressure P.

The temperature dependence of $P(T)$, or the pressure dependence of $T(P)$ can be fitted to various empirical equations. A simple equation is the so-called Antoine equation:

$$P(T)/\mathrm{Pa} = \exp\left\{A(1) - \frac{A(2)}{(T/\mathrm{K}) + A(3)}\right\} \tag{1}$$

or

$$T(P)/\mathrm{K} = \frac{A(2)}{A(1) - \ln(P/\mathrm{Pa})} - A(3) \tag{2}$$

The Antoine equation (1) or (2) is valid between the triple point temperature and a reduced temperature $T_r < 0.85$, where $T_r = T/T_c$, T_c being the critical temperature of the substance. It fails near T_c. This table gives the coefficients $A(i)$ and the valid temperature range for 393 common substances. The equation should not be used for extrapolation beyond this range. The program, Property Code VLPT, permits the calculation of $P(T)$ at any desired set of temperatures in the valid range. The program, Property Code VLTP, permits the calculation of $T(P)$ in the corresponding range of pressures. The program also permits estimation of an approximate value of the molar enthalpy of vaporization $\Delta_{vap}H$ (see Table 2.3.5) along the phase boundary curve, *via* the simplified Clausius-Clapeyron equation. If we assume that the volume change upon vaporization of a mole of liquid equals the molar volume V_{id} of an ideal gas, $V_{id} = R\,T/P$, where R = 8.3145 J $\mathrm{K^{-1}\,mol^{-1}}$ is the molar gas constant, then:

$$\Delta_{vap}H/\mathrm{J\,mol^{-1}} = \frac{R\,A(2)}{\{1 + A(3)/(T/\mathrm{K})\}^2} \tag{3}$$

The table gives also the recommended values of the normal boiling temperature T_b, i.e., T under the pressure $P = 101325$ Pa (1 atm), and of P at $T = 298.15$ K. The values generated by the program may differ slightly from these because of the fitting procedure.

The substances are arranged in a modified Hill order, with compounds not containing carbon preceding those that contain carbon.

Physical quantity	**Symbol**	**SI unit**
Boiling temperature	T	K
Vapor pressure	P	Pa
Molar enthalpy of vaporization	$\Delta_{vap}H$	$\mathrm{J\,mol^{-1}}$

REFERENCES

1. Stull, D. R., "Vapor Pressure of Pure Substances", *Ind. Eng. Chem.*, 39, 517-550, 1947 (a compilation of bibliographical data with tables of selected data for about 1200 organic compounds and about 300 inorganic compounds).
2. Ohe, S., *Computer Aided Data Book of Vapor Pressure*, Data Book Publishing Co., Tokyo, 1976, (a compilation of bibliographical data with graphical presentation and Antoine coefficients for 2000 substances).

TABLE 2.1.3 (VLPT and VLTP)
VAPOR PRESSURE AND BOILING TEMPERATURE OF LIQUIDS (continued)

3. Chao, J.; Lin, C. T.; Chung, T. H., "Vapor Pressure of Coal Chemicals", *J. Phys. Chem. Ref. Data*, 12, 1033-1063, 1983 (a compilation, evaluation and correlation of vapor pressure data for 324 coal chemicals).
4. Boublik, T.; Fried, V.; Hala, E., *The Vapour Pressures of Pure Substances*, Second Revised Edition, Elsevier, Amsterdam, 1984, (a compilation of primary literature data and Antoine coefficients).
5. Dykyj, J.; Repas, M., *The Vapor Pressure of Pure Organic Compounds* (in Slovak, with English Translation of the Guide to the Tables), Slovak Academy of Sciences, Bratislava, 1979 and 1984 (Supplement, with Svoboda, J.) (a compilation of bibliographical data and Antoine coefficients).
6. *TRC Thermodynamic Tables—Hydrocarbons*, Thermodynamic Research Center: The Texas A & M University System, College Station, TX (Loose-leaf data sheets) (recommended data and Antoine coefficients).
7. *TRC Thermodynamic Tables—Non-hydrocarbons*, Thermodynamic Research Center: The Texas A & M University System, College Station, TX (Loose-leaf data sheets) (recommended data and Antoine coefficients).
8. *TRC Databases for Chemistry and Engineering, Vapor Pressure*, Thermodynamic Research Center: The Texas A & M University System, College Station, TX (computerized database with Antoine coefficients for more than 5700 compounds).

TABLE 2.1.3 (VLPT and VLTP)
VAPOR PRESSURE AND BOILING TEMPERATURE OF LIQUIDS (continued)

SN	Molecular formula	Name	T_b/K	P(298.15 K)/kPa	$A(1)$	$A(2)$	$A(3)$	T/K-Range
1	Ar	Argon	87.280		0.22946E+02	0.10325E+04	0.31300E+01	73-90
2	$AsCl_3$	Arsenic trichloride	403.000	1.29	0.19368E+02	0.22988E+04	-0.10986E+03	293-404
3	BBr_3	Boron tribromide	364.000	8.93	0.20816E+02	0.29519E+04	-0.46242E+02	273-365
4	B_2F_4	Tetrafluorodiborane(4)	239.300		0.19841E+02	0.12485E+04	-0.89155E+02	219-242
5	B_2H_6	Diborane(6)	180.770		0.19616E+02	0.12173E+04	-0.30295E+02	118-181
6	$BiCl_3$	Bismuth trichloride	712.500		0.23606E+02	0.83210E+04	-0.23680E+02	423-715
7	Br_2	Bromine	331.900	28.72	0.20729E+02	0.25782E+04	-0.51770E+02	268-354
8	Br_3In	Indium tribromide	698.543		0.18403E+02	0.16566E+04	-0.45765E+03	583-700
9	Br_3Sb	Antimony tribromide	560.260		0.21017E+02	0.43697E+04	-0.99860E+02	399-561
10	ClH	Hydrogen chloride	188.150		0.21402E+02	0.17173E+04	-0.14268E+02	158-200
11	Cl_2	Chlorine	239.181		0.20906E+02	0.19972E+04	-0.26253E+02	206-270
12	Cl_4Si	Silicon tetrachloride	330.800		0.20863E+02	0.27269E+04	-0.38744E+02	275-332
13	FH	Hydrogen fluoride	292.650	122.90	0.22893E+02	0.36178E+04	0.25627E+02	273-303
14	F_2	Fluorine	84.950		0.20470E+02	0.70058E+03	-0.66160E+01	54-88
15	H_2	Hydrogen	20.384		0.18304E+02	0.15544E+03	0.25500E+01	13-23
16	H_2O	Water	373.150		0.23195E+02	0.38140E+04	-0.46290E+02	353-393
17	H_2S	Dihydrogen sulfide	212.840		0.21250E+02	0.18580E+04	-0.21761E+02	160-228
18	H_3N	Ammonia	239.720		0.21841E+02	0.21325E+04	-0.32980E+02	194-255
19	N_2	Nitrogen	77.351		0.19847E+02	0.58873E+03	-0.66000E+01	53-83
20	O_2	Oxygen	90.169		0.20300E+02	0.73455E+03	-0.64530E+01	65-98
21	O_2S	Sulfur dioxide	263.132		0.21661E+02	0.23024E+04	-0.35960E+02	195-280
22	$CBrClF_2$	Bromochlorodifluoro-methane	269.139		0.20646E+02	0.21561E+04	-0.32730E+02	178-283
23	CBrN	Cyanogen bromide	334.600		0.24514E+02	0.36043E+04	-0.55250E+02	273-313
24	CBr_2F_2	Dibromodifluoromethane	295.946	109.51	0.21367E+02	0.27236E+04	-0.19180E+02	247-300
25	CCl_3F	Trichlorofluoromethane	296.780	106.34	0.20761E+02	0.24030E+04	-0.36566E+02	243-334
26	CCl_4	Tetrachloromethane	349.788	15.28	0.20738E+02	0.27923E+04	-0.46667E+02	287-350
27	CFN	Cyanogen fluoride	227.160		0.20478E+02	0.15985E+04	-0.48592E+02	197-230
28	CF_4	Tetrafluoromethane	145.073		0.21066E+02	0.13013E+04	-0.86720E+01	115-146
29	$CHBr_3$	Tribromomethane	422.360	0.73	0.21193E+02	0.35505E+04	-0.55066E+02	298-423
30	$CHCl_2F$	Dichlorofluoromethane	282.050	186.57	0.16828E+02	0.65598E+03	-0.15833E+03	243-317
31	$CHCl_3$	Trichloromethane	334.328	26.24	0.20907E+02	0.26961E+04	-0.46926E+02	263-335
32	CHF_3	Trifluoromethane	191.000		0.21234E+02	0.16302E+04	-0.23084E+02	145-192
33	CHN	Hydrogen cyanide	298.811	98.84	0.22226E+02	0.30606E+04	-0.12773E+02	257-316
34	CH_2BrCl	Bromochloromethane	341.201	19.52	0.19804E+02	0.21474E+04	-0.81774E+02	289-342
35	CH_2Cl_2	Dichloromethane	312.790	58.24	0.21160E+02	0.25955E+04	-0.43386E+02	264-313
36	CH_2F_2	Difluoromethane	221.455		0.21330E+02	0.18919E+04	-0.28492E+02	191-241
37	CH_2O	Methanal	254.050		0.21370E+02	0.22041E+04	-0.30150E+02	185-270
38	CH_2O_2	Methanoic acid	373.710	5.75	0.22321E+02	0.38893E+04	-0.13424E+02	310-374
39	CH_3Br	Bromomethane	276.703	217.70	0.21240E+02	0.24175E+04	-0.27831E+02	203-277
40	CH_3Cl	Chloromethane	248.950	574.60	0.21012E+02	0.20774E+04	-0.29961E+02	183-249
41	CH_3Cl_3Si	Trichloromethylsilane	339.270	22.49	0.21184E+02	0.29450E+04	-0.34330E+02	287-340

TABLE 2.1.3 (VLPT and VLTP)
VAPOR PRESSURE AND BOILING TEMPERATURE OF LIQUIDS (continued)

SN	Molecular formula	Name	T_b/K	P(298.15 K)/kPa	A(1)	A(2)	A(3)	T/K-Range
42	CH_3F	Fluoromethane	194.700		0.18627E+02	0.93661E+03	-0.62797E+02	164-217
43	CH_3I	Iodomethane	315.579	53.97	0.20983E+02	0.26396E+04	-0.36473E+02	259-316
44	CH_3NO_2	Nitromethane	374.350	4.89	0.21658E+02	0.33313E+04	-0.45572E+02	330-410
45	CH_4	Methane	111.651		0.20982E+02	0.11075E+04	0.54690E+01	100-190
46	CH_4Cl_2Si	Dichloromethylsilane	314.039	57.24	0.21065E+02	0.26835E+04	-0.32709E+02	275-315
47	CH_4O	Methanol	337.696	16.94	0.23593E+02	0.36971E+04	-0.31317E+02	275-338
48	CH_4S	Methanethiol	279.100		0.21161E+02	0.23728E+04	-0.32820E+02	222-280
49	CH_5N	Methylamine	266.820	357.30	0.21808E+02	0.23370E+04	-0.39524E+02	190-267
50	CH_6N_2	Methylhydrazine	364.100	6.61	0.20075E+02	0.23356E+04	-0.91069E+02	275-300
51	CH_6OSi	Methoxysilane			0.19807E+02	0.16821E+04	-0.52535E+02	184-215
52	COS	Carbonyl sulfide	223.000		0.20798E+02	0.18536E+04	-0.23094E+02	162-224
53	CS_2	Carbon disulfide	319.375	48.17	0.20801E+02	0.26524E+04	-0.33402E+02	255-320
54	CSe_2	Carbon diselenide	398.000	2.38	0.20472E+02	0.31037E+04	-0.53675E+02	278-323
55	C_2ClF_3	Chlorotrifluoroethene	245.000		0.20764E+02	0.19543E+04	-0.33457E+02	206-262
56	C_2ClF_5	Chloropentafluoroethane	234.014		0.20636E+02	0.18520E+04	-0.30712E+02	178-235
57	$C_2Cl_3F_3$	1,1,1-Trichlorotrifluoro-ethane	318.950		0.15126E+02	0.47376E+03	-0.18879E+03	287-309
58	$C_2Cl_3F_3$	1,1,2-Trichlorotrifluoro-ethane	320.783		0.20043E+02	0.22952E+04	-0.51301E+02	273-321
59	C_2Cl_3N	Trichloroethanenitrile	358.649	9.88	0.21408E+02	0.31362E+04	-0.41274E+02	290-360
60	C_2Cl_4	Tetrachloroethene	394.222	2.42	0.20961E+02	0.31959E+04	-0.55510E+02	298-395
61	$C_2Cl_4F_2$	Tetrachloro-1,2-difluoro-ethane	366.000	7.29	0.29760E+02	0.98069E+04	0.17184E+03	283-367
62	C_2Cl_4O	Trichloroacetyl chloride	391.360	2.77	0.20991E+02	0.32024E+04	-0.53008E+02	298-392
63	C_2F_3N	Trifluoroethanenitrile	205.500		0.21273E+02	0.17726E+04	-0.23633E+02	142-206
64	C_2F_4	Tetrafluoroethene	197.500		0.20775E+02	0.15751E+04	-0.27195E+02	142-208
65	C_2F_6	Hexafluoroethane	194.900		0.20676E+02	0.15591E+04	-0.24509E+02	180-195
66	$C_2HBrClF_3$	Bromochloro-1,1,1-trifluoroethane	323.320	39.85	0.20857E+02	0.25833E+04	-0.46465E+02	222-328
67	C_2HCl_3	Trichloroethene	360.340	6.30	0.19796E+02	0.22898E+04	-0.83445E+02	291-361
68	$C_2HCl_3O_2$	Trichloroethanoic acid	470.750		0.21650E+02	0.36782E+04	-0.10744E+03	386-471
69	C_2HCl_5	Pentachloroethane	433.030	0.48	0.20552E+02	0.32690E+04	-0.70865E+02	298-435
70	$C_2HF_3O_2$	Trifluoroethanoic acid	344.930	15.09	0.22104E+02	0.32436E+04	-0.38289E+02	275-355
71	C_2H_2	Ethyne	188.430		0.21159E+02	0.16068E+04	-0.21688E+02	186-206
72	$C_2H_2Br_2$	*cis*-1,2-Dibromoethene		2.53	0.21305E+02	0.32203E+04	-0.59062E+02	298-351
73	$C_2H_2Br_2$	*trans*-1,2-Dibromoethene		4.43	0.15523E+02	0.93413E+03	-0.16706E+03	277-343
74	$C_2H_2Cl_2$	1,1-Dichloroethene	304.709	80.04	0.20951E+02	0.25338E+04	-0.35876E+02	245-306
75	$C_2H_2Cl_2$	*cis*-1,2-Dichloroethene	333.780	26.84	0.21053E+02	0.27742E+04	-0.42600E+02	274-357
76	$C_2H_2Cl_2$	*trans*-1,2-Dichloroethene	320.821	44.19	0.20933E+02	0.26308E+04	-0.41152E+02	235-358
77	$C_2H_2Cl_4$	1,1,1,2-Tetrachloroethane	403.300		0.20783E+02	0.31474E+04	-0.63289E+02	332-404
78	$C_2H_2Cl_4$	1,1,2,2-Tetrachloroethane	418.300	0.62	0.20276E+02	0.28572E+04	-0.91749E+02	298-419
79	C_2H_2O	Ketene	223.320		0.22622E+02	0.24684E+04	-0.85600E+00	185-224

80	C_2H_3Cl	Chloroethene	259.342	354.60	0.20747E+02	0.20786E+04	-0.33933E+02	208-260
81	C_2H_3ClO	Acetyl chloride	323.900	38.41	0.20848E+02	0.25467E+04	-0.50704E+02	267-324
82	$C_2H_3ClO_2$	Chloroethanoic acid	462.500		0.22272E+02	0.39690E+04	-0.93140E+02	377-463
83	$C_2H_3Cl_3$	1,1,2-Trichloroethane	387.000	3.00	0.20892E+02	0.30247E+04	-0.64044E+02	323-388
84	$C_2H_3F_3$	1,1,1-Trifluoroethane	225.600		0.20799E+02	0.18142E+04	-0.29956E+02	173-226
85	C_2H_3N	Ethanenitrile	354.750	11.86	0.21640E+02	0.32712E+04	-0.31298E+02	288-362
86	C_2H_3NO	Methyl isocyanate	312.200	57.69	0.21219E+02	0.24804E+04	-0.56310E+02	230-340
87	C_2H_4	Ethene	169.380		0.20430E+02	0.13454E+04	-0.18288E+02	149-188
88	$C_2H_4Br_2$	1,2-Dibromoethane	404.510	1.04	0.19878E+02	0.26626E+04	-0.85704E+02	298-405
89	$C_2H_4Cl_2$	1,1-Dichloroethane	330.450	30.49	0.20984E+02	0.27116E+04	-0.43741E+02	234-331
90	$C_2H_4Cl_2$	1,2-Dichloroethane	356.633	10.64	0.21108E+02	0.29434E+04	-0.49456E+02	242-372
91	$C_2H_4F_2$	1,1-Difluoroethane	248.200	601.00	0.23020E+02	0.32766E+04	0.39179E+02	298-387
92	C_2H_4O	Ethanal	293.600	120.02	0.23349E+02	0.36999E+04	0.19332E+02	273-307
93	C_2H_4O	Oxirane	283.600		0.21641E+02	0.25816E+04	-0.28378E+02	238-284
94	$C_2H_4O_2$	Ethanoic acid	391.035	2.08	0.21884E+02	0.35178E+04	-0.51408E+02	300-400
95	$C_2H_4O_2$	Methyl methanoate	304.900	81.93	0.11774E+02	0.36220E+01	-0.29028E+03	294-306
96	C_2H_5Cl	Chloroethane	285.420	159.88	0.20995E+02	0.23789E+04	-0.34194E+02	217-286
97	C_2H_5F	Fluoroethane	235.500		0.21518E+02	0.21824E+04	-0.17076E+02	169-255
98	C_2H_5NO	*N*-Methylmethanamide	472.640		0.22300E+02	0.43648E+04	-0.67519E+02	369-473
99	$C_2H_5NO_3$	Ethyl nitrate		8.56	0.21384E+02	0.30807E+04	-0.48281E+02	273-333
100	C_2H_6O	Dimethyl ether	248.310	593.30	0.20964E+02	0.20503E+04	-0.31066E+02	195-250
101	C_2H_6O	Ethanol	351.447	7.87	0.23584E+02	0.36745E+04	-0.46702E+02	293-366
102	C_2H_6OS	Dimethyl sulfoxide	462.200	0.08	0.22259E+02	0.44263E+04	-0.49797E+02	298-463
103	$C_2H_6O_2$	1,2-Ethanediol	470.490	0.01	0.23352E+02	0.46862E+04	-0.74214E+02	323-473
104	C_2H_6S	Ethanethiol	308.151	70.29	0.20900E+02	0.24970E+04	-0.41776E+02	273-339
105	C_2H_6S	2-Thiapropane	310.483	64.41	0.21364E+02	0.27559E+04	-0.30340E+02	250-311
106	C_2H_7N	Dimethylamine	280.030	196.80	0.21210E+02	0.22151E+04	-0.51298E+02	201-282
107	C_2H_7N	Ethylamine	289.800		0.18713E+02	0.12881E+04	-0.11057E+03	275-290
108	C_2H_7NO	2-Aminoethanol	444.100	0.05	0.21986E+02	0.35786E+04	-0.10198E+03	338-448
109	$C_2H_8N_2$	1,2-Dimethylhydrazine		9.32	0.17865E+02	0.14748E+04	-0.12913E+03	274-299
110	$C_2H_8N_2$	1,2-Ethanediamine	390.070	1.62	0.21443E+02	0.30959E+04	-0.77896E+02	298-392
111	C_3F_8	Octafluoropropane	236.400		0.20853E+02	0.19122E+04	-0.31396E+02	194-237
112	C_3H_3N	Propenenitrile	350.570	14.10	0.20818E+02	0.27822E+04	-0.51150E+02	255-385
113	C_3H_4	Propyne	250.163		0.21047E+02	0.20593E+04	-0.33870E+02	193-251
114	C_3H_4O	Propenal	326.256	36.17	0.20798E+02	0.26065E+04	-0.45150E+02	235-360
115	$C_3H_4O_2$	Propenoic acid	414.200		0.21462E+02	0.33192E+04	-0.80150E+02	315-450
116	$C_3H_4O_2$	Vinyl methanoate	319.610	40.58	0.21546E+02	0.25697E+04	-0.63150E+02	240-350
117	C_3H_5N	Propanenitrile	370.600	6.14	0.17129E+02	0.15598E+04	-0.11260E+03	188-299
118	C_3H_6	Cyclopropane	240.300		0.20769E+02	0.19777E+04	-0.26332E+02	183-241
119	C_3H_6	Propene	225.460	1156.60	0.20613E+02	0.18152E+04	-0.25705E+02	166-226
120	$C_3H_6Cl_2$	1,2-Dichloropropane	369.520	7.08	0.22411E+02	0.39541E+04	-0.62500E+01	288-373
121	C_3H_6O	Methyl vinyl ether	278.152	179.75	0.19353E+02	0.19802E+04	-0.25150E+02	190-315
122	C_3H_6O	Propanal	321.200	42.21	0.21388E+02	0.27881E+04	-0.38492E+02	286-322
123	C_3H_6O	2-Propanone	329.215	30.80	0.21299E+02	0.27958E+04	-0.43148E+02	259-351
124	C_3H_6O	2-Propen-1-ol	370.180	3.14	0.21799E+02	0.29282E+04	-0.85150E+02	286-400
125	$C_3H_6O_2$	1,3-Dioxolane	348.800	14.64	0.20991E+02	0.28250E+04	-0.50338E+02	281-355
126	$C_3H_6O_2$	Ethyl methanoate	327.460	32.27	0.21018E+02	0.25865E+04	-0.54973E+02	277-328
127	$C_3H_6O_2$	Methyl ethanoate	330.018	28.84	0.21309E+02	0.27392E+04	-0.50035E+02	259-351

TABLE 2.1.3 (VLPT and VLTP) VAPOR PRESSURE AND BOILING TEMPERATURE OF LIQUIDS (continued)

SN	Molecular formula	Name	T_b/K	P(298.15 K)/kPa	A(1)	A(2)	A(3)	T/K-Range
128	$C_3H_6O_2$	Propanoic acid	414.317	0.45	0.22277E+02	0.37192E+04	-0.68362E+02	328-437
129	$C_3H_6O_3$	1,3,5-Trioxane	387.180		0.22932E+02	0.41308E+04	-0.25032E+02	329-388
130	C_3H_7Br	1-Bromopropane	344.120	18.58	0.20946E+02	0.28380E+04	-0.42860E+02	273-345
131	C_3H_7Br	2-Bromopropane	332.560	28.93	0.20954E+02	0.27646E+04	-0.39335E+02	273-333
132	C_3H_7Cl	1-Chloropropane	319.670	45.85	0.20887E+02	0.25798E+04	-0.44089E+02	248-320
133	C_3H_7Cl	2-Chloropropane	308.890	68.90	0.22813E+02	0.36694E+04	0.16205E+02	273-309
134	C_3H_7NO	*N,N*-Dimethylmethanamide	426.000	0.44	0.20829E+02	0.32268E+04	-0.79286E+02	298-363
135	C_3H_7NO	*N*-Methylethanamide	479.000		0.10955E+02	0.28039E+03	-0.28247E+03	313-363
136	$C_3H_7NO_2$	1-Nitropropane	404.330		0.21276E+02	0.33764E+04	-0.58029E+02	330-405
137	$C_3H_7NO_2$	2-Nitropropane	393.400	2.27	0.22133E+02	0.38316E+04	-0.32155E+02	255-394
138	C_3H_8	Propane	231.080	948.10	0.20558E+02	0.18513E+04	-0.26110E+02	95-370
139	C_3H_8O	Ethyl methyl ether	280.250		0.11437E+02	0.11283E+00	-0.28152E+03	278-281
140	C_3H_8O	1-Propanol	370.301	2.80	0.22728E+02	0.33125E+04	-0.74598E+02	333-378
141	C_3H_8O	2-Propanol	355.392	5.78	0.22718E+02	0.31319E+04	-0.75557E+02	325-362
142	$C_3H_8O_2$	2-Methoxyethanol	397.800	1.31	0.22985E+02	0.41527E+04	-0.35395E+02	298-398
143	$C_3H_8O_3$	1,2,3-Propanetriol	563.200		0.19170E+02	0.24340E+04	-0.24213E+03	456-533
144	C_3H_8S	1-Propanethiol	340.867	20.56	0.20846E+02	0.27246E+04	-0.48532E+02	297-375
145	C_3H_8S	2-Propanethiol	325.706	36.91	0.20730E+02	0.25655E+04	-0.46959E+02	284-359
146	C_3H_8S	2-Thiabutane	339.801	21.33	0.20868E+02	0.27221E+04	-0.48409E+02	296-374
147	C_3H_9N	Isopropylamine	304.914	78.03	0.20758E+02	0.22695E+04	-0.59082E+02	277-334
148	C_3H_9N	Propylamine	320.377	42.11	0.20842E+02	0.24041E+04	-0.62310E+02	296-351
149	C_3H_9N	Trimethylamine	276.020	226.50	0.20943E+02	0.22305E+04	-0.39150E+02	200-300
150	C_4F_8	Octafluorocyclobutane	267.172		0.20585E+02	0.19859E+04	-0.47965E+02	241-273
151	C_4F_{10}	Decafluorobutane	270.960		0.21091E+02	0.22764E+04	-0.32955E+02	233-271
152	C_4H_4O	Furan	304.505	79.93	0.20954E+02	0.24427E+04	-0.45408E+02	276-334
153	C_4H_5N	Pyrrole	402.920	1.10	0.21693E+02	0.34598E+04	-0.62620E+02	339-439
154	C_4H_6	1,3-Butadiene	268.700		0.20663E+02	0.21460E+04	-0.33821E+02	197-271
155	C_4H_6	1-Butyne	281.200		0.21075E+02	0.23233E+04	-0.37903E+02	194-283
156	$C_4H_6O_2$	Vinyl ethanoate	345.700	15.39	0.21504E+02	0.29861E+04	-0.46419E+02	295-346
157	$C_4H_6O_3$	Ethanoic anhydride	413.200	0.68	0.21298E+02	0.33170E+04	-0.73774E+02	336-415
158	C_4H_7N	Butanenitrile	390.770	2.55	0.21308E+02	0.33438E+04	-0.48950E+02	333-400
159	C_4H_8	1-Butene	267.000		0.19914E+02	0.18578E+04	-0.45514E+02	196-269
160	C_4H_8	*cis*-2-Butene	276.860		0.20690E+02	0.22018E+04	-0.36600E+02	203-296
161	C_4H_8	*trans*-2-Butene	274.000		0.20639E+02	0.21817E+04	-0.34601E+02	205-283
162	C_4H_8	2-Methylpropene	266.050		0.19927E+02	0.18466E+04	-0.46256E+02	216-273
163	C_4H_8O	Butanal	348.000	15.70	0.19645E+02	0.21225E+04	-0.86586E+02	304-350
164	C_4H_8O	2-Butanone	352.730	12.08	0.21147E+02	0.28988E+04	-0.51425E+02	316-361
165	C_4H_8O	2-Methylpropanal	337.260	23.01	0.20408E+02	0.24288E+04	-0.63822E+02	285-338
166	C_4H_8O	Oxolane	339.113	21.62	0.21000E+02	0.27686E+04	-0.46883E+02	296-373
167	$C_4H_8O_2$	Butanoic acid	436.868	0.10	0.22280E+02	0.37824E+04	-0.85137E+02	350-452
168	$C_4H_8O_2$	1,4-Dioxane	374.470	4.95	0.22025E+02	0.35885E+04	-0.32691E+02	293-398
169	$C_4H_8O_2$	Ethyl ethanoate	350.262	12.60	0.21189E+02	0.28380E+04	-0.56563E+02	271-373

170	$C_4H_8O_2$	Isopropyl methanoate	342.570	10.50	0.13078E+02	0.11613E+03	-0.26774E+03	298-345
171	$C_4H_8O_2$	Methyl propanoate	352.520	11.51	0.20865E+02	0.26874E+04	-0.64744E+02	294-355
172	$C_4H_8O_2$	2-Methylpropanoic acid	427.850	0.19	0.16123E+02	0.88924E+03	-0.23442E+03	331-428
173	$C_4H_8O_2$	Propyl methanoate	353.950	10.92	0.20690E+02	0.26146E+04	-0.68632E+02	298-355
174	C_4H_8S	Thiacyclopentane	394.100	2.45	0.21003E+02	0.32266E+04	-0.53619E+02	344-433
175	C_4H_9Br	1-Bromobutane	374.750	5.26	0.17076E+02	0.15848E+04	-0.11188E+03	195-300
176	C_4H_9Br	2-Bromo-2-methylpropane	346.400	17.71	0.21922E+02	0.34912E+04	-0.10561E+02	273-347
177	C_4H_9Cl	1-Chlorobutane	351.580	13.68	0.20612E+02	0.26881E+04	-0.55725E+02	256-352
178	C_4H_9Cl	2-Chlorobutane	341.400	21.17	0.20602E+02	0.26680E+04	-0.47437E+02	273-343
179	C_4H_9Cl	2-Chloro-2-methylpropane	323.900	42.74	0.15681E+02	0.62208E+03	-0.17419E+03	295-335
180	C_4H_9N	Pyrrolidine	359.707	8.41	0.20837E+02	0.27172E+04	-0.67893E+02	298-394
181	C_4H_9NO	*N,N*-Dimethylethanamide	439.300	0.17	0.18981E+02	0.21399E+04	-0.15225E+03	297-440
182	C_4H_9NO	Morpholine	402.090	1.34	0.21370E+02	0.33333E+04	-0.63463E+02	346-403
183	C_4H_{10}	Butane	272.650	243.00	0.20442E+02	0.21494E+04	-0.31591E+02	205-280
184	C_4H_{10}	2-Methylpropane	261.420	349.06	0.20726E+02	0.21818E+04	-0.24280E+02	115-410
185	$C_4H_{10}O$	1-Butanol	390.879	0.91	0.21968E+02	0.30740E+04	-0.96496E+02	352-399
186	$C_4H_{10}O$	2-Butanol	372.664	2.32	0.21531E+02	0.26939E+04	-0.10341E+03	341-380
187	$C_4H_{10}O$	Diethyl ether	307.600	71.61	0.20837E+02	0.24439E+04	-0.45090E+02	250-328
188	$C_4H_{10}O$	2-Methyl-1-propanol	381.040	1.53	0.21856E+02	0.29249E+04	-0.97885E+02	343-389
189	$C_4H_{10}O$	2-Methyl-2-propanol	355.500	5.64	0.21535E+02	0.25448E+04	-0.10126E+03	330-363
190	$C_4H_{10}O_2$	2,5-Dioxahexane	358.000	9.93	0.20917E+02	0.28698E+04	-0.53150E+02	262-393
191	$C_4H_{10}O_2$	2-Ethoxyethanol	408.800	0.71	0.22941E+02	0.42021E+04	-0.40667E+02	336-410
192	$C_4H_{10}O_3$	3-Oxa-1,5-pentanediol	518.840	0.00	0.22340E+02	0.44196E+04	-0.11013E+03	412-519
193	$C_4H_{10}S$	3-Thiapentane	365.250	7.78	0.20844E+02	0.28950E+04	-0.54552E+02	298-395
194	$C_4H_{11}N$	Diethylamine	328.700	30.06	0.18234E+02	0.13365E+04	-0.12947E+03	298-334
195	$C_5H_4O_2$	2-Furaldehyde	435.000	0.33	0.20176E+02	0.28477E+04	-0.10578E+03	329-436
196	C_5H_5N	Pyridine	388.404	2.76	0.21105E+02	0.31634E+04	-0.58168E+02	297-426
197	C_5H_6O	2-Methylfuran	338.000		0.22607E+02	0.37787E+04	0.30140E+01	333-373
198	C_5H_6S	2-Methylthiophene	385.700	3.45	0.22654E+02	0.41819E+04	-0.99130E+01	298-388
199	C_5H_6S	3-Methylthiophene	388.600	2.98	0.22214E+02	0.38977E+04	-0.23919E+02	298-390
200	C_5H_8	2-Methyl-1,3-butadiene	307.183	73.38	0.20846E+02	0.25167E+04	-0.37148E+02	257-308
201	C_5H_8	*cis*-1,3-Pentadiene	317.200		0.20918E+02	0.26098E+04	-0.39850E+02	213-242
202	C_5H_8	*trans*-1,3-Pentadiene	315.100		0.20957E+02	0.26179E+04	-0.37734E+02	213-242
203	C_5H_8O	Cyclopentanone	403.900	1.54	0.21161E+02	0.33368E+04	-0.57582E+02	340-415
204	$C_5H_8O_2$	Methyl 2-methylpropenoate	373.500	5.10	0.21179E+02	0.30294E+04	-0.59660E+02	305-375
205	$C_5H_8O_2$	2,4-Pentanedione	411.080	1.02	0.18990E+02	0.22116E+04	-0.11475E+03	289-412
206	C_5H_9N	Pentanenitrile	414.500	0.94	0.21228E+02	0.34845E+04	-0.55364E+02	342-415
207	C_5H_{10}	2-Methyl-1-butene	304.300	81.37	0.20697E+02	0.24149E+04	-0.40969E+02	274-336
208	C_5H_{10}	3-Methyl-1-butene	293.209	120.27	0.20611E+02	0.23337E+04	-0.36326E+02	273-324
209	C_5H_{10}	2-Methyl-2-butene	311.718	62.14	0.20834E+02	0.25307E+04	-0.39836E+02	276-344
210	C_5H_{10}	1-Pentene	303.116	85.02	0.20658E+02	0.24068E+04	-0.39552E+02	286-304
211	C_5H_{10}	*cis*-2-Pentene	310.072	65.99	0.20724E+02	0.24620E+04	-0.42393E+02	275-342
212	C_5H_{10}	*trans*-2-Pentene	309.494	67.40	0.20794E+02	0.24957E+04	-0.40198E+02	274-341
213	$C_5H_{10}O$	3-Methyl-2-butanone	367.500		0.20930E+02	0.29141E+04	-0.57631E+02	329-376
214	$C_5H_{10}O$	2-Pentanone	375.408	4.72	0.21044E+02	0.30155E+04	-0.58585E+02	336-385
215	$C_5H_{10}O$	3-Pentanone	375.108	4.72	0.21059E+02	0.30116E+04	-0.59182E+02	330-384
216	$C_5H_{10}O_2$	Butyl methanoate	379.300	3.53	0.16352E+02	0.95465E+03	-0.18150E+03	295-385
217	$C_5H_{10}O_2$	Ethyl propanoate	372.250	4.97	0.21007E+02	0.29085E+04	-0.65478E+02	307-373

TABLE 2.1.3 (VLPT and VLTP) VAPOR PRESSURE AND BOILING TEMPERATURE OF LIQUIDS (continued)

SN	Molecular formula	Name	T_b/K	P(298.15 K)/kPa	A(1)	A(2)	A(3)	T/K-Range
218	$C_5H_{10}O_2$	3-Methylbutanoic acid	449.650		0.11462E+02	0.17223E+03	-0.31084E+03	360-377
219	$C_5H_{10}O_2$	Pentanoic acid	458.700	0.02	0.20343E+02	0.20232E+04	-0.17744E+03	354-390
220	$C_5H_{10}O_2$	Propyl ethanoate	374.688	4.49	0.21104E+02	0.29869E+04	-0.62849E+02	290-399
221	C_5H_{12}	2,2-Dimethylpropane	282.650	171.30	0.20077E+02	0.20323E+04	-0.44982E+02	257-300
222	C_5H_{12}	2-Methylbutane	301.030	91.73	0.20569E+02	0.23707E+04	-0.38856E+02	289-302
223	C_5H_{12}	Pentane	309.216	68.33	0.20701E+02	0.24665E+04	-0.40384E+02	269-341
224	$C_5H_{12}O$	Butyl methyl ether	343.290	18.51	0.20747E+02	0.26736E+04	-0.53329E+02	296-345
225	$C_5H_{12}O$	*tert*-Butyl methyl ether	328.300	33.56	0.20918E+02	0.26904E+04	-0.41851E+02	288-363
226	$C_5H_{12}O$	Ethyl propyl ether	336.300	24.15	0.20978E+02	0.27375E+04	-0.46667E+02	293-337
227	$C_5H_{12}O$	2-Methyl-1-butanol	401.825	0.42	0.21166E+02	0.27522E+04	-0.11632E+03	298-423
228	$C_5H_{12}O$	2-Methyl-2-butanol	375.200	2.23	0.21734E+02	0.28815E+04	-0.92928E+02	298-376
229	$C_5H_{12}O$	1-Pentanol	411.133	0.29	0.20729E+02	0.25418E+04	-0.13493E+03	410-514
230	$C_5H_{12}O$	2-Pentanol	392.200	0.78	0.21543E+02	0.28931E+04	-0.10338E+03	298-393
231	$C_5H_{12}O_3$	3,6-Dioxa-1-heptanol	467.300	0.02	0.22129E+02	0.41431E+04	-0.76551E+02	385-468
232	$C_5H_{12}S$	Butyl methyl sulfide	396.572		0.20887E+02	0.31408E+04	-0.61050E+02	347-436
233	$C_5H_{12}S$	1-Pentanethiol	399.785		0.20859E+02	0.31548E+04	-0.61760E+02	349-439
234	C_6ClF_5	Chloropentafluorobenzene	391.100	2.37	0.21168E+02	0.31981E+04	-0.59431E+02	298-417
235	$C_6Cl_3F_3$	1,3,5-Trichloro-2,4,6-trifluorobenzene	471.500		0.21343E+02	0.39451E+04	-0.69651E+02	364-496
236	C_6F_6	Hexafluorobenzene	353.405	11.27	0.21069E+02	0.28183E+04	-0.58066E+02	290-377
237	$C_6H_2F_4$	1,2,3,4-Tetrafluorobenzene	367.500	6.51	0.21090E+02	0.29738E+04	-0.56570E+02	298-392
238	$C_6H_2F_4$	1,2,3,5-Tetrafluorobenzene	357.608	9.80	0.21069E+02	0.28853E+04	-0.55247E+02	287-381
239	$C_6H_2F_4$	1,2,4,5-Tetrafluorobenzene	363.411	7.55	0.21125E+02	0.29425E+04	-0.56861E+02	293-387
240	$C_6H_4Cl_2$	1,2-Dichlorobenzene	453.630	0.17	0.21029E+02	0.36819E+04	-0.66186E+02	360-454
241	$C_6H_4Cl_2$	1,3-Dichlorobenzene	446.143	0.25	0.21117E+02	0.37097E+04	-0.59333E+02	364-447
242	$C_6H_4Cl_2$	1,4-Dichlorobenzene	447.270	0.24	0.21045E+02	0.36548E+04	-0.63321E+02	368-448
243	C_6H_5Br	Bromobenzene	429.058	0.56	0.20710E+02	0.33232E+04	-0.67210E+02	298-430
244	C_6H_5Cl	Chlorobenzene	404.837	1.57	0.20964E+02	0.32969E+04	-0.55515E+02	335-405
245	C_6H_5F	Fluorobenzene	357.884	10.39	0.21459E+02	0.31814E+04	-0.37602E+02	255-358
246	C_6H_5I	Iodobenzene	461.480	0.13	0.21088E+02	0.38136E+04	-0.62654E+02	298-462
247	$C_6H_5NO_2$	Nitrobenzene	483.950	0.04	0.21259E+02	0.40106E+04	-0.71893E+02	407-484
248	C_6H_6	Benzene	353.244	12.69	0.20767E+02	0.27738E+04	-0.53081E+02	294-378
249	C_6H_6O	Phenol	454.989	0.06	0.22337E+02	0.41306E+04	-0.72932E+02	344-455
250	C_6H_7N	Aniline	457.143	0.09	0.21515E+02	0.38266E+04	-0.74048E+02	298-458
251	C_6H_7N	2-Methylpyridine	402.556	1.50	0.21103E+02	0.32725E+04	-0.60864E+02	298-403
252	C_6H_7N	3-Methylpyridine	417.290	0.80	0.21095E+02	0.33909E+04	-0.62936E+02	298-418
253	C_6H_7N	4-Methylpyridine	418.506	0.76	0.21117E+02	0.34159E+04	-0.62336E+02	298-420
254	C_6H_{10}	Cyclohexene	356.130	11.84	0.20715E+02	0.28131E+04	-0.50001E+02	298-364
255	C_6H_{10}	3-Hexyne	354.580		0.18649E+02	0.20680E+04	-0.75084E+02	253-297
256	$C_6H_{10}O$	Cyclohexanone	428.800	640.00	0.20951E+02	0.34404E+04	-0.63751E+02	363-439
257	$C_6H_{11}N$	Hexanenitrile	436.611	0.36	0.21174E+02	0.35769E+04	-0.65854E+02	371-442

258	C_6H_{12}	Cyclohexane	353.880	13.04	0.20682E+02	0.27890E+04	-0.49281E+02	298-358
259	C_6H_{12}	2,3-Dimethyl-2-butene	346.380	16.69	0.20843E+02	0.27706E+04	-0.48998E+02	289-347
260	C_6H_{12}	1-Hexene	336.633	24.79	0.20710E+02	0.26594E+04	-0.47068E+02	289-337
261	C_6H_{12}	Methylcyclopentane	344.962	18.33	0.20704E+02	0.27362E+04	-0.46843E+02	288-346
262	$C_6H_{12}O$	Cyclohexanol	434.250		0.19217E+02	0.20604E+04	-0.16635E+03	367-435
263	$C_6H_{12}O$	3,3-Dimethyl-2-butanone	379.263	4.27	0.20879E+02	0.29986E+04	-0.58642E+02	289-405
264	$C_6H_{12}O$	2-Hexanone	400.731	1.54	0.21074E+02	0.32130E+04	-0.64204E+02	298-428
265	$C_6H_{12}O$	3-Hexanone	396.653		0.21012E+02	0.31455E+04	-0.65057E+02	349-406
266	$C_6H_{12}O_2$	Butyl ethanoate	399.211	1.66	0.21085E+02	0.32978E+04	-0.54214E+02	333-400
267	$C_6H_{12}O_2$	Hexanoic acid	478.170	0.01	0.22445E+02	0.40979E+04	-0.10279E+03	386-442
268	$C_6H_{13}N$	Cyclohexylamine	407.110	1.20	0.20431E+02	0.29095E+04	-0.80369E+02	334-408
269	C_6H_{14}	2,2-Dimethylbutane	322.891	42.54	0.20463E+02	0.24984E+04	-0.43333E+02	288-324
270	C_6H_{14}	2,3-Dimethylbutane	331.134	31.28	0.20594E+02	0.26068E+04	-0.43656E+02	287-332
271	C_6H_{14}	Hexane	341.886	20.17	0.20749E+02	0.27081E+04	-0.48251E+02	286-343
272	C_6H_{14}	2-Methylpentane	333.421	28.24	0.20711E+02	0.26531E+04	-0.44579E+02	286-334
273	C_6H_{14}	3-Methylpentane	336.432	25.31	0.20703E+02	0.26758E+04	-0.44870E+02	288-337
274	$C_6H_{14}O$	Butyl ethyl ether	365.390	7.46	0.20875E+02	0.28880E+04	-0.56482E+02	311-366
275	$C_6H_{14}O$	Diisopropyl ether	341.660	19.84	0.20651E+02	0.26190E+04	-0.54634E+02	297-342
276	$C_6H_{14}O$	Dipropyl ether	363.230	8.35	0.20878E+02	0.28884E+04	-0.54369E+02	298-365
277	$C_6H_{14}O$	1-Hexanol	430.650	0.11	0.21183E+02	0.30071E+04	-0.11925E+03	325-431
278	$C_6H_{14}O$	2-Hexanol	413.050	0.44	0.21800E+02	0.32656E+04	-0.95205E+02	313-415
279	$C_6H_{14}O$	3-Hexanol	408.600	0.96	0.22825E+02	0.39904E+04	-0.55439E+02	313-411
280	$C_6H_{15}N$	Triethylamine	362.020	7.70	0.18382E+02	0.16027E+04	-0.12825E+03	298-368
281	C_7F_{14}	Tetradecafluoromethyl-cyclohexane	349.450	14.10	0.20605E+02	0.26101E+04	-0.61952E+02	298-384
282	C_7F_{16}	Hexadecafluoroheptane	355.634	10.18	0.20868E+02	0.27202E+04	-0.64466E+02	271-379
283	C_7H_6O	Benzaldehyde	451.899	0.17	0.21213E+02	0.37271E+04	-0.67156E+02	311-481
284	C_7H_7Cl	2-Chlorotoluene	432.300	0.48	0.20936E+02	0.34721E+04	-0.63304E+02	346-433
285	C_7H_7Cl	4-Chlorotoluene	435.140		0.20930E+02	0.34812E+04	-0.64951E+02	338-436
286	C_7H_8	Toluene	383.780	3.80	0.21600E+02	0.36266E+04	-0.23778E+02	360-580
287	C_7H_8O	Anisole	426.750	0.47	0.21135E+02	0.34330E+04	-0.69475E+02	383-437
288	C_7H_8O	2-Methylphenol	464.154	0.04	0.20333E+02	0.29933E+04	-0.12426E+03	415-465
289	C_7H_8O	3-Methylphenol	475.382	0.02	0.22159E+02	0.42606E+04	-0.74688E+02	423-476
290	C_7H_8O	4-Methylphenol	475.090	0.02	0.21752E+02	0.39449E+04	-0.89304E+02	419-476
291	C_7H_8O	Phenylmethanol	478.460	0.02	0.20995E+02	0.34048E+04	-0.11890E+03	404-507
292	C_7H_9N	2,4-Dimethylpyridine	431.553	0.40	0.21102E+02	0.34858E+04	-0.67517E+02	298-450
293	C_7H_9N	2,5-Dimethylpyridine	430.150		0.21200E+02	0.35470E+04	-0.63500E+02	358-431
294	C_7H_9N	2,6-Dimethylpyridine	417.191	0.75	0.21145E+02	0.33873E+04	-0.65032E+02	298-418
295	C_7H_9N	3,5-Dimethylpyridine	445.060		0.21772E+02	0.41028E+04	-0.44648E+02	436-446
296	C_7H_9N	2-Methylaniline	473.470	0.04	0.21184E+02	0.37375E+04	-0.86509E+02	391-474
297	C_7H_9N	3-Methylaniline	476.500	0.04	0.21217E+02	0.37500E+04	-0.89545E+02	395-477
298	C_7H_9N	4-Methylaniline	473.650		0.21125E+02	0.36496E+04	-0.93440E+02	393-475
299	C_7H_9N	*N*-Methylaniline	469.400		0.21328E+02	0.38554E+04	-0.76077E+02	323-473
300	C_7H_{12}	1-Heptyne	373.130		0.21351E+02	0.30267E+04	-0.65053E+02	336-374
301	C_7H_{12}	2-Heptyne	385.130		0.21548E+02	0.32876E+04	-0.57100E+02	346-386
302	C_7H_{14}	Cycloheptane	391.950	2.90	0.20682E+02	0.30714E+04	-0.56493E+02	298-395
303	C_7H_{14}	1-Heptene	366.790	7.52	0.20818E+02	0.29162E+04	-0.52948E+02	295-368
304	C_7H_{14}	*cis*-2-Heptene	371.560		0.20913E+02	0.29916E+04	-0.52856E+02	332-372

TABLE 2.1.3 (VLPT and VLTP)
VAPOR PRESSURE AND BOILING TEMPERATURE OF LIQUIDS (continued)

SN	Molecular formula	Name	T_b/K	P(298.15 K)/kPa	A(1)	A(2)	A(3)	T/K-Range
305	C_7H_{14}	*trans*-2-Heptene	371.093		0.20882E+02	0.29709E+04	-0.53542E+02	331-372
306	C_7H_{14}	*cis*-3-Heptene	368.900		0.20889E+02	0.29636E+04	-0.52377E+02	329-370
307	C_7H_{14}	*trans*-3-Heptene	368.820		0.20918E+02	0.29775E+04	-0.51795E+02	329-370
308	C_7H_{14}	Methylcyclohexane	374.084	6.18	0.20617E+02	0.29334E+04	-0.51395E+02	298-375
309	$C_7H_{14}O$	2,4-Dimethyl-3-pentanone	398.400	1.33	0.20954E+02	0.31938E+04	-0.59630E+02	321-400
310	$C_7H_{14}O$	2-Heptanone	424.201	0.51	0.21060E+02	0.33624E+04	-0.71514E+02	328-452
311	C_7H_{16}	2,2-Dimethylpentane	352.341	14.03	0.20581E+02	0.27380E+04	-0.49952E+02	288-353
312	C_7H_{16}	2,3-Dimethylpentane	362.931	9.18	0.20684E+02	0.28561E+04	-0.51056E+02	291-364
313	C_7H_{16}	2,4-Dimethylpentane	353.650	13.12	0.20624E+02	0.27519E+04	-0.51158E+02	287-354
314	C_7H_{16}	3,3-Dimethylpentane	359.210	11.04	0.20616E+02	0.28313E+04	-0.47723E+02	286-360
315	C_7H_{16}	3-Ethylpentane	366.621	7.74	0.20732E+02	0.28865E+04	-0.53069E+02	294-367
316	C_7H_{16}	Heptane	371.574	6.09	0.20785E+02	0.29187E+04	-0.56357E+02	298-372
317	C_7H_{16}	2-Methylhexane	363.202	8.78	0.20730E+02	0.28526E+04	-0.53283E+02	292-364
318	C_7H_{16}	3-Methylhexane	365.000	8.21	0.20711E+02	0.28587E+04	-0.53775E+02	293-366
319	$C_7H_{16}O$	1-Heptanol	449.850		0.20766E+02	0.29445E+04	-0.13075E+03	333-426
320	C_8H_8	Vinylbenzene	418.290	0.84	0.21146E+02	0.34589E+04	-0.58730E+02	300-420
321	C_8H_{10}	1,2-Dimethylbenzene	417.560	0.88	0.21016E+02	0.34003E+04	-0.59239E+02	336-418
322	C_8H_{10}	1,3-Dimethylbenzene	412.270	1.10	0.21027E+02	0.33636E+04	-0.58255E+02	332-413
323	C_8H_{10}	1,4-Dimethylbenzene	411.509	1.20	0.20985E+02	0.33438E+04	-0.57992E+02	331-412
324	C_8H_{10}	Ethylbenzene	409.343	1.30	0.20904E+02	0.32735E+04	-0.60274E+02	330-410
325	$C_8H_{10}O$	2,3-Dimethylphenol	490.020		0.21043E+02	0.36570E+04	-0.10577E+03	422-491
326	$C_8H_{10}O$	2,4-Dimethylphenol	484.080	0.02	0.21141E+02	0.36573E+04	-0.10371E+03	417-485
327	$C_8H_{10}O$	2,5-Dimethylphenol	484.281	0.02	0.21028E+02	0.35931E+04	-0.10613E+03	417-485
328	$C_8H_{10}O$	2,6-Dimethylphenol	474.176		0.21174E+02	0.37496E+04	-0.85537E+02	418-477
329	$C_8H_{10}O$	3,4-Dimethylphenol	500.000		0.21199E+02	0.37357E+04	-0.11378E+03	445-502
330	$C_8H_{10}O$	3,5-Dimethylphenol	494.890		0.21315E+02	0.37790E+04	-0.10884E+03	428-496
331	$C_8H_{11}N$	*N,N*-Dimethylaniline	467.200		0.21869E+02	0.42945E+04	-0.51979E+02	344-473
332	$C_8H_{11}N$	2,4,6-Trimethylpyridine	443.830	0.16	0.20970E+02	0.34466E+04	-0.78881E+02	298-445
333	C_8H_{14}	1-Octyne	399.400		0.21168E+02	0.32868E+04	-0.58525E+02	358-400
334	C_8H_{14}	2-Octyne	411.370		0.21231E+02	0.34077E+04	-0.60233E+02	369-412
335	C_8H_{14}	3-Octyne	406.360		0.21306E+02	0.33959E+04	-0.59149E+02	365-407
336	C_8H_{14}	4-Octyne	404.970		0.21330E+02	0.33996E+04	-0.58198E+02	363-405
337	C_8H_{16}	Cyclooctane	424.310		0.20873E+02	0.34357E+04	-0.56737E+02	373-434
338	C_8H_{16}	1-Octene	394.436	2.30	0.20867E+02	0.31234E+04	-0.60051E+02	318-395
339	$C_8H_{16}O_2$	Octanoic acid	513.100		0.18937E+02	0.23752E+04	-0.19459E+03	398-437
340	C_8H_{18}	2-Methylheptane	390.800		0.20756E+02	0.30398E+04	-0.61454E+02	315-392
341	C_8H_{18}	3-Methylheptane	392.076		0.20761E+02	0.30540E+04	-0.61374E+02	316-393
342	C_8H_{18}	2,2,4-Trimethylpentane	372.388	6.50	0.21347E+02	0.34428E+04	-0.21844E+02	340-540
343	C_8H_{18}	2,3,4-Trimethylpentane	386.622		0.20680E+02	0.30316E+04	-0.55450E+02	310-387
344	$C_8H_{18}O$	Dibutyl ether	413.442	0.90	0.20545E+02	0.29893E+04	-0.82006E+02	362-414
345	$C_8H_{18}O$	Di-*tert*-butyl ether	380.380	4.34	0.20858E+02	0.30772E+04	-0.50634E+02	277-383
346	$C_8H_{18}O$	2-Ethyl-1-hexanol	457.774		0.20266E+02	0.27686E+04	-0.14100E+03	370-480

347	$C_8H_{18}O$	1-Octanol	468.345	0.01	0.20445E+02	0.29025E+04	-0.14292E+03	386-479
348	$C_8H_{18}O$	2-Octanol	453.000	0.00	0.19602E+02	0.24417E+04	-0.15065E+03	318-468
349	$C_8H_{18}O$	3-Octanol	447.820	0.00	0.16916E+02	0.12901E+04	-0.20845E+03	340-467
350	$C_8H_{18}O$	4-Octanol	449.672		0.18109E+02	0.17511E+04	-0.18365E+03	340-467
351	C_9H_7N	Isoquinoline	516.390		0.20809E+02	0.39684E+04	-0.88882E+02	440-517
352	C_9H_7N	Quinoline	510.250	0.01	0.20601E+02	0.38415E+04	-0.86938E+02	438-511
353	C_9H_{10}	Indan	451.100		0.20957E+02	0.36190E+04	-0.67352E+02	355-482
354	C_9H_{12}	Isopropylbenzene	425.560	0.61	0.20854E+02	0.33565E+04	-0.65735E+02	343-426
355	C_9H_{12}	Propylbenzene	432.392		0.20907E+02	0.34399E+04	-0.65723E+02	349-433
356	C_9H_{12}	1,2,3-Trimethylbenzene	449.230		0.21100E+02	0.36667E+04	-0.66245E+02	363-450
357	C_9H_{12}	1,2,4-Trimethylbenzene	442.528		0.21097E+02	0.36129E+04	-0.65061E+02	358-443
358	C_9H_{12}	1,3,5-Trimethylbenzene	437.893	0.33	0.21188E+02	0.36187E+04	-0.63360E+02	354-439
359	C_9H_{18}	1-Nonene	420.033	0.71	0.20918E+02	0.33139E+04	-0.67188E+02	340-421
360	$C_9H_{18}O$	5-Nonanone	461.600		0.21048E+02	0.36027E+04	-0.16324E+03	358-485
361	C_9H_{20}	3,3-Diethylpentane	419.313		0.20774E+02	0.33485E+04	-0.57240E+02	336-420
362	C_9H_{20}	Nonane	423.950	0.57	0.20868E+02	0.32941E+04	-0.71323E+02	298-425
363	C_9H_{20}	2,2,3,3-Tetramethylpentane	413.440		0.20622E+02	0.32213E+04	-0.59272E+02	331-414
364	C_9H_{20}	2,2,3,4-Tetramethylpentane	406.162		0.20631E+02	0.31688E+04	-0.58140E+02	325-407
365	C_9H_{20}	2,2,4,4-Tetramethylpentane	395.431		0.20543E+02	0.30507E+04	-0.57096E+02	316-396
366	C_9H_{20}	2,3,3,4-Tetramethylpentane	414.700		0.20699E+02	0.32712E+04	-0.58082E+02	332-416
367	C_9H_{20}	2,2,4-Trimethylhexane	399.690	2.12	0.21349E+02	0.34717E+04	-0.44554E+02	238-303
368	C_9H_{20}	2,2,5-Trimethylhexane	397.243	2.21	0.20640E+02	0.30539E+04	-0.62159E+02	298-398
369	$C_9H_{20}O$	1-Nonanol	486.650		0.22396E+02	0.41393E+04	-0.10584E+03	425-494
370	$C_{10}H_8$	Naphthalene	491.157	0.01	0.21100E+02	0.40526E+04	-0.67866E+02	353-453
371	$C_{10}H_{14}$	Butylbenzene	456.420	0.15	0.21239E+02	0.38229E+04	-0.62836E+02	374-457
372	$C_{10}H_{18}$	*cis*-Bicyclo[4.4.0]decane	468.924	0.10	0.20724E+02	0.36718E+04	-0.69735E+02	325-470
373	$C_{10}H_{18}$	*trans*-Bicyclo[4.4.0]decane	460.421	0.16	0.20693E+02	0.36119E+04	-0.66424E+02	335-461
374	$C_{10}H_{20}$	Cyclodecane	475.500		0.20753E+02	0.37300E+04	-0.71254E+02	344-489
375	$C_{10}H_{20}$	1-Decene	443.747	0.21	0.20906E+02	0.34491E+04	-0.76048E+02	360-445
376	$C_{10}H_{22}$	Decane	447.305	0.17	0.20914E+02	0.34640E+04	-0.78319E+02	367-448
377	$C_{10}H_{22}O$	1-Decanol	504.267		0.20367E+02	0.31451E+04	-0.14853E+03	400-528
378	$C_{11}H_{10}$	1-Methylnaphthalene	517.832	0.01	0.21090E+02	0.42036E+04	-0.78302E+02	415-518
379	$C_{11}H_{10}$	2-Methylnaphthalene	514.200		0.21175E+02	0.42433E+04	-0.74458E+02	412-515
380	$C_{11}H_{22}O$	2-Undecanone	506.301		0.21075E+02	0.38959E+04	-0.98297E+02	393-538
381	$C_{11}H_{22}O$	6-Undecanone	500.550		0.21104E+02	0.38863E+04	-0.94787E+02	388-532
382	$C_{11}H_{24}$	Undecane	469.038		0.20958E+02	0.36216E+04	-0.85088E+02	377-470
383	$C_{12}H_{10}$	Acenaphthene	552.000		0.21610E+02	0.48412E+04	-0.70026E+02	368-413
384	$C_{12}H_{10}$	Biphenyl	528.400		0.21572E+02	0.45995E+04	-0.70542E+02	342-544
385	$C_{12}H_{26}$	Dodecane	489.473	0.02	0.20969E+02	0.37473E+04	-0.92661E+02	400-490
386	$C_{12}H_{26}O$	1-Dodecanol	537.718		0.20466E+02	0.34591E+04	-0.15080E+03	426-550
387	$C_{13}H_{12}$	Diphenylmethane	537.000		0.22318E+02	0.50690E+04	-0.64082E+02	353-433
388	$C_{13}H_{28}$	Tridecane	508.616	0.01	0.21034E+02	0.39020E+04	-0.98234E+02	412-509
389	$C_{14}H_{10}$	Anthracene	614.000		0.21965E+02	0.58733E+04	-0.51394E+02	496-615
390	$C_{14}H_{10}$	Phenanthrene	612.550		0.22127E+02	0.59709E+04	-0.48748E+02	477-620
391	$C_{14}H_{30}$	Tetradecane	526.660		0.21067E+02	0.40273E+04	-0.10454E+03	428-527
392	$C_{15}H_{32}$	Pentadecane	543.760		0.21089E+02	0.41383E+04	-0.11102E+03	443-544
393	$C_{16}H_{34}$	Hexadecane	560.000		0.21095E+02	0.42291E+04	-0.11803E+03	463-561

TABLE 2.1.4 (VSPT)
SUBLIMATION PRESSURE OF SOLIDS

The sublimation pressure P_{sub} of a solid (see Section 2.1) is a monotonically increasing function of the temperature T. The temperature dependence of $P_{sub}(T)$ can be fitted to various empirical equations. A simple equation is the so-called Antoine equation:

$$P_{sub}(T)/\text{Pa} = \exp\left\{A(1) - \frac{A(2)}{(T/\text{K}) + A(3)}\right\} \qquad (1)$$

This table gives the coefficients $A(i)$ and the valid temperature range for 38 substances. The first and last T represent the lower and the upper limit of the T range of validity of the equation. The upper limit for each substance is the triple point temperature and pressure. The equation should not be used for extrapolation beyond this range. The program, Property Code VSPT, permits the calculation of $P_{sub}(T)$ at any desired set of temperatures in the valid range. The program also permits estimation of an approximate value of the molar enthalpy of sublimation $\Delta_{sub}H$ along the phase boundary curve, via the simplified Clausius-Clapeyron equation. If we assume that the volume change upon sublimation of a mole of solid equals the molar volume V_{id} of an ideal gas, $V_{id} = RT/P$, where $R = 8.3145\ \text{J K}^{-1}\ \text{mol}^{-1}$ is the molar gas constant, then:

$$\Delta_{sub}H/\text{J mol}^{-1} = \frac{R\,A(2)}{\{1 + A(3)/(T/\text{K})\}^2} \qquad (2)$$

For $A(3) = 0$,

$$\Delta_{sub}H/\text{J mol}^{-1} = R\,A(2) \qquad (3)$$

independent of T.

The substances are arranged in a modified Hill order, with compounds not containing carbon preceding those that contain carbon.

Physical quantity	Symbol	SI unit
Temperature	T	K
Sublimation pressure	P_{sub}	Pa
Molar enthalpy of sublimation	$\Delta_{sub}H$	J mol^{-1}

REFERENCES

1. Lide, D. R., Editor, *CRC Handbook of Chemistry and Physics*, 73rd Edition, CRC Press, Boca Raton, FL, 1992 (a collection of Antoine coefficients for ca. 140 substances published before 1960).
2. *TRC Thermodynamic Tables—Hydrocarbons*, Thermodynamic Research Center: The Texas A & M University System, College Station, TX (Loose-leaf data sheets) (recommended data and Antoine coefficients).
3. *TRC Thermodynamic Tables—Non-hydrocarbons*, Thermodynamic Research Center: The Texas A & M University System, College Station, TX (Loose-leaf data sheets) (recommended data and Antoine coefficients).

SN	Molecular formula	Name	A(1)	A(2)	A(3)	T/K range
1	Br_2	Bromine	0.27276E+02	0.47003E+04	-0.13050E+02	169-247
2	ClH	Hydrogen chloride	0.24354E+02	0.23550E+04		134-150
3	Cl_2	Chlorine	0.27240E+02	0.33254E+04	-0.60170E+01	116-172
4	F_2Xe	Xenon difluoride	0.27964E+02	0.61800E+04	-0.11470E+02	253-310
5	F_4Xe	Xenon tetrafluoride	0.30023E+02	0.71266E+04	-0.35940E+01	260-316
6	HI	Hydrogen iodide	0.21451E+02	0.20144E+04	-0.32730E+02	154-175
7	I_2	Iodine	0.27359E+02	0.66016E+04	-0.18970E+02	236-347

TABLE 2.1.4 (VSPT)
SUBLIMATION PRESSURE OF SOLIDS (continued)

SN	Molecular formula	Name	*A*(1)	*A*(2)	*A*(3)	*T*/K range
8	CCl_4	Tetrachloromethane	0.17613E+02	0.16431E+04	-0.95250E+02	232-250
9	CF_4	Tetrafluoromethane	0.17003E+02	0.72633E+03	-0.30710E+02	73-90
10	CH_2O_2	Methanoic acid	0.32090E+02	0.64452E+04	-0.16350E+02	236-277
11	CH_3I	Iodomethane	0.28893E+02	0.48350E+04		176-204
12	C_2Cl_6	Hexachloroethane	0.17435E+02	0.20020E+04	-0.15080E+03	280-347
13	C_2F_6	Hexafluoroethane	0.23472E+02	0.21549E+04	-0.10910E+02	104-143
14	$C_2H_4O_2$	Ethanoic acid	0.32331E+02	0.67852E+04	-0.20440E+02	238-288
15	$C_5H_{10}O_2$	2,2-Dimethylpropanoic acid	0.36925E+02	0.72407E+04	-0.30750E+02	238-257
16	$C_6H_6O_2$	1,3-Benzenediol	0.33807E+02	0.11147E+05		328-380
17	$C_6H_6O_2$	1,4-Benzenediol	0.35137E+02	0.12233E+05		341-400
18	$C_6H_{11}NO$	Hexahydro-2-azepinone	0.34650E+02	0.10741E+05		301-341
19	C_6H_{14}	Hexane	0.31224E+02	0.48186E+04	-0.23150E+02	168-178
20	$C_7H_6O_2$	Benzoic acid	0.14870E+02	0.47196E+04		293-314
21	C_8H_{18}	2,2,3,3-Tetramethylbutane	0.20970E+02	0.29160E+04	-0.74720E+02	253-287
22	C_9H_7NO	2-Hydroxyquinoline	0.35510E+02	0.13856E+05		375-391
23	C_9H_7NO	4-Hydroxyquinoline	0.35830E+02	0.15500E+05		415-434
24	$C_{10}H_8$	Naphthalene	0.31143E+02	0.85750E+04		270-305
25	$C_{12}H_8N_2$	Phenazine	0.30859E+02	0.10869E+05		281-324
26	$C_{12}H_8O$	Dibenzofuran	0.32770E+02	0.10100E+05		298-356
27	$C_{12}H_9N$	Carbazole	0.32330E+02	0.12176E+05		346-364
28	$C_{13}H_9N$	Acridine	0.31966E+02	0.11020E+05		281-324
29	$C_{13}H_{11}N$	9-Methylcarbazole	0.33549E+02	0.11425E+05		312-333
30	$C_{13}H_{12}$	Diphenylmethane	0.22318E+02	0.50690E+04	-0.64080E+02	353-434
31	$C_{14}H_{10}$	Anthracene	0.31620E+02	0.11378E+05		353-400
32	$C_{14}H_{10}$	Phenanthrene	0.31541E+02	0.10484E+05		324-363
33	$C_{14}H_{12}$	9,10-Dihydroanthracene	0.34014E+02	0.11130E+05		315-380
34	$C_{14}H_{12}$	9,10-Dihydrophenanthrene	0.22702E+02	0.63478E+04	-0.21750E+02	380-434
35	$C_{14}H_{12}$	*trans*-1,2-Diphenylethene	0.35755E+02	0.12111E+05		297-317
36	$C_{14}H_{13}N$	9-Ethylcarbazole	0.34953E+02	0.11830E+05		310-329
37	$C_{16}H_{10}$	Pyrene	0.31922E+02	0.11732E+05		365-421
38	$C_{18}H_{12}$	Chrysene	0.34988E+02	0.14598E+05		383-421

TABLE 2.1.5 (SLTP)
INFLUENCE OF PRESSURE ON MELTING TEMPERATURE

The melting temperature T_m of a solid changes with the pressure P, as represented by its melting curve (see Section 2.1).The effect becomes measurable at high pressure increments only, of the order of megapascals. For most 'normal' substances, T_m increases monotonically with P. Several substances, such as Sb and Ga, present falling melting curves, i.e., the melting temperature T_m decreases with P. A few substances, such as alkali metals, are 'abnormal', having maxima in the melting curves.

The melting temperature-pressure dependence of normal substances can be represented by the so-called Simon equation:

$$P = P_o + a\left[\left(\frac{T_m}{T_{m,o}}\right)^c - 1\right] \quad (1)$$

where $T_{m,o}$ is the melting temperature at the pressure P_o (usually at the triple point) and a and c are adjustable parameters. Equation 1 can be written in the form:

$$T_m(P)/\mathrm{K} = (T_{m,1}/\mathrm{K})\left(\frac{(P-P_1)/\mathrm{Pa}}{A(1)} + 1\right)^{A(2)} \quad (2)$$

where $T_{m,1}$ is the melting temperature at the lowest pressure P_1 (usually taken as zero, P_1 being negligible compared to P) and $A(1)$ and $A(2)$ are adjustable parameters. This table gives the coefficients $A(1)$ and $A(2)$, the values of T_m (P) for $P = P_1$, $P = 100$ MPa, and $P = 1000$ MPa, and the pressure range, P-Range, for 110 substances. The equation should not be used for extrapolation beyond this range. The program, Property Code SLTP, permits the calculation of T_m (P) at selected pressure increments within the range of validity of the fitting equation. The auxiliary coefficients displayed by the program have the following meaning:

$$V(1) = P_1 \ ; \ V(2) = T_{m,1} \quad (3)$$

The substances are arranged in the Hill order, with compounds not containing carbon preceding those that contain carbon.

Physical quantity	Symbol	SI unit
Melting temperature	T_m	K
Pressure	P	Pa

REFERENCES

1. Babb Jr., S. E., "Parameters in the Simon Equation Relating Pressure and Melting Temperature", *Rev. Mod. Phys.*, 35, 400-413, 1963 (a compilation of bibliographical data and correlations).
2. Reeves, L. E.; Scott, G. J.; Babb Jr., S. E., "Melting Curves of Pressure-Transmitting Fluids", *J. Chem. Phys.*, 40, 3662-3666, 1964 (melting curves of nine fluids, correlated values).
3. Isaacs, N. S., *Liquid Phase High Pressure Chemistry*, John Wiley, New York, 1981 (freezing temperatures and pressures for ca. 70 liquids).
4. Merrill, L. J., "Behavior of AB-Type Compounds at High Pressures and High Temperatures", *J. Phys. Chem. Ref. Data*, 1205-1252, 1977 (melting curves for 87 inorganic compounds).
5. Merrill, L. J., "Behavior of AB_2 -Type Compounds at High Pressures and High Temperatures", *J. Phys. Chem. Ref. Data*, 1005-1064, 1982 (melting curves for 168 inorganic compounds).

TABLE 2.1.5 (SLTP)
INFLUENCE OF PRESSURE ON MELTING TEMPERATURE (continued)

SN	Molecular formula	Name	$A(1)$	$A(2)$	T/K			(P/MPa)-range
					$P = P_1$	100 MPa	1000MPa	
1	Ag	Silver	0.9000E+09	0.1316	1234.00	1251.2	1361	0 - 5000
2	Ar	Argon	0.2114E+09	0.6277	83.81	106.9	251	0 - 1800
3	Bi	Bismuth	-0.2725E+10	0.1786	544.20	540.6	502	0 - 1800
4	BrNa	Sodium bromide	0.1220E+10	0.3448	1014.00	1041.9	1246	0 - 2000
5	Cd	Cadmium	0.4500E+10	0.4167	594.10	599.6	646	0 - 5000
6	ClK	Potassium chloride (c,I)	0.6900E+09	0.1754	1043.00	1068.1	1220	0 - 3000
7	ClK	Potassium chloride (c,II)	0.1210E+10	0.2500	1315.00			1895 - 3000
8	ClLi	Lithium chloride	0.1450E+10	0.4000	878.00	901.7	1083	0 - 2500
9	ClNa	Sodium chloride	0.1670E+10	0.3704	1073.50	1096.9	1277	0 - 2000
10	ClRb	Rubidium chloride(c,I)	0.6600E+09	0.1667	990.50	1014.1		0 - 800
11	ClRb	Rubidium chloride(c,II)	0.7200E+09	0.2500	1127.00		1205	780 - 3000
12	Cl_4Si	Silicon tetrachloride	0.4620E+09	0.7042	205.00	235.3	461	0 - 1200
13	Cs	Cesium	0.2674E+09	0.2227	302.90	325.1	428	0 - 5000
14	FNa	Sodium fluoride	0.1430E+10	0.1818	1265.00	1280.6	1393	0 - 1200
15	Ga	Gallium	-0.5750E+10	0.4000	303.01	300.9	281	0 - 1200
16	H_2	Hydrogen	0.2742E+08	0.5734	14.15	34.2	113	0 - 1000
17	H_2O	Water	-0.3952E+09	0.1111	273.15	264.4		0 - 200
18	H_3N	Ammonia	0.5270E+09	0.2326	195.30	203.4		0 - 300
19	H_3P	Phosphine	0.3300E+08	0.0667	139.40			0 - 2
20	$^1H\,^2H$	(^1H,^2H)Hydrogen	0.2596E+08	0.4484	16.60			0 - 9
21	$^2H_2$	($^2H_2$)Hydrogen	0.4297E+08	0.5595	18.81	36.9		0 - 350
22	^2H_2O	($^2H_2$)Dihydrogen oxide	-0.4596E+09	0.1290	276.97	268.3		0 - 200
23	$^3H_2$	($^3H_2$)Hydrogen	0.5299E+08	0.5668	20.68	37.7		0 - 300
24	Hg	Mercury	0.3821E+10	0.8496	234.29	239.5	285	0 - 2000
25	INa	Sodium iodide	0.1010E+10	0.3571	928.00	959.8	1187	0 - 3000
26	In	Indium	0.3580E+10	0.4348	429.76	434.9	478	0 - 5000
27	K	Potassium	0.4270E+09	0.2252	335.70	352.0	441	0 - 5000
28	Kr	Krypton	0.2376E+09	0.6185	115.75	143.8	321	0 - 1200
29	Li	Lithium	0.9000E+09	0.0676	453.70	456.9	477	0 - 5000
30	Mg	Magnesium	0.6400E+09	0.1724	923.00	946.4	1086	0 - 5000
31	N_2	Nitrogen	0.1607E+09	0.5583	63.16	82.8	191	0 - 1000
32	Na	Sodium	0.1197E+10	0.2830	370.78	379.3	440	0 - 5000
33	Ne	Neon	0.1038E+09	0.6250	24.62	37.5		0 - 340
34	O_2	Oxygen	0.2733E+09	0.5739	54.38	65.0		0 - 300
35	Pb	Lead	0.3230E+10	0.4158	600.20	607.9	671	0 - 5000
36	Rb	Rubidium	0.3951E+09	0.2674	311.90	331.3	437	0 - 9000
37	S	Sulfur (c,I)	0.2795E+09	0.2392	387.20	416.6		0 - 160
38	S	Sulfur (c,II)	0.2914E+09	0.2320	393.17	415.9		20 - 300
39	Sb	Antimony	-0.2940E+10	0.0167	903.70	903.2	897	0 - 2600
40	Se	Selenium	0.1170E+10	0.4902	495.50	515.8	671	0 - 1000
41	Sn	Tin (c,I)	0.5700E+10	0.2941	505.05	507.6	530	0 - 3350
42	Sn	Tin (c,II)	0.1470E+10	0.1923	591.05			3350 - 8000
43	Tl	Thallium	0.6400E+10	0.7246	576.80	583.3	641	0 - 3000
44	Zn	Zinc	0.6000E+10	0.4167	692.70	697.5	739	0 - 5000
45	CCl_2F_2	Dichlorodifluoromethane	0.3288E+09	0.4482	117.90	132.8	220	0 - 1000
46	CCl_4	Tetrachloromethane	0.2919E+09	0.4717	250.60	288.0		0 - 900
47	$CHBr_3$	Tribromomethane	0.5340E+09	0.4878	280.94	305.5	470	0 - 1000
48	$CHCl_3$	Trichloromethane	0.8330E+09	0.6579	209.70	225.9	352	0 - 2500
49	CHN	Hydrogen cyanide	0.3080E+09	0.2778	259.20	280.3		0 - 300
50	CH_2Cl_2	Dichloromethane	0.1059E+10	0.6689	176.50	187.5	275	0 - 3000
51	CH_2O_2	Methanoic acid	0.4100E+09	0.1923	281.60	293.7		0 - 300
52	CH_3NO	Methanamide	0.1070E+10	0.3704	275.71	285.0		0 - 100
53	CO_2	Carbon dioxide	0.4000E+09	0.3846	216.56	236.0	351	0 - 1200
54	CS_2	Carbon disulfide	0.8250E+09	0.6667	161.60	174.4	274	0 - 3500
55	C_2H_3N	Ethanenitrile	0.5250E+09	0.4854	229.30	249.6	385	0 - 1000
56	C_2H_4	Ethene	0.3275E+09	0.5522	103.80	120.3	225	0 - 1000

TABLE 2.1.5 (SLTP)
INFLUENCE OF PRESSURE ON MELTING TEMPERATURE (continued)

SN	Molecular formula	Name	$A(1)$	$A(2)$	T/K $P = P_1$	T/K 100 MPa	T/K 1000MPa	(P/MPa)-range
57	$C_2H_4O_2$	Ethanoic acid	0.2400E+09	0.2045	289.84	311.2		0 - 400
58	C_2H_5Br	Bromoethane	0.5400E+09	0.4405	154.20	166.2	245	0 - 3000
59	C_2H_5NO	Ethanamide	0.3330E+09	0.1333	354.70	367.3		0 - 500
60	C_2H_6O	Ethanol	0.1060E+10	0.6211	155.90	164.9	236	0 - 3500
61	C_3H_6	Propene (c,I)	0.3196E+09	0.3545	86.00	94.7		0 - 800
62	C_3H_6	Propene (c,II)	0.3064E+09	0.2583	109.60		143	445 - 1000
63	$C_3H_6O_2$	Propanoic acid	0.3300E+09	0.2941	252.40	272.8		0 - 100
64	C_3H_7Br	1-Bromopropane	0.7350E+09	0.5747	163.20	175.6	267	0 - 4000
65	C_3H_8	Propane	0.7180E+09	0.7794	85.30	94.4	168	0 - 1000
66	C_4H_4S	Thiophene	0.1300E+09	0.3030	234.60	278.9		0 - 500
67	C_4H_6O	Cyclobutanone	0.4550E+09	0.5000	222.20	245.4		0 - 450
68	$C_4H_8O_2$	1,4-Dioxane	0.7300E+09	0.3030	284.50	295.8		0 - 600
69	$C_4H_8O_2$	Ethyl ethanoate	0.7600E+09	0.4545	189.60	200.6	278	0 - 3000
70	C_4H_{10}	Butane	0.3634E+09	0.4525	134.50	150.1	245	0 - 1000
71	C_4H_{10}	2-Methylpropane (c,I)	0.4246E+09	0.4036	128.20	139.6		0 - 400
72	C_4H_{10}	2-Methylpropane (c,II)	0.7942E+09	0.6365	160.20		237	326 - 1000
73	$C_4H_{10}O$	1-Butanol	0.5160E+09	0.4167	183.40	197.4	287	0 - 3500
74	$C_4H_{10}O$	2-Methyl-2-propanol	0.1290E+09	0.1786	298.70	330.9		0 - 300
75	C_5H_{12}	Pentane	0.6600E+09	0.6064	143.50	156.3	251	0 - 1000
76	C_5H_{12}	2-Methylbutane	0.5916E+09	0.6398	112.50	124.3	212	0 - 1000
77	$C_5H_{12}O$	2-Methyl-2-butanol	0.6930E+08	0.0781	264.80	283.9		0 - 300
78	C_6H_5Br	Bromobenzene	0.4800E+09	0.4065	242.10	261.5	383	0 - 1200
79	C_6H_5Cl	Chlorobenzene	0.4980E+09	0.4132	227.70	245.6	359	0 - 1200
80	$C_6H_5NO_2$	Nitrobenzene	0.6100E+09	0.5181	278.80	301.6	461	0 - 1200
81	$C_6H_5NO_3$	2-Nitrophenol	0.1030E+10	0.7143	319.20	341.0		0 - 400
82	$C_6H_5NO_3$	3-Nitrophenol	0.9300E+09	0.5000	368.90	388.2		0 - 300
83	$C_6H_5NO_3$	4-Nitrophenol	0.4700E+09	0.3322	385.60	411.1		0 - 400
84	C_6H_6	Benzene	0.3600E+09	0.3846	278.80	306.4	465	0 - 1200
85	C_6H_6O	Phenol (c,I)	0.2670E+09	0.1250	314.03	326.8		0 - 200
86	C_6H_6O	Phenol (c,II)	0.6600E+09	0.3937	337.20		460	208 - 1200
87	$C_6H_6O_2$	1,3-Benzenediol	0.3100E+09	0.1266	383.20	397.0		0 - 300
88	C_6H_7N	Aniline	0.5352E+09	0.4275	266.80	287.1	419	0 - 1200
89	C_6H_{12}	Cyclohexane	0.4425E+09	0.4329	279.87	305.7		0 - 400
90	C_7H_5N	Benzonitrile	0.1080E+10	0.8475	260.00	280.3	453	0 - 1200
91	C_7H_8O	2-Methylphenol	0.2700E+09	0.1613	303.56	319.4		0 - 800
92	C_7H_8O	3-Methylphenol	0.1790E+10	0.8621	285.06	298.7		0 - 100
93	C_7H_8O	4-Methylphenol	0.7000E+09	0.5263	307.00	329.4		0 - 300
94	C_7H_{14}	Methylcyclohexane	0.9600E+10	0.4651	146.80	147.5	154	0 - 1900
95	C_8H_8O	Methyl phenyl ketone	0.3160E+09	0.2703	292.90	315.5		0 - 300
96	C_9H_20	Nonane	0.1030E+10	0.6250	219.70	232.8		0 - 500
97	$C_{10}H_8$	Naphthalene	0.3600E+09	0.4016	353.17			0 - 60
98	$C_{10}H_8O$	1-Naphthol	0.7600E+09	0.5051	369.20	393.0		0 - 300
99	$C_{10}H_{12}O$	1-Methoxy-4-(1-propenyl)benzene	0.4930E+09	0.3571	295.70	315.9		0 - 300
100	$C_{12}H_{11}N$	Diphenylamine	0.4750E+09	0.4065	327.20	353.6		0 - 800
101	$C_{12}H_{24}O_2$	Dodecanoic acid	0.3240E+09	0.2500	321.20	343.5		0 - 300
102	$C_{12}H_{26}$	Dodecane	0.3550E+09	0.3279	263.65	286.0		0 - 800
103	$C_{13}H_{10}O$	Diphenyl ketone	0.4720E+09	0.4255	320.93	348.3	521	0 - 1200
104	$C_{13}H_{12}$	Diphenylmethane	0.9000E+08	0.1818	300.20	343.9		0 - 300
105	$C_{14}H_{14}$	1,2-Diphenylethane	0.2700E+09	0.3030	324.50	357.0		0 - 150
106	$C_{15}H_{32}$	Pentadecane	0.2700E+09	0.3030	283.10	311.5		0 - 140
107	$C_{18}H_{38}$	Octadecane	0.3450E+09	0.2950	300.95	324.4		0 - 600
108	$C_{19}H_{16}$	Triphenylmethane	0.2670E+09	0.2778	365.70	399.5		0 - 300
109	$C_{24}H_{50}$	Tetracosane (c,I)	0.4500E+09	0.3521	323.50			0 - 40
110	$C_{24}H_{50}$	Tetracosane (c,II)	0.4240E+09	0.2584	352.59		471	127 - 4500

2.2. VOLUMETRIC PROPERTIES

The *density* (mass density) ρ of a system is the ratio of its mass m to the volume.

The *specific volume* v of a system is the ratio of its volume to its mass. Hence:

$$v = 1/\rho \tag{1}$$

The *molar volume* V of a system is the ratio of its volume to its amount of substance. Hence:

$$V = M\,v = M/\rho \tag{2}$$

and

$$v = V/M \tag{3}$$

where M is the *absolute molar mass* for single-component systems, or the *absolute average molar mass* for multi-component systems.

The molar volume V of a single-component system is a function of temperature T and pressure P. The P-V-T relationship is called the *equation of state* of the system.

The quantity

$$\alpha = (1/V)(\partial V/\partial T)_P \tag{4}$$

is called the *cubic expansion coefficient* (isobaric expansivity or coefficient of thermal expansion).

The quantity

$$\kappa_T = -(1/V)(\partial V/\partial P)_T \tag{5}$$

is called the *isothermal compressibility* (isothermal coefficient of bulk compressibility).

The quantity

$$\beta = (\partial P/\partial T)_V \tag{6}$$

is called the *pressure coefficient* (isothermal pressure coefficient).

These three quantities are interrelated by

$$\alpha = \kappa_T \beta \tag{7}$$

The equation of state of a single-component ideal gas is the same for all the substances

$$P = RT/V \tag{8}$$

where R is the molar gas constant. Hence, for an ideal gas,

$$\alpha = 1/T$$

$$\kappa_T = 1/P$$

$$\beta = R/V = P/T \tag{9}$$

Table 2.2.1 deals with the volumetric properties of real gases at low densities (or pressures).

Table 2.2.2 covers the volumetric properties of liquids as a function of T. The densities of selected liquids at $T = 298.15$ K are also listed in Table 2.1.1.

TABLE 2.2.1 (VIBT)
VIRIAL COEFFICIENTS OF GASES

The *equation of state*, i.e., the relationship between the *pressure* P, the *temperature* T, and the *molar volume* V, of gases can be represented by a polynomial series (virial equation of state):

$$P = \frac{RT}{V}\left[1 + \frac{B(T)}{V} + \frac{C(T)}{V^2} + \cdots\right] \tag{1}$$

where R is the *molar gas constant*. The parameters $B(T)$, $C(T)$, ..., are called *second, third, ..., molar virial coefficients.* For pure gases they are functions only of T. The exact definition of $B(T)$ is

$$B(T) = \lim_{V\to\infty} (Z - 1) \tag{2}$$

where $Z = PV/RT$ is called the *compressibility factor*. In many applications, Equation (1) is truncated after the second term

$$P = \frac{RT}{V}\left[1 + \frac{B(T)}{V}\right] \tag{3}$$

and $B(T)$ is fitted to best describe the P-V relationship at any given T. Equation (3) is valid for most nonpolar or polar gases at low densities (or pressures), say at reduced volumes $V_r < 0.5$, where $V_r = V/V_c$, V_c being the *critical molar volume* of the gas (see Table 2.1.1). Strongly associated substances such as carboxylic acids are an exception.

$B(T)$ is related to the intermolecular potential energy function. At low T, when attractive forces prevail, $B(T)$ is negative ($Z < 1$). $B(T)$ increases with increasing T and becomes positive ($Z > 1$) at high T, when the repulsive forces prevail over the attractive ones. The temperature at which $B(T) = 0$ ($Z = 1$), i.e., when the gas behaves as a perfect gas, is called the *Boyle temperature*. Most organic substances are chemically unstable at the Boyle temperature and therefore have $B(T) < 0$ in the whole range of experimental T.

The temperature dependence of $B(T)$ can be fitted to various empirical equations. A simple equation is

$$B(T)/\mathrm{cm}^3\,\mathrm{mol}^{-1} = \sum_{i=1}^{n} A(i)\left[(T_o/T) - 1\right]^{i-1} \tag{4}$$

where $T_o = 298.15$ K. The table gives the coefficients $A(i)$ and values of $B(T)$ for a few temperatures T. The first and last T represent the lower and the upper limit of the T range of validity of the equation. The equation should not be used for extrapolation beyond this range. The program, Property Code VIBT, permits the calculation of $B(T)$ at selected temperature increments within the range of validity of the fitting equation.

The substances are arranged in the Hill order, with substances not containing carbon preceding those that contain carbon.

Physical quantity	Symbol	SI unit
Second molar virial coefficient	B	$m^3\ mol^{-1}$
Temperature	T	K

REFERENCES

1. Dymond, J. H.; Smith, E. B., *The Virial Coefficients of Pure Gases and Mixtures*, Clarendon Press, Oxford, 1980 (a comprehensive compilation of experimental data published up to early 1979 on ca. 400 pure substances and 530 binary mixtures).
2. Cholinski, J.; Szafranski, A.; Wyrzykowska-Stankiewicz, D., *Computer-Aided Second Virial Coefficient Data for Organic Individual Compounds and Binary Systems*, PWN-Polish Scientific Publishers, Warsaw, 1986 (experimental and correlated data for 232 pure organic substances and 192 binary mixtures).
3. Reid, R. C.; Prausnitz, J. M.; Poling, B. E. *The Properties of Gases and Liquids*, 4th Ed., McGraw-Hill, New York, 1987 (estimation methods).

TABLE 2.2.1 (VIBT)
VIRIAL COEFFICIENTS OF GASES (continued)

System Number: 1
Argon, Ar

T/K	B/cm³mol⁻¹
100	-184
280	-19
460	3
640	13
820	18
1000	21

$A(1)$ = -0.1598E+02
$A(2)$ = -0.5983E+02
$A(3)$ = -0.9729E+01
$A(4)$ = -0.1512E+01

System Number: 2
Boron trifluoride, BF_3

T/K	B/cm³mol⁻¹
200	-336
250	-181
300	-103
350	-61
400	-36
450	-20

$A(1)$ = -0.1058E+03
$A(2)$ = -0.3404E+03
$A(3)$ = -0.2636E+03
$A(4)$ = -0.1512E+01

System Number: 95
Hydrogen chloride, ClH

T/K	B/cm³mol⁻¹
190	-450
246	-227
302	-140
358	-96
414	-71
470	-54

$A(1)$ = -0.1442E+03
$A(2)$ = -0.3254E+03
$A(3)$ = -0.2772E+03
$A(4)$ = -0.1698E+03

System Number: 109
Chlorine, Cl_2

T/K	B/cm³mol⁻¹
210	-510
348	-224
486	-106
624	-57
762	-36
900	-26

$A(1)$ = -0.3028E+03
$A(2)$ = -0.5553E+03
$A(3)$ = 0.9160E+01
$A(4)$ = 0.3289E+03

System Number: 87
Fluorine, F_2

T/K	B/cm³mol⁻¹
80	-378
114	-153
148	-99
182	-68
216	-42
250	-20

$A(1)$ = 0.8499E+01
$A(2)$ = -0.1632E+03
$A(3)$ = 0.8404E+02
$A(4)$ = -0.2794E+02

System Number: 88
Silicon tetrafluoride, F_4Si

T/K	B/cm³mol⁻¹
210	-268
263	-179
316	-120
369	-77
422	-46
475	-21

$A(1)$ = -0.1376E+03
$A(2)$ = -0.3116E+03
$A(3)$ = 0.0000E+00
$A(4)$ = 0.0000E+00

System Number: 89
Iodine pentafluoride, F_5I

T/K	B/cm³mol⁻¹
320	-2540
338	-2204
356	-1940
374	-1732
392	-1570
410	-1443

$A(1)$ = -0.3077E+04
$A(2)$ = -0.8474E+04
$A(3)$ = -0.9116E+04
$A(4)$ = 0.0000E+00

System Number: 90
Phosphorus pentafluoride, F_5P

T/K	B/cm³mol⁻¹
320	-162
348	-136
376	-114
404	-95
432	-78
460	-64

$A(1)$ = -0.1860E+03
$A(2)$ = -0.3454E+03
$A(3)$ = 0.0000E+00
$A(4)$ = 0.0000E+00

System Number: 91
Molybdenum hexafluoride, F_6Mo

T/K	B/cm³mol⁻¹
295	-945
315	-771
335	-650
355	-568
375	-516
395	-484

$A(1)$ = -0.9142E+03
$A(2)$ = -0.2922E+04
$A(3)$ = -0.4778E+04
$A(4)$ = 0.0000E+00

System Number: 92
Sulfur hexafluoride, F_6S

T/K	B/cm³mol⁻¹
200	-685
265	-365
330	-219
395	-139
460	-89
525	-56

$A(1)$ = -0.2788E+03
$A(2)$ = -0.6468E+03
$A(3)$ = -0.3351E+03
$A(4)$ = -0.7175E+02

System Number: 93
Uranium hexafluoride, F_6U

T/K	B/cm³mol⁻¹
320	-1030
350	-852
380	-724
410	-629
440	-559
470	-507

$A(1)$ = -0.1204E+04
$A(2)$ = -0.2690E+04
$A(3)$ = -0.2144E+04
$A(4)$ = 0.0000E+00

System Number: 94
Tungsten hexafluoride, F_6W

T/K	B/cm³mol⁻¹
320	-640
348	-555
376	-482
404	-419
432	-364
460	-316

$A(1)$ = -0.7190E+03
$A(2)$ = -0.1143E+04
$A(3)$ = 0.0000E+00
$A(4)$ = 0.0000E+00

TABLE 2.2.1 (VIBT)
VIRIAL COEFFICIENTS OF GASES (continued)

System Number: 96
Hydrogen, H_2

T/K	*B*/cm³mol⁻¹
15	-229
92	-5
169	8
246	13
323	16
400	17

$A(1) = 0.1542E+02$
$A(2) = -0.9014E+01$
$A(3) = -0.2109E+00$
$A(4) = 0.0000E+00$

System Number: 97
Water, H_2O

T/K	*B*/cm³mol⁻¹
300	-1122
480	-186
660	-80
840	-46
1020	-23
1200	-2

$A(1) = -0.1151E+04$
$A(2) = -0.4663E+04$
$A(3) = -0.7028E+04$
$A(4) = -0.3802E+04$

System Number: 98
Ammonia, H_3N

T/K	*B*/cm³mol⁻¹
290	-302
316	-221
342	-175
368	-146
394	-122
420	-100

$A(1) = -0.2710E+03$
$A(2) = -0.1022E+04$
$A(3) = -0.2715E+04$
$A(4) = -0.4189E+04$

System Number: 99
Phosphine, H_3P

T/K	*B*/cm³mol⁻¹
190	-457
210	-363
230	-305
250	-258
270	-212
290	-165

$A(1) = -0.1461E+03$
$A(2) = -0.7327E+03$
$A(3) = 0.1022E+04$
$A(4) = -0.1220E+04$

System Number: 100
Helium, He

T/K	*B*/cm³mol⁻¹
2	-172
141	11
281	12
420	12
560	13
700	13

$A(1) = 0.1244E+02$
$A(2) = -0.1248E+01$
$A(3) = 0.0000E+00$
$A(4) = 0.0000E+00$

System Number: 101
Krypton, Kr

T/K	*B*/cm³mol⁻¹
110	-363
228	-90
346	-35
464	-12
582	0
700	8

$A(1) = -0.5128E+02$
$A(2) = -0.1185E+03$
$A(3) = -0.2851E+02$
$A(4) = -0.5209E+01$

System Number: 102
Nitrogen oxide, NO

T/K	*B*/cm³mol⁻¹
120	-232
150	-113
180	-72
210	-51
240	-36
270	-23

$A(1) = -0.1238E+02$
$A(2) = -0.1185E+03$
$A(3) = 0.8913E+02$
$A(4) = -0.7347E+02$

System Number: 103
Nitrogen, N_2

T/K	*B*/cm³mol⁻¹
75	-274
200	-34
325	0
450	13
575	19
700	23

$A(1) = -0.4252E+01$
$A(2) = -0.5570E+02$
$A(3) = -0.1177E+02$
$A(4) = 0.0000E+00$

System Number: 104
Dinitrogen oxide, N_2O

T/K	*B*/cm³mol⁻¹
240	-219
272	-162
304	-124
336	-98
368	-80
400	-68

$A(1) = -0.1302E+03$
$A(2) = -0.3070E+03$
$A(3) = -0.2478E+03$
$A(4) = 0.0000E+00$

System Number: 105
Neon, Ne

T/K	*B*/cm³mol⁻¹
60	-25
168	4
276	10
384	12
492	13
600	14

$A(1) = 0.1082E+02$
$A(2) = -0.7520E+01$
$A(3) = -0.3857E+00$
$A(4) = 0.0000E+00$

System Number: 106
Oxygen, O_2

T/K	*B*/cm³mol⁻¹
90	-240
152	-86
214	-42
276	-21
338	-9
400	0

$A(1) = -0.1633E+02$
$A(2) = -0.6220E+02$
$A(3) = -0.7787E+01$
$A(4) = -0.3137E+01$

System Number: 107
Sulfur dioxide, O_2S

T/K	*B*/cm³mol⁻¹
290	-464
326	-335
362	-251
398	-195
434	-157
470	-131

$A(1) = -0.4303E+03$
$A(2) = -0.1193E+04$
$A(3) = -0.1029E+04$
$A(4) = 0.0000E+00$

TABLE 2.2.1 (VIBT)
VIRIAL COEFFICIENTS OF GASES (continued)

System Number: 108
Xenon, Xe

T/K	*B*/cm³mol⁻¹
160	-421
258	-173
356	-90
454	-50
552	-28
650	-14

A(1) = -0.1303E+03
A(2) = -0.2617E+03
A(3) = -0.8724E+02
A(4) = 0.0000E+00

System Number: 3
Chlorotrifluoromethane, $CClF_3$

T/K	*B*/cm³mol⁻¹
240	-368
300	-219
360	-144
420	-98
480	-65
540	-39

A(1) = -0.2228E+03
A(2) = -0.5036E+03
A(3) = -0.3404E+03
A(4) = -0.2909E+03

System Number: 4
Dichlorodifluoromethane, CCl_2F_2

T/K	*B*/cm³mol⁻¹
250	-769
290	-520
330	-378
370	-289
410	-227
450	-183

A(1) = -0.4856E+03
A(2) = -0.1217E+04
A(3) = -0.1188E+04
A(4) = -0.6985E+03

System Number: 5
Trichlorofluoromethane, CCl_3F

T/K	*B*/cm³mol⁻¹
240	-1140
290	-826
340	-612
390	-457
440	-340
490	-248

A(1) = -0.7859E+03
A(2) = -0.1428E+04
A(3) = -0.1416E+03
A(4) = 0.0000E+00

System Number: 6
Tetrachloromethane, CCl_4

T/K	*B*/cm³mol⁻¹
320	-1344
340	-1171
360	-1040
380	-941
400	-868
420	-814

A(1) = -0.1600E+04
A(2) = -0.4059E+04
A(3) = -0.4653E+04
A(4) = 0.0000E+00

System Number: 7
Tetrafluoromethane, CF_4

T/K	*B*/cm³mol⁻¹
250	-136
350	-54
450	-16
550	5
650	19
750	29

A(1) = -0.8840E+02
A(2) = -0.2377E+03
A(3) = -0.6967E+02
A(4) = 0.0000E+00

System Number: 8
Chlorodifluoromethane, $CHClF_2$

T/K	*B*/cm³mol⁻¹
300	-343
330	-289
360	-242
390	-200
420	-164
450	-131

A(1) = -0.3467E+03
A(2) = -0.5753E+03
A(3) = 0.1872E+03
A(4) = 0.0000E+00

System Number: 9
Dichlorofluoromethane, $CHCl_2F$

T/K	*B*/cm³mol⁻¹
250	-727
290	-586
330	-478
370	-394
410	-326
450	-271

A(1) = -0.5619E+03
A(2) = -0.8616E+03
A(3) = 0.0000E+00
A(4) = 0.0000E+00

System Number: 10
Trichloromethane, $CHCl_3$

T/K	*B*/cm³mol⁻¹
320	-1000
340	-858
360	-740
380	-641
400	-558
420	-488

A(1) = -0.1193E+04
A(2) = -0.2936E+04
A(3) = -0.1751E+04
A(4) = 0.0000E+00

System Number: 11
Trifluoromethane, CHF_3

T/K	*B*/cm³mol⁻¹
200	-433
240	-288
280	-203
320	-150
360	-115
400	-91

A(1) = -0.1769E+03
A(2) = -0.3992E+03
A(3) = -0.2501E+03
A(4) = 0.0000E+00

System Number: 12
Dichloromethane, CH_2Cl_2

T/K	*B*/cm³mol⁻¹
320	-706
340	-574
360	-481
380	-419
400	-379
420	-356

A(1) = -0.9131E+03
A(2) = -0.3371E+04
A(3) = -0.5013E+04
A(4) = 0.0000E+00

System Number: 13
Difluoromethane, CH_2F_2

T/K	*B*/cm³mol⁻¹
280	-375
295	-329
310	-293
325	-267
340	-247
355	-233

A(1) = -0.3208E+03
A(2) = -0.7536E+03
A(3) = -0.1300E+04
A(4) = 0.0000E+00

TABLE 2.2.1 (VIBT)
VIRIAL COEFFICIENTS OF GASES (continued)

System Number: 14
Bromomethane, CH_3Br

T/K	B/cm^{3}mol^{-1}
280	-645
300	-551
320	-468
340	-396
360	-331
380	-273

$A(1) = -0.5592E+03$
$A(2) = -0.1324E+04$
$A(3) = 0.0000E+00$
$A(4) = 0.0000E+00$

System Number: 15
Chloromethane, CH_3Cl

T/K	B/cm^{3}mol^{-1}
280	-466
340	-303
400	-206
460	-142
520	-98
580	-66

$A(1) = -0.4070E+03$
$A(2) = -0.8872E+03$
$A(3) = -0.3850E+03$
$A(4) = 0.0000E+00$

System Number: 16
Fluoromethane, CH_3F

T/K	B/cm^{3}mol^{-1}
280	-244
310	-189
340	-149
370	-120
400	-98
430	-81

$A(1) = -0.2087E+03$
$A(2) = -0.5252E+03$
$A(3) = -0.3650E+03$
$A(4) = 0.0000E+00$

System Number: 17
Iodomethane, CH_3I

T/K	B/cm^{3}mol^{-1}
300	-823
320	-645
340	-531
360	-462
380	-427
400	-417

$A(1) = -0.8439E+03$
$A(2) = -0.3353E+04$
$A(3) = -0.6590E+04$
$A(4) = 0.0000E+00$

System Number: 18
Methane, CH_4

T/K	B/cm^{3}mol^{-1}
110	-327
210	-95
310	-39
410	-13
510	0
610	10

$A(1) = -0.4346E+02$
$A(2) = -0.1135E+03$
$A(3) = -0.1933E+02$
$A(4) = -0.6660E+01$

System Number: 19
Methanol, CH_4O

T/K	B/cm^{3}mol^{-1}
320	-1431
340	-1174
360	-945
380	-740
400	-556
420	-390

$A(1) = -0.1752E+04$
$A(2) = -0.4694E+04$
$A(3) = 0.0000E+00$
$A(4) = 0.0000E+00$

System Number: 20
Methylamine, CH_5N

T/K	B/cm^{3}mol^{-1}
300	-451
350	-304
400	-220
450	-170
500	-140
550	-122

$A(1) = -0.4588E+03$
$A(2) = -0.1191E+04$
$A(3) = -0.9952E+03$
$A(4) = 0.0000E+00$

System Number: 21
Carbon oxide, CO

T/K	B/cm^{3}mol^{-1}
200	-41
260	-17
320	-4
380	3
440	8
500	12

$A(1) = -0.8530E+01$
$A(2) = -0.5823E+02$
$A(3) = -0.1781E+02$
$A(4) = 0.0000E+00$

System Number: 22
Carbon dioxide, CO_2

T/K	B/cm^{3}mol^{-1}
220	-244
396	-63
572	-16
748	2
924	13
1100	19

$A(1) = -0.1273E+03$
$A(2) = -0.2876E+03$
$A(3) = -0.1183E+03$
$A(4) = 0.0000E+00$

System Number: 23
Carbon disulfide, CS_2

T/K	B/cm^{3}mol^{-1}
280	-931
310	-739
340	-603
370	-503
400	-430
430	-375

$A(1) = -0.8075E+03$
$A(2) = -0.1829E+04$
$A(3) = -0.1371E+04$
$A(4) = 0.0000E+00$

System Number: 24
1,2-Dichlorotetrafluoroethane, $C_2Cl_2F_4$

T/K	B/cm^{3}mol^{-1}
300	-800
340	-608
380	-474
420	-378
460	-307
500	-253

$A(1) = -0.8119E+03$
$A(2) = -0.1773E+04$
$A(3) = -0.9633E+03$
$A(4) = 0.0000E+00$

System Number: 25
1,1,2-Trichlorotrifluoroethane, $C_2Cl_3F_3$

T/K	B/cm^{3}mol^{-1}
300	-990
330	-856
360	-745
390	-650
420	-570
450	-500

$A(1) = -0.9992E+03$
$A(2) = -0.1479E+04$
$A(3) = 0.0000E+00$
$A(4) = 0.0000E+00$

TABLE 2.2.1 (VIBT)
VIRIAL COEFFICIENTS OF GASES (continued)

System Number: 26
Trichloroethene, C_2HCl_3

T/K	*B*/cm^{3}mol^{-1}
320	-2107
330	-1696
340	-1401
350	-1206
360	-1095
370	-1056

A(1) = -0.3523E+04
A(2) = -0.2507E+05
A(3) = -0.6372E+05
A(4) = 0.0000E+00

System Number: 27
Ethyne, C_2H_2

T/K	*B*/cm^{3}mol^{-1}
200	-572
214	-474
228	-399
242	-341
256	-297
270	-263

A(1) = -0.2161E+03
A(2) = -0.3750E+03
A(3) = -0.7156E+03
A(4) = 0.0000E+00

System Number: 28
Ethanenitrile, C_2H_3N

T/K	*B*/cm^{3}mol^{-1}
330	-3468
346	-2716
362	-2175
378	-1801
394	-1560
410	-1424

A(1) = -0.5840E+04
A(2) = -0.2918E+05
A(3) = -0.4761E+05
A(4) = 0.0000E+00

System Number: 29
Ethene, C_2H_4

T/K	*B*/cm^{3}mol^{-1}
240	-218
282	-157
324	-117
366	-88
408	-67
450	-51

A(1) = -0.1404E+03
A(2) = -0.2964E+03
A(3) = -0.1012E+03
A(4) = 0.0000E+00

System Number: 30
1,2-Dichloroethane, $C_2H_4Cl_2$

T/K	*B*/cm^{3}mol^{-1}
370	-812
410	-634
450	-508
490	-415
530	-346
570	-294

A(1) = -0.1362E+04
A(2) = -0.3240E+04
A(3) = -0.2100E+04
A(4) = 0.0000E+00

System Number: 31
Ethanal, C_2H_4O

T/K	*B*/cm^{3}mol^{-1}
290	-1352
326	-861
362	-575
398	-411
434	-323
470	-283

A(1) = -0.1217E+04
A(2) = -0.4647E+04
A(3) = -0.5725E+04
A(4) = 0.0000E+00

System Number: 32
Methyl methanoate, $C_2H_4O_2$

T/K	*B*/cm^{3}mol^{-1}
320	-820
336	-702
352	-609
368	-536
384	-479
400	-435

A(1) = -0.1035E+04
A(2) = -0.3425E+04
A(3) = -0.4203E+04
A(4) = 0.0000E+00

System Number: 33
Chloroethane, C_2H_5Cl

T/K	*B*/cm^{3}mol^{-1}
320	-634
376	-395
432	-262
488	-185
544	-140
600	-114

A(1) = -0.7767E+03
A(2) = -0.2205E+04
A(3) = -0.1764E+04
A(4) = 0.0000E+00

System Number: 34
Ethane, C_2H_6

T/K	*B*/cm^{3}mol^{-1}
200	-408
280	-208
360	-123
440	-75
520	-45
600	-24

A(1) = -0.1837E+03
A(2) = -0.3755E+03
A(3) = -0.1426E+03
A(4) = -0.5351E+02

System Number: 35
Ethanol, C_2H_6O

T/K	*B*/cm^{3}mol^{-1}
320	-2710
334	-1938
348	-1381
362	-997
376	-753
390	-621

A(1) = -0.4475E+04
A(2) = -0.2972E+05
A(3) = -0.5672E+05
A(4) = 0.0000E+00

System Number: 36
Dimethyl ether, C_2H_6O

T/K	*B*/cm^{3}mol^{-1}
275	-535
282	-509
289	-485
296	-461
303	-439
310	-417

A(1) = -0.4546E+03
A(2) = -0.9646E+03
A(3) = 0.0000E+00
A(4) = 0.0000E+00

System Number: 37
Dimethylamine, C_2H_7N

T/K	*B*/cm^{3}mol^{-1}
310	-605
328	-530
346	-466
364	-411
382	-364
400	-322

A(1) = -0.6622E+03
A(2) = -0.1504E+04
A(3) = -0.6672E+03
A(4) = 0.0000E+00

TABLE 2.2.1 (VIBT) VIRIAL COEFFICIENTS OF GASES (continued)

System Number: 38
Ethylamine, C_2H_7N

T/K	B/cm^{3}mol^{-1}
300	-772
320	-653
340	-558
360	-480
380	-416
400	-363

$A(1) = -0.7849E+03$
$A(2) = -0.2012E+04$
$A(3) = -0.1397E+04$
$A(4) = 0.0000E+00$

System Number: 39
Propyne, C_3H_4

T/K	B/cm^{3}mol^{-1}
350	-284
370	-248
390	-219
410	-195
430	-175
450	-159

$A(1) = -0.4249E+03$
$A(2) = -0.1075E+04$
$A(3) = -0.8541E+03$
$A(4) = 0.0000E+00$

System Number: 40
Cyclopropane, C_3H_6

T/K	B/cm^{3}mol^{-1}
310	-355
328	-313
346	-279
364	-249
382	-224
400	-203

$A(1) = -0.3878E+03$
$A(2) = -0.8606E+03$
$A(3) = -0.5383E+03$
$A(4) = 0.0000E+00$

System Number: 41
Propene, C_3H_6

T/K	B/cm^{3}mol^{-1}
280	-395
324	-290
368	-220
412	-170
456	-134
500	-106

$A(1) = -0.3469E+03$
$A(2) = -0.7272E+03$
$A(3) = -0.3251E+03$
$A(4) = 0.0000E+00$

System Number: 42
2-Propanone, C_3H_6O

T/K	B/cm^{3}mol^{-1}
300	-1996
336	-1253
372	-866
408	-641
444	-491
480	-374

$A(1) = -0.2051E+04$
$A(2) = -0.8903E+04$
$A(3) = -0.1806E+05$
$A(4) = -0.1645E+05$

System Number: 43
Ethyl methanoate, $C_3H_6O_2$

T/K	B/cm^{3}mol^{-1}
330	-1002
344	-883
358	-784
372	-700
386	-631
400	-573

$A(1) = -0.1371E+04$
$A(2) = -0.4231E+04$
$A(3) = -0.4312E+04$
$A(4) = 0.0000E+00$

System Number: 44
Methyl ethanoate, $C_3H_6O_2$

T/K	B/cm^{3}mol^{-1}
320	-1320
334	-1138
348	-997
362	-889
376	-808
390	-749

$A(1) = -0.1709E+04$
$A(2) = -0.6348E+04$
$A(3) = -0.9650E+04$
$A(4) = 0.0000E+00$

System Number: 45
1-Chloropropane, C_3H_7Cl

T/K	B/cm^{3}mol^{-1}
310	-1001
364	-641
418	-447
472	-330
526	-253
580	-198

$A(1) = -0.1121E+04$
$A(2) = -0.3271E+04$
$A(3) = -0.3786E+04$
$A(4) = -0.1974E+04$

System Number: 46
Propane, C_3H_8

T/K	B/cm^{3}mol^{-1}
240	-640
302	-375
364	-253
426	-181
488	-132
550	-95

$A(1) = -0.3859E+03$
$A(2) = -0.8444E+03$
$A(3) = -0.7198E+03$
$A(4) = -0.5739E+03$

System Number: 47
1-Propanol, C_3H_8O

T/K	B/cm^{3}mol^{-1}
380	-873
390	-782
400	-709
410	-651
420	-605
430	-571

$A(1) = -0.2690E+04$
$A(2) = -0.1204E+05$
$A(3) = -0.1674E+05$
$A(4) = 0.0000E+00$

System Number: 48
2-Propanol, C_3H_8O

T/K	B/cm^{3}mol^{-1}
380	-821
388	-736
396	-665
404	-609
412	-565
420	-532

$A(1) = -0.3165E+04$
$A(2) = -0.1609E+05$
$A(3) = -0.2420E+05$
$A(4) = 0.0000E+00$

System Number: 49
Trimethylamine, C_3H_9N

T/K	B/cm^{3}mol^{-1}
310	-674
324	-610
338	-554
352	-504
366	-461
380	-423

$A(1) = -0.7373E+03$
$A(2) = -0.1669E+04$
$A(3) = -0.9857E+03$
$A(4) = 0.0000E+00$

TABLE 2.2.1 (VIBT) VIRIAL COEFFICIENTS OF GASES (continued)

System Number: 50
1-Butene, C_4H_8

T/K	B/cm^{3}mol^{-1}
300	-624
324	-524
348	-445
372	-383
396	-333
420	-293

$A(1)$ = -0.6333E+03
$A(2)$ = -0.1442E+04
$A(3)$ = -0.9319E+03
$A(4)$ = 0.0000E+00

System Number: 51
2-Butanone, C_4H_8O

T/K	B/cm^{3}mol^{-1}
310	-2055
322	-1844
334	-1647
346	-1464
358	-1294
370	-1134

$A(1)$ = -0.2282E+04
$A(2)$ = -0.5907E+04
$A(3)$ = 0.0000E+00
$A(4)$ = 0.0000E+00

System Number: 52
Propyl methanoate, $C_4H_8O_2$

T/K	B/cm^{3}mol^{-1}
330	-1496
344	-1302
358	-1145
372	-1018
386	-915
400	-833

$A(1)$ = -0.2118E+04
$A(2)$ = -0.7299E+04
$A(3)$ = -0.8851E+04
$A(4)$ = 0.0000E+00

System Number: 53
Ethyl ethanoate, $C_4H_8O_2$

T/K	B/cm^{3}mol^{-1}
330	-1543
344	-1329
358	-1164
372	-1038
386	-945
400	-877

$A(1)$ = -0.2272E+04
$A(2)$ = -0.8818E+04
$A(3)$ = -0.1313E+05
$A(4)$ = 0.0000E+00

System Number: 54
Methyl propanoate, $C_4H_8O_2$

T/K	B/cm^{3}mol^{-1}
330	-1588
344	-1391
358	-1230
372	-1100
386	-994
400	-908

$A(1)$ = -0.2216E+04
$A(2)$ = -0.7339E+04
$A(3)$ = -0.8658E+04
$A(4)$ = 0.0000E+00

System Number: 55
1-Chlorobutane, C_4H_9Cl

T/K	B/cm^{3}mol^{-1}
330	-1224
378	-848
426	-628
474	-486
522	-385
570	-308

$A(1)$ = -0.1643E+04
$A(2)$ = -0.4897E+04
$A(3)$ = -0.6178E+04
$A(4)$ = -0.3718E+04

System Number: 56
Butane, C_4H_{10}

T/K	B/cm^{3}mol^{-1}
250	-1169
310	-667
370	-441
430	-315
490	-231
550	-170

$A(1)$ = -0.7354E+03
$A(2)$ = -0.1835E+04
$A(3)$ = -0.1922E+04
$A(4)$ = -0.1330E+04

System Number: 57
2-Methylpropane, C_4H_{10}

T/K	B/cm^{3}mol^{-1}
270	-900
318	-604
366	-432
414	-326
462	-258
510	-214

$A(1)$ = -0.7073E+03
$A(2)$ = -0.1719E+04
$A(3)$ = -0.1282E+04
$A(4)$ = 0.0000E+00

System Number: 58
1-Butanol, $C_4H_{10}O$

T/K	B/cm^{3}mol^{-1}
350	-1693
368	-1429
386	-1191
404	-973
422	-775
440	-592

$A(1)$ = -0.2629E+04
$A(2)$ = -0.6315E+04
$A(3)$ = 0.0000E+00
$A(4)$ = 0.0000E+00

System Number: 59
2-Methyl-1-propanol, $C_4H_{10}O$

T/K	B/cm^{3}mol^{-1}
390	-1076
400	-979
410	-887
420	-799
430	-715
440	-636

$A(1)$ = -0.2269E+04
$A(2)$ = -0.5065E+04
$A(3)$ = 0.0000E+00
$A(4)$ = 0.0000E+00

System Number: 60
2-Butanol, $C_4H_{10}O$

T/K	B/cm^{3}mol^{-1}
380	-1110
388	-1025
396	-944
404	-867
412	-792
420	-720

$A(1)$ = -0.2232E+04
$A(2)$ = -0.5209E+04
$A(3)$ = 0.0000E+00
$A(4)$ = 0.0000E+00

System Number: 61
2-Methyl-2-propanol, $C_4H_{10}O$

T/K	B/cm^{3}mol^{-1}
375	-973
385	-874
395	-781
405	-692
415	-607
425	-526

$A(1)$ = -0.1952E+04
$A(2)$ = -0.4775E+04
$A(3)$ = 0.0000E+00
$A(4)$ = 0.0000E+00

TABLE 2.2.1 (VIBT) VIRIAL COEFFICIENTS OF GASES (continued)

System Number: 62
Diethyl ether, $C_4H_{10}O$

T/K	*B*/cm³mol⁻¹
280	-1550
304	-1142
328	-876
352	-689
376	-546
400	-427

A(1) = -0.1226E+04
A(2) = -0.4458E+04
A(3) = -0.7746E+04
A(4) = -0.1000E+05

System Number: 63
Diethylamine, $C_4H_{11}N$

T/K	*B*/cm³mol⁻¹
310	-1343
330	-1134
350	-988
370	-868
390	-754
410	-637

A(1) = -0.1522E+04
A(2) = -0.5204E+04
A(3) = -0.1505E+05
A(4) = -0.2883E+05

System Number: 64
Pyridine, C_5H_5N

T/K	*B*/cm³mol⁻¹
340	-1342
360	-1175
380	-1026
400	-891
420	-769
440	-659

A(1) = -0.1765E+04
A(2) = -0.3431E+04
A(3) = 0.0000E+00
A(4) = 0.0000E+00

System Number: 65
Cyclopentane, C_5H_{10}

T/K	*B*/cm³mol⁻¹
295	-1084
301	-1042
307	-1001
313	-961
319	-923
325	-887

A(1) = -0.1062E+04
A(2) = -0.2116E+04
A(3) = 0.0000E+00
A(4) = 0.0000E+00

System Number: 66
1-Pentene, C_5H_{10}

T/K	*B*/cm³mol⁻¹
310	-965
330	-836
350	-728
370	-638
390	-560
410	-494

A(1) = -0.1055E+04
A(2) = -0.2377E+04
A(3) = -0.1189E+04
A(4) = 0.0000E+00

System Number: 67
2-Pentanone, $C_5H_{10}O$

T/K	*B*/cm³mol⁻¹
330	-2849
344	-2273
358	-1853
372	-1560
386	-1369
400	-1263

A(1) = -0.4962E+04
A(2) = -0.2637E+05
A(3) = -0.4654E+05
A(4) = 0.0000E+00

System Number: 68
Pentane, C_5H_{12}

T/K	*B*/cm³mol⁻¹
300	-1233
350	-818
400	-579
450	-435
500	-348
550	-294

A(1) = -0.1254E+04
A(2) = -0.3345E+04
A(3) = -0.2726E+04
A(4) = 0.0000E+00

System Number: 69
2-Methylbutane, C_5H_{12}

T/K	*B*/cm³mol⁻¹
280	-1263
314	-972
348	-767
382	-618
416	-508
450	-424

A(1) = -0.1095E+04
A(2) = -0.2503E+04
A(3) = -0.1534E+04
A(4) = 0.0000E+00

System Number: 70
2,2-Dimethylpropane, C_5H_{12}

T/K	*B*/cm³mol⁻¹
300	-916
350	-629
400	-464
450	-356
500	-278
550	-217

A(1) = -0.9310E+03
A(2) = -0.2387E+04
A(3) = -0.2641E+04
A(4) = -0.1810E+04

System Number: 71
Benzene, C_6H_6

T/K	*B*/cm³mol⁻¹
290	-1588
352	-969
414	-656
476	-478
538	-366
600	-290

A(1) = -0.1477E+04
A(2) = -0.3851E+04
A(3) = -0.3683E+04
A(4) = -0.1423E+04

System Number: 72
2-Methylpyridine, C_6H_7N

T/K	*B*/cm³mol⁻¹
360	-1656
374	-1473
388	-1317
402	-1184
416	-1069
430	-971

A(1) = -0.2940E+04
A(2) = -0.8813E+04
A(3) = -0.7809E+04
A(4) = 0.0000E+00

System Number: 73
3-Methylpyridine, C_6H_7N

T/K	*B*/cm³mol⁻¹
380	-1819
392	-1575
404	-1393
416	-1262
428	-1176
440	-1128

A(1) = -0.6304E+04
A(2) = -0.3042E+05
A(3) = -0.4455E+05
A(4) = 0.0000E+00

TABLE 2.2.1 (VIBT)
VIRIAL COEFFICIENTS OF GASES (continued)

System Number: 74
4-Methylpyridine, C_6H_7N

T/K	*B*/cm^{3}mol^{-1}
375	-1911
387	-1635
399	-1430
411	-1287
423	-1195
435	-1147

A(1) = -0.6553E+04
A(2) = -0.3287E+05
A(3) = -0.4987E+05
A(4) = 0.0000E+00

System Number: 75
Cyclohexane, C_6H_{12}

T/K	*B*/cm^{3}mol^{-1}
300	-1698
352	-1066
404	-768
456	-595
508	-471
560	-368

A(1) = -0.1733E+04
A(2) = -0.5618E+04
A(3) = -0.9486E+04
A(4) = -0.7936E+04

System Number: 76
Methylcyclopentane, C_6H_{12}

T/K	*B*/cm^{3}mol^{-1}
305	-1446
313	-1374
321	-1305
329	-1239
337	-1176
345	-1117

A(1) = -0.1512E+04
A(2) = -0.2910E+04
A(3) = 0.0000E+00
A(4) = 0.0000E+00

System Number: 110
1-Hexene, C_6H_{12}

T/K	*B*/cm^{3}mol^{-1}
313	-1499
333	-1272
353	-1093
373	-951
393	-837
413	-746

A(1) = -0.1707E+04
A(2) = -0.4575E+04
A(3) = -0.4020E+04
A(4) = 0.0000E+00

System Number: 77
Hexane, C_6H_{14}

T/K	*B*/cm^{3}mol^{-1}
300	-1920
330	-1424
360	-1122
390	-916
420	-755
450	-615

A(1) = -0.1961E+04
A(2) = -0.6691E+04
A(3) = -0.1317E+05
A(4) = -0.1527E+05

System Number: 111
2,3-Dimethylbutane, C_6H_{14}

T/K	*B*/cm^{3}mol^{-1}
293	-1599
301	-1519
309	-1443
317	-1372
325	-1303
333	-1238

A(1) = -0.1547E+04
A(2) = -0.2948E+04
A(3) = 0.0000E+00
A(4) = 0.0000E+00

System Number: 78
Triethylamine, $C_6H_{15}N$

T/K	*B*/cm^{3}mol^{-1}
330	-1561
344	-1400
358	-1266
372	-1154
386	-1060
400	-982

A(1) = -0.2061E+04
A(2) = -0.5735E+04
A(3) = -0.5899E+04
A(4) = 0.0000E+00

System Number: 79
Toluene, C_7H_8

T/K	*B*/cm^{3}mol^{-1}
350	-1641
366	-1439
382	-1269
398	-1126
414	-1005
430	-902

A(1) = -0.2620E+04
A(2) = -0.7548E+04
A(3) = -0.6349E+04
A(4) = 0.0000E+00

System Number: 80
1-Heptene, C_7H_{14}

T/K	*B*/cm^{3}mol^{-1}
340	-1781
354	-1602
368	-1444
382	-1305
396	-1182
410	-1072

A(1) = -0.2491E+04
A(2) = -0.6230E+04
A(3) = -0.3780E+04
A(4) = 0.0000E+00

System Number: 81
Heptane, C_7H_{16}

T/K	*B*/cm^{3}mol^{-1}
300	-2781
380	-1414
460	-861
540	-582
620	-416
700	-303

A(1) = -0.2834E+04
A(2) = -0.8523E+04
A(3) = -0.1007E+05
A(4) = -0.5051E+04

System Number: 82
1,2-Dimethylbenzene, C_8H_{10}

T/K	*B*/cm^{3}mol^{-1}
380	-2045
392	-1811
404	-1622
416	-1471
428	-1352
440	-1261

A(1) = -0.5632E+04
A(2) = -0.2287E+05
A(3) = -0.2890E+05
A(4) = 0.0000E+00

System Number: 83
1,3-Dimethylbenzene, C_8H_{10}

T/K	*B*/cm^{3}mol^{-1}
380	-2082
392	-1825
404	-1613
416	-1438
428	-1297
440	-1183

A(1) = -0.5808E+04
A(2) = -0.2324E+05
A(3) = -0.2761E+05
A(4) = 0.0000E+00

TABLE 2.2.1 (VIBT)
VIRIAL COEFFICIENTS OF GASES (continued)

System Number: 84
1,4-Dimethylbenzene, C_8H_{10}

T/K	B/cm^{3}mol^{-1}
380	-2043
392	-1815
404	-1617
416	-1446
428	-1298
440	-1171

$A(1)$ = -0.4921E+04
$A(2)$ = -0.1684E+05
$A(3)$ = -0.1616E+05
$A(4)$ = 0.0000E+00

System Number: 85
1-Octene, C_8H_{16}

T/K	B/cm^{3}mol^{-1}
360	-2146
370	-2000
380	-1861
390	-1729
400	-1603
410	-1484

$A(1)$ = -0.3273E+04
$A(2)$ = -0.6557E+04
$A(3)$ = 0.0000E+00
$A(4)$ = 0.0000E+00

System Number: 86
Octane, C_8H_{18}

T/K	B/cm^{3}mol^{-1}
300	-4042
380	-1972
460	-1164
540	-767
620	-533
700	-375

$A(1)$ = -0.4123E+04
$A(2)$ = -0.1312E+05
$A(3)$ = -0.1641E+05
$A(4)$ = -0.8580E+04

TABLE 2.2.2 (PVTL)
SATURATED DENSITIES AND MOLAR VOLUMES OF LIQUIDS

The *saturated density* ρ and the liquid *molar volume* V (see definitions in Section 2.2), are functions of the temperature T. The temperature dependence of V at constant pressure P and a given T is given by the *cubic expansion coefficient* α:

$$\alpha = (1/V)(\partial V/\partial T)_P \tag{1}$$

The variation of V over a large temperature range can be fitted to various empirical equations. A simple equation is the so-called Rackett equation which is used here in the following form:

$$V(T)/\mathrm{m^3\,mol^{-1}} = A(2) \times A(1)^{\left[1+(1-T/T_c)^{2/7}\right]} \tag{2}$$

where T_c is the critical temperature (or a value close to it, see Table 2.1.1) and $A(1)$ and $A(2)$ are adjustable parameters. The parameter $A(2)$ has a value close to

$$A(2) = RT_c/P_c \tag{3}$$

where P_c is the critical pressure. In this table Equation 2 was fitted, whenever possible, to the experimental density at 298.15 K.

The table gives T_c, the coefficients $A(1)$ and $A(2)$, and the values of ρ and V at $T = 298.15$ K (except for substances which do not exist as liquids at this T). The 298.15 K value is the most accurate value available at that temperature, and is not necessarily generated from Equation 2. The program, Property Code PVTL permits the calculation of ρ and V at selected temperature increments within the range of validity of the fitting equation. Moreover it permits one to calculate the specific volume $v = V/M$, where M is the molar mass, and an approximate value of the cubic expansion coefficient α:

$$\alpha/\mathrm{K^{-1}} = -\frac{2}{7}\ \ln\ A(1)\left[1-(T/T_c)\right]^{-(5/7)}/(T_c/K) \tag{4}$$

The substances are arranged in a modified Hill order, with compounds not containing carbon preceding those that contain carbon.

Physical quantity	Symbol	SI unit
Molar volume	V	$\mathrm{m^3\ mol^{-1}}$
Specific volume	v	$\mathrm{m^3\ kg^{-1}}$
Density	ρ	$\mathrm{kg\ m^{-3}}$
Cubic expansion coefficient	α	$\mathrm{K^{-1}}$
Temperature	T	K

REFERENCES

1. Spencer, C. F.; Adler, S. B., "A Critical Review of Equations for Predicting Saturated Liquid Density", *J. Chem. Eng. Data*, 23, 82-89, 1987 (correlated data for 165 pure organic and inorganic substances).
2. Tekac, V.; Cibulka, I.; Holub, R., "PVT Properties of Liquids and Liquid Mixtures: A Review of the Experimental Methods and the Literature Data", *Fluid Phase Equilib.*, 19, 33-149, 1985 (an annotated bibliography on experimental data for ca. 350 pure substances and 170 binary liquid mixtures).
3. Hankinson, R. W.; Thomson, G. H., "A New Correlation for Saturated Densities of Liquids and their Mixtures", *AIChE J.*, 25, 653-663, 1979 (correlated data for 200 pure organic and inorganic substances).
4. *TRC Thermodynamic Tables—Hydrocarbons*, Thermodynamic Research Center: The Texas A & M University System, College Station, TX (Loose-leaf data sheets) (recommended data).
5. *TRC Thermodynamic Tables—Non-hydrocarbons*, Thermodynamic Research Center: The Texas A & M University System, College Station, TX (Loose-leaf data sheets) (recommended data).
6. Riddick, J. A.; Bunger, W. B.; Sakano, T. K., "Organic Solvents", in: *Techniques of Chemistry*, Vol. II, 4th Edition, John Wiley & Sons, New York, 1986 (recommended density data, mainly at room temperature, for ca. 500 organic substances).

TABLE 2.2.2 (PVTL) SATURATED DENSITIES AND MOLAR VOLUMES OF LIQUIDS (continued)

SN	Molecular formula	Name	ρ (298.15 K)/ g cm^{-3}	V (298.15 K)/ cm^3 mol^{-1}	T_c/K	$A(1)$	$A(2)$	T/K range
1	Ar	Argon			150.76	0.29216E+00	0.25719E-03	86-144
2	ClH	Hydrogen chloride	0.80400	45.349	324.60	0.26568E+00	0.32616E-03	163-323
3	Cl_2	Chlorine	1.39400	50.865	417.15	0.27676E+00	0.45101E-03	173-416
4	FH	Hydrogen fluoride	0.95800	20.883	461.00	0.14514E+00	0.60348E-03	189-443
5	F_2	Fluorine			144.30	0.28867E+00	0.22992E-03	54-143
6	H_2	Hydrogen			33.26	0.31997E+00	0.21322E-03	14-32
7	H_2S	Dihydrogen sulfide	0.76886	44.328	372.54	0.28476E+00	0.34395E-03	282-368
8	H_3N	Ammonia	0.60198	28.291	405.54	0.24658E+00	0.29899E-03	197-403
9	Kr	Krypton			209.43	0.29007E+00	0.31649E-03	118-162
10	NO	Nitrogen oxide	0.50556	59.352	309.59	0.27592E+00	0.35531E-03	243-309
11	N_2	Nitrogen			126.26	0.28971E+00	0.30900E-03	64-125
12	Ne	Neon			44.37	0.30852E+00	0.13386E-03	25-41
13	O_2	Oxygen			154.59	0.28962E+00	0.25473E-03	61-150
14	O_2S	Sulfur dioxide			430.80	0.26729E+00	0.45438E-03	323-423
15	Xe	Xenon			289.71	0.28288E+00	0.41273E-03	164-285
16	$CClF_3$	Chlorotrifluoromethane	0.83000	125.854	302.10	0.27971E+00	0.65076E-03	131-299
17	CCl_2O	Carbonyl dichloride	1.37193	72.100	455.15	0.27931E+00	0.66149E-03	264-333
18	CCl_3F	Trichlorofluoromethane	1.47600	93.067	471.15	0.27560E+00	0.88903E-03	200-464
19	CCl_4	Tetrachloromethane	1.58436	97.088	556.37	0.27222E+00	0.10140E-02	253-554
20	CF_4	Tetrafluoromethane			227.60	0.28008E+00	0.50613E-03	91-226
21	$CHClF_2$	Chlorodifluoromethane	1.1942	72.41	369.20	0.26800E+00	0.78300E-03	273-354
22	$CHCl_3$	Trichloromethane	1.47970	80.676	536.37	0.27498E+00	0.81677E-03	210-343
23	CHF_3	Trifluoromethane	0.67300	104.033	299.07	0.25871E+00	0.52101E-03	118-298
24	CH_2Cl_2	Dichloromethane	1.31678	64.500	510.00	0.26184E+00	0.69873E-03	178-383
25	CH_2F_2	Difluoromethane	0.96000	54.191	351.60	0.24651E+00	0.49790E-03	143-347
26	CH_3Cl	Chloromethane	0.91110	55.414	416.26	0.26793E+00	0.51844E-03	175-414
27	CH_3F	Fluoromethane	0.56600	60.129	317.80	0.24909E+00	0.45214E-03	133-316
28	CH_3NO_2	Nitromethane	1.13128	53.957	587.98	0.23126E+00	0.77176E-03	298-476
29	CH_4	Methane			190.54	0.28941E+00	0.34416E-03	89-190
30	CH_4O	Methanol	0.78664	40.733	512.65	0.23320E+00	0.54345E-03	260-511
31	CO	Carbon oxide			132.93	0.28966E+00	0.31617E-03	68-131
32	CO_2	Carbon dioxide	0.71965	61.155	304.21	0.27275E+00	0.34276E-03	218-303
33	CS_2	Carbon disulfide	1.25550	60.648	552.15	0.28492E+00	0.58194E-03	244-313
34	$C_2Cl_2F_4$	1,2-Dichlorotetrafluoroethane	1.45501	117.471	418.60	0.27526E+00	0.10536E-02	196-415
35	C_2F_3N	Trifluoroethanenitrile	0.85055	111.720	311.11	0.26636E+00	0.71510E-03	193-288
36	C_2H_2	Ethyne	0.37716	69.036	308.32	0.27063E+00	0.41770E-03	192-300
37	$C_2H_3F_3$	1,1,1-Trifluoroethane	0.96330	87.243	346.26	0.25183E+00	0.75926E-03	233-344
38	C_2H_3N	Ethanenitrile	0.77649	52.869	547.87	0.19866E+00	0.96796E-03	297-542
39	C_2H_4	Ethene			282.37	0.28054E+00	0.46668E-03	104-281
40	$C_2H_4F_2$	1,1-Difluoroethane	0.89600	73.717	386.65	0.25335E+00	0.71636E-03	193-385
41	C_2H_4O	Oxirane	0.87172	50.536	468.98	0.25762E+00	0.54202E-03	233-461

42	$C_2H_4O_2$	Ethanoic acid	1.04392	57.526	592.71	0.22253E+00	0.88494E-03	294-589
43	$C_2H_4O_2$	Methyl methanoate	0.96640	62.141	487.15	0.25778E+00	0.67817E-03	273-473
44	C_2H_5Br	Bromoethane	1.45050	75.123	503.93	0.28962E+00	0.67704E-03	162-303
45	C_2H_5Cl	Chloroethane	0.88891	72.577	460.43	0.26540E+00	0.73209E-03	113-458
46	C_2H_6	Ethane	0.31480	95.520	305.43	0.28128E+00	0.52525E-03	89-305
47	C_2H_6O	Dimethyl ether	0.66120	69.675	400.00	0.27420E+00	0.60975E-03	273-398
48	C_2H_6O	Ethanol	0.78493	58.692	516.15	0.25041E+00	0.69186E-03	238-511
49	C_2H_7N	Ethylamine	0.67690	66.604	456.15	0.26419E+00	0.67390E-03	273-421
50	C_3F_6O	Hexafluoro-2-propanone	1.31285	126.460	357.14	0.26641E+00	0.10467E-02	210-300
51	C_3H_4	Propadiene	0.58363	68.648	393.15	0.27283E+00	0.59798E-03	194-303
52	C_3H_4	Propyne	0.60665	66.043	402.37	0.27027E+00	0.59470E-03	218-368
53	C_3H_5N	Propanenitrile	0.77682	70.904	564.37	0.21690E+00	0.11218E-02	273-372
54	C_3H_6	Cyclopropane	0.61757	68.139	398.25	0.27429E+00	0.59407E-03	293-393
55	C_3H_6	Propene	0.50480	83.361	364.76	0.27821E+00	0.65829E-03	200-363
56	C_3H_6O	Methyloxirane	0.82090	70.752	488.15	0.26221E+00	0.75000E-03	273-474
57	C_3H_6O	2-Propanone	0.78440	74.044	508.15	0.24494E+00	0.90169E-03	179-506
58	$C_3H_6O_2$	Ethyl methanoate	0.91530	80.935	508.43	0.25863E+00	0.89503E-03	273-493
59	$C_3H_6O_2$	Methyl ethanoate	0.92790	79.836	506.87	0.25523E+00	0.90269E-03	273-493
60	$C_3H_6O_2$	Propanoic acid	0.98808	74.973	612.04	0.24906E+00	0.94939E-03	273-313
61	C_3H_8	Propane	0.49270	89.500	369.82	0.27664E+00	0.72298E-03	89-369
62	C_3H_8O	Ethyl methyl ether	0.69220	86.819	437.80	0.26728E+00	0.84155E-03	273-427
63	C_3H_8O	1-Propanol	0.79960	75.158	536.78	0.25272E+00	0.88551E-03	256-533
64	C_3H_8O	2-Propanol	0.78126	76.922	508.32	0.24962E+00	0.90589E-03	266-505
65	C_3H_9N	Isopropylamine	0.68210	86.661	471.85	0.26849E+00	0.86722E-03	213-373
66	C_3H_9N	Propylamine	0.71210	83.010	496.95	0.26444E+00	0.87385E-03	213-413
67	C_3H_9N	Trimethylamine	0.62700	94.276	433.26	0.27148E+00	0.88426E-03	273-308
68	C_4F_8	Octafluorocyclobutane	1.50000	133.354	388.37	0.27048E+00	0.11671E-02	233-383
69	C_4F_{10}	Decafluorobutane			386.37	0.26988E+00	0.13845E-02	318-358
70	C_4H_6	1,3-Butadiene	0.61490	87.968	425.37	0.27130E+00	0.81692E-03	164-413
71	C_4H_7N	Butanenitrile	0.78650	87.866	582.21	0.22873E+00	0.12776E-02	273-369
72	C_4H_8	1-Butene	0.58850	95.340	419.56	0.27351E+00	0.86570E-03	195-416
73	C_4H_8	*cis*-2-Butene	0.61650	91.010	435.59	0.27044E+00	0.86198E-03	195-353
74	C_4H_8	*trans*-2-Butene	0.59910	93.653	428.65	0.27212E+00	0.86931E-03	223-353
75	C_4H_8	Cyclobutane	0.68900	81.433	460.37	0.27634E+00	0.76555E-03	233-333
76	C_4H_8	2-Methylpropene	0.58910	95.243	417.93	0.27277E+00	0.86664E-03	203-417
77	$C_4H_8O_2$	Butanoic acid	0.95290	92.461	628.15	0.27426E+00	0.98914E-03	272-528
78	$C_4H_8O_2$	Ethyl ethanoate	0.89455	98.492	523.30	0.25389E+00	0.11393E-02	273-503
79	$C_4H_8O_2$	Methyl propanoate	0.90900	96.927	530.54	0.25656E+00	0.11065E-02	273-523
80	$C_4H_8O_2$	2-Methylpropanoic acid	0.94310	93.422	609.15	0.24027E+00	0.12613E-02	214-508
81	$C_4H_8O_2$	Propyl methanoate	0.89960	97.939	538.04	0.25910E+00	0.11045E-02	273-533
82	C_4H_{10}	Butane	0.57287	101.460	425.15	0.27331E+00	0.93011E-03	143-424
83	C_4H_{10}	2-Methylpropane	0.55092	105.502	408.15	0.27569E+00	0.92808E-03	193-407
84	$C_4H_{10}O$	1-Butanol	0.80575	91.992	563.05	0.25380E+00	0.10949E-02	253-558
85	$C_4H_{10}O$	Diethyl ether	0.70782	104.720	466.71	0.26444E+00	0.10704E-02	273-463
86	$C_4H_{11}N$	Butylamine	0.73686	99.256	524.15	0.26581E+00	0.10585E-02	213-412
87	$C_4H_{11}N$	Diethylamine	0.70160	104.245	496.65	0.25677E+00	0.11558E-02	253-413
88	$C_4H_{11}N$	(2-Methylpropyl)amine	0.72970	100.230	516.00	0.27347E+00	0.10098E-02	213-433
89	C_5H_{10}	Cyclopentane	0.74045	94.719	511.61	0.26824E+00	0.98423E-03	293-508

TABLE 2.2.2 (PVTL)
SATURATED DENSITIES AND MOLAR VOLUMES OF LIQUIDS (continued)

SN	Molecular formula	Name	ρ (298.15 K)/ g cm^{-3}	V (298.15 K)/ cm^3 mol^{-1}	T_c/K	A(1)	A(2)	T/K range
90	C_5H_{10}	1-Pentene	0.63530	110.396	465.04	0.27035E+00	0.10837E-02	273-463
91	$C_5H_{10}O_2$	Ethyl propanoate	0.88400	115.535	546.04	0.25459E+00	0.13522E-02	273-533
92	$C_5H_{10}O_2$	Methyl butanoate	0.89261	114.421	554.43	0.25627E+00	0.13308E-02	273-543
93	$C_5H_{10}O_2$	Methyl 2-methylpropanoate	0.88538	115.355	540.76	0.25848E+00	0.13090E-02	273-533
94	$C_5H_{10}O_2$	Pentanoic acid	0.93450	109.292	650.98	0.24882E+00	0.14120E-02	233-533
95	$C_5H_{10}O_2$	Propyl ethanoate	0.88303	115.662	549.37	0.25264E+00	0.13756E-02	273-533
96	C_5H_{12}	2,2-Dimethylpropane	0.58520	123.292	433.76	0.27570E+00	0.11270E-02	253-432
97	C_5H_{12}	2-Methylbutane	0.61420	117.470	460.43	0.27167E+00	0.11377E-02	150-455
98	C_5H_{12}	Pentane	0.62139	116.111	469.65	0.26853E+00	0.11590E-02	183-469
99	$C_5H_{12}O$	1-Pentanol	0.81080	108.719	588.15	0.25960E+00	0.12606E-02	213-453
100	C_6F_6	Hexafluorobenzene	1.61441	115.247	516.73	0.25667E+00	0.13007E-02	398-498
101	C_6H_5Br	Bromobenzene	1.48820	105.503	670.15	0.26370E+00	0.12344E-02	273-543
102	C_6H_5Cl	Chlorobenzene	1.10090	102.242	632.37	0.26510E+00	0.11662E-02	273-543
103	C_6H_5F	Fluorobenzene	1.01940	94.275	560.09	0.26616E+00	0.10278E-02	273-548
104	C_6H_5I	Iodobenzene	1.82290	111.915	721.15	0.26453E+00	0.13252E-02	273-373
105	C_6H_6	Benzene	0.87360	89.416	562.15	0.26967E+00	0.95325E-03	273-558
106	C_6H_7N	Aniline	1.01750	91.527	699.00	0.26165E+00	0.10946E-02	273-553
107	C_6H_{12}	Cyclohexane	0.77389	108.751	553.54	0.27286E+00	0.11290E-02	280-529
108	$C_6H_{12}O$	4-Methyl-2-pentanone	0.79620	125.798	571.00	0.25892E+00	0.14512E-02	293-373
109	C_6H_{14}	2,3-Dimethylbutane	0.65702	131.164	499.98	0.26937E+00	0.13299E-02	343-473
110	C_6H_{14}	Hexane	0.65484	131.600	507.43	0.26355E+00	0.14062E-02	183-500
111	$C_6H_{14}O$	1-Hexanol	0.81534	125.318	610.37	0.25748E+00	0.14922E-02	253-513
112	$C_6H_{15}N$	Dipropylamine	0.73290	138.070	550.15	0.26909E+00	0.14666E-02	213-453
113	$C_6H_{15}N$	Triethylamine	0.72305	139.951	535.15	0.26934E+00	0.14693E-02	213-433
114	C_7H_8	Toluene	0.86219	106.868	591.82	0.26455E+00	0.11997E-02	178-578
115	C_7H_9N	*N*-Methylaniline	0.98221	109.096	701.21	0.28490E+00	0.11185E-02	253-553
116	C_7H_{14}	Cycloheptane	0.80660	121.731	604.26	0.26957E+00	0.13086E-02	477-580
117	C_7H_{14}	Methylcyclohexane	0.76506	128.340	572.21	0.26986E+00	0.13746E-02	273-533
118	C_7H_{16}	2,2-Dimethylpentane	0.66953	149.663	520.48	0.26733E+00	0.15755E-02	203-513
119	C_7H_{16}	2,3-Dimethylpentane	0.69091	145.032	537.37	0.25987E+00	0.15370E-02	363-533
120	C_7H_{16}	2,4-Dimethylpentane	0.66832	149.934	519.76	0.26539E+00	0.15981E-02	298-513
121	C_7H_{16}	3,3-Dimethylpentane	0.68908	145.417	536.43	0.27348E+00	0.15147E-02	363-533
122	C_7H_{16}	3-Ethylpentane	0.69395	144.397	540.65	0.26641E+00	0.15556E-02	373-538
123	C_7H_{16}	Heptane	0.67946	147.476	540.26	0.26074E+00	0.16469E-02	183-538
124	C_7H_{16}	2-Methylhexane	0.67439	148.585	530.32	0.26288E+00	0.16135E-02	363-523

125	C_7H_{16}	3-Methylhexane	0.68295	146.722	535.26	0.26103E+00	0.15822E-02	373-533
126	C_7H_{16}	2,2,3-Trimethylbutane	0.68588	146.096	531.15	0.26978E+00	0.14957E-02	373-523
127	C_8H_{10}	1,2-Dimethylbenzene	0.87594	121.204	630.37	0.26326E+00	0.13990E-02	293-608
128	C_8H_{10}	1,3-Dimethylbenzene	0.86009	123.438	617.04	0.25919E+00	0.14569E-02	245-603
129	C_8H_{10}	1,4-Dimethylbenzene	0.85661	123.939	616.26	0.25888E+00	0.14655E-02	273-613
130	C_8H_{10}	Ethylbenzene	0.86253	123.088	617.15	0.26186E+00	0.14259E-02	178-391
131	$C_8H_{11}N$	*N,N*-Dimethylaniline	0.95232	127.249	687.10	0.25577E+00	0.15853E-02	293-553
132	C_8H_{16}	Cyclooctane	0.83200	134.874	647.15	0.26672E+00	0.15099E-02	569-638
133	C_8H_{18}	2,2-Dimethylhexane	0.69112	165.284	549.87	0.26393E+00	0.18085E-02	393-548
134	C_8H_{18}	2,3-Dimethylhexane	0.70809	161.323	563.48	0.26220E+00	0.17832E-02	393-558
135	C_8H_{18}	2,4-Dimethylhexane	0.69620	164.078	553.54	0.26579E+00	0.18010E-02	382-543
136	C_8H_{18}	2,5-Dimethylhexane	0.68934	165.711	550.04	0.26143E+00	0.18400E-02	393-543
137	C_8H_{18}	3,3-Dimethylhexane	0.70596	161.809	561.98	0.26007E+00	0.17615E-02	393-553
138	C_8H_{18}	3,4-Dimethylhexane	0.71516	159.728	568.87	0.26322E+00	0.17575E-02	403-563
139	C_8H_{18}	3-Ethylhexane	0.70948	161.007	565.48	0.25853E+00	0.18034E-02	393-563
140	C_8H_{18}	3-Ethyl-2-methylpentane	0.71522	159.714	567.09	0.26122E+00	0.17468E-02	403-563
141	C_8H_{18}	3-Ethyl-3-methylpentane	0.72354	157.878	576.59	0.26657E+00	0.17081E-02	393-573
142	C_8H_{18}	2-Methylheptane	0.69392	164.617	559.65	0.25806E+00	0.18737E-02	393-553
143	C_8H_{18}	3-Methylheptane	0.70175	162.780	563.65	0.25763E+00	0.18412E-02	393-558
144	C_8H_{18}	4-Methylheptane	0.70055	163.059	561.78	0.25884E+00	0.18381E-02	403-553
145	C_8H_{18}	Octane	0.69862	163.509	568.82	0.25678E+00	0.19122E-02	223-567
146	C_8H_{18}	2,2,3,3-Tetramethylbutane			567.93	0.27450E+00	0.16473E-02	367-510
147	C_8H_{18}	2,2,3-Trimethylpentane	0.71207	160.421	563.48	0.26494E+00	0.17170E-02	393-553
148	C_8H_{18}	2,2,4-Trimethylpentane	0.68781	166.079	543.98	0.26719E+00	0.17795E-02	298-533
149	C_8H_{18}	2,3,3-Trimethylpentane	0.72232	158.144	573.54	0.26855E+00	0.16917E-02	393-568
150	C_8H_{18}	2,3,4-Trimethylpentane	0.71503	159.757	566.43	0.26558E+00	0.17260E-02	393-563
151	C_9H_{12}	Isopropylbenzene	0.85743	140.180	631.15	0.26164E+00	0.16370E-02	273-425
152	C_9H_{12}	Propylbenzene	0.85790	140.103	638.37	0.25990E+00	0.16616E-02	178-373
153	C_9H_{20}	Nonane	0.71375	179.696	594.65	0.25456E+00	0.21668E-02	219-533
154	C_9H_{20}	2,2,5-Trimethylhexane	0.70322	182.386	568.04	0.26365E+00	0.20275E-02	393-548
155	$C_{10}H_8$	Naphthalene			748.43	0.26100E+00	0.15365E-02	333-473
156	$C_{10}H_{22}$	Decane	0.72635	195.890	617.65	0.25074E+00	0.24572E-02	243-616
157	$C_{10}H_{22}O$	1-Decanol	0.82630	191.558	700.37	0.26274E+00	0.22813E-02	293-553
158	$C_{11}H_{24}$	Undecane	0.73650	212.236	638.71	0.24990E+00	0.27055E-02	253-573
159	$C_{12}H_{10}$	Biphenyl			789.26	0.27432E+00	0.17061E-02	331-644
160	$C_{12}H_{26}$	Dodecane	0.74518	228.587	658.26	0.24692E+00	0.30046E-02	263-633
161	$C_{13}H_{28}$	Tridecane	0.75271	244.935	675.76	0.24698E+00	0.32411E-02	273-593
162	$C_{14}H_{30}$	Tetradecane	0.75920	261.317	691.87	0.24322E+00	0.35796E-02	283-553
163	$C_{17}H_{36}$	Heptadecane	0.77460	310.448	733.37	0.23431E+00	0.46251E-02	293-613
164	$C_{18}H_{38}$	Octadecane	0.77910	326.659	745.04	0.22917E+00	0.50915E-02	298-613
165	$C_{20}H_{42}$	Icosane	0.78230	361.183	767.04	0.22811E+00	0.57116E-02	310-423

2.3. CALORIMETRIC PROPERTIES OF NON-REACTING SYSTEMS

In the absence of chemical reactions, the *internal energy*, U, and *enthalpy*, $H = U + PV$, are determined by temperature, pressure, and physical state. This section deals with the dependence on temperature and physical state. At moderate pressures (say, up to 10 atmospheres), the pressure dependence of these properties is normally quite small.

Heat capacity is a measure of the change of enthalpy, H, or internal energy, U, of a system with a change in temperature. The most common types of heat capacity are the molar isobaric (constant-pressure) heat capacity, C_p, defined as

$$C_p = (\partial H / \partial T)_P \tag{1}$$

and the molar isochoric (constant-volume) heat capacity, C_v, defined as

$$C_v = (\partial U / \partial T)_V \tag{2}$$

The former is more commonly used in applied calculations, since many processes of interest take place at constant pressure.

Tables 2.3.1, 2.3.2, and 2.3.3 in this section give the molar isobaric heat capacity, C_p. The specific heat capacity, c_p, which refers to a mass rather than an amount-of-substance basis, may be calculated from it by the relation

$$c_p / \mathrm{J\,K^{-1}\,kg^{-1}} = (1000 / M_r)(C_p / \mathrm{J\,K^{-1}\,mol^{-1}}) \tag{3}$$

where M_r is the relative molar mass (molecular weight).

Enthalpy and internal energy (as well as other thermodynamic properties) generally show a change when the physical state of the substance changes. Tables 2.3.4 and 2.3.5 give the change of enthalpy upon fusion (melting) and vaporization (boiling) for a number of common substances.

Physical quantity	Symbol	SI unit
Molar isobaric heat capacity	C_p	$\mathrm{J\,K^{-1}\,mol^{-1}}$
Specific isobaric heat capacity	c_p	$\mathrm{J\,K^{-1}\,kg^{-1}}$
Molar enthalpy of fusion	$\Delta_{fus} H$	$\mathrm{J\,mol^{-1}}$
Molar enthalpy of vaporization	$\Delta_{vap} H$	$\mathrm{J\,mol^{-1}}$
Normal melting point temperature	T_m	K
Normal boiling point temperature	T_b	K

TABLE 2.3.1 (CPGT)
IDEAL GAS HEAT CAPACITY

The molar isobaric heat capacity C_p of an ideal gas may be calculated with high accuracy if the structure and dynamics of the molecules comprising the gas are accurately known; i.e., if the molecular parameters which describe the rotation and internal vibrations of the molecule are available. To a good approximation C_p may be written as the sum of translational, rotational, and vibrational contributions. The translational contribution is simply 5/2 R, where R is the gas constant, or 20.8 J K^{-1} mol^{-1}; the actual C_p for the noble gases is very close to this value, independent of temperature. For a linear molecule the sum of translational and rotational contributions is 7/2 R or 29.1 J K^{-1} mol^{-1}. Diatomic molecules such as H_2 and N_2 exhibit C_p values close to this at ambient conditions, but the heat capacity rises with increasing temperature as higher vibrational states are excited. The translational and rotational contributions amount to 4 R or 33.2 J K^{-1} mol^{-1} for polyatomic molecules. However, the vibrational contribution becomes dominant, especially at higher temperatures, as the size and complexity of the molecule increases.

The molar isochoric heat capacity C_v for an ideal gas is equal to C_p-R.

The ideal gas heat capacity can be expressed as a power series:

$$C_p / \mathrm{J\,K^{-1}\,mol^{-1}} = A(1) + A(2)(T/\mathrm{K}) + A(3)(T/\mathrm{K})^2 + A(4)(T/\mathrm{K})^3 + A(5)(T/\mathrm{K})^4 \qquad (1)$$

This table gives the coefficients $A(i)$ and the valid temperature range for 76 common substances in the gaseous state. The program CPGT permits the calculation of C_p and the specific heat capacity c_p at any desired set of temperatures in the valid range. The table also gives the recommended value of C_p at 298.15 K; the value generated by the program at 298.15 K may differ slightly from this because of the fitting procedure.

The substances are arranged in a modified Hill order, with compounds not containing carbon preceding those that contain carbon.

REFERENCES

1. *DIPPR Data Compilation of Pure Compound Properties*, Design Institute for Physical Property Data, American Institute of Chemical Engineers, New York.
2. Wagman, D. D.; Evans, W. H.; Parker, V. B.; Schumm, R. H.; Halow, I., Bailey, S. M.; Churney, K. L.; Nuttall, R. L., *The NBS Tables of Chemical Thermodynamic Properties*, *J. Phys. Chem. Ref. Data*, Vol. 11, Suppl. 2, 1982.
3. Chase, M. W.; Davies, C. A.; Downey, J. R.; Frurip, D. J.; McDonald, R. A.; Syverud, A. N., *JANAF Thermochemical Tables, Third Edition*, *J. Phys. Chem. Ref. Data*, Vol. 14, Suppl. 1, 1985.
4. Gurvich, L. V.; Veyts, I. V.; Alcock, C. B., *Thermodynamic Properties of Individual Substances, Fourth Edition*, Hemisphere Publishing Corp., New York, 1989.
5. Cox, J. D.; Wagman, D. D.; Medvedev, V. A., *CODATA Key Values for Thermodynamics*, Hemisphere Publishing Corp. New York, 1989.
6. Lide, D. R., Editor, *CRC Handbook of Chemistry and Physics, 74th Edition*, CRC Press, Boca Raton, FL, 1993.

SN	Molecular formula	Name	C_p (298.15 K)/ J K^{-1} mol^{-1}	$A(1)$	$A(2)$	$A(3)$	$A(4)$	$A(5)$	(T/K)-range
1	Ar	Argon	20.79	20.79					100-6000
2	BrH	Hydrogen bromide	29.13	29.72	-0.00416	7.3177E-06			298-800
3	Br_2	Bromine	36.05	30.11	0.03353	-5.5009E-05	3.1711E-08		298-800
4	ClH	Hydrogen chloride	29.14	29.81	-0.00412	6.2231E-06			298-800
5	Cl_2	Chlorine	33.95	22.85	0.06543	-1.2517E-04	1.1484E-07	-4.0946E-11	298-800
6	Cl_4Si	Silicon tetrachloride	90.26	34.28	0.33371	-6.4945E-04	6.0070E-07	-2.1564E-10	298-800
7	FH	Hydrogen fluoride	29.14	28.94	0.00152	-4.0674E-06	3.8970E-09		298-800
8	F_2	Fluorine	31.30	23.06	0.03742	-3.6836E-05	1.3515E-08		298-800
9	F_3N	Nitrogen trifluoride	53.65	15.67	0.16200	-1.1617E-04			200-500
10	H_2	Hydrogen	28.84	22.66	0.04381	-1.0835E-04	1.1710E-07	-4.5660E-11	298-800
11	H_2O	Water	33.59	33.80	-0.00795	2.8228E-05	-1.3115E-08		298-800
12	H_2S	Dihydrogen sulfide	34.19	32.34	-0.00112	2.8857E-05	-1.4441E-08		298-800
13	H_3N	Ammonia	35.62	29.29	0.01103	4.2446E-05	-2.7706E-08		298-800
14	H_4N_2	Hydrazine	50.80	3.03	0.20665	-1.7381E-04	6.0707E-08		298-800
15	He	Helium	20.79	20.79					100-6000

TABLE 2.3.1 (CPGT)
IDEAL GAS HEAT CAPACITY (continued)

SN	Molecular formula	Name	C_p (298.15 K)/ J K^{-1} mol^{-1}	$A(1)$	$A(2)$	$A(3)$	$A(4)$	$A(5)$	(T/K)-range
16	Kr	Krypton	20.79	20.79					100-6000
17	NO	Nitrogen oxide	29.86	33.58	-0.02593	5.3326E-05	-2.7744E-08		298-800
18	NO_2	Nitrogen dioxide	37.18	32.06	-0.00984	1.3807E-04	-1.8157E-07	7.5253E-11	298-800
19	N_2	Nitrogen	29.13	30.81	-0.01187	2.3968E-05	-1.0176E-08		298-800
20	N_2O	Dinitrogen oxide	38.62	20.66	0.07862	-6.9156E-05	2.5297E-08		298-800
21	Ne	Neon	20.79	20.79					100-6000
22	O_2	Oxygen	29.38	32.83	-0.03633	1.1532E-04	-1.2194E-07	4.5412E-11	298-800
23	O_2S	Sulfur dioxide	39.84	26.07	0.05417	-2.6774E-05			298-800
24	Xe	Xenon	20.79	20.79					100-6000
25	$CClF_3$	Chlorotrifluoromethane	66.89	8.31	0.28828	-3.7324E-04	2.3666E-07	-5.9407E-11	298-800
26	CCl_2F_2	Dichlorodifluoromethane	72.48	11.63	0.31680	-4.7414E-04	3.5379E-07	-1.0655E-10	298-800
27	CCl_4	Tetrachloromethane	83.43	33.73	0.24808	-3.1546E-04	1.4222E-07		298-800
28	CF_4	Tetrafluoromethane	61.05	5.66	0.25229	-2.5070E-04	9.2650E-08		298-800
29	CO	Carbon oxide	29.14	31.08	-0.01452	3.1415E-05	-1.4973E-08		298-800
30	CO_2	Carbon dioxide	37.14	18.86	0.07937	-6.7834E-05	2.4426E-08		298-800
31	CS_2	Carbon disulfide	45.67	25.61	0.09321	-9.8938E-05	4.0345E-08		298-800
32	$CHClF_2$	Chlorodifluoromethane	55.85	18.21	0.14523	-4.3440E-05	-8.5459E-08	5.9451E-11	298-800
33	$CHCl_3$	Trichloromethane	65.38	16.99	0.23836	-3.1152E-04	2.0520E-07	-5.3348E-11	298-800
34	CHF_3	Trifluoromethane	51.07	18.91	0.09524	1.0471E-04	-2.4552E-07	1.2157E-10	298-800
35	CH_2O	Methanal	35.39	40.02	-0.08726	3.3666E-04	-3.6336E-07	1.3748E-10	298-800
36	CH_2O_2	Methanoic acid	45.68	22.06	0.05715	1.2443E-04	-1.9288E-07	7.9794E-11	298-800
37	CH_4	Methane	35.67	30.65	-0.01739	1.3903E-04	-8.1395E-08		298-800
38	CH_4O	Methanol	44.07	26.53	0.03703	9.4541E-05	-7.2006E-08		298-800
39	CH_5N	Methylamine	50.05	36.08	-0.03869	4.3004E-04	-5.4852E-07	2.2957E-10	298-800
40	C_2H_2	Ethyne	44.05	10.82	0.15889	-1.8447E-04	8.5291E-08		298-800
41	C_2H_4	Ethene	42.86	8.39	0.12453	-2.5224E-05	-1.5679E-08		298-800
42	C_2H_4O	Ethanal	55.23	16.26	0.14533	-4.9024E-05			298-800
43	$C_2H_4O_2$	Ethanoic acid	63.44	5.08	0.22262	-9.0193E-05			298-800
44	C_2H_5Cl	Chloroethane	62.58	-0.40	0.26404	-1.9668E-04	6.5798E-08		298-800
45	C_2H_6	Ethane	52.38	6.82	0.16840	-5.2347E-05			298-800
46	C_2H_6O	Dimethyl ether	65.54	26.36	0.12604	3.3699E-05	-5.2344E-08		298-800
47	C_2H_6O	Ethanol	65.59	9.53	0.21234	-8.1600E-05			298-800
48	C_2H_7N	Ethylamine	71.53	31.59	0.05325	4.4450E-04	-6.7378E-07	3.0421E-10	298-800
49	C_3H_6O	2-Propanone	74.47	17.85	0.19910	-1.9211E-05	-3.8962E-08		298-800
50	C_3H_8	Propane	73.52	0.56	0.27559	-1.0355E-04			298-800
51	C_3H_8O	1-Propanol	85.55	27.02	0.12010	4.3871E-04	-7.1090E-07	3.2405E-10	298-800
52	C_3H_8O	2-Propanol	89.27	0.63	0.35409	-2.0062E-04	3.4142E-08		298-800
53	C_3H_9N	Propylamine	91.16	33.02	0.08971	5.7678E-04	-8.6439E-07	3.8314E-10	298-800
54	C_3H_9N	Isopropylamine	97.51	6.81	0.35575	-1.8116E-04	2.7973E-08		298-800
55	C_4H_4O	Furan	65.37	-35.45	0.42977	-3.3742E-04	1.0093E-07		298-800
56	C_4H_5N	Pyrrole	71.56	-37.71	0.46553	-3.6530E-04	1.1104E-07		298-800
57	C_4H_8O	Oxolane	76.23	6.42	0.08953	7.9470E-04	-1.1977E-06	5.3368E-10	298-800
58	C_4H_9N	Pyrrolidine	82.10	12.85	0.05899	9.3456E-04	-1.3648E-06	6.0180E-10	298-800
59	C_4H_{10}	Butane	101.01	172.02	-1.08574	4.4887E-03	-6.5539E-06	3.4654E-09	298-540
60	C_4H_{10}	2-Methylpropane	98.96	174.86	-1.20302	5.1042E-03	-7.7053E-06	4.2101E-09	298-540
61	$C_4H_{10}O$	Diethylether	119.45	90.32	-0.11372	1.0583E-03	-1.3366E-06	5.5606E-10	298-800
62	$C_4H_{11}N$	Butylamine	119.46	49.28	0.12250	6.2393E-04	-9.4817E-07	4.2077E-10	298-800
63	$C_4H_{11}N$	2-Aminobutane	120.46	-12.63	0.54979	-3.7819E-04	1.0515E-07		298-800
64	C_5H_5N	Pyridine	77.68	-35.73	0.46626	-3.0915E-04	7.1061E-08		298-800
65	C_5H_{12}	Pentane	120.11	2.02	0.44729	-1.7174E-04			298-800
66	C_5H_{12}	2,2-Dimethylpropane	121.07	-38.49	0.70122	-6.4458E-04	2.9390E-07		298-800
67	C_6H_6	Benzene	82.39	-46.48	0.53735	-3.8303E-04	1.0184E-07		298-800

TABLE 2.3.1 (CPGT)
IDEAL GAS HEAT CAPACITY (continued)

SN	Molecular formula	Name	C_p (298.15 K)/ J K^{-1} mol^{-1}	$A(1)$	$A(2)$	$A(3)$	$A(4)$	$A(5)$	(T/K)-range
68	C_6H_6O	Phenol	103.19	-39.24	0.61377	-5.0501E-04	1.6316E-07		298-800
69	C_6H_7N	Aniline	107.90	-39.01	0.62351	-4.8143E-04	1.4368E-07		298-800
70	C_6H_{12}	Cyclohexane	105.33	13.18	0.07780	1.2238E-03	-1.7215E-06	7.3188E-10	298-800
71	C_6H_{14}	Hexane	142.13	-13.27	0.61995	-3.5408E-04	7.6704E-08		298-800
72	C_7H_{16}	Heptane	164.75	-18.02	0.73364	-4.3377E-04	9.7678E-08		298-800
73	C_8H_{18}	Octane	187.40	-19.49	0.82772	-4.7837E-04	9.9072E-08		298-800
74	C_9H_7N	Quinoline	129.09	-64.38	0.81722	-6.1752E-04	1.7765E-07		298-800
75	C_9H_{20}	Nonane	209.87	-26.18	0.95102	-5.7168E-04	1.2524E-07		298-800
76	$C_{10}H_{22}$	Decane	232.63	-27.48	1.04462	-6.1517E-04	1.2603E-07		298-800

TABLE 2.3.2 (CPLT) HEAT CAPACITY OF LIQUIDS

There are two types of liquid heat capacities, the customary isobaric heat capacity, C_p, and the saturation heat capacity, C_{sat}, defined such that the pressure changes with temperature along the vapor-liquid saturation curve rather than remaining constant. The two quantities are related by

$$C_{sat} = C_p - T(\partial V / \partial T)_p (dp_{vap} / dT)_{sat} \tag{1}$$

where p_{vap} is the vapor pressure of the liquid. The difference is generally quite small (less than 1%) below the normal boiling point, but can become significant at higher temperatures.

Liquid heat capacities are not very sensitive to temperature unless the temperature is fairly high (say, 0.7 times the critical temperature). Likewise, the pressure dependence is slight at normal pressures. The relation of C_p to molecular structure is much more complex than in the case of gases.

This table gives the coefficients in the expression for the temperature dependence of C_p for 85 representative liquids. The equation used is

$$C_p / \mathrm{J\,K^{-1}\,mol^{-1}} = A(1) + A(2)(T/\mathrm{K}) + A(3)(T/\mathrm{K})^2 + A(4)(T/\mathrm{K})^3 + A(5)(T/\mathrm{K})^4 \tag{2}$$

The pressure is taken to be in the ambient range. The table also gives the recommended value of C_p at 298.15 K; this value may be slightly different from that calculated from the equation at 298.15 K because of the fitting procedure.

The program CPLT allows calculation of C_p and c_p at any desired temperature (or set of temperatures) within the range of validity of the equation, which is given in the last column of the table.

The substances are arranged in a modified Hill order, with compounds not containing carbon preceding those that contain carbon.

REFERENCES

1. *DIPPR Data Compilation of Pure Compound Properties*, Design Institute for Physical Property Data, American Institute of Chemical Engineers, New York.
2. Wagman, D. D.; Evans, W. H.; Parker, V. B.; Schumm, R. H.; Halow, I., Bailey, S. M.; Churney, K. L.; Nuttall, R. L., *The NBS Tables of Chemical Thermodynamic Properties, J. Phys. Chem. Ref. Data*, Vol. 11, Suppl. 2, 1982.
3. Chase, M. W.; Davies, C. A.; Downey, J. R.; Frurip, D. J.; McDonald, R. A.; Syverud, A. N., *JANAF Thermochemical Tables, Third Edition, J. Phys. Chem. Ref. Data*, Vol. 14, Suppl. 1, 1985.
4. Gurvich, L. V.; Veyts, I. V.; Alcock, C. B., *Thermodynamic Properties of Individual Substances, Fourth Edition*, Hemisphere Publishing Corp., New York, 1989.
5. Lide, D. R., Editor, *CRC Handbook of Chemistry and Physics, 74th Edition*, CRC Press, Boca Raton, FL, 1993.
6. Liley, P. E.; Makita, T.; Tanaka, Y., *Properties of Inorganic and Organic Fluids*, Hemisphere Publishing Corp., New York, 1988.
7. Dolmalski, E. S.; Evans, W. H.; Hearing, E. D., *Heat Capacities and Entropies of Organic Compounds in the Condensed Phase, J. Phys. Chem. Ref. Data*, Vol. 13, Suppl. 1, 1984; Vol. 19, No. 4, 881-1047, 1990.
8. Zabransky, M.; Ruzicka, V.; Majer, V, "Heat Capacities of Organic Compounds in the Liquid State", *J. Phys. Chem. Ref. Data*, 19, 719-762, 1990; 20, 405-444, 1991.

TABLE 2.3.2 (CPLT) HEAT CAPACITY OF LIQUIDS (continued)

SN	Molecular formula	Name	C_p (298.15 K)/ J K^{-1} mol^{-1}	A(1)	A(2)	A(3)	A(4)	A(5)	(T/K)-range
1	Cl_4Si	Silicon tetrachloride	145.3	188.0	-0.4953	0.0011810			205 - 460
2	H_2O	Water	75.3	917.5	-10.1016	0.0454134	-9.07517E-05	6.80700E-08	273 - 373
3	H_4N_2	Hydrazine	98.9	79.2	0.0662				275 - 385
4	Hg	Mercury	28.0	30.5	-0.0120	0.0000102	1.93667E-09	-1.94110E-12	260 - 600
5	CCl_4	Tetrachloromethane	130.7	339.2	-1.4830	0.0026280			255 - 375
6	CS_2	Carbon disulfide	76.4	85.6	-0.1220	0.0005605	-1.45200E-06	2.00800E-09	165 - 550
7	$CHCl_3$	Trichloromethane	114.2	108.1	-0.0536	0.0002480			210 - 450
8	CH_2Br_2	Dibromomethane	105.0	202.6	-0.7263	0.0013377			240 - 370
9	CH_4O	Methanol	81.1	87.4	-0.1510	0.0001588	9.27250E-07		180 - 380
10	C_2Cl_4	Tetrachloroethene	143.4	73.0	0.2360				250 - 420
11	C_2HCl_3	Trichloroethene	124.4	90.8	0.1126				225 - 420
12	$C_2H_3Cl_3$	1,1,1-Trichloroethane	144.3	117.8	0.0875	0.0000044			245 - 330
13	C_2H_3N	Ethanenitrile	91.4	97.6	-0.1222	0.0003409			230 - 350
14	$C_2H_4O_2$	Ethanoic acid	123.3	50.1	0.2456				290 - 410
15	C_2H_5Cl	Chloroethane	104.3	80.0	0.5359	-0.0051801	1.89610E-05	-2.24630E-08	140 - 300
16	C_2H_5NO	*N*-Methylmethanamide	127.1	87.3	0.1344				298 - 470
17	C_2H_6O	Ethanol	112.3	108.6	-0.2068	0.0002197	1.73529E-06		180 - 370
18	$C_2H_6O_2$	1,2-Ethanediol	148.6	114.8	-0.0750	0.0008020	-5.70000E-07		260 - 550
19	C_2H_7NO	2-Aminoethanol	195.5	154.2	0.1384				292 - 440
20	C_3H_5ClO	(Chloromethyl)oxirane	131.6	162.8	-0.5650	0.0015443			216 - 385
21	C_3H_6O	2-Propanone	126.3	135.6	-0.1770	0.0002837	6.89000E-07		178 - 329
22	$C_3H_6O_2$	Methyl ethanoate	141.9	191.6	-0.6363	0.0015750			175 - 330
23	C_3H_7NO	*N,N*-Dimethyl-methanamide	150.6	32.0	0.8580	-0.0022460	2.35600E-06		250 - 475
24	C_3H_8O	1-Propanol	143.9	-56.8	2.7837	-0.0177120	4.87380E-05	-4.38570E-08	180 - 380
25	C_3H_8O	2-Propanol	156.5	466.4	-4.1086	0.0145060	-1.41260E-05		190 - 460
26	$C_3H_8O_2$	1,2-Propanediol	190.8	58.1	0.4452				215 - 540
27	$C_3H_8O_3$	1,2,3-Propanetriol	218.9	68.2	0.5052				291 - 540
28	C_3H_9N	Trimethylamine	137.9	149.2	-0.4740	0.0018511	-1.30350E-06		160 - 300
29	C_4H_4O	Furan	115.3	121.7	-0.2811	0.0008702			191 - 300
30	C_4H_4S	Thiophene	123.8	77.6	0.1406	0.0000510			240 - 345
31	C_4H_6	2-Butyne	125.2	88.2	0.1242				241 - 300
32	C_4H_8O	2-Butanone	158.7	132.3	0.2009	-0.0009597	1.95330E-06		186 - 373
33	C_4H_8O	Oxolane	124.0	171.7	-0.7984	0.0028812	-2.48130E-06		165 - 320
34	$C_4H_8O_2$	1,4-Dioxane	152.1	91.2	0.2044				285 - 450
35	$C_4H_8O_2$	Ethyl ethanoate	170.7	155.9	0.2370	-0.0019976	4.59200E-06		190 - 325
36	$C_4H_{10}O$	1-Butanol	177.2	-392.3	8.4877	-0.0510570	1.34066E-04	-1.23460E-07	190 - 360
37	$C_4H_{10}O$	Diethyl ether	175.6	44.4	1.3010	-0.0055000	8.76300E-06		157 - 460
38	C_5H_5N	Pyridine	132.7	107.9	-0.0348	0.0003957			232 - 385
39	C_5H_8	Cyclopentene	122.4	125.4	-0.3497	0.0011430			140 - 315
40	C_5H_{10}	1-Pentene	155.3	149.8	-0.3320	0.0014087	-7.84700E-07		110 - 400

41	C_5H_{10}	Cyclopentane	126.8	122.5	-0.4038	0.0017344	-1.09750E-06		182 - 320
42	C_5H_{12}	Pentane	167.2	236.3	-1.3975	0.0069262	-1.33770E-05	1.09450E-08	220 - 310
43	$C_5H_{12}O$	1-Pentanol	208.1	-637.2	12.4614	-0.0728180	1.86572E-04	-1.69790E-07	210 - 380
44	C_6H_5Cl	Chlorobenzene	150.1	-1307.5	15.3380	-0.0539740	6.34830E-05		230 - 320
45	C_6H_6	Benzene	136.3	1201.2	-15.4799	0.0817688	-1.89230E-04	1.64109E-07	280 - 340
46	C_6H_6O	Phenol		101.7	0.3176				314 - 425
47	C_6H_7N	Aniline	191.9	573.6	-4.6836	0.0199610	-3.55780E-05	2.31750E-08	270 - 460
48	C_6H_{10}	Cyclohexene	148.3	105.9	-0.0600	0.0006800			170 - 350
49	$C_6H_{10}O$	4-Methyl-3-penten-2-one	212.5	137.8	0.2505				295 - 400
50	$C_6H_{11}NO$	Hexahydro-2-azepinone		64.4	0.5145				342 - 385
51	C_6H_{12}	Cyclohexane	154.9	-220.6	3.1183	-0.0094216	1.06870E-05		280 - 400
52	C_6H_{12}	1-Hexene	183.3	192.6	-0.5712	0.0024004	-1.97600E-06		135 - 500
53	$C_6H_{12}O$	Cyclohexanol	208.2	85.3	0.4122				300 - 430
54	$C_6H_{12}O$	2-Hexanone	213.3	272.5	-0.7907	0.0025834	-2.00400E-06		220 - 380
55	C_6H_{14}	Hexane	195.6	306.5	-2.1801	0.0119790	-2.74460E-05	2.55240E-08	220 - 370
56	$C_6H_{14}O$	1-Hexanol	240.4	-2510.9	39.1242	-0.2123200	5.08680E-04	-4.45670E-07	230 - 360
57	C_7H_6O	Benzaldehyde	172.0	111.3	0.1693	0.0001210			250 - 450
58	C_7H_8	Toluene	157.3	190.4	-0.7506	0.0029723	-2.77550E-06		180 - 420
59	C_7H_8O	Anisole	197.2	92.8	0.3500				235 - 426
60	C_7H_{16}	Heptane	224.7	460.6	-3.5847	0.0172050	-3.42420E-05	2.67090E-08	220 - 370
61	$C_7H_{16}O$	1-Heptanol	272.1	-7717.7	108.4227	-0.5527915	1.24501E-03	-1.03693E-06	240 - 360
62	C_8H_8	Vinylbenzene	182.0	113.3	0.2902	-0.0006051	1.35670E-06		243 - 415
63	C_8H_{10}	Ethylbenzene	183.2	86.9	0.7080	-0.0026330	4.50000E-06		180 - 420
64	C_8H_{10}	1,2-Dimethylbenzene	186.1	36.5	1.0175	-0.0026300	3.02000E-06		250 - 415
65	C_8H_{10}	1,3-Dimethylbenzene	183.0	175.6	-0.2995	0.0010880			225 - 360
66	C_8H_{10}	1,4-Dimethylbenzene	181.5	-35.5	1.2872	-0.0025990	2.42600E-06		286 - 600
67	C_8H_{18}	Octane	254.6	711.2	-6.1969	0.0282755	-5.41980E-05	3.97160E-08	230 - 340
68	$C_8H_{18}O$	1-Octanol	305.2	-6469.5	92.1991	-0.4728750	1.07215E-03	-8.97820E-07	280 - 360
69	C_9H_{12}	Isopropylbenzene	210.7	1547.6	-12.6530	0.0386760	-3.78240E-05		280 - 380
70	C_9H_{12}	1,3,5-Trimethylbenzene	209.3	148.1	0.0197	0.0006226			230 - 350
71	C_9H_{20}	Nonane	284.4	868.4	-6.6513	0.0236178	-2.71400E-05	2.39615E-09	250 - 440
72	$C_9H_{20}O$	1-Nonanol	336.5	11379.7	-124.6350	0.5136820	-9.17390E-04	6.03372E-07	310 - 460
73	$C_{10}H_{14}$	4-Isopropyltoluene	236.4	145.6	0.2487	0.0001870			205 - 450
74	$C_{10}H_{22}$	Decane	314.4	815.2	-6.3042	0.0266868	-4.69670E-05	3.18082E-08	250 - 440
75	$C_{10}H_{22}O$	1-Decanol	370.6	4449.2	-47.7202	0.1979381	-3.46591E-04	2.20024E-07	290 - 500
76	$C_{11}H_{24}$	Undecane	344.9	996.9	-7.6655	0.0305561	-4.99650E-05	3.05461E-08	260 - 420
77	$C_{12}H_{10}O$	Diphenyl ether	267.6	185.7	-0.0527	0.0017702	-2.70270E-06	1.50448E-09	300 - 570
78	$C_{12}H_{26}$	Dodecane	375.8	545.3	-1.7704	0.0045286	-1.66670E-06		270 - 310
79	$C_{12}H_{26}O$	1-Dodecanol		5526.4	-58.8897	0.2436502	-4.28464E-04	2.74256E-07	300 - 480
80	$C_{13}H_{28}$	Tridecane	406.7	776.0	-2.9975	0.0058993			273 - 300
81	$C_{14}H_{30}$	Tetradecane	438.3	745.3	-3.1778	0.0093058	-6.63000E-06	-1.39710E-09	290 - 430
82	$C_{14}H_{30}O$	1-Tetradecanol		2576.1	-23.9715	0.0946331	-1.50798E-04	8.40150E-08	320 - 470
83	$C_{16}H_{22}O_4$	Dibutyl 1,2-benzene-dicarboxylate	468.4	268.0	0.6720				295 - 610
84	$C_{16}H_{34}$	Hexadecane	501.6	2186.3	-16.4087	0.0529118	-6.06990E-05	1.42749E-08	300 - 410
85	$C_{18}H_{38}$	Octadecane	564.7	544.3	-0.4774	0.0018391	-2.52530E-08		302 - 370

TABLE 2.3.3 (CPST) HEAT CAPACITY OF SOLIDS

The molar isobaric heat capacities C_p of some representative solids are presented in this table. The table gives the coefficients in the expression for the temperature dependence of C_p as represented by the equation

$$C_p / \mathrm{J\,K^{-1}\,mol^{-1}} = A(1) + A(2)(T/\mathrm{K}) + A(3)(T/\mathrm{K})^2 + A(4)(T/\mathrm{K})^3 + A(5)(T/\mathrm{K})^4 \quad (1)$$

The pressure is taken to be in the ambient range. The table also gives the recommended value of C_p at 298.15 K; this value may be slightly different from that calculated from the equation at 298.15 K because of the fitting procedure.

The program CPST allows calculation of C_p and the specific heat capacity c_p for these solids at any desired temperature within the range of validity of the equation, which is given in the last column of the table.

For simple crystalline solids the heat capacity is zero at $T = 0$ and increases as T^3 in the very low temperature range, in accordance with the Debye model of lattice vibrations. In metals, there is an additional contribution, proportional to T, from the conduction electrons. At ambient temperatures most metals have a heat capacity in the neighborhood of 3 R (about 25 J K^{-1} mol^{-1}), the classical value for an ideal lattice, and C_p increases only gradually with further increase in temperature. Inorganic salts tend to have higher C_p because of the additional degrees of freedom within the ions, and complex organic solids exhibit still higher values. In both cases the heat capacity continues to increase with increasing temperature.

The substances in the table are arranged in a modified Hill order, with compounds not containing carbon preceding those that contain carbon.

REFERENCES

1. *DIPPR Data Compilation of Pure Compound Properties*, Design Institute for Physical Property Data, American Institute of Chemical Engineers, New York.
2. Wagman, D. D.; Evans, W. H.; Parker, V. B.; Schumm, R. H.; Halow, I., Bailey, S. M.; Churney, K. L.; Nuttall, R. L., *The NBS Tables of Chemical Thermodynamic Properties, J. Phys. Chem. Ref. Data*, Vol. 11, Suppl. 2, 1982.
3. Chase, M. W.; Davies, C. A.; Downey, J. R.; Frurip, D. J.; McDonald, R. A.; Syverud, A. N., *JANAF Thermochemical Tables, Third Edition, J. Phys. Chem. Ref. Data*, Vol. 14, Suppl. 1, 1985.
4. Gurvich, L. V.; Veyts, I. V.; Alcock, C. B., *Thermodynamic Properties of Individual Substances, Fourth Edition*, Hemisphere Publishing Corp., New York, 1989.
5. Touloukian, Y. S.; Buyco, E. H., Editors, *Thermophysical Properties of Matter, Vol. 4: Specific Heat of Metallic Elements and Alloys*; Vol. 5: *Specific Heat of Nonmetallic Solids*, IFI/Plenum, New York, 1970.
6. Cox, J. D.; Wagman, D. D.; Medvedev, V. A., *CODATA Key Values for Thermodynamics*, Hemisphere Publishing Corp. New York, 1989.

TABLE 2.3.3 (CPST) HEAT CAPACITY OF SOLIDS (continued)

SN	Molecular formula	Name	C_p (298.15 K)/ J K^{-1} mol^{-1}	A(1)	A(2)	A(3)	A(4)	A(5)	(T/K)-range
1	Ag	Silver	25.4	26.12	-0.0110	3.8259E-05	-3.74954E-08	1.39583E-11	290 - 800
2	Al	Aluminum	24.4	6.56	0.1153	-2.4603E-04	1.94148E-07		200 - 450
3	Au	Gold	25.4	34.97	-0.0768	2.1169E-04	-2.34981E-07	9.50000E-11	290 - 800
4	B	Boron	11.1	57.68	-0.7708	3.9339E-03	-8.04611E-06	5.93000E-09	200 - 450
5	Be	Beryllium	16.4	-2.54	0.0034	5.7503E-04	-1.68037E-06	1.43333E-09	200 - 450
6	Ca	Calcium	25.9	13.53	0.0993	-2.7773E-04	2.81481E-07		200 - 450
7	ClCs	Cesium chloride	52.5	43.38	0.0467	-8.9725E-05	1.42059E-07	-8.23669E-11	200 - 600
8	ClLi	Lithium chloride	48.0	17.89	0.2163	-5.8823E-04	7.99275E-07	-4.08397E-10	200 - 600
9	ClNa	Sodium chloride	50.2	25.19	0.1973	-6.0114E-04	8.81505E-07	-4.76500E-10	200 - 600
10	Cu	Copper	24.4	8.11	0.1368	-4.3683E-04	6.49778E-07	-3.60000E-10	200 - 450
11	CuO	Copper oxide	42.3	-6.44	0.3202	-7.3435E-04	7.99788E-07	-3.33540E-10	290 - 600
12	CuO_4S	Copper sulfate	98.9	-13.81	0.7036	-1.6356E-03	2.17557E-06	-1.18191E-09	200 - 600
13	Fe	Iron	25.1	-10.99	0.3353	-1.2384E-03	2.16274E-06	-1.40667E-09	200 - 450
14	Ge	Germanium	23.3	12.71	0.0673	-1.4989E-04	1.62068E-07	-6.70906E-11	290 - 600
15	I_2	Iodine	54.4	-136.00	2.5712	-1.2875E-02	2.78169E-05	-2.13756E-08	200 - 386
16	K	Potassium	29.6	64.89	-0.9436	7.6721E-03	-2.57239E-05	3.10933E-08	100 - 336
17	Li	Lithium	24.8	-24.58	0.5439	-2.3674E-03	4.68111E-06	-3.36667E-09	200 - 450
18	Mg	Magnesium	24.9	6.76	0.1523	-5.0581E-04	8.15185E-07	-5.00000E-10	200 - 450
19	O_2Si	Silicon dioxide (quartz)	44.4	-6.30	0.2544	-3.3497E-04	1.84135E-07	-8.54020E-12	200 - 600
20	Pb	Lead	26.4	24.84	-0.0044	7.7428E-05	-1.77259E-07	1.40000E-10	200 - 450
21	Si	Silicon	20.0	-6.25	0.1681	-3.4374E-04	2.49407E-07	6.66667E-12	200 - 450
22	W	Tungsten	24.3	0.09	0.2361	-8.8412E-04	1.51489E-06	-9.80000E-10	200 - 450
23	Zn	Zinc	25.4	-6.91	0.3695	-1.6116E-03	3.13915E-06	-2.25000E-09	200 - 450
24	C	Carbon (graphite)	8.5	-12.19	0.1126	-1.9469E-04	1.91880E-07	-7.79919E-11	290 - 600
25	C_2H_5NO	Ethanamide	91.3	32.75	0.1233	2.2020E-04			100 - 345
26	C_6Cl_6	Hexachlorobenzene	201.2	-1.80	1.3455	-3.4134E-03	3.99100E-06		100 - 320
27	$C_6H_4Cl_2$	1,4-Dichlorobenzene	147.8	-9.43	1.2446	-7.0730E-03	2.43000E-05	-2.89400E-08	100 - 326
28	C_6H_6O	Phenol	127.4	-5.97	1.0380	-6.4670E-03	2.30400E-05	-2.65800E-08	100 - 314
29	$C_6H_{11}NO$	Hexahydro-2-azepinone	156.8	29.00	0.2800	5.4300E-04			100 - 342
30	$C_7H_6O_2$	Benzoic acid	146.8	-6.50	1.1150	-5.6500E-03	1.77800E-05	-1.87900E-08	100 - 395
31	$C_{10}H_8$	Naphthalene	165.7	-6.16	1.0383	-5.3550E-03	1.89130E-05	-2.05260E-08	100 - 353
32	$C_{12}H_{10}$	Biphenyl	198.4	36.86	0.3426	6.3960E-04			185 - 342
33	$C_{14}H_{10}$	Anthracene	210.5	11.10	0.5816	2.7900E-04			100 - 488
34	$C_{14}H_{10}$	Phenanthrene		-0.42	1.0200	-3.4200E-03	8.60000E-06		100 - 270
35	$C_{18}H_{14}$	1,4-Diphenylbenzene	278.7	32.37	0.8123	8.6890E-05			296 - 348

TABLE 2.3.4
ENTHALPY OF FUSION

The *enthalpy of fusion* (melting), $\Delta_{fus}H$, is defined as the enthalpy of the liquid phase minus the enthalpy of the solid at a given temperature on the solid-liquid phase boundary or melting curve (see Section 2.1). This table gives the molar enthalpy of fusion at the normal melting point temperature for about 500 substances. An entry followed by "t" indicates a value at the solid-liquid-vapor triple point, rather than the normal melting point (see Section 2.1).

Substances are arranged in a modified Hill order in which compounds not containing carbon precede those that contain carbon.

REFERENCES

1. Chase, M. W.; Davies, C. A.; Downey, J. R.; Frurip, D. J.; McDonald, R. A.; Syverud, A. N.; *JANAF Thermochemical Tables, Third Edition, J. Phys. Chem. Ref. Data*, Vol. 14, Suppl. 1, 1985.
2. Dinsdale, A. T., "SGTE Data for Pure Elements", *CALPHAD*, 15, 317-425, 1991.
3. *DIPPR Data Compilation of Pure Compound Properties*, Design Institute for Physical Property Data, American Institute of Chemical Engineers, New York.
4. *Landolt-Börnstein, Numerical Data and Functions for Physics, Chemistry, Astronomy, Geophysics, and Technology, Sixth Edition*, Vol.2, Part 4: Thermal Properties, Springer-Verlag, Heidelberg, 1961.
5. Janz, G. J.; Allen, C. B.; Bansal, N. P.; Murphy, R. M.; Tompkins, R. P. T., *Physical Properties Data Compilations Relevant to Energy Storage. II. Molten Salts: Data on Single and Multi-Component Salt Systems,* Natl. Stand. Ref. Data Sys.— Natl. Bur. Standards (U.S.), No. 61, Part 2, 1969.
6. Tamir, A.; Tamir, E.; Stephan, K, *Heats of Phase Change of Pure Components and Mixtures*, Elsevier, Amsterdam, 1983 (a comprehensive bibliography).

Molecular formula	Name	T_m/K	$\Delta_{fus}H$/ kJ mol^{-1}
Ag	Silver	1234.93	11.3
AgBr	Silver bromide	705	9.12
AgCl	Silver chloride	728	13.2
AgI	Silver iodide	831	9.41
$AgNO_3$	Silver nitrate	485	11.5
Ag_2S	Disilver sulfide	1098	14.1
Al	Aluminum	933.47	10.71
$AlBr_3$	Aluminum tribromide	370.6	11.25
$AlCl_3$	Aluminum trichloride	463	35.4
AlF_3	Aluminum trifluoride	2523	98
AlI_3	Aluminum triiodide	464	15.9
Al_2O_3	Dialuminum trioxide	2327	111.1
Ar	Argon	83.80	1.12
As	Arsenic	1090 t	24.4 t
$AsBr_3$	Arsenic tribromide	304.2	11.7
$AsCl_3$	Arsenic trichloride	257	10.1
AsF_3	Arsenic trifluoride	267.2	10.4
Au	Gold	1337.33	12.55
B	Boron	2348	50.2
BCl_3	Boron trichloride	166	2.1
BF_3	Boron trifluoride	146.3	4.2
$BNaO_2$	Sodium metaborate	1239	36.2
Ba	Barium	1000	7.12
$BaBr_2$	Barium dibromide	1130	31.96
$BaCl_2$	Barium dichloride	1235	16
BaF_2	Barium difluoride	1641	23.36
BaH_2O_2	Barium dihydroxide	681	16.7
BaI_2	Barium diiodide	984	26.53
BaO	Barium oxide	2286	59
BaO_4S	Barium sulfate	1623	40.6
Be	Beryllium	1560	7.9
$BeBr_2$	Beryllium dibromide	781	9.8

TABLE 2.3.4
ENTHALPY OF FUSION (continued)

Molecular formula	Name	T_m/K	$\Delta_{fus}H$/ kJ mol^{-1}
$BeCl_2$	Beryllium dichloride	688	8.66
BeF_2	Beryllium difluoride	825	4.76
BeI_2	Beryllium diiodide	753	21
BeO	Beryllium oxide	2780	85
Bi	Bismuth	544.55	11.3
$BiCl_3$	Bismuth trichloride	503	10.9
BrF_5	Bromine pentafluoride	212.6	5.67
BrH	Hydrogen bromide	186.34	2.41
BrK	Potassium bromide	1007	25.5
BrLi	Lithium bromide	825	17.6
BrNa	Sodium bromide	1014	26.11
$BrNaO_3$	Sodium bromate	1020	28.11
BrRb	Rubidium bromide	955	15.5
BrTl	Thallium bromide	733	25.1
Br_2	Bromine	265.9	10.57
Br_2Ca	Calcium dibromide	1015	29.08
Br_2Fe	Iron dibromide	964	50.2
Br_2Hg	Mercury dibromide	509	17.9
Br_2Mg	Magnesium dibromide	984	39.3
Br_2Pb	Lead dibromide	644	16.44
Br_2Sr	Strontium dibromide	930	10.12
Br_4Sn	Tin tetrabromide	304	12
Br_4Ti	Titanium tetrabromide	312	12.9
Ca	Calcium	1115	8.54
$CaCl_2$	Calcium dichloride	1045	28.54
CaF_2	Calcium difluoride	1691	29.71
CaI_2	Calcium diiodide	1052	41.8
CaO	Calcium oxide	3200	59
CaO_4S	Calcium sulfate	1723	28.03
Cd	Cadmium	594.22	6.19
Ce	Cerium	1072	5.46
ClCs	Cesium chloride	918	15.9
ClCu	Copper chloride	703	10.2
ClH	Hydrogen chloride	158.97	2
ClI	Iodine chloride	300.53	11.6
ClK	Potassium chloride	1043	26.53
ClLi	Lithium chloride	878	19.9
$ClLiO_4$	Lithium perchlorate	509	29
ClNa	Sodium chloride	1073.5	28.16
$ClNaO_3$	Sodium chlorate	521	22.1
ClRb	Rubidium chloride	990	18.4
ClTl	Thallium chloride	703	17.8
Cl_2	Chlorine	171.6	6.4
Cl_2Co	Cobalt dichloride	1013	45
Cl_2Cr	Chromium dichloride	1087	32.2
Cl_2Cu	Copper dichloride	703	20.4
Cl_2Fe	Iron dichloride	950	43.01
Cl_2Hg	Mercury dichloride	549	19.41
Cl_2Mg	Magnesium dichloride	987	43.1
Cl_2Mn	Manganese dichloride	923	30.7
Cl_2Ni	Nickel dichloride	1304	71.2
Cl_2Pb	Lead dichloride	774	21.9
Cl_2Sn	Tin dichloride	520	12.8
Cl_2Sr	Strontium dichloride	1147	16.2
Cl_3Fe	Iron trichloride	577	43.1
Cl_3OP	Phosphoryl trichloride	274	13.1

TABLE 2.3.4
ENTHALPY OF FUSION (continued)

Molecular formula	Name	T_m/K	$\Delta_{fus}H$/ kJ mol^{-1}
Cl_3P	Phosphorus trichloride	161	7.1
Cl_3Sb	Antimony trichloride	346.5	12.7
Cl_4OW	Tungstenosyl tetrachloride	484	45
Cl_4Si	Silicon tetrachloride	205	7.6
Cl_4Sn	Tin tetrachloride	240	9.2
Cl_4Ti	Titanium tetrachloride	248	9.97
Cl_4V	Vanadium tetrachloride	247.4	2.3
Cl_4Zr	Zirconium tetrachloride	710	50
Cl_5Mo	Molybdenum pentachloride	467	19
Cl_5Nb	Niobium pentachloride	477.8	33.9
Cl_5Ta	Tantalum pentachloride	489	35.1
Cl_6W	Tungsten hexachloride	548	6.6
Co	Cobalt	1768	16.20
CoF_2	Cobalt difluoride	1400	59
Cr	Chromium	2180	21.0
Cr_2O_3	Dichromium trioxide	2603	130
Cs	Cesium	302.9	2.1
CsF	Cesium fluoride	976	21.7
Cs_2O_4S	Dicesium sulfate	1278	35.7
Cu	Copper	1357.77	13.26
CuF_2	Copper difluoride	1109	55
CuO	Copper oxide	1719	11.8
Dy	Dysprosium	1684	10.78
Er	Erbium	1802	19.9
Eu	Europium	1095	9.21
FH	Hydrogen fluoride	189.79	4.58
FK	Potassium fluoride	1131	27.2
FLi	Lithium fluoride	1121.3	27.09
FNa	Sodium fluoride	1265	33.35
FRb	Rubidium fluoride	1106	17.3
F_2	Fluorine	53.49	0.51
F_2Fe	Iron difluoride	1373	52
F_2HK	Potassium hydrogen difluoride	512.0	6.62
F_2Mg	Magnesium difluoride	1536	58.7
F_2Pb	Lead difluoride	1103	14.7
F_2Sr	Strontium difluoride	1750	29.7
F_4Zr	Zirconium tetrafluoride	1205	64.2
F_6Mo	Molybdenum hexafluoride	290.6	4.33
F_6W	Tungsten hexafluoride	275.4	4.1
Fe	Iron	1811	13.81
FeI_2	Iron diiodide	860	45
FeO	Iron oxide	1650	24
FeS	Iron sulfide	1463	31.5
Fe_3O_4	Triiron tetraoxide	1870	138
Ga	Gallium	303.0	5.59
Gd	Gadolinium	1587	9.81
Ge	Germanium	1211.40	36.9
HI	Hydrogen iodide	222.38	2.87
HKO	Potassium hydroxide	679	8.6
HLi	Lithium hydride	961.8	22.59
HLiO	Lithium hydroxide	744.3	20.88
HNO_3	Nitric acid	231.5	10.5
HNaO	Sodium hydroxide	596	6.6
H_2	Hydrogen	13.81	0.12
H_2O	Water	273.15	6.010
H_2O_2	Dihydrogen peroxide	272.72	12.5

TABLE 2.3.4
ENTHALPY OF FUSION (continued)

Molecular formula	Name	T_m/K	$\Delta_{fus} H$/ kJ mol^{-1}
H_2O_2Sr	Strontium dihydroxide	783	21
H_2O_4S	Sulfuric acid	283.46	10.71
H_2S	Dihydrogen sulfide	187.6	23.8
H_3N	Ammonia	195.41	5.66
H_3O_2P	Phosphonous acid	299.6	9.7
H_3O_3P	Phosphorous acid	347.5	12.8
H_3O_4P	Phosphoric acid	315.5	13.4
H_4IN	Ammonium iodide	824	21
H_4N_2	Hydrazine	274.5	12.6
$H_4N_2O_3$	Ammonium nitrate	442.7	6.4
Hf	Hafnium	2506	27.2
Hg	Mercury	234.32	2.29
HgI_2	Mercury diiodide	532	18.9
Hg_2I_2	Dimercury diiodide	563	27
Ho	Holmium	1745	11.76
IK	Potassium iodide	954	24
ILi	Lithium iodide	742	14.6
INa	Sodium iodide	928	23.6
IRb	Rubidium iodide	915	12.5
ITl	Thallium iodide	713	13.1
I_2	Iodine	386.8	15.52
I_2Mg	Magnesium diiodide	907	29
I_2Pb	Lead diiodide	683	23.4
I_2Sr	Strontium diiodide	811	19.7
I_4Si	Silicon tetraiodide	393.6	19.7
I_4Ti	Titanium tetraiodide	423	19.8
In	Indium	429.75	3.28
Ir	Iridium	2719	41.12
K	Potassium	336.53	2.32
KNO_3	Potassium nitrate	610	10.1
K_2O_4S	Dipotassium sulfate	1342	36.4
K_2S	Dipotassium sulfide	1221	16.15
Kr	Krypton	115.79	1.37
La	Lanthanum	1193	6.2
Li	Lithium	453.6	3
$LiNO_3$	Lithium nitrate	526	24.9
Li_2O_3Si	Dilithium metasilicate	1474	28
Li_2O_4S	Dilithium sulfate	1132	7.5
Lu	Lutetium	1936	18.65
Mg	Magnesium	923	8.48
MgO	Magnesium oxide	3099	78
MgO_4S	Magnesium sulfate	1400	14.6
Mg_2O_4Si	Dimagnesium silicate	2171	71
Mn	Manganese	1519	12.91
MnO	Manganese oxide	2057	54.4
Mo	Molybdenum	2896	37.48
MoO_3	Molybdenum trioxide	1074	48
$NNaO_3$	Sodium nitrate	580	15.0
NO	Nitrogen oxide	109.5	2.3
NO_3Rb	Rubidium nitrate	578	5.6
NO_3Tl	Thallium nitrate	479	9.6
N_2	Nitrogen	63.15	0.71
N_2O	Dinitrogen oxide	182.3	6.54
N_2O_4	Dinitrogen tetraoxide	263.8	14.65
Na	Sodium	370.87	2.6
Na_2O	Disodium oxide	1405	48

TABLE 2.3.4
ENTHALPY OF FUSION (continued)

Molecular formula	Name	T_m/K	$\Delta_{fus}H$/ kJ mol^{-1}
Na_2O_3Si	Disodium metasilicate	1362	52
Na_2O_4S	Disodium sulfate	1157	23.6
Na_2S	Disodium sulfide	1445	19
Nb	Niobium	2750	30.0
NbO	Niobium oxide	2210	85
NbO_2	Niobium dioxide	2175	92
Nb_2O_5	Diniobium pentaoxide	1785	104.3
Nd	Neodymium	1289	7.14
Ne	Neon	24.56	0.34
Ni	Nickel	1728.3	17.48
NiS	Nickel sulfide	1249	30.1
OSr	Strontium oxide	2938	75
OV	Vanadium oxide	2063	63
O_2	Oxygen	54.36	0.44
O_2Si	Silicon dioxide	1883	8.51
O_2Zr	Zirconium dioxide	2983	87
O_3S	Sulfur trioxide	289.9	
O_3W	Tungsten trioxide	1745	73
O_3Y_2	Diyttrium trioxide	2712	105
O_4Os	Osmium tetraoxide	314	9.8
O_4STl_2	Dithallium sulfate	905	23
O_5P_2	Diphosphorus pentaoxide	693	27.2
O_5Ta_2	Ditantalum pentaoxide	2058	120
O_5V_2	Divanadium pentaoxide	943	64.5
O_7Re_2	Dirhenium heptaoxide	570	64.2
Os	Osmium	3306	57.8
P	Phosphorus	317.30	0.66
Pb	Lead	600.61	4.77
PbS	Lead sulfide	1386.5	19
Pd	Palladium	1828	16.74
Pr	Praseodymium	1204	6.89
Pt	Platinum	2041.5	22.17
Rb	Rubidium	312.46	2.19
Rh	Rhodium	2237	26.59
Ru	Ruthenium	2607	38.59
S	Sulfur	388.36	1.72
STl_2	Dithallium sulfide	721	12
Sb	Antimony	903.78	19.87
Sc	Scandium	1814	14.10
Se	Selenium	495	6.69
Si	Silicon	1687	50.21
Sm	Samarium	1345	8.62
Sn	Tin	505.08	7.03
Sr	Strontium	1050	7.43
Ta	Tantalum	3290	36.57
Tb	Terbium	1632	10.15
Te	Tellurium	722.66	17.49
Th	Thorium	2023	13.81
Ti	Titanium	1941	14.15
Tl	Thallium	577	4.14
Tm	Thulium	1818	16.84
U	Uranium	1408	9.14
V	Vanadium	2183	21.5
W	Tungsten	3695	52.31
Xe	Xenon	161.40	1.81
Y	Yttrium	1799	11.4

TABLE 2.3.4
ENTHALPY OF FUSION (continued)

Molecular formula	Name	T_m/K	$\Delta_{fus}H$/ kJ mol^{-1}
Yb	Ytterbium	1097	7.66
Zn	Zinc	692.68	7.32
Zr	Zirconium	2127.85	21.0
C	Carbon	4247 t	104.6 t
$CBrCl_3$	Bromotrichloromethane	267.4	2.54
$CCaO_3$	Calcium carbonate	1612.2	53.1
CCl_2O	Carbonyl dichloride	145.2	5.74
CCl_4	Tetrachloromethane	250	3.28
$CHCl_3$	Trichloromethane	209.5	8.8
CHN	Hydrogen cyanide	259.7	8.41
CH_2Cl_2	Dichloromethane	176.5	6.0
CH_2N_2	Cyanamide	317	8.76
CH_2O_2	Methanoic acid	281.4	12.7
CH_3Br	Bromomethane	179.4	5.98
CH_4	Methane	90.69	0.94
CH_4O	Methanol	175.47	3.18
CH_4S	Methanethiol	150	5.91
CH_5N	Methylamine	179.71	6.13
CK_2O_3	Dipotassium carbonate	1171	27.6
CLi_2O_3	Dilithium carbonate	996	41
CNa_2O_3	Disodium carbonate	1131.2	29.7
CO	Carbon oxide	68	0.83
CO_2	Carbon dioxide	216.58 t	9.02 t
CO_3Tl_2	Dithallium carbonate	546	18.4
CS_2	Carbon disulfide	161.6	4.4
$C_2Br_2F_4$	1,2-Dibromotetrafluoroethane	162.7	7.04
$C_2Cl_2F_4$	1,2-Dichlorotetrafluoroethane	179	6.32
$C_2Cl_3F_3$	1,1,2-Trichlorotrifluoroethane	238	2.47
$C_2Cl_4F_2$	Tetrachloro-1,2-difluoroethane	299	3.7
$C_2HCl_3O_2$	Trichloroethanoic acid	330.6	5.88
$C_2H_3ClO_2$	Chloroethanoic acid	334.4	12.3
$C_2H_3Cl_3$	1,1,1-Trichloroethane	242.7	2.73
$C_2H_3Cl_3$	1,1,2-Trichloroethane	236.5	11.5
$C_2H_3F_3$	1,1,1-Trifluoroethane	161.8	6.19
$C_2H_4Br_2$	1,2-Dibromoethane	283.08	10.8
$C_2H_4Cl_2$	1,2-Dichloroethane	237.6	8.83
$C_2H_4O_2$	Ethanoic acid	289.7	11.5
C_2H_5Cl	Chloroethane	134.4	4.45
C_2H_6	Ethane	90.3	2.86
C_2H_6O	Dimethyl ether	131.6	4.94
C_2H_6O	Ethanol	159.0	5.02
$C_2H_6O_2$	1,2-Ethanediol	260	11.2
C_2H_6S	2-Thiapropane	174.8	7.99
C_2H_6S	Ethanethiol	125.26	4.98
$C_2H_6S_2$	2,3-Dithiabutane	188	9.19
C_2H_7N	Dimethylamine	180.9	5.94
$C_2H_8N_2$	1,2-Ethanediamine	284.2	22.6
C_3H_3N	Propenenitrile	189.6	6.23
$C_3H_4O_2$	Propenoic acid	285.4	11.2
$C_3H_5N_3O_9$	1,2,3-Propanetriol trinitrate	286	21.9
C_3H_6	Propene	87.90	3.0
C_3H_6	Cyclopropane	145.7	5.44
$C_3H_6Br_2$	1,3-Dibromopropane	238.9	13.6
$C_3H_6Cl_2$	1,2-Dichloropropane	172.6	6.4
C_3H_6O	2-Propanone	178.3	5.69
$C_3H_6O_3$	1,3,5-Trioxane	333.3	15.1

TABLE 2.3.4
ENTHALPY OF FUSION (continued)

Molecular formula	Name	T_m/K	$\Delta_{fus}H$/ kJ mol^{-1}
C_3H_7Cl	2-Chloropropane	155.9	7.39
C_3H_7N	Cyclopropylamine	237.7	13.2
C_3H_8	Propane	85.46	3.53
C_3H_8O	1-Propanol	147.0	5.2
C_3H_8O	2-Propanol	183.6	5.37
$C_3H_8O_3$	1,2,3-Propanetriol	291.3	18.28
C_3H_9N	Propylamine	190	11.0
C_3H_9N	Isopropylamine	178.01	7.33
C_3H_9N	Trimethylamine	156.0	6.55
$C_4H_4N_2$	Butanedinitrile	327.6	3.70
$C_4H_4O_3$	Oxolan-2,5-dione	392	20.4
C_4H_4S	Thiophene	233.7	5.09
C_4H_5N	Pyrrole	249.73	7.91
C_4H_6	1,3-Butadiene	164.2	7.98
C_4H_6	2-Butyne	240.79	9.23
$C_4H_6O_2$	Oxolan-2-one	229.78	9.57
$C_4H_6O_2$	*cis*-2-Butenoic acid	288	12.6
$C_4H_6O_2$	*trans*-2-Butenoic acid	345	12.98
$C_4H_6O_4$	Butanedioic acid	461	33.0
$C_4H_6O_4$	Dimethyl oxalate	327.50	21.1
C_4H_8	*cis*-2-Butene	134.2	7.58
C_4H_8	2-Methylpropene	132.7	5.93
C_4H_8O	2-Butanone	186.48	8.44
C_4H_8O	Oxolane	164.76	8.54
$C_4H_8O_2$	1,4-Dioxane	284.9	12.9
$C_4H_8O_2$	Ethyl ethanoate	189.5	10.5
$C_4H_8O_2$	Butanoic acid	267.4	11.1
C_4H_9Br	2-Bromobutane	160.4	6.89
C_4H_{10}	2-Methylpropane	134.8	4.66
$C_4H_{10}O$	1-Butanol	183.3	9.28
$C_4H_{10}O$	2-Methyl-2-propanol	298.5	6.79
$C_4H_{10}O$	Diethyl ether	156.8	7.27
$C_4H_{12}Si$	Tetramethylsilane	174.11	6.88
C_5H_8	2-Methyl-1,3-butadiene	127.2	4.79
C_5H_8	1,4-Pentadiene	124.3	6.14
C_5H_8	Cyclopentene	138.0	3.36
C_5H_8	Spiropentane	138.5	5.76
$C_5H_8O_3$	4-Oxopentanoic acid	306	9.22
$C_5H_8O_4$	Pentanedioic acid	370.9	20.9
C_5H_{10}	1-Pentene	107.9	5.81
C_5H_{10}	*cis*-2-Pentene	121.7	7.12
C_5H_{10}	*trans*-2-Pentene	132.9	8.36
C_5H_{10}	Cyclopentane	179.3	0.61
$C_5H_{10}O$	2-Pentanone	196.2	10.6
$C_5H_{10}O$	3-Pentanone	234	11.6
$C_5H_{11}N$	Cyclopentylamine	190.4	8.31
C_5H_{12}	Pentane	143.4	8.42
C_5H_{12}	2-Methylbutane	113.2	5.15
C_5H_{12}	2,2-Dimethylpropane	256.5	3.10
$C_5H_{12}O$	1-Pentanol	194.2	9.83
C_6Cl_6	Hexachlorobenzene	504.9	23.9
C_6F_6	Hexafluorobenzene	278.50	11.59
C_6HF_5	Pentafluorobenzene	225.8	10.9
C_6HF_5O	Pentafluorophenol	305.9	12.9
$C_6H_3Cl_3$	1,3,5-Trichlorobenzene	336.6	18.2
$C_6H_4ClNO_2$	1-Chloro-3-nitrobenzene	317.5	19.4

TABLE 2.3.4
ENTHALPY OF FUSION (continued)

Molecular formula	Name	T_m/K	$\Delta_{fus}H$/ kJ mol^{-1}
$C_6H_4ClNO_2$	1-Chloro-4-nitrobenzene	356.6	20.8
$C_6H_4Cl_2$	1,2-Dichlorobenzene	256.4	12.9
$C_6H_4Cl_2$	1,3-Dichlorobenzene	248.3	12.6
$C_6H_4Cl_2$	1,4-Dichlorobenzene	325.8	17.2
$C_6H_4F_2$	1,3-Difluorobenzene	204.06	8.58
$C_6H_4O_2$	1,4-Benzoquinone	388.8	18.5
C_6H_5Br	Bromobenzene	242.5	10.6
C_6H_5Cl	Chlorobenzene	227.9	9.61
C_6H_5ClO	2-Chlorophenol	282.9	12.5
C_6H_5ClO	3-Chlorophenol	305.7	14.9
C_6H_5ClO	4-Chlorophenol	315.8	14.1
C_6H_5F	Fluorobenzene	230.94	11.3
C_6H_5I	Iodobenzene	241.8	9.76
$C_6H_5NO_2$	Nitrobenzene	278.8	11.6
$C_6H_5NO_3$	2-Nitrophenol	317.9	17.4
$C_6H_5NO_3$	3-Nitrophenol	369.9	19.2
$C_6H_5NO_3$	4-Nitrophenol	386.9	18.3
C_6H_6	Benzene	278.68	9.95
$C_6H_6N_2O_2$	2-Nitroaniline	344.3	16.1
$C_6H_6N_2O_2$	3-Nitroaniline	387	23.7
$C_6H_6N_2O_2$	4-Nitroaniline	420	21.1
C_6H_6O	Phenol	314.0	11.3
$C_6H_6O_2$	1,4-Benzenediol	445.4	27.1
C_6H_6S	Benzenethiol	258.2	11.5
C_6H_7N	Aniline	267.13	10.6
$C_6H_8N_2$	Phenylhydrazine	292.7	16.4
C_6H_{10}	Cyclohexene	169.6	3.29
$C_6H_{10}O_4$	Hexanedioic acid	426.3	34.9
C_6H_{12}	Cyclohexane	279.7	2.63
C_6H_{12}	Methylcyclopentane	130.6	6.93
C_6H_{12}	2,3-Dimethyl-2-butene	198.5	5.46
$C_6H_{12}O$	Cyclohexanol	298.61	1.76
$C_6H_{12}O$	2-Hexanone	217.6	14.9
$C_6H_{12}O$	3-Hexanone	217.6	13.5
C_6H_{14}	Hexane	177.8	13.1
C_6H_{14}	2-Methylpentane	119.4	6.27
C_6H_{14}	2,2-Dimethylbutane	174	0.58
C_6H_{14}	2,3-Dimethylbutane	144.3	0.80
$C_6H_{14}O$	Dipropyl ether	147.0	8.83
$C_6H_{14}O$	Diisopropyl ether	186.3	11.0
C_7F_8	Octafluorotoluene	207.5	11.6
$C_7H_3F_5$	2,3,4,5,6-Pentafluorotoluene	243.3	13.0
$C_7H_5ClO_2$	2-Chlorobenzoic acid	413.3	25.7
$C_7H_6O_2$	Benzoic acid	395.5	18.06
C_7H_7NO	Benzamide	402.2	18.5
$C_7H_7NO_2$	4-Nitrotoluene	324.7	16.8
C_7H_8	Toluene	178.16	6.85
C_7H_8O	2-Methylphenol	302.9	13.9
C_7H_8O	3-Methylphenol	284.9	9.41
C_7H_8O	4-Methylphenol	307.8	11.9
C_7H_8O	Phenylmethanol	257.9	8.97
C_7H_9N	4-Methylaniline	316.90	18.2
C_7H_{14}	Cycloheptane	265.12	1.88
C_7H_{14}	Methylcyclohexane	146.5	6.75
C_7H_{14}	1-Heptene	153.4	12.7
C_7H_{16}	Heptane	182.5	14.2

TABLE 2.3.4
ENTHALPY OF FUSION (continued)

Molecular formula	Name	T_m/K	$\Delta_{fus}H$/ kJ mol^{-1}
C_7H_{16}	2-Methylhexane	154.9	8.87
C_7H_{16}	2,2-Dimethylpentane	149.3	5.86
C_7H_{16}	2,4-Dimethylpentane	153.2	6.69
C_7H_{16}	3,3-Dimethylpentane	138.2	7.07
C_7H_{16}	3-Ethylpentane	154.5	9.55
C_7H_{16}	2,2,3-Trimethylbutane	248	2.2
$C_8H_8O_2$	2-Toluic acid	376.8	20.2
$C_8H_8O_2$	3-Toluic acid	381.90	15.7
$C_8H_8O_2$	4-Toluic acid	452.7	22.7
$C_8H_8O_2$	Phenylethanoic acid	349.8	14.5
C_8H_{10}	1,2-Dimethylbenzene	247.9	13.6
C_8H_{10}	1,3-Dimethylbenzene	225.3	11.6
C_8H_{10}	1,4-Dimethylbenzene	286.3	16.8
$C_8H_{10}O$	2,3-Dimethylphenol	345.9	21.0
$C_8H_{10}O$	2,5-Dimethylphenol	347.9	23.4
$C_8H_{10}O$	2,6-Dimethylphenol	318.8	18.9
$C_8H_{10}O$	3,4-Dimethylphenol	333.9	18.1
$C_8H_{10}O$	3,5-Dimethylphenol	336.7	18.0
C_8H_{16}	Cyclooctane	287.98	2.41
C_8H_{16}	Ethylcyclohexane	161.8	8.33
C_8H_{16}	1,1-Dimethylcyclohexane	239.8	2.06
C_8H_{16}	*cis*-1,2-Dimethylcyclohexane	223.2	1.64
C_8H_{16}	*trans*-1,2-Dimethylcyclohexane	183	10.5
C_8H_{16}	*cis*-1,3-Dimethylcyclohexane	197.5	10.8
C_8H_{16}	*trans*-1,3-Dimethylcyclohexane	183.0	9.86
C_8H_{16}	*cis*-1,4-Dimethylcyclohexane	185.7	9.31
C_8H_{16}	*trans*-1,4-Dimethylcyclohexane	236.2	12.3
$C_8H_{16}O_2$	Octanoic acid	289.4	21.4
C_8H_{18}	Octane	216.3	20.7
C_8H_{18}	3-Methylheptane	152.6	11.4
C_8H_{18}	4-Methylheptane	152	10.8
C_8H_{18}	2,2,4-Trimethylpentane	165.8	9.04
C_9H_7N	Quinoline	258.37	10.66
C_9H_{12}	Propylbenzene	173.59	9.27
C_9H_{12}	1,2,3-Trimethylbenzene	247.7	8.37
C_9H_{12}	1,2,4-Trimethylbenzene	229.3	3.76
C_9H_{12}	1,3,5-Trimethylbenzene	228.4	9.51
C_9H_{18}	Propylcyclohexane	178.2	10.4
$C_9H_{18}O$	5-Nonanone	267.2	24.9
$C_9H_{18}O_2$	Nonanoic acid	285.50	20.3
C_9H_{20}	Nonane	219.6	15.5
C_9H_{20}	2,2,3,3-Tetramethylpentane	263.3	2.33
C_9H_{20}	3,3-Diethylpentane	240.0	10.1
C_9H_{20}	2,2,4,4-Tetramethylpentane	206.61	9.75
$C_{10}H_7Br$	1-Bromonaphthalene	271.3	15.2
$C_{10}H_7Cl$	1-Chloronaphthalene	270.6	12.9
$C_{10}H_8$	Naphthalene	353.3	17.87
$C_{10}H_8O$	1-Naphthol	369	23.3
$C_{10}H_8O$	2-Naphthol	396	17.5
$C_{10}H_{14}$	Butylbenzene	185.2	11.2
$C_{10}H_{14}$	4-Isopropyltoluene	204.2	9.6
$C_{10}H_{14}$	1,2,4,5-Tetramethylbenzene	352.4	21.0
$C_{10}H_{14}O$	2-Isopropyl-5-methylphenol	324.6	17.3
$C_{10}H_{18}O_4$	Decanedioic acid	403.9	40.8
$C_{10}H_{20}$	Butylcyclohexane	198.42	14.2
$C_{10}H_{20}O_2$	Decanoic acid	305.14	28.0

TABLE 2.3.4
ENTHALPY OF FUSION (continued)

Molecular formula	Name	T_m/K	$\Delta_{fus}H$/ kJ mol^{-1}
$C_{10}H_{22}$	Decane	243.4	28.8
$C_{11}H_{10}$	2-Methylnaphthalene	307.5	12.0
$C_{11}H_{24}$	Undecane	247.5	22.3
$C_{12}H_9N$	Carbazole	519.3	26.9
$C_{12}H_{10}$	Acenaphthene	366.5	21.5
$C_{12}H_{10}$	Biphenyl	342	18.6
$C_{12}H_{10}N_2$	Azobenzene	340.2	22.0
$C_{12}H_{10}N_2O$	Azoxybenzene	309	17.9
$C_{12}H_{10}O$	Diphenyl ether	300.02	17.22
$C_{12}H_{11}N$	Diphenylamine	326.13	17.9
$C_{12}H_{16}$	Cyclohexylbenzene	280.4	15.3
$C_{12}H_{24}O_2$	Dodecanoic acid	316.37	36.6
$C_{12}H_{26}$	Dodecane	263.5	36.6
$C_{13}H_{10}$	Fluorene	387.9	19.6
$C_{13}H_{10}O$	Diphenyl ketone	321.03	18.2
$C_{14}H_{10}$	Anthracene	488.1	28.8
$C_{14}H_{10}$	Phenanthrene	372.39	16.5
$C_{14}H_{10}O_2$	Benzil	368.01	23.54
$C_{14}H_{12}$	*trans*-1,2-Diphenylethene	396	27.4
$C_{14}H_{12}O_2$	Diphenylethanoic acid	420.44	31.27
$C_{14}H_{28}O_2$	Tetradecanoic acid	327.11	45.4
$C_{16}H_{10}$	Fluoranthene	380.9	18.9
$C_{16}H_{10}$	Pyrene	424.3	17.1
$C_{16}H_{32}O_2$	Hexadecanoic acid	334.97	53.4
$C_{16}H_{34}O$	1-Hexadecanol	322.4	34.3
$C_{18}H_{12}$	Chrysene	531.3	26.2
$C_{18}H_{14}$	1,4-Diphenylbenzene	483.2	35.5
$C_{18}H_{36}O_2$	Octadecanoic acid	341.97	63.0
$C_{18}H_{38}$	Octadecane	301.3	61.4
$C_{19}H_{40}$	Nonadecane	305.2	45.8
$C_{20}H_{12}$	Perylene	546.6	31.8
$C_{20}H_{42}$	Icosane	309.9	69.9
$C_{24}H_{12}$	Coronene	710.4	19.2

TABLE 2.3.5
ENTHALPY OF VAPORIZATION

The *enthalpy of vaporization*, $\Delta_{vap}H$, is defined as the enthalpy of the gas phase minus the enthalpy of the liquid at a given temperature on the vapor pressure curve (see Section 2.1 and Table 2.1.3). This table gives the molar enthalpy of vaporization at the normal boiling point temperature and/or at 298.15 K for about 650 substances. See Reference 1 for a discussion of the accuracy of the data and methods of estimating enthalpy of vaporization of organic compounds at other temperatures.

Substances are arranged in a modified Hill order in which compounds not containing carbon precede those that contain carbon.

REFERENCES

1. Majer, V.; Svoboda, V., *Enthalpies of Vaporization of Organic Compounds,* Blackwell Scientific Publications, Oxford, 1985.
2. Chase, M. W.; Davies, C. A.; Downey, J. R.; Frurip, D. J.; McDonald, R. A.; Syverud, A. N.; *JANAF Thermochemical Tables, Third Edition, J. Phys. Chem. Ref. Data*, Vol. 14, Suppl. 1, 1985.
3. *DIPPR Data Compilation of Pure Compound Properties*, Design Institute for Physical Property Data, American Institute of Chemical Engineers, New York.
4. *Landolt-Börnstein, Numerical Data and Functions for Physics, Chemistry, Astronomy, Geophysics, and Technology, Sixth Edition*, Vol.2, Part 4: Thermal Properties, Springer-Verlag, Heidelberg, 1961.
5. Majer, V.; Svoboda, V.; Pick, J., *Heats of Vaporization of Fluids*, Elsevier, Amsterdam, 1989 (experimental techniques and predictive methods).
6. Tamir, A.; Tamir, E.; Stephan, K., *Heats of Phase Change of Pure Components and Mixtures*, Elsevier, Amsterdam, 1983 (review of correlation and predictive methods and a comprehensive bibliography).

Molecular formula	Name	T_b/K	$\Delta_{vap}H(T_b)$/ kJ mol^{-1}	$\Delta_{vap}H$(298.15 K)/ kJ mol^{-1}
AgBr	Silver bromide	1775	198	
AgCl	Silver chloride	1820	199	
AgI	Silver iodide	1779	143.9	
Al	Aluminum	2792	294	
AlB_3H_{12}	Aluminum trihydride-tris(borane)	317.6	30	
$AlBr_3$	Aluminum tribromide	528	23.5	
AlI_3	Aluminum triiodide	655	32.2	
Ar	Argon	87.30	6.43	
$AsBr_3$	Arsenic tribromide	494	41.8	
$AsCl_3$	Arsenic trichloride	403	35.01	
AsF_3	Arsenic trifluoride	330.9	29.7	
AsF_5	Arsenic pentafluoride	219.9	20.8	
AsH_3	Arsane	210.6	16.69	
AsI_3	Arsenic triiodide	697	59.3	
Au	Gold	3129	324	
B	Boron	4273	480	
BBr_3	Boron tribromide	364	30.5	
BCl_3	Boron trichloride	285.80	23.77	23.1
BF_3	Boron trifluoride	172	19.33	
BI_3	Boron triiodide	483	40.5	
B_2F_4	Tetrafluorodiborane(4)	239.3	28	
B_2H_6	Diborane(6)	180.77	14.28	
B_4H_{10}	Tetraborane(10)	291	27.1	
B_5H_{11}	Pentaborane(11)	336	31.8	
Ba	Barium	2170	140	
$BeCl_2$	Beryllium dichloride	820	105	
BeI_2	Beryllium diiodide	760	70.5	
Bi	Bismuth	1837	151	
$BiBr_3$	Bismuth tribromide	726	75.4	
$BiCl_3$	Bismuth trichloride	712.5	72.61	
BrF_3	Bromine trifluoride	398.9	47.57	

TABLE 2.3.5
ENTHALPY OF VAPORIZATION (continued)

Molecular formula	Name	T_b/K	$\Delta_{vap}H(T_b)$/ kJ mol^{-1}	$\Delta_{vap}H$(298.15 K)/ kJ mol^{-1}
BrF_5	Bromine pentafluoride	313.91	30.6	
BrH	Hydrogen bromide	206.77		12.69
BrH_3Si	Bromosilane	275	24.4	
$BrIn$	Indium bromide	929	92	
$BrTl$	Thallium bromide	1092	99.56	
Br_2	Bromine	331.9	29.96	30.91
Br_2Cd	Cadmium dibromide	1117	115	
Br_2Hg	Mercury dibromide	595	58.89	
Br_2Pb	Lead dibromide	1165	133	
Br_2Sn	Tin dibromide	912	102	
Br_2Zn	Zinc dibromide	970	118	
Br_3Ga	Gallium tribromide	552	38.9	
Br_3HSi	Tribromosilane	382	34.8	
Br_3OP	Phosphoryl tribromide	464.8	38	
Br_3P	Phosphorus tribromide	446.10	38.8	
Br_3Sb	Antimony tribromide	560.26	59	
Br_4Ge	Germanium tetrabromide	459.50	41.4	
Br_4Si	Silicon tetrabromide	427	37.9	
Br_4Sn	Tin tetrabromide	478	43.5	
Br_4Ti	Titanium tetrabromide	503	44.37	
Br_5Ta	Tantalum pentabromide	622	62.3	
Cd	Cadmium	1040	99.87	
$CdCl_2$	Cadmium dichloride	1233	124.3	
CdF_2	Cadmium difluoride	2021	214	
CdI_2	Cadmium diiodide	1015	115	
ClF	Chlorine fluoride	172.0	24	
$ClFO_3$	Perchloryl fluoride	226.40	19.33	
ClF_3	Chlorine trifluoride	284.90	27.53	
ClF_3Si	Chlorotrifluorosilane	203.10	18.7	
ClH	Hydrogen chloride	188.15	16.15	9.08
ClH_3Si	Chlorosilane	242.7	21	
$ClNO$	Nitrosyl chloride	267.60	25.78	
ClO_2	Chlorine dioxide	284	30	
$ClTl$	Thallium chloride	1080	102.2	
Cl_2	Chlorine	239.18	20.41	17.65
Cl_2Cr	Chromium dichloride	1573	197	
Cl_2CrO_2	Chromyl dichloride	390	35.1	
Cl_2H_2Si	Dichlorosilane	281.4	25	24.2
Cl_2Hg	Mercury dichloride	577	58.9	
Cl_2O	Dichlorine oxide	275.3	25.9	
Cl_2OS	Sulfinyl dichloride	348.7	31.7	31
Cl_2O_2S	Sulfonyl dichloride	342.5	31.4	30.1
Cl_2Pb	Lead dichloride	1224	127	
Cl_2Sn	Tin dichloride	896	86.8	
Cl_2Zn	Zinc dichloride	1005	126	
Cl_3Ga	Gallium trichloride	474	23.9	
Cl_3HSi	Trichlorosilane	306		25.7
Cl_3OP	Phosphoryl trichloride	378.6	34.35	38.6
Cl_3OV	Vanadyl trichloride	400	36.78	
Cl_3P	Phosphorus trichloride	349.10	30.5	32.1
Cl_3Sb	Antimony trichloride	493.4	45.19	
Cl_4Ge	Tetrachlorogerman	359.70	27.9	
Cl_4OW	Tungstenosyl tetrachloride	500.70	67.8	
Cl_4Si	Silicon tetrachloride	330.80	28.7	29.7
Cl_4Sn	Tin tetrachloride	387.30	34.9	
Cl_4Te	Tellurium tetrachloride	660	77	

TABLE 2.3.5
ENTHALPY OF VAPORIZATION (continued)

Molecular formula	Name	T_b/K	$\Delta_{vap}H(T_b)$/ kJ mol^{-1}	$\Delta_{vap}H$(298.15 K)/ kJ mol^{-1}
Cl_4Th	Thorium tetrachloride	1194	146.4	
Cl_4Ti	Titanium tetrachloride	409.60	36.2	
Cl_4V	Vanadium tetrachloride	425	41.4	42.5
Cl_5Mo	Molybdenum pentachloride	541	62.8	
Cl_5Nb	Niobium pentachloride	527.20	52.7	
Cl_5Ta	Tantalum pentachloride	512.50	54.8	
Cl_6W	Tungsten hexachloride	619.90	52.7	
FH_3Si	Fluorosilane	174.5	18.8	
FLi	Lithium fluoride	1946	147	
FNO	Nitrosyl fluoride	213.2	19.28	
FNO_2	Nitryl fluoride	200.7	18.05	
F_2	Fluorine	84.95	6.62	
F_2H_2Si	Difluorosilane	195.3	16.3	
F_2O	Oxygen difluoride	128.40	11.09	
F_2OS	Sulfinyl difluoride	229.3	21.8	
F_2Pb	Lead difluoride	1566	160.4	
F_2Zn	Zinc difluoride	1773	190.1	
F_3HSi	Trifluorosilane	178	16.2	
F_3N	Nitrogen trifluoride	144.40	11.56	
F_3P	Phosphorus trifluoride	171.6	16.5	
F_3PS	Thiophosphoryl trifluoride	220.90	19.6	
F_4N_2	Tetrafluorohydrazine	199	13.27	
F_4S	Sulfur tetrafluoride	232.70	26.44	
F_4Se	Selenium tetrafluoride	379	47.2	
F_4Th	Thorium tetrafluoride	1953	258	
F_5I	Iodine pentafluoride	373.6	41.3	
F_5Nb	Niobium pentafluoride	502	52.3	
F_5P	Phosphorus pentafluoride	188.5	17.2	
F_5Ta	Tantalum pentafluoride	502.3	56.9	
F_5V	Vanadium pentafluoride	321.4	44.52	
F_6Ir	Iridium hexafluoride	326	36	
F_6Mo	Molybdenum hexafluoride	307	27.2	28
F_6S	Sulfur hexafluoride	209.3s		8.99
F_6W	Tungsten hexafluoride	290	27	26.6
Ga	Gallium	2477	254	
GaI_3	Gallium triiodide	613	56.5	
Ge	Germanium	3106	334	
GeH_4	Germane	183	14.06	
Ge_2H_6	Digermane	303.9	25.1	
HI	Hydrogen iodide	237.60	19.76	17.36
HLiO	Lithium hydroxide	1899	188	
HNO_3	Nitric acid	356		39.1
HN_3	Hydrogen azide	308.8	30.5	
HNaO	Sodium hydroxide	1661	175	
H_2	Hydrogen	20.38	0.90	
H_2O	Water	373.15	40.65	43.98
H_2O_2	Dihydrogen peroxide	423.3		51.6
H_2S	Dihydrogen sulfide	212.84	18.67	14.08
H_2Se	Dihydrogen selenide	231.90	19.7	
H_2Te	Dihydrogen telluride	271	19.2	
H_3N	Ammonia	239.72	23.33	19.86
H_3P	Phosphine	185.40	14.6	
H_3Sb	Stibine	256	21.3	
H_4N_2	Hydrazine	386.70	41.8	44.7
H_4Si	Silane	161	12.1	
H_4Sn	Stannane	221.3	19.05	

TABLE 2.3.5
ENTHALPY OF VAPORIZATION (continued)

Molecular formula	Name	T_b/K	$\Delta_{vap}H(T_b)$/ kJ mol^{-1}	$\Delta_{vap}H$(298.15 K)/ kJ mol^{-1}
H_6Si_2	Disilane	258.8	21.2	
H_8Si_3	Trisilane	326	28.5	
He	Helium	4.22	0.083	
Hg	Mercury	629.88	59.11	
HgI_2	Mercury diiodide	627	59.2	
IIn	Indium iodide	985	90.8	
ITl	Thallium iodide	1097	104.7	
I_2	Iodine	457.5	41.57	
I_2Pb	Lead diiodide	1145	104	
I_3Sb	Antimony triiodide	674	68.6	
I_4Si	Silicon tetraiodide	560.50	50.2	
I_4Sn	Tin tetraiodide	637.50	56.9	
I_4Ti	Titanium tetraiodide	650	58.4	
Kr	Krypton	119.93	9.08	
MoO_3	Molybdenum trioxide	1428	138	
NO	Nitrogen oxide	121.41	13.83	
N_2	Nitrogen	77.35	5.57	
N_2O	Dinitrogen oxide	184.67	16.53	
N_2O_4	Dinitrogen tetraoxide	294.30	38.12	
Ne	Neon	27.07	1.71	
O_2	Oxygen	90.17	6.82	
O_2S	Sulfur dioxide	263.13	24.94	22.92
O_3S	Sulfur trioxide	318	40.69	43.14
P	Phosphorus	550	12.4	14.2
Pb	Lead	2022	179.5	
S	Sulfur	717.75	45	
STl_2	Dithallium sulfide	1640	154	
Se	Selenium	958	95.48	
Te	Tellurium	1261	114.1	
Xe	Xenon	165.11	12.62	
CBr_4	Tetrabromomethane	462.6	43.5	
$CClF_3$	Chlorotrifluoromethane	191.7	15.75	
CCl_2F_2	Dichlorodifluoromethane	243.3	20.11	
CCl_3F	Trichlorofluoromethane	296.78	25.06	25.02
CCl_4	Tetrachloromethane	349.8	29.82	32.43
CF_4	Tetrafluoromethane	145.07	12.33	
$CHBr_3$	Tribromomethane	422.36	39.66	46.05
$CHClF_2$	Chlorodifluoromethane	232.40	20.24	
$CHCl_2F$	Dichlorofluoromethane	282.05	25.15	24.23
$CHCl_3$	Trichloromethane	334.33	29.24	31.28
CHF_3	Trifluoromethane	191.0	18.4	
CH_2BrCl	Bromochloromethane	341.20	30	32.85
CH_2Br_2	Dibromomethane	370	32.92	36.97
CH_2Cl_2	Dichloromethane	312.79	28.06	28.82
CH_2I_2	Diiodomethane	455	42.49	49.38
CH_2O	Methanal	254.05	23.3	
CH_2O_2	Methanoic acid	373.71	22.69	20.1
CH_3Br	Bromomethane	276.7	23.91	22.81
CH_3Cl	Chloromethane	248.95	21.4	18.92
CH_3I	Iodomethane	315.57	27.34	27.97
CH_3NO	Methanamide	493		60.15
CH_3NO_2	Nitromethane	374.35	33.99	38.27
CH_4	Methane	111.65	8.19	
CH_4O	Methanol	337.7	35.21	37.43
CH_5N	Methylamine	266.82	25.6	23.37
CH_6N_2	Methylhydrazine	364.1	36.12	40.37

TABLE 2.3.5
ENTHALPY OF VAPORIZATION (continued)

Molecular formula	Name	T_b/K	$\Delta_{vap}H(T_b)$/ kJ mol^{-1}	$\Delta_{vap}H$(298.15 K)/ kJ mol^{-1}
CN_4O_8	Tetranitromethane	399.2	40.74	49.93
CO	Carbon oxide	81.6	6.04	
CS_2	Carbon disulfide	319.37	26.74	27.51
$C_2Br_2ClF_3$	1,2-Dibromo-1-chloro-1,2,2-trifluoro-ethane	366	31.17	35.04
$C_2Br_2F_4$	1,2-Dibromotetrafluoroethane	320.50	27.03	28.39
C_2ClF_5	Chloropentafluoroethane	234.0	19.41	
$C_2Cl_2F_4$	1,2-Dichlorotetrafluoroethane	276.9	23.25	
$C_2Cl_3F_3$	1,1,1-Trichlorotrifluoroethane	318.95	26.85	28.08
$C_2Cl_3F_3$	1,1,2-Trichlorotrifluoroethane	320.8	27.04	28.4
C_2Cl_4	Tetrachloroethene	394.2	34.68	39.68
C_2F_4	Tetrafluoroethene	197.50	16.82	
C_2F_6	Hexafluoroethane	194.9	16.15	
C_2HCl_3	Trichloroethene	360.34	31.4	34.54
C_2HCl_5	Pentachloroethane	433.03	36.94	
$C_2HF_3O_2$	Trifluoroethanoic acid	344.9	33.26	
$C_2H_2Br_4$	1,1,2,2-Tetrabromoethane	516.70	48.65	70
$C_2H_2Cl_2$	1,1-Dichloroethene	304.7	26.14	26.48
$C_2H_2Cl_2$	*cis*-1,2-Dichloroethene	333.78	30.23	31.57
$C_2H_2Cl_2$	*trans*-1,2-Dichloroethene	320.82	28.89	30.04
$C_2H_2Cl_4$	1,1,2,2-Tetrachloroethane	418.3	37.64	45.71
C_2H_2O	Ketene	223.34	20.4	
$C_2H_2O_2$	Ethanedial	323.5	38	
C_2H_3Br	Bromoethene	288.9	23.43	22.6
C_2H_3Cl	Chloroethene	259.34	20.8	18.64
$C_2H_3Cl_3$	1,1,1-Trichloroethane	347.24	29.86	32.5
$C_2H_3Cl_3$	1,1,2-Trichloroethane	387.0	34.82	40.24
C_2H_3F	Fluoroethene	201	17.1	
C_2H_3N	Ethanenitrile	354.80	29.75	32.94
C_2H_4	Ethene	169.38	13.53	
$C_2H_4Br_2$	1,2-Dibromoethane	404.5	34.77	41.73
$C_2H_4Cl_2$	1,1-Dichloroethane	330.5	28.85	30.62
$C_2H_4Cl_2$	1,2-Dichloroethane	356.6	31.98	35.16
$C_2H_4F_2$	1,1-Difluoroethane	248.20	21.56	19.08
C_2H_4O	Ethanal	293.6	25.76	25.47
C_2H_4O	Oxirane	283.6	25.54	24.75
$C_2H_4O_2$	Ethanoic acid	391.0	23.7	23.36
$C_2H_4O_2$	Methyl methanoate	304.9	27.92	28.35
C_2H_5Br	Bromoethane	311.6	27.04	28.03
C_2H_5Cl	Chloroethane	285.4	24.65	
C_2H_5ClO	2-Chloroethanol	401.7	41.43	
C_2H_5F	Fluoroethane	235.50	21.07	
C_2H_5I	Iodoethane	345.6	29.44	31.93
C_2H_5NO	*N*-Methylmethanamide	472.66		56.19
$C_2H_5NO_2$	Nitroethane	387.22	37.97	41.59
C_2H_6	Ethane	184.5	14.69	5.16
C_2H_6O	Dimethyl ether	248.3	21.51	18.51
C_2H_6O	Ethanol	351.44	38.56	42.32
C_2H_6OS	Dimethyl sulfoxide	462.2	43.14	52.88
$C_2H_6O_2$	1,2-Ethanediol	470.49	50.46	67.8
C_2H_6S	2-Thiapropane	310.48	27	27.65
C_2H_6S	Ethanethiol	308.2	26.79	27.3
$C_2H_6S_2$	2,3-Dithiabutane	382.9	33.78	37.86
C_2H_7N	Dimethylamine	280.03	26.4	25.05
C_2H_7NO	2-Aminoethanol	444.1	49.83	
$C_2H_8N_2$	1,1-Dimethylhydrazine	337.0	32.55	35

TABLE 2.3.5
ENTHALPY OF VAPORIZATION (continued)

Molecular formula	Name	T_b/K	$\Delta_{vap}H(T_b)$/ kJ mol^{-1}	$\Delta_{vap}H$(298.15 K)/ kJ mol^{-1}
$C_2H_8N_2$	1,2-Ethanediamine	390	37.98	44.98
C_2N_2	Ethanedinitrile	252.0	23.33	19.75
C_3F_8	Octafluoropropane	236.40	20.6	
C_3H_3N	Propenenitrile	350.57	32.55	
C_3H_4O	Propenal	325.26	28.33	
$C_3H_4O_2$	2-Oxetanone	435		47.03
C_3H_5Br	3-Bromopropene	343.2	30.24	32.73
C_3H_5Cl	3-Chloropropene	318.30	29.04	
C_3H_5ClO	(Chloromethyl)oxirane	389.26	37.91	
$C_3H_5ClO_2$	Methyl chloroethanoate	402.6	39.23	46.73
$C_3H_5Cl_3$	1,2,3-Trichloropropane	430	37.12	46.94
C_3H_5N	Propanenitrile	370.6	31.81	36.03
C_3H_6	Propene	225.46	18.42	14.24
C_3H_6	Cyclopropane	240.34	20.05	16.93
$C_3H_6Br_2$	1,2-Dibromopropane	415.0	35.61	41.67
$C_3H_6Cl_2$	1,2-Dichloropropane	369.52	32	36.4
$C_3H_6Cl_2$	1,3-Dichloropropane	394.0	35.18	40.75
C_3H_6O	2-Propanone	329.20	29.1	30.99
C_3H_6O	2-Propen-1-ol	370.23	39.96	
C_3H_6O	Propanal	321.2	28.31	29.62
C_3H_6O	Methyloxirane	308	27.35	27.89
C_3H_6O	Oxetane	320.7	28.67	29.85
$C_3H_6O_2$	Ethyl methanoate	327.5	29.91	31.96
$C_3H_6O_2$	Methyl ethanoate	330.02	30.32	32.29
$C_3H_6O_2$	Propanoic acid	414.30	32.28	32.14
$C_3H_6O_2$	1,3-Dioxolane	348.8		35.6
C_3H_6S	Thiacyclobutane	368	32.32	35.97
C_3H_7Br	1-Bromopropane	344.1	29.84	32.01
C_3H_7Br	2-Bromopropane	332.6	28.33	30.17
C_3H_7Cl	1-Chloropropane	319.67	27.18	28.35
C_3H_7Cl	2-Chloropropane	308.89	26.3	26.9
C_3H_7I	1-Iodopropane	375.7	32.08	36.25
C_3H_7I	2-Iodopropane	362.6	30.68	34.06
C_3H_7NO	*N,N*-Dimethylmethanamide	426	38.44	46.89
$C_3H_7NO_2$	1-Nitropropane	404.33	38.47	43.39
$C_3H_7NO_2$	2-Nitropropane	393.40	36.79	41.34
C_3H_8	Propane	231.08	19.04	14.79
C_3H_8O	1-Propanol	370.3	41.44	47.45
C_3H_8O	2-Propanol	355.4	39.85	45.39
$C_3H_8O_2$	1,2-Propanediol	460.7	52.35	64.4
$C_3H_8O_2$	1,3-Propanediol	487.5	57.86	72.8
$C_3H_8O_2$	2-Methoxyethanol	397.8	37.54	45.17
$C_3H_8O_2$	2,4-Dioxapentane	315		28.89
$C_3H_8O_3$	1,2,3-Propanetriol	563.2	61.04	
C_3H_8S	2-Thiabutane	339.8	29.53	31.85
C_3H_8S	1-Propanethiol	340.9	29.54	31.89
C_3H_8S	2-Propanethiol	325.7	27.91	29.45
$C_3H_8S_2$	1,3-Propanedithiol	446.0		49.66
C_3H_9N	Propylamine	320.37	29.55	31.27
C_3H_9N	Isopropylamine	304.91	27.83	28.36
C_3H_9N	Trimethylamine	276.02	22.94	21.66
C_4F_8	Octafluorocyclobutane	267.16	23.24	
C_4F_{10}	Decafluorobutane	270.96	22.9	
$C_4H_4N_2$	Pyridazine	481		53.47
$C_4H_4N_2$	Pyrimidine	396.9	43.09	49.79
$C_4H_4N_2$	Butanedinitrile	539	48.5	63.97

TABLE 2.3.5
ENTHALPY OF VAPORIZATION (continued)

Molecular formula	Name	T_b/K	$\Delta_{vap}H(T_b)$/ kJ mol^{-1}	$\Delta_{vap}H$(298.15 K)/ kJ mol^{-1}
C_4H_4O	Furan	304.5	27.1	27.45
C_4H_4S	Thiophene	357.1	31.48	34.7
C_4H_5N	Pyrrole	402.94	38.75	45.09
C_4H_6	1,2-Butadiene	284.0	24.02	23.21
C_4H_6	1,3-Butadiene	268.74	22.47	20.86
C_4H_6	1-Butyne	281.23	24.52	23.35
$C_4H_6O_2$	Oxolan-2-one	477	52.22	
$C_4H_6O_2$	Methyl propenoate	353.8	33.1	29.2
$C_4H_6O_2$	Vinyl ethanoate	345.7	34.55	37.2
$C_4H_6O_3$	Ethanoic anhydride	412.20	38.2	
$C_4H_7ClO_2$	Ethyl chloroethanoate	417.4	40.43	49.47
C_4H_7N	Butanenitrile	390.77	33.68	39.33
C_4H_7N	2-Methylpropanenitrile	377.0	32.39	37.13
C_4H_8	1-Butene	267.0	22.07	20.22
C_4H_8	*cis*-2-Butene	276.86	23.34	22.16
C_4H_8	*trans*-2-Butene	274.03	22.72	21.4
C_4H_8	Cyclobutane	285.7	24.19	23.51
$C_4H_8Cl_2$	1,2-Dichlorobutane	397.2	33.9	39.58
$C_4H_8Cl_2O$	Bis(2-chloroethyl) ether	451.6	45.23	
C_4H_8O	Butanal	348.0	31.5	33.68
C_4H_8O	2-Butanone	352.74	31.3	34.79
C_4H_8O	Oxolane	339.1	29.81	31.99
C_4H_8O	Ethyl vinyl ether	308.70	26.2	27.5
$C_4H_8O_2$	1,3-Dioxane	379.2	34.37	39.09
$C_4H_8O_2$	1,4-Dioxane	374.47	34.16	38.6
$C_4H_8O_2$	Ethyl ethanoate	350.26	31.94	35.6
$C_4H_8O_2$	Methyl propanoate	352.5	32.24	35.85
$C_4H_8O_2$	Propyl methanoate	354.0	33.61	37.53
$C_4H_8O_2$	Butanoic acid	436.90		40.45
$C_4H_8O_2$	2-Methylpropanoic acid	427.85		35.3
C_4H_8S	Thiacyclopentane	394.1	34.66	39.43
C_4H_9Br	1-Bromobutane	374.7	32.51	36.64
C_4H_9Br	2-Bromobutane	364.5	30.77	34.41
C_4H_9Cl	1-Chlorobutane	351.58	30.39	33.51
C_4H_9Cl	2-Chlorobutane	341.4	29.17	31.53
C_4H_9I	1-Iodobutane	403.7	34.66	40.63
C_4H_9I	2-Iodobutane	393.2	33.27	38.46
C_4H_9N	Pyrrolidine	359.71	33.01	37.52
C_4H_9NO	*N,N*-Dimethylethanamide	439.3	43.35	50.24
C_4H_9NO	Morpholine	402.1	37.05	43.96
C_4H_{10}	Butane	272.6	22.44	21.02
C_4H_{10}	2-Methylpropane	261.42	21.3	19.23
$C_4H_{10}O$	1-Butanol	390.88	43.29	52.35
$C_4H_{10}O$	2-Butanol	372.66	40.75	49.72
$C_4H_{10}O$	2-Methyl-2-propanol	355.5	39.07	46.69
$C_4H_{10}O$	2-Methyl-1-propanol	381.04	41.82	50.82
$C_4H_{10}O$	Diethyl ether	307.6	26.52	27.1
$C_4H_{10}O$	Methyl propyl ether	312.2	26.75	27.6
$C_4H_{10}O$	Isopropyl methyl ether	303.92	26.05	26.41
$C_4H_{10}O_2$	1,4-Butanediol	508		76.6
$C_4H_{10}O_2$	2-Ethoxyethanol	408.8	39.22	48.21
$C_4H_{10}O_2$	2,5-Dioxahexane	358	32.42	36.39
$C_4H_{10}O_3$	3-Oxa-1,5-pentanediol	518.84	52.26	
$C_4H_{10}S$	1-Butanethiol	371.6	32.23	36.63
$C_4H_{10}S$	3-Thiapentane	365.2	31.77	35.8
$C_4H_{10}S$	Methyl propyl sulfide	368.7	32.08	36.24

TABLE 2.3.5
ENTHALPY OF VAPORIZATION (continued)

Molecular formula	Name	T_b/K	$\Delta_{vap}H(T_b)$/ kJ mol^{-1}	$\Delta_{vap}H$(298.15 K)/ kJ mol^{-1}
$C_4H_{11}N$	Butylamine	350.15	31.81	35.72
$C_4H_{11}N$	(2-Methylpropyl)amine	340.90	30.61	33.85
$C_4H_{11}N$	2-Aminobutane	335.88	29.92	32.85
$C_4H_{11}N$	(1,1-Dimethylethyl)amine	317.19	28.27	29.64
$C_4H_{11}N$	Diethylamine	328.7	29.06	31.31
$C_4H_{11}NO_2$	Bis(2-hydroxyethyl)amine	541.9	65.23	
$C_5H_2F_6O_2$	1,1,1,5,5,5-Hexafluoroacetylacetone	327.30	27.05	30.58
$C_5H_4O_2$	2-Furaldehyde	435.00	43.22	
C_5H_5N	Pyridine	388.38	35.09	40.21
$C_5H_6O_2$	2-Furylmethanol	444	53.64	64.43
C_5H_6S	2-Methylthiophene	385.7	33.9	38.87
C_5H_6S	3-Methylthiophene	388.6	34.24	39.43
C_5H_8	Spiropentane	312	26.76	27.49
C_5H_8O	Cyclopentanone	403.9	36.35	42.72
C_5H_8O	Cyclopropyl methyl ketone	384.4	34.07	39.41
$C_5H_8O_2$	Ethyl propenoate	372.5	34.7	
$C_5H_8O_2$	Methyl 2-methylpropenoate	373.5	36	40.7
C_5H_9N	Pentanenitrile	414.5	36.09	43.6
C_5H_{10}	1-Pentene	303.11	25.2	25.47
C_5H_{10}	*cis*-2-Pentene	310.08	26.11	26.86
C_5H_{10}	*trans*-2-Pentene	309.49	26.07	26.76
C_5H_{10}	2-Methyl-1-butene	304.3	25.5	25.86
C_5H_{10}	2-Methyl-2-butene	311.72	26.31	27.06
C_5H_{10}	Cyclopentane	322.4	27.3	28.52
$C_5H_{10}O$	Cyclopentanol	413.57		57.6
$C_5H_{10}O$	2-Pentanone	375.41	33.44	38.4
$C_5H_{10}O$	3-Pentanone	375.11	33.45	38.52
$C_5H_{10}O$	3-Methyl-2-butanone	367.48	32.35	36.78
$C_5H_{10}O$	3,3-Dimethyloxetane	353.7	30.85	33.94
$C_5H_{10}O$	Oxane	361	31.17	34.58
$C_5H_{10}O_2$	Butyl methanoate	379.3	36.58	41.11
$C_5H_{10}O_2$	Isobutyl methanoate	371.40	33.55	
$C_5H_{10}O_2$	Propyl ethanoate	374.69	33.92	39.72
$C_5H_{10}O_2$	Isopropyl ethanoate	361.7	32.93	37.2
$C_5H_{10}O_2$	Ethyl propanoate	372.2	33.88	39.21
$C_5H_{10}O_2$	Methyl butanoate	375.9	33.79	39.28
$C_5H_{10}O_2$	Methyl 2-methylpropanoate	365.6	32.61	37.32
$C_5H_{10}O_2$	Pentanoic acid	458.70	44.06	69.29
$C_5H_{10}O_2$	2-Tetrahydrofurylmethanol	451	45.19	51.55
$C_5H_{10}O_3$	Diethyl carbonate	399	36.15	43.6
$C_5H_{10}O_3$	2-Methoxyethyl ethanoate	416	43.9	50.27
$C_5H_{11}Br$	1-Bromopentane	402.9	35.01	41.28
$C_5H_{11}Cl$	1-Chloropentane	380.9	33.15	38.24
$C_5H_{11}I$	1-Iodopentane	428		45.27
$C_5H_{11}N$	Piperidine	379.37		39.29
C_5H_{12}	Pentane	309.21	25.79	26.43
C_5H_{12}	2-Methylbutane	301.03	24.69	24.85
C_5H_{12}	2,2-Dimethylpropane	282.63	22.74	21.84
$C_5H_{12}O$	Butyl methyl ether	343.31	29.55	32.37
$C_5H_{12}O$	Isobutyl methyl ether	331.7	28.02	30.13
$C_5H_{12}O$	*tert*-Butyl methyl ether	328.3	27.94	29.82
$C_5H_{12}O$	Ethyl propyl ether	336.36	28.94	31.43
$C_5H_{12}O$	Ethyl isopropyl ether	327.2	28.21	30.08
$C_5H_{12}O$	1-Pentanol	411.13	44.36	57.02
$C_5H_{12}O$	2-Pentanol	392.2	41.4	54.21
$C_5H_{12}O$	3-Pentanol	389.40	43.5	54

TABLE 2.3.5
ENTHALPY OF VAPORIZATION (continued)

Molecular formula	Name	T_b/K	$\Delta_{vap}H(T_b)$/ kJ mol^{-1}	$\Delta_{vap}H$(298.15 K)/ kJ mol^{-1}
$C_5H_{12}O$	2-Methyl-2-butanol	375.3	39.04	50.1
$C_5H_{12}O$	3-Methyl-1-butanol	404.2	44.07	55.61
$C_5H_{12}O_2$	3,5-Dioxaheptane	361	31.33	35.65
$C_5H_{12}S$	1-Pentanethiol	399.8	34.88	41.24
$C_5H_{13}N$	Pentylamine	377.4	34.01	40.08
C_6ClF_5	Chloropentafluorobenzene	391.11	34.76	41.07
C_6F_6	Hexafluorobenzene	353.41	31.66	35.71
C_6HF_5	Pentafluorobenzene	358.89	32.15	36.27
$C_6H_4Cl_2$	1,2-Dichlorobenzene	453.6	39.66	50.21
$C_6H_4Cl_2$	1,3-Dichlorobenzene	446.14	38.62	48.58
$C_6H_4Cl_2$	1,4-Dichlorobenzene	447.27	38.79	49
$C_6H_4F_2$	1,2-Difluorobenzene	367	32.21	36.18
$C_6H_4F_2$	1,3-Difluorobenzene	355.7	31.1	34.59
$C_6H_4F_2$	1,4-Difluorobenzene	362	31.77	35.54
C_6H_5Br	Bromobenzene	429.06		44.54
C_6H_5Cl	Chlorobenzene	404.87	35.19	40.97
C_6H_5F	Fluorobenzene	357.88	31.19	34.58
C_6H_5I	Iodobenzene	461.48	39.5	49.58
$C_6H_5NO_2$	Nitrobenzene	483.9		55.01
C_6H_6	Benzene	353.24	30.72	33.83
C_6H_6ClN	2-Chloroaniline	481.99	44.35	56.75
C_6H_6O	Phenol	455.02	45.69	57.82
C_6H_6S	Benzenethiol	442.2	39.93	47.56
C_6H_7N	Aniline	457.14	42.44	55.83
C_6H_7N	2-Methylpyridine	402.53	36.17	42.48
C_6H_7N	3-Methylpyridine	417.29	37.35	44.44
C_6H_7N	4-Methylpyridine	418.51	37.51	44.56
C_6H_{10}	Cyclohexene	356.13	30.46	33.47
$C_6H_{10}O$	Cyclohexanone	428.80	40.25	45.06
$C_6H_{10}O$	4-Methyl-3-penten-2-one	403	36.11	43.4
$C_6H_{10}O_3$	Propanoic anhydride	440.1	41.7	
$C_6H_{11}N$	Hexanenitrile	436.61		47.91
C_6H_{12}	Cyclohexane	353.88	29.97	33.01
C_6H_{12}	Methylcyclopentane	344.96	29.08	31.64
C_6H_{12}	1-Hexene	336.63	28.28	30.61
C_6H_{12}	*cis*-2-Hexene	341.9		32.19
C_6H_{12}	2,3-Dimethyl-2-butene	346.4	29.64	32.53
$C_6H_{12}O$	Cyclohexanol	434.25		62.01
$C_6H_{12}O$	2-Hexanone	400.7	36.35	43.14
$C_6H_{12}O$	3-Hexanone	396.6	35.36	42.47
$C_6H_{12}O$	2-Methyl-3-pentanone	386.6	33.84	39.79
$C_6H_{12}O$	3-Methyl-2-pentanone	390.6	34.16	40.53
$C_6H_{12}O$	4-Methyl-2-pentanone	389.6	34.49	40.61
$C_6H_{12}O$	3,3-Dimethyl-2-butanone	379.2	33.39	37.91
$C_6H_{12}O$	Butyl vinyl ether	367	31.58	36.17
$C_6H_{12}O_2$	Butyl ethanoate	399.2	36.28	43.86
$C_6H_{12}O_2$	Isobutyl ethanoate	389.70	35.85	39.2
$C_6H_{12}O_2$	Propyl propanoate	395.6	35.54	43.45
$C_6H_{12}O_2$	Ethyl butanoate	394.6	35.47	42.68
$C_6H_{12}O_2$	Ethyl 2-methylpropanoate	383.2	33.67	39.83
$C_6H_{12}O_2$	Methyl pentanoate	400.5	35.36	43.1
$C_6H_{12}O_3$	3-Oxapentyl ethanoate	429.5		52.69
$C_6H_{12}S$	Cyclohexanethiol	432.0	37.06	44.57
$C_6H_{13}Br$	1-Bromohexane	428.4		45.89
$C_6H_{13}Cl$	1-Chlorohexane	408	35.67	42.83
$C_6H_{13}I$	1-Iodohexane	454		49.75

TABLE 2.3.5
ENTHALPY OF VAPORIZATION (continued)

Molecular formula	Name	T_b/K	$\Delta_{vap}H(T_b)$/ kJ mol^{-1}	$\Delta_{vap}H(298.15\ K)$/ kJ mol^{-1}
$C_6H_{13}N$	Cyclohexylamine	407.11	36.14	43.67
C_6H_{14}	Hexane	341.88	28.85	31.56
C_6H_{14}	2-Methylpentane	333.41	27.79	29.89
C_6H_{14}	3-Methylpentane	336.42	28.06	30.28
C_6H_{14}	2,2-Dimethylbutane	322.88	26.31	27.68
C_6H_{14}	2,3-Dimethylbutane	331.13	27.38	29.12
$C_6H_{14}O$	Methyl pentyl ether	372	32.02	36.85
$C_6H_{14}O$	Butyl ethyl ether	365.4	31.63	36.32
$C_6H_{14}O$	Dipropyl ether	363.23	31.31	35.69
$C_6H_{14}O$	Diisopropyl ether	341.66	29.1	32.12
$C_6H_{14}O$	1-Hexanol	430.7	44.5	61.61
$C_6H_{14}O$	2-Hexanol	413.0	41.01	58.46
$C_6H_{14}O$	4-Methyl-1-pentanol	425.0	44.46	60.47
$C_6H_{14}O$	2-Methyl-2-pentanol	394.2	39.59	54.77
$C_6H_{14}O$	4-Methyl-2-pentanol	404.80	44.22	58.7
$C_6H_{14}O_2$	1,1-Diethoxyethane	375.40	36.28	43.2
$C_6H_{14}O_2$	3,6-Dioxaoctane	392.5	36.28	43.2
$C_6H_{14}O_2$	2-Butoxyethanol	441.5		56.59
$C_6H_{14}O_2$	2-Methyl-2,4-pentanediol	470.2	57.3	
$C_6H_{14}O_3$	2,5,8-Trioxanonane	435	36.17	44.69
$C_6H_{14}O_4$	3,6-Dioxa-1,8-octanediol	558	71.4	
$C_6H_{15}N$	Hexylamine	405.9	36.54	45.1
$C_6H_{15}N$	Triethylamine	362	31.01	34.84
$C_6H_{15}N$	Dipropylamine	382.4	33.47	40.04
$C_6H_{15}N$	Diisopropylamine	357.0	30.4	34.61
$C_7H_3F_5$	2,3,4,5,6-Pentafluorotoluene	390.6	34.75	41.12
C_7H_5N	Benzonitrile	464.30	45.94	55.48
C_7H_6O	Benzaldehyde	451.9	42.5	50.3
$C_7H_6O_2$	2-Hydroxybenzaldehyde	470	38.24	
C_7H_7Cl	2-Chlorotoluene	432.3	37.53	45.63
C_7H_7Cl	4-Chlorotoluene	435.14	38.74	
C_7H_7Cl	Benzyl chloride	452		51.5
C_7H_7F	4-Fluorotoluene	389.7	34.08	39.42
C_7H_8	Toluene	383.78	33.18	38.01
C_7H_8O	2-Methylphenol	464.19	45.19	
C_7H_8O	3-Methylphenol	475.42	47.4	61.71
C_7H_8O	4-Methylphenol	475.13	47.45	
C_7H_8O	Phenylmethanol	478.46	50.48	
C_7H_8O	Anisole	426.8	38.97	46.9
C_7H_9N	*N*-Methylaniline	469.40		53.1
C_7H_9N	2-Methylaniline	473.49	44.6	56.74
C_7H_9N	3-Methylaniline	476.52	44.85	57.28
C_7H_9N	4-Methylaniline	473.7	44.27	56.2
C_7H_9N	2,3-Dimethylpyridine	434.41	39.08	47.73
C_7H_9N	2,4-Dimethylpyridine	431.6	38.53	47.48
C_7H_9N	2,6-Dimethylpyridine	417.2	37.46	45.36
C_7H_9N	3,4-Dimethylpyridine	452.28	39.99	50.53
C_7H_9N	3,5-Dimethylpyridine	445.06	39.46	49.48
C_7H_9N	Benzylamine	458		60.16
$C_7H_{10}O$	Dicyclopropyl ketone	434		53.7
C_7H_{14}	Methylcyclohexane	374.08	31.27	35.36
C_7H_{14}	Ethylcyclopentane	376.6	31.96	36.4
C_7H_{14}	1-Heptene	366.79		35.49
$C_7H_{14}O$	2-Heptanone	424.20	38.3	47.24
$C_7H_{14}O_2$	Pentyl ethanoate	422.40	41	
$C_7H_{14}O_2$	3-Methylbutyl ethanoate	415.70	37.53	

TABLE 2.3.5
ENTHALPY OF VAPORIZATION (continued)

Molecular formula	Name	T_b/K	$\Delta_{vap}H(T_b)$/ kJ mol^{-1}	$\Delta_{vap}H$(298.15 K)/ kJ mol^{-1}
$C_7H_{14}O_2$	Methyl hexanoate	422.6	38.55	48.04
$C_7H_{14}O_2$	Ethyl pentanoate	419.2	36.96	47.01
$C_7H_{14}O_2$	Ethyl 3-methylbutanoate	408.20	36.95	47.3
$C_7H_{15}Br$	1-Bromoheptane	452		50.6
$C_7H_{15}Cl$	1-Chloroheptane	432		47.66
C_7H_{16}	Heptane	371.6	31.77	36.57
C_7H_{16}	2-Methylhexane	363.19	30.62	34.87
C_7H_{16}	3-Methylhexane	364.99	30.89	35.06
C_7H_{16}	2,2-Dimethylpentane	352.3	29.23	32.42
C_7H_{16}	2,3-Dimethylpentane	362.93	30.46	34.26
C_7H_{16}	2,4-Dimethylpentane	353.64	29.55	32.88
C_7H_{16}	3,3-Dimethylpentane	359.21	29.62	33.03
C_7H_{16}	3-Ethylpentane	366.6	31.12	35.22
C_7H_{16}	2,2,3-Trimethylbutane	354.01	28.9	32.05
$C_7H_{16}O$	Hexyl methyl ether	399.2	34.93	42.07
$C_7H_{16}O$	Ethyl pentyl ether	390.7	34.41	41.01
$C_7H_{16}O$	Butyl propyl ether	391.2	33.72	40.22
$C_7H_{16}O$	1-Heptanol	449.85		66.81
$C_7H_{17}N$	Heptylamine	429		49.96
C_8F_{18}	Octadecafluorooctane	379.0	33.38	41.13
C_8H_8	Vinylbenzene	418.29	38.7	43.93
C_8H_8O	Methyl phenyl ketone	475	38.81	53.39
$C_8H_8O_2$	Methyl benzoate	472	43.18	55.57
$C_8H_8O_3$	Methyl 2-hydroxybenzoate	496.10	46.67	
C_8H_{10}	Ethylbenzene	409.34	35.57	42.24
C_8H_{10}	1,2-Dimethylbenzene	417.6	36.24	43.43
C_8H_{10}	1,3-Dimethylbenzene	412.27	35.66	42.65
C_8H_{10}	1,4-Dimethylbenzene	411.52	35.67	42.4
$C_8H_{10}O$	2,4-Dimethylphenol	484.13	47.14	64.96
$C_8H_{10}O$	2,5-Dimethylphenol	484.33	46.94	59.96
$C_8H_{10}O$	2,6-Dimethylphenol	474.22	44.52	75.31
$C_8H_{10}O$	3,4-Dimethylphenol	500	49.67	85.03
$C_8H_{10}O$	3,5-Dimethylphenol	494.89	49.31	82.01
$C_8H_{10}O$	Ethyl phenyl ether	442.96	40.7	51.04
$C_8H_{11}N$	*N,N*-Dimethylaniline	467.20		52.83
C_8H_{14}	1-Octyne	399.4	35.83	42.3
$C_8H_{15}N$	Octanenitrile	478.40		56.8
C_8H_{16}	Ethylcyclohexane	405.0	34.04	40.56
C_8H_{16}	1,1-Dimethylcyclohexane	392.7	32.51	37.92
C_8H_{16}	*cis*-1,2-Dimethylcyclohexane	402.9	33.47	39.7
C_8H_{16}	*trans*-1,2-Dimethylcyclohexane	396.6	32.96	38.36
C_8H_{16}	*cis*-1,3-Dimethylcyclohexane	393.2	32.91	38.26
C_8H_{16}	*trans*-1,3-Dimethylcyclohexane	397.6	33.39	39.16
C_8H_{16}	*cis*-1,4-Dimethylcyclohexane	397.5	33.28	39.02
C_8H_{16}	*trans*-1,4-Dimethylcyclohexane	392.5	32.56	37.9
C_8H_{16}	1-Octene	394.44	34.07	40.39
$C_8H_{16}O_2$	Octanoic acid	513.1	58.45	
$C_8H_{16}O_2$	Isobutyl 2-methylpropanoate	421.80	38.2	46.4
$C_8H_{17}Br$	1-Bromooctane	473		55.77
$C_8H_{17}Cl$	1-Chlorooctane	454.6		52.42
C_8H_{18}	Octane	398.82	34.41	41.49
C_8H_{18}	2-Methylheptane	390.81	33.26	39.67
C_8H_{18}	3-Methylheptane	392.09	33.66	39.83
C_8H_{18}	4-Methylheptane	390.87	33.35	39.69
C_8H_{18}	2,2-Dimethylhexane	380.01	32.07	37.28
C_8H_{18}	2,3-Dimethylhexane	388.77	33.17	38.78

TABLE 2.3.5
ENTHALPY OF VAPORIZATION (continued)

Molecular formula	Name	T_b/K	$\Delta_{vap}H(T_b)$/ kJ mol^{-1}	$\Delta_{vap}H(298.15\ K)$/ kJ mol^{-1}
C_8H_{18}	2,4-Dimethylhexane	382.6	32.51	37.76
C_8H_{18}	2,5-Dimethylhexane	382.27	32.54	37.85
C_8H_{18}	3,3-Dimethylhexane	385.12	32.31	37.53
C_8H_{18}	3,4-Dimethylhexane	390.88	33.24	38.97
C_8H_{18}	3-Ethylhexane	391.7	33.59	39.64
C_8H_{18}	3-Ethyl-2-methylpentane	388.81	32.93	38.52
C_8H_{18}	3-Ethyl-3-methylpentane	391.42	32.78	37.99
C_8H_{18}	2,2,3-Trimethylpentane	383	31.94	36.91
C_8H_{18}	2,2,4-Trimethylpentane	372.37	30.79	35.14
C_8H_{18}	2,3,3-Trimethylpentane	387.9	32.12	37.27
C_8H_{18}	2,3,4-Trimethylpentane	386.6	32.36	37.75
C_8H_{18}	2,2,3,3-Tetramethylbutane	379.60		42.9
$C_8H_{18}O$	1-Octanol	468.31	46.9	70.98
$C_8H_{18}O$	(±)-2-Octanol	453.03	44.35	
$C_8H_{18}O$	2-Ethyl-1-hexanol	457.77	54.2	
$C_8H_{18}O$	Dibutyl ether	413.43	36.49	44.97
$C_8H_{18}O$	Di-*tert*-butyl ether	380.38	32.15	37.61
$C_8H_{18}S$	5-Thianonane	462	41.28	52.96
$C_8H_{19}N$	Dibutylamine	432.7	38.44	49.45
C_9H_7N	Quinoline	510.25	49.71	53.9
C_9H_7N	Isoquinoline	516.37	48.96	60.26
C_9H_{10}	Indan	451.12	39.63	48.79
C_9H_{12}	Isopropylbenzene	425.56	37.53	45.13
C_9H_{12}	Propylbenzene	432.39		46.22
C_9H_{12}	1,2,3-Trimethylbenzene	449.27		49.05
C_9H_{12}	1,2,4-Trimethylbenzene	442.53		47.93
C_9H_{12}	1,3,5-Trimethylbenzene	437.89	39.04	47.5
$C_9H_{14}O_6$	1,2,3-Propanetriol tris(ethanoate)	532	57.8	85.74
C_9H_{18}	Butylcyclopentane	429.7	36.16	45.89
$C_9H_{18}O$	5-Nonanone	461.60		53.3
$C_9H_{18}O$	2,6-Dimethyl-4-heptanone	442.5	39.92	50.92
C_9H_{20}	Nonane	423.97	36.91	46.41
C_9H_{20}	2,2,5-Trimethylhexane	397.24	33.65	40.16
C_9H_{20}	3,3-Diethylpentane	419.3	34.61	42
C_9H_{20}	2,2,4,4-Tetramethylpentane	395.44	32.51	38.49
$C_9H_{20}O$	1-Nonanol	486.65		76.86
$C_{10}H_7Br$	1-Bromonaphthalene	554	39.3	
$C_{10}H_7Cl$	1-Chloronaphthalene	532.4	52.05	
$C_{10}H_8$	Naphthalene	491.14	43.18	
$C_{10}H_{12}$	1,2,3,4-Tetrahydronaphthalene	480.77	43.85	55.23
$C_{10}H_{14}$	Butylbenzene	456.46	38.87	51.36
$C_{10}H_{14}$	Isobutylbenzene	445.94	37.82	47.86
$C_{10}H_{14}$	4-Isopropyltoluene	450.28	38.16	50.29
$C_{10}H_{16}O$	1,7,7-Trimethylbicyclo[2.2.1]hepten-2-one	480	59.5	
$C_{10}H_{18}$	*cis*-Bicyclo[4.4.0]decane	468.96	41	51.34
$C_{10}H_{18}$	*trans*-Bicyclo[4.4.0]decane	460.46	40.23	49.87
$C_{10}H_{20}$	1-Decene	443.7	38.66	50.43
$C_{10}H_{22}$	Decane	447.30	38.75	51.38
$C_{10}H_{22}O$	1-Decanol	504.27		81.5
$C_{11}H_{10}$	1-Methylnaphthalene	517.83	45.48	60.07
$C_{11}H_{24}$	Undecane	469.08	42.5	56.43
$C_{12}H_{10}O$	Diphenyl ether	531.20	48.2	66.9
$C_{12}H_{16}$	Cyclohexylbenzene	513.2		59.94
$C_{12}H_{24}$	1-Dodecene	486.9	43.97	60.78
$C_{12}H_{26}$	Dodecane	489.47	44.5	61.51

TABLE 2.3.5
ENTHALPY OF VAPORIZATION (continued)

Molecular formula	Name	T_b/K	$\Delta_{vap} H(T_b)$/ kJ mol^{-1}	$\Delta_{vap} H(298.15\ K)$/ kJ mol^{-1}
$C_{12}H_{26}O$	1-Dodecanol	537.7		91.96
$C_{12}H_{27}N$	Tributylamine	489.6	46.9	
$C_{13}H_{26}O_2$	Methyl dodecanoate	540		77.17
$C_{13}H_{28}$	Tridecane	508.62	45.65	66.43
$C_{14}H_{10}$	Phenanthrene	612.5		75.5
$C_{14}H_{12}O_2$	Benzyl benzoate	596.6	53.6	
$C_{14}H_{30}$	Tetradecane	526.66	47.5	71.3
$C_{14}H_{30}O$	1-Tetradecanol	562		102.2
$C_{15}H_{32}$	Pentadecane	543.76		76.11
$C_{16}H_{22}O_4$	Dibutyl 1,2-benzenedicarboxylate	613	79.2	
$C_{16}H_{32}$	1-Hexadecene	558.0		80.25
$C_{16}H_{34}$	Hexadecane	560.01		81.38
$C_{17}H_{36}$	Heptadecane	575.1		86.02
$C_{18}H_{38}$	Octadecane	589.50	52.8	
$C_{19}H_{40}$	Nonadecane	603.0	56	
$C_{20}H_{42}$	Icosane	617	59	

2.4. CALORIMETRIC PROPERTIES ASSOCIATED WITH CHEMICAL REACTIONS

The changes in enthalpy, H, energy, U, and entropy, S, govern the course of a chemical reaction and the heat that is evolved or absorbed. The basic thermodynamic relations

$$H = U + PV$$

$$G = H - TS$$

apply to the properties themselves and to the changes in these properties which occur in a reaction (including the reaction that leads to formation of a compound from its constituent elements). When all the substances involved in a reaction are in their standard states (see Table 2.4.1), the changes in properties are described by terms such as "standard molar enthalpy of reaction", "standard molar Gibbs energy of formation", etc. A particularly important relation is that between the standard molar Gibbs energy of reaction and the equilibrium constant K_p:

$$\ln K_p = -\Delta G^\circ / RT$$

where R is the gas constant. When the reaction refers to formation of a compound from its elements in their standard states, this relation becomes:

$$\ln K_\mathrm{f} = -\Delta_\mathrm{f} G^\circ / RT$$

where K_f is called the equilibrium constant of formation.

The following tables give the necessary data for calculating these changes in thermodynamic properties for reactions involving many common inorganic and organic substances.

Physical quantity	Symbol	SI unit
Standard molar enthalpy	H°	J mol^{-1}
Standard molar entropy	S°	J K^{-1} mol^{-1}
Standard molar Gibbs energy	G°	J mol^{-1}
Isobaric molar heat capacity	C_p	J K^{-1}mol^{-1}
Standard molar enthalpy of formation	$\Delta_\mathrm{f} H^\circ$	J mol^{-1}
Standard molar Gibbs energy of formation	$\Delta_\mathrm{f} G^\circ$	J mol^{-1}
Equilibrium constant of formation	K_f	1
Reference temperature	T_r	K

TABLE 2.4.1
STANDARD STATE THERMOCHEMICAL PROPERTIES AT 298.15 K

In calculating the changes in thermodynamic quantities associated with a chemical reaction, it is convenient to have the standard enthalpy and Gibbs energy of formation of the reactants and products. The standard molar enthalpy of formation, $\Delta_f H°$, of a substance is defined as the enthalpy change when one mole of the substance is formed in its standard state from the constituent elements, each in its own standard reference state. An analogous definition applies to $\Delta_f G°$. The standard states are defined for different phases by:

- The standard state of a pure gaseous substance is that of the substance as a (hypothetical) ideal gas at the standard-state pressure
- The standard state of a pure liquid substance is that of the liquid under the standard-state pressure
- The standard state of a pure crystalline substance is that of the crystalline substance under the standard-state pressure

In this table the standard-state pressure is taken as 100000 Pa (1 bar), and all data refer to a temperature of 298.15 K.

The table gives the standard enthalpy and Gibbs energy of formation, the standard entropy, and the heat capacity for about 1750 individual substances, many of them in more than one physical state. An entry of 0.0 in the $\Delta_f H°$ column indicates the reference state for an element (see References 1 and 2 for more details on reference states). A blank entry means no value is available.

The standard-state enthalpy change in a reaction, $\Delta H°(298.15\ \mathrm{K})$, may be calculated from the equation

$$\Delta H°(298.15\,\mathrm{K}) = \sum_{\mathrm{p}} \nu_{\mathrm{p}}\, \Delta_f H°(298.15\,\mathrm{K})_{\mathrm{p}} - \sum_{\mathrm{r}} \nu_{\mathrm{r}}\, \Delta_f H°(298.15\,\mathrm{K})_{\mathrm{r}}$$

where the ν_p and ν_r are the stoichiometric coefficients of the products and reactants, respectively. The enthalpy change at a temperature T not too far from 298.15 K may be estimated from the relation:

$$\Delta H°(T) = \Delta H°(298.15\,\mathrm{K}) + \Delta C_p (T - 298.15\,\mathrm{K})$$

where ΔC_p is the sum of the C_p values of the products, each multiplied by its stoichiometric coefficient, minus the corresponding sum for the reactants.

Substances are arranged in a modified Hill order, with substances not containing carbon preceding those that contain carbon.

REFERENCES

1. Cox, J. D.; Wagman, D. D.; Medvedev, V. A., *CODATA Key Values for Thermodynamics*, Hemisphere Publishing Corp. New York, 1989.
2. Wagman, D. D.; Evans, W. H.; Parker, V. B.; Schumm, R. H.; Halow, I.; Bailey, S. M.; Churney, K. L.; Nuttall, R. L., *The NBS Tables of Chemical Thermodynamic Properties, J. Phys. Chem. Ref. Data*, Vol. 11, Suppl. 2, 1982.
3. Pedley, J. B.; Naylor, R. D.; Kirby, S. P., *Thermochemical Data of Organic Compounds, Second Edition*, Chapman and Hall, London, 1986.
4. *TRC Thermodynamic Tables*, Thermodynamic Research Center, Texas A & M University, College Station, TX.
5. Chase, M. W.; Davies, C. A.; Downey, J. R.; Frurip, D. J.; McDonald, R. A.; Syverud, A. N.; *JANAF Thermochemical Tables, Third Edition, J. Phys. Chem. Ref. Data*, Vol. 14, Suppl. 1, 1985.
6. Gurvich, L. V.; Veyts, I. V.; Alcock, C. B., *Thermodynamic Properties of Individual Substances, Fourth Edition*, Hemisphere Publishing Corp., New York, 1989.
7. Dolmalski, E. S.; Evans, W. H.; Hearing, E. D., *Heat Capacities and Entropies of Organic Compounds in the Condensed Phase, J. Phys. Chem. Ref. Data*, Vol. 13, Suppl. 1, 1984; Vol. 19, No. 4, 881-1047, 1990.

Molecular formula	Name	State	$\Delta_f H°$/ kJ mol^{-1}	$\Delta_f G°$/ kJ mol^{-1}	$S°$/ J K^{-1} mol^{-1}	C_p/ J K^{-1} mol^{-1}
Ac	Actinium	cr	0.0	0.0	56.5	27.2
		g	406.0	366.0	188.1	20.8
Ag	Silver	cr	0.0	0.0	42.6	25.4
		g	284.9	246.0	173.0	20.8

TABLE 2.4.1
STANDARD STATE THERMOCHEMICAL PROPERTIES AT 298.15 K (continued)

Molecular formula	Name	State	$\Delta_f H°$/ kJ mol^{-1}	$\Delta_f G°$/ kJ mol^{-1}	$S°$/ J K^{-1} mol^{-1}	C_p/ J K^{-1} mol^{-1}
AgBr	Silver bromide	cr	-100.4	-96.9	107.1	52.4
$AgBrO_3$	Silver bromate	cr	-10.5	71.3	151.9	
AgCl	Silver chloride	cr	-127.0	-109.8	96.3	50.8
$AgClO_3$	Silver chlorate	cr	-30.3	64.5	142.0	
$AgClO_4$	Silver perchlorate	cr	-31.1			
AgF	Silver fluoride	cr	-204.6			
AgF_2	Silver difluoride	cr	-360.0			
AgI	Silver iodide	cr	-61.8	-66.2	115.5	56.8
$AgIO_3$	Silver iodate	cr	-171.1	-93.7	149.4	102.9
$AgNO_3$	Silver nitrate	cr	-124.4	-33.4	140.9	93.1
Ag_2	Disilver	g	410.0	358.8	257.1	37.0
Ag_2CrO_4	Disilver chromate	cr	-731.7	-641.8	217.6	142.3
Ag_2O	Disilver oxide	cr	-31.1	-11.2	121.3	65.9
Ag_2O_2	Disilver peroxide	cr	-24.3	27.6	117.0	88.0
Ag_2O_3	Disilver trioxide	cr	33.9	121.4	100.0	
Ag_2O_4S	Disilver sulfate	cr	-715.9	-618.4	200.4	131.4
Ag_2S	Disilver sulfide (argentite)	cr	-32.6	-40.7	144.0	76.5
Al	Aluminum	cr	0.0	0.0	28.3	24.4
		g	330.0	289.4	164.6	21.4
AlB_3H_{12}	Aluminum trihydride-tris(borane)	l	-16.3	145.0	289.1	194.6
		g	13.0	147.0	379.2	
AlBr	Aluminum bromide	g	-4.0	-42.0	239.5	35.6
$AlBr_3$	Aluminum tribromide	cr	-527.2			101.7
		g	-425.1			
AlCl	Aluminum chloride	g	-47.7	-74.1	228.1	35.0
$AlCl_2$	Aluminum dichloride	g	-331.0			
$AlCl_3$	Aluminum trichloride	cr	-704.2	-628.8	110.7	91.8
		g	-583.2			
AlF	Aluminum fluoride	g	-258.2	-283.7	215.0	31.9
AlF_3	Aluminum trifluoride	cr	-1510.4	-1431.1	66.5	75.1
		g	-1204.6	-1188.2	277.1	62.6
AlF_4Na	Sodium tetrafluoroaluminate	g	-1869.0	-1827.5	345.7	105.9
AlH	Aluminum hydride	g	259.2	231.2	187.9	29.4
AlH_3	Aluminum trihydride	cr	-46.0			
AlH_4K	Potassium tetrahydroaluminate	cr	-183.7			
AlH_4Li	Lithium tetrahydroaluminate	cr	-116.3	-44.7	78.7	83.2
AlI	Aluminum iodide	g	65.5			36.0
AlI_3	Aluminum triiodide	cr	-313.8	-300.8	159.0	98.7
		g	-207.5			
AlN	Aluminum nitride	cr	-318.0	-287.0	20.2	30.1
AlO	Aluminum oxide	g	91.2	65.3	218.4	30.9
AlO_4P	Aluminum phosphate	cr	-1733.8	-1617.9	90.8	93.2
AlP	Aluminum phosphide	cr	-166.5			
AlS	Aluminum sulfide	g	200.9	150.1	230.6	33.4
Al_2	Dialuminum	g	485.9	433.3	233.2	36.4
Al_2Br_6	Dialuminum hexabromide	g	-970.7			
Al_2Cl_6	Dialuminum hexachloride	g	-1290.8	-1220.4	490.0	
Al_2F_6	Dialuminum hexafluoride	g	-2628.0			
Al_2I_6	Dialuminum hexaiodide	g	-516.7			
Al_2O	Dialuminum oxide	g	-130.0	-159.0	259.4	45.7
Al_2O_3	Dialuminum trioxide (corundum)	cr	-1675.7	-1582.3	50.9	79.0
Al_2S_3	Dialuminum trisulfide	cr	-724.0			
Am	Americium	cr	0.0	0.0		
Ar	Argon	g	0.0	0.0	154.8	20.8
As	Arsenic (gray)	cr	0.0	0.0	35.1	24.6
	Arsenic (yellow)	cr	14.6			

TABLE 2.4.1
STANDARD STATE THERMOCHEMICAL PROPERTIES AT 298.15 K (continued)

Molecular formula	Name	State	$\Delta_f H°$/ kJ mol^{-1}	$\Delta_f G°$/ kJ mol^{-1}	$S°$/ J K^{-1} mol^{-1}	C_p/ J K^{-1} mol^{-1}
		g	302.5	261.0	174.2	20.8
$AsBr_3$	Arsenic tribromide	cr	-197.5			
		g	-130.0	-159.0	363.9	79.2
$AsCl_3$	Arsenic trichloride	l	-305.0	-259.4	216.3	
		g	-261.5	-248.9	327.2	75.7
AsF_3	Arsenic trifluoride	l	-821.3	-774.2	181.2	126.6
		g	-785.8	-770.8	289.1	65.6
$AsGa$	Gallium arsenide	cr	-71.0	-67.8	64.2	46.2
AsH_3	Arsane	g	66.4	68.9	222.8	38.1
AsH_3O_4	Arsenic acid	cr	-906.3			
AsI_3	Arsenic triiodide	cr	-58.2	-59.4	213.1	105.8
		g			388.3	80.6
$AsIn$	Indium arsenide	cr	-58.6	-53.6	75.7	47.8
AsO	Arsenic oxide	g	70.0			
As_2	Diarsenic	g	222.2	171.9	239.4	35.0
As_2O_5	Diarsenic pentaoxide	cr	-924.9	-782.3	105.4	116.5
As_2S_3	Diarsenic trisulfide	cr	-169.0	-168.6	163.6	116.3
At	Astatine	cr	0.0	0.0		
Au	Gold	cr	0.0	0.0	47.4	25.4
		g	366.1	326.3	180.5	20.8
$AuBr$	Gold bromide	cr	-14.0			
$AuBr_3$	Gold tribromide	cr	-53.3			
$AuCl$	Gold chloride	cr	-34.7			
$AuCl_3$	Gold trichloride	cr	-117.6			
AuF_3	Gold trifluoride	cr	-363.6			
AuH	Gold hydride	g	295.0	265.7	211.2	29.2
AuI	Gold iodide	cr	0.0			
Au_2	Digold	g	515.1			36.9
B	Boron (rhombic)	cr	0.0	0.0	5.9	11.1
		g	565.0	521.0	153.4	20.8
BBr	Boron bromide	g	238.1	195.4	225.0	32.9
BBr_3	Boron tribromide	l	-239.7	-238.5	229.7	
		g	-205.6	-232.5	324.2	67.8
BCl	Boron chloride	g	149.5	120.9	213.2	31.7
$BClO$	Chloroxyborane	g	-314.0			
BCl_3	Boron trichloride	l	-427.2	-387.4	206.3	106.7
		g	-403.8	-388.7	290.1	62.7
$BCsO_2$	Cesium metaborate	cr	-972.0	-915.0	104.4	80.6
BF	Boron fluoride	g	-122.2	-149.8	200.5	29.6
BFO	Fluorooxyborane	g	-607.0			
BF_3	Boron trifluoride	g	-1136.0	-1119.4	254.4	
BF_3H_3N	Ammonia-boron trifluoride	cr	-1353.9			
BF_3H_3P	Phosphine-boron trifluoride	g	-854.0			
BF_4Na	Sodium tetrafluoroborate	cr	-1844.7	-1750.1	145.3	120.3
BH	Boron hydride	g	449.6	419.6	171.9	29.2
BHO_2	Metaboric acid (monoclinic)	cr	-794.3	-723.4	38.0	
		g	-561.9	-551.0	240.1	42.2
BH_3	Borane	g	100.0			
BH_3O_3	Boric acid	cr	-1094.3	-968.9	88.8	81.4
		g	-994.1			
BH_4K	Potassium hydride-borane	cr	-227.4	-160.3	106.3	96.1
BH_4Li	Lithium hydride-borane	cr	-190.8	-125.0	75.9	82.6
BH_4Na	Sodium hydride-borane	cr	-188.6	-123.9	101.3	86.8
BI_3	Boron triiodide	g	71.1	20.7	349.2	70.8
BKO_2	Potassium metaborate	cr	-981.6	-923.4	80.0	66.7
$BLiO_2$	Lithium metaborate	cr	-1032.2	-976.1	51.5	59.8

TABLE 2.4.1
STANDARD STATE THERMOCHEMICAL PROPERTIES AT 298.15 K (continued)

Molecular formula	Name	State	$\Delta_f H°$/ kJ mol^{-1}	$\Delta_f G°$/ kJ mol^{-1}	$S°$/ J K^{-1} mol^{-1}	C_p/ J K^{-1} mol^{-1}
BN	Boron nitride	cr	-254.4	-228.4	14.8	19.7
		g	647.5	614.5	212.3	29.5
$BNaO_2$	Sodium metaborate	cr	-977.0	-920.7	73.5	65.9
BO	Boron oxide	g	25.0	-4.0	203.5	29.2
BO_2	Boron dioxide	g	-300.4	-305.9	229.6	43.0
BO_2Rb	Rubidium metaborate	cr	-971.0	-913.0	94.3	74.1
BS	Boron sulfide	g	342.0	288.8	216.2	30.0
B_2	Diboron	g	830.5	774.0	201.9	30.5
B_2Cl_4	Tetrachlorodiborane(4)	l	-523.0	-464.8	262.3	137.7
		g	-490.4	-460.6	357.4	95.4
B_2F_4	Tetrafluorodiborane(4)	g	-1440.1	-1410.4	317.3	79.1
B_2H_6	Diborane(6)	g	35.6	86.7	232.1	56.9
B_2O_2	Diboron dioxide	g	-454.8	-462.3	242.5	57.3
		g	-843.8	-832.0	279.8	66.9
B_2O_3	Diboron trioxide	cr	-1273.5	-1194.3	54.0	
B_2S_3	Diboron trisulfide	cr	-240.6			
		g	67.0			
$B_3H_6N_3$	Cyclotriborazane	l	-541.0	-392.7	199.6	
B_4H_{10}	Tetraborane(10)	g	66.1			
$B_4Na_2O_7$	Disodium tetraborate	cr	-3291.1	-3096.0	189.5	186.8
B_5H_9	Pentaborane(9)	l	42.7	171.8	184.2	151.1
		g	73.2	175.0	275.9	96.8
B_5H_{11}	Pentaborane(11)	l	73.2			
		g	103.3			
B_6H_{10}	Hexaborane(10)	l	56.3			
		g	94.6			
Ba	Barium	cr	0.0	0.0	62.8	28.1
		g	180.0	146.0	170.2	20.8
$BaBr_2$	Barium dibromide	cr	-757.3	-736.8	146.0	
$BaCl_2$	Barium dichloride	cr	-858.6	-810.4	123.7	75.1
BaF_2	Barium difluoride	cr	-1207.1	-1156.8	96.4	71.2
BaH_2	Barium dihydride	cr	-178.7			
BaH_2O_2	Barium dihydroxide	cr	-944.7			
BaI_2	Barium diiodide	cr	-602.1			
BaN_2O_4	Barium dinitrite	cr	-768.2			
BaN_2O_6	Barium dinitrate	cr	-992.1	-796.6	213.8	151.4
BaO	Barium oxide	cr	-553.5	-525.1	70.4	47.8
BaO_4S	Barium sulfate	cr	-1473.2	-1362.2	132.2	101.8
BaS	Barium sulfide	cr	-460.0	-456.0	78.2	49.4
Be	Beryllium	cr	0.0	0.0	9.5	16.4
		g	324.0	286.6	136.3	20.8
$BeBr_2$	Beryllium dibromide	cr	-353.5			
$BeCl_2$	Beryllium dichloride	cr	-490.4	-445.6	82.7	64.8
BeF_2	Beryllium difluoride	cr	-1026.8	-979.4	53.4	51.8
BeH_2O_2	Beryllium dihydroxide	cr	-902.5	-815.0	51.9	
BeI_2	Beryllium diiodide	cr	-192.5			
BeO	Beryllium oxide	cr	-609.4	-580.1	13.8	
BeO_4S	Beryllium sulfate	cr	-1205.2	-1093.8	77.9	85.7
BeS	Beryllium sulfide	cr	-234.3			
Bi	Bismuth	cr	0.0	0.0	56.7	25.5
		g	207.1	168.2	187.0	20.8
BiClO	Bismuth oxychloride	cr	-366.9	-322.1	120.5	
$BiCl_3$	Bismuth trichloride	cr	-379.1	-315.0	177.0	105.0
		g	-265.7	-256.0	358.9	79.7
BiH_3O_3	Bismuth trihydroxide	cr	-711.3			
BiI_3	Bismuth triiodide	cr		-175.3		

TABLE 2.4.1
STANDARD STATE THERMOCHEMICAL PROPERTIES AT 298.15 K (continued)

Molecular formula	Name	State	$\Delta_f H°$/ kJ mol^{-1}	$\Delta_f G°$/ kJ mol^{-1}	$S°$/ J K^{-1} mol^{-1}	C_p/ J K^{-1} mol^{-1}
Bi_2	Dibismuth	g	219.7			36.9
$Bi_2O_{12}S_3$	Dibismuth trisulfate	cr	-2544.3			
Bi_2O_3	Dibismuth trioxide	cr	-573.9	-493.7	151.5	113.5
Bi_2S_3	Dibismuth trisulfide	cr	-143.1	-140.6	200.4	122.2
Bk	Berkelium	cr	0.0	0.0		
Br	Bromine (atomic)	g	111.9	82.4	175.0	20.8
BrCl	Bromine chloride	g	14.6	-1.0	240.1	35.0
$BrCl_3Si$	Bromotrichlorosilane	g			350.1	90.9
BrCs	Cesium bromide	cr	-405.8	-391.4	113.1	52.9
BrCu	Copper bromide	cr	-104.6	-100.8	96.1	54.7
BrF	Bromine fluoride	g	-93.8	-109.2	229.0	33.0
BrF_3	Bromine trifluoride	l	-300.8	-240.5	178.2	124.6
		g	-255.6	-229.4	292.5	66.6
BrF_5	Bromine pentafluoride	l	-458.6	-351.8	225.1	
		g	-428.9	-350.6	320.2	99.6
BrGe	Germanium bromide	g	235.6			37.1
$BrGeH_3$	Bromogermane	g			274.8	56.4
BrH	Hydrogen bromide	g	-36.3	-53.4	198.7	29.1
BrHSi	Bromosilylene	cr	-464.4			
BrH_3Si	Bromosilane	g			262.4	52.8
BrH_4N	Ammonium bromide	cr	-270.8	-175.2	113.0	96.0
BrI	Iodine bromide	g	40.8	3.7	258.8	36.4
BrIn	Indium bromide	cr	-175.3	-169.0	113.0	
		g	-56.9	-94.3	259.5	36.7
BrK	Potassium bromide	cr	-393.8	-380.7	95.9	52.3
$BrKO_3$	Potassium bromate	cr	-360.2	-271.2	149.2	105.2
$BrKO_4$	Potassium perbromate	cr	-287.9	-174.4	170.1	120.2
BrLi	Lithium bromide	cr	-351.2	-342.0	74.3	
BrNO	Nitrosyl bromide	g	82.2	82.4	273.7	45.5
BrNa	Sodium bromide	cr	-361.1	-349.0	86.8	51.4
		g	-143.1	-177.1	241.2	36.3
$BrNaO_3$	Sodium bromate	cr	-334.1	-242.6	128.9	
BrO	Bromine oxide	g	125.8	108.2	237.6	32.1
BrO_2	Bromine superoxide	cr	48.5			
BrRb	Rubidium bromide	cr	-394.6	-381.8	110.0	52.8
BrSi	Bromosilyldyne	g	209.0			38.6
BrTl	Thallium bromide	cr	-173.2	-167.4	120.5	
		g	-37.7			
Br_2	Bromine	l	0.0	0.0	152.2	75.7
		g	30.9	3.1	245.5	36.0
Br_2Ca	Calcium dibromide	cr	-682.8	-663.6	130.0	
Br_2Cd	Cadmium dibromide	cr	-316.2	-296.3	137.2	76.7
Br_2Co	Cobalt dibromide	cr	-220.9			79.5
Br_2Cr	Chromium dibromide	cr	-302.1			
Br_2Cu	Copper dibromide	cr	-141.8			
Br_2Fe	Iron dibromide	cr	-249.8	-238.1	140.6	
Br_2H_2Si	Dibromosilane	g			309.7	65.5
Br_2Hg	Mercury dibromide	cr	-170.7	-153.1	172.0	
Br_2Hg_2	Dimercury dibromide	cr	-206.9	-181.1	218.0	
Br_2Mg	Magnesium dibromide	cr	-524.3	-503.8	117.2	
Br_2Mn	Manganese dibromide	cr	-384.9			
Br_2Ni	Nickel dibromide	cr	-212.1			
Br_2Pb	Lead dibromide	cr	-278.7	-261.9	161.5	80.1
Br_2Pt	Platinum dibromide	cr	-82.0			
Br_2S_2	Disulfur dibromide	l	-13.0			
Br_2Se	Selenium dibromide	g	-21.0			

TABLE 2.4.1
STANDARD STATE THERMOCHEMICAL PROPERTIES AT 298.15 K (continued)

Molecular formula	Name	State	$\Delta_f H°$/ kJ mol^{-1}	$\Delta_f G°$/ kJ mol^{-1}	$S°$/ J K^{-1} mol^{-1}	C_p/ J K^{-1} mol^{-1}
Br_2Sn	Tin dibromide	cr	-243.5			
Br_2Sr	Strontium dibromide	cr	-717.6	-697.1	135.1	75.3
Br_2Ti	Titanium dibromide	cr	-402.0			
Br_2Zn	Zinc dibromide	cr	-328.7	-312.1	138.5	
Br_3ClSi	Tribromochlorosilane	g			377.1	95.3
Br_3Fe	Iron tribromide	cr	-268.2			
Br_3Ga	Gallium tribromide	cr	-386.6	-359.8	180.0	
Br_3HSi	Tribromosilane	l	-355.6	-336.4	248.1	
		g	-317.6	-328.5	348.6	80.8
Br_3In	Indium tribromide	cr	-428.9			
		g	-282.0			
Br_3OP	Phosphoryl tribromide	cr	-458.6			
		g			359.8	89.9
Br_3P	Phosphorus tribromide	l	-184.5	-175.7	240.2	
		g	-139.3	-162.8	348.1	76.0
Br_3Pt	Platinum tribromide	cr	-120.9			
Br_3Re	Rhenium tribromide	cr	-167.0			
Br_3Ru	Ruthenium tribromide	cr	-138.0			
Br_3Sb	Antimony tribromide	cr	-259.4	-239.3	207.1	
		g	-194.6	-223.9	372.9	80.2
Br_3Sc	Scandium tribromide	cr	-743.1			
Br_3Ti	Titanium tribromide	cr	-548.5	-523.8	176.6	101.7
Br_4Ge	Germanium tetrabromide	l	-347.7	-331.4	280.7	
		g	-300.0	-318.0	396.2	101.8
Br_4Pa	Protactinium tetrabromide	cr	-824.0	-787.8	234.0	
Br_4Pt	Platinum tetrabromide	cr	-156.5			
Br_4Si	Silicon tetrabromide	l	-457.3	-443.9	277.8	
		g	-415.5	-431.8	377.9	97.1
Br_4Sn	Tin tetrabromide	cr	-377.4	-350.2	264.4	
		g	-314.6	-331.4	411.9	103.4
Br_4Te	Tellurium tetrabromide	cr	-190.4			
Br_4Ti	Titanium tetrabromide	cr	-616.7	-589.5	243.5	131.5
		g	-549.4	-568.2	398.4	100.8
Br_4V	Vanadium tetrabromide	g	-336.8			
Br_4Zr	Zirconium tetrabromide	cr	-760.7			
Br_5P	Phosphorus pentabromide	cr	-269.9			
Br_5Ta	Tantalum pentabromide	cr	-598.3			
Br_6W	Tungsten hexabromide	cr	-348.5			
Ca	Calcium	cr	0.0	0.0	41.6	25.9
		g	177.8	144.0	154.9	20.8
$CaCl_2$	Calcium dichloride	cr	-795.4	-748.8	108.4	72.9
CaF_2	Calcium difluoride	cr	-1228.0	-1175.6	68.5	67.0
CaH_2	Calcium dihydride	cr	-181.5	-142.5	41.4	41.0
CaH_2O_2	Calcium dihydroxide	cr	-985.2	-897.5	83.4	87.5
CaI_2	Calcium diiodide	cr	-533.5	-528.9	142.0	
CaN_2O_6	Calcium dinitrate	cr	-938.2	-742.8	193.2	149.4
CaO	Calcium oxide	cr	-634.9	-603.3	38.1	42.0
CaO_4S	Calcium sulfate	cr	-1434.5	-1322.0	106.5	99.7
CaS	Calcium sulfide	cr	-482.4	-477.4	56.5	47.4
$Ca_3O_8P_2$	Tricalcium bisphosphate	cr	-4120.8	-3884.7	236.0	227.8
Cd	Cadmium	cr	0.0	0.0	51.8	26.0
		g	111.8		167.7	20.8
$CdCl_2$	Cadmium dichloride	cr	-391.5	-343.9	115.3	74.7
CdF_2	Cadmium difluoride	cr	-700.4	-647.7	77.4	
CdH_2O_2	Cadmium dihydroxide	cr	-560.7	-473.6	96.0	
CdI_2	Cadmium diiodide	cr	-203.3	-201.4	161.1	80.0

TABLE 2.4.1
STANDARD STATE THERMOCHEMICAL PROPERTIES AT 298.15 K (continued)

Molecular formula	Name	State	$\Delta_f H°$/ kJ mol^{-1}	$\Delta_f G°$/ kJ mol^{-1}	$S°$/ J K^{-1} mol^{-1}	C_p/ J K^{-1} mol^{-1}
CdO	Cadmium oxide	cr	-258.4	-228.7	54.8	43.4
CdO_4S	Cadmium sulfate	cr	-933.3	-822.7	123.0	99.6
CdS	Cadmium sulfide	cr	-161.9	-156.5	64.9	
CdTe	Cadmium telluride	cr	-92.5	-92.0	100.0	
Ce	Cerium	cr	0.0	0.0	72.0	26.9
		g	423.0	385.0	191.8	23.1
$CeCl_3$	Cerium trichloride	cr	-1053.5	-977.8	151.0	87.4
CeO_2	Cerium dioxide	cr	-1088.7	-1024.6	62.3	61.6
CeS	Cerium sulfide	cr	-459.4	-451.5	78.2	50.0
Ce_2O_3	Dicerium trioxide	cr	-1796.2	-1706.2	150.6	114.6
Cf	Californium	cr	0.0	0.0		
Cl	Chlorine (atomic)	g	121.3	105.3	165.2	21.8
ClCs	Cesium chloride	cr	-443.0	-414.5	101.2	52.5
$ClCsO_4$	Cesium perchlorate	cr	-443.1	-314.3	175.1	108.3
ClCu	Copper chloride	cr	-137.2	-119.9	86.2	48.5
ClF	Chlorine fluoride	g	-50.3	-51.8	217.9	32.1
$ClFO_3$	Perchloryl fluoride	g	-23.8	48.2	279.0	64.9
ClGe	Germanium chloride	g	155.2	124.2	247.0	36.9
ClF_3	Chlorine trifluoride	l	-189.5			
		g	-163.2	-123.0	281.6	63.9
ClF_5S	Sulfur chloride pentafluoride	l	-1065.7			
$ClGeH_3$	Chlorogermane	g			263.7	54.7
ClH	Hydrogen chloride	g	-92.3	-95.3	186.9	29.1
ClHO	Hypochlorous acid	g	-78.7	-66.1	236.7	37.2
$ClHO_4$	Perchloric acid	l	-40.6			
ClH_3Si	Chlorosilane	g			250.7	51.0
ClH_4N	Ammonium chloride	cr	-314.4	-202.9	94.6	84.1
ClH_4NO_4	Ammonium perchlorate	cr	-295.3	-88.8	186.2	
ClH_4P	Phosphonium chloride	cr	-145.2			
ClI	Iodine chloride	l	-23.9	-13.6	135.1	
		g	17.8	-5.5	247.6	35.6
ClIn	Indium chloride	cr	-186.2			
		g	-75.0			
ClK	Potassium chloride	cr	-436.5	-408.5	82.6	51.3
$ClKO_3$	Potassium chlorate	cr	-397.7	-296.3	143.1	100.3
$ClKO_4$	Potassium perchlorate	cr	-432.8	-303.1	151.0	112.4
ClLi	Lithium chloride	cr	-408.6	-384.4	59.3	48.0
$ClLiO_4$	Lithium perchlorate	cr	-381.0			
ClNO	Nitrosyl chloride	g	51.7	66.1	261.7	44.7
$ClNO_2$	Nitryl chloride	g	12.6	54.4	272.2	53.2
ClNa	Sodium chloride	cr	-411.2	-384.1	72.1	50.5
$ClNaO_2$	Sodium chlorite	cr	-307.0			
$ClNaO_3$	Sodium chlorate	cr	-365.8	-262.3	123.4	
$ClNaO_4$	Sodium perchlorate	cr	-383.3	-254.9	142.3	
ClO	Chlorine oxide	g	101.8	98.1	226.6	31.5
ClOV	Vanadium oxychloride	cr	-607.0	-556.0	75.0	
ClO_2	Chlorine dioxide	g	102.5	120.5	256.8	42.0
ClO_2	Chlorine superoxide	g	89.1	105.0	263.7	46.0
ClO_4Rb	Rubidium perchlorate	cr	-437.2	-306.9	161.1	
ClRb	Rubidium chloride	cr	-435.4	-407.8	95.9	52.4
ClSi	Chlorosilylidyne	g	189.9			36.9
ClTl	Thallium chloride	cr	-204.1	-184.9	111.3	50.9
		g	-67.8			
Cl_2	Chlorine	g	0.0	0.0	223.1	33.9
Cl_2Co	Cobalt dichloride	cr	-312.5	-269.8	109.2	78.5
Cl_2Cr	Chromium dichloride	cr	-395.4	-356.0	115.3	71.2

TABLE 2.4.1
STANDARD STATE THERMOCHEMICAL PROPERTIES AT 298.15 K (continued)

Molecular formula	Name	State	$\Delta_f H°$/ kJ mol^{-1}	$\Delta_f G°$/ kJ mol^{-1}	$S°$/ J K^{-1} mol^{-1}	C_p/ J K^{-1} mol^{-1}
Cl_2CrO_2	Chromyl dichloride	l	-579.5	-510.8	221.8	
		g	-538.1	-501.6	329.8	84.5
Cl_2Cu	Copper dichloride	cr	-220.1	-175.7	108.1	71.9
Cl_2Fe	Iron dichloride	cr	-341.8	-302.3	118.0	76.7
Cl_2H_2Si	Dichlorosilane	g			285.7	60.5
Cl_2Hg	Mercury dichloride	cr	-224.3	-178.6	146.0	
Cl_2Hg_2	Dimercury dichloride	cr	-265.4	-210.7	191.6	
Cl_2Mg	Magnesium dichloride	cr	-641.3	-591.8	89.6	71.4
Cl_2Mn	Manganese dichloride	cr	-481.3	-440.5	118.2	72.9
Cl_2Ni	Nickel dichloride	cr	-305.3	-259.0	97.7	71.7
Cl_2O	Dichlorine oxide	g	80.3	97.9	266.2	45.4
Cl_2OS	Sulfinyl dichloride	l	-245.6			121.0
		g	-212.5	-198.3	309.8	66.5
Cl_2O_2S	Sulfonyl dichloride	l	-394.1			134.0
		g	-364.0	-320.0	311.9	77.0
Cl_2O_2U	Uranyl dichloride	cr	-1243.9	-1146.4	150.5	107.9
Cl_2Pb	Lead dichloride	cr	-359.4	-314.1	136.0	
Cl_2Pt	Platinum dichloride	cr	-123.4			
Cl_2S	Sulfur dichloride	l	-50.0			
Cl_2S_2	Disulfur dichloride	l	-59.4			
Cl_2Sn	Tin dichloride	cr	-325.1			
Cl_2Sr	Strontium dichloride	cr	-828.9	-781.1	114.9	75.6
Cl_2Ti	Titanium dichloride	cr	-513.8	-464.4	87.4	69.8
Cl_2Zn	Zinc dichloride	cr	-415.1	-369.4	111.5	71.3
		g	-266.1			
Cl_2Zr	Zirconium dichloride	cr	-502.0			
Cl_3Cr	Chromium trichloride	cr	-556.5	-486.1	123.0	91.8
Cl_3Dy	Dysprosium trichloride	cr	-1000.0			
Cl_3Er	Erbium trichloride	cr	-998.7			100.0
Cl_3Eu	Europium trichloride	cr	-936.0			
Cl_3Fe	Iron trichloride	cr	-399.5	-334.0	142.3	96.7
Cl_3Ga	Gallium trichloride	cr	-524.7	-454.8	142.0	
Cl_3Gd	Gadolinium trichloride	cr	-1008.0			88.0
Cl_3HSi	Trichlorosilane	l	-539.3	-482.5	227.6	
		g	-513.0	-482.0	313.9	75.8
Cl_3Ho	Holmium trichloride	cr	-1005.4			88.0
Cl_3In	Indium trichloride	cr	-537.2			
		g	-374.0			
Cl_3Ir	Iridium trichloride	cr	-245.6			
Cl_3La	Lanthanum trichloride	cr	-1071.1			108.8
Cl_3Lu	Lutetium trichloride	cr	-945.6			
		g	-649.0			
Cl_3N	Nitrogen trichloride	l	230.0			
Cl_3Nd	Neodymium trichloride	cr	-1041.0			113.0
Cl_3OP	Phosphoryl trichloride	l	-597.1	-520.8	222.5	138.8
		g	-558.5	-512.9	325.5	84.9
Cl_3OV	Vanadyl trichloride	l	-734.7	-668.5	244.3	
		g	-695.6	-659.3	344.3	89.9
Cl_3Os	Osmium trichloride	cr	-190.4			
Cl_3P	Phosphorus trichloride	l	-319.7	-272.3	217.1	
		g	-287.0	-267.8	311.8	71.8
Cl_3Pr	Praseodymium trichloride	cr	-1056.9			100.0
Cl_3Pt	Platinum trichloride	cr	-182.0			
Cl_3Re	Rhenium trichloride	cr	-264.0	-188.0	123.8	92.4
Cl_3Rh	Rhodium trichloride	cr	-299.2			
Cl_3Ru	Ruthenium trichloride	cr	-205.0			

TABLE 2.4.1
STANDARD STATE THERMOCHEMICAL PROPERTIES AT 298.15 K (continued)

Molecular formula	Name	State	$\Delta_f H°$/ kJ mol^{-1}	$\Delta_f G°$/ kJ mol^{-1}	$S°$/ J K^{-1} mol^{-1}	C_p/ J K^{-1} mol^{-1}
Cl_3Sb	Antimony trichloride	cr	-382.2	-323.7	184.1	107.9
Cl_3Sc	Scandium trichloride	cr	-925.1			
Cl_3Sm	Samarium trichloride	cr	-1025.9			
Cl_3Tb	Terbium trichloride	cr	-997.0			
Cl_3Ti	Titanium trichloride	cr	-720.9	-653.5	139.7	97.2
Cl_3Tl	Thallium trichloride	cr	-315.1			
Cl_3Tm	Thullium trichloride	cr	-986.6			
Cl_3U	Uranium trichloride	cr	-866.5	-799.1	159.0	102.5
Cl_3V	Vanadium trichloride	cr	-580.7	-511.2	131.0	93.2
Cl_3Y	Yttrium trichloride	cr	-1000.0			
		g	-750.2			75.0
Cl_3Yb	Ytterbium trichloride	cr	-959.8			
Cl_4Ge	Tetrachlorogermane	l	-531.8	-462.7	245.6	
		g	-495.8	-457.3	347.7	96.1
Cl_4Hf	Hafnium tetrachloride	cr	-990.4	-901.3	190.8	120.5
		g	-884.5			
Cl_4Pa	Protactinium tetrachloride	cr	-1043.0	-953.0	192.0	
Cl_4Pb	Lead tetrachloride	l	-329.3			
Cl_4Pt	Platinum tetrachloride	cr	-231.8			
Cl_4Si	Silicon tetrachloride	l	-687.0	-619.8	239.7	145.3
		g	-657.0	-617.0	330.7	90.3
Cl_4Sn	Tin tetrachloride	l	-511.3	-440.1	258.6	165.3
		g	-471.5	-432.2	365.8	98.3
Cl_4Te	Tellurium tetrachloride	cr	-326.4			138.5
Cl_4Th	Thorium tetrachloride	cr	-1186.6	-1094.5	190.4	
Cl_4Ti	Titanium tetrachloride	l	-804.2	-737.2	252.3	145.2
		g	-763.2	-726.3	353.2	95.4
Cl_4U	Uranium tetrachloride	cr	-1019.2	-930.0	197.1	122.0
		g	-809.6	-786.6	419.0	
Cl_4V	Vanadium tetrachloride	l	-569.4	-503.7	255.0	
		g	-525.5	-492.0	362.4	96.2
Cl_4Zr	Zirconium tetrachloride	cr	-980.5	-889.9	181.6	119.8
Cl_5Nb	Niobium pentachloride	cr	-797.5	-683.2	210.5	148.1
		g	-703.7	-646.0	400.6	120.8
Cl_5P	Phosphorus pentachloride	cr	-443.5			
		g	-374.9	-305.0	364.6	112.8
Cl_5Pa	Protactinium pentachloride	cr	-1145.0	-1034.0	238.0	
Cl_5Ta	Tantalum pentachloride	cr	-859.0			
Cl_6U	Uranium hexachloride	cr	-1092.0	-962.0	285.8	175.7
		g	-1013.0	-928.0	431.0	
Cl_6W	Tungsten hexachloride	cr	-602.5			
		g	-513.8			
Cm	Curium	cr	0.0	0.0		
Co	Cobalt	cr	0.0	0.0	30.0	24.8
		g	424.7	380.3	179.5	23.0
CoF_2	Cobalt difluoride	cr	-692.0	-647.2	82.0	68.8
CoH_2O_2	Cobalt dihydroxide	cr	-539.7	-454.3	79.0	
CoI_2	Cobalt diiodide	cr	-88.7			
CoN_2O_6	Cobalt dinitrate	cr	-420.5			
CoO	Cobalt oxide	cr	-237.9	-214.2	53.0	55.2
CoO_4S	Cobalt sulfate	cr	-888.3	-782.3	118.0	
CoS	Cobalt sulfide	cr	-82.8			
Co_2S_3	Dicobalt trisulfide	cr	-147.3			
Co_3O_4	Tricobalt tetraoxide	cr	-891.0	-774.0	102.5	123.4
Cr	Chromium	cr	0.0	0.0	23.8	23.4
		g	396.6	351.8	174.5	20.8

TABLE 2.4.1
STANDARD STATE THERMOCHEMICAL PROPERTIES AT 298.15 K (continued)

Molecular formula	Name	State	$\Delta_f H°$/ kJ mol^{-1}	$\Delta_f G°$/ kJ mol^{-1}	$S°$/ J K^{-1} mol^{-1}	C_p/ J K^{-1} mol^{-1}
CrF_2	Chromium difluoride	cr	-778.0			
CrF_3	Chromium trifluoride	cr	-1159.0	-1088.0	93.9	78.7
CrI_2	Chromium diiodide	cr	-156.9			
CrI_3	Chromium triiodide	cr	-205.0			
CrO_2	Chromium dioxide	cr	-598.0			
CrO_4Pb	Lead chromate	cr	-930.9			
Cr_2FeO_4	Dichromium iron tetraoxide	cr	-1444.7	-1343.8	146.0	133.6
Cr_2O_3	Dichromium trioxide	cr	-1139.7	-1058.1	81.2	118.7
Cr_3O_4	Trichromium tetraoxide	cr	-1531.0			
Cs	Cesium	cr	0.0	0.0	85.2	32.2
		g	76.5	49.6	175.6	20.8
CsF	Cesium fluoride	cr	-553.5	-525.5	92.8	51.1
CsF_2H	Cesium hydrogen difluoride	cr	-923.8	-858.9	135.2	87.3
CsH	Cesium hydride	cr	-54.2			
CsHO	Cesium hydroxide	cr	-417.2			
$CsHO_4S$	Cesium hydrogen sulfate	cr	-1158.1			
CsH_2N	Cesium dihydronitride	cr	-118.4			
CsI	Cesium iodide	cr	-346.6	-340.6	123.1	52.8
$CsNO_3$	Cesium nitrate	cr	-506.0	-406.5	155.2	
CsO_2	Cesium hyperoxide	cr	-286.2			
Cs_2O	Dicesium oxide	cr	-345.8	-308.1	146.9	76.0
Cs_2O_3S	Dicesium sulfite	cr	-1134.7			
Cs_2O_4S	Dicesium sulfate	cr	-1443.0	-1323.6	211.9	134.9
Cs_2S	Dicesium sulfide	cr	-359.8			
Cu	Copper	cr	0.0	0.0	33.2	24.4
		g	337.4	297.7	166.4	20.8
CuF_2	Copper difluoride	cr	-542.7			
CuH_2O_2	Copper dihydroxide	cr	-449.8			
CuI	Copper iodide	cr	-67.8	-69.5	96.7	54.1
CuN_2O_6	Copper dinitrate	cr	-302.9			
CuO	Copper oxide	cr	-157.3	-129.7	42.6	42.3
CuO_4S	Copper sulfate	cr	-771.4	-662.2	109.2	
CuO_4W	Copper tungstate	cr	-1105.0			
CuS	Copper sulfide	cr	-53.1	-53.6	66.5	47.8
CuSe	Copper selenide	cr	-39.5			
Cu_2	Dicopper	g	484.2	431.9	241.6	36.6
Cu_2O	Dicopper oxide	cr	-168.6	-146.0	93.1	63.6
Cu_2S	Dicopper sulfide	cr	-79.5	-86.2	120.9	76.3
Dy	Dysprosium	cr	0.0	0.0	74.8	28.2
		g	290.4	254.4	196.6	20.8
Dy_2O_3	Didysprosium trioxide	cr	-1863.1	-1771.5	149.8	116.3
Er	Erbium	cr	0.0	0.0	73.2	28.1
		g	317.1	280.7	195.6	20.8
ErF_3	Erbium trifluoride	cr	-1711.0			
Er_2O_3	Dierbium trioxide	cr	-1897.9	-1808.7	155.6	108.5
Es	Einsteinium	cr	0.0	0.0		
Eu	Europium	cr	0.0	0.0	77.8	27.7
		g	175.3	142.2	188.8	20.8
Eu_2O_3	Dieuropium trioxide	cr	-1651.4	-1556.8	146.0	122.2
Eu_3O_4	Trieuropium tetraoxide	cr	-2272.0	-2142.0	205.0	
F	Fluorine (atomic)	g	79.4	62.3	158.8	22.7
FGa	Gallium fluoride	g	-251.9			33.3
FGe	Germanium fluoride	g	-33.4			34.7
$FGeH_3$	Fluorogermane	g			252.8	51.6
FH	Hydrogen fluoride	l	-299.8			
		g	-273.3	-275.4	173.8	

TABLE 2.4.1
STANDARD STATE THERMOCHEMICAL PROPERTIES AT 298.15 K (continued)

Molecular formula	Name	State	$\Delta_f H°$/ kJ mol^{-1}	$\Delta_f G°$/ kJ mol^{-1}	$S°$/ J K^{-1} mol^{-1}	C_p/ J K^{-1} mol^{-1}
FH_3Si	Fluorosilane	g			238.4	47.4
FH_4N	Ammonium fluoride	cr	-464.0	-348.7	72.0	65.3
FI	Iodine fluoride	g	-95.7	-118.5	236.2	33.4
FIn	Indium fluoride	g	-203.4			
FK	Potassium fluoride	cr	-567.3	-537.8	66.6	49.0
FLi	Lithium fluoride	cr	-616.0	-587.7	35.7	41.6
FNO	Nitrosyl fluoride	g	-66.5	-51.0	248.1	41.3
FNO_2	Nitryl fluoride	g			260.4	49.8
FNS	Thionitrosyl fluoride	g			259.8	44.1
FNa	Sodium fluoride	cr	-576.6	-546.3	51.1	46.9
FO	Fluorine oxide	g	109.0	105.0	216.8	30.5
FRb	Rubidium fluoride	cr	-557.7			
FSi	Fluorosilylidyne	g	7.1	-24.3	225.8	32.6
FTl	Thallium fluoride	cr	-324.7			
		g	-182.4			
F_2	Fluorine	g	0.0	0.0	202.8	31.3
F_2Fe	Iron difluoride	cr	-711.3	-668.6	87.0	68.1
F_2HK	Potassium hydrogen difluoride	cr	-927.7	-859.7	104.3	76.9
F_2HN	Difluoroazane	g			252.8	43.4
F_2HNa	Sodium hydrogen difluoride	cr	-920.3	-852.2	90.9	75.0
F_2HRb	Rubidium hydrogen difluoride	cr	-922.6	-855.6	120.1	79.4
F_2Mg	Magnesium difluoride	cr	-1124.2	-1071.1	57.2	61.6
F_2N	Difluoroamidogen	g	43.1	57.8	249.9	41.0
F_2N_2	*cis*-Difluorodiazene	g	69.5			
F_2N_2	*trans*-Difluorodiazene	g	82.0			
F_2Ni	Nickel difluoride	cr	-651.4	-604.1	73.6	64.1
F_2O	Oxygen difluoride	g	24.7	41.9	247.4	43.3
F_2OS	Sulfinyl difluoride	g			278.7	56.8
F_2O_2	Dioxygen difluoride	g	18.0			
F_2O_2S	Sulfonyl difluoride	g			284.0	66.0
F_2O_2U	Uranyl difluoride	cr	-1648.1	-1551.8	135.6	103.2
F_2Pb	Lead difluoride	cr	-664.0	-617.1	110.5	
F_2Si	Difluorosilylene	g	-619.0	-628.0	252.7	43.9
F_2Sr	Strontium difluoride	cr	-1216.3	-1164.8	82.1	70.0
F_2Zn	Zinc difluoride	cr	-764.4	-713.3	73.7	65.7
F_3Ga	Gallium trifluoride	cr	-1163.0	-1085.3	84.0	
F_3Gd	Gadolinium trifluoride	g	-1297.0			
F_3HSi	Trifluorosilane	g			271.9	60.5
F_3Ho	Holmium trifluoride	cr	-1707.0			
F_3N	Nitrogen trifluoride	g	-132.1	-90.6	260.8	53.4
F_3Nd	Neodymium trifluoride	cr	-1657.0			
F_3OP	Phosphoryl trifluoride	g	-1254.3	-1205.8	285.4	68.8
F_3P	Phosphorus trifluoride	g	-958.4	-936.9	273.1	58.7
F_3Sb	Antimony trifluoride	cr	-915.5			
		g	-1247.0	-1234.0	300.5	67.8
F_3Sc	Scandium trifluoride	cr	-1629.2	-1555.6	92.0	
F_3Sm	Samarium trifluoride	cr	-1778.0			
F_3Th	Thorium trifluoride	g	-1182.0	-1176.5	339.4	73.2
F_3U	Uranium trifluoride	cr	-1502.1	-1433.4	123.4	95.1
F_3Y	Yttrium trifluoride	cr	-1718.8	-1644.7	100.0	
		g	-1288.7	-1277.8	311.8	70.3
F_4Ge	Tetrafluorogerman	g	-1190.2	-1150.0	301.9	
F_4Hf	Hafnium tetrafluoride	cr	-1930.5	-1830.4	113.0	
		g	-1669.8			
F_4N_2	Tetrafluorohydrazine	g	-8.4	79.9	301.2	79.2
F_4Pb	Lead tetrafluoride	cr	-941.8			

TABLE 2.4.1
STANDARD STATE THERMOCHEMICAL PROPERTIES AT 298.15 K (continued)

Molecular formula	Name	State	$\Delta_f H°$/ kJ mol^{-1}	$\Delta_f G°$/ kJ mol^{-1}	$S°$/ J K^{-1} mol^{-1}	C_p/ J K^{-1} mol^{-1}
F_4S	Sulfur tetrafluoride	g	-763.2	-722.0	299.6	77.6
F_4Si	Silicon tetrafluoride	g	-1615.0	-1572.8	282.8	73.6
F_4Th	Thorium tetrafluoride	cr	-2091.6	-1997.0	142.1	110.5
F_4U	Uranium tetrafluoride	cr	-1914.2	-1823.3	151.7	116.0
		g	-1598.7	-1572.7	368.0	91.2
F_4V	Vanadium tetrafluoride	cr	-1403.3			
F_4Xe	Xenon tetrafluoride	cr	-261.5			
F_4Zr	Zirconium tetrafluoride	cr	-1911.3	-1809.9	104.6	103.7
F_5I	Iodine pentafluoride	l	-864.8			
		g	-822.5	-751.7	327.7	99.2
F_5Nb	Niobium pentafluoride	cr	-1813.8	-1699.0	160.2	134.7
		g	-1739.7	-1673.6	321.9	97.1
F_5P	Phosphorus pentafluoride	g	-1594.4	-1520.7	300.8	84.8
F_5Ta	Tantalum pentafluoride	cr	-1903.6			
F_5V	Vanadium pentafluoride	l	-1480.3	-1373.1	175.7	
		g	-1433.9	-1369.8	320.9	98.6
$F_6H_8N_2Si$	Diammonium hexafluorosilicate	cr	-2681.7	-2365.3	280.2	228.1
F_6Ir	Iridium hexafluoride	cr	-579.7	-461.6	247.7	
		g	-544.0	-460.0	357.8	121.1
F_6K_2Si	Dipotassium hexafluorosilicate	cr	-2956.0	-2798.6	226.0	
F_6Mo	Molybdenum hexafluoride	l	-1585.5	-1473.0	259.7	169.8
		g	-1557.7	-1472.2	350.5	120.6
F_6Na_2Si	Disodium hexafluorosilicate	cr	-2909.6	-2754.2	207.1	187.1
F_6Os	Osmium hexafluoride	cr			246.0	
		g			358.1	120.8
F_6Pt	Platinum hexafluoride	cr			235.6	
		g			348.3	122.8
F_6S	Sulfur hexafluoride	g	-1220.5	-1116.5	291.5	97.0
F_6Se	Selenium hexafluoride	g	-1117.0	-1017.0	313.9	110.5
F_6Te	Tellurium hexafluoride	g	-1318.0			
F_6U	Uranium hexafluoride	cr	-2197.0	-2068.5	227.6	166.8
		g	-2147.4	-2063.7	377.9	129.6
F_6W	Tungsten hexafluoride	l	-1747.7	-1631.4	251.5	
		g	-1721.7	-1632.1	341.1	119.0
Fe	Iron	cr	0.0	0.0	27.3	25.1
		g	416.3	370.7	180.5	25.7
FeI_2	Iron diiodide	cr	-113.0			
FeI_3	Iron triiodide	g	71.0			
$FeMoO_4$	Iron molybdate	cr	-1075.0	-975.0	129.3	118.5
FeO	Iron oxide	cr	-272.0			
FeO_4S	Iron sulfate	cr	-928.4	-820.8	107.5	100.6
FeO_4W	Iron tungstate	cr	-1155.0	-1054.0	131.8	114.6
FeS	Iron sulfide	cr	-100.0	-100.4	60.3	50.5
FeS_2	Iron disulfide	cr	-178.2	-166.9	52.9	62.2
Fe_2O_3	Diiron trioxide	cr	-824.2	-742.2	87.4	103.9
Fe_2O_4Si	Diiron silicate	cr	-1479.9	-1379.0	145.2	132.9
Fe_3O_4	Triiron tetraoxide	cr	-1118.4	-1015.4	146.4	143.4
Fm	Fermium	cr	0.0	0.0		
Fr	Francium	cr	0.0	0.0	95.4	
Ga	Gallium	cr	0.0	0.0	40.9	25.9
		l	5.6			
		g	277.0	238.9	169.1	25.4
GaH_3O_3	Gallium trihydroxide	cr	-964.4	-831.3	100.0	
GaI_3	Gallium triiodide	cr	-238.9			
GaN	Gallium nitride	cr	-110.5			
GaO	Gallium oxide	g	279.5	253.5	231.1	32.1

TABLE 2.4.1
STANDARD STATE THERMOCHEMICAL PROPERTIES AT 298.15 K (continued)

Molecular formula	Name	State	$\Delta_f H°$/ kJ mol^{-1}	$\Delta_f G°$/ kJ mol^{-1}	$S°$/ J K^{-1} mol^{-1}	C_p/ J K^{-1} mol^{-1}
GaP	Gallium phosphide	cr	-88.0			
GaSb	Gallium antimonide	cr	-41.8	-38.9	76.1	48.5
Ga_2	Digallium	g	438.5			
Ga_2O	Digallium oxide	cr	-356.0			
Ga_2O_3	Digallium trioxide	cr	-1089.1	-998.3	85.0	92.1
Gd	Gadolinium	cr	0.0		68.1	37.0
		g	397.5	359.8	194.3	27.5
Gd_2O_3	Digadolinium trioxide	cr	-1819.6			106.7
Ge	Germanium	cr	0.0	0.0	31.1	23.3
		g	372.0	331.2	167.9	30.7
GeH_3I	Iodogermane	g			283.2	57.5
GeH_4	Germane	g	90.8	113.4	217.1	45.0
GeI_4	Germanium tetraiodide	cr	-141.8	-144.3	271.1	
		g	-56.9	-106.3	428.9	104.1
GeO	Germanium oxide (brown)	cr	-261.9	-237.2	50.0	
		g	-46.2	-73.2	224.3	30.9
GeO_2	Germanium dioxide (tetragonal)	cr	-580.0	-521.4	39.7	52.1
GeP	Germanium phosphide	cr	-21.0	-17.0	63.0	
GeS	Germanium sulfide	cr	-69.0	-71.5	71.0	
		g	92.0	42.0	234.0	33.7
Ge_2	Digermanium	g	473.1	416.3	252.8	35.6
Ge_2H_6	Digermane	l	137.3			
		g	162.3			
Ge_3H_8	Trigermane	l	193.7			
		g	226.8			
H	Hydrogen (atomic)	g	218.0	203.3	114.7	20.8
HI	Hydrogen iodide	g	26.5	1.7	206.6	29.2
HIO_3	Iodic acid	cr	-230.1			
HK	Potassium hydride	cr	-57.7			
HKO	Potassium hydroxide	cr	-424.8	-379.1	78.9	64.9
HKO_4S	Potassium hydrogen sulfate	cr	-1160.6	-1031.3	138.1	
HLi	Lithium hydride	cr	-90.5	-68.3	20.0	27.9
HLiO	Lithium hydroxide	cr	-484.9	-439.0	42.8	49.7
HN	Imidogen	g	351.5	345.6	181.2	29.2
HNO_2	Nitrous acid	g	-79.5	-46.0	254.1	45.6
HNO_3	Nitric acid	l	-174.1	-80.7	155.6	109.9
		g	-135.1	-74.7	266.4	53.4
HN_3	Hydrogen azide	l	264.0	327.3	140.6	
		g	294.1	328.1	239.0	43.7
HNa	Sodium hydride	cr	-56.3	-33.5	40.0	36.4
HNaO	Sodium hydroxide	cr	-425.6	-379.5	64.5	59.5
$HNaO_4S$	Sodium hydrogen sulfate	cr	-1125.5	-992.8	113.0	
HNa_2O_4P	Disodium hydrogen phosphate	cr	-1748.1	-1608.2	150.5	135.3
HO	Hydroxyl	g	39.0	34.2	183.7	29.9
HORb	Rubidium hydroxide	cr	-418.2			
HOTl	Thallium hydroxide	cr	-238.9	-195.8	88.0	
HO_2	Hydroperoxy	g	10.5	22.6	229.0	34.9
HO_3P	Metaphosphoric acid	cr	-948.5			
HO_4RbS	Rubidium hydrogen sulfate	cr	-1159.0			
HO_4Re	Perrhenic acid	cr	-762.3	-656.4	158.2	
HRb	Rubidium hydride	cr	-52.3			
HS	Mercapto	g	142.7	113.3	195.7	32.3
HSi	Silylidyne	g	361.0			
HTa_2	Ditantalum hydride	cr	-32.6	-69.0	79.1	90.8
H_2	Hydrogen	g	0.0	0.0	130.7	28.8
H_2KN	Potassium dihydronitride	cr	-128.9			

TABLE 2.4.1
STANDARD STATE THERMOCHEMICAL PROPERTIES AT 298.15 K (continued)

Molecular formula	Name	State	$\Delta_f H°$/ kJ mol^{-1}	$\Delta_f G°$/ kJ mol^{-1}	$S°$/ J K^{-1} mol^{-1}	C_p/ J K^{-1} mol^{-1}
H_2KO_4P	Potassium dihydrogen phosphate	cr	-1568.3	-1415.9	134.9	116.6
H_2LiN	Lithium dihydronitride	cr	-179.5			
H_2Mg	Magnesium dihydride	cr	-75.3	-35.9	31.1	35.4
H_2MgO_2	Magnesium dihydroxide	cr	-924.5	-833.5	63.2	77.0
H_2N	Amidogen	g	184.9	194.6	195.0	33.9
H_2NNa	Sodium dihydronitride	cr	-123.8	-64.0	76.9	66.2
H_2NRb	Rubidium dihydronitride	cr	-113.0			
$H_2N_2O_2$	Nitroazane	cr	-89.5			
H_2NiO_2	Nickel dihydroxide	cr	-529.7	-447.2	88.0	
H_2O	Water	l	-285.8	-237.1	70.0	75.3
		g	-241.8	-228.6	188.8	33.6
H_2O_2	Dihydrogen peroxide	l	-187.8	-120.4	109.6	89.1
		g	-136.3	-105.6	232.7	43.1
H_2O_2Sn	Tin dihydroxide	cr	-561.1	-491.6	155.0	
H_2O_2Sr	Strontium dihydroxide	cr	-959.0			
H_2O_2Zn	Zinc dihydroxide	cr	-641.9	-553.5	81.2	
H_2O_3Si	Metasilicic acid	cr	-1188.7	-1092.4	134.0	
H_2O_4S	Sulfuric acid	l	-814.0	-690.0	156.9	138.9
H_2O_4Se	Selenic acid	cr	-530.1			
H_2S	Dihydrogen sulfide	g	-20.6	-33.4	205.8	34.2
H_2S_2	Dihydrogen disulfide	l	-18.1			84.1
		g	15.5			51.5
H_2Se	Dihydrogen selenide	g	29.7	15.9	219.0	34.7
H_2Sr	Strontium dihydride	cr	-180.3			
H_2Te	Dihydrogen telluride	g	99.6			
H_2Th	Thorium dihydride	cr	-139.7	-100.0	50.7	36.7
H_2Zr	Zirconium dihydride	cr	-169.0	-128.8	35.0	31.0
H_3ISi	Iodosilane	g			270.9	54.4
H_3N	Ammonia	g	-45.9	-16.4	192.8	35.1
H_3NO	Hydroxylamine	cr	-114.2			
H_3O_2P	Phosphonous acid	cr	-604.6			
		l	-595.4			
H_3O_3P	Phosphorous acid	cr	-964.4			
H_3O_4P	Phosphoric acid	cr	-1284.4	-1124.3	110.5	106.1
		l	-1271.7	-1123.6	150.8	145.0
H_3P	Phosphine	g	5.4	13.4	210.2	37.1
H_3Sb	Stibine	g	145.1	147.8	232.8	41.1
H_3U	Uranium trihydride	cr	-127.2	-72.8	63.7	49.3
H_4IN	Ammonium iodide	cr	-201.4	-112.5	117.0	
H_4N_2	Hydrazine	l	50.6	149.3	121.2	98.9
		g	95.4	159.4	238.5	49.6
$H_4N_2O_2$	Ammonium nitrite	cr	-256.5			
$H_4N_2O_3$	Ammonium nitrate	cr	-365.6	-183.9	151.1	139.3
H_4N_4	Ammonium azide	cr	115.5	274.2	112.5	
H_4O_4Si	Silicic acid	cr	-1481.1	-1332.9	192.0	
$H_4O_7P_2$	Diphosphoric acid	cr	-2241.0			
		l	-2231.7			
H_4P_2	Diphosphane	l	-5.0			
		g	20.9			
H_4Si	Silane	g	34.3	56.9	204.6	42.8
H_4Sn	Stannane	g	162.8	188.3	227.7	49.0
H_5NO	Ammonium hydroxide	l	-361.2	-254.0	165.6	154.9
H_5NO_3S	Ammonium hydrogen sulfite	cr	-768.6			
H_5NO_4S	Ammonium hydrogen sulfate	cr	-1027.0			
H_6Si_2	Disilane	g	80.3	127.3	272.7	80.8
$H_8N_2O_4S$	Diammonium sulfate	cr	-1180.9	-901.7	220.1	187.5

TABLE 2.4.1
STANDARD STATE THERMOCHEMICAL PROPERTIES AT 298.15 K (continued)

Molecular formula	Name	State	$\Delta_f H°$/ kJ mol^{-1}	$\Delta_f G°$/ kJ mol^{-1}	$S°$/ J K^{-1} mol^{-1}	C_p/ J K^{-1} mol^{-1}
H_8Si_3	Trisilane	l	92.5			
		g	120.9			
$H_9N_2O_4P$	Diammonium hydrogen phosphate	cr	-1566.9			188.0
$H_{12}N_3O_4P$	Triammonium phosphate	cr	-1671.9			
He	Helium	g	0.0	0.0	126.2	20.8
Hf	Hafnium	cr	0.0	0.0	43.6	25.7
		g	619.2	576.5	186.9	20.8
HfO_2	Hafnium dioxide	cr	-1144.7	-1088.2	59.3	60.3
Hg	Mercury	l	0.0	0.0	75.9	28.0
		g	61.4	31.8	175.0	20.8
HgI_2	Mercury diiodide (red)	cr	-105.4	-101.7	180.0	
HgO	Mercury oxide (red)	cr	-90.8	-58.5	70.3	44.1
HgO_4S	Mercury sulfate	cr	-707.5			
HgS	Mercury sulfide	cr	-58.2	-50.6	82.4	48.4
HgTe	Mercury telluride	cr	-42.0			
Hg_2	Dimercury	g	108.8	68.2	288.1	37.4
Hg_2I_2	Dimercury diiodide	cr	-121.3	-111.0	233.5	
Hg_2O_4S	Dimercury sulfate	cr	-743.1	-625.8	200.7	132.0
Ho	Holmium	cr	0.0	0.0	75.3	27.2
		g	300.8	264.8	195.6	20.8
Ho_2O_3	Diholmium trioxide	cr	-1880.7	-1791.1	158.2	115.0
I	Iodine (atomic)	g	106.8	70.2	180.8	20.8
IIn	Indium iodide	cr	-116.3	-120.5	130.0	
		g	7.5	-37.7	267.3	36.8
IK	Potassium iodide	cr	-327.9	-324.9	106.3	52.9
IKO_3	Potassium iodate	cr	-501.4	-418.4	151.5	106.5
IKO_4	Potassium periodate	cr	-467.2	-361.4	175.7	
ILi	Lithium iodide	cr	-270.4	-270.3	86.8	51.0
INa	Sodium iodide	cr	-287.8	-286.1	98.5	52.1
$INaO_3$	Sodium iodate	cr	-481.8			92.0
$INaO_4$	Sodium periodate	cr	-429.3	-323.0	163.0	
IO	Iodine oxide	g	175.1	149.8	245.5	32.9
IRb	Rubidium iodide	cr	-333.8	-328.9	118.4	53.2
ITl	Thallium iodide	cr	-123.8	-125.4	127.6	
		g	7.1			
I_2	Iodine (rhombic)	cr	0.0	0.0	116.1	54.4
		g	62.4	19.3	260.7	36.9
I_2Mg	Magnesium diiodide	cr	-364.0	-358.2	129.7	
I_2Ni	Nickel diiodide	cr	-78.2			
I_2Pb	Lead diiodide	cr	-175.5	-173.6	174.9	77.4
I_2Sn	Tin diiodide	cr	-143.5			
I_2Sr	Strontium diiodide	cr	-558.1			81.6
I_2Zn	Zinc diiodide	cr	-208.0	-209.0	161.1	
I_3In	Indium triiodide	cr	-238.0			
		g	-120.5			
I_3Lu	Lutetium triiodide	cr	-548.0			
I_3P	Phosphorus triiodide	cr	-45.6			
		g			374.4	78.4
I_3Ru	Ruthenium triiodide	cr	-65.7			
I_3Sb	Antimony triiodide	cr	-100.4			
I_4Pt	Platinum tetraiodide	cr	-72.8			
I_4Si	Silicon tetraiodide	cr	-189.5			
I_4Sn	Tin tetraiodide	cr				84.9
		g			446.1	105.4
I_4Ti	Titanium tetraiodide	cr	-375.7	-371.5	249.4	125.7
		g	-277.8			

TABLE 2.4.1
STANDARD STATE THERMOCHEMICAL PROPERTIES AT 298.15 K (continued)

Molecular formula	Name	State	$\Delta_f H°$/ kJ mol^{-1}	$\Delta_f G°$/ kJ mol^{-1}	$S°$/ J K^{-1} mol^{-1}	C_p/ J K^{-1} mol^{-1}
I_4V	Vanadium tetraiodide	g	-122.6			
I_4Zr	Zirconium tetraiodide	cr	-481.6			
In	Indium	cr	0.0	0.0	57.8	26.7
		g	243.3	208.7	173.8	20.8
InO	Indium oxide	g	387.0	364.4	236.5	32.6
InP	Indium phosphide	cr	-88.7	-77.0	59.8	45.4
InS	Indium sulfide	cr	-138.1	-131.8	67.0	
		g	238.0			
InSb	Indium antimonide	cr	-30.5	-25.5	86.2	49.5
		g	344.3			
In_2	Diindium	g	380.9			
In_2O_3	Diindium trioxide	cr	-925.8	-830.7	104.2	92.0
In_2S_3	Diindium trisulfide	cr	-427.0	-412.5	163.6	118.0
Ir	Iridium	cr	0.0	0.0	35.5	25.1
		g	665.3	617.9	193.6	20.8
IrO_2	Iridium dioxide	cr	-274.1			57.3
IrS_2	Iridium disulfide	cr	-138.0			
Ir_2S_3	Diiridium trisulfide	cr	-234.0			
K	Potassium	cr	0.0	0.0	64.7	29.6
		g	89.0	60.5	160.3	20.8
$KMnO_4$	Potassium permanganate	cr	-837.2	-737.6	171.7	117.6
KNO_2	Potassium nitrite	cr	-369.8	-306.6	152.1	107.4
KNO_3	Potassium nitrate	cr	-494.6	-394.9	133.1	96.4
KNa	Potassium sodium	l	6.3			
KO_2	Potassium hyperoxide	cr	-284.9	-239.4	116.7	77.5
K_2	Dipotassium	g	123.7	87.5	249.7	37.9
K_2O	Dipotassium oxide	cr	-361.5			
K_2O_2	Dipotassium peroxide	cr	-494.1	-425.1	102.1	
K_2O_4S	Dipotassium sulfate	cr	-1437.8	-1321.4	175.6	131.5
K_2S	Dipotassium sulfide	cr	-380.7	-364.0	105.0	
K_3O_4P	Tripotassium phosphate	cr	-1950.2			
Kr	Krypton	g	0.0	0.0	164.1	20.8
La	Lanthanum	cr	0.0	0.0	56.9	27.1
		g	431.0	393.6	182.4	22.8
LaS	Lanthanum sulfide	cr	-456.0	-451.5	73.2	59.0
La_2O_3	Dilanthanum trioxide	cr	-1793.7	-1705.8	127.3	108.8
Li	Lithium	cr	0.0	0.0	29.1	24.8
		g	159.3	126.6	138.8	20.8
$LiNO_2$	Lithium nitrite	cr	-372.4	-302.0	96.0	
$LiNO_3$	Lithium nitrate	cr	-483.1	-381.1	90.0	
Li_2	Dilithium	g	215.9	174.4	197.0	36.1
Li_2O	Dilithium oxide	cr	-597.9	-561.2	37.6	54.1
Li_2O_2	Dilithium peroxide	cr	-634.3			
Li_2O_3Si	Dilithium metasilicate	cr	-1648.1	-1557.2	79.8	99.1
Li_2O_4S	Dilithium sulfate	cr	-1436.5	-1321.7	115.1	117.6
Li_2S	Dilithium sulfide	cr	-441.4			
Li_3O_4P	Trilithium phosphate	cr	-2095.8			
Lr	Lawrencium	cr	0.0	0.0		
Lu	Lutetium	cr	0.0	0.0	51.0	26.9
		g	427.6	387.8	184.8	20.9
Lu_2O_3	Dilutetium trioxide	cr	-1878.2	-1789.0	110.0	101.8
Md	Mendelevium	cr	0.0	0.0		
Mg	Magnesium	cr	0.0	0.0	32.7	24.9
		g	147.1	112.5	148.6	20.8
MgN_2O_6	Magnesium dinitrate	cr	-790.7	-589.4	164.0	141.9
MgO	Magnesium oxide	cr	-601.6	-569.3	27.0	37.2

TABLE 2.4.1
STANDARD STATE THERMOCHEMICAL PROPERTIES AT 298.15 K (continued)

Molecular formula	Name	State	$\Delta_f H°$/ kJ mol^{-1}	$\Delta_f G°$/ kJ mol^{-1}	$S°$/ J K^{-1} mol^{-1}	C_p/ J K^{-1} mol^{-1}
MgO_4S	Magnesium sulfate	cr	-1284.9	-1170.6	91.6	96.5
MgO_4Se	Magnesium selenate	cr	-968.5			
MgS	Magnesium sulfide	cr	-346.0	-341.8	50.3	45.6
Mg_2	Dimagnesium	g	287.7			
Mg_2O_4Si	Dimagnesium silicate	cr	-2174.0	-2055.1	95.1	118.5
Mn	Manganese	cr	0.0	0.0	32.0	26.3
		g	280.7	238.5	173.7	20.8
MnN_2O_6	Manganese dinitrate	cr	-576.3			
$MnNa_2O_4$	Disodium manganate	cr	-1156.0			
MnO	Manganese oxide	cr	-385.2	-362.9	59.7	45.4
MnO_2	Manganese dioxide	cr	-520.0	-465.1	53.1	54.1
MnO_3Si	Manganese metasilicate	cr	-1320.9	-1240.5	89.1	86.4
MnS	Manganese sulfide	cr	-214.2	-218.4	78.2	50.0
MnSe	Manganese selenide	cr	-106.7	-111.7	90.8	51.0
Mn_2O_3	Dimanganese trioxide	cr	-959.0	-881.1	110.5	107.7
Mn_2O_4Si	Dimanganese silicate	cr	-1730.5	-1632.1	163.2	129.9
Mn_3O_4	Trimanganese tetraoxide	cr	-1387.8	-1283.2	155.6	139.7
Mo	Molybdenum	cr	0.0	0.0	28.7	24.1
		g	658.1	612.5	182.0	20.8
$MoNa_2O_4$	Disodium molybdate	cr	-1468.1	-1354.3	159.7	141.7
MoO_2	Molybdenum dioxide	cr	-588.9	-533.0	46.3	56.0
MoO_3	Molybdenum trioxide	cr	-745.1	-668.0	77.7	75.0
MoO_4Pb	Lead molybdate	cr	-1051.9	-951.4	166.1	119.7
MoS_2	Molybdenum disulfide	cr	-235.1	-225.9	62.6	63.6
N	Nitrogen (atomic)	g	472.7	455.5	153.3	20.8
$NNaO_2$	Sodium nitrite	cr	-358.7	-284.6	103.8	
$NNaO_3$	Sodium nitrate	cr	-467.9	-367.0	116.5	92.9
NO	Nitrogen oxide	g	90.3	86.6	210.8	29.8
NO_2	Nitrogen dioxide	g	33.2	51.3	240.1	37.2
NO_2Rb	Rubidium nitrite	cr	-367.4	-306.2	172.0	
NO_3Rb	Rubidium nitrate	cr	-495.1	-395.8	147.3	102.1
NO_3Tl	Thallium nitrate	cr	-243.9	-152.4	160.7	99.5
NP	Phosphorous nitride	cr	-63.0			
		g	109.9	87.7	211.2	29.7
N_2	Nitrogen	g	0.0	0.0	191.6	29.1
N_2O	Dinitrogen oxide	g	82.1	104.2	219.9	38.5
N_2O_3	Dinitrogen trioxide	l	50.3			
		g	83.7	139.5	312.3	65.6
N_2O_4	Dinitrogen tetraoxide	l	-19.5	97.5	209.2	142.7
		g	9.2	97.9	304.3	77.3
N_2O_4Sr	Strontium dinitrite	cr	-762.3			
N_2O_5	Dinitrogen pentaoxide	cr	-43.1	113.9	178.2	143.1
		g	11.3	115.1	355.7	84.5
N_2O_6Pb	Lead dinitrate	cr	-451.9			
N_2O_6Ra	Radium dinitrate	cr	-992.0	-796.1	222.0	
N_2O_6Sr	Strontium dinitrate	cr	-978.2	-780.0	194.6	149.9
N_2O_6Zn	Zinc dinitrate	cr	-483.7			
N_3Na	Sodium azide	cr	21.7	93.8	96.9	76.6
N_4Si_3	Trisilicon tetranitride	cr	-743.5	-642.6	101.3	
Na	Sodium	cr	0.0	0.0	51.3	28.2
		g	107.5	77.0	153.7	20.8
NaO_2	Sodium hyperoxide	cr	-260.2	-218.4	115.9	72.1
Na_2	Disodium	g	142.1	103.9	230.2	37.6
Na_2O	Disodium oxide	cr	-414.2	-375.5	75.1	69.1
Na_2O_2	Disodium peroxide	cr	-510.9	-447.7	95.0	89.2
Na_2O_3S	Disodium sulfite	cr	-1100.8	-1012.5	145.9	120.3

TABLE 2.4.1
STANDARD STATE THERMOCHEMICAL PROPERTIES AT 298.15 K (continued)

Molecular formula	Name	State	$\Delta_f H°$/ kJ mol^{-1}	$\Delta_f G°$/ kJ mol^{-1}	$S°$/ J K^{-1} mol^{-1}	C_p/ J K^{-1} mol^{-1}
Na_2O_3Si	Disodium metasilicate	cr	-1554.9	-1462.8	113.9	
Na_2O_4S	Disodium sulfate	cr	-1387.1	-1270.2	149.6	128.2
Na_2S	Disodium sulfide	cr	-364.8	-349.8	83.7	
Nb	Niobium	cr	0.0	0.0	36.4	24.6
		g	725.9	681.1	186.3	30.2
NbO	Niobium oxide	cr	-405.8	-378.6	48.1	41.3
NbO_2	Niobium dioxide	cr	-796.2	-740.5	54.5	57.5
Nb_2O_5	Diniobium pentaoxide	cr	-1899.5	-1766.0	137.2	132.1
Nd	Neodymium	cr	0.0	0.0	71.5	27.5
		g	327.6	292.4	189.4	22.1
Nd_2O_3	Dineodymium trioxide	cr	-1807.9	-1720.8	158.6	111.3
Ne	Neon	g	0.0	0.0	146.3	20.8
Ni	Nickel	cr	0.0	0.0	29.9	26.1
		g	429.7	384.5	182.2	23.4
NiO_4S	Nickel sulfate	cr	-872.9	-759.7	92.0	138.0
NiS	Nickel sulfide	cr	-82.0	-79.5	53.0	47.1
Ni_2O_3	Dinickel trioxide	cr	-489.5			
No	Nobelium	cr	0.0	0.0		
O	Oxygen (atomic)	g	249.2	231.7	161.1	21.9
OP	Phosphorous oxide	g	-28.5	-51.9	222.8	31.8
OPb	Lead oxide (yellow)	cr	-217.3	-187.9	68.7	45.8
	Leads oxide (red)	cr	-219.0	-188.9	66.5	45.8
OPd	Palladium oxide	cr	-85.4			31.4
		g	348.9	325.9	218.0	
ORa	Radium oxide	cr	-523.0			
ORb_2	Dirubidium oxide	cr	-339.0			
ORh	Rhodium oxide	g	385.0			
OS	Sulfur oxide	g	6.3	-19.9	222.0	30.2
OSe	Selenium oxide	g	53.4	26.8	234.0	31.3
OSi	Silicon oxide	g	-99.6	-126.4	211.6	29.9
OSn	Tin oxide (tetragonal)	cr	-280.7	-251.9	57.2	44.3
		g	15.1	-8.4	232.1	31.6
OSr	Strontium oxide	cr	-592.0	-561.9	54.4	45.0
OTi	Titanium oxide	cr	-519.7	-495.0	50.0	40.0
OTl_2	Dithallium oxide	cr	-178.7	-147.3	126.0	
OU	Uranium oxide	g	21.0			
OV	Vanadium oxide	cr	-431.8	-404.2	38.9	45.4
OZn	Zinc oxide	cr	-350.5	-320.5	43.7	40.3
O_2	Oxygen	g	0.0	0.0	205.2	29.4
O_2P	Phosphorous dioxide	g	-279.9	-281.6	252.1	39.5
O_2Pb	Lead dioxide	cr	-277.4	-217.3	68.6	64.6
O_2Rb	Rubidium hyperoxide	cr	-278.7			
O_2Rb_2	Dirubidium peroxide	cr	-472.0			
O_2Ru	Ruthenium dioxide	cr	-305.0			
O_2S	Sulfur dioxide	l	-320.5			
		g	-296.8	-300.1	248.2	39.9
O_2Se	Selenium dioxide	cr	-225.4			
O_2Si	Silicon dioxide (α-quartz)	cr	-910.7	-856.3	41.5	44.4
		g	-322.0			
O_2Sn	Tin dioxide (tetragonal)	cr	-577.6	-515.8	49.0	52.6
O_2Te	Tellurium dioxide	cr	-322.6	-270.3	79.5	
O_2Th	Thorium dioxide	cr	-1226.4	-1169.2	65.2	61.8
O_2Ti	Titanium dioxide (rutile)	cr	-944.0	-888.8	50.6	55.0
O_2U	Uranium dioxide	cr	-1085.0	-1031.8	77.0	63.6
		g	-465.7	-471.5	274.6	51.4
O_2W	Tungsten dioxide	cr	-589.7	-533.9	50.5	56.1

TABLE 2.4.1
STANDARD STATE THERMOCHEMICAL PROPERTIES AT 298.15 K (continued)

Molecular formula	Name	State	$\Delta_f H°$/ kJ mol^{-1}	$\Delta_f G°$/ kJ mol^{-1}	$S°$/ J K^{-1} mol^{-1}	C_p/ J K^{-1} mol^{-1}
O_2Zr	Zirconium dioxide	cr	-1100.6	-1042.8	50.4	56.2
O_3	Ozone	g	142.7	163.2	238.9	39.2
O_3PbS	Lead sulfite	cr	-669.9			
O_3PbSi	Lead metasilicate	cr	-1145.7	-1062.1	109.6	90.0
O_3Pr_2	Dipraseodymium trioxide	cr	-1809.6			117.4
O_3Rh_2	Dirhodium trioxide	cr	-343.0			103.8
O_3S	Sulfur trioxide	cr	-454.5	-374.2	70.7	
		l	-441.0	-373.8	113.8	
		g	-395.7	-371.1	256.8	50.7
O_3Sc_2	Discandium trioxide	cr	-1908.8	-1819.4	77.0	94.2
O_3SiSr	Strontium metasilicate	cr	-1633.9	-1549.7	96.7	88.5
O_3Sm_2	Disamarium trioxide	cr	-1823.0	-1734.6	151.0	114.5
O_3Tb_2	Diterbium trioxide	cr	-1865.2			115.9
O_3Ti_2	Dititanium trioxide	cr	-1520.9	-1434.2	78.8	97.4
O_3Tm_2	Dithullium trioxide	cr	-1888.7	-1794.5	139.7	116.7
O_3U	Uranium trioxide	cr	-1223.8	-1145.7	96.1	81.7
O_3V_2	Divanadium trioxide	cr	-1218.8	-1139.3	98.3	103.2
O_3W	Tungsten trioxide	cr	-842.9	-764.0	75.9	73.8
O_3Y_2	Diyttrium trioxide	cr	-1905.3	-1816.6	99.1	102.5
O_3Yb_2	Diytterbium trioxide	cr	-1814.6	-1726.7	133.1	115.4
O_4Os	Osmium tetraoxide	cr	-394.1	-304.9	143.9	
		g	-337.2	-292.8	293.8	74.1
O_4PbS	Lead sulfate	cr	-920.0	-813.0	148.5	103.2
O_4PbSe	Lead selenate	cr	-609.2	-504.9	167.8	
O_4Pb_2Si	Dilead silicate	cr	-1363.1	-1252.6	186.6	137.2
O_4Pb_3	Trilead tetraoxide	cr	-718.4	-601.2	211.3	146.9
O_4RaS	Radium sulfate	cr	-1471.1	-1365.6	138.0	
O_4Rb_2S	Dirubidium sulfate	cr	-1435.6	-1316.9	197.4	134.1
O_4Ru	Ruthenium tetraoxide	cr	-239.3	-152.2	146.4	
O_4SSr	Strontium sulfate	cr	-1453.1	-1340.9	117.0	
O_4STl_2	Dithallium sulfate	cr	-931.8	-830.4	230.5	
O_4SZn	Zinc sulfate	cr	-982.8	-871.5	110.5	99.2
O_4SiSr_2	Distrontium silicate	cr	-2304.5	-2191.1	153.1	134.3
O_4SiZn_2	Dizinc silicate	cr	-1636.7	-1523.2	131.4	123.3
O_4SiZr	Zirconium silicate	cr	-2033.4	-1919.1	84.1	98.7
O_5Sb_2	Diantimony pentaoxide	cr	-971.9	-829.2	125.1	
O_5Ta_2	Ditantalum pentaoxide	cr	-2046.0	-1911.2	143.1	135.1
O_5Ti_3	Trititanium pentaoxide	cr	-2459.4	-2317.4	129.3	154.8
O_5V_2	Divanadium pentaoxide	cr	-1550.6	-1419.5	131.0	127.7
O_5V_3	Trivanadium pentaoxide	cr	-1933.0	-1803.0	163.0	
O_7Re_2	Dirhenium heptaoxide	cr	-1240.1	-1066.0	207.1	166.1
		g	-1100.0	-994.0	452.0	
O_7U_3	Triuranium heptaoxide	cr	-3427.1	-3242.9	250.5	215.5
O_8S_2Zr	Zirconium disulfate	cr	-2217.1			172.0
O_8U_3	Triuranium octaoxide	cr	-3574.8	-3369.5	282.6	238.4
O_9U_4	Tetrauranium nonaoxide	cr	-4510.4	-4275.1	334.1	293.3
Os	Osmium	cr	0.0	0.0	32.6	24.7
		g	791.0	745.0	192.6	20.8
P	Phosphorus (white)	cr	0.0	0.0	41.1	23.8
	Phosphorus (red)	cr	-17.6		22.8	21.2
	Phosphorus (black)	cr	-39.3			
		g	316.5	280.1	163.2	20.8
P_2	Diphosphorous	g	144.0	103.5	218.1	32.1
P_4	Tetraphosphorous	g	58.9	24.4	280.0	67.2
Pa	Protactinium	cr	0.0	0.0	51.9	
		g	607.0	563.0	198.1	22.9

TABLE 2.4.1
STANDARD STATE THERMOCHEMICAL PROPERTIES AT 298.15 K (continued)

Molecular formula	Name	State	$\Delta_f H°$/ kJ mol^{-1}	$\Delta_f G°$/ kJ mol^{-1}	$S°$/ J K^{-1} mol^{-1}	C_p/ J K^{-1} mol^{-1}
Pb	Lead	cr	0.0	0.0	64.8	26.4
		g	195.2	162.2	175.4	20.8
PbS	Lead sulfide	cr	-100.4	-98.7	91.2	49.5
PbSe	Lead selenide	cr	-102.9	-101.7	102.5	50.2
PbTe	Lead telluride	cr	-70.7	-69.5	110.0	50.5
Pd	Palladium	cr	0.0	0.0	37.6	26.0
		g	378.2	339.7	167.1	20.8
PdS	Palladium sulfide	cr	-75.0	-67.0	46.0	
Pm	Promethium	cr	0.0	0.0		
		g			187.1	24.3
Po	Polonium	cr	0.0	0.0		
Pr	Praseodymium	cr	0.0	0.0	73.2	27.2
		g	355.6	320.9	189.8	21.4
Pt	Platinum	cr	0.0	0.0	41.6	25.9
		g	565.3	520.5	192.4	25.5
PtS	Platinum sulfide	cr	-81.6	-76.1	55.1	43.4
PtS_2	Platinum disulfide	cr	-108.8	-99.6	74.7	65.9
Pu	Plutonium	cr	0.0	0.0		
Ra	Radium	cr	0.0	0.0	71.0	
		g	159.0	130.0	176.5	20.8
Rb	Rubidium	cr	0.0	0.0	76.8	31.1
		g	80.9	53.1	170.1	20.8
Re	Rhenium	cr	0.0	0.0	36.9	25.5
		g	769.9	724.6	188.9	20.8
Rh	Rhodium	cr	0.0	0.0	31.5	25.0
		g	556.9	510.8	185.8	21.0
Rn	Radon	g	0.0	0.0	176.2	20.8
Ru	Ruthenium	cr	0.0	0.0	28.5	24.1
		g	642.7	595.8	186.5	21.5
S	Sulfur (rhombic)	cr	0.0	0.0	32.1	22.6
	Sulfur (monoclinic)	cr	0.3			
		g	277.2	236.7	167.8	23.7
SSi	Silicon sulfide	g	112.5	60.9	223.7	32.3
SSn	Tin sulfide	cr	-100.0	-98.3	77.0	49.3
SSr	Strontium sulfide	cr	-472.4	-467.8	68.2	48.7
STl_2	Dithallium sulfide	cr	-97.1	-93.7	151.0	
SZn	Zinc sulfide (wurtzite)	cr	-192.6			
	Zinc sulfide (sphalerite)	cr	-206.0	-201.3	57.7	46.0
S_2	Disulfur	g	128.6	79.7	228.2	32.5
Sb	Antimony	cr	0.0	0.0	45.7	25.2
		g	262.3	222.1	180.3	20.8
Sb_2	Diantimony	g	235.6	187.0	254.9	36.4
Sc	Scandium	cr	0.0	0.0	34.6	25.5
		g	377.8	336.0	174.8	22.1
Se	Selenium	cr	0.0	0.0	42.4	25.4
		g	227.1	187.0	176.7	20.8
SeSr	Strontium selenide	cr	-385.8			
$SeTl_2$	Dithallium selenide	cr	-59.0	-59.0	172.0	
SeZn	Zinc selenide	cr	-163.0	-163.0	84.0	
Se_2	Diselenium	g	146.0	96.2	252.0	35.4
Si	Silicon	cr	0.0	0.0	18.8	20.0
		g	450.0	405.5	168.0	22.3
Si_2	Disilicon	g	594.0	536.0	229.9	34.4
Sm	Samarium	cr	0.0	0.0	69.6	29.5
		g	206.7	172.8	183.0	30.4
Sn	Tin (white)	cr	0.0	0.0	51.2	27.0

TABLE 2.4.1
STANDARD STATE THERMOCHEMICAL PROPERTIES AT 298.15 K (continued)

Molecular formula	Name	State	$\Delta_f H°$/ kJ mol^{-1}	$\Delta_f G°$/ kJ mol^{-1}	$S°$/ J K^{-1} mol^{-1}	C_p/ J K^{-1} mol^{-1}
	Tin (gray)	cr	-2.1	0.1	44.1	25.8
		g	301.2	266.2	168.5	21.3
Sr	Strontium	cr	0.0	0.0	52.3	26.4
		g	164.4	130.9	164.6	20.8
Ta	Tantalum	cr	0.0	0.0	41.5	25.4
		g	782.0	739.3	185.2	20.9
Tb	Terbium	cr	0.0	0.0	73.2	28.9
		g	388.7	349.7	203.6	24.6
Tc	Technetium	cr	0.0	0.0		
		g	678.0		181.1	20.8
Te	Tellurium	cr	0.0	0.0	49.7	25.7
		g	196.7	157.1	182.7	20.8
Te_2	Ditellurium	g	168.2	118.0	268.1	36.7
Th	Thorium	cr	0.0	0.0	51.8	27.3
		g	602.0	560.7	190.2	20.8
Ti	Titanium	cr	0.0	0.0	30.7	25.0
		g	473.0	428.4	180.3	24.4
Tl	Thallium	cr	0.0	0.0	64.2	26.3
		g	182.2	147.4	181.0	20.8
Tm	Thulium	cr	0.0	0.0	74.0	27.0
		g	232.2	197.5	190.1	20.8
U	Uranium	cr	0.0	0.0	50.2	27.7
		g	533.0	488.4	199.8	23.7
V	Vanadium	cr	0.0	0.0	28.9	24.9
		g	514.2	754.4	182.3	26.0
W	Tungsten	cr	0.0	0.0	32.6	24.3
		g	849.4	807.1	174.0	21.3
Xe	Xenon	g	0.0	0.0	169.7	20.8
Y	Yttrium	cr	0.0	0.0	44.4	26.5
		g	421.3	381.1	179.5	25.9
Yb	Ytterbium	cr	0.0	0.0	59.9	26.7
		g	152.3	118.4	173.1	20.8
Zn	Zinc	cr	0.0	0.0	41.6	25.4
		g	130.4	94.8	161.0	20.8
Zr	Zirconium	cr	0.0	0.0	39.0	25.4
		g	608.8	566.5	181.4	26.7
C	Carbon (graphite)	cr	0.0	0.0	5.7	8.5
	Carbon (diamond)	cr	1.9	2.9	2.4	6.1
		g	716.7	671.3	158.1	20.8
CAgN	Silver cyanide	cr	146.0	156.9	107.2	66.7
CAg_2O_3	Disilver carbonate	cr	-505.8	-436.8	167.4	112.3
$CBaO_3$	Barium carbonate	cr	-1216.3	-1137.6	112.1	85.3
$CBeO_3$	Beryllium carbonate	cr	-1025.0			
$CBrClF_2$	Bromochlorodifluoromethane	g			318.5	74.6
$CBrCl_2F$	Bromodichlorofluoromethane	g			330.6	80.0
$CBrCl_3$	Bromotrichloromethane	g	-41.8			85.3
$CBrF_3$	Bromotrifluoromethane	g	-648.3			69.3
CBrN	Cyanogen bromide	cr	140.5			
		g	186.2	165.3	248.3	46.9
CBr_2ClF	Dibromochlorofluoromethane	g			342.8	82.4
CBr_2Cl_2	Dibromodichloromethane	g			347.8	87.1
CBr_2F_2	Dibromodifluoromethane	g			325.3	77.0
CBr_2O	Carbonyl dibromide	l	-127.2			
		g	-96.2	-110.9	309.1	61.8
CBr_3Cl	Tribromochloromethane	g			357.8	89.4
CBr_3F	Tribromofluoromethane	g			345.9	84.4

TABLE 2.4.1
STANDARD STATE THERMOCHEMICAL PROPERTIES AT 298.15 K (continued)

Molecular formula	Name	State	$\Delta_f H°$/ kJ mol^{-1}	$\Delta_f G°$/ kJ mol^{-1}	$S°$/ J K^{-1} mol^{-1}	C_p/ J K^{-1} mol^{-1}
CBr_4	Tetrabromomethane	cr	18.8	47.7	212.5	144.3
		g	79.0	67.0	358.1	91.2
$CCaO_3$	Calcium carbonate (calcite)	cr	-1207.6	-1129.1	91.7	83.5
	Calcium carbonate (aragonite)	cr	-1207.8	-1128.2	88.0	82.3
$CCdO_3$	Cadmium carbonate	cr	-750.6	-669.4	92.5	
CClFO	Carbonyl chloride fluoride	g			276.7	52.4
$CClF_3$	Chlorotrifluoromethane	g	-706.3			66.9
CClN	Cyanogen chloride	l	112.1			
		g	138.0	131.0	236.2	45.0
CCl_2F_2	Dichlorodifluoromethane	g	-477.4	-439.4	300.8	72.3
CCl_2O	Carbonyl dichloride	g	-219.1	-204.9	283.5	57.7
CCl_3	Trichloromethyl	g	59.0			
CCl_3F	Trichlorofluoromethane	l	-301.3	-236.8	225.4	121.6
		g	-268.3			78.1
CCl_4	Tetrachloromethane	l	-128.2			130.7
		g	-95.8			83.3
$CCoO_3$	Cobalt carbonate	cr	-713.0			
$CCsHO_3$	Cesium hydrogen carbonate	cr	-966.1			
CCs_2O_3	Dicesium carbonate	cr	-1139.7	-1054.3	204.5	123.9
CCuN	Copper cyanide	cr	96.2	111.3	84.5	
CFN	Cyanogen fluoride	g			224.7	41.8
CF_2O	Carbonyl difluoride	g	-639.8			46.8
CF_3	Trifluoromethyl	g	-477.0	-464.0	264.5	49.6
CF_3I	Trifluoroiodomethane	g	-587.8		307.4	70.9
CF_4	Tetrafluoromethane	g	-933.6		261.6	61.1
CFe_3	Triiron carbide	cr	25.1	20.1	104.6	105.9
$CFeO_3$	Iron carbonate	cr	-740.6	-666.7	92.9	82.1
CH	Methylidyne	g	595.8			
CHBrClF	Bromochlorofluoromethane	g			304.3	63.2
$CHBrCl_2$	Bromodichloromethane	g			316.4	67.4
$CHBrF_2$	Bromodifluoromethane	g			295.1	58.7
$CHBr_2Cl$	Chlorodibromomethane	g			327.7	69.2
$CHBr_2F$	Dibromofluoromethane	g			316.8	65.1
$CHBr_3$	Tribromomethane	l	-28.5	-5.0	220.9	130.7
		g	17.0	8.0	330.9	71.2
$CHClF_2$	Chlorodifluoromethane	g	-482.6		280.9	55.9
$CHCl_2F$	Dichlorofluoromethane	g			293.1	60.9
$CHCl_3$	Trichloromethane	l	-134.5	-73.7	201.7	114.2
		g	-103.1	6.0	295.7	65.7
CHFO	Formyl fluoride	g			246.6	39.9
CHF_3	Trifluoromethane	g	-695.4		259.7	51.0
CHI_3	Triiodomethane	cr	141.0			
		g			356.2	75.0
$CHKO_2$	Potassium methanoate	cr	-679.7			
$CHKO_3$	Potassium hydrogen carbonate	cr	-963.2	-863.5	115.5	
CHN	Hydrogen cyanide	l	108.9	125.0	112.8	70.6
		g	135.1	124.7	201.8	35.9
CHNO	Isocyanic acid	g			238.0	44.9
CHNS	Isothiocyanic acid	g	127.6	113.0	247.8	46.9
$CHNaO_2$	Sodium methanoate	cr	-666.5	-599.9	103.8	82.7
$CHNaO_3$	Sodium hydrogen carbonate	cr	-950.8	-851.0	101.7	87.6
CHO	Oxomethyl	g	43.1	28.0	224.7	34.6
CH_2	Methylene	g	390.4	372.9	194.9	33.8
CH_2BrCl	Bromochloromethane	g			287.6	52.7
CH_2BrF	Bromofluoromethane	g			276.3	49.2
CH_2Br_2	Dibromomethane	g			293.2	54.7

TABLE 2.4.1
STANDARD STATE THERMOCHEMICAL PROPERTIES AT 298.15 K (continued)

Molecular formula	Name	State	$\Delta_f H°$/ kJ mol^{-1}	$\Delta_f G°$/ kJ mol^{-1}	$S°$/ J K^{-1} mol^{-1}	C_p/ J K^{-1} mol^{-1}
CH_2ClF	Chlorofluoromethane	g			264.4	47.0
CH_2Cl_2	Dichloromethane	l	-124.1		177.8	101.2
		g	-95.6		270.2	51.0
CH_2F_2	Difluoromethane	g	-452.2		246.7	42.9
CH_2I_2	Diiodomethane	l	66.9	90.4	174.1	134.0
		g	113.0	95.8	309.7	57.7
CH_2N_2	Cyanamide	cr	58.8			
CH_2N_2	Diazomethane	g			242.9	52.5
CH_2O	Methanal	g	-108.6	-102.5	218.8	35.4
CH_2O_2	Methanoic acid	l	-424.7	-361.4	129.0	99.0
		g	-378.6			
CH_3	Methyl	g	145.7	147.9	194.2	38.7
CH_3BO	Carbonyltrihydroboron	g	-111.2	-92.9	249.4	59.5
CH_3Br	Bromomethane	l	-59.4			
		g	-35.5	-26.3	246.4	42.4
CH_3Cl	Chloromethane	g	-81.9		234.6	40.8
CH_3F	Fluoromethane	g			222.9	37.5
CH_3I	Iodomethane	l	-12.3		163.2	126.0
		g	14.7		254.1	44.1
CH_3NO	Methanamide	l	-254.0			
CH_3NO_2	Nitromethane	l	-113.1	-14.4	171.8	106.6
		g	-74.7	-6.8	275.0	57.3
CH_3NO_3	Methyl nitrate	l	-159.0	-43.4	217.1	157.3
		g	-124.7	-39.2	318.5	
CH_4	Methane	g	-74.4	-50.3	186.3	35.3
CH_4N_2	Ammonium cyanide	cr	0.4			134.0
CH_4N_2O	Urea	cr	-333.6			
CH_4O	Methanol	l	-239.1	-166.6	126.8	81.1
		g	-201.5	-162.6	239.8	43.9
CH_4S	Methanethiol	l	-46.4	-7.7	169.2	90.5
		g	-22.3	-9.3	255.2	50.3
CH_5N	Methylamine	l	-47.3	35.7	150.2	102.1
		g	-22.5	32.7	242.9	50.1
CH_5NO_3	Ammonium hydrogen carbonate	cr	-849.4	-665.9	120.9	
CH_6N_2	Methylhydrazine	l	54.0	180.0	165.9	134.9
		g	94.3	187.0	278.8	71.1
CH_6Si	Methylsilane	g			256.5	65.9
CHg_2O_3	Dimercury carbonate	cr	-553.5	-468.1	180.0	
CIN	Cyanogen iodide	cr	166.2	185.0	96.2	
		g	225.5	196.6	256.8	48.3
CI_4	Tetraiodomethane	g			391.9	95.9
CKN	Potassium cyanide	cr	-113.0	-101.9	128.5	66.3
$CKNS$	Potassium thiocyanate	cr	-200.2	-178.3	124.3	88.5
CK_2O_3	Dipotassium carbonate	cr	-1151.0	-1063.5	155.5	114.4
CLi_2O_3	Dilithium carbonate	cr	-1215.9	-1132.1	90.4	99.1
$CMgO_3$	Magnesium carbonate	cr	-1095.8	-1012.1	65.7	75.5
$CMnO_3$	Manganese carbonate	cr	-894.1	-816.7	85.8	81.5
CN	Cyanide	g	437.6	407.5	202.6	29.2
$CNNa$	Sodium cyanide	cr	-87.5	-76.4	115.6	70.4
$CNNaO$	Sodium cyanate	cr	-405.4	-358.1	96.7	86.6
CN_4O_8	Tetranitromethane	l	38.4			
CNa_2O_3	Disodium carbonate	cr	-1130.7	-1044.4	135.0	112.3
CO	Carbon oxide	g	-110.5	-137.2	197.7	29.1
COS	Carbonyl sulfide	g	-142.0	-169.2	231.6	41.5
CO_2	Carbon dioxide	g	-393.5	-394.4	213.8	37.1
CO_3Pb	Lead carbonate	cr	-699.1	-625.5	131.0	87.4

TABLE 2.4.1
STANDARD STATE THERMOCHEMICAL PROPERTIES AT 298.15 K (continued)

Molecular formula	Name	State	$\Delta_f H°$/ kJ mol^{-1}	$\Delta_f G°$/ kJ mol^{-1}	$S°$/ J K^{-1} mol^{-1}	C_p/ J K^{-1} mol^{-1}
CO_3Rb_2	Dirubidium carbonate	cr	-1136.0	-1051.0	181.3	117.6
CO_3Sr	Strontium carbonate	cr	-1220.1	-1140.1	97.1	81.4
CO_3Tl_2	Dithallium carbonate	cr	-700.0	-614.6	155.2	
CO_3Zn	Zinc carbonate	cr	-812.8	-731.5	82.4	79.7
CS	Carbon sulfide	g	234.0	184.0	210.6	29.8
CS_2	Carbon disulfide	l	89.0	64.6	151.3	76.4
		g	116.6	67.1	237.8	45.4
CSe_2	Carbon diselenide	l	164.8			
CSi	Silicon carbide (cubic)	cr	-65.3	-62.8	16.6	26.9
	Silicon carbide (hexagonal)	cr	-62.8	-60.2	16.5	26.7
C_2	Dicarbon	g	831.9	775.9	199.4	43.2
$C_2Br_2ClF_3$	1,2-Dibromo-1-chloro-1,2,2-trifluoroethane	g	-656.6			
$C_2Br_2F_4$	1,2-Dibromotetrafluoroethane	g	-789.1			
C_2Br_4	Tetrabromoethene	g			387.1	102.7
C_2Br_6	Hexabromoethane	g			441.9	139.3
C_2Ca	Calcium acetylide	cr	-59.8	-64.9	70.0	62.7
C_2CaN_2	Calcium dicyanide	cr	-184.5			
C_2CaO_4	Calcium ethanedioate	cr	-1360.6			
C_2ClF_3	Chlorotrifluoroethene	g	-555.2	-523.8	322.1	83.9
$C_2Cl_2F_4$	1,2-Dichlorotetrafluoroethane	l	-939.7			111.7
		g	-916.3			
$C_2Cl_3F_3$	1,1,2-Trichlorotrifluoroethane	l	-805.8			170.1
		g	-777.3			
C_2Cl_3N	Trichloroethanenitrile	g			336.6	96.1
C_2Cl_4	Tetrachloroethene	l	-50.6	3.0	266.9	143.4
$C_2Cl_4F_2$	Tetrachloro-1,1-difluoroethane	g	-489.9	-407.0	382.9	123.4
C_2Cl_4O	Trichloroacetyl chloride	l	-280.8			
C_2Cl_6	Hexachloroethane	cr	-202.8		237.3	198.2
C_2F_3N	Trifluoroethanenitrile	g	-497.9		298.1	77.9
C_2F_4	Tetrafluoroethene	cr	-820.5			
		g	-658.9		300.1	80.5
C_2F_6	Hexafluoroethane	g	-1344.2		332.3	106.7
C_2HBr	Bromoethyne	g			253.7	55.7
C_2HCl	Chloroethyne	g			242.0	54.3
C_2HClF_2	1-Chloro-2,2-difluoroethene	g	-315.5	-289.1	303.0	72.1
C_2HCl_2F	1,1-Dichloro-2-fluoroethene	g			313.9	76.5
$C_2HCl_2F_3$	2,2-Dichloro-1,1,1-trifluoroethane	g			352.8	102.5
C_2HCl_3	Trichloroethene	l	-43.6		228.4	124.4
		g	-8.1		324.8	80.3
C_2HCl_3O	Trichloroethanal	l	-236.2			151.0
		g	-196.6			
C_2HCl_3O	Dichloroacetyl chloride	l	-280.4			
$C_2HCl_3O_2$	Trichloroethanoic acid	cr	-503.3			
C_2HCl_5	Pentachloroethane	l	-187.6			173.8
C_2HF	Fluoroethyne	g			231.7	52.4
C_2HF_3	Trifluoroethene	g	-490.4			
$C_2HF_3O_2$	Trifluoroethanoic acid	l	-1069.9			
		g	-1031.4			
C_2H_2	Ethyne	g	228.2	210.7	200.9	43.9
$C_2H_2Br_2$	*cis*-1,2-Dibromoethene	g			311.3	68.8
$C_2H_2Br_2$	*trans*-1,2-Dibromoethene	g			313.5	70.3
$C_2H_2ClF_3$	2-Chloro-1,1,1-trifluoroethane	g			326.5	89.1
$C_2H_2Cl_2$	1,1-Dichloroethene	l	-23.9	24.1	201.5	111.3
		g	2.6	25.4	289.0	67.1
$C_2H_2Cl_2$	*cis*-1,2-Dichloroethene	l	-26.4		198.4	116.4

TABLE 2.4.1
STANDARD STATE THERMOCHEMICAL PROPERTIES AT 298.15 K (continued)

Molecular formula	Name	State	$\Delta_f H°$/ kJ mol^{-1}	$\Delta_f G°$/ kJ mol^{-1}	$S°$/ J K^{-1} mol^{-1}	C_p/ J K^{-1} mol^{-1}
		g	4.6		289.6	65.1
$C_2H_2Cl_2$	*trans*-1,2-Dichloroethene	l	-23.1	27.3	195.9	116.8
		g	6.2	28.6	290.0	66.7
$C_2H_2Cl_2O$	Chloroacetyl chloride	l	-283.7			
$C_2H_2Cl_4$	1,1,1,2-Tetrachloroethane	g			356.0	102.7
$C_2H_2Cl_4$	1,1,2,2-Tetrachloroethane	l	-195.0		246.9	162.3
		g	-149.2		362.8	100.8
$C_2H_2F_2$	1,1-Difluoroethene	g	-335.0		266.2	60.1
$C_2H_2F_2$	*cis*-1,2-Difluoroethene	g			268.3	58.2
C_2H_2O	Ketene	l	-67.9			
		g	-47.5	-48.3	247.6	51.8
$C_2H_2O_2$	Ethanedial	g	-212.0			
$C_2H_2O_4$	Ethanedioic acid	cr	-821.7		109.8	91.0
		g	-723.7			
$C_2H_2O_4Sr$	Strontium dimethanoate	cr	-1393.3			
C_2H_3Br	Bromoethene	g	79.2	81.8	275.8	55.5
C_2H_3BrO	Acetyl bromide	l	-223.4			
C_2H_3Cl	Chloroethene	cr	-94.1			59.4
		l	14.6			
		g	37.3	53.6	264.0	53.7
$C_2H_3ClF_2$	1-Chloro-1,1-difluoroethane	g			307.2	82.5
C_2H_3ClO	Acetyl chloride	l	-273.8	-208.0	200.8	117.0
		g	-243.5	-205.8	295.1	67.8
$C_2H_3ClO_2$	Chloroethanoic acid	cr	-510.5			
$C_2H_3Cl_2F$	1,1-Dichloro-1-fluoroethane	g			320.2	88.7
$C_2H_3Cl_3$	1,1,1-Trichloroethane	l	-177.4		227.4	144.3
		g	-144.6		323.1	93.3
$C_2H_3Cl_3$	1,1,2-Trichloroethane	l	-191.5		232.6	150.9
		g	-151.2		337.2	89.0
C_2H_3F	Fluoroethene	g	-138.8			
C_2H_3FO	Acetyl fluoride	l	-467.2			
		g	-442.1			
$C_2H_3F_3$	1,1,1-Trifluoroethane	g	-744.6		279.9	78.2
$C_2H_3F_3$	1,1,2-Trifluoroethane	g	-730.9			
C_2H_3I	Iodoethene	g			285.0	57.9
C_2H_3IO	Acetyl iodide	l	-162.5			
$C_2H_3KO_2$	Potassium ethanoate	cr	-723.0			
C_2H_3N	Ethanenitrile	l	31.4	77.2	149.6	91.4
		g	64.3	81.7	245.1	52.2
C_2H_3N	Methyl isocyanide	l	117.2	159.5	159.0	
		g	149.0	165.7	246.9	52.9
C_2H_3NO	Methyl isocyanate	l	-92.0			
$C_2H_3NaO_2$	Sodium ethanoate	cr	-708.8	-607.2	123.0	79.9
C_2H_4	Ethene	g	52.5	68.4	219.6	43.6
$C_2H_4Br_2$	1,1-Dibromoethane	g			327.7	80.8
$C_2H_4Br_2$	1,2-Dibromoethane	l	-79.2		223.3	136.0
C_2H_4ClF	1-Chloro-1-fluoroethane	g	-313.4			
$C_2H_4Cl_2$	1,1-Dichloroethane	l	-158.4	-73.8	211.8	126.3
		g	-127.7	-70.8	305.1	76.2
$C_2H_4Cl_2$	1,2-Dichloroethane	l	-167.4			128.4
		g	-126.9		308.4	78.7
$C_2H_4F_2$	1,1-Difluoroethane	g	-497.0		282.5	67.8
C_2H_4O	Ethanal	l	-191.8	-127.6	160.2	89.0
		g	-166.2	-132.8	263.7	55.3
C_2H_4O	Oxirane	l	-77.8	-11.8	153.9	88.0
		g	-52.6	-13.0	242.5	47.9

TABLE 2.4.1
STANDARD STATE THERMOCHEMICAL PROPERTIES AT 298.15 K (continued)

Molecular formula	Name	State	$\Delta_f H°$/ kJ mol^{-1}	$\Delta_f G°$/ kJ mol^{-1}	$S°$/ J K^{-1} mol^{-1}	C_p/ J K^{-1} mol^{-1}
$C_2H_4O_2$	Ethanoic acid	l	-484.5	-389.9	159.8	123.3
		g	-432.8	-374.5	282.5	66.5
$C_2H_4O_2$	Methyl methanoate	l	-386.1			119.1
		g	-355.5		285.3	64.4
C_2H_4Si	Ethynylsilane	g			269.4	72.6
C_2H_5Br	Bromoethane	l	-90.1	-25.8	198.7	100.8
		g	-61.9	-23.9	286.7	64.5
C_2H_5Cl	Chloroethane	l	-136.5	-59.3	190.8	104.3
		g	-112.2	-60.4	276.0	62.8
C_2H_5F	Fluoroethane	g			264.5	58.6
C_2H_5I	Iodoethane	l	-40.2	14.7	211.7	115.1
		g	-7.7	19.2	306.0	66.9
C_2H_5N	Aziridine	l	91.9			
		g	126.5			
C_2H_5NO	Ethanamide	cr	-317.0		115.0	91.3
$C_2H_5NO_2$	Nitroethane	l	-143.9			134.4
$C_2H_5NO_2$	Aminoethanoic acid	cr	-528.5			
		g	-392.1			
C_2H_6	Ethane	g	-83.8	-31.9	229.6	52.6
C_2H_6Cd	Dimethylcadmium	l	63.6	139.0	201.9	132.0
		g	101.6	146.9	303.0	
C_2H_6Hg	Dimethylmercury	l	59.8	140.3	209.0	
		g	94.4	146.1	306.0	83.3
C_2H_6O	Dimethyl ether	g	-184.1	-112.6	266.4	64.4
C_2H_6O	Ethanol	l	-277.7	-174.8	160.7	112.3
		g	-235.1	-168.5	282.7	65.4
C_2H_6OS	Dimethyl sulfoxide	l	-204.2	-99.9	188.3	153.0
$C_2H_6O_2$	1,2-Ethanediol	l	-455.3		163.2	148.6
		g	-387.5		303.8	82.7
$C_2H_6O_2S$	Dimethyl sulfone	cr	-451.0	-302.4	142.0	
		g	-371.1	-272.7	310.6	100.0
C_2H_6S	2-Thiapropane	l	-65.4		196.4	118.1
		g	-37.5		286.0	74.1
C_2H_6S	Ethanethiol	l	-73.6	-5.5	207.0	117.9
		g	-45.3	-4.8	296.2	72.7
$C_2H_6S_2$	2,3-Dithiabutane	l	-62.6		235.4	146.1
C_2H_6Zn	Dimethylzinc	l	23.4		201.6	129.2
		g	53.0			
C_2H_7N	Dimethylamine	l	-43.9	70.0	182.3	137.7
		g	-18.5	68.5	273.1	70.7
C_2H_7N	Ethylamine	l	-74.1			130.0
		g	-47.5	36.3	283.8	71.5
$C_2H_8N_2$	1,1-Dimethylhydrazine	l	48.9	206.4	198.0	164.1
		g	83.9			
$C_2H_8N_2$	1,2-Ethanediamine	l	-63.0			172.6
$C_2H_8N_2O_4$	Diammonium ethanedioate	cr	-1123.0			226.0
C_2HgO_4	Mercury ethanedioate	cr	-678.2			
C_2I_2	Diiodoethyne	g			313.1	70.3
C_2I_4	Tetraiodoethene	cr	305.0			
$C_2K_2O_4$	Dipotassium ethanedioate	g	-1346.0			
C_2MgO_4	Magnesium ethanedioate	cr	-1269.0			
C_2N_2	Ethanedinitrile	g	306.7		241.9	56.8
		l	285.9			
$C_2Na_2O_4$	Disodium ethanedioate	g	-1318.0			
C_2O_4Pb	Lead ethanedioate	cr	-851.4	-750.1	146.0	105.4
C_3F_8	Octafluoropropane	g	-1783.2			

TABLE 2.4.1
STANDARD STATE THERMOCHEMICAL PROPERTIES AT 298.15 K (continued)

Molecular formula	Name	State	$\Delta_f H°$/ kJ mol^{-1}	$\Delta_f G°$/ kJ mol^{-1}	$S°$/ J K^{-1} mol^{-1}	C_p/ J K^{-1} mol^{-1}
$C_3H_3F_3$	3,3,3-Trifluoropropene	g	-614.2			
C_3H_3N	Propenenitrile	l	147.1			
		g	180.6			
C_3H_3NO	Oxazole	l	-48.0			
		g	-15.5			
C_3H_3NO	Isoxazole	l	42.1			
		g	78.6			
C_3H_4	Propyne	g	184.9			
C_3H_4	Propadiene	g	190.5			
C_3H_4	Cyclopropene	g	277.1			
$C_3H_4Cl_2$	2,3-Dichloropropene	l	-73.3			
$C_3H_4N_2$	Imidazole	cr	58.5			
$C_3H_4O_2$	Propenoic acid	l	-383.8			145.7
$C_3H_4O_2$	2-Oxetanone	l	-329.9		175.3	122.1
C_3H_5Br	3-Bromopropene	l	12.2			
		g	45.2			
C_3H_5Cl	2-Chloropropene	g	-21.0			
C_3H_5ClO	(Chloromethyl)oxirane	l	-148.4			131.6
		g	-107.8			
$C_3H_5Cl_3$	1,2,3-Trichloropropane	l	-230.6			183.6
C_3H_5N	Propanenitrile	l	15.5			119.3
		g	51.5			
$C_3H_5N_3O_9$	1,2,3-Propanetriol trinitrate	l	-370.9			
		g	-270.9			
C_3H_6	Propene	l	1.7			
		g	20.0			
C_3H_6	Cyclopropane	g	53.3			
$C_3H_6Br_2$	1,2-Dibromopropane	g	-71.5			
$C_3H_6Cl_2$	1,2-Dichloropropane	l	-198.8			149.1
		g	-162.8			
$C_3H_6Cl_2$	1,3-Dichloropropane	l	-200.0			
C_3H_6O	2-Propanone	l	-248.1		199.8	126.3
		g	-217.3		297.6	75.0
C_3H_6O	2-Propen-1-ol	l	-171.8			138.9
		g	-124.5			
C_3H_6O	Propanal	l	-215.3			
		g	-185.6		304.5	80.7
C_3H_6O	Methyloxirane	l	-122.6		196.5	120.4
		g	-94.7		286.9	72.6
$C_3H_6O_2$	Methyl ethanoate	l	-445.8			141.9
		g	-411.9		324.4	86.0
$C_3H_6O_2$	Propanoic acid	l	-510.7		191.0	152.8
$C_3H_6O_2$	1,3-Dioxolane	l	-333.5			118.0
		g	-298.0			
$C_3H_6O_3$	1,3,5-Trioxane	cr	-522.5		133.0	111.4
C_3H_6S	Thiacyclobutane	l	24.7		184.9	
		g	60.6			
C_3H_7Br	1-Bromopropane	l	-121.8			
		g	-87.0			
C_3H_7Br	2-Bromopropane	l	-130.5			
		g	-99.4			
C_3H_7Cl	1-Chloropropane	l	-160.6			
		g	-131.9			
C_3H_7Cl	2-Chloropropane	l	-172.1			
		g	-144.9			
C_3H_7F	1-Fluoropropane	g	-285.9			

TABLE 2.4.1
STANDARD STATE THERMOCHEMICAL PROPERTIES AT 298.15 K (continued)

Molecular formula	Name	State	$\Delta_f H°$/ kJ mol^{-1}	$\Delta_f G°$/ kJ mol^{-1}	$S°$/ J K^{-1} mol^{-1}	C_p/ J K^{-1} mol^{-1}
C_3H_7F	2-Fluoropropane	g	-293.5			
C_3H_7I	1-Iodopropane	l	-66.0			
C_3H_7I	2-Iodopropane	l	-74.8			
		g	-40.3			
C_3H_7N	Cyclopropylamine	l	45.8		187.7	147.1
		g	77.0			
C_3H_7NO	*N,N*-Dimethylmethanamide	l	-239.3			150.6
		g	-191.7			
$C_3H_7NO_2$	1-Nitropropane	l	-167.2			
$C_3H_7NO_2$	2-Nitropropane	l	-180.3			170.3
$C_3H_7NO_2$	(*S*)-(+)-Alanine	cr	-604.0			
		g	-465.9			
$C_3H_7NO_2$	(*R*)-(-)-Alanine	cr	-561.2			
$C_3H_7NO_2S$	(*R*)-(+)-Cysteine	cr	-515.5			
$C_3H_7NO_3$	(*S*)-(+)-Serine	cr	-732.7			
C_3H_8	Propane	g	-104.7			
C_3H_8O	1-Propanol	l	-302.6		193.6	143.9
		g	-255.1		322.6	85.6
C_3H_8O	2-Propanol	l	-318.1		181.1	156.5
		g	-272.8		309.2	89.3
C_3H_8O	Ethyl methyl ether	g	-216.4		309.2	93.3
$C_3H_8O_2$	1,2-Propanediol	l	-485.7			190.8
$C_3H_8O_2$	1,3-Propanediol	l	-464.9			
$C_3H_8O_2$	2,4-Dioxapentane	l	-377.7		244.0	162.0
		g	-348.4			
$C_3H_8O_3$	1,2,3-Propanetriol	l	-668.5		206.3	218.9
		g	-582.7			
C_3H_8S	2-Thiabutane	l	-91.6		239.1	144.6
		g	-59.6			
C_3H_8S	1-Propanethiol	l	-99.9		242.5	144.6
		g	-67.9			
C_3H_8S	2-Propanethiol	l	-105.9		233.5	145.3
		g	-76.2			
C_3H_9Al	Trimethylaluminum	l	-136.4	-9.9	209.4	155.6
		g	-74.1			
C_3H_9B	Trimethylborane	l	-143.1	-32.1	238.9	
		g	-124.3	-35.9	314.7	88.5
C_3H_9ClSi	Chlorotrimethylsilane	l	-382.8	-246.4	278.2	
		g	-352.8	-243.5	369.1	
C_3H_9N	Propylamine	l	-101.5			164.1
		g	-70.1	39.9	325.4	91.2
C_3H_9N	Isopropylamine	l	-112.3		218.3	163.8
		g	-83.7	32.2	312.2	97.5
C_3H_9N	Trimethylamine	l	-45.7		208.5	137.9
		g	-23.7		287.1	91.8
$C_3H_{10}Si$	Trimethylsilane	g			331.0	117.9
$C_3H_{12}BN$	Trimethylamine-borane	cr	-142.5	70.7	187.0	
$C_3H_{12}BN$	Ammonia-trimethylborane	cr	-284.1	-79.3	218.0	
C_4F_8	Octafluorocyclobutane	g	-1542.6			
$C_4H_2O_3$	1-Oxa-3-cyclopenten-2,5-dione	cr	-469.8			
$C_4H_4N_2$	Pyridazine	l	224.8			
		g	278.3			
$C_4H_4N_2$	Pyrimidine	l	145.9			
$C_4H_4N_2$	Butanedinitrile	l	139.7		191.6	145.6
		g	209.7			
$C_4H_4N_2O_2$	Uracil	cr	-429.4			120.5

TABLE 2.4.1
STANDARD STATE THERMOCHEMICAL PROPERTIES AT 298.15 K (continued)

Molecular formula	Name	State	$\Delta_f H°$/ kJ mol⁻¹	$\Delta_f G°$/ kJ mol⁻¹	$S°$/ J K⁻¹ mol⁻¹	C_p/ J K⁻¹ mol⁻¹
		g	-302.9			
C_4H_4O	Furan	l	-62.3		177.0	115.3
		g	-34.9		267.2	65.4
$C_4H_4O_2$	4-Methyleneoxetan-2-one	l	-233.1			
$C_4H_4O_3$	Oxolan-2,5-dione	cr	-607.8			
$C_4H_4O_4$	Fumaric acid	cr	-811.7		168.0	142.0
		g	-675.8			
$C_4H_4O_4$	*cis*-Butanedioic acid	cr	-789.4		160.8	137.0
		g	-679.4			
C_4H_4S	Thiophene	l	80.2		181.2	123.8
		g	114.9			
C_4H_5N	Cyclopropanecarbonitrile	l	140.8			
		g	181.8			
C_4H_5N	Pyrrole	l	63.1		156.4	127.7
C_4H_5NS	4-Methylthiazole	l	67.9			
$C_4H_5N_3O$	Cytosine	cr	-221.3			132.6
C_4H_6	1,2-Butadiene	l	139.0			
		g	162.3			
C_4H_6	1,3-Butadiene	l	87.9		199.0	123.6
		g	110.0			
C_4H_6	1-Butyne	l	141.9			
		g	165.2			
C_4H_6	2-Butyne	l	119.1			
		g	145.7			
C_4H_6	Cyclobutene	g	156.7			
C_4H_6O	*trans*-2-Butenal	l	-138.7			
		g	-100.6			
C_4H_6O	Divinyl ether	l	-39.8			
		g	-13.6			
$C_4H_6O_2$	Methyl propenoate	l	-362.2		239.5	158.8
		g	-333.0			
$C_4H_6O_2$	Vinyl ethanoate	l	-280.1			
		g	-314.9			
$C_4H_6O_3$	Ethanoic anhydride	l	-624.4			
		g	-572.5			
$C_4H_6O_4$	Butanedioic acid	cr	-940.5		167.3	153.1
		g	-823.0			
$C_4H_6O_4$	Dimethyl oxalate	cr	-756.3			
		g	-708.9			
C_4H_6S	2,3-Dihydrothiophene	l	52.9			
C_4H_6S	2,5-Dihydrothiophene	l	47.0			
C_4H_7N	Butanenitrile	l	-5.8			
C_4H_7N	2-Methylpropanenitrile	l	-13.8			
C_4H_7NO	2-Pyrrolidinone	l	-286.2			
$C_4H_7NO_4$	(*S*)-(+)-Aspartic acid	cr	-973.3			
C_4H_8	1-Butene	l	-20.5		227.0	118.0
		g	0.1			
C_4H_8	*cis*-2-Butene	l	-29.7		219.9	127.0
		g	-7.1			
C_4H_8	*trans*-2-Butene	l	-33.0			
		g	-11.4			
C_4H_8	2-Methylpropene	l	-37.5			
		g	-16.9			
C_4H_8	Cyclobutane	l	3.7			
		g	28.4			
C_4H_8	Methylcyclopropane	l	1.7			

TABLE 2.4.1
STANDARD STATE THERMOCHEMICAL PROPERTIES AT 298.15 K (continued)

Molecular formula	Name	State	$\Delta_f H°$/ kJ mol^{-1}	$\Delta_f G°$/ kJ mol^{-1}	$S°$/ J K^{-1} mol^{-1}	C_p/ J K^{-1} mol^{-1}
$C_4H_8Br_2$	1,4-Dibromobutane	l	-140.1			
		g	-87.0			
$C_4H_8N_2O_3$	*L*-Asparagine	cr	-789.4			
C_4H_8O	Ethyloxirane	l	-168.9		230.9	147.0
C_4H_8O	Butanal	l	-239.2		246.6	163.7
		g	-204.8		343.7	103.4
C_4H_8O	2-Methylpropanal	l	-247.4			
		g	-215.8			
C_4H_8O	2-Butanone	l	-273.3		239.1	158.7
		g	-238.7		339.9	101.7
C_4H_8O	Oxolane	l	-216.2		204.3	124.0
		g	-184.2		302.4	76.3
C_4H_8O	Ethyl vinyl ether	l	-167.4			
		g	-140.8			
$C_4H_8O_2$	1,3-Dioxane	l	-379.7			143.9
$C_4H_8O_2$	1,4-Dioxane	l	-353.9		270.2	152.1
		g	-315.8			
$C_4H_8O_2$	Ethyl ethanoate	l	-479.3		257.7	170.7
		g	-444.1			
$C_4H_8O_2$	Propyl methanoate	l	-500.3			
$C_4H_8O_2$	Butanoic acid	l	-533.8		222.2	178.6
C_4H_8S	Thiacyclopentane	l	-72.9			
C_4H_9Br	1-Bromobutane	l	-143.8			
C_4H_9Br	2-Bromobutane	l	-154.8			
		g	-120.3			
C_4H_9Br	2-Bromo-2-methylpropane	l	-163.8			
		g	-132.4			
C_4H_9Cl	1-Chlorobutane	l	-188.1			
		g	-154.6			
C_4H_9Cl	1-Chloro-2-methylpropane	l	-191.1			
		g	-159.4			
C_4H_9Cl	2-Chloro-2-methylpropane	l	-211.2			
		g	-182.2			
C_4H_9I	2-Iodo-2-methylpropane	l	-107.4			
C_4H_9N	Pyrrolidine	l	-41.0		204.1	156.6
		g	-3.4			
C_4H_9NO	*N,N*-Dimethylethanamide	l	-278.3			175.6
$C_4H_9NO_3$	*L*-Threonine	cr	-807.2			
C_4H_{10}	Butane	l	-146.6			140.9
		g	-125.6			
C_4H_{10}	2-Methylpropane	l	-153.5			
		g	-134.2			
$C_4H_{10}Hg$	Diethyl mercury	l	30.1			182.8
		g	75.3			
$C_4H_{10}O$	1-Butanol	l	-327.3		225.8	177.2
$C_4H_{10}O$	2-Butanol	l	-342.6		214.9	196.9
		g	-292.9		359.5	112.7
$C_4H_{10}O$	2-Methyl-2-propanol	l	-359.2		193.3	218.6
		g	-312.5		326.7	113.6
$C_4H_{10}O$	2-Methyl-1-propanol	l	-334.7		214.7	181.5
$C_4H_{10}O$	Diethyl ether	l	-279.3		172.4	175.6
		g	-252.1		342.7	119.5
$C_4H_{10}O$	Methyl propyl ether	l	-266.0		262.9	165.4
		g	-238.2			
$C_4H_{10}O$	Isopropyl methyl ether	l	-278.7		253.8	161.9
		g	-252.0			

TABLE 2.4.1
STANDARD STATE THERMOCHEMICAL PROPERTIES AT 298.15 K (continued)

Molecular formula	Name	State	$\Delta_f H°$/ kJ mol^{-1}	$\Delta_f G°$/ kJ mol^{-1}	$S°$/ J K^{-1} mol^{-1}	C_p/ J K^{-1} mol^{-1}
$C_4H_{10}O_2$	(±)-1,3-Butanediol	l	-501.0			
		g	-433.2			
$C_4H_{10}O_2$	1,4-Butanediol	l	-503.3		223.4	200.1
		g	-426.7			
$C_4H_{10}O_3$	3-Oxa-1,5-pentanediol	l	-628.5			244.8
		g	-571.2			
$C_4H_{10}S$	1-Butanethiol	l	-124.7			171.2
		g	-88.1			
$C_4H_{10}S$	2-Butanethiol	l	-131.0			
		g	-96.9			
$C_4H_{10}S$	2-Methyl-1-propanethiol	l	-132.0			
		g	-97.3			
$C_4H_{10}S$	2-Methyl-2-propanethiol	l	-140.5			
		g	-109.6			
$C_4H_{10}S$	3-Thiapentane	l	-119.4		269.3	171.4
		g	-83.6		368.1	117.0
$C_4H_{10}S$	Methyl propyl sulfide	l	-118.5		272.5	171.6
		g	-82.3			
$C_4H_{10}S$	Isopropyl methyl sulfide	l	-124.7		263.1	172.4
		g	-90.5			
$C_4H_{10}S_2$	3,4-Dithiahexane	l	-120.1		269.3	171.4
$C_4H_{11}N$	Butylamine	l	-127.7			179.2
		g	-92.0			
$C_4H_{11}N$	(2-Methylpropyl)amine	l	-132.6			183.2
		g	-98.7			
$C_4H_{11}N$	2-Aminobutane	l	-137.5			
		g	-104.9			
$C_4H_{11}N$	(1,1-Dimethylethyl)amine	l	-150.6			192.1
		g	-120.9			
$C_4H_{11}N$	Diethylamine	l	-103.7			169.2
		g	-72.5			
$C_4H_{12}Pb$	Tetramethyllead	l	97.9			
		g	135.9			
$C_4H_{12}Si$	Tetramethylsilane	l	-264.0	-100.0	277.3	204.1
		g	-239.1	-99.9	359.0	143.9
$C_4H_{12}Sn$	Tetramethylstannane	l	-52.3			
		g	-18.8			
C_4NiO_4	Nickel tetracarbonyl	l	-633.0	-588.2	313.4	204.6
		g	-602.9	-587.2	410.6	145.2
C_5FeO_5	Iron pentacarbonyl	l	-774.0	-705.3	338.1	240.6
$C_5H_4N_4O$	Hypoxanthine	cr	-110.8		145.6	134.5
$C_5H_4N_4O_2$	Xanthine	cr	-379.6		161.1	151.3
$C_5H_4N_4O_3$	Uric acid	cr	-618.8		173.2	166.1
$C_5H_4O_2$	2-Furaldehyde	l	-201.6			163.2
$C_5H_5N_5$	Adenine	cr	96.0			147.0
		g	204.8			
C_5H_5N	Pyridine	l	100.2			132.7
		g	140.4			
$C_5H_5N_5O$	Guanine	cr	-183.9			
C_5H_6	*cis*-3-Penten-1-yne	g	81.4			
C_5H_6	*trans*-3-Penten-1-yne	l	228.2			
C_5H_6	1,3-Cyclopentadiene	l	105.9			
		g	134.3			
$C_5H_6N_2O_2$	Thymine	cr	-462.8			150.8
		g	-328.7			
$C_5H_6O_2$	2-Furylmethanol	l	-276.2			204.0

TABLE 2.4.1
STANDARD STATE THERMOCHEMICAL PROPERTIES AT 298.15 K (continued)

Molecular formula	Name	State	$\Delta_f H°$/ kJ mol^{-1}	$\Delta_f G°$/ kJ mol^{-1}	$S°$/ J K^{-1} mol^{-1}	C_p/ J K^{-1} mol^{-1}
C_5H_6S	2-Methylthiophene	l	44.6		218.5	149.8
C_5H_6S	3-Methylthiophene	l	43.1			
C_5H_7N	Cyclobutanecarbonitrile	l	103.0			
C_5H_8	2-Methyl-1,3-butadiene	l	48.2		229.3	152.6
		g	75.5			
C_5H_8	*cis*-1,3-Pentadiene	g	81.4			
C_5H_8	*trans*-1,3-Pentadiene	g	76.1			
C_5H_8	1,4-Pentadiene	g	105.6			
C_5H_8	Cyclopentene	l	4.4		201.2	122.4
		g	33.9			
C_5H_8	Spiropentane	l	157.7		193.7	134.5
		g	185.2			
C_5H_8O	Cyclopentanone	l	-235.7			
$C_5H_8O_2$	2,4-Pentanedione	l	-423.8			
$C_5H_8O_4$	Pentanedioic acid	cr	-960.0			
C_5H_9N	Pentanenitrile	l	-33.1			
C_5H_9N	2,2-Dimethylpropanenitrile	l	-39.8		232.0	179.4
C_5H_9NO	1-Methyl-2-pyrrolidinone	l	-262.2			307.8
$C_5H_9NO_2$	(*S*)-(-)-2-Pyrrolidinecarboxylic acid	cr	-512.2			
		g	-366.2			
$C_5H_9NO_4$	(*S*)-(+)-2-Aminopentanedioic acid	cr	-1009.7			
C_5H_{10}	1-Pentene	l	-46.9		262.6	154.0
		g	-21.3			
C_5H_{10}	*cis*-2-Pentene	l	-53.7		258.6	151.7
		g	-27.6			
C_5H_{10}	*trans*-2-Pentene	l	-58.2		256.5	157.0
		g	-31.9			
C_5H_{10}	2-Methyl-1-butene	l	-61.0		254.0	157.2
		g	-35.3			
C_5H_{10}	2-Methyl-2-butene	l	-68.6		251.0	152.8
		g	-41.8			
C_5H_{10}	3-Methyl-1-butene	l	-51.5		253.3	156.1
		g	-27.6			
C_5H_{10}	Cyclopentane	l	-105.1		204.5	128.8
		g	-76.4			
$C_5H_{10}N_2O_3$	*L*-Glutamine	cr	-826.4			
$C_5H_{10}O$	Cyclopentanol	l	-300.1		206.3	184.1
$C_5H_{10}O$	2-Pentanone	l	-297.3			184.1
$C_5H_{10}O$	3-Pentanone	l	-296.5		266.0	190.9
$C_5H_{10}O$	3-Methyl-2-butanone	l	-299.4		268.5	179.9
		g	-262.5			
$C_5H_{10}O$	3,3-Dimethyloxetane	l	-182.2			
		g	-148.2			
$C_5H_{10}O$	Oxane	l	-258.3			
		g	-223.4			
$C_5H_{10}O$	Pentanal	l	-267.3			
		g	-288.5			
$C_5H_{10}O_2$	Isopropyl ethanoate	l	-518.9			199.4
		g	-481.7			
$C_5H_{10}O_2$	Ethyl propanoate	l	-502.7			
		g	-463.6			
$C_5H_{10}O_2$	Pentanoic acid	l	-559.4		259.8	210.3
$C_5H_{10}O_2$	3-Methylbutanoic acid	l	-561.6			
$C_5H_{10}O_2$	2-Methylbutanoic acid	l	-554.5			
$C_5H_{10}O_2$	2-Tetrahydrofurylmethanol	l	-435.7			
		g	-369.2			

TABLE 2.4.1
STANDARD STATE THERMOCHEMICAL PROPERTIES AT 298.15 K (continued)

Molecular formula	Name	State	$\Delta_f H°$/ kJ mol^{-1}	$\Delta_f G°$/ kJ mol^{-1}	$S°$/ J K^{-1} mol^{-1}	C_p/ J K^{-1} mol^{-1}
$C_5H_{10}O_3$	Diethyl carbonate	l	-681.5			
		g	-637.9			
$C_5H_{10}S$	Cyclopentanethiol	l	-89.5		256.9	165.2
$C_5H_{10}S$	Thiacyclohexane	l	-106.3		218.2	163.3
$C_5H_{11}Br$	1-Bromopentane	l	-170.2			
$C_5H_{11}Cl$	1-Chloropentane	l	-213.2			
$C_5H_{11}Cl$	1-Chloro-3-methylbutane	l	-216.0			
		g	-179.2			
$C_5H_{11}N$	Piperidine	l	-86.4		210.0	179.9
$C_5H_{11}N$	Cyclopentylamine	l	-95.1		241.0	181.2
		g	-54.9			
$C_5H_{11}NO_2$	*L*-Valine	cr	-617.9			
		g	-455.1			
C_5H_{12}	Pentane	l	-173.5			167.2
		g	-146.9			
C_5H_{12}	2-Methylbutane	l	-178.5		260.4	164.8
		g	-153.7			
C_5H_{12}	2,2-Dimethylpropane	l	-190.2			
		g	-168.1			
$C_5H_{12}O$	Butyl methyl ether	l	-290.6		295.3	192.7
		g	-258.1			
$C_5H_{12}O$	*tert*-Butyl methyl ether	l	-313.6		265.3	187.5
		g	-283.5			
$C_5H_{12}O$	Ethyl propyl ether	l	-303.6		295.0	197.2
		g	-272.2			
$C_5H_{12}O$	1-Pentanol	l	-351.6			208.1
$C_5H_{12}O$	2-Pentanol	l	-365.2			
$C_5H_{12}O$	3-Pentanol	l	-368.9			239.7
$C_5H_{12}O$	2-Methyl-1-butanol	l	-356.6			
		g	-302.0			
$C_5H_{12}O$	2-Methyl-2-butanol	l	-379.5			247.1
$C_5H_{12}O$	3-Methyl-1-butanol	l	-356.4			
$C_5H_{12}O$	3-Methyl-2-butanol	l	-366.6			
		g	-315.2			
$C_5H_{12}O$	2,2-Dimethyl-1-propanol	l	-399.4			
$C_5H_{12}O_2$	1,5-Pentanediol	l	-531.5			
		g	-449.1			
$C_5H_{12}O_2$	3,5-Dioxaheptane	l	-450.4			
		g	-414.8			
$C_5H_{12}O_4$	Pentaerythritol	cr	-920.6			
		g	-776.7			
$C_5H_{12}S$	Butyl methyl sulfide	l	-142.9		307.5	200.9
$C_5H_{12}S$	*tert*-Butyl methyl sulfide	l	-157.1		276.1	199.9
		g	-121.3			
$C_5H_{12}S$	Ethyl propyl sulfide	l	-144.8		309.5	198.4
$C_5H_{12}S$	Ethyl isopropyl sulfide	l	-156.1			
$C_5H_{12}S$	1-Pentanethiol	l	-151.3			
$C_5H_{12}S$	2-Methyl-1-butanethiol	l	-154.4			
$C_5H_{12}S$	3-Methyl-1-butanethiol	l	-154.3			
$C_5H_{12}S$	2-Methyl-2-butanethiol	l	-162.8		290.1	198.1
		g	-127.1			
C_6ClF_5	Chloropentafluorobenzene	cr	-858.7			
C_6Cl_6	Hexachlorobenzene	cr	-127.6		260.2	201.2
		g	-35.5			
C_6F_6	Hexafluorobenzene	l	-991.3		280.8	221.6
		g	-955.4			

TABLE 2.4.1
STANDARD STATE THERMOCHEMICAL PROPERTIES AT 298.15 K (continued)

Molecular formula	Name	State	$\Delta_f H°$/ kJ mol^{-1}	$\Delta_f G°$/ kJ mol^{-1}	$S°$/ J K^{-1} mol^{-1}	C_p/ J K^{-1} mol^{-1}
C_6F_{10}	Decafluorocyclohexene	l	-1963.5			
		g	-1932.7			
C_6F_{12}	Dodecafluorocyclohexane	l	-2406.3			
		g	-2370.4			
C_6HF_5	Pentafluorobenzene	cr	-852.7			
		l	-841.8			
		g	-806.5			
C_6HF_5O	Pentafluorophenol	cr	-1024.1			
		l	-1007.7			
$C_6H_2F_4$	1,2,4,5-Tetrafluorobenzene	l	-683.7			
$C_6H_4Cl_2$	1,2-Dichlorobenzene	l	-17.5			162.4
		g	30.2			
$C_6H_4Cl_2$	1,3-Dichlorobenzene	l	-20.7			
		g	25.7			
$C_6H_4Cl_2$	1,4-Dichlorobenzene	cr	-42.3		175.4	147.8
		g	22.5			
$C_6H_4F_2$	1,2-Difluorobenzene	l	-330.0		222.6	159.0
		g	-293.8			
$C_6H_4F_2$	1,3-Difluorobenzene	l	-343.9		223.8	159.1
		g	-309.2			
$C_6H_4F_2$	1,4-Difluorobenzene	l	-342.3			157.5
		g	-306.7			
$C_6H_4O_2$	1,4-Benzoquinone	cr	-185.7			129.0
		g	-122.9			
C_6H_5Br	Bromobenzene	l	60.9		219.2	154.3
C_6H_5Cl	Chlorobenzene	l	11.0			150.1
C_6H_5ClO	3-Chlorophenol	cr	-206.4			
		l	-189.3			
C_6H_5ClO	4-Chlorophenol	cr	-197.7			
		l	-181.3			
C_6H_5F	Fluorobenzene	l	-150.6		205.9	146.4
C_6H_5I	Iodobenzene	l	117.2		205.4	158.7
$C_6H_5NO_2$	Nitrobenzene	l	12.5			185.8
		g	67.5			
C_6H_6	Benzene	l	49.0			136.3
		g	82.6			
$C_6H_6N_2O_2$	2-Nitroaniline	cr	-26.1			166.0
		l	-9.4			
		g	63.8			
$C_6H_6N_2O_2$	3-Nitroaniline	cr	-38.3			158.8
		l	-14.4			
		g	58.4			
$C_6H_6N_2O_2$	4-Nitroaniline	cr	-42.0			167.0
		l	-20.7			
		g	58.8			
C_6H_6O	Phenol	cr	-165.1		144.0	127.4
		g	-96.4			
$C_6H_6O_2$	1,4-Benzenediol	cr	-364.5			136.0
		g	-265.3			
C_6H_6S	Benzenethiol	l	63.7		222.8	173.2
C_6H_7N	Aniline	l	31.3			191.9
		g	87.5	-7.0	317.9	107.9
C_6H_7N	2-Methylpyridine	l	56.7			158.6
C_6H_7N	3-Methylpyridine	l	61.9		216.3	158.7
C_6H_7N	4-Methylpyridine	l	59.2		209.1	159.0
C_6H_7N	1-Cyclopentenecarbonitrile	l	111.5			

TABLE 2.4.1
STANDARD STATE THERMOCHEMICAL PROPERTIES AT 298.15 K (continued)

Molecular formula	Name	State	$\Delta_f H°$/ kJ mol^{-1}	$\Delta_f G°$/ kJ mol^{-1}	$S°$/ J K^{-1} mol^{-1}	C_p/ J K^{-1} mol^{-1}
		g	156.5			
$C_6H_8N_2$	Hexanedinitrile	l	85.1			128.7
$C_6H_8N_2$	Phenylhydrazine	l	141.0			217.0
$C_6H_8N_2$	1,2-Phenylenediamine	cr	-0.3			
$C_6H_8N_2$	1,3-Phenylenediamine	cr	-7.8		154.5	159.6
$C_6H_8N_2$	1,4-Phenylenediamine	cr	3.1			
C_6H_9N	Cyclopentanecarbonitrile	l	0.7			
		g	43.0			
$C_6H_9NO_3$	Tris(acetyl)amine	l	-610.5			
		g	-550.1			
$C_6H_9N_3O_2$	(*S*)-(+)-2-Amino-3-(4-imidazolyl)-propanoic acid	cr	-466.7			
C_6H_{10}	Cyclohexene	l	-38.5		214.6	148.3
		g	-5.0			
C_6H_{10}	1,5-Hexadiene	l	54.1			
		g	84.1			
C_6H_{10}	3,3-Dimethyl-1-butyne	l	78.4			
$C_6H_{10}O$	Cyclohexanone	l	-271.2			182.2
$C_6H_{10}O_2$	Methyl cyclobutanecarboxylate	l	-395.0			
$C_6H_{10}O_3$	Propanoic anhydride	l	-679.1			
		g	-626.5			
$C_6H_{10}O_4$	Hexanedioic acid	cr	-994.3			
$C_6H_{10}O_4$	Diethyl ethanedioate	l	-805.5			
		g	-742.0			
$C_6H_{11}Cl$	Chlorocyclohexane	l	-207.2			
		g	-163.7			
$C_6H_{11}NO$	Hexahydro-2-azepinone	cr	-329.4			156.8
		g	-246.2			
C_6H_{12}	Cyclohexane	l	-156.4			154.9
		g	-123.4			
C_6H_{12}	Methylcyclopentane	l	-137.9			
		g	-106.2			
C_6H_{12}	Ethylcyclobutane	l	-59.0			
		g	-26.3			
C_6H_{12}	1-Hexene	l	-74.2		295.2	183.3
		g	-43.5			
C_6H_{12}	*cis*-2-Hexene	l	-83.9			
		g	-52.3			
C_6H_{12}	*trans*-2-Hexene	l	-85.5			
		g	-53.9			
C_6H_{12}	2-Methyl-1-pentene	l	-90.0			
		g	-59.4			
C_6H_{12}	2-Methyl-2-pentene	l	-98.5			
		g	-66.9			
C_6H_{12}	4-Methyl-1-pentene	l	-80.0			
		g	-51.3			
C_6H_{12}	*cis*-4-Methyl-2-pentene	l	-87.0			
		g	-57.5			
C_6H_{12}	*trans*-4-Methyl-2-pentene	l	-91.5			
		g	-61.5			
C_6H_{12}	2,3-Dimethyl-1-butene	l	-93.3			
		g	-62.6			
C_6H_{12}	2,3-Dimethyl-2-butene	l	-101.5		270.2	174.7
		g	-68.2			
C_6H_{12}	2-Ethyl-1-butene	l	-87.1			
		g	-56.0			

TABLE 2.4.1
STANDARD STATE THERMOCHEMICAL PROPERTIES AT 298.15 K (continued)

Molecular formula	Name	State	$\Delta_f H°$/ kJ mol^{-1}	$\Delta_f G°$/ kJ mol^{-1}	$S°$/ J K^{-1} mol^{-1}	C_p/ J K^{-1} mol^{-1}
$C_6H_{12}O$	Cyclohexanol	l	-348.2			208.2
$C_6H_{12}O$	2-Hexanone	l	-322.0			213.3
$C_6H_{12}O$	3-Hexanone	l	-320.2		305.3	216.9
$C_6H_{12}O$	2-Methyl-3-pentanone	l	-325.9			
$C_6H_{12}O$	3,3-Dimethyl-2-butanone	l	-328.6			
$C_6H_{12}O_2$	Butyl ethanoate	l	-529.2			227.8
$C_6H_{12}O_2$	Methyl pentanoate	l	-514.2			229.3
$C_6H_{12}O_2$	Methyl 2,2-dimethylpropanoate	l	-530.0			257.9
$C_6H_{12}O_2$	Hexanoic acid	l	-583.8			
$C_6H_{12}S$	Cyclohexanethiol	l	-140.7		255.6	192.6
$C_6H_{13}Br$	1-Bromohexane	l	-194.2		453.0	203.5
		g	-148.1			
$C_6H_{13}N$	Cyclohexylamine	l	-147.7			
$C_6H_{13}NO_2$	*L*-Leucine	cr	-637.4			200.1
		g	-486.8			
$C_6H_{13}NO_2$	*D*-Leucine	cr	-637.3			
$C_6H_{13}NO_2$	*L*-Isoleucine	cr	-637.9			
C_6H_{14}	Hexane	l	-198.7			195.6
		g	-167.1			
C_6H_{14}	2-Methylpentane	l	-204.6		290.6	193.7
		g	-174.8			
C_6H_{14}	3-Methylpentane	l	-202.4		292.5	190.7
		g	-172.1			
C_6H_{14}	2,2-Dimethylbutane	l	-213.8		272.5	191.9
		g	-186.1			
C_6H_{14}	2,3-Dimethylbutane	l	-207.4		287.8	189.7
		g	-178.3			
$C_6H_{14}N_2$	Azopropane	l	11.5			
		g	51.5			
$C_6H_{14}N_2O_2$	*DL*-2,6-Diaminohexanoic acid	cr	-678.7			
$C_6H_{14}N_4O_2$	(*R*)-(-)-6-Aza-2,7-diamino-7-iminoheptanoic acid	cr	-623.5		250.6	232.0
$C_6H_{14}O$	Dipropyl ether	l	-328.8		323.9	221.6
		g	-292.9			
$C_6H_{14}O$	Diisopropyl ether	l	-351.5			216.8
		g	-319.2			
$C_6H_{14}O$	1-Hexanol	l	-377.5		287.4	240.4
$C_6H_{14}O$	2-Hexanol	l	-392.0			
$C_6H_{14}O$	4-Methyl-2-pentanol	l	-394.7			273.0
$C_6H_{14}O_2$	1,1-Diethoxyethane	l	-491.4			
$C_6H_{14}O_2$	3,6-Dioxaoctane	l	-451.4			259.4
$C_6H_{14}O_2$	1,6-Hexanediol	cr	-569.9			
		l	-544.4			
		g	-461.2			
$C_6H_{14}O_3$	2-Ethyl-2-(hydroxymethyl)-1,3-propanediol	cr	-750.9			
$C_6H_{14}O_4$	3,6-Dioxa-1,8-octanediol	l	-804.2			
		g	-725.0			
$C_6H_{14}S$	2-Thiaheptane	l	-167.1			
$C_6H_{14}S$	3-Thiaheptane	l	-172.3			
$C_6H_{14}S$	2,4-Dimethyl-3-thiapentane	l	-181.6		313.0	232.0
$C_6H_{15}B$	Triethylborane	l	-194.6	9.4	336.7	241.2
		g	-157.7	16.1	437.8	
$C_6H_{15}N$	Triethylamine	l	-127.7			219.9
		g	-92.8			
$C_6H_{15}N$	Dipropylamine	l	-156.1			

TABLE 2.4.1
STANDARD STATE THERMOCHEMICAL PROPERTIES AT 298.15 K (continued)

Molecular formula	Name	State	$\Delta_f H°$/ kJ mol^{-1}	$\Delta_f G°$/ kJ mol^{-1}	$S°$/ J K^{-1} mol^{-1}	C_p/ J K^{-1} mol^{-1}
$C_6H_{15}N$	Diisopropylamine	l	-178.5			
		g	-144.0			
$C_6H_{18}OSi_2$	Hexamethyldisiloxane	l	-815.0	-541.5	433.8	311.4
		g	-777.7	-534.5	535.0	238.5
C_6MoO_6	Molybdenum hexacarbonyl	cr	-982.8	-877.7	325.9	242.3
		g	-912.1	-856.0	490.0	205.0
C_7F_8	Octafluorotoluene	l	-1311.1		355.5	262.3
C_7F_{14}	Tetradecafluoromethylcyclohexane	l	-2931.1			353.1
		g	-2897.2			
C_7F_{16}	Hexadecafluoroheptane	l	-3420.0		561.8	419.0
		g	-3383.6			
$C_7H_3F_5$	2,3,4,5,6-Pentafluorotoluene	l	-883.8		306.4	225.8
		g	-842.9			
$C_7H_4Cl_2O_2$	3-Chlorobenzoyl chloride	l	-189.7			
C_7H_5ClO	Benzoyl chloride	l	-158.0			
		g	-103.2			
	4-Chlorobenzoic acid	cr	-428.9			163.2
		g	-341.0			
C_7H_5N	Benzonitrile	l	163.2		209.1	165.2
		g	215.7			
C_7H_6O	Benzaldehyde	l	-87.0		221.2	172.0
		g	-36.7			
$C_7H_6O_2$	Benzoic acid	cr	-385.2		167.6	146.8
		g	-294.1			
$C_7H_6O_3$	2-Hydroxybenzoic acid	cr	-589.9			
		g	-494.8			
C_7H_7Cl	Benzyl chloride	l	-32.6			
		g	18.9			
C_7H_7NO	Benzamide	cr	-202.6			
$C_7H_7NO_2$	2-Nitrotoluene	l	-9.7			
$C_7H_7NO_2$	3-Nitrotoluene	l	-31.5			
$C_7H_7NO_2$	4-Nitrotoluene	cr	-48.1			172.3
		g	31.0			
C_7H_8	Toluene	l	12.4			157.3
		g	50.4			
C_7H_8O	2-Methylphenol	cr	-204.6		165.4	154.6
		g	-128.6			
C_7H_8O	3-Methylphenol	l	-194.0		212.6	224.9
		g	-132.3			
C_7H_8O	4-Methylphenol	cr	-199.3		167.3	150.2
		g	-125.4			
C_7H_8O	Phenylmethanol	l	-160.7		216.7	217.9
		g	-100.4			
C_7H_8O	Anisole	l	-114.8			
		g	-67.9			
C_7H_9N	2-Methylaniline	l	-6.3			
		g	56.4	167.6	351.0	130.2
C_7H_9N	3-Methylaniline	l	-8.1			
		g	54.6	165.4	352.5	125.5
C_7H_9N	4-Methylaniline	cr	-23.5			
		g	55.3	167.7	347.0	126.2
C_7H_9N	1-Cyclohexenecarbonitrile	l	48.1			
		g	101.6			
C_7H_9N	2,3-Dimethylpyridine	l	19.4		243.7	189.5
		g	68.3			
C_7H_9N	2,4-Dimethylpyridine	l	16.2		248.5	184.8

TABLE 2.4.1
STANDARD STATE THERMOCHEMICAL PROPERTIES AT 298.15 K (continued)

Molecular formula	Name	State	$\Delta_f H°$/ kJ mol^{-1}	$\Delta_f G°$/ kJ mol^{-1}	$S°$/ J K^{-1} mol^{-1}	C_p/ J K^{-1} mol^{-1}
		g	63.9			
C_7H_9N	2,5-Dimethylpyridine	l	18.7		248.8	184.7
		g	66.5			
C_7H_9N	2,6-Dimethylpyridine	l	12.7		244.2	185.2
		g	58.7			
C_7H_9N	3,4-Dimethylpyridine	l	18.3		240.7	191.8
		g	70.1			
C_7H_9N	3,5-Dimethylpyridine	l	22.5		241.7	184.5
		g	72.8			
$C_7H_{11}N$	Cyclohexanecarbonitrile	l	-47.2			
		g	4.8			
C_7H_{12}	1-Methylbicyclo[3.1.0]hexane	l	-33.2			
		g	1.5			
C_7H_{14}	Cycloheptane	l	-156.6			
		g	-118.1			
C_7H_{14}	Methylcyclohexane	l	-190.1			184.8
		g	-154.7			
C_7H_{14}	Ethylcyclopentane	l	-163.4		279.9	
		g	-126.9			
C_7H_{14}	*cis*-1,3-Dimethylcyclopentane	l	-170.1			
		g	-135.9			
C_7H_{14}	*cis*-1,2-Dimethylcyclopentane	l	-165.3		269.2	
		g	-129.5			
C_7H_{14}	*trans*-1,2-Dimethylcyclopentane	l	-171.2			
		g	-136.6			
C_7H_{14}	1-Heptene	l	-97.9		327.6	211.8
		g	-62.3			
C_7H_{14}	*cis*-2-Heptene	l	-105.1			
C_7H_{14}	*cis*-3-Heptene	l	-104.3			
$C_7H_{14}O$	2,4-Dimethyl-3-pentanone	l	-352.9		318.0	233.7
		g	-311.3			
$C_7H_{14}O$	2,2-Dimethyl-3-pentanone	l	-356.1			
		g	-313.7			
$C_7H_{14}O$	Heptanal	l	-311.5		335.4	230.1
		g	-263.8			
$C_7H_{14}O$	(±)-*cis*-2-Methylcyclohexanol	l	-390.2			
		g	-327.0			
$C_7H_{14}O$	(±)-*trans*-2-Methylcyclohexanol	l	-415.7			
		g	-352.5			
$C_7H_{14}O$	*cis*-3-Methylcyclohexanol	l	-416.1			
		g	-350.9			
$C_7H_{14}O$	*trans*-3-Methylcyclohexanol	l	-394.4			
		g	-329.1			
$C_7H_{14}O$	*cis*-4-Methylcyclohexanol	l	-413.2			
		g	-347.5			
$C_7H_{14}O$	*trans*-4-Methylcyclohexanol	l	-433.3			
		g	-367.2			
$C_7H_{14}O_2$	Ethyl 2,2-dimethylpropanoate	l	-577.2			
		g	-536.0			
$C_7H_{14}O_2$	Methyl hexanoate	l	-540.2			
		g	-492.6			
$C_7H_{14}O_2$	Ethyl pentanoate	l	-553.0			
		g	-506.9			
$C_7H_{14}O_2$	Ethyl 3-methylbutanoate	l	-570.9			
		g	-527.0			
$C_7H_{14}O_2$	Heptanoic acid	l	-610.2			265.4

TABLE 2.4.1
STANDARD STATE THERMOCHEMICAL PROPERTIES AT 298.15 K (continued)

Molecular formula	Name	State	$\Delta_f H°$/ kJ mol^{-1}	$\Delta_f G°$/ kJ mol^{-1}	$S°$/ J K^{-1} mol^{-1}	C_p/ J K^{-1} mol^{-1}
		g	-536.2			
$C_7H_{15}Br$	1-Bromoheptane	l	-218.4			
		g	-167.8			
C_7H_{16}	Heptane	l	-224.2			224.7
		g	-187.7			
C_7H_{16}	2-Methylhexane	l	-229.5		323.3	222.9
		g	-194.6			
C_7H_{16}	3-Methylhexane	l	-226.4			
		g	-191.3			
C_7H_{16}	2,2-Dimethylpentane	l	-238.3		300.3	221.1
		g	-205.9			
C_7H_{16}	2,3-Dimethylpentane	l	-233.1			
		g	-198.9			
C_7H_{16}	2,4-Dimethylpentane	l	-234.6		303.2	224.2
		g	-201.7			
C_7H_{16}	3,3-Dimethylpentane	l	-234.2			
		g	-201.2			
C_7H_{16}	3-Ethylpentane	l	-224.8		314.5	219.6
		g	-189.6			
C_7H_{16}	2,2,3-Trimethylbutane	l	-236.5		292.2	213.5
		g	-204.5			
$C_7H_{16}O$	1-Heptanol	l	-403.3			272.1
		g	-336.4			
$C_8H_4O_3$	1,2-Benzenedicarboxylic anhydride	cr	-460.1		180.0	160.0
		g	-371.4			
$C_8H_6O_4$	Phthalic acid	cr	-782.0		207.9	188.1
$C_8H_6O_4$	Isophthalic acid	cr	-803.0			
		g	-696.3			
$C_8H_6O_4$	Terephthalic acid	cr	-816.1			
		g	-717.9			
C_8H_8	Vinylbenzene	l	103.8			182.0
		g	147.9			
C_8H_8O	Methyl phenyl ketone	l	-142.5			
		g	-86.7			
$C_8H_8O_2$	2-Toluic acid	cr	-416.5			174.9
$C_8H_8O_2$	3-Toluic acid	cr	-426.1			163.6
$C_8H_8O_2$	4-Toluic acid	cr	-429.2			169.0
$C_8H_8O_2$	Methyl benzoate	l	-343.5			221.3
		g	-287.9			
C_8H_{10}	Ethylbenzene	l	-12.3			183.2
		g	29.9			
C_8H_{10}	1,2-Dimethylbenzene	l	-24.4			186.1
		g	19.1			
C_8H_{10}	1,3-Dimethylbenzene	l	-25.4			183.0
		g	17.3			
C_8H_{10}	1,4-Dimethylbenzene	l	-24.4			181.5
		g	18.0			
$C_8H_{10}O$	2-Ethylphenol	l	-208.8			
		g	-145.2			
$C_8H_{10}O$	3-Ethylphenol	l	-214.3			
		g	-146.1			
$C_8H_{10}O$	4-Ethylphenol	cr	-224.4			206.9
		g	-144.1			
$C_8H_{10}O$	2,3-Dimethylphenol	cr	-241.1			
		g	-157.2			
$C_8H_{10}O$	2,4-Dimethylphenol	l	-228.7			

TABLE 2.4.1
STANDARD STATE THERMOCHEMICAL PROPERTIES AT 298.15 K (continued)

Molecular formula	Name	State	$\Delta_f H°$/ kJ mol^{-1}	$\Delta_f G°$/ kJ mol^{-1}	$S°$/ J K^{-1} mol^{-1}	C_p/ J K^{-1} mol^{-1}
		g	-162.9			
$C_8H_{10}O$	2,5-Dimethylphenol	cr	-246.6			
		g	-161.6			
$C_8H_{10}O$	2,6-Dimethylphenol	cr	-237.4			
		g	-161.8			
$C_8H_{10}O$	3,4-Dimethylphenol	cr	-242.3			
		g	-156.6			
$C_8H_{10}O$	3,5-Dimethylphenol	cr	-244.4			
		g	-161.5			
$C_8H_{10}O$	Ethyl phenyl ether	l	-152.6			228.5
		g	-101.6			
$C_8H_{11}N$	*N*,*N*-Dimethylaniline	l	47.7			
		g	100.5			
$C_8H_{11}N$	*N*-Ethylaniline	l	4.0			
		g	56.3			
$C_8H_{15}N$	Octanenitrile	l	-107.3			
		g	-50.5			
C_8H_{16}	Cyclooctane	l	-167.7			
		g	-124.4			
C_8H_{16}	Propylcyclopentane	l	-188.8		310.8	216.3
		g	-147.7			
C_8H_{16}	1-Ethyl-1-methylcyclopentane	l	-193.8			
C_8H_{16}	Ethylcyclohexane	l	-211.9		280.9	211.8
		g	-171.7			
C_8H_{16}	1,1-Dimethylcyclohexane	l	-218.7		267.2	209.2
		g	-180.9			
C_8H_{16}	*cis*-1,2-Dimethylcyclohexane	l	-211.8		274.1	210.2
		g	-172.1			
C_8H_{16}	*trans*-1,2-Dimethylcyclohexane	l	-218.2		273.2	209.4
		g	-179.9			
C_8H_{16}	*cis*-1,3-Dimethylcyclohexane	l	-222.9		272.6	209.4
		g	-184.6			
C_8H_{16}	*trans*-1,3-Dimethylcyclohexane	l	-215.7		276.3	212.8
		g	-176.5			
C_8H_{16}	*cis*-1,4-Dimethylcyclohexane	l	-215.6		271.1	212.1
		g	-176.6			
C_8H_{16}	*trans*-1,4-Dimethylcyclohexane	l	-222.4		268.0	210.2
		g	-184.5			
$C_8H_{16}O$	2,2,4-Trimethyl-3-pentanone	l	-381.6			
		g	-338.3			
$C_8H_{16}O$	2-Ethylhexanal	l	-348.5			
		g	-299.6			
$C_8H_{16}O_2$	Octanoic acid	l	-636.0			297.9
		g	-554.3			
$C_8H_{16}O_2$	(±)-2-Ethylhexanoic acid	l	-635.1			
		g	-559.5			
$C_8H_{16}O_2$	Methyl heptanoate	l	-567.1			285.1
		g	-515.9			
$C_8H_{17}Br$	1-Bromooctane	l	-245.1			
		g	-189.7			
$C_8H_{17}Cl$	1-Chlorooctane	l	-291.3			
		g	-238.9			
C_8H_{18}	Octane	l	-250.1			254.6
		g	-208.6			
C_8H_{18}	2-Methylheptane	l	-255.0		356.4	252.0
		g	-215.4			
C_8H_{18}	3-Methylheptane	l	-252.3		362.6	250.2

TABLE 2.4.1
STANDARD STATE THERMOCHEMICAL PROPERTIES AT 298.15 K (continued)

Molecular formula	Name	State	$\Delta_f H°$/ kJ mol^{-1}	$\Delta_f G°$/ kJ mol^{-1}	$S°$/ J K^{-1} mol^{-1}	C_p/ J K^{-1} mol^{-1}
		g	-212.5			
C_8H_{18}	4-Methylheptane	l	-251.6			251.1
		g	-212.0			
C_8H_{18}	2,2-Dimethylhexane	l	-261.9			
		g	-224.6			
C_8H_{18}	2,3-Dimethylhexane	l	-252.6			
		g	-213.8			
C_8H_{18}	2,4-Dimethylhexane	l	-257.0			
		g	-219.2			
C_8H_{18}	2,5-Dimethylhexane	l	-260.4			249.2
		g	-222.5			
C_8H_{18}	3,3-Dimethylhexane	l	-257.5			246.6
		g	-220.0			
C_8H_{18}	3,4-Dimethylhexane	l	-251.8			
		g	-212.8			
C_8H_{18}	3-Ethylhexane	l	-250.4			
		g	-210.7			
C_8H_{18}	3-Ethyl-2-methylpentane	l	-249.6			
		g	-211.0			
C_8H_{18}	3-Ethyl-3-methylpentane	l	-252.8			
		g	-214.8			
C_8H_{18}	2,2,3-Trimethylpentane	l	-256.9			
		g	-220.0			
C_8H_{18}	2,2,4-Trimethylpentane	l	-259.2			239.1
		g	-224.0			
C_8H_{18}	2,3,3-Trimethylpentane	l	-253.5			245.6
		g	-216.3			
C_8H_{18}	2,3,4-Trimethylpentane	l	-255.0		329.3	247.3
		g	-217.3			
C_8H_{18}	2,2,3,3-Tetramethylbutane	cr	-268.9		273.7	239.2
		g	-225.6			
$C_8H_{18}N_2$	Azobutane	l	-40.1			
		g	9.2			
$C_8H_{18}O$	1-Octanol	l	-426.5			305.2
		g	-355.5			
$C_8H_{18}O$	2-Ethyl-1-hexanol	l	-432.8		347.0	317.5
		g	-365.3			
$C_8H_{18}O$	Dibutyl ether	l	-377.9			278.2
		g	-333.4			
$C_8H_{18}O$	Di-*sec*-butyl ether	l	-401.5			
		g	-360.9			
$C_8H_{18}O$	Di-*tert*-butyl ether	l	-399.6			276.1
		g	-362.0			
$C_8H_{18}O_5$	3,6,9-Trioxa-1,11-undecanediol	l	-981.7			428.8
		g	-883.0			
$C_8H_{18}S$	5-Thianonane	l	-220.7		405.1	284.3
		g	-167.4			
$C_8H_{18}S$	Diisobutyl sulfide	l	-229.2			
		g	-179.5			
$C_8H_{18}S$	Bis(1,1-dimethylethyl) sulfide	l	-232.6			
		g	-188.9			
$C_8H_{19}N$	Dibutylamine	l	-206.0			292.9
		g	-156.6			
$C_8H_{19}N$	Bis(2-methylpropyl)amine	l	-218.5			
		g	-179.2			
$C_8H_{20}Pb$	Tetraethyllead	l	52.7		464.6	307.4
		g	109.6			

TABLE 2.4.1
STANDARD STATE THERMOCHEMICAL PROPERTIES AT 298.15 K (continued)

Molecular formula	Name	State	$\Delta_f H°$/ kJ mol^{-1}	$\Delta_f G°$/ kJ mol^{-1}	$S°$/ J K^{-1} mol^{-1}	C_p/ J K^{-1} mol^{-1}
C_9H_7N	Isoquinoline	l	144.5		216.0	196.2
C_9H_8	Indene	l	110.6		215.3	186.9
		g	163.4			
C_9H_{10}	Cyclopropylbenzene	l	100.3			
		g	150.5			
C_9H_{10}	Indan	l	11.5		56.0	190.2
		g	60.7			
$C_9H_{11}NO_2$	*L*-Phenylalanine	cr	-466.9		213.6	203.0
		g	-312.9			
$C_9H_{11}NO_3$	*L*-Tyrosine	cr	-685.1		214.0	216.4
C_9H_{12}	Isopropylbenzene	l	-41.1			210.7
		g	4.0			
C_9H_{12}	2-Ethyltoluene	l	-46.4			
		g	1.3			
C_9H_{12}	3-Ethyltoluene	l	-48.7			
		g	-1.8			
C_9H_{12}	4-Ethyltoluene	l	-49.8			
		g	-3.2			
C_9H_{12}	Propylbenzene	l	-38.3		287.8	214.7
		g	7.9			
C_9H_{12}	1,2,3-Trimethylbenzene	l	-58.5		267.9	216.4
		g	-9.5			
C_9H_{12}	1,2,4-Trimethylbenzene	l	-61.8			215.0
		g	-13.8			
C_9H_{12}	1,3,5-Trimethylbenzene	l	-63.4			209.3
		g	-15.9			
$C_9H_{14}O_6$	1,2,3-Propanetriol tris(ethanoate)	l	-1330.8		458.3	384.7
		g	-1248.8			
C_9H_{18}	Propylcyclohexane	l	-237.4		311.9	242.0
		g	-192.5			
$C_9H_{18}O$	2-Nonanone	l	-397.2			
		g	-340.7			
$C_9H_{18}O$	5-Nonanone	l	-398.2		401.4	303.6
		g	-344.9			
$C_9H_{18}O$	2,6-Dimethyl-4-heptanone	l	-408.5			297.3
		g	-357.6			
$C_9H_{18}O_2$	Nonanoic acid	l	-659.7			362.4
		g	-577.3			
$C_9H_{18}O_2$	Methyl octanoate	l	-590.3			
		g	-533.8			
C_9H_{20}	Nonane	l	-274.7			284.4
		g	-228.2			
C_9H_{20}	2,2-Dimethylheptane	l	-288.2			
C_9H_{20}	2,2,5-Trimethylhexane	l	-293.3			
C_9H_{20}	2,3,5-Trimethylhexane	l	-284.0			
C_9H_{20}	2,2,3,3-Tetramethylpentane	l	-278.3			271.5
		g	-237.1			
C_9H_{20}	2,2,3,4-Tetramethylpentane	l	-277.7			
		g	-236.9			
C_9H_{20}	3,3-Diethylpentane	l	-275.4			278.2
		g	-232.3			
C_9H_{20}	2,2,4,4-Tetramethylpentane	l	-280.0			266.3
		g	-241.6			
C_9H_{20}	2,3,3,4-Tetramethylpentane	l	-277.9			
		g	-236.1			
$C_9H_{20}O$	1-Nonanol	l	-456.5			
$C_{10}H_7Cl$	1-Chloronaphthalene	l	54.6			212.6

TABLE 2.4.1
STANDARD STATE THERMOCHEMICAL PROPERTIES AT 298.15 K (continued)

Molecular formula	Name	State	$\Delta_f H°$/ kJ mol^{-1}	$\Delta_f G°$/ kJ mol^{-1}	$S°$/ J K^{-1} mol^{-1}	C_p/ J K^{-1} mol^{-1}
		g	119.8			
$C_{10}H_8$	Azulene	cr	212.3			
		g	289.1			
$C_{10}H_8$	Naphthalene	cr	77.9		167.4	165.7
		g	150.3			
$C_{10}H_8O$	1-Naphthol	cr	-121.0			166.9
		g	-29.9			
$C_{10}H_8O$	2-Naphthol	l	-124.2			
		g	-30.0			
$C_{10}H_{10}O_4$	Dimethyl 1,3-benzenedicarboxylate	cr	-730.9			
$C_{10}H_{10}O_4$	Dimethyl 1,4-benzenedicarboxylate	cr	-732.6			261.1
$C_{10}H_{12}$	1,2,3,4-Tetrahydronaphthalene	l	-29.2			217.5
		g	26.0			
$C_{10}H_{14}$	Butylbenzene	l	-63.2		321.2	243.4
		g	-13.1			
$C_{10}H_{14}$	Isobutylbenzene	l	-69.8			
		g	-21.5			
$C_{10}H_{14}$	(1-Methylpropyl)benzene	l	-66.4			
		g	-17.4			
$C_{10}H_{14}$	*tert*-Butylbenzene	l	-70.7			
		g	-22.6			
$C_{10}H_{14}$	2-Isopropyltoluene	l	-73.3			
$C_{10}H_{14}$	3-Isopropyltoluene	l	-78.6			
$C_{10}H_{14}$	4-Isopropyltoluene	l	-78.0			236.4
$C_{10}H_{14}$	1,2-Diethylbenzene	l	-68.5			
$C_{10}H_{14}$	1,3-Diethylbenzene	l	-73.5			
$C_{10}H_{14}$	1,4-Diethylbenzene	l	-72.8			
$C_{10}H_{14}$	1,2,4,5-Tetramethylbenzene	cr	-119.9		245.6	215.1
$C_{10}H_{14}$	1-Ethyl-2,3-dimethylbenzene	l	-80.5			
$C_{10}H_{14}O$	2-Isopropyl-5-methylphenol	cr	-309.7			
		g	-218.5			
$C_{10}H_{16}$	2,7,7-Trimethylbicyclo[3.1.1]-hept-2-ene	l	-16.4			
		g	28.3			
$C_{10}H_{16}$	7,7-Dimethyl-2-methylenebicyclo-[3.1.1]heptane	l	-7.7			
		g	38.7			
$C_{10}H_{16}O$	1,7,7-Trimethylbicyclo[2.2.1]-hepten-2-one	cr	-319.4			271.2
		g	-267.5			
$C_{10}H_{18}$	*cis*-Bicyclo[4.4.0]decane	l	-219.4		265.0	232.0
		g	-169.2			
$C_{10}H_{18}$	*trans*-Bicyclo[4.4.0]decane	l	-230.6		264.9	228.5
		g	-182.1			
$C_{10}H_{18}O_4$	Decanedioic acid	cr	-1082.6			
		g	-921.9			
$C_{10}H_{19}N$	Decanenitrile	l	-158.4			
		g	-91.5			
$C_{10}H_{20}$	Butylcyclohexane	l	-263.1		345.0	271.0
		g	-213.3			
$C_{10}H_{20}$	1-Decene	l	-173.8		425.0	300.8
		g	-123.4			
$C_{10}H_{20}O_2$	Decanoic acid	cr	-713.7			
		l	-684.3			
		g	-594.9			
$C_{10}H_{22}$	Decane	l	-300.9			314.4
		g	-249.5			

TABLE 2.4.1
STANDARD STATE THERMOCHEMICAL PROPERTIES AT 298.15 K (continued)

Molecular formula	Name	State	$\Delta_f H°$/ kJ mol^{-1}	$\Delta_f G°$/ kJ mol^{-1}	$S°$/ J K^{-1} mol^{-1}	C_p/ J K^{-1} mol^{-1}
$C_{10}H_{22}$	2-Methylnonane	l	-309.8		420.1	313.3
		g	-259.9			
$C_{10}H_{22}$	5-Methylnonane	l	-307.9		423.8	314.4
		g	-258.6			
$C_{10}H_{22}O$	1-Decanol	l	-478.1			370.6
		g	-396.4			
$C_{10}H_{22}S$	1-Decanethiol	cr	-309.9			
		l	-276.5		476.1	350.4
		g	-211.5			
$C_{10}H_{22}S$	Bis(3-methylbutyl) sulfide	l	-281.8			
		g	-221.5			
$C_{11}H_{10}$	1-Methylnaphthalene	l	56.3		254.8	224.4
$C_{11}H_{10}$	2-Methylnaphthalene	cr	44.9		220.0	196.0
		g	106.7			
$C_{11}H_{12}N_2O_2$	(*S*)-(-)-Tryptophan	cr	-415.3		251.0	238.1
$C_{11}H_{24}$	Undecane	l	-327.2			344.9
		g	-270.9			
$C_{12}H_8$	Acenaphthylene	cr	186.7			166.4
		g	259.7			
$C_{12}H_9N$	Carbazole	cr	125.1			
		g	209.6			
$C_{12}H_{10}$	Acenaphthene	cr	70.3		188.9	190.4
		g	156.0			
$C_{12}H_{10}$	Biphenyl	cr	99.4		209.4	198.4
		g	181.4			
$C_{12}H_{10}O$	Diphenyl ether	cr	-32.1		233.9	216.6
		g	52.0			
$C_{12}H_{11}N$	Diphenylamine	cr	130.2			
		g	219.3			
$C_{12}H_{12}N_2$	4,4'-Biphenyldiamine	cr	70.7			
$C_{12}H_{14}O_4$	Diethyl 1,2-benzenedicarboxylate	l	-776.6		425.1	366.1
		g	-688.4			
$C_{12}H_{16}$	Cyclohexylbenzene	l	-76.6			
		g	-16.7			
$C_{12}H_{18}$	5,7-Dodecadiyne	l	181.5			
$C_{12}H_{18}$	3,9-Dodecadiyne	l	197.8			
$C_{12}H_{18}$	Hexamethylbenzene	cr	-161.5		306.3	245.6
		g	-86.8			
$C_{12}H_{22}$	Bicyclohexyl	l	-273.7			
		g	-215.7			
$C_{12}H_{24}$	1-Dodecene	l	-226.2		484.8	360.7
		g	-165.4			
$C_{12}H_{24}O_2$	Dodecanoic acid	cr	-774.6			404.3
		l	-737.9			
		g	-642.0			
$C_{12}H_{26}$	Dodecane	l	-350.9			375.8
		g	-289.7			
$C_{12}H_{26}O$	1-Dodecanol	l	-528.5			438.1
		g	-436.6			
$C_{12}H_{27}N$	Tributylamine	l	-281.6			
$C_{13}H_{10}O$	Diphenyl ketone	cr	-34.5			224.8
		g	54.9			
$C_{13}H_{12}$	Diphenylmethane	cr	71.5		239.3	
		l	89.7			
		g	139.0			
$C_{13}H_{26}O_2$	Methyl dodecanoate	l	-693.0			
		g	-614.8			

TABLE 2.4.1
STANDARD STATE THERMOCHEMICAL PROPERTIES AT 298.15 K (continued)

Molecular formula	Name	State	$\Delta_f H°$/ kJ mol^{-1}	$\Delta_f G°$/ kJ mol^{-1}	$S°$/ J K^{-1} mol^{-1}	C_p/ J K^{-1} mol^{-1}
$C_{13}H_{28}O$	1-Tridecanol	cr	-599.4			
$C_{14}H_{10}$	Anthracene	cr	129.2		207.5	210.5
		g	230.9			
$C_{14}H_{10}$	Phenanthrene	cr	116.2		215.1	220.6
		g	207.5			
$C_{14}H_{10}$	Diphenylethyne	cr	312.4			225.9
$C_{14}H_{12}$	*cis*-1,2-Diphenylethene	l	183.3			
		g	252.3			
$C_{14}H_{12}$	*trans*-1,2-Diphenylethene	cr	136.9			
		g	236.1			
$C_{14}H_{14}$	1,1-Diphenylethane	l	48.7			
$C_{14}H_{14}$	1,2-Diphenylethane	cr	51.5			
		g	142.9			
$C_{14}H_{27}N$	Tetradecanenitrile	l	-260.2			
		g	-174.9			
$C_{14}H_{28}O_2$	Tetradecanoic acid	cr	-833.5			432.0
		l	-788.8			
		g	-693.7			
$C_{14}H_{30}O$	1-Tetradecanol	cr	-629.6			388.0
		l	-580.6			
$C_{15}H_{30}O_2$	Pentadecanoic acid	cr	-861.7			443.3
		l	-811.7			
		g	-699.0			
$C_{16}H_{10}$	Fluoranthene	cr	189.9		230.6	230.2
		g	289.0			
$C_{16}H_{10}$	Pyrene	cr	125.5		224.9	229.7
		g	225.7			
$C_{16}H_{22}O_4$	Dibutyl 1,2-benzenedicarboxylate	l	-842.6			
		g	-750.9			
$C_{16}H_{26}$	Decylbenzene	l	-218.3			
		g	-138.6			
$C_{16}H_{32}$	1-Hexadecene	l	-328.7		587.9	488.9
		g	-248.5			
$C_{16}H_{32}O_2$	Hexadecanoic acid	cr	-891.5		452.4	460.7
		l	-838.1			
		g	-737.1			
$C_{16}H_{34}$	Hexadecane	l	-456.1			501.6
		g	-374.8			
$C_{16}H_{34}O$	1-Hexadecanol	cr	-686.5			422.0
		g	-517.0			
$C_{17}H_{34}O_2$	Heptadecanoic acid	cr	-924.4			475.7
		l	-865.6			
$C_{18}H_{12}$	Chrysene	cr	145.3			
		g	269.8			
$C_{18}H_{36}O_2$	Octadecanoic acid	cr	-947.7			501.5
		l	-884.7			
		g	-781.2			
$C_{18}H_{38}$	Octadecane	cr	-567.4		480.2	485.6
		g	-414.6			
$C_{19}H_{36}O_2$	Methyl *cis*-9-octanedecenoate	l	-734.5			
		g	-649.9			
$C_{20}H_{12}$	Perylene	cr	182.8		264.6	274.9
$C_{20}H_{40}O_2$	Icosanoic acid	cr	-1011.9			545.1
		l	-940.0			
		g	-812.4			
$C_{22}H_{42}O_2$	*trans*-13-Docosenoic acid	cr	-960.7			

TABLE 2.4.2
THERMOCHEMICAL PROPERTIES AT HIGH TEMPERATURES

This table gives standard-state thermodynamic properties for 80 substances as a function of temperature from 298.15 K to 1500 K. The standard-state pressure is 100000 Pa, and the reference temperature T_r is 298.15 K. See Table 2.4.1 and References 1 and 2 for a discussion of reference states.

REFERENCES

1. Gurvich, L. V.; Veyts, I. V.; Alcock, C. B., *Thermodynamic Properties of Individual Substances, Fourth Edition*, Hemisphere Publishing Corp., New York, 1989.
2. Chase, M. W.; Davies, C. A.; Downey, J. R.; Frurip, D. J.; McDonald, R. A.; Syverud, A. N.; *JANAF Thermochemical Tables, Third Edition*, *J. Phys. Chem. Ref. Data*, Vol. 14, Suppl. 1, 1985.

ORDER OF LISTING OF TABLES

No.	Formula	Name	State
1	Ar	Argon	g
2	Br	Bromine	g
3	Br_2	Dibromine	g
4	BrH	Hydrogen bromide	g
5	C	Carbon (graphite)	cr
6	C	Carbon (diamond)	cr
7	C_2	Dicarbon	g
8	C_3	Tricarbon	g
9	CO	Carbon oxide	g
10	CO_2	Carbon dioxide	g
11	CH_4	Methane	g
12	C_2H_2	Acetylene	g
13	C_2H_4	Ethylene	g
14	C_2H_6	Ethane	g
15	C_3H_6	Cyclopropane	g
16	C_3H_8	Propane	g
17	C_6H_6	Benzene	l
18	C_6H_6	Benzene	g
19	$C_{10}H_8$	Naphthalene	cr, l
20	$C_{10}H_8$	Naphthalene	g
21	CH_2O	Formaldehyde	g
22	CH_4O	Methanol	g
23	C_2H_4O	Acetaldehyde	g
24	C_2H_6O	Ethanol	g
25	$C_2H_4O_2$	Acetic acid	g
26	C_3H_6O	Acetone	g
27	C_6H_6O	Phenol	g
28	CF_4	Carbon tetrafluoride	g
29	CHF_3	Trifluoromethane	g
30	$CClF_3$	Chlorotrifluoromethane	g
31	CCl_2F_2	Dichlorodifluoromethane	g
32	$CHClF_2$	Chlorodifluoromethane	g
33	CH_5N	Methylamine	g
34	Cl	Chlorine	g
35	Cl_2	Dichlorine	g
36	ClH	Hydrogen chloride	g
37	Cu	Copper	cr, l
38	Cu	Copper	g
39	CuO	Copper oxide	cr
40	Cu_2O	Dicopper oxide	cr
41	$CuCl_2$	Copper dichloride	cr, l
42	$CuCl_2$	Copper dichloride	g
43	F	Fluorine	g
44	F_2	Difluorine	g
45	FH	Hydrogen fluoride	g
46	Ge	Germanium	cr, l
47	Ge	Germanium	g
48	GeO_2	Germanium dioxide	cr, l
49	$GeCl_4$	Germanium tetrachloride	g
50	H	Hydrogen	g
51	H_2	Dihydrogen	g
52	HO	Hydroxyl	g
53	H_2O	Water	l
54	H_2O	Water	g
55	I	Iodine	g
56	I_2	Diiodine	cr, l
57	I_2	Diiodine	g
58	IH	Hydrogen iodide	g
59	K	Potassium	cr, l
60	K	Potassium	g
61	K_2O	Dipotassium oxide	cr, l
62	KOH	Potassium hydroxide	cr, l
63	KOH	Potassium hydroxide	g
64	KCl	Potassium chloride	cr, l
65	KCl	Potassium chloride	g
66	N_2	Dinitrogen	g
67	NO	Nitric oxide	g
68	NO_2	Nitrogen dioxide	g
69	NH_3	Ammonia	g
70	O	Oxygen	g
71	O_2	Dioxygen	g
72	S	Sulfur	cr, l
73	S	Sulfur	g
74	S_2	Disulfur	g
75	S_8	Octasulfur	g
76	SO_2	Sulfur dioxide	g
77	Si	Silicon	cr
78	Si	Silicon	g
79	SiO_2	Silicon dioxide	cr
80	$SiCl_4$	Silicon tetrachloride	g

TABLE 2.4.2
THERMOCHEMICAL PROPERTIES AT HIGH TEMPERATURES (continued)

	J K^{-1} mol^{-1}			kJ mol^{-1}			
T/K	C_p°	S°	$-(G^\circ-H^\circ(T_r))/T$	$H^\circ-H^\circ(T_r)$	$\Delta_f H^\circ$	$\Delta_f G^\circ$	Log K_f
1. ARGON Ar (g)							
298.15	20.786	154.845	154.845	0.000	0.000	0.000	0.000
300	20.786	154.973	154.845	0.038	0.000	0.000	0.000
400	20.786	160.953	155.660	2.117	0.000	0.000	0.000
500	20.786	165.591	157.200	4.196	0.000	0.000	0.000
600	20.786	169.381	158.924	6.274	0.000	0.000	0.000
700	20.786	172.585	160.653	8.353	0.000	0.000	0.000
800	20.786	175.361	162.322	10.431	0.000	0.000	0.000
900	20.786	177.809	163.909	12.510	0.000	0.000	0.000
1000	20.786	179.999	165.410	14.589	0.000	0.000	0.000
1100	20.786	181.980	166.828	16.667	0.000	0.000	0.000
1200	20.786	183.789	168.167	18.746	0.000	0.000	0.000
1300	20.786	185.453	169.434	20.824	0.000	0.000	0.000
1400	20.786	186.993	170.634	22.903	0.000	0.000	0.000
1500	20.786	188.427	171.773	24.982	0.000	0.000	0.000
2. BROMINE Br (g)							
298.15	20.786	175.017	175.017	0.000	111.870	82.379	–14.432
300	20.786	175.146	175.018	0.038	111.838	82.196	–14.311
400	20.787	181.126	175.833	2.117	96.677	75.460	–9.854
500	20.798	185.765	177.373	4.196	96.910	70.129	–7.326
600	20.833	189.559	179.097	6.277	97.131	64.752	–5.637
700	20.908	192.776	180.827	8.364	97.348	59.338	–4.428
800	21.027	195.575	182.499	10.461	97.568	53.893	–3.519
900	21.184	198.061	184.093	12.571	97.796	48.420	–2.810
1000	21.365	200.302	185.604	14.698	98.036	42.921	–2.242
1100	21.559	202.347	187.034	16.844	98.291	37.397	–1.776
1200	21.752	204.231	188.390	19.010	98.560	31.850	–1.386
1300	21.937	205.980	189.676	21.195	98.844	26.279	–1.056
1400	22.107	207.612	190.900	23.397	99.141	20.686	–0.772
1500	22.258	209.142	192.065	25.615	99.449	15.072	–0.525
3. DIBROMINE Br_2 (g)							
298.15	36.057	245.467	245.467	0.000	30.910	3.105	–0.544
300	36.074	245.690	245.468	0.067	30.836	2.933	–0.511
332.25	36.340	249.387	245.671	1.235		pressure = 1 bar	
400	36.729	256.169	246.892	3.711	0.000	0.000	0.000
500	37.082	264.406	249.600	7.403	0.000	0.000	0.000
600	37.305	271.188	252.650	11.123	0.000	0.000	0.000
700	37.464	276.951	255.720	14.862	0.000	0.000	0.000
800	37.590	281.962	258.694	18.615	0.000	0.000	0.000
900	37.697	286.396	261.530	22.379	0.000	0.000	0.000
1000	37.793	290.373	264.219	26.154	0.000	0.000	0.000
1100	37.883	293.979	266.763	29.938	0.000	0.000	0.000
1200	37.970	297.279	269.170	33.730	0.000	0.000	0.000
1300	38.060	300.322	271.451	37.532	0.000	0.000	0.000
1400	38.158	303.146	273.615	41.343	0.000	0.000	0.000
1500	38.264	305.782	275.673	45.164	0.000	0.000	0.000
4. HYDROGEN BROMIDE HBr (g)							
298.15	29.141	198.697	198.697	0.000	–36.290	–53.360	9.348
300	29.141	198.878	198.698	0.054	–36.333	–53.466	9.309
400	29.220	207.269	199.842	2.971	–52.109	–55.940	7.305
500	29.454	213.811	202.005	5.903	–52.484	–56.854	5.939

TABLE 2.4.2
THERMOCHEMICAL PROPERTIES AT HIGH TEMPERATURES (continued)

	J K^{-1} mol^{-1}			kJ mol^{-1}			
T/K	C_p°	S°	-(G°-H° (T_r))/T	H°-H° (T_r)	$\Delta_f H^\circ$	$\Delta_f G^\circ$	Log K_f
4. HYDROGEN BROMIDE HBr (g) (continued)							
600	29.872	219.216	204.436	8.868	–52.844	–57.694	5.023
700	30.431	223.861	206.886	11.882	–53.168	–58.476	4.363
800	31.063	227.965	209.269	14.957	–53.446	–59.214	3.866
900	31.709	231.661	211.555	18.095	–53.677	–59.921	3.478
1000	32.335	235.035	213.737	21.298	–53.864	–60.604	3.166
1100	32.919	238.145	215.816	24.561	–54.012	–61.271	2.909
1200	33.454	241.032	217.799	27.880	–54.129	–61.925	2.696
1300	33.938	243.729	219.691	31.250	–54.220	–62.571	2.514
1400	34.374	246.261	221.499	34.666	–54.291	–63.211	2.358
1500	34.766	248.646	223.230	38.123	–54.348	–63.846	2.223
5. CARBON (GRAPHITE) C (cr; graphite)							
298.15	8.536	5.740	5.740	0.000	0.000	0.000	0.000
300	8.610	5.793	5.740	0.016	0.000	0.000	0.000
400	11.974	8.757	6.122	1.054	0.000	0.000	0.000
500	14.537	11.715	6.946	2.385	0.000	0.000	0.000
600	16.607	14.555	7.979	3.945	0.000	0.000	0.000
700	18.306	17.247	9.113	5.694	0.000	0.000	0.000
800	19.699	19.785	10.290	7.596	0.000	0.000	0.000
900	20.832	22.173	11.479	9.625	0.000	0.000	0.000
1000	21.739	24.417	12.662	11.755	0.000	0.000	0.000
1100	22.452	26.524	13.827	13.966	0.000	0.000	0.000
1200	23.000	28.502	14.968	16.240	0.000	0.000	0.000
1300	23.409	30.360	16.082	18.562	0.000	0.000	0.000
1400	23.707	32.106	17.164	20.918	0.000	0.000	0.000
1500	23.919	33.749	18.216	23.300	0.000	0.000	0.000
6. CARBON (DIAMOND) C (cr; diamond)							
298.15	6.109	2.362	2.362	0.000	1.850	2.857	–0.501
300	6.201	2.400	2.362	0.011	1.846	2.863	–0.499
400	10.321	4.783	2.659	0.850	1.645	3.235	–0.422
500	13.404	7.431	3.347	2.042	1.507	3.649	–0.381
600	15.885	10.102	4.251	3.511	1.415	4.087	–0.356
700	17.930	12.709	5.274	5.205	1.361	4.537	–0.339
800	19.619	15.217	6.361	7.085	1.338	4.993	–0.326
900	21.006	17.611	7.479	9.118	1.343	5.450	–0.316
1000	22.129	19.884	8.607	11.277	1.372	5.905	–0.308
1100	23.020	22.037	9.731	13.536	1.420	6.356	–0.302
1200	23.709	24.071	10.842	15.874	1.484	6.802	–0.296
1300	24.222	25.990	11.934	18.272	1.561	7.242	–0.291
1400	24.585	27.799	13.003	20.714	1.646	7.675	–0.286
1500	24.824	29.504	14.047	23.185	1.735	8.103	–0.282
7. DICARBON C_2 (g)							
298.15	43.548	197.095	197.095	0.000	830.457	775.116	–135.795
300	43.575	197.365	197.096	0.081	830.506	774.772	–134.898
400	42.169	209.809	198.802	4.403	832.751	755.833	–98.700
500	39.529	218.924	201.959	8.483	834.170	736.423	–76.933
600	37.837	225.966	205.395	12.342	834.909	716.795	–62.402
700	36.984	231.726	208.758	16.078	835.148	697.085	–52.016
800	36.621	236.637	211.943	19.755	835.020	677.366	–44.227
900	36.524	240.943	214.931	23.411	834.618	657.681	–38.170
1000	36.569	244.793	217.728	27.065	834.012	638.052	–33.328

TABLE 2.4.2
THERMOCHEMICAL PROPERTIES AT HIGH TEMPERATURES (continued)

	J K⁻¹ mol⁻¹			kJ mol⁻¹			
T/K	C_p°	S°	$-(G^\circ-H^\circ(T_r))/T$	$H^\circ-H^\circ(T_r)$	$\Delta_f H^\circ$	$\Delta_f G^\circ$	Log K_f
7. DICARBON C_2 (g) (continued)							
1100	36.696	248.284	220.349	30.728	833.252	618.492	–29.369
1200	36.874	251.484	222.812	34.406	832.383	599.006	–26.074
1300	37.089	254.444	225.133	38.104	831.437	579.596	–23.288
1400	37.329	257.201	227.326	41.824	830.445	560.261	–20.903
1500	37.589	259.785	229.405	45.570	829.427	540.997	–18.839
8. TRICARBON C_3 (g)							
298.15	42.202	237.611	237.611	0.000	839.958	774.249	–135.643
300	42.218	237.872	237.611	0.078	839.989	773.841	–134.736
400	43.383	250.164	239.280	4.354	841.149	751.592	–98.147
500	44.883	260.003	242.471	8.766	841.570	729.141	–76.172
600	46.406	268.322	246.104	13.331	841.453	706.659	–61.519
700	47.796	275.582	249.807	18.042	840.919	684.230	–51.057
800	48.997	282.045	253.440	22.884	840.053	661.901	–43.217
900	50.006	287.876	256.948	27.835	838.919	639.698	–37.127
1000	50.844	293.189	260.310	32.879	837.572	617.633	–32.261
1100	51.535	298.069	263.524	37.999	836.059	595.711	–28.288
1200	52.106	302.578	266.593	43.182	834.420	573.933	–24.982
1300	52.579	306.768	269.524	48.417	832.690	552.295	–22.191
1400	52.974	310.679	272.326	53.695	830.899	530.793	–19.804
1500	53.307	314.346	275.006	59.010	829.068	509.421	–17.739
9. CARBON OXIDE CO (g)							
298.15	29.141	197.658	197.658	0.000	–110.530	–137.168	24.031
300	29.142	197.838	197.659	0.054	–110.519	–137.333	23.912
400	29.340	206.243	198.803	2.976	–110.121	–146.341	19.110
500	29.792	212.834	200.973	5.930	–110.027	–155.412	16.236
600	30.440	218.321	203.419	8.941	–110.157	–164.480	14.319
700	31.170	223.067	205.895	12.021	–110.453	–173.513	12.948
800	31.898	227.277	208.309	15.175	–110.870	–182.494	11.915
900	32.573	231.074	210.631	18.399	–111.378	–191.417	11.109
1000	33.178	234.538	212.851	21.687	–111.952	–200.281	10.461
1100	33.709	237.726	214.969	25.032	–112.573	–209.084	9.928
1200	34.169	240.679	216.990	28.426	–113.228	–217.829	9.482
1300	34.568	243.430	218.920	31.864	–113.904	–226.518	9.101
1400	34.914	246.005	220.763	35.338	–114.594	–235.155	8.774
1500	35.213	248.424	222.527	38.845	–115.291	–243.742	8.488
10. CARBON DIOXIDE CO_2 (g)							
298.15	37.135	213.783	213.783	0.000	–393.510	–394.373	69.092
300	37.220	214.013	213.784	0.069	–393.511	–394.379	68.667
400	41.328	225.305	215.296	4.004	–393.586	–394.656	51.536
500	44.627	234.895	218.280	8.307	–393.672	–394.914	41.256
600	47.327	243.278	221.762	12.909	–393.791	–395.152	34.401
700	49.569	250.747	225.379	17.758	–393.946	–395.367	29.502
800	51.442	257.492	228.978	22.811	–394.133	–395.558	25.827
900	53.008	263.644	232.493	28.036	–394.343	–395.724	22.967
1000	54.320	269.299	235.895	33.404	–394.568	–395.865	20.678
1100	55.423	274.529	239.172	38.893	–394.801	–395.984	18.803
1200	56.354	279.393	242.324	44.483	–395.035	–396.081	17.241
1300	57.144	283.936	245.352	50.159	–395.265	–396.159	15.918
1400	57.818	288.196	248.261	55.908	–395.488	–396.219	14.783
1500	58.397	292.205	251.059	61.719	–395.702	–396.264	13.799

TABLE 2.4.2
THERMOCHEMICAL PROPERTIES AT HIGH TEMPERATURES (continued)

	J K⁻¹ mol⁻¹			kJ mol⁻¹			
T/K	C_p°	S°	-(G°-H° (T_r))/T	H°-H° (T_r)	$\Delta_f H^\circ$	$\Delta_f G^\circ$	Log K_f
11. METHANE CH_4 (g)							
298.15	35.695	186.369	186.369	0.000	–74.600	–50.530	8.853
300	35.765	186.590	186.370	0.066	–74.656	–50.381	8.772
400	40.631	197.501	187.825	3.871	–77.703	–41.827	5.462
500	46.627	207.202	190.744	8.229	–80.520	–32.525	3.398
600	52.742	216.246	194.248	13.199	–82.969	–22.690	1.975
700	58.603	224.821	198.008	18.769	–85.023	–12.476	0.931
800	64.084	233.008	201.875	24.907	–86.693	–1.993	0.130
900	69.137	240.852	205.773	31.571	–88.006	8.677	–0.504
1000	73.746	248.379	209.660	38.719	–88.996	19.475	–1.017
1100	77.919	255.607	213.511	46.306	–89.698	30.358	–1.442
1200	81.682	262.551	217.310	54.289	–90.145	41.294	–1.797
1300	85.067	269.225	221.048	62.630	–90.367	52.258	–2.100
1400	88.112	275.643	224.720	71.291	–90.390	63.231	–2.359
1500	90.856	281.817	228.322	80.242	–90.237	74.200	–2.584
12. ACETYLENE C_2H_2 (g)							
298.15	44.036	200.927	200.927	0.000	227.400	209.879	–36.769
300	44.174	201.199	200.927	0.082	227.397	209.770	–36.524
400	50.388	214.814	202.741	4.829	227.161	203.928	–26.630
500	54.751	226.552	206.357	10.097	226.846	198.154	–20.701
600	58.121	236.842	210.598	15.747	226.445	192.452	–16.754
700	60.970	246.021	215.014	21.704	225.968	186.823	–13.941
800	63.511	254.331	219.418	27.931	225.436	181.267	–11.835
900	65.831	261.947	223.726	34.399	224.873	175.779	–10.202
1000	67.960	268.995	227.905	41.090	224.300	170.355	–8.898
1100	69.909	275.565	231.942	47.985	223.734	164.988	–7.835
1200	71.686	281.725	235.837	55.067	223.189	159.672	–6.950
1300	73.299	287.528	239.592	62.317	222.676	154.400	–6.204
1400	74.758	293.014	243.214	69.721	222.203	149.166	–5.565
1500	76.077	298.218	246.709	77.264	221.774	143.964	–5.013
13. ETHYLENE C_2H_4 (g)							
298.15	42.883	219.316	219.316	0.000	52.400	68.358	–11.976
300	43.059	219.582	219.317	0.079	52.341	68.457	–11.919
400	53.045	233.327	221.124	4.881	49.254	74.302	–9.703
500	62.479	246.198	224.864	10.667	46.533	80.887	–8.450
600	70.673	258.332	229.441	17.335	44.221	87.982	–7.659
700	77.733	269.770	234.393	24.764	42.278	95.434	–7.121
800	83.868	280.559	239.496	32.851	40.655	103.142	–6.734
900	89.234	290.754	244.630	41.512	39.310	111.036	–6.444
1000	93.939	300.405	249.730	50.675	38.205	119.067	–6.219
1100	98.061	309.556	254.756	60.280	37.310	127.198	–6.040
1200	101.670	318.247	259.688	70.271	36.596	135.402	–5.894
1300	104.829	326.512	264.513	80.599	36.041	143.660	–5.772
1400	107.594	334.384	269.225	91.223	35.623	151.955	–5.669
1500	110.018	341.892	273.821	102.107	35.327	160.275	–5.581
14. ETHANE C_2H_6 (g)							
298.15	52.487	229.161	229.161	0.000	–84.000	–32.015	5.609
300	52.711	229.487	229.162	0.097	–84.094	–31.692	5.518
400	65.459	246.378	231.379	5.999	–88.988	–13.473	1.759
500	77.941	262.344	235.989	13.177	–93.238	5.912	–0.618
600	89.188	277.568	241.660	21.545	–96.779	26.086	–2.271

TABLE 2.4.2
THERMOCHEMICAL PROPERTIES AT HIGH TEMPERATURES (continued)

T/K	C_p° (J K^{-1} mol^{-1})	S° (J K^{-1} mol^{-1})	$-(G^\circ-H^\circ(T_r))/T$ (J K^{-1} mol^{-1})	$H^\circ-H^\circ(T_r)$ (kJ mol^{-1})	$\Delta_f H^\circ$ (kJ mol^{-1})	$\Delta_f G^\circ$ (kJ mol^{-1})	Log K_f
14. ETHANE C_2H_6 (g) (continued)							
700	99.136	292.080	247.835	30.972	–99.663	46.800	–3.492
800	107.936	305.904	254.236	41.334	–101.963	67.887	–4.433
900	115.709	319.075	260.715	52.525	–103.754	89.231	–5.179
1000	122.552	331.628	267.183	64.445	–105.105	110.750	–5.785
1100	128.553	343.597	273.590	77.007	–106.082	132.385	–6.286
1200	133.804	355.012	279.904	90.131	–106.741	154.096	–6.708
1300	138.391	365.908	286.103	103.746	–107.131	175.850	–7.066
1400	142.399	376.314	292.178	117.790	–107.292	197.625	–7.373
1500	145.905	386.260	298.121	132.209	–107.260	219.404	–7.640
15. CYCLOPROPANE C_3H_6 (g)							
298.15	55.571	237.488	237.488	0.000	53.300	104.514	–18.310
300	55.941	237.832	237.489	0.103	53.195	104.832	–18.253
400	76.052	256.695	239.924	6.708	47.967	122.857	–16.043
500	93.859	275.637	245.177	15.230	43.730	142.091	–14.844
600	108.542	294.092	251.801	25.374	40.405	162.089	–14.111
700	120.682	311.763	259.115	36.854	37.825	182.583	–13.624
800	130.910	328.564	266.755	49.447	35.854	203.404	–13.281
900	139.658	344.501	274.516	62.987	34.384	224.441	–13.026
1000	147.207	359.616	282.277	77.339	33.334	245.618	–12.830
1100	153.749	373.961	289.965	92.395	32.640	266.883	–12.673
1200	159.432	387.588	297.538	108.060	32.249	288.197	–12.545
1300	164.378	400.549	304.967	124.257	32.119	309.533	–12.437
1400	168.689	412.892	312.239	140.915	32.215	330.870	–12.345
1500	172.453	424.662	319.344	157.976	32.507	352.193	–12.264
16. PROPANE C_3H_8 (g)							
298.15	73.597	270.313	270.313	0.000	–103.847	–23.458	4.110
300	73.931	270.769	270.314	0.136	–103.972	–22.959	3.997
400	94.014	294.739	273.447	8.517	–110.33	15.029	–0.657
500	112.591	317.768	280.025	18.872	–115.658	34.507	–3.605
600	128.700	339.753	288.162	30.955	–119.973	64.961	–5.655
700	142.674	360.668	297.039	44.540	–123.384	96.065	–7.168
800	154.766	380.528	306.245	59.427	–126.016	127.603	–8.331
900	165.352	399.381	315.555	75.444	–127.982	159.430	–9.253
1000	174.598	417.293	324.841	92.452	–129.380	191.444	–10.000
1100	182.673	434.321	334.026	110.325	–130.296	223.574	–10.617
1200	189.745	450.526	343.064	128.954	–130.802	255.770	–11.133
1300	195.853	465.961	351.929	148.241	–130.961	287.993	–11.572
1400	201.209	480.675	360.604	168.100	–130.829	320.217	–11.947
1500	205.895	494.721	369.080	188.460	–130.445	352.422	–12.272
17. BENZENE C_6H_6 (l)							
298.15	135.950	173.450	173.450	0.000	49.080	124.521	–21.815
300	136.312	174.292	173.453	.252	49.077	124.989	–21.762
400	161.793	216.837	179.082	15.102	48.978	150.320	–19.630
500	207.599	257.048	190.639	33.204	50.330	175.559	–18.340
18. BENZENE C_6H_6 (g)							
298.15	82.430	269.190	269.190	0.000	82.880	129.750	–22.731
300	83.020	269.700	269.190	0.153	82.780	130.040	–22.641
400	113.510	297.840	272.823	10.007	77.780	146.570	–19.140

TABLE 2.4.2
THERMOCHEMICAL PROPERTIES AT HIGH TEMPERATURES (continued)

	J K⁻¹ mol⁻¹			kJ mol⁻¹			
T/K	C_p°	S°	$-(G^\circ - H^\circ(T_r))/T$	$H^\circ - H^\circ(T_r)$	$\Delta_f H^\circ$	$\Delta_f G^\circ$	Log K_f
18. BENZENE C_6H_6 (g) (continued)							
500	139.340	326.050	280.658	22.696	73.740	164.260	–17.160
600	160.090	353.360	290.517	37.706	70.490	182.680	–15.903
700	176.790	379.330	301.360	54.579	67.910	201.590	–15.042
800	190.460	403.860	312.658	72.962	65.910	220.820	–14.418
900	201.840	426.970	324.084	92.597	64.410	240.280	–13.945
1000	211.430	448.740	335.473	113.267	63.340	259.890	–13.575
1100	219.580	469.280	346.710	134.827	62.620	277.640	–13.184
1200	226.540	488.690	357.743	157.137	62.200	299.320	–13.029
1300	232.520	507.070	368.534	180.097	62.000	319.090	–12.821
1400	237.680	524.490	379.056	203.607	61.990	338.870	–12.643
1500	242.140	541.040	389.302	227.607	62.110	358.640	–12.489
19. NAPHTHALENE $C_{10}H_8$ (cr, l)							
298.15	165.720	167.390	167.390	0.000	78.530	201.585	–35.316
300	167.001	168.419	167.393	0.308	78.466	202.349	–35.232
353.43	208.722	198.948	169.833	10.290	96.099	224.543	–33.186
PHASE TRANSITION: $\Delta_{trs}H$ = 18.980 kJ/mol, $\Delta_{trs}S$ = 53.702 J/K·mol, cr–l							
353.43	217.200	252.650	169.833	29.270	96.099	224.543	–33.186
400	241.577	280.916	181.124	39.917	96.067	241.475	–31.533
470	276.409	322.712	199.114	58.091	97.012	266.859	–29.658
20. NAPHTHALENE $C_{10}H_8$ (g)							
298.15	131.920	333.150	333.150	0.000	150.580	224.100	–39.260
300	132.840	333.970	333.157	0.244	150.450	224.560	–39.098
400	180.070	378.800	338.950	15.940	144.190	250.270	–32.681
500	219.740	423.400	351.400	36.000	139.220	277.340	–28.973
600	251.530	466.380	367.007	59.624	135.350	305.330	–26.581
700	277.010	507.140	384.146	86.096	132.330	333.950	–24.919
800	297.730	545.520	401.935	114.868	130.050	362.920	–23.696
900	314.850	581.610	419.918	145.523	128.430	392.150	–22.759
1000	329.170	615.550	437.806	177.744	127.510	421.700	–22.027
1100	341.240	647.500	455.426	211.281	127.100	450.630	–21.398
1200	351.500	677.650	472.707	245.932	126.960	480.450	–20.913
1300	360.260	706.130	489.568	281.531	127.060	509.770	–20.482
1400	367.780	733.110	506.009	317.941	127.390	539.740	–20.137
1500	374.270	758.720	522.019	355.051	127.920	568.940	–19.812
21. FORMALDEHYDE H_2CO (g)							
298.15	35.387	218.760	218.760	0.000	–108.700	–102.667	17.987
300	35.443	218.979	218.761	0.066	–108.731	–102.630	17.869
400	39.240	229.665	220.192	3.789	–110.438	–100.340	13.103
500	43.736	238.900	223.028	7.936	–112.073	–97.623	10.198
600	48.181	247.270	226.381	12.534	–113.545	–94.592	8.235
700	52.280	255.011	229.924	17.560	–114.833	–91.328	6.815
800	55.941	262.236	233.517	22.975	–115.942	–87.893	5.739
900	59.156	269.014	237.088	28.734	–116.889	–84.328	4.894
1000	61.951	275.395	240.603	34.792	–117.696	–80.666	4.213
1100	64.368	281.416	244.042	41.111	–118.382	–76.929	3.653
1200	66.453	287.108	247.396	47.655	–118.966	–73.134	3.183
1300	68.251	292.500	250.660	54.392	–119.463	–69.294	2.784
1400	69.803	297.616	253.833	61.297	–119.887	–65.418	2.441
1500	71.146	302.479	256.915	68.346	–120.249	–61.514	2.142

TABLE 2.4.2
THERMOCHEMICAL PROPERTIES AT HIGH TEMPERATURES (continued)

T/K	J K^{-1} mol^{-1}			kJ mol^{-1}			
	C_p°	S°	$-(G^\circ-H^\circ(T_r))/T$	$H^\circ-H^\circ(T_r)$	$\Delta_f H^\circ$	$\Delta_f G^\circ$	Log K_f
22. METHANOL CH_3OH (g)							
298.15	44.101	239.865	239.865	0.000	–201.000	–162.298	28.434
300	44.219	240.139	239.866	0.082	–201.068	–162.057	28.216
400	51.713	253.845	241.685	4.864	–204.622	–148.509	19.393
500	59.800	266.257	245.374	10.442	–207.750	–134.109	14.010
600	67.294	277.835	249.830	16.803	–210.387	–119.125	10.371
700	73.958	288.719	254.616	23.873	–212.570	–103.737	7.741
800	79.838	298.987	259.526	31.569	–214.350	–88.063	5.750
900	85.025	308.696	264.455	39.817	–215.782	–72.188	4.190
1000	89.597	317.896	269.343	48.553	–216.916	–56.170	2.934
1100	93.624	326.629	274.158	57.718	–217.794	–40.050	1.902
1200	97.165	334.930	278.879	67.262	–218.457	–23.861	1.039
1300	100.277	342.833	283.497	77.137	–218.936	–7.624	0.306
1400	103.014	350.367	288.007	87.304	–219.261	8.644	–0.322
1500	105.422	357.558	292.405	97.729	–219.456	24.930	–0.868
23. ACETALDEHYDE C_2H_4O (g)							
298.15	55.318	263.840	263.840	0.000	–166.190	–133.010	23.302
300	55.510	264.180	263.837	0.103	–166.250	–132.800	23.122
400	66.282	281.620	266.147	6.189	–169.530	–121.130	15.818
500	76.675	297.540	270.850	13.345	–172.420	–108.700	11.356
600	85.942	312.360	276.550	21.486	–174.870	–95.720	8.334
700	94.035	326.230	282.667	30.494	–176.910	–82.350	6.145
800	101.070	339.260	288.938	40.258	–178.570	–68.730	4.487
900	107.190	351.520	295.189	50.698	–179.880	–54.920	3.187
1000	112.490	363.100	301.431	61.669	–180.850	–40.930	2.138
1100	117.080	374.040	307.537	73.153	–181.560	–27.010	1.283
1200	121.060	384.400	313.512	85.065	–182.070	–12.860	0.560
1300	124.500	394.230	319.350	97.344	–182.420	1.240	–0.050
1400	127.490	403.570	325.031	109.954	–182.640	15.470	–0.577
1500	130.090	412.460	330.571	122.834	–182.750	29.580	–1.030
24. ETHANOL C_2H_5OH (g)							
298.15	65.652	281.622	281.622	0.000	–234.800	–167.874	29.410
300	65.926	282.029	281.623	0.122	–234.897	–167.458	29.157
400	81.169	303.076	284.390	7.474	–239.826	–144.216	18.832
500	95.400	322.750	290.115	16.318	–243.940	–119.820	12.517
600	107.656	341.257	297.112	26.487	–247.260	–94.672	8.242
700	118.129	358.659	304.674	37.790	–249.895	–69.023	5.151
800	127.171	375.038	312.456	50.065	–251.951	–43.038	2.810
900	135.049	390.482	320.276	63.185	–253.515	–16.825	0.976
1000	141.934	405.075	328.033	77.042	–254.662	9.539	–0.498
1100	147.958	418.892	335.670	91.543	–255.454	36.000	–1.709
1200	153.232	431.997	343.156	106.609	–255.947	62.520	–2.721
1300	157.849	444.448	350.473	122.168	–256.184	89.070	–3.579
1400	161.896	456.298	357.612	138.160	–256.206	115.630	–4.314
1500	165.447	467.591	364.571	154.531	–256.044	142.185	–4.951
25. ACETIC ACID $C_2H_4O_2$ (g)							
298.15	63.438	283.470	283.470	0.000	–432.249	–374.254	65.567
300	63.739	283.863	283.471	0.118	–432.324	–373.893	65.100
400	79.665	304.404	286.164	7.296	–436.006	–353.840	46.206
500	93.926	323.751	291.765	15.993	–438.875	–332.950	34.783

TABLE 2.4.2
THERMOCHEMICAL PROPERTIES AT HIGH TEMPERATURES (continued)

	J K^{-1} mol^{-1}			kJ mol^{-1}			
T/K	C_p°	S°	$-(G^\circ-H^\circ(T_r))/T$	$H^\circ-H^\circ(T_r)$	$\Delta_f H^\circ$	$\Delta_f G^\circ$	Log K_f
25. ACETIC ACID $C_2H_4O_2$ (g) (continued)							
600	106.181	341.988	298.631	26.014	–440.993	–311.554	27.123
700	116.627	359.162	306.064	37.169	–442.466	–289.856	21.629
800	125.501	375.331	313.722	49.287	–443.395	–267.985	17.497
900	132.989	390.558	321.422	62.223	–443.873	–246.026	14.279
1000	139.257	404.904	329.060	75.844	–443.982	–224.034	11.702
1100	144.462	418.429	336.576	90.039	–443.798	–202.046	9.594
1200	148.760	431.189	343.933	104.707	–443.385	–180.086	7.839
1300	152.302	443.240	351.113	119.765	–442.795	–158.167	6.355
1400	155.220	454.637	358.105	135.146	–442.071	–136.299	5.085
1500	157.631	465.432	364.903	150.793	–441.247	–114.486	3.987
26. ACETONE C_3H_6O (g)							
298.15	74.517	295.349	295.349	0.000	–217.150	–152.716	26.757
300	74.810	295.809	295.349	0.138	–217.233	–152.339	26.521
400	91.755	319.658	298.498	8.464	–222.212	–129.913	16.962
500	107.864	341.916	304.988	18.464	–226.522	–106.315	11.107
600	122.047	362.836	312.873	29.978	–230.120	–81.923	7.133
700	134.306	382.627	321.470	42.810	–233.049	–56.986	4.252
800	144.934	401.246	330.265	56.785	–235.350	–31.673	2.068
900	154.097	418.860	339.141	71.747	–237.149	–6.109	0.353
1000	162.046	435.513	347.950	87.563	–238.404	19.707	–1.030
1100	168.908	451.286	356.617	104.136	–239.283	45.396	–2.157
1200	174.891	466.265	365.155	121.332	–239.827	71.463	–3.110
1300	180.079	480.491	373.513	139.072	–240.120	97.362	–3.912
1400	184.556	493.963	381.596	157.314	–240.203	123.470	–4.607
1500	188.447	506.850	389.533	175.975	–240.120	149.369	–5.202
27. PHENOL C_6H_6O (g)							
298.15	103.220	314.810	314.810	0.000	–96.400	–32.630	5.720
300	103.860	315.450	314.810	0.192	–96.490	–32.230	5.610
400	135.790	349.820	319.278	12.217	–100.870	–10.180	1.330
500	161.910	383.040	328.736	27.152	–104.240	12.970	–1.360
600	182.480	414.450	340.430	44.412	–106.810	36.650	–3.190
700	198.840	443.860	353.134	63.508	–108.800	60.750	–4.530
800	212.140	471.310	366.211	84.079	–110.300	85.020	–5.550
900	223.190	496.950	379.327	105.861	–111.370	109.590	–6.360
1000	232.490	520.960	392.302	128.658	–111.990	134.280	–7.010
1100	240.410	543.500	405.033	152.314	–112.280	158.620	–7.530
1200	247.200	564.720	417.468	176.703	–112.390	183.350	–7.980
1300	253.060	584.740	429.568	201.723	–112.330	208.070	–8.360
1400	258.120	603.680	441.331	227.288	–112.120	233.050	–8.700
1500	262.520	621.650	452.767	253.325	–111.780	257.540	–8.970
28. CARBON TETRAFLUORIDE CF_4 (g)							
298.15	61.050	261.455	261.455	0.000	–933.200	–888.518	155.663
300	61.284	261.833	261.456	0.113	–933.219	–888.240	154.654
400	72.399	281.057	264.001	6.822	–933.986	–873.120	114.016
500	80.713	298.153	269.155	14.499	–934.372	–857.852	89.618
600	86.783	313.434	275.284	22.890	–934.490	–842.533	73.348
700	91.212	327.162	281.732	31.801	–934.431	–827.210	61.726
800	94.479	339.566	288.199	41.094	–934.261	–811.903	53.011
900	96.929	350.842	294.542	50.670	–934.024	–796.622	46.234

TABLE 2.4.2
THERMOCHEMICAL PROPERTIES AT HIGH TEMPERATURES (continued)

T/K	J K^{-1} mol^{-1}			kJ mol^{-1}			
	C_p°	S°	$-(G^\circ-H^\circ(T_r))/T$	$H^\circ-H^\circ(T_r)$	$\Delta_f H^\circ$	$\Delta_f G^\circ$	Log K_f
28. CARBON TETRAFLUORIDE CF_4 (g) (continued)							
1000	98.798	361.156	300.695	60.460	–933.745	–781.369	40.814
1100	100.250	370.643	306.629	70.416	–933.442	–766.146	36.381
1200	101.396	379.417	312.334	80.500	–933.125	–750.952	32.688
1300	102.314	387.571	317.811	90.687	–932.800	–735.784	29.564
1400	103.059	395.181	323.069	100.957	–932.470	–720.641	26.887
1500	103.671	402.313	328.116	111.295	–932.137	–705.522	24.568
29. TRIFLUOROMETHANE CHF_3 (g)							
298.15	51.069	259.675	259.675	0.000	–696.700	–662.237	116.020
300	51.258	259.991	259.676	0.095	–696.735	–662.023	115.267
400	61.148	276.113	261.807	5.722	–698.427	–650.186	84.905
500	69.631	290.700	266.149	12.275	–699.715	–637.969	66.647
600	76.453	304.022	271.368	19.593	–700.634	–625.528	54.456
700	81.868	316.230	276.917	27.519	–701.253	–612.957	45.739
800	86.201	327.455	282.542	35.930	–701.636	–600.315	39.196
900	89.719	337.818	288.116	44.732	–701.832	–587.636	34.105
1000	92.617	347.426	293.572	53.854	–701.879	–574.944	30.032
1100	95.038	356.370	298.879	63.240	–701.805	–562.253	26.699
1200	97.084	364.730	304.022	72.849	–701.629	–549.574	23.922
1300	98.833	372.571	308.997	82.647	–701.368	–536.913	21.573
1400	100.344	379.952	313.804	92.607	–701.033	–524.274	19.561
1500	101.660	386.921	318.449	102.709	–700.635	–511.662	17.817
30. CHLOROTRIFLUOROMETHANE $CClF_3$ (g)							
298.15	66.886	285.419	285.419	0.000	–707.800	–667.238	116.896
300	67.111	285.834	285.421	0.124	–707.810	–666.986	116.131
400	77.528	306.646	288.187	7.383	–708.153	–653.316	85.313
500	85.013	324.797	293.734	15.532	–708.170	–639.599	66.818
600	90.329	340.794	300.271	24.314	–707.975	–625.901	54.489
700	94.132	355.020	307.096	33.547	–707.654	–612.246	45.686
800	96.899	367.780	313.897	43.106	–707.264	–598.642	39.087
900	98.951	379.317	320.536	52.903	–706.837	–585.090	33.957
1000	100.507	389.827	326.947	62.880	–706.396	–571.586	29.856
1100	101.708	399.465	333.108	72.993	–705.950	–558.126	26.503
1200	102.651	408.357	339.013	83.213	–705.505	–544.707	23.710
1300	103.404	416.604	344.668	93.517	–705.064	–531.326	21.349
1400	104.012	424.290	350.084	103.889	–704.628	–517.977	19.326
1500	104.512	431.484	355.273	114.316	–704.196	–504.660	17.574
31. DICHLORODIFLUOROMETHANE CCl_2F_2 (g)							
298.15	72.476	300.903	300.903	0.000	–486.000	–447.030	78.317
300	72.691	301.352	300.905	0.134	–486.002	–446.788	77.792
400	82.408	323.682	303.883	7.919	–485.945	–433.716	56.637
500	89.063	342.833	309.804	16.514	–485.618	–420.692	43.949
600	93.635	359.500	316.729	25.663	–485.136	–407.751	35.497
700	96.832	374.189	323.909	35.196	–484.576	–394.897	29.467
800	99.121	387.276	331.027	44.999	–483.984	–382.126	24.950
900	100.801	399.053	337.942	55.000	–483.388	–369.429	21.441
1000	102.062	409.742	344.596	65.146	–482.800	–356.799	18.637
1100	103.030	419.517	350.969	75.402	–482.226	–344.227	16.346
1200	103.786	428.515	357.061	85.745	–481.667	–331.706	14.439

TABLE 2.4.2
THERMOCHEMICAL PROPERTIES AT HIGH TEMPERATURES (continued)

T/K	J K^{-1} mol^{-1}			kJ mol^{-1}			
	C_p°	S°	$-(G^\circ-H^\circ(T_r))/T$	$H^\circ-H^\circ(T_r)$	$\Delta_f H^\circ$	$\Delta_f G^\circ$	Log K_f
31. DICHLORODIFLUOROMETHANE CCl_2F_2 (g) (continued)							
1300	104.388	436.847	362.882	96.154	–481.121	–319.232	12.827
1400	104.874	444.602	368.445	106.618	–480.588	–306.799	11.447
1500	105.270	451.851	373.767	117.126	–480.065	–294.404	10.252
32. CHLORODIFLUOROMETHANE $CHClF_2$ (g)							
298.15	55.853	280.915	280.915	0.000	–475.000	–443.845	77.759
300	56.039	281.261	280.916	0.104	–475.028	–443.652	77.246
400	65.395	298.701	283.231	6.188	–476.390	–432.978	56.540
500	73.008	314.145	287.898	13.123	–477.398	–422.001	44.086
600	78.940	328.003	293.448	20.733	–478.103	–410.851	35.767
700	83.551	340.533	299.294	28.867	–478.574	–399.603	29.818
800	87.185	351.936	305.172	37.411	–478.870	–388.299	25.353
900	90.100	362.379	310.956	46.280	–479.031	–376.967	21.878
1000	92.475	371.999	316.586	55.413	–479.090	–365.622	19.098
1100	94.433	380.908	322.033	64.761	–479.068	–354.276	16.823
1200	96.066	389.196	327.289	74.289	–478.982	–342.935	14.927
1300	97.438	396.941	332.352	83.966	–478.843	–331.603	13.324
1400	98.601	404.206	337.228	93.769	–478.661	–320.283	11.950
1500	99.593	411.044	341.923	103.681	–478.443	–308.978	10.759
33. METHYLAMINE CH_5N (g)							
298.15	50.053	242.881	242.881	0.000	–22.529	32.734	–5.735
300	50.227	243.196	242.893	0.091	–22.614	33.077	–5.759
400	60.171	258.986	244.975	5.604	–26.846	52.294	–6.829
500	70.057	273.486	249.244	12.121	–30.431	72.510	–7.575
600	78.929	287.063	254.431	19.579	–33.364	93.382	–8.129
700	86.711	299.826	260.008	27.873	–35.712	114.702	–8.559
800	93.545	311.865	265.749	36.893	–37.548	136.316	–8.900
900	99.573	323.239	271.511	46.555	–38.949	158.138	–9.178
1000	104.886	334.006	277.220	56.786	–39.967	180.098	–9.407
1100	109.576	344.233	282.861	67.509	–40.681	201.822	–9.584
1200	113.708	353.944	288.374	78.685	–41.136	224.240	–9.761
1300	117.341	363.190	293.775	90.239	–41.376	246.364	–9.899
1400	120.542	372.012	299.061	102.131	–41.451	268.504	–10.018
1500	123.353	380.426	304.209	114.326	–41.381	290.639	–10.121
34. CHLORINE Cl (g)							
298.15	21.838	165.190	165.190	0.000	121.302	105.306	–18.449
300	21.852	165.325	165.190	0.040	121.311	105.207	–18.318
400	22.467	171.703	166.055	2.259	121.795	99.766	–13.028
500	22.744	176.752	167.708	4.522	122.272	94.203	–9.841
600	22.781	180.905	169.571	6.800	122.734	88.546	–7.709
700	22.692	184.411	171.448	9.074	123.172	82.813	–6.179
800	22.549	187.432	173.261	11.337	123.585	77.019	–5.029
900	22.389	190.079	174.986	13.584	123.971	71.175	–4.131
1000	22.233	192.430	176.615	15.815	124.334	65.289	–3.410
1100	22.089	194.542	178.150	18.031	124.675	59.368	–2.819
1200	21.959	196.458	179.597	20.233	124.996	53.416	–2.325
1300	21.843	198.211	180.963	22.423	125.299	47.439	–1.906
1400	21.742	199.826	182.253	24.602	125.587	41.439	–1.546
1500	21.652	201.323	183.475	26.772	125.861	35.418	–1.233

TABLE 2.4.2
THERMOCHEMICAL PROPERTIES AT HIGH TEMPERATURES (continued)

T/K	C_p°	S°	$-(G^\circ-H^\circ(T_r))/T$	$H^\circ-H^\circ(T_r)$	$\Delta_f H^\circ$	$\Delta_f G^\circ$	Log K_f
	J K^{-1} mol^{-1}			kJ mol^{-1}			
35. DICHLORINE Cl_2 (g)							
298.15	33.949	223.079	223.079	0.000	0.000	0.000	0.000
300	33.981	223.290	223.080	0.063	0.000	0.000	0.000
400	35.296	233.263	224.431	3.533	0.000	0.000	0.000
500	36.064	241.229	227.021	7.104	0.000	0.000	0.000
600	36.547	247.850	229.956	10.736	0.000	0.000	0.000
700	36.874	253.510	232.926	14.408	0.000	0.000	0.000
800	37.111	258.450	235.815	18.108	0.000	0.000	0.000
900	37.294	262.832	238.578	21.829	0.000	0.000	0.000
1000	37.442	266.769	241.203	25.566	0.000	0.000	0.000
1100	37.567	270.343	243.692	29.316	0.000	0.000	0.000
1200	37.678	273.617	246.052	33.079	0.000	0.000	0.000
1300	37.778	276.637	248.290	36.851	0.000	0.000	0.000
1400	37.872	279.440	250.416	40.634	0.000	0.000	0.000
1500	37.961	282.056	252.439	44.426	0.000	0.000	0.000
36. HYDROGEN CHLORIDE HCl (g)							
298.15	29.136	186.902	186.902	0.000	–92.310	–95.298	16.696
300	29.137	187.082	186.902	0.054	–92.314	–95.317	16.596
400	29.175	195.468	188.045	2.969	–92.587	–96.278	12.573
500	29.304	201.990	190.206	5.892	–92.911	–97.164	10.151
600	29.576	207.354	192.630	8.835	–93.249	–97.983	8.530
700	29.988	211.943	195.069	11.812	–93.577	–98.746	7.368
800	30.500	215.980	197.435	14.836	–93.879	–99.464	6.494
900	31.063	219.604	199.700	17.913	–94.149	–100.145	5.812
1000	31.639	222.907	201.858	21.049	–94.384	–100.798	5.265
1100	32.201	225.949	203.912	24.241	–94.587	–101.430	4.816
1200	32.734	228.774	205.867	27.488	–94.760	–102.044	4.442
1300	33.229	231.414	207.732	30.786	–94.908	–102.645	4.124
1400	33.684	233.893	209.513	34.132	–95.035	–103.235	3.852
1500	34.100	236.232	211.217	37.522	–95.146	–103.817	3.615
37. COPPER Cu (cr, l)							
298.15	24.440	33.150	33.150	0.000	0.000	0.000	0.000
300	24.460	33.301	33.150	0.045	0.000	0.000	0.000
400	25.339	40.467	34.122	2.538	0.000	0.000	0.000
500	25.966	46.192	35.982	5.105	0.000	0.000	0.000
600	26.479	50.973	38.093	7.728	0.000	0.000	0.000
700	26.953	55.090	40.234	10.399	0.000	0.000	0.000
800	27.448	58.721	42.322	13.119	0.000	0.000	0.000
900	28.014	61.986	44.328	15.891	0.000	0.000	0.000
1000	28.700	64.971	46.245	18.726	0.000	0.000	0.000
1100	29.553	67.745	48.075	21.637	0.000	0.000	0.000
1200	30.617	70.361	49.824	24.644	0.000	0.000	0.000
1300	31.940	72.862	51.501	27.769	0.000	0.000	0.000
1358	32.844	74.275	52.443	29.647	0.000	0.000	0.000
PHASE TRANSITION: $\Delta_{trs}H$ = 13.141 kJ/mol, $\Delta_{trs}S$ = 9.676 J/K·mol, cr–l							
1358	32.800	83.951	52.443	42.788	0.000	0.000	0.000
1400	32.800	84.950	53.403	44.166	0.000	0.000	0.000
1500	32.800	87.213	55.583	47.446	0.000	0.000	0.000
38. COPPER Cu (g)							
298.15	20.786	166.397	166.397	0.000	337.600	297.873	–52.185
300	20.786	166.525	166.397	0.038	337.594	297.626	–51.821

TABLE 2.4.2
THERMOCHEMICAL PROPERTIES AT HIGH TEMPERATURES (continued)

	J K⁻¹ mol⁻¹			kJ mol⁻¹			
T/K	C_p°	S°	$-(G^\circ-H^\circ(T_r))/T$	$H^\circ-H^\circ(T_r)$	Δ_fH°	Δ_fG°	Log K_f
38. COPPER Cu (g)							
400	20.786	172.505	167.213	2.117	337.179	284.364	–37.134
500	20.786	177.143	168.752	4.196	336.691	271.215	–28.333
600	20.786	180.933	170.476	6.274	336.147	258.170	–22.475
700	20.786	184.137	172.205	8.353	335.554	245.221	–18.298
800	20.786	186.913	173.874	10.431	334.913	232.359	–15.171
900	20.786	189.361	175.461	12.510	334.219	219.581	–12.744
1000	20.786	191.551	176.963	14.589	333.463	206.883	–10.806
1100	20.788	193.532	178.380	16.667	332.631	194.265	–9.225
1200	20.793	195.341	179.719	18.746	331.703	181.726	–7.910
1300	20.803	197.006	180.986	20.826	330.657	169.270	–6.801
1400	20.823	198.548	182.186	22.907	316.342	157.305	–5.869
1500	20.856	199.986	183.325	24.991	315.146	145.987	–5.084
39. COPPER OXIDE CuO (cr)							
298.15	42.300	42.740	42.740	0.000	–162.000	–134.277	23.524
300	42.417	43.002	42.741	0.078	–161.994	–134.105	23.349
400	46.783	55.878	44.467	4.564	–161.487	–124.876	16.307
500	49.190	66.596	47.852	9.372	–160.775	–115.803	12.098
600	50.827	75.717	51.755	14.377	–159.973	–106.883	9.305
700	52.099	83.651	55.757	19.526	–159.124	–98.102	7.320
800	53.178	90.680	59.691	24.791	–158.247	–89.444	5.840
900	54.144	97.000	63.491	30.158	–157.356	–80.897	4.695
1000	55.040	102.751	67.134	35.617	–156.462	–72.450	3.784
1100	55.890	108.037	70.615	41.164	–155.582	–64.091	3.043
1200	56.709	112.936	73.941	46.794	–154.733	–55.812	2.429
1300	57.507	117.507	77.118	52.505	–153.940	–47.601	1.913
1400	58.288	121.797	80.158	58.295	–166.354	–39.043	1.457
1500	59.057	125.845	83.070	64.163	–165.589	–29.975	1.044
40. DICOPPER OXIDE Cu_2O (cr)							
298.15	62.600	92.550	92.550	0.000	–173.100	–150.344	26.339
300	62.721	92.938	92.551	0.116	–173.102	–150.203	26.152
400	67.587	111.712	95.078	6.654	–173.036	–142.572	18.618
500	70.784	127.155	99.995	13.580	–172.772	–134.984	14.101
600	73.323	140.291	105.643	20.789	–172.389	–127.460	11.096
700	75.552	151.764	111.429	28.235	–171.914	–120.009	8.955
800	77.616	161.989	117.121	35.894	–171.363	–112.631	7.354
900	79.584	171.245	122.629	43.755	–170.750	–105.325	6.113
1000	81.492	179.729	127.920	51.809	–170.097	–98.091	5.124
1100	83.360	187.584	132.992	60.052	–169.431	–90.922	4.317
1200	85.202	194.917	137.850	68.480	–168.791	–83.814	3.648
1300	87.026	201.808	142.507	77.092	–168.223	–76.756	3.084
1400	88.836	208.324	146.978	85.885	–194.030	–68.926	2.572
1500	90.636	214.515	151.276	94.858	–193.438	–60.010	2.090
41. COPPER DICHLORIDE $CuCl_2$ (cr, l)							
298.15	71.880	108.070	108.070	0.000	–218.000	–173.826	30.453
300	71.998	108.515	108.071	0.133	–217.975	–173.552	30.218
400	76.338	129.899	110.957	7.577	–216.494	–158.962	20.758
500	78.654	147.204	116.532	15.336	–214.873	–144.765	15.123
600	80.175	161.687	122.884	23.282	–213.182	–130.901	11.396
675	81.056	171.183	127.732	29.329	–211.185	–120.693	9.340

TABLE 2.4.2
THERMOCHEMICAL PROPERTIES AT HIGH TEMPERATURES (continued)

T/K	C_p° (J K^{-1} mol^{-1})	S° (J K^{-1} mol^{-1})	$-(G^\circ-H^\circ(T_r))/T$ (J K^{-1} mol^{-1})	$H^\circ-H^\circ(T_r)$ (kJ mol^{-1})	$\Delta_f H^\circ$ (kJ mol^{-1})	$\Delta_f G^\circ$ (kJ mol^{-1})	Log K_f
41. COPPER DICHLORIDE $CuCl_2$ (cr, l) (continued)							
PHASE TRANSITION: $\Delta_{trs}H$ = 0.700 kJ/mol, $\Delta_{trs}S$ = 1.037 J/K·mol, crII–crI							
675	82.400	172.220	127.732	30.029	–211.185	–120.693	9.340
700	82.400	175.216	129.375	32.089	–210.719	–117.350	8.757
800	82.400	186.219	135.808	40.329	–208.898	–104.137	6.799
871	82.400	193.226	140.207	46.179	–192.649	–94.893	5.691
PHASE TRANSITION: $\Delta_{trs}H$ = 15.001 kJ/mol, $\Delta_{trs}S$ = 17.221 J/K·mol, crI–l							
871	100.000	210.447	140.207	61.180	–192.649	–94.893	5.691
900	100.000	213.723	142.523	64.080	–191.640	–91.655	5.319
1000	100.000	224.259	150.179	74.080	–188.212	–80.730	4.217
1100	100.000	233.790	157.353	84.080	–184.873	–70.144	3.331
1130.75	100.000	236.547	159.470	87.155	–183.867	–66.951	3.093
42. COPPER DICHLORIDE $CuCl_2$ (g)							
298.15	56.814	278.418	278.418	0.000	–43.268	–49.883	8.739
300	56.869	278.769	278.419	0.105	–43.271	–49.924	8.692
400	58.992	295.456	280.679	5.911	–43.428	–52.119	6.806
500	60.111	308.752	285.010	11.871	–43.606	–54.271	5.670
600	60.761	319.774	289.911	17.918	–43.814	–56.385	4.909
700	61.168	329.173	294.865	24.015	–44.060	–58.462	4.362
800	61.439	337.360	299.677	30.147	–44.349	–60.500	3.950
900	61.630	344.608	304.274	36.301	–44.688	–62.499	3.627
1000	61.776	351.109	308.638	42.471	–45.088	–64.457	3.367
1100	61.900	357.003	312.771	48.655	–45.566	–66.372	3.152
1200	62.022	362.394	316.685	54.851	–46.139	–68.239	2.970
1300	62.159	367.364	320.395	61.060	–46.829	–70.053	2.815
1400	62.325	371.976	323.916	67.284	–60.784	–71.404	2.664
1500	62.531	376.283	327.265	73.526	–61.613	–72.133	2.512
43. FLUORINE F (g)							
298.15	22.746	158.750	158.750	0.000	79.380	62.280	–10.911
300	22.742	158.891	158.750	0.042	79.393	62.173	–10.825
400	22.432	165.394	159.639	2.302	80.043	56.332	–7.356
500	22.100	170.363	161.307	4.528	80.587	50.340	–5.259
600	21.832	174.368	163.161	6.724	81.046	44.246	–3.852
700	21.629	177.717	165.008	8.897	81.442	38.081	–2.842
800	21.475	180.595	166.780	11.052	81.792	31.862	–2.080
900	21.357	183.117	168.458	13.193	82.106	25.601	–1.486
1000	21.266	185.362	170.039	15.324	82.391	19.308	–1.009
1100	21.194	187.386	171.525	17.447	82.654	12.986	–0.617
1200	21.137	189.227	172.925	19.563	82.897	6.642	–0.289
1300	21.091	190.917	174.245	21.675	83.123	0.278	–0.011
1400	21.054	192.479	175.492	23.782	83.335	–6.103	0.228
1500	21.022	193.930	176.673	25.886	83.533	–12.498	0.435
44. DIFLUORINE F_2 (g)							
298.15	31.304	202.790	202.790	0.000	0.000	0.000	0.000
300	31.337	202.984	202.790	0.058	0.000	0.000	0.000
400	32.995	212.233	204.040	3.277	0.000	0.000	0.000
500	34.258	219.739	206.453	6.643	0.000	0.000	0.000
600	35.171	226.070	209.208	10.117	0.000	0.000	0.000
700	35.839	231.545	212.017	13.669	0.000	0.000	0.000
800	36.343	236.365	214.765	17.279	0.000	0.000	0.000

TABLE 2.4.2
THERMOCHEMICAL PROPERTIES AT HIGH TEMPERATURES (continued)

	J K⁻¹ mol⁻¹			kJ mol⁻¹			
T/K	C_p°	S°	$-(G^\circ - H^\circ(T_r))/T$	$H^\circ - H^\circ(T_r)$	$\Delta_f H^\circ$	$\Delta_f G^\circ$	Log K_f
44. DIFLUORINE F_2 (g) (continued)							
900	36.740	240.669	217.409	20.934	0.000	0.000	0.000
1000	37.065	244.557	219.932	24.625	0.000	0.000	0.000
1100	37.342	248.103	222.334	28.346	0.000	0.000	0.000
1200	37.588	251.363	224.619	32.093	0.000	0.000	0.000
1300	37.811	254.381	226.794	35.863	0.000	0.000	0.000
1400	38.019	257.191	228.866	39.654	0.000	0.000	0.000
1500	38.214	259.820	230.843	43.466	0.000	0.000	0.000
45. HYDROGEN FLUORIDE HF (g)							
298.15	29.137	173.776	173.776	0.000	–273.300	–275.399	48.248
300	29.137	173.956	173.776	0.054	–273.302	–275.412	47.953
400	29.149	182.340	174.919	2.968	–273.450	–276.096	36.054
500	29.172	188.846	177.078	5.884	–273.679	–276.733	28.910
600	29.230	194.169	179.496	8.804	–273.961	–277.318	24.142
700	29.350	198.683	181.923	11.732	–274.277	–277.852	20.733
800	29.549	202.614	184.269	14.676	–274.614	–278.340	18.174
900	29.827	206.110	186.505	17.645	–274.961	–278.785	16.180
1000	30.169	209.270	188.626	20.644	–275.309	–279.191	14.583
1100	30.558	212.163	190.636	23.680	–275.652	–279.563	13.275
1200	30.974	214.840	192.543	26.756	–275.988	–279.904	12.184
1300	31.403	217.336	194.355	29.875	–276.315	–280.217	11.259
1400	31.831	219.679	196.081	33.037	–276.631	–280.505	10.466
1500	32.250	221.889	197.729	36.241	–276.937	–280.771	9.777
46. GERMANIUM Ge (cr, l)							
298.15	23.222	31.090	31.090	0.000	0.000	0.000	0.000
300	23.249	31.234	31.090	0.043	0.000	0.000	0.000
400	24.310	38.083	32.017	2.426	0.000	0.000	0.000
500	24.962	43.582	33.798	4.892	0.000	0.000	0.000
600	25.452	48.178	35.822	7.414	0.000	0.000	0.000
700	25.867	52.133	37.876	9.980	0.000	0.000	0.000
800	26.240	55.612	39.880	12.586	0.000	0.000	0.000
900	26.591	58.723	41.804	15.227	0.000	0.000	0.000
1000	26.926	61.542	43.639	17.903	0.000	0.000	0.000
1100	27.252	64.124	45.386	20.612	0.000	0.000	0.000
1200	27.571	66.509	47.048	23.353	0.000	0.000	0.000
1211.4	27.608	66.770	47.232	23.668	0.000	0.000	0.000
		PHASE TRANSITION: $\Delta_{trs} H$ = 37.030 kJ/mol, $\Delta_{trs} S$ = 30.568 J/K·mol, cr–l					
1211.4	27.600	97.338	47.232	60.698	0.000	0.000	0.000
1300	27.600	99.286	50.714	63.143	0.000	0.000	0.000
1400	27.600	101.331	54.258	65.903	0.000	0.000	0.000
1500	27.600	103.236	57.460	68.663	0.000	0.000	0.000
47. GERMANIUM Ge (g)							
298.15	30.733	167.903	167.903	0.000	367.800	327.009	–57.290
300	30.757	168.094	167.904	0.057	367.814	326.756	–56.893
400	31.071	177.025	169.119	3.162	368.536	312.959	–40.868
500	30.360	183.893	171.415	6.239	369.147	298.991	–31.235
600	29.265	189.334	173.965	9.222	369.608	284.914	–24.804
700	28.102	193.758	176.487	12.090	369.910	270.773	–20.205
800	27.029	197.439	178.882	14.845	370.060	256.598	–16.754
900	26.108	200.567	181.122	17.501	370.073	242.414	–14.069

TABLE 2.4.2
THERMOCHEMICAL PROPERTIES AT HIGH TEMPERATURES (continued)

T/K	C_p° (J K^{-1} mol^{-1})	S° (J K^{-1} mol^{-1})	$-(G^\circ - H^\circ(T_r))/T$ (J K^{-1} mol^{-1})	$H^\circ - H^\circ(T_r)$ (kJ mol^{-1})	$\Delta_f H^\circ$ (kJ mol^{-1})	$\Delta_f G^\circ$ (kJ mol^{-1})	Log K_f
47. GERMANIUM Ge (g) (continued)							
1000	25.349	203.277	183.205	20.072	369.969	228.234	–11.922
1100	24.741	205.664	185.141	22.575	369.763	214.069	–10.165
1200	24.264	207.795	186.941	25.025	369.471	199.928	–8.703
1300	23.898	209.722	188.621	27.432	332.088	188.521	–7.575
1400	23.624	211.483	190.192	29.807	331.704	177.492	–6.622
1500	23.426	213.105	191.666	32.159	331.296	166.491	–5.798
48. GERMANIUM DIOXIDE GeO_2 (cr, l)							
298.15	50.166	39.710	39.710	0.000	–580.200	–521.605	91.382
300	50.475	40.021	39.711	0.093	–580.204	–521.242	90.755
400	61.281	56.248	41.850	5.759	–579.893	–501.610	65.503
500	66.273	70.519	46.191	12.164	–579.013	–482.134	50.368
600	69.089	82.872	51.299	18.943	–577.915	–462.859	40.295
700	70.974	93.671	56.597	25.952	–576.729	–443.776	33.115
800	72.449	103.247	61.841	33.125	–575.498	–424.866	27.741
900	73.764	111.857	66.928	40.436	–574.235	–406.113	23.570
1000	75.049	119.696	71.819	47.877	–572.934	–387.502	20.241
1100	76.378	126.910	76.504	55.447	–571.582	–369.024	17.523
1200	77.796	133.616	80.987	63.155	–570.166	–350.671	15.264
1300	79.332	139.903	85.279	71.010	–605.685	–329.732	13.249
1308	79.460	140.390	85.615	71.646	–584.059	–328.034	13.100
PHASE TRANSITION: $\Delta_{trs}H$ = 21.500 kJ/mol, $\Delta_{trs}S$ = 16.437 J/K·mol, crII–crI							
1308	80.075	156.827	85.615	93.146	–584.059	–328.034	13.100
1388	81.297	161.617	89.858	99.601	–565.504	–312.415	11.757
PHASE TRANSITION: $\Delta_{trs}H$ = 17.200 kJ/mol, $\Delta_{trs}S$ = 12.392 J/K·mol, crI–l							
1388	78.500	174.009	89.858	116.801	–565.504	–312.415	11.757
1400	78.500	174.685	90.582	117.743	–565.328	–310.228	11.575
1500	78.500	180.100	96.372	125.593	–563.882	–292.057	10.170
49. GERMANIUM TETRACHLORIDE $GeCl_4$ (g)							
298.15	95.918	348.393	348.393	0.000	–500.000	–461.582	80.866
300	96.041	348.987	348.395	0.178	–499.991	–461.343	80.326
400	100.750	377.342	352.229	10.045	–499.447	–448.540	58.573
500	103.206	400.114	359.604	20.255	–498.845	–435.882	45.536
600	104.624	419.067	367.980	30.652	–498.234	–423.347	36.855
700	105.509	435.266	376.463	41.162	–497.634	–410.914	30.662
800	106.096	449.396	384.715	51.744	–497.057	–398.565	26.023
900	106.504	461.917	392.611	62.375	–496.509	–386.287	22.419
1000	106.799	473.155	400.113	73.041	–495.993	–374.068	19.539
1100	107.020	483.344	407.224	83.733	–495.512	–361.899	17.185
1200	107.189	492.664	413.961	94.444	–495.067	–349.772	15.225
1300	107.320	501.249	420.349	105.169	–531.677	–334.973	13.459
1400	107.425	509.206	426.416	115.907	–531.265	–319.857	11.934
1500	107.509	516.621	432.185	126.654	–530.861	–304.771	10.613
50. HYDROGEN H (g)							
298.15	20.786	114.716	114.716	0.000	217.998	203.276	–35.613
300	20.786	114.845	114.716	0.038	218.010	203.185	–35.377
400	20.786	120.824	115.532	2.117	218.635	198.149	–25.875
500	20.786	125.463	117.071	4.196	219.253	192.956	–20.158
600	20.786	129.252	118.795	6.274	219.867	187.639	–16.335
700	20.786	132.457	120.524	8.353	220.476	182.219	–13.597

TABLE 2.4.2
THERMOCHEMICAL PROPERTIES AT HIGH TEMPERATURES (continued)

T/K	J K^{-1} mol^{-1}			kJ mol^{-1}			
	C_p°	S°	$-(G^\circ - H^\circ(T_r))/T$	$H^\circ - H^\circ(T_r)$	$\Delta_f H^\circ$	$\Delta_f G^\circ$	Log K_f
50. HYDROGEN H (g) (continued)							
800	20.786	135.232	122.193	10.431	221.079	176.712	–11.538
900	20.786	137.680	123.780	12.510	221.670	171.131	–9.932
1000	20.786	139.870	125.282	14.589	222.247	165.485	–8.644
1100	20.786	141.852	126.700	16.667	222.806	159.781	–7.587
1200	20.786	143.660	128.039	18.746	223.345	154.028	–6.705
1300	20.786	145.324	129.305	20.824	223.864	148.230	–5.956
1400	20.786	146.864	130.505	22.903	224.360	142.393	–5.313
1500	20.786	148.298	131.644	24.982	224.835	136.522	–4.754
51. DIHYDROGEN H_2 (g)							
298.15	28.836	130.680	130.680	0.000	0.000	0.000	0.000
300	28.849	130.858	130.680	0.053	0.000	0.000	0.000
400	29.181	139.217	131.818	2.960	0.000	0.000	0.000
500	29.260	145.738	133.974	5.882	0.000	0.000	0.000
600	29.327	151.078	136.393	8.811	0.000	0.000	0.000
700	29.440	155.607	138.822	11.749	0.000	0.000	0.000
800	29.623	159.549	141.172	14.702	0.000	0.000	0.000
900	29.880	163.052	143.412	17.676	0.000	0.000	0.000
1000	30.204	166.217	145.537	20.680	0.000	0.000	0.000
1100	30.580	169.113	147.550	23.719	0.000	0.000	0.000
1200	30.991	171.791	149.460	26.797	0.000	0.000	0.000
1300	31.422	174.288	151.275	29.918	0.000	0.000	0.000
1400	31.860	176.633	153.003	33.082	0.000	0.000	0.000
1500	32.296	178.846	154.653	36.290	0.000	0.000	0.000
52. HYDROXYL OH (g)							
298.15	29.886	183.737	183.737	0.000	39.349	34.631	–6.067
300	29.879	183.922	183.738	0.055	39.350	34.602	–6.025
400	29.604	192.476	184.906	3.028	39.384	33.012	–4.311
500	29.495	199.067	187.104	5.982	39.347	31.422	–3.283
600	29.513	204.445	189.560	8.931	39.252	29.845	–2.598
700	29.655	209.003	192.020	11.888	39.113	28.287	–2.111
800	29.914	212.979	194.396	14.866	38.945	26.752	–1.747
900	30.265	216.522	196.661	17.874	38.763	25.239	–1.465
1000	30.682	219.731	198.810	20.921	38.577	23.746	–1.240
1100	31.135	222.677	200.848	24.012	38.393	22.272	–1.058
1200	31.603	225.406	202.782	27.149	38.215	20.814	–0.906
1300	32.069	227.954	204.621	30.332	38.046	19.371	–0.778
1400	32.522	230.347	206.374	33.562	37.886	17.941	–0.669
1500	32.956	232.606	208.048	36.836	37.735	16.521	–0.575
53. WATER H_2O (l)							
298.15	75.300	69.950	69.950	0.000	–285.830	–237.141	41.546
300	75.281	70.416	69.951	0.139	–285.771	–236.839	41.237
373.21	76.079	86.896	71.715	5.666	–283.454	–225.160	31.513
54. WATER H_2O (g)							
298.15	33.598	188.832	188.832	0.000	–241.826	–228.582	40.046
300	33.606	189.040	188.833	0.062	–241.844	–228.500	39.785
400	34.283	198.791	190.158	3.453	–242.845	–223.900	29.238
500	35.259	206.542	192.685	6.929	–243.822	–219.050	22.884
600	36.371	213.067	195.552	10.509	–244.751	–214.008	18.631

TABLE 2.4.2
THERMOCHEMICAL PROPERTIES AT HIGH TEMPERATURES (continued)

	J K^{-1} mol^{-1}			kJ mol^{-1}			
T/K	C_p°	S°	$-(G^\circ - H^\circ(T_r))/T$	$H^\circ - H^\circ(T_r)$	$\Delta_f H^\circ$	$\Delta_f G^\circ$	Log K_f
54. WATER H_2O (g) (continued)							
700	37.557	218.762	198.469	14.205	–245.620	–208.814	15.582
800	38.800	223.858	201.329	18.023	–246.424	–203.501	13.287
900	40.084	228.501	204.094	21.966	–247.158	–198.091	11.497
1000	41.385	232.792	206.752	26.040	–247.820	–192.603	10.060
1100	42.675	236.797	209.303	30.243	–248.410	–187.052	8.882
1200	43.932	240.565	211.753	34.574	–248.933	–181.450	7.898
1300	45.138	244.129	214.108	39.028	–249.392	–175.807	7.064
1400	46.281	247.516	216.374	43.599	–249.792	–170.132	6.348
1500	47.356	250.746	218.559	48.282	–250.139	–164.429	5.726
55. IODINE I (g)							
298.15	20.786	180.787	180.787	0.000	106.760	70.172	–12.294
300	20.786	180.915	180.787	0.038	106.748	69.945	–12.178
400	20.786	186.895	181.602	2.117	97.974	58.060	–7.582
500	20.786	191.533	183.142	4.196	75.988	50.202	–5.244
600	20.786	195.323	184.866	6.274	76.190	45.025	–3.920
700	20.786	198.527	186.594	8.353	76.385	39.816	–2.971
800	20.787	201.303	188.263	10.432	76.574	34.579	–2.258
900	20.789	203.751	189.851	12.510	76.757	29.319	–1.702
1000	20.795	205.942	191.352	14.589	76.936	24.038	–1.256
1100	20.806	207.924	192.770	16.669	77.109	18.740	–0.890
1200	20.824	209.735	194.110	18.751	77.277	13.426	–0.584
1300	20.851	211.403	195.377	20.835	77.440	8.098	–0.325
1400	20.889	212.950	196.577	22.921	77.596	2.758	–0.103
1500	20.936	214.392	197.717	25.013	77.745	–2.592	0.090
56. DIIODINE I_2 (cr, l)							
298.15	54.440	116.139	116.139	0.000	0.000	0.000	0.000
300	54.518	116.476	116.140	0.101	0.000	0.000	0.000
386.75	61.531	131.039	117.884	5.088	0.000	0.000	0.000
		PHASE TRANSITION: $\Delta_{trs}H$ = 15.665 kJ/mol, $\Delta_{trs}S$ = 40.504 J/K·mol, cr–l					
386.75	79.555	171.543	117.884	20.753	0.000	0.000	0.000
400	79.555	174.223	119.706	21.807	0.000	0.000	0.000
457.67	79.555	184.938	127.266	26.395	0.000	0.000	0.000
57. DIIODINE I_2 (g)							
298.15	36.887	260.685	260.685	0.000	62.420	19.324	–3.385
300	36.897	260.913	260.685	0.068	62.387	19.056	–3.318
400	37.256	271.584	262.138	3.778	44.391	5.447	–0.711
457.67	37.385	276.610	263.652	5.931		pressure = 1 bar	
500	37.464	279.921	264.891	7.515	0.000	0.000	0.000
600	37.613	286.765	267.983	11.269	0.000	0.000	0.000
700	37.735	292.573	271.092	15.037	0.000	0.000	0.000
800	37.847	297.619	274.099	18.816	0.000	0.000	0.000
900	37.956	302.083	276.965	22.606	0.000	0.000	0.000
1000	38.070	306.088	279.681	26.407	0.000	0.000	0.000
1100	38.196	309.722	282.249	30.220	0.000	0.000	0.000
1200	38.341	313.052	284.679	34.047	0.000	0.000	0.000
1300	38.514	316.127	286.981	37.890	0.000	0.000	0.000
1400	38.719	318.989	289.166	41.751	0.000	0.000	0.000
1500	38.959	321.668	291.245	45.635	0.000	0.000	0.000

TABLE 2.4.2
THERMOCHEMICAL PROPERTIES AT HIGH TEMPERATURES (continued)

T/K	C_p° (J K^{-1} mol^{-1})	S° (J K^{-1} mol^{-1})	$-(G^\circ - H^\circ(T_r))/T$ (J K^{-1} mol^{-1})	$H^\circ - H^\circ(T_r)$ (kJ mol^{-1})	$\Delta_f H^\circ$ (kJ mol^{-1})	$\Delta_f G^\circ$ (kJ mol^{-1})	Log K_f
58. HYDROGEN IODIDE HI (g)							
298.15	29.157	206.589	206.589	0.000	26.500	1.700	–0.298
300	29.158	206.769	206.589	0.054	26.477	1.546	–0.269
400	29.329	215.176	207.734	2.977	17.093	–6.289	0.821
500	29.738	221.760	209.904	5.928	–5.481	–9.946	1.039
600	30.351	227.233	212.348	8.931	–5.819	–10.806	0.941
700	31.070	231.965	214.820	12.002	–6.101	–11.614	0.867
800	31.807	236.162	217.230	15.145	–6.323	–12.386	0.809
900	32.511	239.950	219.548	18.362	–6.489	–13.133	0.762
1000	33.156	243.409	221.763	21.646	–6.608	–13.865	0.724
1100	33.735	246.597	223.878	24.991	–6.689	–14.586	0.693
1200	34.249	249.555	225.896	28.391	–6.741	–15.302	0.666
1300	34.703	252.314	227.823	31.839	–6.775	–16.014	0.643
1400	35.106	254.901	229.666	35.330	–6.797	–16.723	0.624
1500	35.463	257.336	231.430	38.858	–6.814	–17.432	0.607
59. POTASSIUM K (cr, l)							
298.15	29.600	64.680	64.680	0.000	0.000	0.000	0.000
300	29.671	64.863	64.681	0.055	0.000	0.000	0.000
336.86	32.130	68.422	64.896	1.188	0.000	0.000	0.000
PHASE TRANSITION: $\Delta_{trs}H$ = 2.321 kJ/mol, $\Delta_{trs}S$ = 6.891 J/K·mol, cr–l							
336.86	32.129	75.313	64.896	3.509	0.000	0.000	0.000
400	31.552	80.784	66.986	5.519	0.000	0.000	0.000
500	30.741	87.734	70.469	8.632	0.000	0.000	0.000
600	30.158	93.283	73.824	11.675	0.000	0.000	0.000
700	29.851	97.905	76.943	14.673	0.000	0.000	0.000
800	29.838	101.887	79.818	17.655	0.000	0.000	0.000
900	30.130	105.415	82.470	20.651	0.000	0.000	0.000
1000	30.730	108.618	84.927	23.691	0.000	0.000	0.000
1039.4	31.053	109.812	85.847	24.908	0.000	0.000	0.000
60. POTASSIUM K (g)							
298.15	20.786	160.340	160.340	0.000	89.000	60.479	–10.596
300	20.786	160.468	160.340	0.038	88.984	60.302	–10.499
400	20.786	166.448	161.155	2.117	85.598	51.332	–6.703
500	20.786	171.086	162.695	4.196	84.563	42.887	–4.480
600	20.786	174.876	164.419	6.274	83.599	34.643	–3.016
700	20.786	178.080	166.148	8.353	82.680	26.557	–1.982
800	20.786	180.856	167.817	10.431	81.776	18.601	–1.215
900	20.786	183.304	169.404	12.510	80.859	10.759	–0.624
1000	20.786	185.494	170.905	14.589	79.897	3.021	–0.158
1039.4	20.786	186.297	171.474	15.408		pressure = 1 bar	
1100	20.786	187.475	172.323	16.667	0.000	0.000	0.000
1200	20.786	189.284	173.662	18.746	0.000	0.000	0.000
1300	20.789	190.948	174.929	20.825	0.000	0.000	0.000
1400	20.793	192.489	176.129	22.904	0.000	0.000	0.000
1500	20.801	193.923	177.268	24.983	0.000	0.000	0.000
61. DIPOTASSIUM OXIDE K_2O (cr, l)							
298.15	72.000	96.000	96.000	0.000	–361.700	–321.171	56.267
300	72.130	96.446	96.001	0.133	–361.704	–320.920	55.876
400	79.154	118.158	98.914	7.698	–366.554	–306.416	40.013

TABLE 2.4.2
THERMOCHEMICAL PROPERTIES AT HIGH TEMPERATURES (continued)

61. DIPOTASSIUM OXIDE K_2O (cr, l) (continued)

	J K^{-1} mol^{-1}			kJ mol^{-1}			
T/K	C_p°	S°	$-(G^\circ - H^\circ(T_r))/T$	$H^\circ - H^\circ(T_r)$	$\Delta_f H^\circ$	$\Delta_f G^\circ$	Log K_f
500	86.178	136.575	104.647	15.964	–366.043	–291.423	30.444
590	92.500	151.348	110.662	24.005	–364.204	–278.079	24.619
PHASE TRANSITION: $\Delta_{trs}H$ = 0.700 kJ/mol, $\Delta_{trs}S$ = 1.186 J/K·mol, crIII–crII							
590	100.000	152.534	110.662	24.705	–364.204	–278.079	24.619
600	100.000	154.215	111.374	25.705	–363.968	–276.621	24.082
645	100.000	161.447	114.618	30.205	–358.901	–270.109	21.874
PHASE TRANSITION: $\Delta_{trs}H$ = 4.000 kJ/mol, $\Delta_{trs}S$ = 6.202 J/K·mol, crII–crI							
645	100.000	167.649	114.618	34.205	–358.901	–270.109	21.874
700	100.000	175.832	119.111	39.705	–357.592	–262.592	19.595
800	100.000	189.185	127.054	49.705	–355.224	–249.183	16.270
900	100.000	200.963	134.625	59.705	–352.919	–236.067	13.701
1000	100.000	211.499	141.794	69.705	–350.732	–223.202	11.659
1013	100.000	212.791	142.697	71.005	–323.459	–221.546	11.424
PHASE TRANSITION: $\Delta_{trs}H$ = 27.000 kJ/mol, $\Delta_{trs}S$ =26.654 J/K·mol, crI–l							
1013	100.000	239.444	142.697	98.005	–323.459	–221.546	11.424
1100	100.000	247.684	150.679	106.705	–479.439	–203.633	9.670
1200	100.000	256.385	159.131	116.705	–475.371	–178.740	7.780
1300	100.000	264.389	166.924	126.705	–471.321	–154.185	6.195
1400	100.000	271.800	174.154	136.705	–467.287	–129.941	4.848
1500	100.000	278.699	180.896	146.705	–463.268	–105.986	3.691

62. POTASSIUM HYDROXIDE KOH (cr, l)

T/K	C_p°	S°	$-(G^\circ - H^\circ(T_r))/T$	$H^\circ - H^\circ(T_r)$	$\Delta_f H^\circ$	$\Delta_f G^\circ$	Log K_f
298.15	64.900	78.870	78.870	0.000	–424.580	–378.747	66.354
300	65.038	79.272	78.871	0.120	–424.569	–378.463	65.895
400	72.519	99.007	81.512	6.998	–426.094	–362.765	47.372
500	80.000	115.993	86.745	14.624	–424.572	–347.093	36.260
520	81.496	119.159	87.931	16.239	–417.725	–344.002	34.555
PHASE TRANSITION: $\Delta_{trs}H$ = 6.450 kJ/mol, $\Delta_{trs}S$ = 12.404 J/K·mol, crII–crI							
520	79.000	131.563	87.931	22.689	–417.725	–344.002	34.555
600	79.000	142.868	94.520	29.009	–416.274	–332.766	28.969
678	79.000	152.523	100.649	35.171	–405.464	–321.998	24.807
PHASE TRANSITION: $\Delta_{trs}H$ = 9.400 kJ/mol, $\Delta_{trs}S$ = 13.865 J/K·mol, crI–l							
678	83.000	166.388	100.649	44.571	–405.464	–321.998	24.807
700	83.000	169.038	102.757	46.397	–404.981	–319.297	23.826
800	83.000	180.121	111.750	54.697	–402.808	–307.206	20.058
900	83.000	189.897	119.901	62.997	–400.694	–295.383	17.143
1000	83.000	198.642	127.345	71.297	–398.668	–283.791	14.824
1100	83.000	206.553	134.192	79.597	–475.618	–267.780	12.716
1200	83.000	213.775	140.527	87.897	–472.711	–249.014	10.839
1300	83.000	220.418	146.421	96.197	–469.843	–230.490	9.261
1400	83.000	226.569	151.929	104.497	–467.011	–212.184	7.917
1500	83.000	232.296	157.098	112.797	–464.217	–194.080	6.758

63. POTASSIUM HYDROXIDE KOH (g)

T/K	C_p°	S°	$-(G^\circ - H^\circ(T_r))/T$	$H^\circ - H^\circ(T_r)$	$\Delta_f H^\circ$	$\Delta_f G^\circ$	Log K_f
298.15	49.184	238.283	238.283	0.000	–227.989	–229.685	40.239
300	49.236	238.588	238.284	0.091	–228.007	–229.696	39.993
400	51.178	253.053	240.243	5.124	–231.377	–229.667	29.991
500	52.178	264.591	243.998	10.296	–232.309	–229.129	23.937
600	52.804	274.163	248.251	15.547	–233.145	–228.413	19.885
700	53.296	282.340	252.551	20.853	–233.934	–227.562	16.981
800	53.758	289.487	256.730	26.206	–234.708	–226.599	14.795
900	54.229	295.846	260.730	31.605	–235.495	–225.538	13.090
1000	54.713	301.585	264.533	37.052	–236.322	–224.388	11.721

TABLE 2.4.2
THERMOCHEMICAL PROPERTIES AT HIGH TEMPERATURES (continued)

	J K⁻¹ mol⁻¹			kJ mol⁻¹			
T/K	C_p°	S°	$-(G^\circ-H^\circ(T_r))/T$	$H^\circ-H^\circ(T_r)$	$\Delta_f H^\circ$	$\Delta_f G^\circ$	Log K_f
63. POTASSIUM HYDROXIDE KOH (g) (continued)							
1100	55.203	306.823	268.143	42.548	–316.077	–218.535	10.377
1200	55.686	311.647	271.570	48.092	–315.925	–209.674	9.127
1300	56.153	316.122	274.827	53.684	–315.764	–200.826	8.069
1400	56.598	320.300	277.927	59.322	–315.595	–191.991	7.163
1500	57.016	324.220	280.884	65.003	–315.420	–183.169	6.378
64. POTASSIUM CHLORIDE KCl (cr, l)							
298.15	51.300	82.570	82.570	0.000	–436.490	–408.568	71.579
300	51.333	82.887	82.571	0.095	–436.481	–408.395	71.107
400	52.977	97.886	84.605	5.312	–438.463	–398.651	52.058
500	54.448	109.867	88.498	10.685	–437.990	–388.749	40.612
600	55.885	119.921	92.919	16.201	–437.332	–378.960	32.991
700	57.425	128.649	97.413	21.865	–436.502	–369.295	27.557
800	59.205	136.430	101.812	27.694	–435.505	–359.760	23.490
900	61.361	143.523	106.058	33.719	–434.337	–350.360	20.334
1000	64.032	150.121	110.138	39.983	–432.981	–341.100	17.817
1044	65.405	152.908	111.882	42.830	–485.450	–336.720	16.847
PHASE TRANSITION: $\Delta_{trs}H$ = 26.320 kJ/mol, $\Delta_{trs}S$ = 25.210 J/K·mol, cr–l							
1044	72.000	178.118	111.882	69.150	–485.450	–336.720	16.847
1100	72.000	181.880	115.351	73.182	–483.633	–328.790	15.613
1200	72.000	188.145	121.160	80.382	–480.393	–314.856	13.705
1300	72.000	193.908	126.537	87.582	–477.158	–301.192	12.102
1400	72.000	199.244	131.542	94.782	–473.928	–287.778	10.737
1500	72.000	204.211	136.223	101.982	–470.704	–274.594	9.562
65. POTASSIUM CHLORIDE KCl (g)							
298.15	36.505	239.091	239.091	0.000	–214.575	–233.320	40.876
300	36.518	239.317	239.092	0.068	–214.594	–233.436	40.644
400	37.066	249.904	240.532	3.749	–218.112	–239.107	31.224
500	37.384	258.212	243.267	7.473	–219.287	–244.219	25.513
600	37.597	265.048	246.344	11.222	–220.396	–249.100	21.686
700	37.769	270.857	249.441	14.991	–221.461	–253.799	18.938
800	37.907	275.910	252.441	18.775	–222.509	–258.347	16.868
900	38.041	280.382	255.302	22.572	–223.568	–262.764	15.250
1000	38.162	284.397	258.014	26.383	–224.667	–267.061	13.950
1100	38.279	288.039	260.581	30.205	–304.696	–266.627	12.661
1200	38.401	291.375	263.010	34.039	–304.821	–263.161	11.455
1300	38.518	294.454	265.312	37.885	–304.941	–259.684	10.434
1400	38.639	297.313	267.496	41.743	–305.053	–256.199	9.559
1500	38.761	299.983	269.574	45.613	–305.159	–252.706	8.800
66. DINITROGEN N_2 (g)							
298.15	29.124	191.608	191.608	0.000	0.000	0.000	0.000
300	29.125	191.788	191.608	0.054	0.000	0.000	0.000
400	29.249	200.180	192.752	2.971	0.000	0.000	0.000
500	29.580	206.738	194.916	5.911	0.000	0.000	0.000
600	30.109	212.175	197.352	8.894	0.000	0.000	0.000
700	30.754	216.864	199.812	11.936	0.000	0.000	0.000
800	31.433	221.015	202.208	15.046	0.000	0.000	0.000
900	32.090	224.756	204.509	18.222	0.000	0.000	0.000
1000	32.696	228.169	206.706	21.462	0.000	0.000	0.000
1100	33.241	231.311	208.802	24.759	0.000	0.000	0.000

TABLE 2.4.2
THERMOCHEMICAL PROPERTIES AT HIGH TEMPERATURES (continued)

T/K	J K⁻¹ mol⁻¹			kJ mol⁻¹			
	C_p°	S°	$-(G^\circ-H^\circ(T_r))/T$	$H^\circ-H^\circ(T_r)$	$\Delta_f H^\circ$	$\Delta_f G^\circ$	Log K_f
66. DINITROGEN N_2 (g) (continued)							
1200	33.723	234.224	210.801	28.108	0.000	0.000	0.000
1300	34.147	236.941	212.708	31.502	0.000	0.000	0.000
1400	34.517	239.485	214.531	34.936	0.000	0.000	0.000
1500	34.842	241.878	216.275	38.404	0.000	0.000	0.000
67. NITRIC OXIDE NO (g)							
298.15	29.862	210.745	210.745	0.000	91.277	87.590	–15.345
300	29.858	210.930	210.746	0.055	91.278	87.567	–15.247
400	29.954	219.519	211.916	3.041	91.320	86.323	–11.272
500	30.493	226.255	214.133	6.061	91.340	85.071	–8.887
600	31.243	231.879	216.635	9.147	91.354	83.816	–7.297
700	32.031	236.754	219.168	12.310	91.369	82.558	–6.160
800	32.770	241.081	221.642	15.551	91.386	81.298	–5.308
900	33.425	244.979	224.022	18.862	91.405	80.036	–4.645
1000	33.990	248.531	226.298	22.233	91.426	78.772	–4.115
1100	34.473	251.794	228.469	25.657	91.445	77.505	–3.680
1200	34.883	254.811	230.540	29.125	91.464	76.237	–3.318
1300	35.234	257.618	232.516	32.632	91.481	74.967	–3.012
1400	35.533	260.240	234.404	36.170	91.495	73.697	–2.750
1500	35.792	262.700	236.209	39.737	91.506	72.425	–2.522
68. NITROGEN DIOXIDE NO_2 (g)							
298.15	37.178	240.166	240.166	0.000	34.193	52.316	–9.165
300	37.236	240.397	240.167	0.069	34.181	52.429	–9.129
400	40.513	251.554	241.666	3.955	33.637	58.600	–7.652
500	43.664	260.939	244.605	8.167	33.319	64.882	–6.778
600	46.383	269.147	248.026	12.673	33.174	71.211	–6.199
700	48.612	276.471	251.575	17.427	33.151	77.553	–5.787
800	50.405	283.083	255.107	22.381	33.213	83.893	–5.478
900	51.844	289.106	258.555	27.496	33.334	90.221	–5.236
1000	53.007	294.631	261.891	32.741	33.495	96.534	–5.042
1100	53.956	299.729	265.102	38.090	33.686	102.828	–4.883
1200	54.741	304.459	268.187	43.526	33.898	109.105	–4.749
1300	55.399	308.867	271.148	49.034	34.124	115.363	–4.635
1400	55.960	312.994	273.992	54.603	34.360	121.603	–4.537
1500	56.446	316.871	276.722	60.224	34.604	127.827	–4.451
69. AMMONIA NH_3 (g)							
298.15	35.630	192.768	192.768	0.000	–45.940	–16.407	2.874
300	35.678	192.989	192.769	0.066	–45.981	–16.223	2.825
400	38.674	203.647	194.202	3.778	–48.087	–5.980	0.781
500	41.994	212.633	197.011	7.811	–49.908	4.764	–0.498
600	45.229	220.578	200.289	12.174	–51.430	15.846	–1.379
700	48.269	227.781	203.709	16.850	–52.682	27.161	–2.027
800	51.112	234.414	207.138	21.821	–53.695	38.639	–2.523
900	53.769	240.589	210.516	27.066	–54.499	50.231	–2.915
1000	56.244	246.384	213.816	32.569	–55.122	61.903	–3.233
1100	58.535	251.854	217.027	38.309	–55.589	73.629	–3.496
1200	60.644	257.039	220.147	44.270	–55.920	85.392	–3.717
1300	62.576	261.970	223.176	50.432	–56.136	97.177	–3.905
1400	64.339	266.673	226.117	56.779	–56.251	108.975	–4.066
1500	65.945	271.168	228.971	63.295	–56.282	120.779	–4.206

TABLE 2.4.2
THERMOCHEMICAL PROPERTIES AT HIGH TEMPERATURES (continued)

	J K^{-1} mol^{-1}			kJ mol^{-1}			
T/K	C_p°	S°	$-(G^\circ - H^\circ(T_r))/T$	$H^\circ - H^\circ(T_r)$	$\Delta_f H^\circ$	$\Delta_f G^\circ$	Log K_f
70. OXYGEN O (g)							
298.15	21.911	161.058	161.058	0.000	249.180	231.743	–40.600
300	21.901	161.194	161.059	0.041	249.193	231.635	–40.331
400	21.482	167.430	161.912	2.207	249.874	225.677	–29.470
500	21.257	172.197	163.511	4.343	250.481	219.556	–22.937
600	21.124	176.060	165.290	6.462	251.019	213.319	–18.571
700	21.040	179.310	167.067	8.570	251.500	206.997	–15.446
800	20.984	182.115	168.777	10.671	251.932	200.610	–13.098
900	20.944	184.584	170.399	12.767	252.325	194.171	–11.269
1000	20.915	186.789	171.930	14.860	252.686	187.689	–9.804
1100	20.893	188.782	173.372	16.950	253.022	181.173	–8.603
1200	20.877	190.599	174.733	19.039	253.335	174.628	–7.601
1300	20.864	192.270	176.019	21.126	253.630	168.057	–6.753
1400	20.853	193.815	177.236	23.212	253.908	161.463	–6.024
1500	20.845	195.254	178.389	25.296	254.171	154.851	–5.392
71. DIOXYGEN O_2 (g)							
298.15	29.378	205.148	205.148	0.000	0.000	0.000	0.000
300	29.387	205.330	205.148	0.054	0.000	0.000	0.000
400	30.109	213.873	206.308	3.026	0.000	0.000	0.000
500	31.094	220.695	208.525	6.085	0.000	0.000	0.000
600	32.095	226.454	211.045	9.245	0.000	0.000	0.000
700	32.987	231.470	213.612	12.500	0.000	0.000	0.000
800	33.741	235.925	216.128	15.838	0.000	0.000	0.000
900	34.365	239.937	218.554	19.244	0.000	0.000	0.000
1000	34.881	243.585	220.878	22.707	0.000	0.000	0.000
1100	35.314	246.930	223.096	26.217	0.000	0.000	0.000
1200	35.683	250.019	225.213	29.768	0.000	0.000	0.000
1300	36.006	252.888	227.233	33.352	0.000	0.000	0.000
1400	36.297	255.568	229.162	36.968	0.000	0.000	0.000
1500	36.567	258.081	231.007	40.611	0.000	0.000	0.000
72. SULFUR S (cr, l)							
298.15	22.690	32.070	32.070	0.000	0.000	0.000	0.000
300	22.737	32.210	32.070	0.042	0.000	0.000	0.000
368.3	24.237	37.030	32.554	1.649	0.000	0.000	0.000
PHASE TRANSITION: $\Delta_{trs} H$ = 0.401 kJ/mol, $\Delta_{trs} S$ = 1.089 J/K·mol, crII–crI							
368.3	24.773	38.119	32.553	2.050	0.000	0.000	0.000
388.36	25.180	39.444	32.875	2.551	0.000	0.000	0.000
PHASE TRANSITION: $\Delta_{trs} H$ = 1.722 kJ/mol, $\Delta_{trs} S$ = 4.431 J/K·mol, crI–l							
388.36	31.710	43.875	32.872	4.273	0.000	0.000	0.000
400	32.369	44.824	33.206	4.647	0.000	0.000	0.000
500	38.026	53.578	36.411	8.584	0.000	0.000	0.000
600	34.371	60.116	39.842	12.164	0.000	0.000	0.000
700	32.451	65.278	43.120	15.511	0.000	0.000	0.000
800	32.000	69.557	46.163	18.715	0.000	0.000	0.000
882.38	32.000	72.693	48.496	21.351	0.000	0.000	0.000
73. SULFUR S (g)							
298.15	23.673	167.828	167.828	0.000	277.180	236.704	–41.469
300	23.669	167.974	167.828	0.044	277.182	236.453	–41.170
400	23.233	174.730	168.752	2.391	274.924	222.962	–29.115
500	22.741	179.860	170.482	4.689	273.286	210.145	–21.953

TABLE 2.4.2
THERMOCHEMICAL PROPERTIES AT HIGH TEMPERATURES (continued)

T/K	J K^{-1} mol^{-1}			kJ mol^{-1}			
	C_p°	S°	$-(G^\circ - H^\circ(T_r))/T$	$H^\circ - H^\circ(T_r)$	$\Delta_f H^\circ$	$\Delta_f G^\circ$	Log K_f
73. SULFUR S (g) (continued)							
600	22.338	183.969	172.398	6.942	271.958	197.646	–17.206
700	22.031	187.388	174.302	9.160	270.829	185.352	–13.831
800	21.800	190.314	176.125	11.351	269.816	173.210	–11.309
900	21.624	192.871	177.847	13.522	215.723	162.258	–9.417
1000	21.489	195.142	179.465	15.677	216.018	156.301	–8.164
1100	21.386	197.185	180.985	17.821	216.284	150.317	–7.138
1200	21.307	199.043	182.413	19.955	216.525	144.309	–6.282
1300	21.249	200.746	183.759	22.083	216.743	138.282	–5.556
1400	21.209	202.319	185.029	24.206	216.940	132.239	–4.934
1500	21.186	203.781	186.231	26.325	217.119	126.182	–4.394
74. DISULFUR S_2 (g)							
298.15	32.505	228.165	228.165	0.000	128.600	79.696	–13.962
300	32.540	228.366	228.165	0.060	128.576	79.393	–13.823
400	34.108	237.956	229.462	3.398	122.703	63.380	–8.276
500	35.133	245.686	231.959	6.863	118.296	49.031	–5.122
600	35.815	252.156	234.800	10.413	114.685	35.530	–3.093
700	36.305	257.715	237.686	14.020	111.599	22.588	–1.685
800	36.697	262.589	240.501	17.671	108.841	10.060	–0.657
882.38	36.985	266.200	242.734	20.706		pressure = 1 bar	
900	37.045	266.932	243.201	21.358	0.000	0.000	0.000
1000	37.377	270.852	245.773	25.079	0.000	0.000	0.000
1100	37.704	274.430	248.218	28.833	0.000	0.000	0.000
1200	38.030	277.725	250.541	32.620	0.000	0.000	0.000
1300	38.353	280.781	252.751	36.439	0.000	0.000	0.000
1400	38.669	283.635	254.856	40.290	0.000	0.000	0.000
1500	38.976	286.314	256.865	44.173	0.000	0.000	0.000
75. OCTASULFUR S_8 (g)							
298.15	156.500	432.536	432.536	0.000	101.277	48.810	–8.551
300	156.768	433.505	432.539	0.290	101.231	48.484	–8.442
400	167.125	480.190	438.834	16.542	80.642	32.003	–4.179
500	173.181	518.176	451.022	33.577	66.185	21.409	–2.237
600	177.936	550.180	464.951	51.137	55.101	13.549	–1.180
700	182.441	577.948	479.152	69.157	46.349	7.343	–0.548
800	186.764	602.596	493.071	87.620	39.177	2.263	–0.148
900	190.595	624.821	506.495	106.494	–392.062	6.554	–0.380
1000	193.618	645.067	519.355	125.712	–387.728	50.614	–2.644
1100	195.684	663.625	531.639	145.185	–383.272	94.233	–4.475
1200	196.825	680.707	543.359	164.817	–378.786	137.444	–5.983
1300	197.195	696.480	554.539	184.524	–374.356	180.283	–7.244
1400	196.988	711.089	565.206	204.237	–370.048	222.785	–8.312
1500	196.396	724.662	575.389	223.909	–365.905	264.984	–9.227
76. SULFUR DIOXIDE SO_2 (g)							
298.15	39.842	248.219	248.219	0.000	–296.810	–300.090	52.574
300	39.909	248.466	248.220	0.074	–296.833	–300.110	52.253
400	43.427	260.435	249.828	4.243	–300.240	–300.935	39.298
500	46.490	270.465	252.978	8.744	–302.735	–300.831	31.427
600	48.938	279.167	256.634	13.520	–304.699	–300.258	26.139
700	50.829	286.859	260.413	18.513	–306.308	–299.386	22.340

TABLE 2.4.2
THERMOCHEMICAL PROPERTIES AT HIGH TEMPERATURES (continued)

T/K	C_p° (J K⁻¹ mol⁻¹)	S° (J K⁻¹ mol⁻¹)	$-(G^\circ-H^\circ(T_r))/T$ (J K⁻¹ mol⁻¹)	$H^\circ-H^\circ(T_r)$ (kJ mol⁻¹)	$\Delta_f H^\circ$ (kJ mol⁻¹)	$\Delta_f G^\circ$ (kJ mol⁻¹)	Log K_f
76. SULFUR DIOXIDE SO_2 (g) (continued)							
800	52.282	293.746	264.157	23.671	–307.691	–298.302	19.477
900	53.407	299.971	267.796	28.958	–362.075	–295.987	17.178
1000	54.290	305.646	271.301	34.345	–362.012	–288.647	15.077
1100	54.993	310.855	274.664	39.810	–361.934	–281.314	13.358
1200	55.564	315.665	277.882	45.339	–361.849	–273.989	11.926
1300	56.033	320.131	280.963	50.920	–361.763	–266.671	10.715
1400	56.426	324.299	283.911	56.543	–361.680	–259.359	9.677
1500	56.759	328.203	286.735	62.203	–361.605	–252.053	8.777
77. SILICON Si (cr)							
298.15	19.789	18.810	18.810	0.000	0.000	0.000	0.000
300	19.855	18.933	18.810	0.037	0.000	0.000	0.000
400	22.301	25.023	19.624	2.160	0.000	0.000	0.000
500	23.610	30.152	21.231	4.461	0.000	0.000	0.000
600	24.472	34.537	23.092	6.867	0.000	0.000	0.000
700	25.124	38.361	25.006	9.348	0.000	0.000	0.000
800	25.662	41.752	26.891	11.888	0.000	0.000	0.000
900	26.135	44.802	28.715	14.478	0.000	0.000	0.000
1000	26.568	47.578	30.464	17.114	0.000	0.000	0.000
1100	26.974	50.130	32.138	19.791	0.000	0.000	0.000
1200	27.362	52.493	33.737	22.508	0.000	0.000	0.000
1300	27.737	54.698	35.265	25.263	0.000	0.000	0.000
1400	28.103	56.767	36.728	28.055	0.000	0.000	0.000
1500	28.462	58.719	38.130	30.883	0.000	0.000	0.000
78. SILICON Si (g)							
298.15	22.251	167.980	167.980	0.000	450.000	405.525	–71.045
300	22.234	168.117	167.980	0.041	450.004	405.249	–70.559
400	21.613	174.416	168.843	2.229	450.070	390.312	–50.969
500	21.316	179.204	170.456	4.374	449.913	375.388	–39.216
600	21.153	183.074	172.246	6.497	449.630	360.508	–31.385
700	21.057	186.327	174.032	8.607	449.259	345.682	–25.795
800	21.000	189.135	175.748	10.709	448.821	330.915	–21.606
900	20.971	191.606	177.375	12.808	448.329	316.205	–18.352
1000	20.968	193.815	178.911	14.904	447.791	301.553	–15.751
1100	20.989	195.815	180.358	17.002	447.211	286.957	–13.626
1200	21.033	197.643	181.723	19.103	446.595	272.416	–11.858
1300	21.099	199.329	183.014	21.209	445.946	257.927	–10.364
1400	21.183	200.895	184.236	23.323	445.268	243.489	–9.085
1500	21.282	202.360	185.396	25.446	444.563	229.101	–7.978
79. SILICON DIOXIDE SiO_2 (cr)							
298.15	44.602	41.460	41.460	0.000	–910.700	–856.288	150.016
300	44.712	41.736	41.461	0.083	–910.708	–855.951	149.032
400	53.477	55.744	43.311	4.973	–910.912	–837.651	109.385
500	60.533	68.505	47.094	10.705	–910.540	–819.369	85.598
600	64.452	79.919	51.633	16.971	–909.841	–801.197	69.749
700	68.234	90.114	56.414	23.590	–908.958	–783.157	58.439
800	76.224	99.674	61.226	30.758	–907.668	–765.265	49.966
848	82.967	104.298	63.533	34.569	–906.310	–756.747	46.613
PHASE TRANSITION: $\Delta_{trs} H$ = 0.411 kJ/mol, $\Delta_{trs} S$ = 0.484 J/K·mol, crII–crII′							
848	67.446	104.782	63.532	34.980	–906.310	–756.747	46.613

TABLE 2.4.2
THERMOCHEMICAL PROPERTIES AT HIGH TEMPERATURES (continued)

T/K	J K⁻¹ mol⁻¹			kJ mol⁻¹			
	C_p°	S°	$-(G^\circ-H^\circ(T_r))/T$	$H^\circ-H^\circ(T_r)$	Δ_fH°	Δ_fG°	Log K_f
79. SILICON DIOXIDE SiO_2 (cr) (continued)							
900	67.953	108.811	66.033	38.500	–905.922	–747.587	43.388
1000	68.941	116.021	70.676	45.345	–905.176	–730.034	38.133
1100	69.940	122.639	75.104	52.289	–904.420	–712.557	33.836
1200	70.947	128.768	79.323	59.333	–901.382	–695.148	30.259
PHASE TRANSITION: $\Delta_{trs}H$ = 2.261 kJ/mol, $\Delta_{trs}S$ = 1.883 J/K·mol, crII′–crI							
1200	71.199	130.651	79.323	61.594	–901.382	–695.148	30.259
1300	71.743	136.372	83.494	68.742	–900.574	–677.994	27.242
1400	72.249	141.707	87.463	75.941	–899.782	–660.903	24.658
1500	72.739	146.709	91.248	83.191	–899.004	–643.867	22.421
80. SILICON TETRACHLORIDE $SiCl_4$ (g)							
298.15	90.404	331.446	331.446	0.000	–662.200	–622.390	109.039
300	90.562	332.006	331.448	0.167	–662.195	–622.143	108.323
400	96.893	359.019	335.088	9.572	–661.853	–608.841	79.505
500	100.449	381.058	342.147	19.456	–661.413	–595.637	62.225
600	102.587	399.576	350.216	29.616	–660.924	–582.527	50.713
700	103.954	415.500	358.432	39.948	–660.417	–569.501	42.496
800	104.875	429.445	366.455	50.392	–659.912	–556.548	36.338
900	105.523	441.837	374.155	60.914	–659.422	–543.657	31.553
1000	105.995	452.981	381.490	71.491	–658.954	–530.819	27.727
1100	106.349	463.101	388.456	82.109	–658.515	–518.027	24.599
1200	106.620	472.366	395.068	92.758	–658.107	–505.274	21.994
1300	106.834	480.909	401.347	103.431	–657.735	–492.553	19.791
1400	107.003	488.833	407.316	114.123	–657.400	–479.860	17.904
1500	107.141	496.220	413.000	124.830	–657.104	–467.189	16.269

TABLE 2.4.3
CODATA KEY VALUES

The Committee on Data for Science and Technology of the International Council of Scientific Unions (CODATA) has published recommended values of the standard state thermodynamic properties of about 150 key substances that frequently appear in thermodynamic calculations. The use of these internationally recommended, internally consistent values is encouraged in the analysis of thermodynamic measurements, data reduction, and preparation of other thermodynamic tables.

The standard state pressure adopted for this table is $P = 100000$ Pa (1 bar). If a standard state pressure of 1 atmosphere (101325 Pa) is used, all values of $S°$ for gaseous species must be increased by 0.109 J K^{-1} mol^{-1}. No other data in the table are affected within the stated uncertainty.

An entry of 0 for an element in the $\Delta_f H°$(298.15 K) column indicates the reference state of the element.

Substances are arranged in the "standard order" adopted in the reference.

REFERENCE

1. Cox, J. D.; Wagman, D. D.; Medvedev, V. A., *CODATA Key Values for Thermodynamics*, Hemisphere Publishing Corp. New York, 1989.

Substance	State	$\Delta_f H°$ (298.15 K)/ kJ mol^{-1}	$S°$ (298.15 K)/ J K^{-1} mol^{-1}	[$H°$ (298.15 K)–$H°$ (0)]/ kJ mol^{-1}
O	g	249.18 ± 0.10	161.059 ± 0.003	6.725 ± 0.001
O_2	g	0	205.152 ± 0.005	8.680 ± 0.002
H	g	217.998 ± 0.006	114.717 ± 0.002	6.197 ± 0.001
H^+	aq	0	0	
H_2	g	0	130.680 ± 0.003	8.468 ± 0.001
OH^-	aq	-230.015 ± 0.040	-10.90 ± 0.20	
H_2O	l	-285.830 ± 0.040	69.95 ± 0.03	13.273 ± 0.020
H_2O	g	-241.826 ± 0.040	188.835 ± 0.010	9.905 ± 0.005
He	g	0	126.153 ± 0.002	6.197 ± 0.001
Ne	g	0	146.328 ± 0.003	6.197 ± 0.001
Ar	g	0	154.846 ± 0.003	6.197 ± 0.001
Kr	g	0	164.085 ± 0.003	6.197 ± 0.001
Xe	g	0	169.685 ± 0.003	6.197 ± 0.001
F	g	79.38 ± 0.30	158.751 ± 0.004	6.518 ± 0.001
F^-	aq	-335.35 ± 0.65	-13.8 ± 0.8	
F_2	g	0	202.791 ± 0.005	8.825 ± 0.001
HF	g	-273.30 ± 0.70	173.779 ± 0.003	8.599 ± 0.001
Cl	g	121.301 ± 0.008	165.190 ± 0.004	6.272 ± 0.001
Cl^-	aq	-167.080 ± 0.10	56.60 ± 0.20	
Cl_2	g	0	223.081 ± 0.010	9.181 ± 0.001
ClO_4^-	aq	-128.10 ± 0.40	184.0 ± 1.5	
HCl	g	-92.31 ± 0.10	186.902 ± 0.005	8.640 ± 0.001
Br	g	111.87 ± 0.12	175.018 ± 0.004	6.197 ± 0.001
Br^-	aq	-121.41 ± 0.15	82.55 ± 0.20	
Br_2	l	0	152.21 ± 0.30	24.52 ± 0.01
Br_2	g	30.91 ± 0.11	245.468 ± 0.005	9.725 ± 0.001
HBr	g	-36.29 ± 0.16	198.700 ± 0.004	8.648 ± 0.001
I	g	106.76 ± 0.04	180.787 ± 0.004	6.197 ± 0.001
I^-	aq	-56.78 ± 0.05	106.45 ± 0.30	
I_2	cr	0	116.14 ± 0.30	13.196 ± 0.040
I_2	g	62.42 ± 0.08	260.687 ± 0.005	10.116 ± 0.001
HI	g	26.50 ± 0.10	206.590 ± 0.004	8.657 ± 0.001
S	cr, rhombic	0	32.054 ± 0.050	4.412 ± 0.006
S	g	277.17 ± 0.15	167.829 ± 0.006	6.657 ± 0.001
S_2	g	128.60 ± 0.30	228.167 ± 0.010	9.132 ± 0.002
SO_2	g	-296.81 ± 0.20	248.223 ± 0.050	10.549 ± 0.010
SO_4^{-2}	aq	-909.34 ± 0.40	18.50 ± 0.40	
HS^-	aq	-16.3 ± 1.5	67 ± 5	

TABLE 2.4.3
CODATA KEY VALUES (continued)

Substance	State	$\Delta_f H°$ (298.15 K)/ kJ mol^{-1}	$S°$ (298.15 K)/ J K^{-1} mol^{-1}	[$H°$ (298.15 K)–$H°$ (0)]/ kJ mol^{-1}
H_2S	g	-20.6 ± 0.5	205.81 ± 0.05	9.957 ± 0.010
H_2S	aq, undissoc.	-38.6 ± 1.5	126 ± 5	
HSO_4^-	aq	-886.9 ± 1.0	131.7 ± 3.0	
N	g	472.68 ± 0.40	153.301 ± 0.003	6.197 ± 0.001
N_2	g	0	191.609 ± 0.004	8.670 ± 0.001
NO_3^-	aq	-206.85 ± 0.40	146.70 ± 0.40	
NH_3	g	-45.94 ± 0.35	192.77 ± 0.05	10.043 ± 0.010
NH_4^+	aq	-133.26 ± 0.25	111.17 ± 0.40	
P	cr, white	0	41.09 ± 0.25	5.360 ± 0.015
P	g	316.5 ± 1.0	163.199 ± 0.003	6.197 ± 0.001
P_2	g	144.0 ± 2.0	218.123 ± 0.004	8.904 ± 0.001
P_4	g	58.9 ± 0.3	280.01 ± 0.50	14.10 ± 0.20
HPO_4^{-2}	aq	-1299.0 ± 1.5	-33.5 ± 1.5	
$H_2PO_4^-$	aq	-1302.6 ± 1.5	92.5 ± 1.5	
C	cr, graphite	0	5.74 ± 0.10	1.050 ± 0.020
C	g	716.68 ± 0.45	158.100 ± 0.003	6.536 ± 0.001
CO	g	-110.53 ± 0.17	197.660 ± 0.004	8.671 ± 0.001
CO_2	g	-393.51 ± 0.13	213.785 ± 0.010	9.365 ± 0.003
CO_2	aq, undissoc.	-413.26 ± 0.20	119.36 ± 0.60	
CO_3^{-2}	aq	-675.23 ± 0.25	-50.0 ± 1.0	
HCO_3^-	aq	-689.93 ± 0.20	98.4 ± 0.5	
Si	cr	0	18.81 ± 0.08	3.217 ± 0.008
Si	g	450 ± 8	167.981 ± 0.004	7.550 ± 0.001
SiO_2	cr, alpha quartz	-910.7 ± 1.0	41.46 ± 0.20	6.916 ± 0.020
SiF_4	g	-1615.0 ± 0.8	282.76 ± 0.50	15.36 ± 0.05
Ge	cr	0	31.09 ± 0.15	4.636 ± 0.020
Ge	g	372 ± 3	167.904 ± 0.005	7.398 ± 0.001
GeO_2	cr, tetragonal	-580.0 ± 1.0	39.71 ± 0.15	7.230 ± 0.020
GeF_4	g	-1190.20 ± 0.50	301.9 ± 1.0	17.29 ± 0.10
Sn	cr, white	0	51.18 ± 0.08	6.323 ± 0.008
Sn	g	301.2 ± 1.5	168.492 ± 0.004	6.215 ± 0.001
Sn^{+2}	aq	-8.9 ± 1.0	-16.7 ± 4.0	
SnO	cr, tetragonal	-280.71 ± 0.20	57.17 ± 0.30	8.736 ± 0.020
SnO_2	cr, tetragonal	-577.63 ± 0.20	49.04 ± 0.10	8.384 ± 0.020
Pb	cr	0	64.80 ± 0.30	6.870 ± 0.030
Pb	g	195.2 ± 0.8	175.375 ± 0.005	6.197 ± 0.001
Pb^{+2}	aq	0.92 ± 0.25	18.5 ± 1.0	
$PbSO_4$	cr	-919.97 ± 0.40	148.50 ± 0.60	20.050 ± 0.040
B	cr, rhombic	0	5.90 ± 0.08	1.222 ± 0.008
B	g	565 ± 5	153.436 ± 0.015	6.316 ± 0.002
B_2O_3	cr	-1273.5 ± 1.4	53.97 ± 0.30	9.301 ± 0.040
H_3BO_3	cr	-1094.8 ± 0.8	89.95 ± 0.60	13.52 ± 0.04
H_3BO_3	aq, undissoc.	-1072.8 ± 0.8	162.4 ± 0.6	
BF_3	g	-1136.0 ± 0.8	254.42 ± 0.20	11.650 ± 0.020
Al	cr	0	28.30 ± 0.10	4.540 ± 0.020
Al	g	330.0 ± 4.0	164.554 ± 0.004	6.919 ± 0.001
Al^{+3}	aq	-538.4 ± 1.5	-325 ± 10	
Al_2O_3	cr, corundum	-1675.7 ± 1.3	50.92 ± 0.10	10.016 ± 0.020
AlF_3	cr	-1510.4 ± 1.3	66.5 ± 0.5	11.62 ± 0.04
Zn	cr	0	41.63 ± 0.15	5.657 ± 0.020
Zn	g	130.40 ± 0.40	160.990 ± 0.004	6.197 ± 0.001
Zn^{+2}	aq	-153.39 ± 0.20	-109.8 ± 0.5	
ZnO	cr	-350.46 ± 0.27	43.65 ± 0.40	6.933 ± 0.040
Cd	cr	0	51.80 ± 0.15	6.247 ± 0.015
Cd	g	111.80 ± 0.20	167.749 ± 0.004	6.197 ± 0.001
Cd^{+2}	aq	-75.92 ± 0.60	-72.8 ± 1.5	

TABLE 2.4.3
CODATA KEY VALUES (continued)

Substance	State	$\Delta_f H°$ (298.15 K)/ kJ mol^{-1}	$S°$ (298.15 K)/ J K^{-1} mol^{-1}	[$H°$ (298.15 K)–$H°$ (0)]/ kJ mol^{-1}
CdO	cr	-258.35 ± 0.40	54.8 ± 1.5	8.41 ± 0.08
$CdSO_4 \cdot 8/3\,H_2O$	cr	-1729.30 ± 0.80	229.65 ± 0.40	35.56 ± 0.04
Hg	l	0	75.90 ± 0.12	9.342 ± 0.008
Hg	g	61.38 ± 0.04	174.971 ± 0.005	6.197 ± 0.001
Hg^{+2}	aq	170.21 ± 0.20	-36.19 ± 0.80	
Hg_2^{+2}	aq	166.87 ± 0.50	65.74 ± 0.80	
HgO	cr, red	-90.79 ± 0.12	70.25 ± 0.30	9.117 ± 0.025
Hg_2Cl_2	cr	-265.37 ± 0.40	191.6 ± 0.8	23.35 ± 0.20
Hg_2SO_4	cr	-743.09 ± 0.40	200.70 ± 0.20	26.070 ± 0.030
Cu	cr	0	33.15 ± 0.08	5.004 ± 0.008
Cu	g	337.4 ± 1.2	166.398 ± 0.004	6.197 ± 0.001
Cu^{+2}	aq	64.9 ± 1.0	-98 ± 4	
$CuSO_4$	cr	-771.4 ± 1.2	109.2 ± 0.4	16.86 ± 0.08
Ag	cr	0	42.55 ± 0.20	5.745 ± 0.020
Ag	g	284.9 ± 0.8	172.997 ± 0.004	6.197 ± 0.001
Ag^+	aq	105.79 ± 0.08	73.45 ± 0.40	
AgCl	cr	-127.01 ± 0.05	96.25 ± 0.20	12.033 ± 0.020
Ti	cr	0	30.72 ± 0.10	4.824 ± 0.015
Ti	g	473 ± 3	180.298 ± 0.010	7.539 ± 0.002
TiO_2	cr, rutile	-944.0 ± 0.8	50.62 ± 0.30	8.68 ± 0.05
$TiCl_4$	g	-763.2 ± 3.0	353.2 ± 4.0	21.5 ± 0.5
U	cr	0	50.20 ± 0.20	6.364 ± 0.020
U	g	533 ± 8	199.79 ± 0.10	6.499 ± 0.020
UO_2	cr	-1085.0 ± 1.0	77.03 ± 0.20	11.280 ± 0.020
UO_2^{+2}	aq	-1019.0 ± 1.5	-98.2 ± 3.0	
UO_3	cr, gamma	-1223.8 ± 1.2	96.11 ± 0.40	14.585 ± 0.050
U_3O_8	cr	-3574.8 ± 2.5	282.55 ± 0.50	42.74 ± 0.10
Th	cr	0	51.8 ± 0.5	6.35 ± 0.05
Th	g	602 ± 6	190.17 ± 0.05	6.197 ± 0.003
ThO_2	cr	-1226.4 ± 3.5	65.23 ± 0.20	10.560 ± 0.020
Be	cr	0	9.50 ± 0.08	1.950 ± 0.020
Be	g	324 ± 5	136.275 ± 0.003	6.197 ± 0.001
BeO	cr	-609.4 ± 2.5	13.77 ± 0.04	2.837 ± 0.008
Mg	cr	0	32.67 ± 0.10	4.998 ± 0.030
Mg	g	147.1 ± 0.8	148.648 ± 0.003	6.197 ± 0.001
Mg^{+2}	aq	-467.0 ± 0.6	-137 ± 4	
MgO	cr	-601.60 ± 0.30	26.95 ± 0.15	5.160 ± 0.020
MgF_2	cr	-1124.2 ± 1.2	57.2 ± 0.5	9.91 ± 0.06
Ca	cr	0	41.59 ± 0.40	5.736 ± 0.040
Ca	g	177.8 ± 0.8	154.887 ± 0.004	6.197 ± 0.001
Ca^{+2}	aq	-543.0 ± 1.0	-56.2 ± 1.0	
CaO	cr	-634.92 ± 0.90	38.1 ± 0.4	6.75 ± 0.06
Li	cr	0	29.12 ± 0.20	4.632 ± 0.040
Li	g	159.3 ± 1.0	138.782 ± 0.010	6.197 ± 0.001
Li^+	aq	-278.47 ± 0.08	12.24 ± 0.15	
Na	cr	0	51.30 ± 0.20	6.460 ± 0.020
Na	g	107.5 ± 0.7	153.718 ± 0.003	6.197 ± 0.001
Na^+	aq	-240.34 ± 0.06	58.45 ± 0.15	
K	cr	0	64.68 ± 0.20	7.088 ± 0.020
K	g	89.0 ± 0.8	160.341 ± 0.003	6.197 ± 0.001
K^+	aq	-252.14 ± 0.08	101.20 ± 0.20	
Rb	cr	0	76.78 ± 0.30	7.489 ± 0.020
Rb	g	80.9 ± 0.8	170.094 ± 0.003	6.197 ± 0.001
Rb^+	aq	-251.12 ± 0.10	121.75 ± 0.25	
Cs	cr	0	85.23 ± 0.40	7.711 ± 0.020
Cs	g	76.5 ± 1.0	175.601 ± 0.003	6.197 ± 0.001
Cs^+	aq	-258.00 ± 0.50	132.1 ± 0.5	

2.5. SURFACE TENSION

The surface tension σ is a measure of the force per unit length in the plane of the interface between a liquid and gas which resists an increase in the area of that surface. This force exists because increasing the area requires molecules to be transported from the bulk liquid, where the attractive forces exerted by surrounding molecules are balanced, to the surface where the forces exerted by molecules in the gas phase are much lower than those due to the liquid phase. The surface tension (dimensions force per unit length, SI unit $N\ m^{-1}$) can be equated to the surface Gibbs energy per unit area (SI unit $J\ m^{-2} = N\ m^{-1}$). Thus surface tension can be considered as a thermodynamic property of the two-phase system.

The value of the surface tension depends on whether the liquid is in equilibrium with its own vapor or surrounded by air (or some other gas). These differences are small but often measurable. Surface tension decreases with increasing temperature and vanishes at the critical temperature. Over a limited temperature range σ can be expressed as a linear function of temperature to a good approximation.

Physical quantity	Symbol	SI unit
Surface tension	σ	$N\ m^{-1}$

REFERENCE

1. Jasper, J. J., *J. Phys. Chem. Ref. Data*, 1, 841-1010, 1972.

TABLE 2.5.1 (STLT)
SURFACE TENSION OF COMMON LIQUIDS

This table gives the coefficients in the equation

$$\sigma / \mathrm{N\,m^{-1}} = A(1) + A(2)(T/\mathrm{K})$$

and the valid temperature range for 201 liquids. The value of σ(298.15 K) is also tabulated in units of mN m^{-1}, which is identical to the cgs unit dyn cm^{-1}. Substances are arranged in a modified Hill order with compounds not containing carbon preceding those that contain carbon.

The program STLT permits calculation of the surface tension of these liquids at any temperature in the valid range.

SN	Molecular formula	Name	σ (298.15 K)/ mN m^{-1}	$A(1)$	$A(2)$	(T/K) range
1	Br_2	Bromine	40.95	0.09521	-0.1820E-03	278 - 323
2	Cl_2O_2S	Sulfonyl dichloride	28.78	0.06837	-0.1328E-03	288 - 318
3	Cl_3OP	Phosphoryl trichloride	32.03	0.07005	-0.1275E-03	288 - 358
4	Cl_3P	Phosphorus trichloride	27.98	0.06572	-0.1266E-03	288 - 338
5	Cl_4Si	Silicon tetrachloride	18.29	0.04799	-0.9962E-04	278 - 323
6	H_2O	Water	71.99	0.12196	-0.1676E-03	283 - 373
7	H_4N_2	Hydrazine	66.39	0.13816	-0.2407E-03	288 - 313
8	Hg	Mercury	485.48	0.54657	-0.2049E-03	278 - 473
9	CCl_4	Tetrachloromethane	26.43	0.06292	-0.1224E-03	288 - 378
11	$CHBr_3$	Tribromomethane	44.87	0.08387	-0.1308E-03	288 - 368
12	$CHCl_3$	Trichloromethane	26.67	0.06528	-0.1295E-03	288 - 348
13	CH_2Br_2	Dibromomethane	39.05	0.08341	-0.1488E-03	293 - 358
14	CH_2Cl_2	Dichloromethane	27.20	0.06548	-0.1284E-03	293 - 313
15	CH_2O_2	Methanoic acid	37.13	0.06986	-0.1098E-03	288 - 363
16	CH_3I	Iodomethane	30.34	0.06713	-0.1234E-03	283 - 313
17	CH_3NO	Methanamide	57.03	0.08213	-0.8420E-04	298 - 393
18	CH_3NO_2	Nitromethane	36.53	0.08655	-0.1678E-03	283 - 343
19	CH_4O	Methanol	22.07	0.04511	-0.7730E-04	283 - 333
20	CH_5N	Methylamine	19.15	0.06351	-0.1488E-03	288 - 313
10	CS_2	Carbon disulfide	31.58	0.07583	-0.1484E-03	283 - 323
21	C_2HCl_5	Pentachloroethane	34.15	0.06927	-0.1178E-03	288 - 358
22	$C_2HF_3O_2$	Trifluoroethanoic acid	13.53	0.03870	-0.8444E-04	297 - 341
23	$C_2H_2Cl_4$	1,1,2,2-Tetrachloroethane	35.58	0.07339	-0.1268E-03	288 - 378
24	$C_2H_3Cl_3$	1,1,1-Trichloroethane	25.18	0.06221	-0.1242E-03	288 - 338
25	$C_2H_3Cl_3$	1,1,2-Trichloroethane	34.02	0.07430	-0.1351E-03	288 - 378
26	C_2H_3N	Ethanenitrile	28.66	0.06632	-0.1263E-03	293 - 333
27	$C_2H_4Br_2$	1,2-Dibromoethane	39.55	0.07891	-0.1320E-03	293 - 358
28	$C_2H_4Cl_2$	1,1-Dichloroethane	24.07	0.05943	-0.1186E-03	293 - 313
29	$C_2H_4Cl_2$	1,2-Dichloroethane	31.86	0.07444	-0.1428E-03	293 - 358
30	C_2H_4O	Ethanal	20.50	0.06105	-0.1360E-03	283 - 323
31	$C_2H_4O_2$	Ethanoic acid	27.10	0.05673	-0.9940E-04	293 - 363
32	$C_2H_4O_2$	Methyl methanoate	24.36	0.07123	-0.1572E-03	283 - 373
33	C_2H_5Br	Bromoethane	23.62	0.05818	-0.1159E-03	283 - 303
34	C_2H_5I	Iodoethane	28.46	0.06680	-0.1286E-03	283 - 343
35	$C_2H_5NO_2$	Nitroethane	32.13	0.06955	-0.1255E-03	283 - 343
36	C_2H_6O	Ethanol	21.97	0.04678	-0.8320E-04	283 - 343
37	C_2H_6OS	Dimethyl sulfoxide	42.92	0.07706	-0.1145E-03	293 - 333
38	$C_2H_6O_2$	1,2-Ethanediol	47.99	0.07452	-0.8900E-04	293 - 413
39	C_2H_6S	2-Thiapropane	24.06	0.04806	-0.8050E-04	283 - 298
40	C_2H_6S	Ethanethiol	23.08	0.04672	-0.7930E-04	288 - 303
41	$C_2H_6S_2$	2,3-Dithiabutane	33.39	0.07343	-0.1343E-03	288 - 333
42	C_2H_7N	Dimethylamine	26.34	0.06405	-0.1265E-03	288 - 313
43	C_2H_7N	Ethylamine	19.20	0.06011	-0.1372E-03	288 - 313
44	C_2H_7NO	2-Aminoethanol	48.32	0.08162	-0.1117E-03	288 - 358
45	C_3H_5Br	3-Bromopropene	26.31	0.06378	-0.1257E-03	288 - 333

TABLE 2.5.1 (STLT)
SURFACE TENSION OF COMMON LIQUIDS (continued)

SN	Molecular formula	Name	σ (298.15 K)/ mN m^{-1}	$A(1)$	$A(2)$	(T/K) range
46	C_3H_5Cl	3-Chloropropene	23.14	0.05134	-0.9460E-04	288 - 303
47	C_3H_5ClO	(Chloromethyl)oxirane	36.36	0.07691	-0.1360E-03	283 - 373
48	C_3H_5N	Propanenitrile	26.75	0.06112	-0.1153E-03	293 - 333
49	$C_3H_6Cl_2$	1,2-Dichloropropane	28.32	0.06529	-0.1240E-03	293 - 358
50	C_3H_6O	2-Propanone	23.46	0.05685	-0.1120E-03	298 - 323
51	C_3H_6O	2-Propen-1-ol	25.28	0.05217	-0.9020E-04	283 - 363
52	$C_3H_6O_2$	Ethyl methanoate	23.18	0.06239	-0.1315E-03	283 - 313
53	$C_3H_6O_2$	Methyl ethanoate	24.73	0.06316	-0.1289E-03	283 - 333
54	$C_3H_6O_2$	Propanoic acid	26.20	0.05580	-0.9930E-04	288 - 363
55	C_3H_7Br	1-Bromopropane	25.26	0.06157	-0.1218E-03	283 - 333
56	C_3H_7Br	2-Bromopropane	23.25	0.05852	-0.1183E-03	283 - 323
57	C_3H_7Cl	1-Chloropropane	21.30	0.05844	-0.1246E-03	283 - 313
58	C_3H_7Cl	2-Chloropropane	19.16	0.04549	-0.8830E-04	283 - 313
59	$C_3H_7NO_2$	2-Nitropropane	29.29	0.06381	-0.1158E-03	283 - 343
60	C_3H_8O	1-Propanol	23.32	0.04648	-0.7770E-04	283 - 363
61	C_3H_8O	2-Propanol	20.93	0.04445	-0.7890E-04	283 - 353
62	$C_3H_8O_2$	2-Methoxyethanol	30.84	0.06018	-0.9840E-04	283 - 373
63	C_3H_8S	1-Propanethiol	24.20	0.06212	-0.1272E-03	288 - 333
64	C_3H_8S	2-Propanethiol	21.33	0.05633	-0.1174E-03	288 - 323
65	C_3H_9N	Propylamine	21.75	0.05881	-0.1243E-03	288 - 313
66	C_3H_9N	Trimethylamine	13.41	0.04719	-0.1133E-03	288 - 313
67	$C_4H_4N_2$	Pyridazine	47.96	0.07885	-0.1036E-03	283 - 373
68	$C_4H_4N_2$	Pyrimidine	30.33	0.06044	-0.1010E-03	293 - 373
69	C_4H_4S	Thiophene	30.68	0.07027	-0.1328E-03	293 - 333
70	C_4H_5N	Pyrrole	37.06	0.06986	-0.1100E-03	283 - 343
71	$C_4H_6O_3$	Ethanoic anhydride	31.93	0.07474	-0.1436E-03	253 - 383
72	C_4H_7N	Butanenitrile	26.92	0.05784	-0.1037E-03	293 - 363
73	C_4H_8O	2-Butanone	23.97	0.05742	-0.1122E-03	298 - 323
74	$C_4H_8O_2$	1,4-Dioxane	32.75	0.07423	-0.1391E-03	293 - 373
75	$C_4H_8O_2$	Ethyl ethanoate	23.39	0.05800	-0.1161E-03	283 - 373
76	$C_4H_8O_2$	Methyl propanoate	24.44	0.06194	-0.1258E-03	283 - 343
77	$C_4H_8O_2$	Butanoic acid	26.05	0.05348	-0.9200E-04	293 - 363
78	C_4H_9Br	1-Bromobutane	25.90	0.05947	-0.1126E-03	283 - 373
79	C_4H_9Cl	1-Chlorobutane	23.18	0.05648	-0.1117E-03	283 - 343
80	C_4H_9I	1-Iodobutane	28.24	0.05898	-0.1031E-03	283 - 373
81	C_4H_9N	Pyrrolidine	29.23	0.05606	-0.9000E-04	283 - 343
82	$C_4H_{10}O$	1-Butanol	24.93	0.05172	-0.8983E-04	283 - 373
83	$C_4H_{10}O$	2-Butanol	22.54	0.04625	-0.7950E-04	283 - 373
84	$C_4H_{10}O$	Diethyl ether	16.65	0.04372	-0.9080E-04	288 - 303
85	$C_4H_{10}O_2$	2-Ethoxyethanol	28.35	0.05509	-0.8970E-04	293 - 373
86	$C_4H_{10}O_3$	3-Oxa-1,5-pentanediol	44.77	0.07101	-0.8800E-04	293 - 413
87	$C_4H_{10}S$	3-Thiapentane	24.57	0.05754	-0.1106E-03	283 - 333
88	$C_4H_{11}N$	Butylamine	23.44	0.05689	-0.1122E-03	288 - 333
89	$C_4H_{11}N$	(2-Methylpropyl)amine	21.75	0.05431	-0.1092E-03	293 - 333
90	$C_4H_{11}N$	(1,1-Dimethylethyl)amine	16.87	0.04752	-0.1028E-03	288 - 313
91	$C_4H_{11}N$	Diethylamine	19.85	0.05393	-0.1143E-03	288 - 313
92	$C_5H_4O_2$	2-Furaldehyde	43.09	0.08266	-0.1327E-03	283 - 373
93	C_5H_5N	Pyridine	36.56	0.07549	-0.1306E-03	293 - 358
94	C_5H_8	Cyclopentene	22.20	0.06678	-0.1495E-03	278 - 303
95	C_5H_8O	Cyclopentanone	32.80	0.06560	-0.1100E-03	278 - 373
96	C_5H_{10}	1-Pentene	15.45	0.04822	-0.1099E-03	243 - 298
97	C_5H_{10}	2-Methyl-2-butene	17.15	0.04612	-0.9715E-04	283 - 303
98	C_5H_{10}	Cyclopentane	21.88	0.06546	-0.1462E-03	278 - 323
99	$C_5H_{10}O$	2-Pentanone	23.25	0.04277	-0.6547E-04	298 - 323
100	$C_5H_{10}O$	3-Pentanone	24.74	0.05596	-0.1047E-03	298 - 323
101	$C_5H_{10}O$	Pentanal	25.44	0.05555	-0.1010E-03	283 - 343

TABLE 2.5.1 (STLT)
SURFACE TENSION OF COMMON LIQUIDS (continued)

SN	Molecular formula	Name	σ (298.15 K)/ mN m^{-1}	$A(1)$	$A(2)$	(T/K) range
102	$C_5H_{10}O_2$	Butyl methanoate	24.52	0.05511	-0.1026E-03	283 - 373
103	$C_5H_{10}O_2$	Propyl ethanoate	23.80	0.05719	-0.1120E-03	283 - 373
104	$C_5H_{10}O_2$	Isopropyl ethanoate	21.76	0.05372	-0.1072E-03	283 - 353
105	$C_5H_{10}O_2$	Ethyl propanoate	23.80	0.05862	-0.1168E-03	283 - 353
106	$C_5H_{10}O_2$	Methyl butanoate	24.62	0.05876	-0.1145E-03	283 - 373
107	$C_5H_{11}Cl$	1-Chloropentane	24.40	0.05648	-0.1076E-03	283 - 373
108	$C_5H_{11}N$	Piperidine	28.91	0.06328	-0.1153E-03	283 - 373
109	C_5H_{12}	Pentane	15.49	0.04835	-0.1102E-03	283 - 303
110	$C_5H_{12}O$	1-Pentanol	25.36	0.05141	-0.8740E-04	283 - 373
111	$C_5H_{12}O$	2-Pentanol	23.45	0.05338	-0.1004E-03	283 - 373
112	$C_5H_{12}O$	3-Methyl-1-butanol	23.71	0.04816	-0.8200E-04	283 - 373
113	$C_5H_{13}N$	Pentylamine	24.69	0.05519	-0.1023E-03	293 - 363
114	$C_6H_4Cl_2$	1,3-Dichlorobenzene	35.43	0.06963	-0.1147E-03	283 - 433
115	C_6H_5Br	Bromobenzene	35.24	0.06983	-0.1160E-03	283 - 423
116	C_6H_5Cl	Chlorobenzene	32.99	0.06850	-0.1191E-03	283 - 403
117	C_6H_5ClO	2-Chlorophenol	39.70	0.07315	-0.1122E-03	298 - 443
118	C_6H_5ClO	3-Chlorophenol	41.18	0.07126	-0.1009E-03	298 - 453
119	C_6H_5F	Fluorobenzene	26.66	0.06256	-0.1204E-03	283 - 353
120	C_6H_5I	Iodobenzene	38.71	0.07219	-0.1123E-03	283 - 433
121	$C_6H_5NO_2$	Nitrobenzene		0.07794	-0.1157E-03	313 - 473
122	C_6H_6	Benzene	28.22	0.06671	-0.1291E-03	288 - 353
123	C_6H_6O	Phenol		0.07271	-0.1068E-03	313 - 413
124	C_6H_7N	Aniline	42.12	0.07447	-0.1085E-03	288 - 363
125	C_6H_7N	2-Methylpyridine	33.00	0.07006	-0.1243E-03	293 - 358
126	$C_6H_8N_2$	Hexanedinitrile	45.45	0.07446	-0.9730E-04	293 - 358
127	C_6H_{10}	Cyclohexene	26.17	0.06264	-0.1223E-03	278 - 343
128	$C_6H_{10}O$	Cyclohexanone	34.57	0.07160	-0.1242E-03	278 - 373
129	$C_6H_{11}N$	Hexanenitrile	27.37	0.05441	-0.9070E-04	293 - 363
130	C_6H_{12}	Cyclohexane	24.65	0.06007	-0.1188E-03	283 - 343
131	C_6H_{12}	Methylcyclopentane	21.72	0.05640	-0.1163E-03	283 - 333
132	C_6H_{12}	1-Hexene	17.90	0.04853	-0.1027E-03	283 - 333
133	$C_6H_{12}O$	Cyclohexanol	32.92	0.06172	-0.9660E-04	293 - 373
134	$C_6H_{12}O$	2-Hexanone	25.45	0.05801	-0.1092E-03	298 - 323
135	$C_6H_{12}O_2$	Butyl ethanoate	24.88	0.05672	-0.1068E-03	283 - 373
136	$C_6H_{12}O_2$	Isobutyl ethanoate	23.06	0.05326	-0.1013E-03	283 - 373
137	$C_6H_{12}O_2$	Ethyl butanoate	23.94	0.05509	-0.1045E-03	283 - 373
138	$C_6H_{12}O_3$	2,4,6-Trimethyl-1,3,5-trioxane	25.63	0.05729	-0.1062E-03	283 - 373
139	$C_6H_{13}Cl$	1-Chlorohexane	25.73	0.05667	-0.1038E-03	283 - 373
140	$C_6H_{13}N$	Cyclohexylamine	31.22	0.06664	-0.1188E-03	288 - 363
141	C_6H_{14}	Hexane	17.89	0.04836	-0.1022E-03	283 - 333
142	C_6H_{14}	2-Methylpentane	16.88	0.04659	-0.9967E-04	283 - 333
143	C_6H_{14}	3-Methylpentane	17.61	0.04921	-0.1060E-03	283 - 333
144	$C_6H_{14}O$	Diisopropyl ether	17.27	0.04852	-0.1048E-03	288 - 333
145	$C_6H_{14}O$	1-Hexanol	25.81	0.04969	-0.8010E-04	288 - 388
146	$C_6H_{14}O_2$	1,1-Diethoxyethane	20.89	0.05159	-0.1030E-03	288 - 363
147	$C_6H_{14}O_2$	2-Butoxyethanol	26.14	0.05047	-0.8160E-04	283 - 373
148	$C_6H_{15}N$	Triethylamine	20.22	0.04980	-0.9920E-04	293 - 333
149	$C_6H_{15}N$	Dipropylamine	22.31	0.05278	-0.1022E-03	288 - 363
150	$C_6H_{15}N$	Diisopropylamine	19.14	0.05125	-0.1077E-03	288 - 333
151	C_7H_5N	Benzonitrile	38.79	0.07335	-0.1159E-03	288 - 363
152	C_7H_6O	Benzaldehyde	38.00	0.07049	-0.1090E-03	283 - 373
153	C_7H_8	Toluene	27.93	0.06338	-0.1189E-03	283 - 373
154	C_7H_8O	2-Methylphenol	36.90	0.06705	-0.1011E-03	298 - 453
155	C_7H_8O	3-Methylphenol	35.69	0.06323	-0.9237E-04	298 - 453
156	C_7H_8O	Phenylmethanol		0.07597	-0.1381E-03	343 - 453
157	C_7H_8O	Anisole	35.10	0.07100	-0.1204E-03	288 - 363

TABLE 2.5.1 (STLT)
SURFACE TENSION OF COMMON LIQUIDS (continued)

SN	Molecular formula	Name	σ (298.15 K)/ mN m^{-1}	*A*(1)	*A*(2)	(*T*/K) range
158	C_7H_9N	*N*-Methylaniline	36.90	0.06581	-0.9698E-04	288 - 363
159	C_7H_9N	2,3-Dimethylpyridine	32.71	0.06458	-0.1069E-03	293 - 358
160	C_7H_9N	Benzylamine	39.30	0.07546	-0.1213E-03	293 - 363
161	C_7H_{14}	Methylcyclohexane	23.29	0.05698	-0.1130E-03	278 - 333
162	C_7H_{14}	1-Heptene	19.80	0.04934	-0.9908E-04	283 - 353
163	$C_7H_{14}O$	2-Heptanone	26.12	0.05762	-0.1056E-03	298 - 323
164	$C_7H_{14}O_2$	Pentyl ethanoate	25.17	0.05482	-0.9943E-04	283 - 373
165	$C_7H_{14}O_2$	Heptanoic acid	27.76	0.05304	-0.8480E-04	288 - 343
166	C_7H_{16}	Heptane	19.65	0.04887	-0.9800E-04	283 - 363
167	C_7H_{16}	3-Methylhexane	19.31	0.04822	-0.9699E-04	253 - 363
168	C_8H_8O	Methyl phenyl ketone	39.04	0.07344	-0.1154E-03	293 - 358
169	$C_8H_8O_2$	Methyl benzoate	37.17	0.07209	-0.1171E-03	288 - 363
170	$C_8H_8O_3$	Methyl 2-hydroxybenzoate	39.22	0.07422	-0.1174E-03	283 - 443
171	C_8H_{10}	Ethylbenzene	28.75	0.06136	-0.1094E-03	283 - 373
172	C_8H_{10}	1,2-Dimethylbenzene	29.76	0.06258	-0.1101E-03	283 - 373
173	C_8H_{10}	1,3-Dimethylbenzene	28.47	0.06139	-0.1104E-03	283 - 373
174	C_8H_{10}	1,4-Dimethylbenzene	28.01	0.06003	-0.1074E-03	293 - 373
175	$C_8H_{10}O$	Ethyl phenyl ether	32.41	0.06533	-0.1104E-03	288 - 363
176	$C_8H_{11}N$	*N,N*-Dimethylaniline	35.52	0.06679	-0.1049E-03	293 - 363
177	$C_8H_{11}N$	*N*-Ethylaniline	36.33	0.06823	-0.1070E-03	293 - 363
178	C_8H_{16}	Ethylcyclohexane	25.15	0.05657	-0.1054E-03	278 - 333
179	C_8H_{18}	Octane	21.14	0.04949	-0.9509E-04	283 - 393
180	C_8H_{18}	2,5-Dimethylhexane	19.40	0.04659	-0.9121E-04	283 - 373
181	$C_8H_{18}O$	1-Octanol	27.10	0.05081	-0.7950E-04	283 - 343
182	$C_8H_{19}N$	Dibutylamine	24.12	0.05250	-0.9520E-04	288 - 363
183	$C_8H_{19}N$	Bis(2-methylpropyl)amine	21.72	0.04891	-0.9120E-04	288 - 363
184	C_9H_7N	Quinoline	42.59	0.07429	-0.1063E-03	283 - 373
185	C_9H_{12}	Isopropylbenzene	27.69	0.05911	-0.1054E-03	283 - 373
186	C_9H_{12}	1,2,4-Trimethylbenzene	29.20	0.05976	-0.1025E-03	283 - 373
187	C_9H_{12}	1,3,5-Trimethylbenzene	27.55	0.05428	-0.8966E-04	283 - 373
188	$C_9H_{18}O$	5-Nonanone	26.28	0.05535	-0.9750E-04	298 - 323
189	C_9H_{20}	Nonane	22.38	0.05025	-0.9347E-04	283 - 393
190	$C_9H_{20}O$	1-Nonanol	27.89	0.05052	-0.7589E-04	283 - 373
191	$C_{10}H_{12}$	1,2,3,4-Tetrahydronaphthalene	33.17	0.06161	-0.9540E-04	293 - 358
192	$C_{10}H_{22}$	Decane	23.37	0.05079	-0.9197E-04	283 - 393
193	$C_{10}H_{22}O$	1-Decanol	28.51	0.05035	-0.7324E-04	283 - 373
194	$C_{11}H_{24}$	Undecane	24.21	0.05107	-0.9010E-04	283 - 393
195	$C_{12}H_{10}O$	Diphenyl ether	26.75	0.05001	-0.7800E-04	288 - 343
196	$C_{12}H_{27}N$	Tributylamine	24.39	0.04917	-0.8310E-04	293 - 363
197	$C_{13}H_{28}$	Tridecane	25.55	0.05155	-0.8719E-04	283 - 393
198	$C_{14}H_{12}O_2$	Benzyl benzoate	42.82	0.07567	-0.1102E-03	283 - 423
199	$C_{14}H_{30}$	Tetradecane	26.13	0.05203	-0.8688E-04	283 - 393
200	$C_{16}H_{34}$	Hexadecane	27.05	0.05251	-0.8540E-04	293 - 393
201	$C_{18}H_{38}$	Octadecane	27.87	0.05300	-0.8428E-04	298 - 393

Section 3
Thermodynamic Properties of Mixtures

3.1. PHASE EQUILIBRIA AND RELATED TOPICS

The phase behavior of a mixture is more complicated than that of a pure substance, since composition variables must be specified, in addition to temperature and pressure, in order to describe the system completely. In a binary mixture, which is the only case considered in this book, the composition in each phase is usually described by the mole fraction of one of the components. By convention, x_1 is generally used to specify the mole fraction of component 1 in the liquid phase (with $x_2 = 1 - x_1$), while y_1 is used for the mole fraction of component 1 in the gas phase.

The most important aspects of the phase behavior of a binary mixture are the vapor-liquid equilibrium, including azeotropic data, the liquid-liquid equilibrium, and the liquid-gas critical line. These topics are covered for representative binary mixtures in the tables of this section.

TABLE 3.1.1 (VLEG)
ISOTHERMAL VAPOR-LIQUID EQUILIBRIA AND EXCESS GIBBS ENERGIES OF BINARY MIXTURES

The *vapor pressure P* and the *vapor phase molar composition* y_i of a binary liquid mixture, at a given temperature T, can be calculated as a function of the molar composition x_i of the liquid phase from: (a) the *vapor pressures*, P_i^o, and *molar volumes* V_i of the pure liquid components, (b) the *partial molar excess Gibbs energies* of the mixture components in the liquid phase, μ_i^E, and (c) the fugacity coefficients of the components in the vapor phase. The molar volumes account for the pressure dependence of the molar Gibbs energies, and the fugacity coefficients account for the nonideality of the vapor-phase.

The partial molar excess Gibbs energies are related to the activity coefficients γ_i through the relation:

$$\mu_i^{\mathrm{E}} = RT \ln \gamma_i \tag{1}$$

At low vapor pressures, the vapor-phase nonideality may be well described in terms of the virial equation of state (see Table 2.2.1), i.e., the second molar virial coefficients of the pure components, B_{11} and B_{22}, and the second molar cross virial coefficient, B_{12}. The latter is often estimated as $B_{12} = (B_{11} + B_{22})/2$.

The vapor pressure P and the vapor-phase mole fraction of component i in a binary mixture, y_i, are then given by the system of equations:

$$P = \sum_{i=1}^{2} x_i P_i^o \exp\left[\frac{\mu_i^{\mathrm{E}} - (B_{ii} - V_i)(P - P_i^o) - 2PB_{12}^{\mathrm{E}}(1 - y_i)^2}{RT}\right] \tag{2}$$

$$y_i = \frac{x_i P_i^o}{P} \exp\left[\frac{\mu_i^{\mathrm{E}} - (B_{ii} - V_i)(P - P_i^o) - 2PB_{12}^{\mathrm{E}}(1 - y_i)^2}{RT}\right] \tag{3}$$

where

$$B_{12}^{\mathrm{E}} = B_{12} - (B_{11} + B_{12})/2 \tag{4}$$

The partial molar excess Gibbs energies μ_i^E of the mixture components can be calculated from the *molar excess Gibbs energy*, G^E. The latter can be represented by various empirical equations. A simple equation is the so-called Redlich-Kister equation:

$$G^{\mathrm{E}}/RT = x_1 x_2 \sum_{i=1}^{n} A(i)(x_1 - x_2)^{i-1} \tag{5}$$

$$\mu_1^{\mathrm{E}}/RT = x_2^2 \left\{ A(1) + \sum_{i=2}^{n} A(i)[(2i-1)x_1 - x_2](x_1 - x_2)^{i-2} \right\} \tag{6}$$

$$\mu_2^{\mathrm{E}}/RT = x_1^2 \left\{ A(1) + \sum_{i=2}^{n} A(i)[x_1 - (2i-1)x_2](x_1 - x_2)^{i-2} \right\} \tag{7}$$

The table gives the coefficients $A(i)$, at the temperature T, obtained by fitting Equations 2 to 7 to experimental $P(x_1)$ data, as well as the molar virial coefficients B_{11}, B_{22}, B_{12}, and the molar volumes V_1 and V_2, used in the calculations. A zero indicates that the corresponding value of B_{ij} or V_i was neglected. The table also lists the calculated values of P, y_1, G^E, μ_1^E, and μ_2^E for increments of $x_1 = 0.1$

The program, Property Code VLEG, permits the calculation of P, y_1, G^E, μ_1^E. μ_2^E, γ_1, and γ_2 at selected increments of x_1 in the whole range of composition, from $x_1 = 0$ to $x_1 = 1$, or at any given value of x_1.

TABLE 3.1.1 (VLEG)
ISOTHERMAL VAPOR-LIQUID EQUILIBRIA AND EXCESS GIBBS ENERGIES OF BINARY MIXTURES (continued)

The auxiliary values displayed by the program have the following meaning:

$V(1) = P_1^\circ$ $V(5) = B_{12}$
$V(2) = P_2^\circ$ $V(6) = V_1$
$V(3) = B_{11}$ $V(7) = V_2$
$V(4) = B_{22}$

The components of each system, and the systems themselves, are arranged in a modified Hill order, with substances that do not contain carbon preceding those that do contain carbon.

Physical quantity	Symbol	SI unit
Vapor pressure	P	Pa
Liquid-phase mole fraction of component *i*	x_i	—
Vapor-phase mole fraction of component *i*	y_i	—
Temperature	T	K
Molar excess Gibbs energy	G^E	J mol^{-1}
Partial molar excess Gibbs energy of component *i*	μ_i^E	J mol^{-1}
Activity coefficient of component *i*	γ_i	—

REFERENCES

1. Van Ness, H. C.; Abbott, M. M., *Classical Thermodynamics of Non-electrolyte Solutions*, McGraw-Hill, New York, 1982 (general theory).
2. Prausnitz, J. M.; Lichtenthaler, R. N.; Azevedo, E. G., *Molecular Thermodynamics of Fluid-Phase Equilibria, 2nd Edition*, Prentice-Hall, Englewood Cliffs, N. J., 1986 (general theory).
3. Reid, R. C.; Prausnitz, J. M.; Poling, B. E. *The Properties of Gases and Liquids*, 4th Ed., McGraw-Hill, New York, 1987 (general theory).
4. H. V. Kehiaian; C. S. Kehiaian, Organic Systems (Mixtures), in: *Bulletin of Chemical Thermodynamics*, Freeman, R. D., Editor, Thermochemistry Inc., 1969-1985 (17 volumes) (a compilation of bibliographical vapor-liquid equilibrium and excess Gibbs energy data).
5. Wisniak, J.; Tamir, A., *Mixing and Excess Properties. A Literature Source Book*, Elsevier, Amsterdam, 1978; *Supplement 1*, 1982; *Supplement 2*, 1986 (a compilation of bibliographical excess Gibbs energy data).
6. Wichterle, I.; Linek, J.; Hala, E., *Vapor-Liquid Equilibrium Bibliography*, 1973; *Supplement 1*, 1976; *Supplement 2*, 1979; *Supplement 3*, 1982; *Supplement 4*, 1985, Elsevier, Amsterdam (a compilation of bibliographical vapor-liquid equilibrium data).
7. Wichterle, I.; Linek, J.; Wagner, Z.; Kehiaian, H. V., *Vapor-Liquid Equilibrium Bibliographic Database, ELDATA*, Montreuil, 1993 (a computerized compilation of bibliographical vapor-liquid equilibrium data covering the literature from 1900 to 1991).
8. Hirata, M.; Ohe, S.; Nagahama, K., *Computer-Aided Data Book of Vapor-Liquid Equilibria*, Kodansha, Tokyo, 1975 (a compilation of numerical vapor-liquid equilibrium data and correlation coefficients for ca. 1000 binary systems).
9. Gmehling, J.; Onken, U.; Arlt, W.; Grenzheuser, P.; Weidlich U.; Kolbe, B.; Rarey-Nies, J., *Vapor-Liquid Equilibrium Data Collection*, DECHEMA, Frankfurt/Main, Germany, 1978-1992 (30 parts) (an extensive compilation of numerical vapor-liquid equilibrium data and correlation coefficients).
10. Knapp, H.; Doering, R.; Oellrich, L. Ploecker, U.; Prausnitz, J. M., *Vapor-Liquid Equilibria for Mixtures of Low Boiling Substances*, DECHEMA, Frankfurt/Main, Germany (numerical data and correlations using equations of state).
11. Ohe, S., *Vapor-Liquid Equilibrium Data at High Pressure*, 1990, Elsevier, Amsterdam (numerical data and correlations using equations of state for 700 binary systems).
12. *Int. DATA Ser., Sel. Data Mixtures, Ser. A* (Ed. H. V. Kehiaian), Thermodynamic Research Center, Texas Engineering Experiment Station, College Station, USA (primary vapor-liquid equilibrium data, correlations, and calculated excess Gibbs energies).

TABLE 3.1.1 (VLEG)
ISOTHERMAL VAPOR-LIQUID EQUILIBRIA AND EXCESS GIBBS ENERGIES OF BINARY MIXTURES (continued)

1. Silicon tetrachloride, Cl_4Si
2. Tetrachloromethane, CCl_4

***T*/K = 323.15**
***System Number*: 390**

x_1	y_1	P/kPa	G^E/J mol^{-1}	μ^E_1/J mol^{-1}	μ^E_2/J mol^{-1}
0.0	0.000	41.60	0	352	0
0.1	0.188	46.31	31	281	4
0.2	0.337	50.71	56	219	15
0.3	0.459	54.86	72	165	33
0.4	0.562	58.80	82	119	57
0.5	0.653	62.57	85	82	88
0.6	0.733	66.21	81	51	125
0.7	0.807	69.74	70	28	168
0.8	0.875	73.19	53	12	216
0.9	0.939	76.58	30	3	269
1.0	1.000	79.93	0	0	327

B_{11}/cm^{3}mol^{-1}	B_{22}/cm^{3}mol^{-1}	B_{12}/cm^{3}mol^{-1}	V_1/cm^{3}mol^{-1}	V_2/cm^{3}mol^{-1}
-1220	-1300	-1260	118	101

$A(1)$ = 0.1264E+00 $A(2)$ = -0.4804E-02 $A(3)$ = 0.0000E+00

---

1. Water, H_2O
2. Dihydrogen peroxide, H_2O_2

***T*/K = 333.15**
***System Number*: 462**

x_1	y_1	P/kPa	G^E/J mol^{-1}	μ^E_1/J mol^{-1}	μ^E_2/J mol^{-1}
0.0	0.000	2.35	0	-2510	0
0.1	0.301	3.00	-234	-2173	-18
0.2	0.530	3.88	-429	-1821	-81
0.3	0.696	5.03	-580	-1468	-199
0.4	0.811	6.51	-681	-1129	-383
0.5	0.888	8.33	-728	-816	-639
0.6	0.937	10.47	-715	-541	-976
0.7	0.967	12.86	-639	-314	-1399
0.8	0.985	15.35	-497	-143	-1911
0.9	0.995	17.78	-285	-37	-2517
1.0	1.000	19.92	0	0	-3216

B_{11}/cm^{3}mol^{-1}	B_{22}/cm^{3}mol^{-1}	B_{12}/cm^{3}mol^{-1}	V_1/cm^{3}mol^{-1}	V_2/cm^{3}mol^{-1}
-720	-1350	-1000	18	25

$A(1)$ = -0.1051E+01 $A(2)$ = -0.1275E+00 $A(3)$ = 0.1695E-01

---

1. Water, H_2O
2. Methanol, CH_4O

***T*/K = 323.15**
***System Number*: 456**

x_1	y_1	P/kPa	G^E/J mol^{-1}	μ^E_1/J mol^{-1}	μ^E_2/J mol^{-1}
0.0	0.000	55.65	0	1118	0
0.1	0.036	51.91	112	1108	2
0.2	0.075	48.34	217	996	23
0.3	0.114	44.94	305	841	75
0.4	0.153	41.67	368	670	168
0.5	0.195	38.42	403	497	310
0.6	0.240	35.08	404	332	512
0.7	0.297	31.36	366	188	781
0.8	0.381	26.77	285	78	1109
0.9	0.544	20.59	160	16	1458
1.0	1.000	12.33	0	0	1737

B_{11}/cm^{3}mol^{-1}	B_{22}/cm^{3}mol^{-1}	B_{12}/cm^{3}mol^{-1}	V_1/cm^{3}mol^{-1}	V_2/cm^{3}mol^{-1}
-810	-1380	-1110	18	42

$A(1)$ = 0.6005E+00 $A(2)$ = 0.1393E+00 $A(3)$ = -0.3455E-01 $A(4)$ = -0.2416E-01 $A(5)$ = -0.3469E-01

TABLE 3.1.1 (VLEG) ISOTHERMAL VAPOR-LIQUID EQUILIBRIA AND EXCESS GIBBS ENERGIES OF BINARY MIXTURES (continued)

1. Water, H_2O
2. Ethanenitrile, C_2H_3N

***T*/K = 323.15**
***System Number*: 451**

x_1	y_1	P/kPa	G^E/J mol^{-1}	μ^E_1/J mol^{-1}	μ^E_2/J mol^{-1}
0.0	0.000	33.86	0	5325	0
0.1	0.163	37.23	483	4362	52
0.2	0.228	38.10	860	3457	211
0.3	0.257	38.09	1133	2678	471
0.4	0.272	37.86	1304	2034	816
0.5	0.282	37.54	1377	1507	1248
0.6	0.290	37.15	1354	1063	1792
0.7	0.297	36.72	1228	679	2509
0.8	0.307	35.94	981	350	3505
0.9	0.351	32.39	585	103	4928
1.0	1.000	12.35	0	0	6970

B_{11}/cm^{3}mol^{-1}	B_{22}/cm^{3}mol^{-1}	B_{12}/cm^{3}mol^{-1}	V_1/cm^{3}mol^{-1}	V_2/cm^{3}mol^{-1}
-810	-4040	-290	18	55

$A(1) = 0.2051E+01$ $A(2) = 0.1926E+00$ $A(3) = 0.2619E+00$ $A(4) = 0.1135E+00$ $A(5) = -0.2448E-01$

1. Water, H_2O
2. Ethanol, C_2H_6O

***T*/K = 323.15**
***System Number*: 453**

x_1	y_1	P/kPa	G^E/J mol^{-1}	μ^E_1/J mol^{-1}	μ^E_2/J mol^{-1}
0.0	0.000	29.48	0	2383	0
0.1	0.098	29.52	234	2265	9
0.2	0.178	29.33	445	1987	59
0.3	0.240	28.90	617	1670	165
0.4	0.292	28.27	744	1357	335
0.5	0.334	27.51	819	1053	585
0.6	0.370	26.65	834	756	950
0.7	0.404	25.68	775	473	1479
0.8	0.446	24.24	627	229	2217
0.9	0.543	21.00	372	60	3182
1.0	1.000	12.35	0	0	4324

B_{11}/cm^{3}mol^{-1}	B_{22}/cm^{3}mol^{-1}	B_{12}/cm^{3}mol^{-1}	V_1/cm^{3}mol^{-1}	V_2/cm^{3}mol^{-1}
-810	-1400	-1300	18	61

$A(1) = 0.1219E+01$ $A(2) = 0.3485E+00$ $A(3) = 0.1010E+00$ $A(4) = 0.1266E-01$ $A(5) = -0.7177E-01$

1. Water, H_2O
2. Dimethyl sulfoxide, C_2H_6OS

***T*/K = 298.15**
***System Number*: 463**

x_1	y_1	P/kPa	G^E/J mol^{-1}	μ^E_1/J mol^{-1}	μ^E_2/J mol^{-1}
0.0	0.000	0.080	0	-2188	0
0.1	0.574	0.171	-266	-2911	28
0.2	0.757	0.264	-561	-2858	13
0.3	0.863	0.393	-829	-2556	-90
0.4	0.925	0.567	-1049	-2189	-289
0.5	0.961	0.804	-1203	-1777	-629
0.6	0.983	1.141	-1269	-1308	-1210
0.7	0.994	1.612	-1210	-807	-2149
0.8	0.998	2.197	-988	-360	-3500
0.9	1.000	2.768	-578	-76	-5101
1.0	1.000	3.170	0	0	-6383

B_{11}/cm^{3}mol^{-1}	B_{22}/cm^{3}mol^{-1}	B_{12}/cm^{3}mol^{-1}	V_1/cm^{3}mol^{-1}	V_2/cm^{3}mol^{-1}
-1180	-5970	-2590	18	71

$A(1) = -0.1942E+01$ $A(2) = -0.9261E+00$ $A(3) = -0.1673E+00$ $A(4) = 0.8010E-01$ $A(5) = 0.3802E+00$

TABLE 3.1.1 (VLEG)
ISOTHERMAL VAPOR-LIQUID EQUILIBRIA AND EXCESS GIBBS ENERGIES OF BINARY MIXTURES (continued)

1. Water, H_2O
2. 1,2-Ethanediol, $C_2H_6O_2$

***T*/K = 333.15**
***System Number*: 468**

x_1	y_1	P/kPa	G^E/J mol^{-1}	μ^E_1/J mol^{-1}	μ^E_2/J mol^{-1}
0.0	0.000	0.21	0	-412	0
0.1	0.903	1.98	-35	-292	-6
0.2	0.956	3.85	-58	-209	-20
0.3	0.974	5.79	-73	-152	-39
0.4	0.984	7.76	-81	-112	-61
0.5	0.989	9.75	-84	-83	-85
0.6	0.993	11.77	-81	-59	-114
0.7	0.996	13.80	-72	-38	-153
0.8	0.997	15.86	-57	-20	-208
0.9	0.999	17.91	-34	-6	-288
1.0	1.000	19.93	0	0	-403

B_{11}/cm^{3}mol^{-1}	B_{22}/cm^{3}mol^{-1}	B_{12}/cm^{3}mol^{-1}	V_1/cm^{3}mol^{-1}	V_2/cm^{3}mol^{-1}
-720	-1700	-1050	18	58

$A(1) = -0.1209E+00$ $A(2) = 0.1667E-02$ $A(3) = -0.2620E-01$

1. Water, H_2O
2. 2-Aminoethanol, C_2H_7NO

***T*/K = 298.15**
***System Number*: 465**

x_1	y_1	P/kPa	G^E/J mol^{-1}	μ^E_1/J mol^{-1}	μ^E_2/J mol^{-1}
0.0	0.000	0.065	0	-1946	0
0.1	0.718	0.207	-190	-1870	-3
0.2	0.855	0.356	-374	-1811	-14
0.3	0.916	0.528	-545	-1671	-63
0.4	0.952	0.749	-689	-1424	-199
0.5	0.974	1.045	-783	-1092	-473
0.6	0.988	1.433	-806	-729	-921
0.7	0.994	1.899	-740	-397	-1539
0.8	0.998	2.386	-577	-154	-2267
0.9	0.999	2.821	-322	-28	-2969
1.0	1.000	3.168	0	0	-3417

B_{11}/cm^{3}mol^{-1}	B_{22}/cm^{3}mol^{-1}	B_{12}/cm^{3}mol^{-1}	V_1/cm^{3}mol^{-1}	V_2/cm^{3}mol^{-1}
-1180	0	0	18	0

$A(1) = -0.1263E+01$ $A(2) = -0.4993E+00$ $A(3) = 0.1816E+00$ $A(4) = 0.2025E+00$ $A(5) = 0.0000E+00$

1. Water, H_2O
2. 2-Propanone, C_3H_6O

***T*/K = 323.15**
***System Number*: 452**

x_1	y_1	P/kPa	G^E/J mol^{-1}	μ^E_1/J mol^{-1}	μ^E_2/J mol^{-1}
0.0	0.000	82.03	0	4655	0
0.1	0.063	80.06	417	3748	47
0.2	0.098	77.56	742	2995	179
0.3	0.119	75.17	981	2353	393
0.4	0.133	73.03	1135	1802	690
0.5	0.144	71.04	1203	1331	1076
0.6	0.153	68.92	1185	930	1567
0.7	0.164	66.24	1073	588	2204
0.8	0.180	61.88	856	304	3065
0.9	0.225	51.38	511	91	4291
1.0	1.000	12.35	0	0	6121

B_{11}/cm^{3}mol^{-1}	B_{22}/cm^{3}mol^{-1}	B_{12}/cm^{3}mol^{-1}	V_1/cm^{3}mol^{-1}	V_2/cm^{3}mol^{-1}
-810	-1430	-870	18	77

$A(1) = 0.1791E+01$ $A(2) = 0.1903E+00$ $A(3) = 0.1733E+00$ $A(4) = 0.8250E-01$ $A(5) = 0.4040E-01$

TABLE 3.1.1 (VLEG) ISOTHERMAL VAPOR-LIQUID EQUILIBRIA AND EXCESS GIBBS ENERGIES OF BINARY MIXTURES (continued)

1. Water, H_2O
2. 2-Propanol, C_3H_8O

***T*/K = 298.15**
***System Number*: 464**

x_1	y_1	P/kPa	G^E/J mol^{-1}	μ^E_1/J mol^{-1}	μ^E_2/J mol^{-1}
0.0	0.000	5.866	0	3334	0
0.1	0.151	6.287	297	2711	28
0.2	0.254	6.516	545	2374	87
0.3	0.326	6.616	751	2024	206
0.4	0.367	6.619	902	1608	432
0.5	0.389	6.577	982	1185	780
0.6	0.407	6.500	982	817	1229
0.7	0.434	6.333	900	531	1761
0.8	0.476	6.021	733	303	2451
0.9	0.544	5.481	457	108	3596
1.0	1.000	3.161	0	0	5896

B_{11}/cm^{3}mol^{-1}	B_{22}/cm^{3}mol^{-1}	B_{12}/cm^{3}mol^{-1}	V_1/cm^{3}mol^{-1}	V_2/cm^{3}mol^{-1}
-1180	-3280	-1770	18	77

$A(1) = 0.1585E+01$ $A(2) = 0.3265E+00$ $A(3) = -0.4479E-01$ $A(4) = 0.1901E+00$ $A(5) = 0.3215E+00$

1. Water, H_2O
2. 2-(Methylamino)ethanol, C_3H_9NO

***T*/K = 298.15**
***System Number*: 466**

x_1	y_1	P/kPa	G^E/J mol^{-1}	μ^E_1/J mol^{-1}	μ^E_2/J mol^{-1}
0.0	0.000	0.123	0	-1921	0
0.1	0.612	0.283	-167	-1490	-20
0.2	0.795	0.469	-302	-1311	-50
0.3	0.876	0.668	-421	-1202	-86
0.4	0.923	0.893	-523	-1065	-161
0.5	0.954	1.168	-597	-869	-325
0.6	0.975	1.514	-626	-626	-626
0.7	0.988	1.929	-591	-375	-1095
0.8	0.995	2.381	-478	-166	-1724
0.9	0.998	2.812	-279	-38	-2448
1.0	1.000	3.168	0	0	-3129

B_{11}/cm^{3}mol^{-1}	B_{22}/cm^{3}mol^{-1}	B_{12}/cm^{3}mol^{-1}	V_1/cm^{3}mol^{-1}	V_2/cm^{3}mol^{-1}
-1180	0	0	18	0

$A(1) = -0.9629E+00$ $A(2) = -0.4393E+00$ $A(3) = -0.5563E-01$ $A(4) = 0.1955E+00$ $A(5) = 0.0000E+00$

1. Water, H_2O
2. 4-Methyl-1,3-dioxolan-2-one, $C_4H_6O_3$

***T*/K = 298.15**
***System Number*: 469**

x_1	y_1	P/kPa	G^E/J mol^{-1}	μ^E_1/J mol^{-1}	μ^E_2/J mol^{-1}
0.0	0.000	0.007	0	5586	0
0.1	0.997	1.987	504	4542	55
0.2	0.998	2.770	899	3650	212
0.3	0.998	3.096	1194	2921	454
0.4	0.998	3.251	1395	2330	772
0.5	0.998	3.323	1506	1831	1180
0.6	0.998	3.321	1521	1377	1738
0.7	0.998	3.245	1425	937	2563
0.8	0.998	3.126	1181	513	3854
0.9	0.997	3.050	734	159	5903
1.0	1.000	3.170	0	0	9120

B_{11}/cm^{3}mol^{-1}	B_{22}/cm^{3}mol^{-1}	B_{12}/cm^{3}mol^{-1}	V_1/cm^{3}mol^{-1}	V_2/cm^{3}mol^{-1}
-1180	0	0	18	0

$A(1) = 0.2429E+01$ $A(2) = 0.5247E+00$ $A(3) = 0.5370E+00$ $A(4) = 0.1881E+00$ $A(5) = 0.0000E+00$

TABLE 3.1.1 (VLEG) ISOTHERMAL VAPOR-LIQUID EQUILIBRIA AND EXCESS GIBBS ENERGIES OF BINARY MIXTURES (continued)

1. Water, H_2O
2. Oxolane, C_4H_8O

***T*/K = 298.15**
***System Number*: 454**

x_1	y_1	P/kPa	G^E/J mol^{-1}	μ^E_1/J mol^{-1}	μ^E_2/J mol^{-1}
0.0	0.000	21.61	0	5899	0
0.1	0.094	22.05	522	4623	67
0.2	0.122	21.85	914	3540	257
0.3	0.130	21.67	1183	2671	545
0.4	0.133	21.55	1343	1998	906
0.5	0.136	21.39	1404	1477	1331
0.6	0.139	21.13	1373	1058	1845
0.7	0.142	20.87	1243	696	2520
0.8	0.144	20.68	999	375	3496
0.9	0.157	19.20	604	116	4995
1.0	1.000	3.17	0	0	7337

B_{11}/cm^{3}mol^{-1}	B_{22}/cm^{3}mol^{-1}	B_{12}/cm^{3}mol^{-1}	V_1/cm^{3}mol^{-1}	V_2/cm^{3}mol^{-1}
-1180	-1360	-1130	18	82

$A(1)$ = 0.2266E+01 $A(2)$ = 0.1178E+00 $A(3)$ = 0.4037E+00 $A(4)$ = 0.1722E+00 $A(5)$ = 0.0000E+00

1. Water, H_2O
2. 1,4-Dioxane, $C_4H_8O_2$

***T*/K = 323.15**
***System Number*: 450**

x_1	y_1	P/kPa	G^E/J mol^{-1}	μ^E_1/J mol^{-1}	μ^E_2/J mol^{-1}
0.0	0.000	15.71	0	5073	0
0.1	0.274	19.91	449	3986	56
0.2	0.363	21.44	789	3082	215
0.3	0.403	21.95	1025	2337	463
0.4	0.426	22.10	1165	1729	790
0.5	0.443	22.07	1214	1232	1196
0.6	0.460	21.92	1172	826	1693
0.7	0.483	21.56	1039	497	2305
0.8	0.523	20.71	808	241	3077
0.9	0.615	18.56	467	67	4073
1.0	1.000	12.35	0	0	5383

B_{11}/cm^{3}mol^{-1}	B_{22}/cm^{3}mol^{-1}	B_{12}/cm^{3}mol^{-1}	V_1/cm^{3}mol^{-1}	V_2/cm^{3}mol^{-1}
-810	-1390	-880	18	88

$A(1)$ = 0.1807E+01 $A(2)$ = 0.2642E-01 $A(3)$ = 0.1384E+00 $A(4)$ = 0.3130E-01 $A(5)$ = 0.0000E+00

1. Water, H_2O
2. 2-(Dimethylamino)ethanol, $C_4H_{11}NO$

***T*/K = 298.15**
***System Number*: 467**

x_1	y_1	P/kPa	G^E/J mol^{-1}	μ^E_1/J mol^{-1}	μ^E_2/J mol^{-1}
0.0	0.000	0.852	0	359	0
0.1	0.287	1.084	13	-41	20
0.2	0.450	1.267	-7	-259	57
0.3	0.572	1.443	-45	-346	84
0.4	0.676	1.631	-89	-342	81
0.5	0.766	1.842	-126	-286	33
0.6	0.841	2.081	-150	-205	-67
0.7	0.900	2.343	-152	-122	-221
0.8	0.945	2.621	-129	-56	-422
0.9	0.978	2.900	-78	-14	-660
1.0	1.000	3.168	0	0	-915

B_{11}/cm^{3}mol^{-1}	B_{22}/cm^{3}mol^{-1}	B_{12}/cm^{3}mol^{-1}	V_1/cm^{3}mol^{-1}	V_2/cm^{3}mol^{-1}
-1180	0	0	18	0

$A(1)$ = -0.2039E+00 $A(2)$ = -0.2570E+00 $A(3)$ = 0.9177E-01

TABLE 3.1.1 (VLEG)
ISOTHERMAL VAPOR-LIQUID EQUILIBRIA AND EXCESS GIBBS ENERGIES OF BINARY MIXTURES (continued)

1. Water, H_2O
2. Pyridine, C_5H_5N

***T*/K = 298.15**
***System Number*: 457**

x_1	y_1	*P*/kPa	G^E/J mol^{-1}	μ^E_1/J mol^{-1}	μ^E_2/J mol^{-1}
0.0	0.000	2.801	0	1271	0
0.1	0.199	3.131	155	1680	-14
0.2	0.342	3.430	319	1527	17
0.3	0.431	3.615	450	1223	119
0.4	0.501	3.737	540	969	254
0.5	0.574	3.835	597	816	377
0.6	0.650	3.912	630	721	495
0.7	0.715	3.954	638	603	720
0.8	0.756	3.946	591	405	1335
0.9	0.774	3.915	420	151	2845
1.0	1.000	3.168	0	0	6037

B_{11}/cm^{3}mol^{-1}	B_{22}/cm^{3}mol^{-1}	B_{12}/cm^{3}mol^{-1}	V_1/cm^{3}mol^{-1}	V_2/cm^{3}mol^{-1}
-1180	-2460	-1740	18	81

A(1) = 0.9632E+00   *A*(2) = 0.3541E+00   *A*(3) = 0.5109E+00   *A*(4) = 0.6072E+00   *A*(5) = 0.0000E+00

1. Water, H_2O
2. 2-Methylpyridine, C_6H_7N

***T*/K = 298.15**
***System Number*: 458**

x_1	y_1	*P*/kPa	G^E/J mol^{-1}	μ^E_1/J mol^{-1}	μ^E_2/J mol^{-1}
0.0	0.000	1.573	0	1652	0
0.1	0.306	2.043	171	1685	2
0.2	0.469	2.418	326	1446	46
0.3	0.569	2.699	449	1191	131
0.4	0.649	2.946	541	1020	222
0.5	0.721	3.190	612	926	297
0.6	0.783	3.414	667	846	397
0.7	0.826	3.566	694	704	671
0.8	0.845	3.614	654	462	1423
0.9	0.844	3.616	467	167	3167
1.0	1.000	3.168	0	0	6665

B_{11}/cm^{3}mol^{-1}	B_{22}/cm^{3}mol^{-1}	B_{12}/cm^{3}mol^{-1}	V_1/cm^{3}mol^{-1}	V_2/cm^{3}mol^{-1}
-1180	-2900	-1900	18	99

A(1) = 0.9872E+00   *A*(2) = 0.5075E+00   *A*(3) = 0.6903E+00   *A*(4) = 0.5035E+00   *A*(5) = 0.0000E+00

1. Water, H_2O
2. 3-Methylpyridine, C_6H_7N

***T*/K = 298.15**
***System Number*: 459**

x_1	y_1	*P*/kPa	G^E/J mol^{-1}	μ^E_1/J mol^{-1}	μ^E_2/J mol^{-1}
0.0	0.000	0.921	0	844	0
0.1	0.418	1.411	130	1545	-27
0.2	0.612	1.892	288	1496	-14
0.3	0.701	2.226	422	1229	77
0.4	0.760	2.499	519	1005	196
0.5	0.814	2.786	589	890	288
0.6	0.861	3.084	643	830	361
0.7	0.895	3.318	677	724	567
0.8	0.910	3.409	654	500	1268
0.9	0.904	3.403	482	189	3119
1.0	1.000	3.168	0	0	7136

B_{11}/cm^{3}mol^{-1}	B_{22}/cm^{3}mol^{-1}	B_{12}/cm^{3}mol^{-1}	V_1/cm^{3}mol^{-1}	V_2/cm^{3}mol^{-1}
-1180	-3300	-2060	18	98

A(1) = 0.9502E+00   *A*(2) = 0.4861E+00   *A*(3) = 0.6595E+00   *A*(4) = 0.7830E+00   *A*(5) = 0.0000E+00

TABLE 3.1.1 (VLEG)
ISOTHERMAL VAPOR-LIQUID EQUILIBRIA AND EXCESS GIBBS ENERGIES OF BINARY MIXTURES (continued)

1. Water, H_2O
2. 4-Methylpyridine, C_6H_7N

***T*/K = 298.15**
***System Number*: 460**

x_1	y_1	P/kPa	G^E/J mol^{-1}	μ^E_1/J mol^{-1}	μ^E_2/J mol^{-1}
0.0	0.000	0.817	0	939	0
0.1	0.446	1.317	134	1535	-22
0.2	0.632	1.781	287	1427	2
0.3	0.718	2.107	411	1152	94
0.4	0.779	2.398	501	962	194
0.5	0.835	2.729	571	901	241
0.6	0.882	3.085	636	890	256
0.7	0.913	3.361	691	805	423
0.8	0.924	3.449	688	567	1172
0.9	0.911	3.418	521	217	3263
1.0	1.000	3.168	0	0	7866

B_{11}/cm^{3}mol^{-1}	B_{22}/cm^{3}mol^{-1}	B_{12}/cm^{3}mol^{-1}	V_1/cm^{3}mol^{-1}	V_2/cm^{3}mol^{-1}
-1180	-3300	-2060	18	98

$A(1)$ = 0.9215E+00 $A(2)$ = 0.5329E+00 $A(3)$ = 0.8543E+00 $A(4)$ = 0.8644E+00 $A(5)$ = 0.0000E+00

---

1. Water, H_2O
2. Triethylamine, $C_6H_{15}N$

***T*/K = 278.15**
***System Number*: 455**

x_1	y_1	P/kPa	G^E/J mol^{-1}	μ^E_1/J mol^{-1}	μ^E_2/J mol^{-1}
0.0	0.000	3.180	0	6759	0
0.1	0.150	3.544	523	4178	117
0.2	0.185	3.563	858	3075	304
0.3	0.212	3.535	1091	2429	518
0.4	0.224	3.504	1240	1875	817
0.5	0.226	3.497	1301	1372	1229
0.6	0.230	3.473	1270	971	1719
0.7	0.244	3.356	1153	679	2260
0.8	0.265	3.177	949	434	3007
0.9	0.270	3.135	612	173	4563
1.0	1.000	0.872	0	0	8358

B_{11}/cm^{3}mol^{-1}	B_{22}/cm^{3}mol^{-1}	B_{12}/cm^{3}mol^{-1}	V_1/cm^{3}mol^{-1}	V_2/cm^{3}mol^{-1}
-1900	-2180	-2040	18	140

$A(1)$ = 0.2249E+01 $A(2)$ = 0.1243E+00 $A(3)$ = 0.2610E+00 $A(4)$ = 0.2214E+00 $A(5)$ = 0.7579E+00

---

1. Water, H_2O
2. 2,6-Dimethylpyridine, C_7H_9N

***T*/K = 298.15**
***System Number*: 461**

x_1	y_1	P/kPa	G^E/J mol^{-1}	μ^E_1/J mol^{-1}	μ^E_2/J mol^{-1}
0.0	0.000	0.769	0	2456	0
0.1	0.513	1.436	228	2095	20
0.2	0.665	1.907	413	1724	86
0.3	0.746	2.281	558	1445	177
0.4	0.804	2.631	670	1271	270
0.5	0.849	2.968	760	1152	368
0.6	0.882	3.248	824	1017	536
0.7	0.901	3.407	846	806	939
0.8	0.904	3.433	777	503	1875
0.9	0.895	3.417	537	174	3808
1.0	1.000	3.168	0	0	7406

B_{11}/cm^{3}mol^{-1}	B_{22}/cm^{3}mol^{-1}	B_{12}/cm^{3}mol^{-1}	V_1/cm^{3}mol^{-1}	V_2/cm^{3}mol^{-1}
-1180	-3430	-2110	18	117

$A(1)$ = 0.1226E+01 $A(2)$ = 0.6327E+00 $A(3)$ = 0.7633E+00 $A(4)$ = 0.3657E+00 $A(5)$ = 0.0000E+00

TABLE 3.1.1 (VLEG) ISOTHERMAL VAPOR-LIQUID EQUILIBRIA AND EXCESS GIBBS ENERGIES OF BINARY MIXTURES (continued)

1. Dichlorodifluoromethane, CCl_2F_2
2. Chlorodifluoromethane, $CHClF_2$

***T*/K = 283.15**
***System Number*: 380**

x_1	y_1	P/kPa	G^E/J mol^{-1}	μ^E_1/J mol^{-1}	μ^E_2/J mol^{-1}
0.0	0.000	680.9	0	1045	0
0.1	0.092	679.0	89	753	15
0.2	0.170	669.5	149	538	52
0.3	0.243	654.6	187	383	104
0.4	0.317	635.2	206	270	164
0.5	0.395	611.7	209	187	232
0.6	0.479	584.2	198	123	309
0.7	0.573	552.7	172	74	401
0.8	0.683	516.7	132	36	515
0.9	0.818	474.6	76	10	664
1.0	1.000	423.4	0	0	863

B_{11}/cm^{3}mol^{-1}	B_{22}/cm^{3}mol^{-1}	B_{12}/cm^{3}mol^{-1}	V_1/cm^{3}mol^{-1}	V_2/cm^{3}mol^{-1}
-560	-400	-480	89	90

$A(1)$ = 0.3555E+00 $A(2)$ = -0.3865E-01 $A(3)$ = 0.4979E-01

1. Tetrachloromethane, CCl_4
2. Trichloromethane, $CHCl_3$

***T*/K = 298.15**
***System Number*: 357**

x_1	y_1	P/kPa	G^E/J mol^{-1}	μ^E_1/J mol^{-1}	μ^E_2/J mol^{-1}
0.0	0.000	26.24	0	420	0
0.1	0.069	25.40	38	337	4
0.2	0.139	24.53	67	264	17
0.3	0.211	23.61	87	201	38
0.4	0.286	22.64	99	146	68
0.5	0.368	21.61	103	101	105
0.6	0.458	20.51	98	64	150
0.7	0.560	19.34	86	36	202
0.8	0.678	18.07	65	16	262
0.9	0.821	16.69	36	4	329
1.0	1.000	15.19	0	0	403

B_{11}/cm^{3}mol^{-1}	B_{22}/cm^{3}mol^{-1}	B_{12}/cm^{3}mol^{-1}	V_1/cm^{3}mol^{-1}	V_2/cm^{3}mol^{-1}
-1800	-1200	-1460	97	81

$A(1)$ = 0.1659E+00 $A(2)$ = -0.3486E-02 $A(3)$ = 0.0000E+00

1. Tetrachloromethane, CCl_4
2. Iodomethane, CH_3I

***T*/K = 298.15**
***System Number*: 361**

x_1	y_1	P/kPa	G^E/J mol^{-1}	μ^E_1/J mol^{-1}	μ^E_2/J mol^{-1}
0.0	0.000	53.79	0	903	0
0.1	0.041	50.61	80	717	10
0.2	0.081	47.46	142	555	38
0.3	0.124	44.29	184	417	84
0.4	0.169	41.03	208	300	147
0.5	0.222	37.61	215	204	226
0.6	0.285	33.95	204	128	319
0.7	0.367	29.97	177	70	425
0.8	0.482	25.58	133	30	544
0.9	0.662	20.65	74	7	674
1.0	1.000	15.09	0	0	815

B_{11}/cm^{3}mol^{-1}	B_{22}/cm^{3}mol^{-1}	B_{12}/cm^{3}mol^{-1}	V_1/cm^{3}mol^{-1}	V_2/cm^{3}mol^{-1}
-1800	-870	-1300	97	63

$A(1)$ = 0.3465E+00 $A(2)$ = -0.1777E-01 $A(3)$ = 0.0000E+00

TABLE 3.1.1 (VLEG) ISOTHERMAL VAPOR-LIQUID EQUILIBRIA AND EXCESS GIBBS ENERGIES OF BINARY MIXTURES (continued)

1. Tetrachloromethane, CCl_4
2. Nitromethane, CH_3NO_2

***T*/K = 318.15**
***System Number*: 373**

x_1	y_1	P/kPa	G^E/J mol^{-1}	μ^E_1/J mol^{-1}	μ^E_2/J mol^{-1}
0.0	0.000	12.49	0	5351	0
0.1	0.596	29.01	478	4274	57
0.2	0.686	35.51	844	3334	222
0.3	0.716	38.01	1103	2551	482
0.4	0.730	39.10	1260	1916	823
0.5	0.741	39.73	1323	1404	1242
0.6	0.750	40.14	1293	983	1757
0.7	0.761	40.38	1165	627	2422
0.8	0.775	40.42	927	325	3336
0.9	0.812	39.79	552	97	4651
1.0	1.000	34.51	0	0	6591

B_{11}/cm^{3}mol^{-1}	B_{22}/cm^{3}mol^{-1}	B_{12}/cm^{3}mol^{-1}	V_1/cm^{3}mol^{-1}	V_2/cm^{3}mol^{-1}
-1450	-3100	-2170	100	55

$A(1) = 0.2001E+01$ $A(2) = 0.1231E+00$ $A(3) = 0.2567E+00$ $A(4) = 0.1115E+00$ $A(5) = 0.0000E+00$

1. Tetrachloromethane, CCl_4
2. Methylamine, CH_5N

***T*/K = 293.15**
***System Number*: 362**

x_1	y_1	P/kPa	G^E/J mol^{-1}	μ^E_1/J mol^{-1}	μ^E_2/J mol^{-1}
0.0	0.000	292.2	0	1210	0
0.1	0.008	264.9	109	982	12
0.2	0.017	239.6	194	784	47
0.3	0.025	215.5	256	611	104
0.4	0.035	191.9	295	460	186
0.5	0.046	168.2	311	330	292
0.6	0.061	143.7	303	219	428
0.7	0.082	117.5	269	129	596
0.8	0.119	88.2	209	60	803
0.9	0.209	54.1	120	16	1055
1.0	1.000	12.1	0	0	1360

B_{11}/cm^{3}mol^{-1}	B_{22}/cm^{3}mol^{-1}	B_{12}/cm^{3}mol^{-1}	V_1/cm^{3}mol^{-1}	V_2/cm^{3}mol^{-1}
-1850	-480	-1160	96	47

$A(1) = 0.5105E+00$ $A(2) = 0.3090E-01$ $A(3) = 0.1663E-01$

1. Tetrachloromethane, CCl_4
2. Carbon disulfide, CS_2

***T*/K = 298.15**
***System Number*: 388**

x_1	y_1	P/kPa	G^E/J mol^{-1}	μ^E_1/J mol^{-1}	μ^E_2/J mol^{-1}
0.0	0.000	48.17	0	857	0
0.1	0.044	45.53	74	644	11
0.2	0.088	42.89	127	465	43
0.3	0.132	40.19	159	318	91
0.4	0.180	37.36	173	203	153
0.5	0.237	34.34	170	118	222
0.6	0.306	31.05	153	60	293
0.7	0.397	27.47	124	24	359
0.8	0.522	23.60	87	6	412
0.9	0.708	19.50	44	0	442
1.0	1.000	15.28	0	0	441

B_{11}/cm^{3}mol^{-1}	B_{22}/cm^{3}mol^{-1}	B_{12}/cm^{3}mol^{-1}	V_1/cm^{3}mol^{-1}	V_2/cm^{3}mol^{-1}
-1800	-810	-1240	97	61

$A(1) = 0.2746E+00$ $A(2) = -0.8387E-01$ $A(3) = -0.1273E-01$

TABLE 3.1.1 (VLEG)
ISOTHERMAL VAPOR-LIQUID EQUILIBRIA AND EXCESS GIBBS ENERGIES OF BINARY MIXTURES (continued)

1. Tetrachloromethane, CCl_4
2. 1,1,1-Trichloroethane, $C_2H_3Cl_3$

***T*/K = 298.15**
***System Number*: 358**

x_1	y_1	P/kPa	G^E/J mol^{-1}	μ^E_1/J mol^{-1}	μ^E_2/J mol^{-1}
0.0	0.000	16.59	0	189	0
0.1	0.100	16.58	20	199	0
0.2	0.199	16.57	39	194	1
0.3	0.297	16.55	58	177	7
0.4	0.393	16.52	73	150	21
0.5	0.486	16.45	83	119	47
0.6	0.580	16.35	87	85	89
0.7	0.674	16.19	82	53	149
0.8	0.772	15.97	67	26	231
0.9	0.879	15.66	40	7	339
1.0	1.000	15.25	0	0	475

B_{11}/cm^{3}mol^{-1}	B_{22}/cm^{3}mol^{-1}	B_{12}/cm^{3}mol^{-1}	V_1/cm^{3}mol^{-1}	V_2/cm^{3}mol^{-1}
-1800	-1520	-1600	97	100

$A(1)$ = 0.1339E+00 $A(2)$ = 0.5775E-01 $A(3)$ = 0.0000E+00

---

1. Tetrachloromethane, CCl_4
2. Ethanenitrile, C_2H_3N

***T*/K = 318.15**
***System Number*: 368**

x_1	y_1	P/kPa	G^E/J mol^{-1}	μ^E_1/J mol^{-1}	μ^E_2/J mol^{-1}
0.0	0.000	27.84	0	4631	0
0.1	0.354	39.78	420	3806	44
0.2	0.469	45.27	751	3032	181
0.3	0.521	47.66	991	2359	405
0.4	0.551	48.75	1143	1798	707
0.5	0.574	49.25	1210	1336	1084
0.6	0.594	49.39	1193	948	1559
0.7	0.614	49.21	1085	613	2185
0.8	0.638	48.59	871	323	3065
0.9	0.693	46.38	524	98	4366
1.0	1.000	34.51	0	0	6326

B_{11}/cm^{3}mol^{-1}	B_{22}/cm^{3}mol^{-1}	B_{12}/cm^{3}mol^{-1}	V_1/cm^{3}mol^{-1}	V_2/cm^{3}mol^{-1}
-1450	-4500	-12	100	54

$A(1)$ = 0.1830E+01 $A(2)$ = 0.1899E+00 $A(3)$ = 0.2412E+00 $A(4)$ = 0.1305E+00 $A(5)$ = 0.0000E+00

---

1. Tetrachloromethane, CCl_4
2. 1,2-Dichloroethane, $C_2H_4Cl_2$

***T*/K = 313.15**
***System Number*: 356**

x_1	y_1	P/kPa	G^E/J mol^{-1}	μ^E_1/J mol^{-1}	μ^E_2/J mol^{-1}
0.0	0.000	20.74	0	1521	0
0.1	0.194	23.34	137	1226	16
0.2	0.325	25.24	242	960	63
0.3	0.423	26.63	316	725	141
0.4	0.503	27.64	359	523	249
0.5	0.574	28.38	371	355	387
0.6	0.644	28.87	353	221	550
0.7	0.715	29.15	305	120	737
0.8	0.795	29.20	230	51	943
0.9	0.888	28.98	127	12	1164
1.0	1.000	28.44	0	0	1395

B_{11}/cm^{3}mol^{-1}	B_{22}/cm^{3}mol^{-1}	B_{12}/cm^{3}mol^{-1}	V_1/cm^{3}mol^{-1}	V_2/cm^{3}mol^{-1}
-1600	-1290	-1450	98	80

$A(1)$ = 0.5698E+00 $A(2)$ = -0.2430E-01 $A(3)$ = -0.9871E-02

TABLE 3.1.1 (VLEG) ISOTHERMAL VAPOR-LIQUID EQUILIBRIA AND EXCESS GIBBS ENERGIES OF BINARY MIXTURES (continued)

1. Tetrachloromethane, CCl_4
2. Nitroethane, $C_2H_5NO_2$

***T*/K = 298.15**
***System Number*: 374**

x_1	y_1	P/kPa	G^E/J mol^{-1}	μ^E_1/J mol^{-1}	μ^E_2/J mol^{-1}
0.0	0.000	2.91	0	2684	0
0.1	0.589	6.47	247	2278	21
0.2	0.730	9.04	452	1919	85
0.3	0.795	10.92	613	1588	195
0.4	0.832	12.27	729	1275	365
0.5	0.856	13.22	793	975	611
0.6	0.874	13.87	799	690	961
0.7	0.888	14.34	735	431	1446
0.8	0.906	14.73	591	213	2105
0.9	0.934	15.10	352	59	2984
1.0	1.000	15.31	0	0	4137

B_{11}/cm^{3}mol^{-1}	B_{22}/cm^{3}mol^{-1}	B_{12}/cm^{3}mol^{-1}	V_1/cm^{3}mol^{-1}	V_2/cm^{3}mol^{-1}
-1800	-4900	-3100	97	72

$A(1) = 0.1280E+01$ $A(2) = 0.2931E+00$ $A(3) = 0.9627E-01$

---

1. Tetrachloromethane, CCl_4
2. Dimethyl sulfoxide, C_2H_6OS

***T*/K = 313.15**
***System Number*: 10**

x_1	y_1	P/kPa	G^E/J mol^{-1}	μ^E_1/J mol^{-1}	μ^E_2/J mol^{-1}
0.0	0.000	0.22	0	4155	0
0.1	0.978	9.42	353	3095	49
0.2	0.988	15.30	625	2567	140
0.3	0.991	19.63	841	2163	275
0.4	0.993	22.39	1000	1756	496
0.5	0.993	23.98	1090	1354	826
0.6	0.994	25.05	1103	992	1269
0.7	0.995	25.94	1032	682	1849
0.8	0.995	26.62	862	402	2703
0.9	0.996	27.11	551	143	4223
1.0	1.000	28.42	0	0	7245

B_{11}/cm^{3}mol^{-1}	B_{22}/cm^{3}mol^{-1}	B_{12}/cm^{3}mol^{-1}	V_1/cm^{3}mol^{-1}	V_2/cm^{3}mol^{-1}
-1600	-4650	-2860	99	72

$A(1) = 0.1674E+01$ $A(2) = 0.4056E+00$ $A(3) = 0.1893E+00$ $A(4) = 0.1879E+00$ $A(5) = 0.3256E+00$

---

1. Tetrachloromethane, CCl_4
2. N,N-Dimethylmethanamide, C_3H_7NO

***T*/K = 303.15**
***System Number*: 371**

x_1	y_1	P/kPa	G^E/J mol^{-1}	μ^E_1/J mol^{-1}	μ^E_2/J mol^{-1}
0.0	0.000	0.72	0	1773	0
0.1	0.834	3.94	156	1428	15
0.2	0.914	6.85	289	1300	37
0.3	0.945	9.53	409	1188	75
0.4	0.960	11.79	509	1038	157
0.5	0.970	13.62	582	860	304
0.6	0.976	15.11	618	675	531
0.7	0.981	16.32	607	491	877
0.8	0.985	17.22	532	298	1468
0.9	0.988	17.89	355	106	2596
1.0	1.000	18.85	0	0	4843

B_{11}/cm^{3}mol^{-1}	B_{22}/cm^{3}mol^{-1}	B_{12}/cm^{3}mol^{-1}	V_1/cm^{3}mol^{-1}	V_2/cm^{3}mol^{-1}
-1750	-4000	-2720	97	78

$A(1) = 0.9233E+00$ $A(2) = 0.4409E+00$ $A(3) = 0.1945E+00$ $A(4) = 0.1681E+00$ $A(5) = 0.1945E+00$

TABLE 3.1.1 (VLEG)
ISOTHERMAL VAPOR-LIQUID EQUILIBRIA AND EXCESS GIBBS ENERGIES OF BINARY MIXTURES (continued)

1. Tetrachloromethane, CCl_4
2. Trimethylamine, C_3H_9N

***T*/K = 293.15**
***System Number*: 363**

x_1	y_1	*P*/kPa	G^E/J mol^{-1}	μ^E_1/J mol^{-1}	μ^E_2/J mol^{-1}
0.0	0.000	185.1	0	-1139	0
0.1	0.006	165.0	-102	-909	-12
0.2	0.014	144.2	-180	-707	-48
0.3	0.026	123.4	-234	-533	-105
0.4	0.043	103.1	-265	-386	-185
0.5	0.068	83.9	-274	-264	-285
0.6	0.107	66.1	-261	-166	-404
0.7	0.168	49.9	-227	-92	-542
0.8	0.273	35.4	-172	-40	-697
0.9	0.478	22.8	-96	-10	-868
1.0	1.000	12.1	0	0	-1055

B_{11}/cm^{3}mol^{-1}	B_{22}/cm^{3}mol^{-1}	B_{12}/cm^{3}mol^{-1}	V_1/cm^{3}mol^{-1}	V_2/cm^{3}mol^{-1}
-1850	-1160	-1500	96	95

A(1) = -0.4501E+00 *A*(2) = 0.1709E-01 *A*(3) = 0.0000E-01

1. Tetrachloromethane, CCl_4
2. Furan, C_4H_4O

***T*/K = 303.15**
***System Number*: 42**

x_1	y_1	*P*/kPa	G^E/J mol^{-1}	μ^E_1/J mol^{-1}	μ^E_2/J mol^{-1}
0.0	0.000	96.69	0	823	0
0.1	0.028	89.63	73	641	9
0.2	0.057	82.78	127	490	36
0.3	0.088	75.98	164	364	78
0.4	0.124	69.11	184	261	133
0.5	0.167	62.03	190	178	201
0.6	0.221	54.63	180	113	281
0.7	0.295	46.78	156	63	373
0.8	0.404	38.34	118	28	478
0.9	0.591	29.14	66	7	597
1.0	1.000	18.97	0	0	733

B_{11}/cm^{3}mol^{-1}	B_{22}/cm^{3}mol^{-1}	B_{12}/cm^{3}mol^{-1}	V_1/cm^{3}mol^{-1}	V_2/cm^{3}mol^{-1}
-1750	-830	-850	97	73

A(1) = 0.3009E+00 *A*(2) = -0.1791E-01 *A*(3) = 0.7768E-02

1. Tetrachloromethane, CCl_4
2. Thiophene, C_4H_4S

***T*/K = 343.15**
***System Number*: 378**

x_1	y_1	*P*/kPa	G^E/J mol^{-1}	μ^E_1/J mol^{-1}	μ^E_2/J mol^{-1}
0.0	0.000	64.49	0	598	0
0.1	0.143	67.94	54	486	6
0.2	0.264	70.90	96	384	24
0.3	0.371	73.42	126	292	55
0.4	0.468	75.57	143	211	98
0.5	0.558	77.39	149	144	153
0.6	0.645	78.91	142	90	219
0.7	0.731	80.15	123	49	295
0.8	0.817	81.13	92	21	379
0.9	0.907	81.86	51	5	468
1.0	1.000	82.33	0	0	560

B_{11}/cm^{3}mol^{-1}	B_{22}/cm^{3}mol^{-1}	B_{12}/cm^{3}mol^{-1}	V_1/cm^{3}mol^{-1}	V_2/cm^{3}mol^{-1}
-1150	-990	-1070	103	84

A(1) = 0.2082E+00 *A*(2) = -0.6609E-02 *A*(3) = -0.5386E-02

TABLE 3.1.1 (VLEG) ISOTHERMAL VAPOR-LIQUID EQUILIBRIA AND EXCESS GIBBS ENERGIES OF BINARY MIXTURES (continued)

1. Tetrachloromethane, CCl_4
2. Oxolane, C_4H_8O

***T*/K = 303.15**
***System Number*: 11**

x_1	y_1	P/kPa	G^E/J mol^{-1}	μ^E_1/J mol^{-1}	μ^E_2/J mol^{-1}
0.0	0.000	26.88	0	-598	0
0.1	0.061	25.68	-54	-488	-6
0.2	0.133	24.52	-96	-374	-26
0.3	0.218	23.43	-124	-270	-61
0.4	0.314	22.45	-138	-183	-108
0.5	0.419	21.60	-139	-115	-163
0.6	0.531	20.88	-129	-66	-223
0.7	0.647	20.27	-108	-33	-283
0.8	0.764	19.76	-79	-13	-342
0.9	0.882	19.33	-43	-3	-400
1.0	1.000	18.97	0	0	-461

B_{11}/cm^{3}mol^{-1}	B_{22}/cm^{3}mol^{-1}	B_{12}/cm^{3}mol^{-1}	V_1/cm^{3}mol^{-1}	V_2/cm^{3}mol^{-1}
-1750	-1280	-1040	97	82

$A(1)$ = -0.2207E+00 $A(2)$ = 0.3812E-01 $A(3)$ = 0.1064E-01 $A(4)$ = -0.1078E-01 $A(5)$ = 0.0000E+00

1. Tetrachloromethane, CCl_4
2. 1,4-Dioxane, $C_4H_8O_2$

***T*/K = 298.15**
***System Number*: 49**

x_1	y_1	P/kPa	G^E/J mol^{-1}	μ^E_1/J mol^{-1}	μ^E_2/J mol^{-1}
0.0	0.000	4.89	0	238	0
0.1	0.285	6.15	33	385	-7
0.2	0.476	7.45	74	412	-10
0.3	0.603	8.67	114	364	7
0.4	0.692	9.74	143	278	53
0.5	0.759	10.70	157	183	131
0.6	0.814	11.58	153	100	233
0.7	0.864	12.43	131	40	343
0.8	0.912	13.30	94	8	438
0.9	0.958	14.20	48	-1	486
1.0	1.000	15.13	0	0	446

B_{11}/cm^{3}mol^{-1}	B_{22}/cm^{3}mol^{-1}	B_{12}/cm^{3}mol^{-1}	V_1/cm^{3}mol^{-1}	V_2/cm^{3}mol^{-1}
-1800	-2160	-1320	97	86

$A(1)$ = 0.2540E+00 $A(2)$ = 0.4200E-01 $A(3)$ = -0.1160E+00

1. Tetrachloromethane, CCl_4
2. 1-Chlorobutane, C_4H_9Cl

***T*/K = 313.15**
***System Number*: 355**

x_1	y_1	P/kPa	G^E/J mol^{-1}	μ^E_1/J mol^{-1}	μ^E_2/J mol^{-1}
0.0	0.000	25.87	0	228	0
0.1	0.116	26.37	21	201	1
0.2	0.226	26.83	40	176	6
0.3	0.331	27.25	55	151	14
0.4	0.432	27.63	67	125	28
0.5	0.528	27.96	74	98	51
0.6	0.621	28.23	76	71	84
0.7	0.712	28.42	71	45	132
0.8	0.804	28.54	58	23	200
0.9	0.899	28.55	35	6	293
1.0	1.000	28.44	0	0	417

B_{11}/cm^{3}mol^{-1}	B_{22}/cm^{3}mol^{-1}	B_{12}/cm^{3}mol^{-1}	V_1/cm^{3}mol^{-1}	V_2/cm^{3}mol^{-1}
-1600	-1400	-1242	98	105

$A(1)$ = 0.1141E+00 $A(2)$ = 0.3637E-01 $A(3)$ = 0.9800E-02

TABLE 3.1.1 (VLEG)
ISOTHERMAL VAPOR-LIQUID EQUILIBRIA AND EXCESS GIBBS ENERGIES OF BINARY MIXTURES (continued)

1. Tetrachloromethane, CCl_4
2. Diethylamine, $C_4H_{11}N$

***T*/K = 293.15**
***System Number*: 366**

x_1	y_1	*P*/kPa	G^E/J mol^{-1}	μ^E_1/J mol^{-1}	μ^E_2/J mol^{-1}
0.0	0.000	25.06	0	-515	0
0.1	0.042	23.52	-53	-522	0
0.2	0.092	21.95	-102	-469	-10
0.3	0.155	20.35	-143	-383	-40
0.4	0.233	18.77	-169	-284	-93
0.5	0.328	17.28	-180	-188	-171
0.6	0.440	15.94	-173	-109	-269
0.7	0.567	14.79	-148	-51	-376
0.8	0.703	13.81	-109	-16	-478
0.9	0.847	12.97	-57	-2	-555
1.0	1.000	12.23	0	0	-584

B_{11}/cm^{3}mol^{-1}	B_{22}/cm^{3}mol^{-1}	B_{12}/cm^{3}mol^{-1}	V_1/cm^{3}mol^{-1}	V_2/cm^{3}mol^{-1}
-1850	-1500	-1330	96	103

A(1) = -0.2952E+00 *A*(2) = -0.1400E-01 *A*(3) = 0.6975E-01

1. Tetrachloromethane, CCl_4
2. Pyridine, C_5H_5N

***T*/K = 313.15**
***System Number*: 369**

x_1	y_1	*P*/kPa	G^E/J mol^{-1}	μ^E_1/J mol^{-1}	μ^E_2/J mol^{-1}
0.0	0.000	6.0	0	1127	0
0.1	0.427	9.4	103	941	9
0.2	0.611	12.5	187	797	35
0.3	0.717	15.3	255	676	75
0.4	0.787	17.9	307	563	136
0.5	0.836	20.1	340	450	230
0.6	0.874	22.1	349	333	373
0.7	0.905	23.9	329	218	590
0.8	0.932	25.5	272	112	910
0.9	0.962	27.0	166	32	1369
1.0	1.000	28.5	0	0	2007

B_{11}/cm^{3}mol^{-1}	B_{22}/cm^{3}mol^{-1}	B_{12}/cm^{3}mol^{-1}	V_1/cm^{3}mol^{-1}	V_2/cm^{3}mol^{-1}
-1600	-2030	-1800	99	82

A(1) = 0.5220E+00 *A*(2) = 0.1691E+00 *A*(3) = 0.7983E-01

1. Tetrachloromethane, CCl_4
2. Cyclopentane, C_5H_{10}

***T*/K = 298.15**
***System Number*: 348**

x_1	y_1	*P*/kPa	G^E/J mol^{-1}	μ^E_1/J mol^{-1}	μ^E_2/J mol^{-1}
0.0	0.000	42.26	0	179	0
0.1	0.041	39.66	15	129	3
0.2	0.086	37.05	25	89	10
0.3	0.137	34.43	31	59	20
0.4	0.196	31.78	33	36	32
0.5	0.265	29.09	32	20	45
0.6	0.349	26.37	29	10	57
0.7	0.452	23.60	23	4	68
0.8	0.585	20.79	16	1	76
0.9	0.760	17.97	8	0	81
1.0	1.000	15.13	0	0	80

B_{11}/cm^{3}mol^{-1}	B_{22}/cm^{3}mol^{-1}	B_{12}/cm^{3}mol^{-1}	V_1/cm^{3}mol^{-1}	V_2/cm^{3}mol^{-1}
-1800	-1060	-1430	97	95

A(1) = 0.5215E-01 *A*(2) = -0.1990E-01 *A*(3) = 0.0000E+00

TABLE 3.1.1 (VLEG)
ISOTHERMAL VAPOR-LIQUID EQUILIBRIA AND EXCESS GIBBS ENERGIES OF BINARY MIXTURES (continued)

1. Tetrachloromethane, CCl_4
2. Diethyl carbonate, $C_5H_{10}O_3$

***T*/K = 298.65**
***System Number*: 376**

x_1	y_1	P/kPa	G^E/J mol^{-1}	μ^E_1/J mol^{-1}	μ^E_2/J mol^{-1}
0.0	0.000	1.48	0	-53	0
0.1	0.540	2.90	0	42	-5
0.2	0.731	4.40	8	101	-15
0.3	0.825	5.91	22	129	-24
0.4	0.881	7.41	38	132	-25
0.5	0.916	8.86	52	118	-13
0.6	0.941	10.26	63	92	19
0.7	0.960	11.60	66	61	77
0.8	0.975	12.91	59	32	168
0.9	0.988	14.22	38	9	298
1.0	1.000	15.57	0	0	473

B_{11}/cm^{3}mol^{-1}	B_{22}/cm^{3}mol^{-1}	B_{12}/cm^{3}mol^{-1}	V_1/cm^{3}mol^{-1}	V_2/cm^{3}mol^{-1}
-1800	-3200	-2100	97	122

$A(1)$ = 0.8446E-01 $A(2)$ = 0.1059E+00 $A(3)$ = 0.0000E+00

---

1. Tetrachloromethane, CCl_4
2. tert-Butyl methyl ether, $C_5H_{12}O$

***T*/K = 313.15**
***System Number*: 1**

x_1	y_1	P/kPa	G^E/J mol^{-1}	μ^E_1/J mol^{-1}	μ^E_2/J mol^{-1}
0.0	0.000	59.85	0	209	0
0.1	0.051	56.75	15	95	6
0.2	0.106	53.62	19	28	17
0.3	0.167	50.44	18	-5	28
0.4	0.238	47.19	14	-17	34
0.5	0.321	43.93	8	-16	33
0.6	0.418	40.69	4	-10	25
0.7	0.530	37.51	2	-4	14
0.8	0.662	34.42	1	0	3
0.9	0.816	31.41	1	1	1
1.0	1.000	28.44	0	0	16

B_{11}/cm^{3}mol^{-1}	B_{22}/cm^{3}mol^{-1}	B_{12}/cm^{3}mol^{-1}	V_1/cm^{3}mol^{-1}	V_2/cm^{3}mol^{-1}
-1600	-1940	-930	99	121

$A(1)$ = 0.1304E-01 $A(2)$ = -0.3717E-01 $A(3)$ = 0.3024E-01

---

1. Tetrachloromethane, CCl_4
2. Hexafluorobenzene, C_6F_6

***T*/K = 278.68**
***System Number*: 50**

x_1	y_1	P/kPa	G^E/J mol^{-1}	μ^E_1/J mol^{-1}	μ^E_2/J mol^{-1}
0.0	0.000	3.988	0	799	0
0.1	0.186	4.417	77	734	4
0.2	0.331	4.808	146	663	16
0.3	0.447	5.156	205	584	43
0.4	0.542	5.455	253	493	92
0.5	0.622	5.699	284	393	175
0.6	0.690	5.888	295	288	305
0.7	0.753	6.023	279	184	498
0.8	0.817	6.100	229	93	775
0.9	0.892	6.105	139	26	1158
1.0	1.000	5.982	0	0	1672

B_{11}/cm^{3}mol^{-1}	B_{22}/cm^{3}mol^{-1}	B_{12}/cm^{3}mol^{-1}	V_1/cm^{3}mol^{-1}	V_2/cm^{3}mol^{-1}
-2000	-3550	-2780	95	113

$A(1)$ = 0.4902E+00 $A(2)$ = 0.1883E+00 $A(3)$ = 0.4302E-01

TABLE 3.1.1 (VLEG)
ISOTHERMAL VAPOR-LIQUID EQUILIBRIA AND EXCESS GIBBS ENERGIES OF BINARY MIXTURES (continued)

1. Tetrachloromethane, CCl_4
2. Chlorobenzene, C_6H_5Cl

***T*/K = 303.15**
***System Number*: 360**

x_1	y_1	P/kPa	G^E/J mol^{-1}	μ^E_1/J mol^{-1}	μ^E_2/J mol^{-1}
0.0	0.000	2.09	0	441	0
0.1	0.529	4.02	37	319	6
0.2	0.710	5.83	64	251	18
0.3	0.804	7.62	85	212	30
0.4	0.863	9.38	101	187	44
0.5	0.902	11.12	113	163	64
0.6	0.931	12.79	120	132	102
0.7	0.953	14.38	118	94	174
0.8	0.970	15.90	102	52	301
0.9	0.985	17.38	66	16	511
1.0	1.000	18.92	0	0	836

B_{11}/cm^{3}mol^{-1}	B_{22}/cm^{3}mol^{-1}	B_{12}/cm^{3}mol^{-1}	V_1/cm^{3}mol^{-1}	V_2/cm^{3}mol^{-1}
-1750	-3030	-2330	97	103

$A(1)$ = 0.1799E+00 $A(2)$ = 0.7833E-01 $A(3)$ = 0.7341E-01

1. Tetrachloromethane, CCl_4
2. Benzene, C_6H_6

***T*/K = 313.15**
***System Number*: 48**

x_1	y_1	P/kPa	G^E/J mol^{-1}	μ^E_1/J mol^{-1}	μ^E_2/J mol^{-1}
0.0	0.000	24.36	0	318	0
0.1	0.125	25.09	29	258	3
0.2	0.239	25.74	51	203	13
0.3	0.344	26.30	67	156	29
0.4	0.443	26.80	76	114	51
0.5	0.538	27.23	79	79	79
0.6	0.630	27.59	76	51	114
0.7	0.721	27.90	67	29	156
0.8	0.812	28.14	51	13	203
0.9	0.905	28.32	29	3	258
1.0	1.000	28.44	0	0	318

B_{11}/cm^{3}mol^{-1}	B_{22}/cm^{3}mol^{-1}	B_{12}/cm^{3}mol^{-1}	V_1/cm^{3}mol^{-1}	V_2/cm^{3}mol^{-1}
-1600	-1300	-1450	98	91

$A(1)$ = 0.1221E+00 $A(2)$ = 0.0000E+00 $A(3)$ = 0.0000E+00

1. Tetrachloromethane, CCl_4
2. Aniline, C_6H_7N

***T*/K = 298.15**
***System Number*: 367**

x_1	y_1	P/kPa	G^E/J mol^{-1}	μ^E_1/J mol^{-1}	μ^E_2/J mol^{-1}
0.0	0.000	0.08	0	3732	0
0.1	0.984	4.91	329	2911	42
0.2	0.991	7.62	578	2292	150
0.3	0.993	9.41	760	1810	310
0.4	0.994	10.69	880	1415	522
0.5	0.995	11.64	938	1074	803
0.6	0.996	12.34	932	765	1181
0.7	0.996	12.86	851	486	1704
0.8	0.997	13.33	682	245	2431
0.9	0.998	13.97	407	70	3438
1.0	1.000	15.07	0	0	4816

B_{11}/cm^{3}mol^{-1}	B_{22}/cm^{3}mol^{-1}	B_{12}/cm^{3}mol^{-1}	V_1/cm^{3}mol^{-1}	V_2/cm^{3}mol^{-1}
-1800	-4080	-2900	97	92

$A(1)$ = 0.1514E+01 $A(2)$ = 0.2186E+00 $A(3)$ = 0.2103E+00

TABLE 3.1.1 (VLEG)
ISOTHERMAL VAPOR-LIQUID EQUILIBRIA AND EXCESS GIBBS ENERGIES OF BINARY MIXTURES (continued)

1. Tetrachloromethane, CCl_4
2. Cyclohexane, C_6H_{12}

***T*/K = 343.15**
***System Number*: 75**

x_1	y_1	P/kPa	G^E/J mol^{-1}	μ^E_1/J mol^{-1}	μ^E_2/J mol^{-1}
0.0	0.000	72.56	0	230	0
0.1	0.118	74.15	20	183	2
0.2	0.228	75.58	36	145	9
0.3	0.333	76.87	48	113	20
0.4	0.434	78.04	55	86	34
0.5	0.531	79.08	58	63	53
0.6	0.626	80.01	57	43	78
0.7	0.719	80.81	51	26	109
0.8	0.812	81.47	40	12	149
0.9	0.905	81.98	23	3	201
1.0	1.000	82.32	0	0	267

B_{11}/cm^{3}mol^{-1}	B_{22}/cm^{3}mol^{-1}	B_{12}/cm^{3}mol^{-1}	V_1/cm^{3}mol^{-1}	V_2/cm^{3}mol^{-1}
-1150	-1140	-1145	103	115

$A(1)$ = 0.8125E-01 $A(2)$ = 0.6550E-02 $A(3)$ = 0.5900E-02

1. Tetrachloromethane, CCl_4
2. Dipropyl ether, $C_6H_{14}O$

***T*/K = 298.15**
***System Number*: 12**

x_1	y_1	P/kPa	G^E/J mol^{-1}	μ^E_1/J mol^{-1}	μ^E_2/J mol^{-1}
0.0	0.000	8.40	0	-226	0
0.1	0.156	8.96	-22	-202	-1
0.2	0.297	9.54	-39	-163	-9
0.3	0.426	10.17	-52	-118	-24
0.4	0.543	10.85	-58	-75	-47
0.5	0.646	11.56	-58	-40	-76
0.6	0.737	12.29	-51	-14	-106
0.7	0.816	13.04	-40	0	-132
0.8	0.884	13.78	-26	4	-145
0.9	0.945	14.51	-11	2	-132
1.0	1.000	15.22	0	0	-82

B_{11}/cm^{3}mol^{-1}	B_{22}/cm^{3}mol^{-1}	B_{12}/cm^{3}mol^{-1}	V_1/cm^{3}mol^{-1}	V_2/cm^{3}mol^{-1}
-1800	-2700	-2200	97	139

$A(1)$ = -0.9320E-01 $A(2)$ = 0.2900E-01 $A(3)$ = 0.3100E-01

1. Tetrachloromethane, CCl_4
2. Toluene, C_7H_8

***T*/K = 313.15**
***System Number*: 349**

x_1	y_1	P/kPa	G^E/J mol^{-1}	μ^E_1/J mol^{-1}	μ^E_2/J mol^{-1}
0.0	0.000	7.89	0	51	0
0.1	0.286	9.97	5	41	1
0.2	0.473	12.04	8	33	2
0.3	0.605	14.10	11	25	5
0.4	0.704	16.15	12	18	8
0.5	0.780	18.20	13	13	13
0.6	0.841	20.24	12	8	18
0.7	0.891	22.28	11	5	25
0.8	0.933	24.33	8	2	33
0.9	0.969	26.38	5	1	41
1.0	1.000	28.44	0	0	51

B_{11}/cm^{3}mol^{-1}	B_{22}/cm^{3}mol^{-1}	B_{12}/cm^{3}mol^{-1}	V_1/cm^{3}mol^{-1}	V_2/cm^{3}mol^{-1}
-1600	-2270	-1930	98	108

$A(1)$ = 0.1966E-01 $A(2)$ = 0.0000E+00 $A(3)$ = 0.0000E+00

TABLE 3.1.1 (VLEG)
ISOTHERMAL VAPOR-LIQUID EQUILIBRIA AND EXCESS GIBBS ENERGIES OF BINARY MIXTURES (continued)

1. Tetrachloromethane, CCl_4
2. N-Methylaniline, C_7H_9N

***T*/K = 298.15**
***System Number*: 364**

x_1	y_1	*P*/kPa	G^E/J mol^{-1}	μ^E_1/J mol^{-1}	μ^E_2/J mol^{-1}
0.0	0.000	0.05	0	2114	0
0.1	0.984	3.04	191	1724	20
0.2	0.992	5.31	342	1402	77
0.3	0.994	7.12	456	1126	169
0.4	0.996	8.60	533	883	300
0.5	0.997	9.84	572	664	480
0.6	0.997	10.90	568	464	724
0.7	0.998	11.84	518	288	1054
0.8	0.998	12.76	412	142	1496
0.9	0.999	13.77	243	39	2081
1.0	1.000	15.07	0	0	2849

B_{11}/cm^{3}mol^{-1}	B_{22}/cm^{3}mol^{-1}	B_{12}/cm^{3}mol^{-1}	V_1/cm^{3}mol^{-1}	V_2/cm^{3}mol^{-1}
-1800	0	0	97	0

A(1) = 0.9225E+00 *A*(2) = 0.1482E+00 *A*(3) = 0.7847E-01

1. Tetrachloromethane, CCl_4
2. Heptane, C_7H_{16}

***T*/K = 298.15**
***System Number*: 13**

x_1	y_1	*P*/kPa	G^E/J mol^{-1}	μ^E_1/J mol^{-1}	μ^E_2/J mol^{-1}
0.0	0.000	6.04	0	372	0
0.1	0.241	7.17	36	339	2
0.2	0.412	8.27	67	297	9
0.3	0.539	9.30	93	251	25
0.4	0.638	10.28	112	201	52
0.5	0.718	11.20	122	151	93
0.6	0.786	12.06	123	104	151
0.7	0.845	12.88	112	63	228
0.8	0.898	13.67	89	30	327
0.9	0.949	14.43	52	8	452
1.0	1.000	15.19	0	0	604

B_{11}/cm^{3}mol^{-1}	B_{22}/cm^{3}mol^{-1}	B_{12}/cm^{3}mol^{-1}	V_1/cm^{3}mol^{-1}	V_2/cm^{3}mol^{-1}
-1800	-2840	-2320	97	147

A(1) = 0.1968E+00 *A*(2) = 0.4691E-01 *A*(3) = 0.0000E+00

1. Tetrachloromethane, CCl_4
2. Dibutyl ether, $C_8H_{18}O$

***T*/K = 308.15**
***System Number*: 375**

x_1	y_1	*P*/kPa	G^E/J mol^{-1}	μ^E_1/J mol^{-1}	μ^E_2/J mol^{-1}
0.0	0.000	1.54	0	-435	0
0.1	0.587	3.36	-41	-387	-3
0.2	0.767	5.28	-76	-323	-14
0.3	0.853	7.31	-103	-258	-36
0.4	0.904	9.44	-120	-199	-68
0.5	0.936	11.65	-129	-146	-111
0.6	0.958	13.93	-127	-98	-170
0.7	0.974	16.27	-114	-57	-246
0.8	0.985	18.64	-89	-25	-342
0.9	0.994	20.99	-51	-6	-453
1.0	1.000	23.26	0	0	-560

B_{11}/cm^{3}mol^{-1}	B_{22}/cm^{3}mol^{-1}	B_{12}/cm^{3}mol^{-1}	V_1/cm^{3}mol^{-1}	V_2/cm^{3}mol^{-1}
-1670	-3000	-2280	98	172

A(1) = -0.2009E+00 *A*(2) = -0.2707E-01 *A*(3) = -0.4100E-02 *A*(4) = 0.2600E-02 *A*(5) = 0.1080E-01

TABLE 3.1.1 (VLEG)
ISOTHERMAL VAPOR-LIQUID EQUILIBRIA AND EXCESS GIBBS ENERGIES OF BINARY MIXTURES (continued)

1. Tetrachloromethane, CCl_4
2. Octamethylcyclotetrasiloxane, $C_8H_{24}O_4Si_4$

***T*/K = 298.15**
***System Number*: 377**

x_1	y_1	P/kPa	G^E/J mol^{-1}	μ^E_1/J mol^{-1}	μ^E_2/J mol^{-1}
0.0	0.000	0.13	0	-483	0
0.1	0.915	1.40	-44	-404	-4
0.2	0.962	2.74	-80	-333	-17
0.3	0.978	4.16	-107	-266	-39
0.4	0.986	5.65	-125	-202	-73
0.5	0.991	7.21	-133	-143	-122
0.6	0.994	8.81	-129	-91	-186
0.7	0.997	10.45	-114	-49	-264
0.8	0.998	12.08	-86	-20	-351
0.9	0.999	13.68	-48	-4	-439
1.0	1.000	15.23	0	0	-514

B_{11}/cm^{3}mol^{-1}	B_{22}/cm^{3}mol^{-1}	B_{12}/cm^{3}mol^{-1}	V_1/cm^{3}mol^{-1}	V_2/cm^{3}mol^{-1}
-1800	0	0	97	0

$A(1)$ = -0.2141E+00 $A(2)$ = -0.1695E-01 $A(3)$ = 0.1288E-01 $A(4)$ = 0.1060E-01 $A(5)$ = 0.0000E+00

1. Tribromomethane, $CHBr_3$
2. Dimethyl sulfoxide, C_2H_6OS

***T*/K = 298.15**
***System Number*: 335**

x_1	y_1	P/kPa	G^E/J mol^{-1}	μ^E_1/J mol^{-1}	μ^E_2/J mol^{-1}
0.0	0.000	0.077	0	-1492	0
0.1	0.398	0.115	-141	-1324	-9
0.2	0.619	0.159	-262	-1138	-43
0.3	0.756	0.211	-358	-941	-109
0.4	0.845	0.273	-427	-743	-216
0.5	0.904	0.344	-462	-552	-373
0.6	0.943	0.425	-461	-376	-589
0.7	0.968	0.514	-418	-224	-871
0.8	0.984	0.606	-330	-105	-1229
0.9	0.994	0.696	-192	-28	-1671
1.0	1.000	0.778	0	0	-2206

B_{11}/cm^{3}mol^{-1}	B_{22}/cm^{3}mol^{-1}	B_{12}/cm^{3}mol^{-1}	V_1/cm^{3}mol^{-1}	V_2/cm^{3}mol^{-1}
-2510	-5970	-4000	88	72

$A(1)$ = -0.7460E+00 $A(2)$ = -0.1440E+00 $A(3)$ = 0.0000E+00

1. Trichloromethane, $CHCl_3$
2. Methanol, CH_4O

***T*/K = 313.15**
***System Number*: 278**

x_1	y_1	P/kPa	G^E/J mol^{-1}	μ^E_1/J mol^{-1}	μ^E_2/J mol^{-1}
0.0	0.000	35.45	0	1975	0
0.1	0.251	42.72	209	2136	-5
0.2	0.419	49.43	418	2021	17
0.3	0.522	54.32	603	1772	102
0.4	0.588	57.48	748	1473	265
0.5	0.634	59.34	842	1165	518
0.6	0.668	60.28	874	865	888
0.7	0.695	60.53	832	576	1428
0.8	0.721	60.19	695	309	2239
0.9	0.770	58.46	433	95	3479
1.0	1.000	48.09	0	0	5385

B_{11}/cm^{3}mol^{-1}	B_{22}/cm^{3}mol^{-1}	B_{12}/cm^{3}mol^{-1}	V_1/cm^{3}mol^{-1}	V_2/cm^{3}mol^{-1}
-1060	-2030	-760	82	41

$A(1)$ = 0.1293E+01 $A(2)$ = 0.4970E+00 $A(3)$ = 0.1206E+00 $A(4)$ = 0.1580E+00 $A(5)$ = 0.0000E+00

TABLE 3.1.1 (VLEG) ISOTHERMAL VAPOR-LIQUID EQUILIBRIA AND EXCESS GIBBS ENERGIES OF BINARY MIXTURES (continued)

1. Trichloromethane, $CHCl_3$
2. Ethanenitrile, C_2H_3N

T/K = 328.15
System Number: 401

x_1	y_1	*P*/kPa	G^E/J mol^{-1}	μ^E_1/J mol^{-1}	μ^E_2/J mol^{-1}
0.0	0.000	40.93	0	1301	0
0.1	0.232	48.76	109	913	20
0.2	0.374	54.53	181	622	71
0.3	0.479	59.17	221	409	141
0.4	0.568	63.17	236	258	221
0.5	0.648	66.80	230	155	305
0.6	0.723	70.20	207	86	388
0.7	0.795	73.43	171	43	468
0.8	0.865	76.54	123	18	544
0.9	0.933	79.54	66	4	620
1.0	1.000	82.45	0	0	699

B_{11}/cm^{3}mol^{-1}	B_{22}/cm^{3}mol^{-1}	B_{12}/cm^{3}mol^{-1}	V_1/cm^{3}mol^{-1}	V_2/cm^{3}mol^{-1}
-940	-3560	-1900	83	55

A(1) = 0.3374E+00 *A*(2) = -0.1104E+00 *A*(3) = 0.2919E-01

1. Trichloromethane, $CHCl_3$
2. Ethanol, C_2H_6O

T/K = 323.15
System Number: 279

x_1	y_1	*P*/kPa	G^E/J mol^{-1}	μ^E_1/J mol^{-1}	μ^E_2/J mol^{-1}
0.0	0.000	29.51	0	1406	0
0.1	0.319	39.00	154	1620	-9
0.2	0.515	48.75	319	1625	-8
0.3	0.629	56.92	474	1479	43
0.4	0.698	62.88	600	1246	170
0.5	0.744	66.95	685	979	390
0.6	0.778	69.70	715	713	717
0.7	0.807	71.54	680	467	1177
0.8	0.838	72.65	567	249	1837
0.9	0.884	72.79	353	78	2832
1.0	1.000	69.36	0	0	4409

B_{11}/cm^{3}mol^{-1}	B_{22}/cm^{3}mol^{-1}	B_{12}/cm^{3}mol^{-1}	V_1/cm^{3}mol^{-1}	V_2/cm^{3}mol^{-1}
-980	-1400	-670	83	60

A(1) = 0.1019E+01 *A*(2) = 0.4382E+00 *A*(3) = 0.1123E-01 *A*(4) = 0.1206E+00 *A*(5) = 0.5153E-01

1. Trichloromethane, $CHCl_3$
2. Dimethyl sulfoxide, C_2H_6OS

T/K = 298.15
System Number: 337

x_1	y_1	*P*/kPa	G^E/J mol^{-1}	μ^E_1/J mol^{-1}	μ^E_2/J mol^{-1}
0.0	0.000	0.08	0	-776	0
0.1	0.947	1.35	-134	-1753	46
0.2	0.971	2.25	-341	-2144	110
0.3	0.983	3.36	-568	-2124	98
0.4	0.991	4.98	-772	-1841	-58
0.5	0.995	7.35	-914	-1424	-405
0.6	0.998	10.57	-967	-971	-961
0.7	0.999	14.56	-911	-562	-1724
0.8	1.000	18.92	-732	-249	-2666
0.9	1.000	23.01	-427	-60	-3734
1.0	1.000	26.24	0	0	-4851

B_{11}/cm^{3}mol^{-1}	B_{22}/cm^{3}mol^{-1}	B_{12}/cm^{3}mol^{-1}	V_1/cm^{3}mol^{-1}	V_2/cm^{3}mol^{-1}
-1200	-5970	-2980	81	73

A(1) = -0.1475E+01 *A*(2) = -0.8220E+00 *A*(3) = 0.3400E+00

TABLE 3.1.1 (VLEG)
ISOTHERMAL VAPOR-LIQUID EQUILIBRIA AND EXCESS GIBBS ENERGIES OF BINARY MIXTURES (continued)

1. Trichloromethane, $CHCl_3$
2. 2-Propanone, C_3H_6O

***T*/K = 323.15**
***System Number*: 280**

x_1	y_1	P/kPa	G^E/J mol^{-1}	μ^E_1/J mol^{-1}	μ^E_2/J mol^{-1}
0.0	0.000	81.98	0	-1475	0
0.1	0.050	77.35	-145	-1418	-4
0.2	0.114	72.72	-279	-1269	-31
0.3	0.198	68.25	-390	-1065	-101
0.4	0.308	64.37	-469	-835	-225
0.5	0.441	61.59	-510	-606	-414
0.6	0.588	60.31	-506	-397	-669
0.7	0.730	60.72	-454	-225	-989
0.8	0.850	62.66	-353	-99	-1368
0.9	0.939	65.71	-201	-24	-1792
1.0	1.000	69.32	0	0	-2243

B_{11}/cm^{3}mol^{-1}	B_{22}/cm^{3}mol^{-1}	B_{12}/cm^{3}mol^{-1}	V_1/cm^{3}mol^{-1}	V_2/cm^{3}mol^{-1}
-980	-1430	-1600	83	77

$A(1)$ = -0.7588E+00 $A(2)$ = -0.1430E+00 $A(3)$ = 0.6680E-01

---

1. Trichloromethane, $CHCl_3$
2. Methyl ethanoate, $C_3H_6O_2$

***T*/K = 313.15**
***System Number*: 70**

x_1	y_1	P/kPa	G^E/J mol^{-1}	μ^E_1/J mol^{-1}	μ^E_2/J mol^{-1}
0.0	0.000	54.10	0	-1668	0
0.1	0.054	51.13	-155	-1444	-12
0.2	0.123	48.28	-287	-1245	-47
0.3	0.209	45.67	-394	-1057	-110
0.4	0.316	43.41	-475	-870	-211
0.5	0.442	41.71	-524	-680	-367
0.6	0.583	40.79	-535	-492	-599
0.7	0.726	40.93	-499	-313	-934
0.8	0.853	42.33	-407	-157	-1406
0.9	0.946	44.89	-245	-44	-2053
1.0	1.000	48.05	0	0	-2921

B_{11}/cm^{3}mol^{-1}	B_{22}/cm^{3}mol^{-1}	B_{12}/cm^{3}mol^{-1}	V_1/cm^{3}mol^{-1}	V_2/cm^{3}mol^{-1}
-1060	-1510	-1110	82	80

$A(1)$ = -0.8047E+00 $A(2)$ = -0.2407E+00 $A(3)$ = -0.7650E-01

---

1. Trichloromethane, $CHCl_3$
2. 1,3-Dioxolane, $C_3H_6O_2$

***T*/K = 308.15**
***System Number*: 277**

x_1	y_1	P/kPa	G^E/J mol^{-1}	μ^E_1/J mol^{-1}	μ^E_2/J mol^{-1}
0.0	0.000	21.32	0	-1522	0
0.1	0.109	21.51	-146	-1406	-6
0.2	0.225	21.82	-279	-1275	-30
0.3	0.349	22.33	-393	-1107	-87
0.4	0.483	23.20	-480	-899	-200
0.5	0.619	24.62	-529	-669	-390
0.6	0.745	26.75	-532	-441	-670
0.7	0.848	29.60	-482	-243	-1037
0.8	0.922	32.93	-373	-99	-1469
0.9	0.970	36.36	-210	-21	-1910
1.0	1.000	39.54	0	0	-2272

B_{11}/cm^{3}mol^{-1}	B_{22}/cm^{3}mol^{-1}	B_{12}/cm^{3}mol^{-1}	V_1/cm^{3}mol^{-1}	V_2/cm^{3}mol^{-1}
-1100	-2050	-4410	82	71

$A(1)$ = -0.8264E+00 $A(2)$ = -0.2180E+00 $A(3)$ = 0.8600E-01 $A(4)$ = 0.7180E-01 $A(5)$ = 0.0000E+00

TABLE 3.1.1 (VLEG)
ISOTHERMAL VAPOR-LIQUID EQUILIBRIA AND EXCESS GIBBS ENERGIES OF BINARY MIXTURES (continued)

1. Trichloromethane, $CHCl_3$
2. Furan, C_4H_4O

***T*/K = 303.15**
***System Number*: 47**

x_1	y_1	*P*/kPa	G^E/J mol^{-1}	μ^E_1/J mol^{-1}	μ^E_2/J mol^{-1}
0.0	0.000	96.91	0	-181	0
0.1	0.035	90.13	-18	-170	-1
0.2	0.076	83.35	-33	-151	-4
0.3	0.125	76.55	-47	-126	-12
0.4	0.183	69.77	-56	-100	-27
0.5	0.255	63.05	-61	-73	-48
0.6	0.343	56.45	-61	-49	-78
0.7	0.454	50.03	-55	-29	-116
0.8	0.593	43.86	-43	-13	-163
0.9	0.770	37.99	-25	-3	-218
1.0	1.000	32.46	0	0	-281

B_{11}/cm^{3}mol^{-1}	B_{22}/cm^{3}mol^{-1}	B_{12}/cm^{3}mol^{-1}	V_1/cm^{3}mol^{-1}	V_2/cm^{3}mol^{-1}
-1150	-830	-1140	81	73

A(1) = -0.9665E-01 *A*(2) = -0.1971E-01 *A*(3) = 0.5014E-02

1. Trichloromethane, $CHCl_3$
2. Oxolane, C_4H_8O

***T*/K = 303.15**
***System Number*: 46**

x_1	y_1	*P*/kPa	G^E/J mol^{-1}	μ^E_1/J mol^{-1}	μ^E_2/J mol^{-1}
0.0	0.000	26.88	0	-3819	0
0.1	0.037	24.80	-353	-3260	-30
0.2	0.102	22.65	-643	-2673	-135
0.3	0.207	20.80	-858	-2093	-329
0.4	0.359	19.67	-993	-1553	-621
0.5	0.539	19.64	-1043	-1075	-1012
0.6	0.711	20.84	-1006	-677	-1498
0.7	0.842	23.15	-879	-370	-2069
0.8	0.927	26.18	-667	-157	-2705
0.9	0.975	29.42	-372	-37	-3383
1.0	1.000	32.38	0	0	-4072

B_{11}/cm^{3}mol^{-1}	B_{22}/cm^{3}mol^{-1}	B_{12}/cm^{3}mol^{-1}	V_1/cm^{3}mol^{-1}	V_2/cm^{3}mol^{-1}
-1150	-1280	-3500	81	82

A(1) = -0.1656E+01 *A*(2) = -0.5011E-01 *A*(3) = 0.9054E-01

1. Trichloromethane, $CHCl_3$
2. 2-Butanone, C_4H_8O

***T*/K = 318.15**
***System Number*: 69**

x_1	y_1	*P*/kPa	G^E/J mol^{-1}	μ^E_1/J mol^{-1}	μ^E_2/J mol^{-1}
0.0	0.000	29.73	0	-2534	0
0.1	0.083	28.99	-242	-2277	-16
0.2	0.195	28.58	-446	-1889	-85
0.3	0.339	28.89	-599	-1478	-223
0.4	0.498	30.23	-696	-1103	-425
0.5	0.649	32.71	-735	-790	-681
0.6	0.775	36.27	-718	-538	-989
0.7	0.870	40.78	-645	-335	-1366
0.8	0.936	46.15	-510	-173	-1857
0.9	0.978	52.13	-302	-52	-2554
1.0	1.000	57.93	0	0	-3601

B_{11}/cm^{3}mol^{-1}	B_{22}/cm^{3}mol^{-1}	B_{12}/cm^{3}mol^{-1}	V_1/cm^{3}mol^{-1}	V_2/cm^{3}mol^{-1}
-1020	-1900	-1100	83	93

A(1) = -0.1112E+01 *A*(2) = -0.8225E-01 *A*(3) = -0.4760E-01 *A*(4) = -0.1194E+00 *A*(5) = 0.0000E+00

TABLE 3.1.1 (VLEG)
ISOTHERMAL VAPOR-LIQUID EQUILIBRIA AND EXCESS GIBBS ENERGIES OF BINARY MIXTURES (continued)

1. Trichloromethane, $CHCl_3$
2. 1,4-Dioxane, $C_4H_8O_2$

***T*/K = 303.15**
***System Number*: 52**

x_1	y_1	*P*/kPa	G^E/J mol^{-1}	μ^E_1/J mol^{-1}	μ^E_2/J mol^{-1}
0.0	0.000	6.18	0	-2104	0
0.1	0.204	6.99	-207	-2018	-6
0.2	0.384	7.95	-399	-1842	-38
0.3	0.547	9.19	-564	-1600	-120
0.4	0.689	10.88	-691	-1315	-275
0.5	0.804	13.15	-768	-1010	-526
0.6	0.888	16.11	-784	-708	-897
0.7	0.943	19.77	-726	-433	-1410
0.8	0.976	23.92	-585	-208	-2090
0.9	0.993	28.16	-346	-56	-2959
1.0	1.000	31.81	0	0	-4040

B_{11}/cm^{3}mol^{-1}	B_{22}/cm^{3}mol^{-1}	B_{12}/cm^{3}mol^{-1}	V_1/cm^{3}mol^{-1}	V_2/cm^{3}mol^{-1}
-1150	-2020	-4860	81	86

$A(1)$ = -0.1219E+01 $A(2)$ = -0.3842E+00 $A(3)$ = 0.0000E+00

---

1. Trichloromethane, $CHCl_3$
2. Ethyl ethanoate, $C_4H_8O_2$

***T*/K = 313.15**
***System Number*: 71**

x_1	y_1	*P*/kPa	G^E/J mol^{-1}	μ^E_1/J mol^{-1}	μ^E_2/J mol^{-1}
0.0	0.000	25.04	0	-2359	0
0.1	0.086	24.51	-225	-2128	-14
0.2	0.197	24.23	-418	-1798	-73
0.3	0.332	24.45	-569	-1466	-184
0.4	0.479	25.30	-674	-1175	-341
0.5	0.621	26.84	-736	-929	-542
0.6	0.748	29.13	-750	-709	-812
0.7	0.854	32.36	-711	-495	-1213
0.8	0.934	36.80	-598	-281	-1868
0.9	0.981	42.47	-379	-91	-2975
1.0	1.000	48.05	0	0	-4826

B_{11}/cm^{3}mol^{-1}	B_{22}/cm^{3}mol^{-1}	B_{12}/cm^{3}mol^{-1}	V_1/cm^{3}mol^{-1}	V_2/cm^{3}mol^{-1}
-1060	-1800	-1430	82	101

$A(1)$ = -0.1130E+01 $A(2)$ = -0.2968E+00 $A(3)$ = -0.2499E+00 $A(4)$ = -0.1769E+00 $A(5)$ = 0.0000E+00

---

1. Trichloromethane, $CHCl_3$
2. Pyridine, C_5H_5N

***T*/K = 303.15**
***System Number*: 408**

x_1	y_1	*P*/kPa	G^E/J mol^{-1}	μ^E_1/J mol^{-1}	μ^E_2/J mol^{-1}
0.0	0.000	3.67	0	-1771	0
0.1	0.328	4.91	-176	-1724	-4
0.2	0.542	6.34	-339	-1562	-34
0.3	0.697	8.13	-477	-1328	-113
0.4	0.808	10.42	-579	-1056	-261
0.5	0.885	13.29	-634	-778	-489
0.6	0.935	16.75	-634	-519	-806
0.7	0.966	20.67	-574	-300	-1214
0.8	0.985	24.81	-450	-135	-1710
0.9	0.995	28.81	-259	-34	-2285
1.0	1.000	32.33	0	0	-2925

B_{11}/cm^{3}mol^{-1}	B_{22}/cm^{3}mol^{-1}	B_{12}/cm^{3}mol^{-1}	V_1/cm^{3}mol^{-1}	V_2/cm^{3}mol^{-1}
-1150	-2300	-1930	81	81

$A(1)$ = -0.1005E+01 $A(2)$ = -0.2290E+00 $A(3)$ = 0.7397E-01

TABLE 3.1.1 (VLEG)
ISOTHERMAL VAPOR-LIQUID EQUILIBRIA AND EXCESS GIBBS ENERGIES OF BINARY MIXTURES (continued)

1. Trichloromethane, $CHCl_3$
2. tert-Butyl methyl ether, $C_5H_{12}O$

***T*/K = 313.15**
***System Number*: 2**

x_1	y_1	P/kPa	G^E/J mol^{-1}	μ^E_1/J mol^{-1}	μ^E_2/J mol^{-1}
0.0	0.000	59.85	0	-2146	0
0.1	0.038	55.69	-215	-2098	-5
0.2	0.092	51.24	-409	-1823	-56
0.3	0.175	46.82	-562	-1446	-183
0.4	0.296	43.15	-659	-1055	-395
0.5	0.448	40.91	-693	-704	-682
0.6	0.609	40.34	-664	-423	-1025
0.7	0.753	41.25	-574	-219	-1401
0.8	0.865	43.21	-429	-89	-1789
0.9	0.945	45.73	-236	-21	-2176
1.0	1.000	48.46	0	0	-2562

B_{11}/cm^{3}mol^{-1}	B_{22}/cm^{3}mol^{-1}	B_{12}/cm^{3}mol^{-1}	V_1/cm^{3}mol^{-1}	V_2/cm^{3}mol^{-1}
-1060	-1940	-840	82	121

A(1) = -0.1065E+01 *A*(2) = -0.1735E-01 *A*(3) = 0.1604E+00 *A*(4) = -0.6245E-01 *A*(5) = 0.0000E+00

1. Trichloromethane, $CHCl_3$
2. Benzene, C_6H_6

***T*/K = 323.15**
***System Number*: 117**

x_1	y_1	P/kPa	G^E/J mol^{-1}	μ^E_1/J mol^{-1}	μ^E_2/J mol^{-1}
0.0	0.000	36.18	0	-659	0
0.1	0.148	38.16	-58	-512	-7
0.2	0.291	40.52	-102	-399	-27
0.3	0.424	43.22	-133	-310	-57
0.4	0.544	46.20	-153	-238	-96
0.5	0.651	49.44	-161	-176	-146
0.6	0.745	52.96	-158	-123	-212
0.7	0.827	56.74	-143	-76	-299
0.8	0.897	60.77	-113	-38	-415
0.9	0.954	65.00	-67	-11	-571
1.0	1.000	69.31	0	0	-779

B_{11}/cm^{3}mol^{-1}	B_{22}/cm^{3}mol^{-1}	B_{12}/cm^{3}mol^{-1}	V_1/cm^{3}mol^{-1}	V_2/cm^{3}mol^{-1}
-980	-1200	-1090	83	92

A(1) = -0.2400E+00 *A*(2) = -0.2229E-01 *A*(3) = -0.2745E-01

1. Trichloromethane, $CHCl_3$
2. Dipropyl ether, $C_6H_{14}O$

***T*/K = 298.15**
***System Number*: 14**

x_1	y_1	P/kPa	G^E/J mol^{-1}	μ^E_1/J mol^{-1}	μ^E_2/J mol^{-1}
0.0	0.000	8.38	0	-1818	0
0.1	0.152	8.86	-173	-1627	-11
0.2	0.313	9.55	-320	-1375	-56
0.3	0.475	10.55	-434	-1098	-149
0.4	0.626	11.96	-508	-822	-299
0.5	0.751	13.81	-537	-569	-506
0.6	0.846	16.06	-520	-355	-767
0.7	0.911	18.60	-455	-190	-1073
0.8	0.955	21.26	-344	-78	-1407
0.9	0.982	23.85	-191	-18	-1749
1.0	1.000	26.25	0	0	-2069

B_{11}/cm^{3}mol^{-1}	B_{22}/cm^{3}mol^{-1}	B_{12}/cm^{3}mol^{-1}	V_1/cm^{3}mol^{-1}	V_2/cm^{3}mol^{-1}
-1200	-2700	-1950	81	139

A(1) = -0.8670E+00 *A*(2) = -0.5060E-01 *A*(3) = 0.8300E-01

TABLE 3.1.1 (VLEG)
ISOTHERMAL VAPOR-LIQUID EQUILIBRIA AND EXCESS GIBBS ENERGIES OF BINARY MIXTURES (continued)

1. Trichloromethane, $CHCl_3$
2. Triethylamine, $C_6H_{15}N$

***T*/K = 283.14**
***System Number*: 176**

x_1	y_1	P/kPa	G^E/J mol^{-1}	μ^E_1/J mol^{-1}	μ^E_2/J mol^{-1}
0.0	0.000	4.19	0	-3591	0
0.1	0.080	4.07	-348	-3324	-17
0.2	0.200	4.00	-651	-2831	-106
0.3	0.374	4.12	-885	-2230	-309
0.4	0.580	4.59	-1031	-1616	-641
0.5	0.760	5.55	-1078	-1058	-1098
0.6	0.878	6.98	-1023	-605	-1650
0.7	0.943	8.71	-872	-283	-2246
0.8	0.975	10.45	-636	-92	-2811
0.9	0.991	11.97	-336	-13	-3246
1.0	1.000	13.26	0	0	-3431

B_{11}/cm^{3}mol^{-1}	B_{22}/cm^{3}mol^{-1}	B_{12}/cm^{3}mol^{-1}	V_1/cm^{3}mol^{-1}	V_2/cm^{3}mol^{-1}
-1310	-2310	-5940	79	137

$A(1)$ = -0.1832E+01 $A(2)$ = 0.3390E-01 $A(3)$ = 0.3400E+00

---

1. Trichloromethane, $CHCl_3$
2. Heptane, C_7H_{16}

***T*/K = 298.15**
***System Number*: 116**

x_1	y_1	P/kPa	G^E/J mol^{-1}	μ^E_1/J mol^{-1}	μ^E_2/J mol^{-1}
0.0	0.000	6.07	0	1090	0
0.1	0.406	9.26	100	917	9
0.2	0.588	12.03	182	763	36
0.3	0.693	14.47	245	623	83
0.4	0.764	16.63	289	493	154
0.5	0.816	18.55	312	371	253
0.6	0.857	20.27	312	259	391
0.7	0.893	21.83	285	159	577
0.8	0.926	23.30	227	78	823
0.9	0.960	24.72	134	21	1145
1.0	1.000	26.17	0	0	1560

B_{11}/cm^{3}mol^{-1}	B_{22}/cm^{3}mol^{-1}	B_{12}/cm^{3}mol^{-1}	V_1/cm^{3}mol^{-1}	V_2/cm^{3}mol^{-1}
-1200	-2840	-1900	81	147

$A(1)$ = 0.5037E+00 $A(2)$ = 0.9478E-01 $A(3)$ = 0.3080E-01

---

1. Dibromomethane, CH_2Br_2
2. Dimethyl sulfoxide, C_2H_6OS

***T*/K = 298.15**
***System Number*: 334**

x_1	y_1	P/kPa	G^E/J mol^{-1}	μ^E_1/J mol^{-1}	μ^E_2/J mol^{-1}
0.0	0.000	0.077	0	-414	0
0.1	0.882	0.588	-41	-404	-1
0.2	0.945	1.111	-80	-374	-6
0.3	0.968	1.658	-114	-328	-22
0.4	0.980	2.234	-140	-272	-53
0.5	0.987	2.843	-157	-210	-103
0.6	0.992	3.482	-161	-148	-180
0.7	0.995	4.145	-150	-91	-286
0.8	0.997	4.819	-121	-44	-429
0.9	0.999	5.485	-72	-12	-612
1.0	1.000	6.119	0	0	-840

B_{11}/cm^{3}mol^{-1}	B_{22}/cm^{3}mol^{-1}	B_{12}/cm^{3}mol^{-1}	V_1/cm^{3}mol^{-1}	V_2/cm^{3}mol^{-1}
-1340	-5970	-3100	70	72

$A(1)$ = -0.2530E+00 $A(2)$ = -0.8600E-01 $A(3)$ = 0.0000E+00

TABLE 3.1.1 (VLEG)
ISOTHERMAL VAPOR-LIQUID EQUILIBRIA AND EXCESS GIBBS ENERGIES OF BINARY MIXTURES (continued)

1. Dichloromethane, CH_2Cl_2
2. Nitromethane, CH_3NO_2

***T*/K = 298.05**
***System Number*: 199**

x_1	y_1	*P*/kPa	G^E/J mol^{-1}	μ^E_1/J mol^{-1}	μ^E_2/J mol^{-1}
0.0	0.000	4.82	0	1379	0
0.1	0.669	13.27	127	1160	13
0.2	0.802	20.19	228	924	54
0.3	0.862	26.09	302	726	120
0.4	0.898	31.30	350	559	210
0.5	0.922	35.98	371	413	330
0.6	0.941	40.35	366	291	478
0.7	0.957	44.63	334	197	653
0.8	0.971	48.83	274	122	882
0.9	0.984	52.72	175	49	1316
1.0	1.000	56.60	0	0	2400

B_{11}/cm^{3}mol^{-1}	B_{22}/cm^{3}mol^{-1}	B_{12}/cm^{3}mol^{-1}	V_1/cm^{3}mol^{-1}	V_2/cm^{3}mol^{-1}
-860	-5080	-1165	65	54

A(1) = 0.5992E+00 *A*(2) = 0.6730E-01 *A*(3) = 0.5420E-01 *A*(4) = 0.4700E-01 *A*(5) = 0.1090E+00

1. Dichloromethane, CH_2Cl_2
2. Methanol, CH_4O

***T*/K = 298.18**
***System Number*: 275**

x_1	y_1	*P*/kPa	G^E/J mol^{-1}	μ^E_1/J mol^{-1}	μ^E_2/J mol^{-1}
0.0	0.000	17.00	0	2703	0
0.1	0.496	30.74	255	2432	14
0.2	0.664	41.84	483	2185	58
0.3	0.744	49.97	677	1890	158
0.4	0.786	55.22	826	1555	339
0.5	0.812	58.40	917	1216	618
0.6	0.830	60.39	942	898	1008
0.7	0.845	61.65	891	610	1548
0.8	0.858	62.33	747	345	2355
0.9	0.879	62.45	473	115	3695
1.0	1.000	58.19	0	0	6075

B_{11}/cm^{3}mol^{-1}	B_{22}/cm^{3}mol^{-1}	B_{12}/cm^{3}mol^{-1}	V_1/cm^{3}mol^{-1}	V_2/cm^{3}mol^{-1}
-860	-2810	-540	65	41

A(1) = 0.1480E+01 *A*(2) = 0.4819E+00 *A*(3) = 0.1434E+00 *A*(4) = 0.1983E+00 *A*(5) = 0.1473E+00

1. Dichloromethane, CH_2Cl_2
2. Ethanenitrile, C_2H_3N

***T*/K = 348.15**
***System Number*: 410**

x_1	y_1	*P*/kPa	G^E/J mol^{-1}	μ^E_1/J mol^{-1}	μ^E_2/J mol^{-1}
0.0	0.000	82.1	0	737	0
0.1	0.313	110.0	64	562	9
0.2	0.495	135.2	111	427	33
0.3	0.618	158.7	144	322	67
0.4	0.709	181.2	163	239	112
0.5	0.780	203.0	169	172	167
0.6	0.838	224.3	164	116	235
0.7	0.886	245.2	145	70	320
0.8	0.928	265.9	113	34	429
0.9	0.965	286.7	66	9	571
1.0	1.000	308.0	0	0	756

B_{11}/cm^{3}mol^{-1}	B_{22}/cm^{3}mol^{-1}	B_{12}/cm^{3}mol^{-1}	V_1/cm^{3}mol^{-1}	V_2/cm^{3}mol^{-1}
-590	-2740	-600	69	57

A(1) = 0.2340E+00 *A*(2) = 0.3271E-02 *A*(3) = 0.2397E-01

TABLE 3.1.1 (VLEG)
ISOTHERMAL VAPOR-LIQUID EQUILIBRIA AND EXCESS GIBBS ENERGIES OF BINARY MIXTURES (continued)

1. Dichloromethane, CH_2Cl_2
2. Dimethyl sulfoxide, C_2H_6OS

***T*/K = 298.15**
***System Number*: 336**

x_1	y_1	P/kPa	G^E/J mol^{-1}	μ^E_1/J mol^{-1}	μ^E_2/J mol^{-1}
0.0	0.000	0.08	0	134	0
0.1	0.985	4.67	-24	-524	31
0.2	0.992	8.29	-99	-805	78
0.3	0.995	12.27	-191	-833	84
0.4	0.997	17.19	-275	-710	16
0.5	0.998	23.24	-331	-519	-143
0.6	0.999	30.25	-347	-322	-385
0.7	1.000	37.77	-318	-160	-685
0.8	1.000	45.15	-244	-55	-998
0.9	1.000	51.88	-133	-8	-1256
1.0	1.000	57.96	0	0	-1373

B_{11}/cm^{3}mol^{-1}	B_{22}/cm^{3}mol^{-1}	B_{12}/cm^{3}mol^{-1}	V_1/cm^{3}mol^{-1}	V_2/cm^{3}mol^{-1}
-860	-5970	-2640	65	72

$A(1)$ = -0.5340E+00 $A(2)$ = -0.3040E+00 $A(3)$ = 0.2840E+00

1. Dichloromethane, CH_2Cl_2
2. 2-Propanone, C_3H_6O

***T*/K = 303.15**
***System Number*: 51**

x_1	y_1	P/kPa	G^E/J mol^{-1}	μ^E_1/J mol^{-1}	μ^E_2/J mol^{-1}
0.0	0.000	38.15	0	-1533	0
0.1	0.112	38.42	-138	-1244	-15
0.2	0.240	39.29	-246	-1002	-57
0.3	0.376	40.81	-327	-797	-126
0.4	0.510	43.02	-381	-617	-223
0.5	0.636	45.97	-406	-458	-353
0.6	0.748	49.70	-401	-316	-527
0.7	0.841	54.21	-362	-194	-756
0.8	0.914	59.42	-287	-94	-1056
0.9	0.966	65.08	-168	-26	-1447
1.0	1.000	70.72	0	0	-1952

B_{11}/cm^{3}mol^{-1}	B_{22}/cm^{3}mol^{-1}	B_{12}/cm^{3}mol^{-1}	V_1/cm^{3}mol^{-1}	V_2/cm^{3}mol^{-1}
-850	-1850	-1540	65	74

$A(1)$ = -0.6438E+00 $A(2)$ = -0.8307E-01 $A(3)$ = -0.4757E-01

1. Dichloromethane, CH_2Cl_2
2. Methyl ethanoate, $C_3H_6O_2$

***T*/K = 303.15**
***System Number*: 15**

x_1	y_1	P/kPa	G^E/J mol^{-1}	μ^E_1/J mol^{-1}	μ^E_2/J mol^{-1}
0.0	0.000	35.84	0	-1041	0
0.1	0.127	36.93	-104	-1020	-2
0.2	0.254	38.22	-201	-922	-20
0.3	0.386	39.93	-282	-776	-70
0.4	0.519	42.29	-340	-608	-161
0.5	0.646	45.45	-369	-439	-299
0.6	0.757	49.48	-367	-286	-487
0.7	0.848	54.27	-329	-160	-721
0.8	0.916	59.58	-254	-69	-993
0.9	0.966	65.07	-144	-17	-1292
1.0	1.000	70.43	0	0	-1601

B_{11}/cm^{3}mol^{-1}	B_{22}/cm^{3}mol^{-1}	B_{12}/cm^{3}mol^{-1}	V_1/cm^{3}mol^{-1}	V_2/cm^{3}mol^{-1}
-850	-1600	-1560	65	80

$A(1)$ = -0.5862E+00 $A(2)$ = -0.1111E+00 $A(3)$ = 0.6197E-01

TABLE 3.1.1 (VLEG)
ISOTHERMAL VAPOR-LIQUID EQUILIBRIA AND EXCESS GIBBS ENERGIES OF BINARY MIXTURES (continued)

1. Dichloromethane, CH_2Cl_2
2. Furan, C_4H_4O

T/K =303.15
System Number: 45

x_1	y_1	*P*/kPa	G^E/J mol^{-1}	μ^E_1/J mol^{-1}	μ^E_2/J mol^{-1}
0.0	0.000	96.72	0	-20	0
0.1	0.076	94.11	-2	-18	0
0.2	0.156	91.50	-4	-16	0
0.3	0.240	88.88	-5	-14	-1
0.4	0.330	86.26	-6	-11	-3
0.5	0.425	83.64	-7	-8	-5
0.6	0.526	81.03	-7	-6	-8
0.7	0.633	78.43	-6	-3	-12
0.8	0.748	75.84	-5	-2	-18
0.9	0.870	73.28	-3	0	-25
1.0	1.000	70.75	0	0	-33

B_{11}/cm^{3}mol^{-1}	B_{22}/cm^{3}mol^{-1}	B_{12}/cm^{3}mol^{-1}	V_1/cm^{3}mol^{-1}	V_2/cm^{3}mol^{-1}
-850	-830	-970	65	73

A(1) = -0.1048E-01 A(2) = -0.2649E-02 A(3) = 0.0000E+00

1. Dichloromethane, CH_2Cl_2
2. Oxolane, C_4H_8O

T/K = 303.15
System Number: 44

x_1	y_1	*P*/kPa	G^E/J mol^{-1}	μ^E_1/J mol^{-1}	μ^E_2/J mol^{-1}
0.0	0.000	26.81	0	-2306	0
0.1	0.119	27.16	-211	-1929	-20
0.2	0.266	28.23	-380	-1544	-89
0.3	0.430	30.33	-501	-1177	-212
0.4	0.590	33.70	-572	-844	-391
0.5	0.725	38.38	-592	-561	-622
0.6	0.828	44.19	-561	-336	-897
0.7	0.900	50.74	-481	-172	-1200
0.8	0.948	57.56	-357	-67	-1513
0.9	0.979	64.18	-194	-14	-1811
1.0	1.000	70.38	0	0	-2063

B_{11}/cm^{3}mol^{-1}	B_{22}/cm^{3}mol^{-1}	B_{12}/cm^{3}mol^{-1}	V_1/cm^{3}mol^{-1}	V_2/cm^{3}mol^{-1}
-850	-1280	-1060	65	82

A(1) = -0.9391E+00 A(2) = 0.4812E-01 A(3) = 0.7241E-01

1. Dichloromethane, CH_2Cl_2
2. Ethyl ethanoate, $C_4H_8O_2$

T/K = 348.15
System Number: 16

x_1	y_1	*P*/kPa	G^E/J mol^{-1}	μ^E_1/J mol^{-1}	μ^E_2/J mol^{-1}
0.0	0.000	94.7	0	-1448	0
0.1	0.182	104.4	-137	-1287	-9
0.2	0.351	116.3	-254	-1091	-44
0.3	0.504	130.9	-344	-879	-115
0.4	0.637	148.9	-405	-670	-229
0.5	0.747	170.5	-432	-475	-388
0.6	0.833	195.3	-422	-306	-595
0.7	0.898	222.8	-374	-171	-845
0.8	0.944	251.7	-287	-75	-1135
0.9	0.977	280.7	-162	-18	-1456
1.0	1.000	308.5	0	0	-1795

B_{11}/cm^{3}mol^{-1}	B_{22}/cm^{3}mol^{-1}	B_{12}/cm^{3}mol^{-1}	V_1/cm^{3}mol^{-1}	V_2/cm^{3}mol^{-1}
-530	-1230	-980	70	106

A(1) = -0.5964E+00 A(2) = -0.5997E-01 A(3) = 0.3610E-01

TABLE 3.1.1 (VLEG) ISOTHERMAL VAPOR-LIQUID EQUILIBRIA AND EXCESS GIBBS ENERGIES OF BINARY MIXTURES (continued)

1. Dichloromethane, CH_2Cl_2
2. 1,4-Dioxane, $C_4H_8O_2$

***T*/K = 303.15**
***System Number*: 43**

x_1	y_1	P/kPa	G^E/J mol^{-1}	μ^E_1/J mol^{-1}	μ^E_2/J mol^{-1}
0.0	0.000	6.35	0	-1485	0
0.1	0.416	9.79	-140	-1325	-9
0.2	0.635	13.80	-262	-1144	-41
0.3	0.768	18.53	-359	-950	-107
0.4	0.853	24.11	-429	-752	-213
0.5	0.909	30.59	-466	-560	-371
0.6	0.946	37.94	-465	-383	-589
0.7	0.970	46.02	-423	-229	-876
0.8	0.985	54.48	-334	-108	-1240
0.9	0.995	62.84	-195	-28	-1692
1.0	1.000	70.43	0	0	-2240

B_{11}/cm^{3}mol^{-1}	B_{22}/cm^{3}mol^{-1}	B_{12}/cm^{3}mol^{-1}	V_1/cm^{3}mol^{-1}	V_2/cm^{3}mol^{-1}
-850	-2020	-1400	65	86

$A(1)$ = -0.7390E+00 $A(2)$ = -0.1497E+00 $A(3)$ = 0.0000E+00

1. Dichloromethane, CH_2Cl_2
2. Pyridine, C_5H_5N

***T*/K = 303.15**
***System Number*: 407**

x_1	y_1	P/kPa	G^E/J mol^{-1}	μ^E_1/J mol^{-1}	μ^E_2/J mol^{-1}
0.0	0.000	3.68	0	-1095	0
0.1	0.588	8.05	-102	-946	-8
0.2	0.775	13.02	-187	-788	-36
0.3	0.865	18.67	-252	-631	-89
0.4	0.916	25.01	-294	-481	-170
0.5	0.947	31.98	-313	-344	-282
0.6	0.967	39.49	-306	-225	-428
0.7	0.981	47.34	-272	-129	-607
0.8	0.990	55.29	-210	-58	-820
0.9	0.996	63.06	-120	-15	-1066
1.0	1.000	70.36	0	0	-1343

B_{11}/cm^{3}mol^{-1}	B_{22}/cm^{3}mol^{-1}	B_{12}/cm^{3}mol^{-1}	V_1/cm^{3}mol^{-1}	V_2/cm^{3}mol^{-1}
-850	-2300	-1600	65	81

$A(1)$ = -0.4969E+00 $A(2)$ = -0.4906E-01 $A(3)$ = 0.1331E-01

1. Dichloromethane, CH_2Cl_2
2. Pentane, C_5H_{12}

***T*/K = 298.19**
***System Number*: 400**

x_1	y_1	P/kPa	G^E/J mol^{-1}	μ^E_1/J mol^{-1}	μ^E_2/J mol^{-1}
0.0	0.000	68.4	0	2450	0
0.1	0.172	75.4	220	1979	24
0.2	0.280	79.3	392	1589	93
0.3	0.359	81.5	520	1259	203
0.4	0.422	82.4	603	972	357
0.5	0.476	82.4	642	720	564
0.6	0.528	81.6	633	496	838
0.7	0.582	79.8	572	304	1198
0.8	0.650	76.7	452	148	1669
0.9	0.758	70.7	265	41	2281
1.0	1.000	58.3	0	0	3071

B_{11}/cm^{3}mol^{-1}	B_{22}/cm^{3}mol^{-1}	B_{12}/cm^{3}mol^{-1}	V_1/cm^{3}mol^{-1}	V_2/cm^{3}mol^{-1}
-860	-1260	-920	65	116

$A(1)$ = 0.1036E+01 $A(2)$ = 0.1252E+00 $A(3)$ = 0.7744E-01

TABLE 3.1.1 (VLEG)
ISOTHERMAL VAPOR-LIQUID EQUILIBRIA AND EXCESS GIBBS ENERGIES OF BINARY MIXTURES (continued)

1. Dichloromethane, CH_2Cl_2
2. tert-Butyl methyl ether, $C_5H_{12}O$

***T*/K = 308.15**
***System Number*: 3**

x_1	y_1	*P*/kPa	G^E/J mol^{-1}	μ^E_1/J mol^{-1}	μ^E_2/J mol^{-1}
0.0	0.000	49.62	0	-1248	0
0.1	0.111	50.00	-114	-1035	-12
0.2	0.238	51.00	-203	-811	-51
0.3	0.376	52.92	-265	-596	-123
0.4	0.513	55.84	-298	-405	-226
0.5	0.639	59.74	-301	-249	-354
0.6	0.747	64.42	-277	-132	-496
0.7	0.834	69.57	-230	-55	-637
0.8	0.902	74.89	-163	-15	-757
0.9	0.955	80.13	-84	-1	-831
1.0	1.000	85.27	0	0	-828

B_{11}/cm^{3}mol^{-1}	B_{22}/cm^{3}mol^{-1}	B_{12}/cm^{3}mol^{-1}	V_1/cm^{3}mol^{-1}	V_2/cm^{3}mol^{-1}
-810	-2060	-800	65	120

A(1) = -0.4703E+00 *A*(2) = 0.8189E-01 *A*(3) = 0.6513E-01

1. Dichloromethane, CH_2Cl_2
2. Chlorobenzene, C_6H_5Cl

***T*/K = 298.03**
***System Number*: 139**

x_1	y_1	*P*/kPa	G^E/J mol^{-1}	μ^E_1/J mol^{-1}	μ^E_2/J mol^{-1}
0.0	0.000	1.68	0	178	0
0.1	0.798	7.53	16	144	2
0.2	0.897	13.29	29	119	6
0.3	0.937	18.98	39	98	13
0.4	0.958	24.60	46	79	24
0.5	0.971	30.17	49	60	39
0.6	0.980	35.68	50	42	61
0.7	0.987	41.16	45	26	92
0.8	0.992	46.63	36	12	132
0.9	0.996	52.15	21	3	183
1.0	1.000	57.78	0	0	245

B_{11}/cm^{3}mol^{-1}	B_{22}/cm^{3}mol^{-1}	B_{12}/cm^{3}mol^{-1}	V_1/cm^{3}mol^{-1}	V_2/cm^{3}mol^{-1}
-860	-3130	-1530	65	102

A(1) = 0.7983E-01 *A*(2) = 0.1696E-01 *A*(3) = 0.5660E-02 *A*(4) = -0.3470E-02 *A*(5) = 0.0000E+00

1. Dichloromethane, CH_2Cl_2
2. Benzene, C_6H_6

***T*/K = 298.15**
***System Number*: 39**

x_1	y_1	*P*/kPa	G^E/J mol^{-1}	μ^E_1/J mol^{-1}	μ^E_2/J mol^{-1}
0.0	0.000	12.7	0	-215	0
0.1	0.314	16.7	-22	-217	0
0.2	0.511	20.8	-42	-188	-6
0.3	0.647	25.1	-58	-145	-20
0.4	0.745	29.6	-67	-100	-44
0.5	0.819	34.3	-69	-61	-76
0.6	0.874	39.0	-64	-32	-111
0.7	0.917	43.8	-53	-14	-145
0.8	0.950	48.6	-38	-4	-174
0.9	0.978	53.3	-20	-1	-193
1.0	1.000	58.0	0	0	-202

B_{11}/cm^{3}mol^{-1}	B_{22}/cm^{3}mol^{-1}	B_{12}/cm^{3}mol^{-1}	V_1/cm^{3}mol^{-1}	V_2/cm^{3}mol^{-1}
-860	-1480	-1020	65	89

A(1) = -0.1107E+00 *A*(2) = 0.1176E-01 *A*(3) = 0.2669E-01 *A*(4) = -0.9080E-02 *A*(5) = 0.0000E+00

TABLE 3.1.1 (VLEG)
ISOTHERMAL VAPOR-LIQUID EQUILIBRIA AND EXCESS GIBBS ENERGIES OF BINARY MIXTURES (continued)

1. Dichloromethane, CH_2Cl_2
2. Triethylamine, $C_6H_{15}N$

***T*/K = 283.15**
***System Number*: 175**

x_1	y_1	P/kPa	G^E/J mol^{-1}	μ^E_1/J mol^{-1}	μ^E_2/J mol^{-1}
0.0	0.000	4.20	0	-642	0
0.1	0.389	6.18	-59	-543	-6
0.2	0.604	8.42	-106	-412	-29
0.3	0.738	10.96	-135	-275	-75
0.4	0.826	13.77	-146	-153	-140
0.5	0.884	16.76	-138	-59	-218
0.6	0.924	19.79	-115	2	-291
0.7	0.951	22.71	-81	28	-338
0.8	0.971	25.44	-44	27	-331
0.9	0.986	28.02	-14	11	-234
1.0	1.000	30.60	0	0	-7

B_{11}/cm^{3}mol^{-1}	B_{22}/cm^{3}mol^{-1}	B_{12}/cm^{3}mol^{-1}	V_1/cm^{3}mol^{-1}	V_2/cm^{3}mol^{-1}
-1030	-2310	-1646	63	137

$A(1)$ = -0.2347E+00 $A(2)$ = 0.1349E+00 $A(3)$ = 0.9780E-01

---

1. Dichloromethane, CH_2Cl_2
2. Toluene, C_7H_8

***T*/K = 298.15**
***System Number*: 125**

x_1	y_1	P/kPa	G^E/J mol^{-1}	μ^E_1/J mol^{-1}	μ^E_2/J mol^{-1}
0.0	0.000	3.92	0	-488	0
0.1	0.580	8.42	-43	-381	-6
0.2	0.763	13.28	-75	-292	-21
0.3	0.852	18.45	-97	-216	-47
0.4	0.903	23.89	-109	-150	-82
0.5	0.935	29.53	-111	-96	-127
0.6	0.957	35.29	-103	-53	-179
0.7	0.973	41.10	-86	-23	-234
0.8	0.984	46.84	-61	-6	-284
0.9	0.993	52.46	-32	0	-315
1.0	1.000	58.01	0	0	-308

B_{11}/cm^{3}mol^{-1}	B_{22}/cm^{3}mol^{-1}	B_{12}/cm^{3}mol^{-1}	V_1/cm^{3}mol^{-1}	V_2/cm^{3}mol^{-1}
-860	-2620	-1580	65	107

$A(1)$ = -0.1793E+00 $A(2)$ = 0.2508E-01 $A(3)$ = 0.1885E-01 $A(4)$ = 0.1128E-01 $A(5)$ = 0.0000E+00

---

1. Nitromethane, CH_3NO_2
2. Methanol, CH_4O

***T*/K = 348.17**
***System Number*: 403**

x_1	y_1	P/kPa	G^E/J mol^{-1}	μ^E_1/J mol^{-1}	μ^E_2/J mol^{-1}
0.0	0.000	151.3	0	4124	0
0.1	0.086	151.4	362	3187	48
0.2	0.136	148.6	631	2439	179
0.3	0.170	144.8	817	1837	379
0.4	0.197	140.3	925	1348	642
0.5	0.223	135.0	959	948	970
0.6	0.252	128.2	920	621	1369
0.7	0.289	119.1	809	362	1852
0.8	0.348	105.2	621	168	2435
0.9	0.475	82.4	353	44	3139
1.0	1.000	42.0	0	0	3984

B_{11}/cm^{3}mol^{-1}	B_{22}/cm^{3}mol^{-1}	B_{12}/cm^{3}mol^{-1}	V_1/cm^{3}mol^{-1}	V_2/cm^{3}mol^{-1}
-2050	-1010	-940	57	41

$A(1)$ = 0.1325E+01 $A(2)$ = -0.1496E-01 $A(3)$ = 0.7583E-01 $A(4)$ = -0.9160E-02 $A(5)$ = 0.0000E+00

TABLE 3.1.1 (VLEG) ISOTHERMAL VAPOR-LIQUID EQUILIBRIA AND EXCESS GIBBS ENERGIES OF BINARY MIXTURES (continued)

1. Nitromethane, CH_3NO_2
2. Ethanenitrile, C_2H_3N

***T*/K = 348.17**
***System Number*: 202**

x_1	y_1	P/kPa	G^E/J mol^{-1}	μ^E_1/J mol^{-1}	μ^E_2/J mol^{-1}
0.0	0.000	82.36	0	2	0
0.1	0.057	78.28	-1	-16	1
0.2	0.118	74.19	-3	-18	1
0.3	0.187	70.10	-4	-11	-1
0.4	0.264	66.01	-5	-1	-7
0.5	0.350	61.97	-3	7	-14
0.6	0.446	57.99	-1	12	-20
0.7	0.555	54.07	3	12	-20
0.8	0.680	50.18	5	8	-7
0.9	0.824	46.26	5	3	25
1.0	1.000	42.18	0	0	87

B_{11}/cm^{3}mol^{-1}	B_{22}/cm^{3}mol^{-1}	B_{12}/cm^{3}mol^{-1}	V_1/cm^{3}mol^{-1}	V_2/cm^{3}mol^{-1}
-2050	-2740	-2960	57	56

$A(1)$ = -0.4521E-02 $A(2)$ = 0.1474E-01 $A(3)$ = 0.1988E-01

1. Nitromethane, CH_3NO_2
2. Ethanol, C_2H_6O

***T*/K = 348.15**
***System Number*: 306**

x_1	y_1	P/kPa	G^E/J mol^{-1}	μ^E_1/J mol^{-1}	μ^E_2/J mol^{-1}
0.0	0.000	89.00	0	4498	0
0.1	0.146	95.76	392	3425	55
0.2	0.216	97.51	679	2597	200
0.3	0.260	97.31	875	1947	416
0.4	0.294	96.13	988	1426	695
0.5	0.324	94.18	1022	1003	1041
0.6	0.356	91.32	980	658	1463
0.7	0.396	86.97	860	382	1976
0.8	0.458	79.78	660	176	2595
0.9	0.585	66.83	375	46	3338
1.0	1.000	42.13	0	0	4218

B_{11}/cm^{3}mol^{-1}	B_{22}/cm^{3}mol^{-1}	B_{12}/cm^{3}mol^{-1}	V_1/cm^{3}mol^{-1}	V_2/cm^{3}mol^{-1}
-2050	-980	-1040	57	64

$A(1)$ = 0.1412E+01 $A(2)$ = -0.2654E-01 $A(3)$ = 0.9305E-01 $A(4)$ = -0.2177E-01 $A(5)$ = 0.0000E+00

1. Nitromethane, CH_3NO_2
2. Thiophene, C_4H_4S

***T*/K = 318.15**
***System Number*: 342**

x_1	y_1	P/kPa	G^E/J mol^{-1}	μ^E_1/J mol^{-1}	μ^E_2/J mol^{-1}
0.0	0.000	25.46	0	2989	0
0.1	0.118	26.32	264	2337	34
0.2	0.191	26.45	462	1797	129
0.3	0.244	26.20	599	1350	277
0.4	0.289	25.70	678	982	475
0.5	0.331	24.98	701	682	720
0.6	0.377	23.99	671	440	1016
0.7	0.435	22.60	586	252	1365
0.8	0.517	20.57	448	115	1777
0.9	0.659	17.50	253	30	2262
1.0	1.000	12.61	0	0	2835

B_{11}/cm^{3}mol^{-1}	B_{22}/cm^{3}mol^{-1}	B_{12}/cm^{3}mol^{-1}	V_1/cm^{3}mol^{-1}	V_2/cm^{3}mol^{-1}
-3100	-1320	-2090	55	81

$A(1)$ = 0.1060E+01 $A(2)$ = -0.2906E-01 $A(3)$ = 0.4035E-01

TABLE 3.1.1 (VLEG) ISOTHERMAL VAPOR-LIQUID EQUILIBRIA AND EXCESS GIBBS ENERGIES OF BINARY MIXTURES (continued)

1. Nitromethane, CH_3NO_2
2. Ethyl ethanoate, $C_4H_8O_2$

***T*/K = 348.22**
***System Number*: 307**

x_1	y_1	P/kPa	G^E/J mol^{-1}	μ^E_1/J mol^{-1}	μ^E_2/J mol^{-1}
0.0	0.000	95.06	0	673	0
0.1	0.059	90.93	62	572	5
0.2	0.120	86.69	114	492	19
0.3	0.184	82.35	156	421	43
0.4	0.253	77.89	189	353	80
0.5	0.326	73.31	210	282	138
0.6	0.406	68.55	217	209	228
0.7	0.497	63.49	205	136	365
0.8	0.606	57.85	169	70	566
0.9	0.755	51.12	103	20	852
1.0	1.000	42.29	0	0	1248

B_{11}/cm^{3}mol^{-1}	B_{22}/cm^{3}mol^{-1}	B_{12}/cm^{3}mol^{-1}	V_1/cm^{3}mol^{-1}	V_2/cm^{3}mol^{-1}
-2050	-1230	-1780	57	106

$A(1)$ = 0.2904E+00 $A(2)$ = 0.9931E-01 $A(3)$ = 0.4144E-01

1. Nitromethane, CH_3NO_2
2. Chlorobenzene, C_6H_5Cl

***T*/K = 298.15**
***System Number*: 200**

x_1	y_1	P/kPa	G^E/J mol^{-1}	μ^E_1/J mol^{-1}	μ^E_2/J mol^{-1}
0.0	0.000	1.629	0	3682	0
0.1	0.506	3.019	326	2892	41
0.2	0.631	3.754	573	2256	152
0.3	0.689	4.184	748	1736	325
0.4	0.726	4.457	856	1306	557
0.5	0.753	4.644	898	944	853
0.6	0.777	4.777	874	638	1227
0.7	0.801	4.876	779	384	1701
0.8	0.834	4.944	608	184	2302
0.9	0.887	4.953	352	50	3069
1.0	1.000	4.782	0	0	4045

B_{11}/cm^{3}mol^{-1}	B_{22}/cm^{3}mol^{-1}	B_{12}/cm^{3}mol^{-1}	V_1/cm^{3}mol^{-1}	V_2/cm^{3}mol^{-1}
-5080	-3130	-1790	54	102

$A(1)$ = 0.1450E+01 $A(2)$ = 0.7331E-01 $A(3)$ = 0.1090E+00

1. Nitromethane, CH_3NO_2
2. Benzene, C_6H_6

***T*/K = 318.15**
***System Number*: 191**

x_1	y_1	P/kPa	G^E/J mol^{-1}	μ^E_1/J mol^{-1}	μ^E_2/J mol^{-1}
0.0	0.000	29.8	0	3075	0
0.1	0.105	30.4	273	2422	34
0.2	0.172	30.2	480	1892	127
0.3	0.222	29.8	627	1454	272
0.4	0.264	29.0	717	1088	470
0.5	0.303	28.1	751	777	724
0.6	0.344	26.9	727	517	1043
0.7	0.393	25.3	644	304	1439
0.8	0.464	23.0	499	142	1927
0.9	0.599	19.2	286	37	2522
1.0	1.000	12.5	0	0	3241

B_{11}/cm^{3}mol^{-1}	B_{22}/cm^{3}mol^{-1}	B_{12}/cm^{3}mol^{-1}	V_1/cm^{3}mol^{-1}	V_2/cm^{3}mol^{-1}
-3100	-1250	-2040	55	92

$A(1)$ = 0.1135E+01 $A(2)$ = 0.4071E-01 $A(3)$ = 0.5902E-01 $A(4)$ = -0.9275E-02 $A(5)$ = 0.0000E+00

TABLE 3.1.1 (VLEG)
ISOTHERMAL VAPOR-LIQUID EQUILIBRIA AND EXCESS GIBBS ENERGIES OF BINARY MIXTURES (continued)

1. Methanol, CH_4O
2. Methanethiol, CH_4S

***T*/K = 288.15**
***System Number*: 343**

x_1	y_1	*P*/kPa	G^E/J mol^{-1}	μ^E_1/J mol^{-1}	μ^E_2/J mol^{-1}
0.0	0.000	142.7	0	5127	0
0.1	0.039	137.3	446	3829	70
0.2	0.048	134.2	752	2655	276
0.3	0.051	132.8	929	1774	567
0.4	0.054	130.5	1001	1170	889
0.5	0.059	125.5	991	759	1223
0.6	0.066	117.8	911	456	1594
0.7	0.077	106.9	764	220	2031
0.8	0.098	88.7	550	59	2513
0.9	0.171	55.0	280	-6	2855
1.0	1.000	10.3	0	0	2580

B_{11}/cm^{3}mol^{-1}	B_{22}/cm^{3}mol^{-1}	B_{12}/cm^{3}mol^{-1}	V_1/cm^{3}mol^{-1}	V_2/cm^{3}mol^{-1}
-3400	-720	-750	41	55

A(1) = 0.1655E+01 *A*(2) = -0.3877E+00 *A*(3) = 0.2098E+00 *A*(4) = -0.1439E+00 *A*(5) = -0.2563E+00

1. Methanol, CH_4O
2. 1,2-Dichloroethane, $C_2H_4Cl_2$

***T*/K = 323.15**
***System Number*: 276**

x_1	y_1	*P*/kPa	G^E/J mol^{-1}	μ^E_1/J mol^{-1}	μ^E_2/J mol^{-1}
0.0	0.000	31.10	0	5971	0
0.1	0.462	54.48	491	4061	94
0.2	0.531	60.31	812	2839	306
0.3	0.563	62.77	1009	2010	579
0.4	0.587	64.21	1104	1408	902
0.5	0.610	65.15	1113	948	1278
0.6	0.636	65.68	1040	593	1711
0.7	0.673	65.69	890	329	2201
0.8	0.729	64.72	666	146	2747
0.9	0.822	61.91	370	38	3363
1.0	1.000	55.55	0	0	4097

B_{11}/cm^{3}mol^{-1}	B_{22}/cm^{3}mol^{-1}	B_{12}/cm^{3}mol^{-1}	V_1/cm^{3}mol^{-1}	V_2/cm^{3}mol^{-1}
-1380	-1200	-1290	42	82

A(1) = 0.1657E+01 *A*(2) = -0.2456E+00 *A*(3) = 0.1535E+00 *A*(4) = -0.1032E+00 *A*(5) = 0.6348E-01

1. Methanol, CH_4O
2. Ethanol, C_2H_6O

***T*/K = 313.15**
***System Number*: 17**

x_1	y_1	*P*/kPa	G^E/J mol^{-1}	μ^E_1/J mol^{-1}	μ^E_2/J mol^{-1}
0.0	0.000	17.90	0	-37	0
0.1	0.177	19.62	-3	-26	-1
0.2	0.327	21.35	-5	-17	-2
0.3	0.455	23.11	-6	-10	-4
0.4	0.566	24.87	-6	-5	-7
0.5	0.662	26.63	-6	-2	-9
0.6	0.746	28.40	-5	-1	-11
0.7	0.821	30.17	-4	0	-13
0.8	0.887	31.93	-2	0	-13
0.9	0.946	33.69	-1	0	-12
1.0	1.000	35.45	0	0	-9

B_{11}/cm^{3}mol^{-1}	B_{22}/cm^{3}mol^{-1}	B_{12}/cm^{3}mol^{-1}	V_1/cm^{3}mol^{-1}	V_2/cm^{3}mol^{-1}
-1480	-2100	-2060	41	60

A(1) = -0.8983E-02 *A*(2) = 0.5360E-02 *A*(3) = 0.0000E+00

TABLE 3.1.1 (VLEG) ISOTHERMAL VAPOR-LIQUID EQUILIBRIA AND EXCESS GIBBS ENERGIES OF BINARY MIXTURES (continued)

1. Methanol, CH_4O
2. 2-Propanone, C_3H_6O

***T*/K = 323.15**
***System Number*: 321**

x_1	y_1	P/kPa	G^E/J mol^{-1}	μ^E_1/J mol^{-1}	μ^E_2/J mol^{-1}
0.0	0.000	81.96	0	1697	0
0.1	0.111	83.55	153	1386	16
0.2	0.200	84.01	274	1103	66
0.3	0.275	83.61	361	851	150
0.4	0.343	82.50	414	630	269
0.5	0.408	80.75	433	441	424
0.6	0.477	78.29	417	284	616
0.7	0.555	74.95	366	161	845
0.8	0.653	70.42	280	72	1112
0.9	0.788	64.22	158	18	1418
1.0	1.000	55.65	0	0	1764

B_{11}/cm^{3}mol^{-1}	B_{22}/cm^{3}mol^{-1}	B_{12}/cm^{3}mol^{-1}	V_1/cm^{3}mol^{-1}	V_2/cm^{3}mol^{-1}
-1380	-1425	-1030	42	77

$A(1)$ = 0.6441E+00 $A(2)$ = 0.1246E-01 $A(3)$ = 0.0000E+00

---

1. Methanol, CH_4O
2. 1-Propanol, C_3H_8O

***T*/K = 313.15**
***System Number*: 19**

x_1	y_1	P/kPa	G^E/J mol^{-1}	μ^E_1/J mol^{-1}	μ^E_2/J mol^{-1}
0.0	0.000	7.01	0	174	0
0.1	0.368	10.01	15	137	2
0.2	0.564	12.96	27	112	6
0.3	0.687	15.87	37	93	12
0.4	0.771	18.76	44	78	20
0.5	0.834	21.62	48	64	32
0.6	0.881	24.44	50	48	51
0.7	0.919	27.22	47	33	81
0.8	0.950	29.97	39	17	128
0.9	0.976	32.69	24	5	198
1.0	1.000	35.45	0	0	299

B_{11}/cm^{3}mol^{-1}	B_{22}/cm^{3}mol^{-1}	B_{12}/cm^{3}mol^{-1}	V_1/cm^{3}mol^{-1}	V_2/cm^{3}mol^{-1}
-1480	-2680	-2020	41	76

$A(1)$ = 0.7388E-01 $A(2)$ = 0.2403E-01 $A(3)$ = 0.1693E-01

---

1. Methanol, CH_4O
2. 2-Propanol, C_3H_8O

***T*/K = 313.15**
***System Number*: 311**

x_1	y_1	P/kPa	G^E/J mol^{-1}	μ^E_1/J mol^{-1}	μ^E_2/J mol^{-1}
0.0	0.000	13.91	0	-450	0
0.1	0.197	15.58	-39	-343	-6
0.2	0.365	17.44	-68	-254	-21
0.3	0.505	19.45	-86	-181	-45
0.4	0.622	21.58	-95	-123	-76
0.5	0.718	23.80	-96	-79	-113
0.6	0.797	26.09	-89	-46	-152
0.7	0.862	28.42	-75	-24	-194
0.8	0.916	30.76	-55	-9	-236
0.9	0.962	33.11	-30	-2	-277
1.0	1.000	35.45	0	0	-315

B_{11}/cm^{3}mol^{-1}	B_{22}/cm^{3}mol^{-1}	B_{12}/cm^{3}mol^{-1}	V_1/cm^{3}mol^{-1}	V_2/cm^{3}mol^{-1}
-1480	-2360	-1890	41	78

$A(1)$ = -0.1470E+00 $A(2)$ = 0.2593E-01 $A(3)$ = 0.0000E+00

TABLE 3.1.1 (VLEG) ISOTHERMAL VAPOR-LIQUID EQUILIBRIA AND EXCESS GIBBS ENERGIES OF BINARY MIXTURES (continued)

1. Methanol, CH_4O
2. Ethyl ethanoate, $C_4H_8O_2$

***T*/K = 328.15**
***System Number*: 68**

x_1	y_1	*P*/kPa	G^E/J mol^{-1}	μ^E_1/J mol^{-1}	μ^E_2/J mol^{-1}
0.0	0.000	46.05	0	2955	0
0.1	0.268	57.71	258	2260	36
0.2	0.393	64.36	447	1702	134
0.3	0.472	68.47	574	1257	281
0.4	0.533	71.18	644	902	472
0.5	0.587	72.99	661	620	702
0.6	0.641	74.08	628	399	972
0.7	0.699	74.46	547	229	1289
0.8	0.769	73.97	416	105	1660
0.9	0.861	72.19	235	27	2101
1.0	1.000	68.29	0	0	2629

B_{11}/cm^{3}mol^{-1}	B_{22}/cm^{3}mol^{-1}	B_{12}/cm^{3}mol^{-1}	V_1/cm^{3}mol^{-1}	V_2/cm^{3}mol^{-1}
-1380	-1550	-1200	42	102

A(1) = 0.9692E+00 *A*(2) = -0.5980E-01 *A*(3) = 0.5420E-01

1. Methanol, CH_4O
2. Diethyl ether, $C_4H_{10}O$

***T*/K = 298.16**
***System Number*: 309**

x_1	y_1	*P*/kPa	G^E/J mol^{-1}	μ^E_1/J mol^{-1}	μ^E_2/J mol^{-1}
0.0	0.000	71.54	0	3914	0
0.1	0.076	71.17	334	2852	54
0.2	0.114	69.58	567	2079	188
0.3	0.140	67.60	717	1508	378
0.4	0.162	65.25	796	1074	610
0.5	0.185	62.37	811	736	886
0.6	0.212	58.69	766	470	1211
0.7	0.249	53.69	663	265	1592
0.8	0.311	46.37	501	118	2034
0.9	0.445	35.01	280	29	2536
1.0	1.000	16.97	0	0	3085

B_{11}/cm^{3}mol^{-1}	B_{22}/cm^{3}mol^{-1}	B_{12}/cm^{3}mol^{-1}	V_1/cm^{3}mol^{-1}	V_2/cm^{3}mol^{-1}
-2200	-1240	-870	41	105

A(1) = 0.1309E+01 *A*(2) = -0.1210E+00 *A*(3) = 0.1029E+00 *A*(4) = -0.4625E-01 *A*(5) = 0.0000E+00

1. Methanol, CH_4O
2. 1-Butanol, $C_4H_{10}O$

***T*/K = 313.15**
***System Number*: 313**

x_1	y_1	*P*/kPa	G^E/J mol^{-1}	μ^E_1/J mol^{-1}	μ^E_2/J mol^{-1}
0.0	0.000	2.52	0	288	0
0.1	0.627	6.10	26	242	2
0.2	0.788	9.61	48	209	8
0.3	0.862	13.06	67	183	17
0.4	0.905	16.45	81	157	31
0.5	0.934	19.77	91	129	54
0.6	0.953	23.00	96	98	92
0.7	0.968	26.14	92	66	153
0.8	0.980	29.21	77	34	248
0.9	0.991	32.28	48	10	388
1.0	1.000	35.45	0	0	588

B_{11}/cm^{3}mol^{-1}	B_{22}/cm^{3}mol^{-1}	B_{12}/cm^{3}mol^{-1}	V_1/cm^{3}mol^{-1}	V_2/cm^{3}mol^{-1}
-1480	-3560	-2400	41	95

A(1) = 0.1406E+00 *A*(2) = 0.5765E-01 *A*(3) = 0.2764E-01

TABLE 3.1.1 (VLEG) ISOTHERMAL VAPOR-LIQUID EQUILIBRIA AND EXCESS GIBBS ENERGIES OF BINARY MIXTURES (continued)

1. Methanol, CH_4O
2. 2-Butanol, $C_4H_{10}O$

***T*/K = 298.15**
***System Number*: 314**

x_1	y_1	P/kPa	G^E/J mol^{-1}	μ^E_1/J mol^{-1}	μ^E_2/J mol^{-1}
0.0	0.000	2.41	0	-495	0
0.1	0.402	3.63	-42	-348	-8
0.2	0.614	4.98	-69	-230	-28
0.3	0.741	6.43	-83	-138	-59
0.4	0.822	7.95	-85	-72	-94
0.5	0.877	9.49	-79	-28	-130
0.6	0.916	11.03	-66	-2	-160
0.7	0.945	12.54	-48	8	-179
0.8	0.967	14.03	-29	8	-178
0.9	0.985	15.48	-12	3	-149
1.0	1.000	16.94	0	0	-85

B_{11}/cm^{3}mol^{-1}	B_{22}/cm^{3}mol^{-1}	B_{12}/cm^{3}mol^{-1}	V_1/cm^{3}mol^{-1}	V_2/cm^{3}mol^{-1}
-2200	-5700	-4200	41	92

$A(1)$ = -0.1272E+00 $A(2)$ = 0.8270E-01 $A(3)$ = 0.1030E-01

---

1. Methanol, CH_4O
2. 2-Methyl-2-propanol, $C_4H_{10}O$

***T*/K = 313.15**
***System Number*: 316**

x_1	y_1	P/kPa	G^E/J mol^{-1}	μ^E_1/J mol^{-1}	μ^E_2/J mol^{-1}
0.0	0.000	13.79	0	-1047	0
0.1	0.166	14.85	-98	-913	-8
0.2	0.326	16.17	-179	-741	-39
0.3	0.476	17.81	-238	-561	-99
0.4	0.610	19.82	-271	-392	-190
0.5	0.721	22.16	-278	-250	-306
0.6	0.810	24.73	-260	-141	-439
0.7	0.878	27.44	-220	-67	-577
0.8	0.929	30.16	-160	-23	-706
0.9	0.969	32.84	-85	-4	-811
1.0	1.000	35.45	0	0	-877

B_{11}/cm^{3}mol^{-1}	B_{22}/cm^{3}mol^{-1}	B_{12}/cm^{3}mol^{-1}	V_1/cm^{3}mol^{-1}	V_2/cm^{3}mol^{-1}
-1480	-2530	-1980	41	98

$A(1)$ = -0.4272E+00 $A(2)$ = 0.4268E-01 $A(3)$ = 0.5763E-01 $A(4)$ = -0.9996E-02 $A(5)$ = 0.0000E+00

---

1. Methanol, CH_4O
2. Diethylamine, $C_4H_{11}N$

***T*/K = 297.97**
***System Number*: 286**

x_1	y_1	P/kPa	G^E/J mol^{-1}	μ^E_1/J mol^{-1}	μ^E_2/J mol^{-1}
0.0	0.000	31.48	0	-1545	0
0.1	0.027	29.23	-171	-1812	12
0.2	0.058	26.88	-358	-1870	20
0.3	0.101	24.28	-543	-1785	-10
0.4	0.165	21.38	-706	-1579	-124
0.5	0.272	18.40	-826	-1274	-377
0.6	0.440	15.87	-876	-905	-833
0.7	0.655	14.50	-832	-532	-1532
0.8	0.843	14.59	-671	-226	-2453
0.9	0.950	15.65	-388	-46	-3462
1.0	1.000	16.88	0	0	-4235

B_{11}/cm^{3}mol^{-1}	B_{22}/cm^{3}mol^{-1}	B_{12}/cm^{3}mol^{-1}	V_1/cm^{3}mol^{-1}	V_2/cm^{3}mol^{-1}
-2200	-1430	-685	41	104

$A(1)$ = -0.1333E+01 $A(2)$ = -0.7239E+00 $A(3)$ = 0.5960E-01 $A(4)$ = 0.1810E+00 $A(5)$ = 0.1070E-01

TABLE 3.1.1 (VLEG)
ISOTHERMAL VAPOR-LIQUID EQUILIBRIA AND EXCESS GIBBS ENERGIES OF BINARY MIXTURES (continued)

1. Methanol, CH_4O
2. Chlorobenzene, C_6H_5Cl

***T*/K = 293.15**
***System Number*: 281**

x_1	y_1	*P*/kPa	G^E/J mol^{-1}	μ^E_1/J mol^{-1}	μ^E_2/J mol^{-1}
0.0	0.000	1.21	0	7754	0
0.1	0.883	9.87	595	4637	146
0.2	0.891	10.47	947	3110	406
0.3	0.898	10.98	1157	2256	687
0.4	0.904	11.37	1267	1651	1011
0.5	0.907	11.55	1286	1155	1417
0.6	0.910	11.71	1215	752	1909
0.7	0.917	11.99	1056	451	2468
0.8	0.930	12.38	812	237	3113
0.9	0.951	12.78	474	80	4024
1.0	1.000	13.07	0	0	5713

B_{11}/cm^{3}mol^{-1}	B_{22}/cm^{3}mol^{-1}	B_{12}/cm^{3}mol^{-1}	V_1/cm^{3}mol^{-1}	V_2/cm^{3}mol^{-1}
-3200	-3330	-900	41	101

A(1) = 0.2111E+01 *A*(2) = -0.2151E+00 *A*(3) = 0.2586E+00 *A*(4) = -0.2034E+00 *A*(5) = 0.3934E+00

---

1. Methanol, CH_4O
2. Aniline, C_6H_7N

***T*/K = 385.15**
***System Number*: 290**

x_1	y_1	*P*/kPa	G^E/J mol^{-1}	μ^E_1/J mol^{-1}	μ^E_2/J mol^{-1}
0.0	0.000	9.8	0	2166	0
0.1	0.899	89.1	195	1767	20
0.2	0.947	156.2	351	1461	74
0.3	0.965	214.5	472	1195	163
0.4	0.975	265.3	557	948	296
0.5	0.981	309.9	601	717	486
0.6	0.985	349.9	602	507	744
0.7	0.989	387.2	552	322	1090
0.8	0.992	423.7	445	166	1562
0.9	0.995	462.2	268	49	2231
1.0	1.000	508.9	0	0	3225

B_{11}/cm^{3}mol^{-1}	B_{22}/cm^{3}mol^{-1}	B_{12}/cm^{3}mol^{-1}	V_1/cm^{3}mol^{-1}	V_2/cm^{3}mol^{-1}
-700	-1730	-550	46	99

A(1) = 0.7513E+00 *A*(2) = 0.1446E+00 *A*(3) = 0.6040E-01 *A*(4) = 0.2071E-01 *A*(5) = 0.3018E-01

---

1. Methanol, CH_4O
2. 1-Decanol, $C_{10}H_{22}O$

***T*/K = 298.15**
***System Number*: 318**

x_1	y_1	*P*/kPa	G^E/J mol^{-1}	μ^E_1/J mol^{-1}	μ^E_2/J mol^{-1}
0.0	0.000	0.00	0	1008	0
0.1	0.999	2.66	111	1149	-5
0.2	1.000	5.18	223	1079	9
0.3	1.000	7.37	321	941	56
0.4	1.000	9.30	399	802	131
0.5	1.000	11.08	455	679	232
0.6	1.000	12.68	487	559	380
0.7	1.000	14.00	487	420	643
0.8	1.000	14.98	434	254	1153
0.9	1.000	15.76	291	86	2135
1.0	1.000	16.93	0	0	3918

B_{11}/cm^{3}mol^{-1}	B_{22}/cm^{3}mol^{-1}	B_{12}/cm^{3}mol^{-1}	V_1/cm^{3}mol^{-1}	V_2/cm^{3}mol^{-1}
-2200	0	0	41	0

A(1) = 0.7347E+00 *A*(2) = 0.3610E+00 *A*(3) = 0.2590E+00 *A*(4) = 0.2260E+00 *A*(5) = 0.0000E+00

TABLE 3.1.1 (VLEG) ISOTHERMAL VAPOR-LIQUID EQUILIBRIA AND EXCESS GIBBS ENERGIES OF BINARY MIXTURES (continued)

1. Methanethiol, CH_4S
2. 2-Thiapropane, C_2H_6S

T/K = 288.15
System Number: 344

x_1	y_1	P/kPa	G^E/J mol^{-1}	μ^E_1/J mol^{-1}	μ^E_2/J mol^{-1}
0.0	0.000	43.8	0	241	0
0.1	0.305	59.2	21	191	2
0.2	0.465	73.6	38	164	7
0.3	0.569	87.1	53	148	12
0.4	0.646	99.9	66	135	20
0.5	0.708	111.4	76	118	34
0.6	0.761	121.6	82	95	62
0.7	0.811	130.0	81	66	115
0.8	0.862	136.6	71	36	207
0.9	0.920	141.1	45	11	354
1.0	1.000	142.9	0	0	575

B_{11}/cm^{3}mol^{-1}	B_{22}/cm^{3}mol^{-1}	B_{12}/cm^{3}mol^{-1}	V_1/cm^{3}mol^{-1}	V_2/cm^{3}mol^{-1}
-720	-960	-840	55	72

$A(1)$ = 0.1263E+00 $A(2)$ = 0.6986E-01 $A(3)$ = 0.4398E-01

---

1. Methylamine, CH_5N
2. Trimethylamine, C_3H_9N

T/K = 293.15
System Number: 181

x_1	y_1	P/kPa	G^E/J mol^{-1}	μ^E_1/J mol^{-1}	μ^E_2/J mol^{-1}
0.0	0.000	184.9	0	1173	0
0.1	0.200	211.8	106	962	11
0.2	0.339	233.2	190	775	44
0.3	0.444	250.4	252	610	99
0.4	0.531	264.2	292	465	178
0.5	0.606	275.1	309	336	283
0.6	0.676	283.6	303	226	418
0.7	0.744	289.9	270	134	589
0.8	0.816	293.8	211	63	803
0.9	0.898	294.9	122	17	1066
1.0	1.000	292.2	0	0	1387

B_{11}/cm^{3}mol^{-1}	B_{22}/cm^{3}mol^{-1}	B_{12}/cm^{3}mol^{-1}	V_1/cm^{3}mol^{-1}	V_2/cm^{3}mol^{-1}
-480	-1160	-770	47	95

$A(1)$ = 0.5079E+00 $A(2)$ = 0.4389E-01 $A(3)$ = 0.1728E-01

---

1. Methylamine, CH_5N
2. Butane, C_4H_{10}

T/K = 288.15
System Number: 150

x_1	y_1	P/kPa	G^E/J mol^{-1}	μ^E_1/J mol^{-1}	μ^E_2/J mol^{-1}
0.0	0.000	175.9	0	3447	0
0.1	0.324	243.6	312	2822	34
0.2	0.445	279.3	556	2228	138
0.3	0.507	297.4	730	1707	312
0.4	0.547	307.1	836	1273	545
0.5	0.581	312.5	877	920	833
0.6	0.615	314.8	854	634	1183
0.7	0.652	313.7	766	399	1622
0.8	0.699	308.1	606	205	2208
0.9	0.777	292.5	359	61	3039
1.0	1.000	243.5	0	0	4259

B_{11}/cm^{3}mol^{-1}	B_{22}/cm^{3}mol^{-1}	B_{12}/cm^{3}mol^{-1}	V_1/cm^{3}mol^{-1}	V_2/cm^{3}mol^{-1}
-500	-800	-640	46	99

$A(1)$ = 0.1464E+01 $A(2)$ = 0.7251E-01 $A(3)$ = 0.1443E+00 $A(4)$ = 0.9689E-01 $A(5)$ = 0.0000E+00

TABLE 3.1.1 (VLEG)
ISOTHERMAL VAPOR-LIQUID EQUILIBRIA AND EXCESS GIBBS ENERGIES OF BINARY MIXTURES (continued)

1. Methylamine, CH_5N
2. Nonane, C_9H_{20}

***T*/K = 293.15**
***System Number*: 151**

x_1	y_1	*P*/kPa	G^E/J mol^{-1}	μ^E_1/J mol^{-1}	μ^E_2/J mol^{-1}
0.0	0.000	0.5	0	3011	0
0.1	0.994	78.7	274	2512	25
0.2	0.997	136.0	500	2132	92
0.3	0.997	177.8	681	1777	211
0.4	0.998	206.8	811	1430	398
0.5	0.998	227.0	885	1103	667
0.6	0.998	242.1	897	809	1028
0.7	0.998	254.2	838	546	1520
0.8	0.999	263.8	695	307	2248
0.9	0.999	273.3	436	101	3443
1.0	1.000	292.2	0	0	5542

B_{11}/cm^{3}mol^{-1}	B_{22}/cm^{3}mol^{-1}	B_{12}/cm^{3}mol^{-1}	V_1/cm^{3}mol^{-1}	V_2/cm^{3}mol^{-1}
-480	-5730	-2200	46	179

A(1) = 0.1452E+01 *A*(2) = 0.3585E+00 *A*(3) = 0.1742E+00 *A*(4) = 0.1608E+00 *A*(5) = 0.1279E+00

1. Carbon disulfide, CS_2
2. Tetrachloroethene, C_2Cl_4

***T*/K = 298.15**
***System Number*: 389**

x_1	y_1	*P*/kPa	G^E/J mol^{-1}	μ^E_1/J mol^{-1}	μ^E_2/J mol^{-1}
0.0	0.000	2.42	0	826	0
0.1	0.751	8.80	83	818	1
0.2	0.867	14.85	161	749	14
0.3	0.913	20.28	228	640	51
0.4	0.938	25.03	277	510	121
0.5	0.953	29.21	304	376	231
0.6	0.965	32.97	305	251	385
0.7	0.974	36.55	276	145	583
0.8	0.983	40.14	217	65	823
0.9	0.991	43.96	125	16	1099
1.0	1.000	48.17	0	0	1405

B_{11}/cm^{3}mol^{-1}	B_{22}/cm^{3}mol^{-1}	B_{12}/cm^{3}mol^{-1}	V_1/cm^{3}mol^{-1}	V_2/cm^{3}mol^{-1}
-810	-2570	-1520	61	103

A(1) = 0.4903E+00 *A*(2) = 0.1169E+00 *A*(3) = -0.4031E-01

1. Carbon disulfide, CS_2
2. Benzene, C_6H_6

***T*/K = 303.15**
***System Number*: 387**

x_1	y_1	*P*/kPa	G^E/J mol^{-1}	μ^E_1/J mol^{-1}	μ^E_2/J mol^{-1}
0.0	0.000	16.27	0	929	0
0.1	0.343	22.43	87	806	7
0.2	0.522	27.81	158	655	34
0.3	0.635	32.52	211	515	81
0.4	0.716	36.81	245	402	141
0.5	0.780	40.86	263	315	212
0.6	0.833	44.69	266	243	300
0.7	0.878	48.22	251	174	430
0.8	0.916	51.35	211	102	651
0.9	0.953	54.10	135	34	1047
1.0	1.000	56.60	0	0	1742

B_{11}/cm^{3}mol^{-1}	B_{22}/cm^{3}mol^{-1}	B_{12}/cm^{3}mol^{-1}	V_1/cm^{3}mol^{-1}	V_2/cm^{3}mol^{-1}
-780	-1420	-1100	61	90

A(1) = 0.4181E+00 *A*(2) = 0.8106E-01 *A*(3) = 0.1116E+00 *A*(4) = 0.8028E-01 *A*(5) = 0.0000E+00

TABLE 3.1.1 (VLEG)
ISOTHERMAL VAPOR-LIQUID EQUILIBRIA AND EXCESS GIBBS ENERGIES OF BINARY MIXTURES (continued)

1. 1,1,2-Trichlorotrifluoroethane, $C_2Cl_3F_3$
2. Benzene, C_6H_6

***T*/K = 308.15**
***System Number*: 142**

x_1	y_1	P/kPa	G^E/J mol^{-1}	μ^E_1/J mol^{-1}	μ^E_2/J mol^{-1}
0.0	0.000	19.77	0	1966	0
0.1	0.390	29.62	169	1447	27
0.2	0.543	36.32	286	1039	98
0.3	0.634	41.39	359	724	203
0.4	0.700	45.61	392	485	330
0.5	0.755	49.40	391	308	474
0.6	0.806	52.97	360	182	628
0.7	0.854	56.44	303	95	789
0.8	0.902	59.85	223	40	954
0.9	0.950	63.25	121	9	1125
1.0	1.000	66.63	0	0	1303

B_{11}/cm^{3}mol^{-1}	B_{22}/cm^{3}mol^{-1}	B_{12}/cm^{3}mol^{-1}	V_1/cm^{3}mol^{-1}	V_2/cm^{3}mol^{-1}
-950	-1360	-1140	124	90

$A(1)$ = 0.6109E+00 $A(2)$ = -0.1293E+00 $A(3)$ = 0.2701E-01

---

1. Tetrachloroethene, C_2Cl_4
2. Pyridine, C_5H_5N

***T*/K = 323.15**
***System Number*: 370**

x_1	y_1	P/kPa	G^E/J mol^{-1}	μ^E_1/J mol^{-1}	μ^E_2/J mol^{-1}
0.0	0.000	9.58	0	2272	0
0.1	0.152	10.29	197	1718	28
0.2	0.252	10.65	341	1300	101
0.3	0.330	10.83	439	983	206
0.4	0.400	10.88	498	735	339
0.5	0.466	10.83	519	535	503
0.6	0.533	10.68	504	369	707
0.7	0.604	10.43	450	227	970
0.8	0.688	10.02	353	113	1317
0.9	0.802	9.38	207	32	1781
1.0	1.000	8.25	0	0	2402

B_{11}/cm^{3}mol^{-1}	B_{22}/cm^{3}mol^{-1}	B_{12}/cm^{3}mol^{-1}	V_1/cm^{3}mol^{-1}	V_2/cm^{3}mol^{-1}
-1910	-1620	-1760	105	83

$A(1)$ = 0.7727E+00 $A(2)$ = 0.2428E-01 $A(3)$ = 0.9720E-01

---

1. Tetrachloroethene, C_2Cl_4
2. Cyclopentane, C_5H_{10}

***T*/K = 298.15**
***System Number*: 353**

x_1	y_1	P/kPa	G^E/J mol^{-1}	μ^E_1/J mol^{-1}	μ^E_2/J mol^{-1}
0.0	0.000	42.3	0	628	0
0.1	0.008	38.4	57	522	6
0.2	0.017	34.7	103	423	23
0.3	0.028	31.0	137	332	54
0.4	0.041	27.3	159	250	98
0.5	0.057	23.7	167	177	157
0.6	0.079	20.0	162	116	232
0.7	0.111	16.1	144	67	324
0.8	0.168	11.9	111	30	433
0.9	0.299	7.5	63	8	562
1.0	1.000	2.5	0	0	710

B_{11}/cm^{3}mol^{-1}	B_{22}/cm^{3}mol^{-1}	B_{12}/cm^{3}mol^{-1}	V_1/cm^{3}mol^{-1}	V_2/cm^{3}mol^{-1}
-2570	-1050	-1690	103	95

$A(1)$ = 0.2698E+00 $A(2)$ = 0.1660E-01 $A(3)$ = 0.0000E+00

TABLE 3.1.1 (VLEG) ISOTHERMAL VAPOR-LIQUID EQUILIBRIA AND EXCESS GIBBS ENERGIES OF BINARY MIXTURES (continued)

1. 1,1,2,2-Tetrachloroethane, $C_2H_2Cl_4$
2. Hexane, C_6H_{14}

***T*/K = 333.15**
***System Number*: 123**

x_1	y_1	P/kPa	G^E/J mol^{-1}	μ^E_1/J mol^{-1}	μ^E_2/J mol^{-1}
0.0	0.000	76.45	0	2580	0
0.1	0.015	70.23	233	2103	25
0.2	0.027	64.75	416	1682	99
0.3	0.039	59.78	549	1311	223
0.4	0.050	55.07	633	986	398
0.5	0.061	50.35	667	705	629
0.6	0.074	45.27	648	466	920
0.7	0.093	39.33	575	272	1281
0.8	0.124	31.68	445	126	1721
0.9	0.202	20.93	255	33	2251
1.0	1.000	4.57	0	0	2884

B_{11}/cm^{3}mol^{-1}	B_{22}/cm^{3}mol^{-1}	B_{12}/cm^{3}mol^{-1}	V_1/cm^{3}mol^{-1}	V_2/cm^{3}mol^{-1}
-2480	-1390	-1880	109	139

$A(1) = 0.9626E+00$ $A(2) = 0.5493E-01$ $A(3) = 0.2380E-01$

1. 1,1,1-Trichloroethane, $C_2H_3Cl_3$
2. Hexane, C_6H_{14}

***T*/K = 333.15**
***System Number*: 122**

x_1	y_1	P/kPa	G^E/J mol^{-1}	μ^E_1/J mol^{-1}	μ^E_2/J mol^{-1}
0.0	0.000	76.45	0	698	0
0.1	0.106	77.13	66	621	5
0.2	0.203	77.48	122	512	24
0.3	0.292	77.42	163	391	65
0.4	0.376	76.95	186	274	128
0.5	0.458	76.05	192	172	211
0.6	0.543	74.70	179	93	308
0.7	0.636	72.89	150	39	407
0.8	0.742	70.61	107	10	492
0.9	0.864	67.94	55	0	545
1.0	1.000	65.11	0	0	542

B_{11}/cm^{3}mol^{-1}	B_{22}/cm^{3}mol^{-1}	B_{12}/cm^{3}mol^{-1}	V_1/cm^{3}mol^{-1}	V_2/cm^{3}mol^{-1}
-1110	-1390	-1240	105	139

$A(1) = 0.2770E+00$ $A(2) = -0.2817E-01$ $A(3) = -0.5316E-01$

1. Ethanenitrile, C_2H_3N
2. Ethanol, C_2H_6O

***T*/K = 343.15**
***System Number*: 296**

x_1	y_1	P/kPa	G^E/J mol^{-1}	μ^E_1/J mol^{-1}	μ^E_2/J mol^{-1}
0.0	0.000	72.34	0	3319	0
0.1	0.198	82.68	285	2456	44
0.2	0.300	87.69	487	1812	156
0.3	0.371	90.27	620	1326	317
0.4	0.430	91.47	692	952	518
0.5	0.485	91.63	708	658	758
0.6	0.542	90.82	672	426	1042
0.7	0.606	88.89	585	245	1379
0.8	0.687	85.41	445	112	1778
0.9	0.803	79.52	251	29	2250
1.0	1.000	69.59	0	0	2798

B_{11}/cm^{3}mol^{-1}	B_{22}/cm^{3}mol^{-1}	B_{12}/cm^{3}mol^{-1}	V_1/cm^{3}mol^{-1}	V_2/cm^{3}mol^{-1}
-2960	-1150	-1520	56	63

$A(1) = 0.9927E+00$ $A(2) = -0.6948E-01$ $A(3) = 0.7936E-01$ $A(4) = -0.2178E-01$ $A(5) = 0.0000E+00$

TABLE 3.1.1 (VLEG)
ISOTHERMAL VAPOR-LIQUID EQUILIBRIA AND EXCESS GIBBS ENERGIES OF BINARY MIXTURES (continued)

1. Ethanenitrile, C_2H_3N
2. 1,2-Ethanediol, $C_2H_6O_2$

***T*/K = 323.15**
***System Number*: 297**

x_1	y_1	P/kPa	G^E/J mol^{-1}	μ^E_1/J mol^{-1}	μ^E_2/J mol^{-1}
0.0	0.000	0.10	0	4456	0
0.1	0.993	13.07	405	3694	40
0.2	0.996	20.42	730	3007	161
0.3	0.997	24.48	974	2390	367
0.4	0.997	26.73	1135	1845	661
0.5	0.997	28.06	1210	1371	1049
0.6	0.997	28.97	1198	964	1548
0.7	0.998	29.70	1090	614	2201
0.8	0.998	30.44	874	319	3094
0.9	0.998	31.56	525	96	4380
1.0	1.000	33.90	0	0	6314

B_{11}/cm^{3}mol^{-1}	B_{22}/cm^{3}mol^{-1}	B_{12}/cm^{3}mol^{-1}	V_1/cm^{3}mol^{-1}	V_2/cm^{3}mol^{-1}
-4100	-1970	-2350	55	57

$A(1)$ = 0.1802E+01 $A(2)$ = 0.2402E+00 $A(3)$ = 0.1664E+00 $A(4)$ = 0.1055E+00 $A(5)$ = 0.3625E-01

---

1. Ethanenitrile, C_2H_3N
2. Methyl ethanoate, $C_3H_6O_2$

***T*/K = 323.15**
***System Number*: 298**

x_1	y_1	P/kPa	G^E/J mol^{-1}	μ^E_1/J mol^{-1}	μ^E_2/J mol^{-1}
0.0	0.000	79.36	0	718	0
0.1	0.058	75.87	64	570	8
0.2	0.115	72.28	113	442	30
0.3	0.174	68.55	146	332	67
0.4	0.237	64.64	166	239	117
0.5	0.307	60.50	171	162	180
0.6	0.387	56.06	162	102	254
0.7	0.483	51.26	141	56	338
0.8	0.604	46.01	106	24	433
0.9	0.766	40.23	59	6	537
1.0	1.000	33.83	0	0	649

B_{11}/cm^{3}mol^{-1}	B_{22}/cm^{3}mol^{-1}	B_{12}/cm^{3}mol^{-1}	V_1/cm^{3}mol^{-1}	V_2/cm^{3}mol^{-1}
-4100	-1280	-2280	55	83

$A(1)$ = 0.2544E+00 $A(2)$ = -0.1290E-01 $A(3)$ = 0.0000E+00

---

1. Ethanenitrile, C_2H_3N
2. Ethyl ethanoate, $C_4H_8O_2$

***T*/K = 353.15**
***System Number*: 299**

x_1	y_1	P/kPa	G^E/J mol^{-1}	μ^E_1/J mol^{-1}	μ^E_2/J mol^{-1}
0.0	0.000	111.4	0	1243	0
0.1	0.117	114.1	110	978	14
0.2	0.215	115.6	194	760	52
0.3	0.304	116.2	252	580	112
0.4	0.387	116.0	288	431	192
0.5	0.468	115.1	300	307	294
0.6	0.551	113.5	290	204	420
0.7	0.638	111.0	257	120	575
0.8	0.736	107.6	199	56	767
0.9	0.851	102.9	114	15	1003
1.0	1.000	96.4	0	0	1294

B_{11}/cm^{3}mol^{-1}	B_{22}/cm^{3}mol^{-1}	B_{12}/cm^{3}mol^{-1}	V_1/cm^{3}mol^{-1}	V_2/cm^{3}mol^{-1}
-2340	-1210	-1400	57	107

$A(1)$ = 0.4090E+00 $A(2)$ = 0.8702E-02 $A(3)$ = 0.2310E-01

TABLE 3.1.1 (VLEG)
ISOTHERMAL VAPOR-LIQUID EQUILIBRIA AND EXCESS GIBBS ENERGIES OF BINARY MIXTURES (continued)

1. Ethanenitrile, C_2H_3N
2. 1-Chlorobutane, C_4H_9Cl

***T*/K = 348.18**
***System Number*: 180**

x_1	y_1	P/kPa	G^E/J mol^{-1}	μ^E_1/J mol^{-1}	μ^E_2/J mol^{-1}
0.0	0.000	91.3	0	3524	0
0.1	0.200	104.3	312	2771	39
0.2	0.308	111.1	549	2158	147
0.3	0.381	114.6	716	1654	314
0.4	0.436	116.2	818	1236	539
0.5	0.484	116.5	856	886	825
0.6	0.531	115.5	830	593	1184
0.7	0.583	113.2	737	353	1631
0.8	0.652	108.7	572	168	2189
0.9	0.762	100.0	330	45	2889
1.0	1.000	82.3	0	0	3767

B_{11}/cm^{3}mol^{-1}	B_{22}/cm^{3}mol^{-1}	B_{12}/cm^{3}mol^{-1}	V_1/cm^{3}mol^{-1}	V_2/cm^{3}mol^{-1}
-2740	-1060	-820	57	112

$A(1)$ = 0.1182E+01 $A(2)$ = 0.4197E-01 $A(3)$ = 0.7679E-01

1. Ethanenitrile, C_2H_3N
2. Diethyl ether, $C_4H_{10}O$

***T*/K = 298.14**
***System Number*: 295**

x_1	y_1	P/kPa	G^E/J mol^{-1}	μ^E_1/J mol^{-1}	μ^E_2/J mol^{-1}
0.0	0.000	71.51	0	3605	0
0.1	0.052	69.13	306	2621	49
0.2	0.083	66.44	522	1949	166
0.3	0.106	63.65	667	1448	332
0.4	0.126	60.72	747	1051	545
0.5	0.146	57.45	769	729	808
0.6	0.169	53.50	732	471	1124
0.7	0.201	48.33	638	270	1498
0.8	0.253	40.99	486	123	1939
0.9	0.373	29.83	274	31	2458
1.0	1.000	11.84	0	0	3051

B_{11}/cm^{3}mol^{-1}	B_{22}/cm^{3}mol^{-1}	B_{12}/cm^{3}mol^{-1}	V_1/cm^{3}mol^{-1}	V_2/cm^{3}mol^{-1}
-6950	-1240	-1900	53	105

$A(1)$ = 0.1240E+01 $A(2)$ = -0.6390E-01 $A(3)$ = 0.7640E-01 $A(4)$ = -0.2890E-01 $A(5)$ = 0.2590E-01

1. Ethanenitrile, C_2H_3N
2. Diethylamine, $C_4H_{11}N$

***T*/K = 298.00**
***System Number*: 184**

x_1	y_1	P/kPa	G^E/J mol^{-1}	μ^E_1/J mol^{-1}	μ^E_2/J mol^{-1}
0.0	0.000	31.44	0	3320	0
0.1	0.105	32.14	287	2500	42
0.2	0.165	31.96	497	1893	148
0.3	0.209	31.39	640	1429	301
0.4	0.247	30.56	724	1061	499
0.5	0.282	29.51	752	758	747
0.6	0.319	28.19	726	504	1058
0.7	0.364	26.42	641	296	1445
0.8	0.431	23.82	494	137	1923
0.9	0.563	19.51	282	36	2501
1.0	1.000	11.81	0	0	3182

B_{11}/cm^{3}mol^{-1}	B_{22}/cm^{3}mol^{-1}	B_{12}/cm^{3}mol^{-1}	V_1/cm^{3}mol^{-1}	V_2/cm^{3}mol^{-1}
-6950	-1430	-3200	53	104

$A(1)$ = 0.1215E+01 $A(2)$ = 0.8300E-02 $A(3)$ = 0.9730E-01 $A(4)$ = -0.3610E-01 $A(5)$ = 0.0000E+00

TABLE 3.1.1 (VLEG)
ISOTHERMAL VAPOR-LIQUID EQUILIBRIA AND EXCESS GIBBS ENERGIES OF BINARY MIXTURES (continued)

1. Ethanenitrile, C_2H_3N
2. Chlorobenzene, C_6H_5Cl

***T*/K = 343.15**
***System Number*: 402**

x_1	y_1	P/kPa	G^E/J mol^{-1}	μ^E_1/J mol^{-1}	μ^E_2/J mol^{-1}
0.0	0.000	13.42	0	3308	0
0.1	0.572	28.69	292	2582	37
0.2	0.706	38.46	512	2010	138
0.3	0.771	45.39	668	1550	290
0.4	0.811	50.68	765	1173	493
0.5	0.841	54.93	804	856	752
0.6	0.866	58.47	785	586	1083
0.7	0.889	61.53	703	358	1509
0.8	0.915	64.33	552	174	2063
0.9	0.947	67.05	322	48	2785
1.0	1.000	69.60	0	0	3725

B_{11}/cm^{3}mol^{-1}	B_{22}/cm^{3}mol^{-1}	B_{12}/cm^{3}mol^{-1}	V_1/cm^{3}mol^{-1}	V_2/cm^{3}mol^{-1}
-2960	-1860	-1300	56	107

$A(1)$ = 0.1128E+01 $A(2)$ = 0.7301E-01 $A(3)$ = 0.1049E+00

---

1. Ethanenitrile, C_2H_3N
2. Benzene, C_6H_6

***T*/K = 313.15**
***System Number*: 167**

x_1	y_1	P/kPa	G^E/J mol^{-1}	μ^E_1/J mol^{-1}	μ^E_2/J mol^{-1}
0.0	0.000	24.38	0	3154	0
0.1	0.193	27.65	266	2266	44
0.2	0.293	29.16	452	1671	148
0.3	0.367	29.97	576	1254	285
0.4	0.430	30.36	648	942	453
0.5	0.488	30.42	674	687	661
0.6	0.543	30.19	653	468	930
0.7	0.602	29.62	580	281	1279
0.8	0.675	28.56	451	132	1728
0.9	0.785	26.57	259	34	2283
1.0	1.000	22.75	0	0	2938

B_{11}/cm^{3}mol^{-1}	B_{22}/cm^{3}mol^{-1}	B_{12}/cm^{3}mol^{-1}	V_1/cm^{3}mol^{-1}	V_2/cm^{3}mol^{-1}
-4920	-1304	-1040	54	91

$A(1)$ = 0.1036E+01 $A(2)$ = 0.1975E-01 $A(3)$ = 0.1341E+00 $A(4)$ = -0.6120E-01 $A(5)$ = 0.0000E+00

---

1. Ethanenitrile, C_2H_3N
2. Aniline, C_6H_7N

***T*/K = 343.15**
***System Number*: 185**

x_1	y_1	P/kPa	G^E/J mol^{-1}	μ^E_1/J mol^{-1}	μ^E_2/J mol^{-1}
0.0	0.000	1.42	0	-275	0
0.1	0.832	7.59	-17	-92	-9
0.2	0.922	14.40	-18	31	-30
0.3	0.955	21.59	-6	109	-56
0.4	0.971	28.94	13	150	-78
0.5	0.981	36.24	38	162	-86
0.6	0.987	43.33	61	148	-68
0.7	0.991	50.08	79	113	-2
0.8	0.995	56.50	81	66	142
0.9	0.997	62.82	60	21	404
1.0	1.000	69.59	0	0	840

B_{11}/cm^{3}mol^{-1}	B_{22}/cm^{3}mol^{-1}	B_{12}/cm^{3}mol^{-1}	V_1/cm^{3}mol^{-1}	V_2/cm^{3}mol^{-1}
-2960	-2580	-1750	56	95

$A(1)$ = 0.5285E-01 $A(2)$ = 0.1741E+00 $A(3)$ = 0.4605E-01 $A(4)$ = 0.2133E-01 $A(5)$ = 0.0000E+00

TABLE 3.1.1 (VLEG) ISOTHERMAL VAPOR-LIQUID EQUILIBRIA AND EXCESS GIBBS ENERGIES OF BINARY MIXTURES (continued)

1. Ethanenitrile, C_2H_3N
2. 1,4-Cyclohexadiene, C_6H_8

***T*/K = 293.15**
***System Number*: 166**

x_1	y_1	*P*/kPa	G^E/J mol^{-1}	μ^E_1/J mol^{-1}	μ^E_2/J mol^{-1}
0.0	0.000	6.94	0	5876	0
0.1	0.423	11.22	480	3962	93
0.2	0.490	12.20	796	2823	289
0.3	0.523	12.63	998	2074	537
0.4	0.546	12.84	1108	1515	837
0.5	0.563	12.94	1135	1063	1206
0.6	0.580	12.96	1079	694	1657
0.7	0.605	12.87	942	405	2195
0.8	0.650	12.56	721	193	2832
0.9	0.741	11.70	411	55	3622
1.0	1.000	9.38	0	0	4711

B_{11}/cm^{3}mol^{-1}	B_{22}/cm^{3}mol^{-1}	B_{12}/cm^{3}mol^{-1}	V_1/cm^{3}mol^{-1}	V_2/cm^{3}mol^{-1}
-7454	-1722	-1380	52	94

A(1) = 0.1862E+01 *A*(2) = -0.1176E+00 *A*(3) = 0.1836E+00 *A*(4) = -0.1213E+00 *A*(5) = 0.1263E+00

1. Ethanenitrile, C_2H_3N
2. Cyclohexene, C_6H_{10}

***T*/K = 313.15**
***System Number*: 165**

x_1	y_1	*P*/kPa	G^E/J mol^{-1}	μ^E_1/J mol^{-1}	μ^E_2/J mol^{-1}
0.0	0.000	22.62	0	7346	0
0.1	0.431	37.25	608	5069	112
0.2	0.472	39.21	1014	3619	363
0.3	0.484	39.63	1273	2651	683
0.4	0.491	39.79	1415	1950	1058
0.5	0.496	39.83	1454	1398	1509
0.6	0.500	39.80	1392	940	2070
0.7	0.506	39.64	1225	561	2777
0.8	0.528	38.81	945	265	3666
0.9	0.605	35.30	541	71	4775
1.0	1.000	22.75	0	0	6149

B_{11}/cm^{3}mol^{-1}	B_{22}/cm^{3}mol^{-1}	B_{12}/cm^{3}mol^{-1}	V_1/cm^{3}mol^{-1}	V_2/cm^{3}mol^{-1}
-4920	-1466	-975	54	104

A(1) = 0.2233E+01 *A*(2) = -0.8568E-01 *A*(3) = 0.3141E+00 *A*(4) = -0.1441E+00 *A*(5) = 0.4428E-01

1. Ethanenitrile, C_2H_3N
2. Toluene, C_7H_8

***T*/K = 343.15**
***System Number*: 168**

x_1	y_1	*P*/kPa	G^E/J mol^{-1}	μ^E_1/J mol^{-1}	μ^E_2/J mol^{-1}
0.0	0.000	27.18	0	3755	0
0.1	0.416	42.72	323	2792	48
0.2	0.552	51.50	556	2118	166
0.3	0.627	57.30	717	1606	335
0.4	0.679	61.45	812	1193	557
0.5	0.720	64.57	844	853	836
0.6	0.758	67.03	815	574	1177
0.7	0.796	68.99	723	350	1594
0.8	0.841	70.42	564	175	2121
0.9	0.899	71.03	330	52	2830
1.0	1.000	69.66	0	0	3868

B_{11}/cm^{3}mol^{-1}	B_{22}/cm^{3}mol^{-1}	B_{12}/cm^{3}mol^{-1}	V_1/cm^{3}mol^{-1}	V_2/cm^{3}mol^{-1}
-2960	-1740	-1000	56	112

A(1) = 0.1184E+01 *A*(2) = 0.1216E-01 *A*(3) = 0.1024E+00 *A*(4) = 0.7620E-02 *A*(5) = 0.4982E-01

TABLE 3.1.1 (VLEG)
ISOTHERMAL VAPOR-LIQUID EQUILIBRIA AND EXCESS GIBBS ENERGIES OF BINARY MIXTURES (continued)

1. 1,2-Dibromoethane, $C_2H_4Br_2$
2. 1-Nitropropane, $C_3H_7NO_2$

***T*/K = 348.15**
***System Number*: 197**

x_1	y_1	P/kPa	G^E/J mol^{-1}	μ^E_1/J mol^{-1}	μ^E_2/J mol^{-1}
0.0	0.000	15.33	0	1394	0
0.1	0.145	16.23	125	1117	14
0.2	0.257	16.86	221	890	54
0.3	0.352	17.30	292	699	118
0.4	0.437	17.58	338	535	206
0.5	0.516	17.74	358	393	322
0.6	0.592	17.76	351	269	474
0.7	0.671	17.64	316	164	671
0.8	0.756	17.36	248	79	926
0.9	0.859	16.84	145	22	1255
1.0	1.000	15.96	0	0	1676

B_{11}/cm^{3}mol^{-1}	B_{22}/cm^{3}mol^{-1}	B_{12}/cm^{3}mol^{-1}	V_1/cm^{3}mol^{-1}	V_2/cm^{3}mol^{-1}
-1520	-3150	-2240	91	94

$A(1)$ = 0.4944E+00 $A(2)$ = 0.4875E-01 $A(3)$ = 0.3583E-01

---

1. 1,2-Dibromoethane, $C_2H_4Br_2$
2. 1,4-Dioxane, $C_4H_8O_2$

***T*/K = 293.15**
***System Number*: 272**

x_1	y_1	P/kPa	G^E/J mol^{-1}	μ^E_1/J mol^{-1}	μ^E_2/J mol^{-1}
0.0	0.000	3.488	0	33	0
0.1	0.036	3.244	13	210	-9
0.2	0.081	3.004	42	311	-26
0.3	0.134	2.775	78	349	-38
0.4	0.193	2.558	117	338	-31
0.5	0.257	2.354	150	291	8
0.6	0.328	2.154	171	223	93
0.7	0.409	1.950	173	146	238
0.8	0.513	1.725	150	74	456
0.9	0.672	1.450	95	21	760
1.0	1.000	1.073	0	0	1165

B_{11}/cm^{3}mol^{-1}	B_{22}/cm^{3}mol^{-1}	B_{12}/cm^{3}mol^{-1}	V_1/cm^{3}mol^{-1}	V_2/cm^{3}mol^{-1}
-3290	-2360	-2800	86	84

$A(1)$ = 0.2458E+00 $A(2)$ = 0.2322E+00 $A(3)$ = 0.0000E+00

---

1. 1,2-Dibromoethane, $C_2H_4Br_2$
2. Benzene, C_6H_6

***T*/K = 303.15**
***System Number*: 106**

x_1	y_1	P/kPa	G^E/J mol^{-1}	μ^E_1/J mol^{-1}	μ^E_2/J mol^{-1}
0.0	0.000	15.73	0	635	0
0.1	0.018	14.40	65	654	0
0.2	0.038	13.12	127	582	14
0.3	0.059	11.93	176	460	55
0.4	0.082	10.79	207	323	129
0.5	0.109	9.65	215	196	233
0.6	0.144	8.42	200	96	355
0.7	0.196	7.04	164	31	474
0.8	0.284	5.46	113	1	563
0.9	0.469	3.71	55	-4	582
1.0	1.000	1.93	0	0	486

B_{11}/cm^{3}mol^{-1}	B_{22}/cm^{3}mol^{-1}	B_{12}/cm^{3}mol^{-1}	V_1/cm^{3}mol^{-1}	V_2/cm^{3}mol^{-1}
-2800	-1420	-2110	87	92

$A(1)$ = 0.3405E+00 $A(2)$ = -0.2952E-01 $A(3)$ = -0.1181E+00

TABLE 3.1.1 (VLEG)
ISOTHERMAL VAPOR-LIQUID EQUILIBRIA AND EXCESS GIBBS ENERGIES OF BINARY MIXTURES (continued)

1. 1,2-Dibromoethane, $C_2H_4Br_2$
2. Cyclohexane, C_6H_{12}

***T*/K = 293.15**
***System Number*: 104**

x_1	y_1	P/kPa	G^E/J mol^{-1}	μ^E_1/J mol^{-1}	μ^E_2/J mol^{-1}
0.0	0.000	10.26	0	3571	0
0.1	0.034	9.72	310	2700	44
0.2	0.054	9.25	536	2047	158
0.3	0.070	8.80	691	1551	322
0.4	0.084	8.33	784	1165	530
0.5	0.098	7.83	819	853	785
0.6	0.114	7.27	797	590	1107
0.7	0.133	6.61	714	366	1525
0.8	0.162	5.73	562	182	2081
0.9	0.233	4.25	329	51	2831
1.0	1.000	1.07	0	0	3841

B_{11}/cm^{3}mol^{-1}	B_{22}/cm^{3}mol^{-1}	B_{12}/cm^{3}mol^{-1}	V_1/cm^{3}mol^{-1}	V_2/cm^{3}mol^{-1}
-3290	-1830	-2030	86	108

$A(1)$ = 0.1344E+01 $A(2)$ = 0.5544E-01 $A(3)$ = 0.1762E+00

---

1. 1,2-Dichloroethane, $C_2H_4Cl_2$
2. 1-Chlorobutane, C_4H_9Cl

***T*/K = 313.15**
***System Number*: 141**

x_1	y_1	P/kPa	G^E/J mol^{-1}	μ^E_1/J mol^{-1}	μ^E_2/J mol^{-1}
0.0	0.000	25.87	0	506	0
0.1	0.096	25.79	48	454	3
0.2	0.189	25.64	90	393	14
0.3	0.278	25.43	123	328	36
0.4	0.365	25.14	148	260	72
0.5	0.452	24.77	160	194	127
0.6	0.540	24.29	160	133	202
0.7	0.633	23.69	146	80	301
0.8	0.735	22.94	116	38	428
0.9	0.853	21.98	67	10	586
1.0	1.000	20.74	0	0	777

B_{11}/cm^{3}mol^{-1}	B_{22}/cm^{3}mol^{-1}	B_{12}/cm^{3}mol^{-1}	V_1/cm^{3}mol^{-1}	V_2/cm^{3}mol^{-1}
-1290	-1420	-1350	80	107

$A(1)$ = 0.2465E+00 $A(2)$ = 0.5200E-01 $A(3)$ = 0.0000E+00

---

1. 1,2-Dichloroethane, $C_2H_4Cl_2$
2. Benzene, C_6H_6

***T*/K = 343.15**
***System Number*: 113**

x_1	y_1	P/kPa	G^E/J mol^{-1}	μ^E_1/J mol^{-1}	μ^E_2/J mol^{-1}
0.0	0.000	73.35	0	43	0
0.1	0.091	72.63	4	38	0
0.2	0.184	71.89	8	33	1
0.3	0.278	71.14	10	27	3
0.4	0.374	70.37	12	22	6
0.5	0.472	69.58	13	16	11
0.6	0.572	68.77	13	11	17
0.7	0.674	67.93	12	7	25
0.8	0.779	67.05	10	3	36
0.9	0.887	66.13	6	1	49
1.0	1.000	65.17	0	0	65

B_{11}/cm^{3}mol^{-1}	B_{22}/cm^{3}mol^{-1}	B_{12}/cm^{3}mol^{-1}	V_1/cm^{3}mol^{-1}	V_2/cm^{3}mol^{-1}
-1070	-1030	-1050	84	94

$A(1)$ = 0.1883E-01 $A(2)$ = 0.3927E-02 $A(3)$ = 0.0000E+00

TABLE 3.1.1 (VLEG) ISOTHERMAL VAPOR-LIQUID EQUILIBRIA AND EXCESS GIBBS ENERGIES OF BINARY MIXTURES (continued)

1. 1,2-Dichloroethane, $C_2H_4Cl_2$
2. Heptane, C_7H_{16}

***T*/K = 343.15**
***System Number*: 124**

x_1	y_1	*P*/kPa	G^E/J mol^{-1}	μ^E_1/J mol^{-1}	μ^E_2/J mol^{-1}
0.0	0.000	40.38	0	2484	0
0.1	0.268	50.28	229	2113	20
0.2	0.417	57.55	418	1772	80
0.3	0.513	62.82	567	1452	187
0.4	0.582	66.55	670	1150	350
0.5	0.636	69.06	725	865	585
0.6	0.681	70.64	724	601	908
0.7	0.724	71.42	661	368	1343
0.8	0.775	71.37	526	178	1915
0.9	0.852	69.84	310	49	2658
1.0	1.000	64.85	0	0	3604

B_{11}/cm^{3}mol^{-1}	B_{22}/cm^{3}mol^{-1}	B_{12}/cm^{3}mol^{-1}	V_1/cm^{3}mol^{-1}	V_2/cm^{3}mol^{-1}
-1070	-1880	-1210	84	157

A(1) = 0.1016E+01 *A*(2) = 0.1962E+00 *A*(3) = 0.5100E-01

---

1. Iodoethane, C_2H_5I
2. Hexane, C_6H_{14}

***T*/K = 332.85**
***System Number*: 145**

x_1	y_1	*P*/kPa	G^E/J mol^{-1}	μ^E_1/J mol^{-1}	μ^E_2/J mol^{-1}
0.0	0.000	75.59	0	1778	0
0.1	0.143	80.08	162	1467	16
0.2	0.250	82.93	290	1181	67
0.3	0.336	84.55	384	920	154
0.4	0.409	85.20	443	688	279
0.5	0.477	85.04	465	486	444
0.6	0.544	84.09	451	317	652
0.7	0.618	82.27	398	181	904
0.8	0.706	79.29	306	82	1202
0.9	0.824	74.71	174	21	1548
1.0	1.000	67.73	0	0	1945

B_{11}/cm^{3}mol^{-1}	B_{22}/cm^{3}mol^{-1}	B_{12}/cm^{3}mol^{-1}	V_1/cm^{3}mol^{-1}	V_2/cm^{3}mol^{-1}
-970	-1390	-1020	86	139

A(1) = 0.6726E+00 *A*(2) = 0.3020E-01 *A*(3) = 0.0000E+00

---

1. Iodoethane, C_2H_5I
2. Heptane, C_7H_{16}

***T*/K = 303.15**
***System Number*: 146**

x_1	y_1	*P*/kPa	G^E/J mol^{-1}	μ^E_1/J mol^{-1}	μ^E_2/J mol^{-1}
0.0	0.000	7.76	0	2047	0
0.1	0.359	11.04	178	1550	25
0.2	0.516	13.34	308	1191	88
0.3	0.613	15.17	401	925	176
0.4	0.685	16.70	459	719	287
0.5	0.742	18.00	487	546	428
0.6	0.791	19.10	482	393	616
0.7	0.833	20.00	440	252	879
0.8	0.876	20.73	354	129	1252
0.9	0.925	21.29	212	37	1780
1.0	1.000	21.57	0	0	2521

B_{11}/cm^{3}mol^{-1}	B_{22}/cm^{3}mol^{-1}	B_{12}/cm^{3}mol^{-1}	V_1/cm^{3}mol^{-1}	V_2/cm^{3}mol^{-1}
-1310	-2700	-1650	82	148

A(1) = 0.7730E+00 *A*(2) = 0.9400E-01 *A*(3) = 0.1330E+00

TABLE 3.1.1 (VLEG)
ISOTHERMAL VAPOR-LIQUID EQUILIBRIA AND EXCESS GIBBS ENERGIES OF BINARY MIXTURES (continued)

1. N-Methylmethanamide, C_2H_5NO
2. Benzene, C_6H_6

***T*/K = 318.15**
***System Number*: 187**

x_1	y_1	*P*/kPa	G^E/J mol^{-1}	μ^E_1/J mol^{-1}	μ^E_2/J mol^{-1}
0.0	0.000	29.76	0	8621	0
0.1	0.003	28.75	640	4788	179
0.2	0.003	28.73	983	2958	490
0.3	0.004	28.29	1163	2004	802
0.4	0.004	27.49	1234	1383	1134
0.5	0.004	26.59	1215	901	1528
0.6	0.005	25.26	1110	527	1984
0.7	0.005	22.62	927	272	2454
0.8	0.007	17.80	678	124	2894
0.9	0.013	10.70	374	40	3374
1.0	1.000	0.16	0	0	4247

B_{11}/cm^{3}mol^{-1}	B_{22}/cm^{3}mol^{-1}	B_{12}/cm^{3}mol^{-1}	V_1/cm^{3}mol^{-1}	V_2/cm^{3}mol^{-1}
-2930	-1250	-1970	60	92

A(1) = 0.1837E+01 *A*(2) = -0.4745E+00 *A*(3) = 0.2125E+00 *A*(4) = -0.3522E+00 *A*(5) = 0.3832E+00

1. Nitroethane, $C_2H_5NO_2$
2. Benzene, C_6H_6

***T*/K = 298.15**
***System Number*: 192**

x_1	y_1	*P*/kPa	G^E/J mol^{-1}	μ^E_1/J mol^{-1}	μ^E_2/J mol^{-1}
0.0	0.000	12.81	0	1815	0
0.1	0.044	12.16	166	1457	22
0.2	0.076	11.56	282	991	104
0.3	0.106	10.92	347	652	216
0.4	0.140	10.14	374	460	317
0.5	0.183	9.23	378	347	409
0.6	0.235	8.29	361	244	537
0.7	0.294	7.36	316	127	756
0.8	0.377	6.28	233	26	1060
0.9	0.553	4.74	114	-13	1261
1.0	1.000	2.91	0	0	839

B_{11}/cm^{3}mol^{-1}	B_{22}/cm^{3}mol^{-1}	B_{12}/cm^{3}mol^{-1}	V_1/cm^{3}mol^{-1}	V_2/cm^{3}mol^{-1}
-4900	-1480	-2860	72	92

A(1) = 0.6103E+00 *A*(2) = -0.5017E-01 *A*(3) = 0.2082E+00 *A*(4) = -0.1468E+00 *A*(5) = -0.2832E+00

1. Ethanol, C_2H_6O
2. 1,2-Ethanediol, $C_2H_6O_2$

***T*/K = 323.15**
***System Number*: 319**

x_1	y_1	*P*/kPa	G^E/J mol^{-1}	μ^E_1/J mol^{-1}	μ^E_2/J mol^{-1}
0.0	0.000	0.12	0	1894	0
0.1	0.980	5.31	172	1562	18
0.2	0.990	9.39	308	1253	72
0.3	0.993	12.69	408	976	165
0.4	0.995	15.45	471	734	295
0.5	0.996	17.89	496	527	464
0.6	0.997	20.14	483	354	676
0.7	0.998	22.29	431	212	940
0.8	0.998	24.47	336	102	1271
0.9	0.999	26.79	195	28	1695
1.0	1.000	29.48	0	0	2247

B_{11}/cm^{3}mol^{-1}	B_{22}/cm^{3}mol^{-1}	B_{12}/cm^{3}mol^{-1}	V_1/cm^{3}mol^{-1}	V_2/cm^{3}mol^{-1}
-1400	-1970	-1730	61	57

A(1) = 0.7380E+00 *A*(2) = 0.4699E-01 *A*(3) = 0.3276E-01 *A*(4) = 0.1872E-01 *A*(5) = 0.0000E+00

TABLE 3.1.1 (VLEG)
ISOTHERMAL VAPOR-LIQUID EQUILIBRIA AND EXCESS GIBBS ENERGIES OF BINARY MIXTURES (continued)

1. Ethanol, C_2H_6O
2. 2-Propanone, C_3H_6O

T/K = 323.15
System Number: 322

x_1	y_1	P/kPa	G^E/J mol^{-1}	μ^E_1/J mol^{-1}	μ^E_2/J mol^{-1}
0.0	0.000	82.03	0	1888	0
0.1	0.066	79.49	170	1533	19
0.2	0.122	76.60	303	1216	74
0.3	0.173	73.44	399	940	166
0.4	0.221	69.97	458	703	294
0.5	0.272	66.11	480	502	459
0.6	0.328	61.69	466	334	664
0.7	0.397	56.46	414	198	917
0.8	0.492	50.00	321	94	1230
0.9	0.648	41.53	185	26	1623
1.0	1.000	29.48	0	0	2119

B_{11}/cm^{3}mol^{-1}	B_{22}/cm^{3}mol^{-1}	B_{12}/cm^{3}mol^{-1}	V_1/cm^{3}mol^{-1}	V_2/cm^{3}mol^{-1}
-1400	-1430	-1040	61	77

$A(1) = 0.7151E+00$ $A(2) = 0.3222E-01$ $A(3) = 0.3061E-01$ $A(4) = 0.1076E-01$ $A(5) = 0.0000E+00$

---

1. Ethanol, C_2H_6O
2. 1,3-Dioxolane, $C_3H_6O_2$

T/K = 313.15
System Number: 310

x_1	y_1	P/kPa	G^E/J mol^{-1}	μ^E_1/J mol^{-1}	μ^E_2/J mol^{-1}
0.0	0.000	26.56	0	3252	0
0.1	0.160	28.91	285	2514	38
0.2	0.250	29.90	499	1944	137
0.3	0.314	30.23	649	1496	286
0.4	0.366	30.17	742	1135	481
0.5	0.413	29.81	780	833	728
0.6	0.459	29.17	762	575	1044
0.7	0.509	28.20	684	354	1455
0.8	0.575	26.67	539	174	1998
0.9	0.688	23.88	316	49	2718
1.0	1.000	17.88	0	0	3672

B_{11}/cm^{3}mol^{-1}	B_{22}/cm^{3}mol^{-1}	B_{12}/cm^{3}mol^{-1}	V_1/cm^{3}mol^{-1}	V_2/cm^{3}mol^{-1}
-2100	-1910	-660	60	71

$A(1) = 0.1199E+01$ $A(2) = 0.8060E-01$ $A(3) = 0.1309E+00$

---

1. Ethanol, C_2H_6O
2. 2-Propanol, C_3H_8O

T/K = 313.15
System Number: 312

x_1	y_1	P/kPa	G^E/J mol^{-1}	μ^E_1/J mol^{-1}	μ^E_2/J mol^{-1}
0.0	0.000	13.91	0	-139	0
0.1	0.120	14.22	-13	-113	-1
0.2	0.237	14.56	-22	-89	-6
0.3	0.350	14.92	-29	-68	-13
0.4	0.458	15.29	-33	-50	-22
0.5	0.562	15.69	-35	-35	-35
0.6	0.660	16.10	-33	-22	-50
0.7	0.753	16.53	-29	-13	-68
0.8	0.841	16.97	-22	-6	-89
0.9	0.923	17.43	-13	-1	-113
1.0	1.000	17.90	0	0	-139

B_{11}/cm^{3}mol^{-1}	B_{22}/cm^{3}mol^{-1}	B_{12}/cm^{3}mol^{-1}	V_1/cm^{3}mol^{-1}	V_2/cm^{3}mol^{-1}
-2100	-2360	-2200	60	78

$A(1) = -0.5337E-01$ $A(2) = 0.0000E+00$ $A(3) = 0.0000E+00$

TABLE 3.1.1 (VLEG)
ISOTHERMAL VAPOR-LIQUID EQUILIBRIA AND EXCESS GIBBS ENERGIES OF BINARY MIXTURES (continued)

1. Ethanol, C_2H_6O
2. 1,4-Dioxane, $C_4H_8O_2$

***T*/K = 323.15**
***System Number*: 102**

x_1	y_1	P/kPa	G^E/J mol^{-1}	μ^E_1/J mol^{-1}	μ^E_2/J mol^{-1}
0.0	0.000	15.70	0	2325	0
0.1	0.294	20.22	210	1907	22
0.2	0.445	23.42	377	1543	86
0.3	0.540	25.73	502	1221	194
0.4	0.608	27.42	583	935	347
0.5	0.662	28.67	619	682	555
0.6	0.711	29.57	607	462	825
0.7	0.758	30.18	545	276	1171
0.8	0.813	30.51	426	131	1608
0.9	0.885	30.42	247	35	2156
1.0	1.000	29.51	0	0	2836

B_{11}/cm^{3}mol^{-1}	B_{22}/cm^{3}mol^{-1}	B_{12}/cm^{3}mol^{-1}	V_1/cm^{3}mol^{-1}	V_2/cm^{3}mol^{-1}
-1400	-1390	-930	61	88

$A(1) = 0.9208E+00$ $A(2) = 0.9505E-01$ $A(3) = 0.3947E-01$

---

1. Ethanol, C_2H_6O
2. Ethyl ethanoate, $C_4H_8O_2$

***T*/K = 328.15**
***System Number*: 67**

x_1	y_1	P/kPa	G^E/J mol^{-1}	μ^E_1/J mol^{-1}	μ^E_2/J mol^{-1}
0.0	0.000	46.05	0	2565	0
0.1	0.160	49.92	231	2077	26
0.2	0.262	52.02	410	1641	103
0.3	0.336	53.00	539	1257	231
0.4	0.395	53.23	616	923	410
0.5	0.448	52.89	641	641	641
0.6	0.503	52.02	616	410	923
0.7	0.566	50.44	539	231	1257
0.8	0.650	47.87	410	103	1641
0.9	0.776	43.77	231	26	2077
1.0	1.000	37.29	0	0	2565

B_{11}/cm^{3}mol^{-1}	B_{22}/cm^{3}mol^{-1}	B_{12}/cm^{3}mol^{-1}	V_1/cm^{3}mol^{-1}	V_2/cm^{3}mol^{-1}
-1300	-1550	-1200	61	102

$A(1) = 0.9400E+00$ $A(2) = 0.0000E+00$ $A(3) = 0.0000E+00$

---

1. Ethanol, C_2H_6O
2. Pyrrolidine, C_4H_9N

***T*/K = 333.35**
***System Number*: 294**

x_1	y_1	P/kPa	G^E/J mol^{-1}	μ^E_1/J mol^{-1}	μ^E_2/J mol^{-1}
0.0	0.000	40.23	0	-2376	0
0.1	0.053	38.15	-237	-2341	-3
0.2	0.118	36.01	-463	-2185	-32
0.3	0.208	33.89	-661	-1915	-124
0.4	0.332	32.12	-815	-1561	-317
0.5	0.493	31.27	-904	-1165	-644
0.6	0.667	31.97	-914	-774	-1124
0.7	0.816	34.51	-831	-435	-1755
0.8	0.916	38.52	-649	-185	-2507
0.9	0.972	43.10	-369	-41	-3313
1.0	1.000	47.33	0	0	-4063

B_{11}/cm^{3}mol^{-1}	B_{22}/cm^{3}mol^{-1}	B_{12}/cm^{3}mol^{-1}	V_1/cm^{3}mol^{-1}	V_2/cm^{3}mol^{-1}
-1200	-1050	-840	61	87

$A(1) = -0.1305E+01$ $A(2) = -0.3763E+00$ $A(3) = 0.1436E+00$ $A(4) = 0.7209E-01$ $A(5) = 0.0000E+00$

TABLE 3.1.1 (VLEG)
ISOTHERMAL VAPOR-LIQUID EQUILIBRIA AND EXCESS GIBBS ENERGIES OF BINARY MIXTURES (continued)

1. Ethanol, C_2H_6O
2. 1-Butanol, $C_4H_{10}O$

T/K = 313.15
***System Number*: 315**

x_1	y_1	P/kPa	G^E/J mol^{-1}	μ^E_1/J mol^{-1}	μ^E_2/J mol^{-1}
0.0	0.000	2.52	0	125	0
0.1	0.449	4.12	12	119	0
0.2	0.646	5.72	24	108	2
0.3	0.756	7.30	33	94	7
0.4	0.827	8.85	41	77	16
0.5	0.876	10.38	45	59	31
0.6	0.913	11.90	46	41	53
0.7	0.941	13.39	43	25	83
0.8	0.964	14.88	34	12	123
0.9	0.983	16.38	20	3	173
1.0	1.000	17.90	0	0	236

B_{11}/cm^{3}mol^{-1}	B_{22}/cm^{3}mol^{-1}	B_{12}/cm^{3}mol^{-1}	V_1/cm^{3}mol^{-1}	V_2/cm^{3}mol^{-1}
-2100	-3560	-2800	60	95

$A(1)$ = 0.6935E-01 $A(2)$ = 0.2134E-01 $A(3)$ = 0.0000E+00

---

1. Ethanol, C_2H_6O
2. 2-Methyl-2-propanol, $C_4H_{10}O$

T/K = 313.15
***System Number*: 317**

x_1	y_1	P/kPa	G^E/J mol^{-1}	μ^E_1/J mol^{-1}	μ^E_2/J mol^{-1}
0.0	0.000	13.79	0	-517	0
0.1	0.108	13.89	-49	-455	-4
0.2	0.220	14.04	-90	-380	-17
0.3	0.334	14.26	-121	-302	-43
0.4	0.450	14.56	-141	-225	-84
0.5	0.563	14.94	-149	-156	-141
0.6	0.670	15.41	-144	-98	-212
0.7	0.768	15.96	-126	-53	-296
0.8	0.856	16.56	-95	-22	-388
0.9	0.933	17.22	-53	-5	-484
1.0	1.000	17.90	0	0	-578

B_{11}/cm^{3}mol^{-1}	B_{22}/cm^{3}mol^{-1}	B_{12}/cm^{3}mol^{-1}	V_1/cm^{3}mol^{-1}	V_2/cm^{3}mol^{-1}
-2100	-2530	-2300	60	98

$A(1)$ = -0.2284E+00 $A(2)$ = -0.1162E-01 $A(3)$ = 0.1810E-01

---

1. Ethanol, C_2H_6O
2. Chlorobenzene, C_6H_5Cl

T/K = 298.15
***System Number*: 282**

x_1	y_1	P/kPa	G^E/J mol^{-1}	μ^E_1/J mol^{-1}	μ^E_2/J mol^{-1}
0.0	0.000	1.644	0	6852	0
0.1	0.736	5.895	535	4222	125
0.2	0.764	6.468	856	2827	364
0.3	0.781	6.803	1045	2000	636
0.4	0.795	7.051	1136	1418	948
0.5	0.805	7.221	1141	957	1324
0.6	0.818	7.373	1063	594	1768
0.7	0.837	7.550	908	329	2259
0.8	0.869	7.745	679	153	2783
0.9	0.918	7.902	381	45	3400
1.0	1.000	7.914	0	0	4337

B_{11}/cm^{3}mol^{-1}	B_{22}/cm^{3}mol^{-1}	B_{12}/cm^{3}mol^{-1}	V_1/cm^{3}mol^{-1}	V_2/cm^{3}mol^{-1}
-2170	-3130	-1190	59	102

$A(1)$ = 0.1841E+01 $A(2)$ = -0.2962E+00 $A(3)$ = 0.1789E+00 $A(4)$ = -0.2110E+00 $A(5)$ = 0.2372E+00

TABLE 3.1.1 (VLEG)
ISOTHERMAL VAPOR-LIQUID EQUILIBRIA AND EXCESS GIBBS ENERGIES OF BINARY MIXTURES (continued)

1. Ethanol, C_2H_6O
2. Aniline, C_6H_7N

***T*/K = 313.15**
***System Number*: 291**

x_1	y_1	*P*/kPa	G^E/J mol^{-1}	μ^E_1/J mol^{-1}	μ^E_2/J mol^{-1}
0.0	0.000	0.25	0	2478	0
0.1	0.944	4.06	223	2008	24
0.2	0.970	6.81	398	1614	94
0.3	0.979	8.90	527	1273	207
0.4	0.984	10.53	611	973	369
0.5	0.987	11.86	647	707	587
0.6	0.989	13.01	634	476	870
0.7	0.992	14.07	567	283	1229
0.8	0.994	15.16	443	134	1679
0.9	0.996	16.40	256	36	2238
1.0	1.000	17.93	0	0	2931

B_{11}/cm^{3}mol^{-1}	B_{22}/cm^{3}mol^{-1}	B_{12}/cm^{3}mol^{-1}	V_1/cm^{3}mol^{-1}	V_2/cm^{3}mol^{-1}
-2100	-3350	-1200	60	93

A(1) = 0.9940E+00 *A*(2) = 0.9195E-01 *A*(3) = 0.3908E-01 *A*(4) = -0.5005E-02 *A*(5) = 0.5744E-02

1. Ethanol, C_2H_6O
2. Hexane, C_6H_{14}

***T*/K = 313.15**
***System Number*: 218**

x_1	y_1	*P*/kPa	G^E/J mol^{-1}	μ^E_1/J mol^{-1}	μ^E_2/J mol^{-1}
0.0	0.000	37.27	0	7829	0
0.1	0.272	48.68	636	5193	129
0.2	0.288	49.29	1041	3567	410
0.3	0.290	49.30	1285	2527	752
0.4	0.294	49.23	1406	1812	1135
0.5	0.301	48.98	1424	1278	1571
0.6	0.310	48.44	1348	854	2089
0.7	0.325	47.35	1176	511	2727
0.8	0.355	44.78	902	246	3525
0.9	0.440	37.91	516	68	4544
1.0	1.000	17.90	0	0	5883

B_{11}/cm^{3}mol^{-1}	B_{22}/cm^{3}mol^{-1}	B_{12}/cm^{3}mol^{-1}	V_1/cm^{3}mol^{-1}	V_2/cm^{3}mol^{-1}
-2100	-1670	-940	60	135

A(1) = 0.2188E+01 *A*(2) = -0.2246E+00 *A*(3) = 0.3760E+00 *A*(4) = -0.1490E+00 *A*(5) = 0.6900E-01

1. Ethanol, C_2H_6O
2. Nonane, C_9H_{20}

***T*/K = 343.21**
***System Number*: 220**

x_1	y_1	*P*/kPa	G^E/J mol^{-1}	μ^E_1/J mol^{-1}	μ^E_2/J mol^{-1}
0.0	0.000	6.33	0	7793	0
0.1	0.884	52.48	641	5325	120
0.2	0.903	61.25	1069	3838	377
0.3	0.909	64.72	1347	2856	701
0.4	0.914	66.73	1506	2133	1089
0.5	0.917	67.96	1558	1556	1560
0.6	0.919	68.86	1505	1081	2142
0.7	0.923	69.74	1344	686	2877
0.8	0.928	70.73	1061	361	3860
0.9	0.942	71.94	631	113	5298
1.0	1.000	72.39	0	0	7598

B_{11}/cm^{3}mol^{-1}	B_{22}/cm^{3}mol^{-1}	B_{12}/cm^{3}mol^{-1}	V_1/cm^{3}mol^{-1}	V_2/cm^{3}mol^{-1}
-1150	-3750	-1200	62	189

A(1) = 0.2184E+01 *A*(2) = -0.3068E-02 *A*(3) = 0.3560E+00 *A*(4) = -0.3106E-01 *A*(5) = 0.1565E+00

TABLE 3.1.1 (VLEG) ISOTHERMAL VAPOR-LIQUID EQUILIBRIA AND EXCESS GIBBS ENERGIES OF BINARY MIXTURES (continued)

1. Ethanol, C_2H_6O
2. 1-Decanol, $C_{10}H_{22}O$

***T*/K = 298.15**
***System Number*: 20**

x_1	y_1	P/kPa	G^E/J mol^{-1}	μ^E_1/J mol^{-1}	μ^E_2/J mol^{-1}
0.0	0.000	0.003	0	603	0
0.1	0.997	1.005	62	612	0
0.2	0.999	1.965	120	557	11
0.3	0.999	2.859	170	481	36
0.4	0.999	3.697	208	404	78
0.5	1.000	4.490	234	331	138
0.6	1.000	5.237	246	259	226
0.7	1.000	5.931	239	184	368
0.8	1.000	6.569	206	105	608
0.9	1.000	7.185	133	34	1022
1.0	1.000	7.879	0	0	1718

B_{11}/cm^{3}mol^{-1}	B_{22}/cm^{3}mol^{-1}	B_{12}/cm^{3}mol^{-1}	V_1/cm^{3}mol^{-1}	V_2/cm^{3}mol^{-1}
-2170	0	0	59	0

$A(1)$ = 0.3782E+00 $A(2)$ = 0.1555E+00 $A(3)$ = 0.9010E-01 $A(4)$ = 0.6940E-01 $A(5)$ = 0.0000E+00

---

1. Dimethyl sulfoxide, C_2H_6OS
2. N,N-Dimethylmethanamide, C_3H_7NO

***T*/K = 373.15**
***System Number*: 345**

x_1	y_1	P/kPa	G^E/J mol^{-1}	μ^E_1/J mol^{-1}	μ^E_2/J mol^{-1}
0.0	0.000	18.69	0	739	0
0.1	0.034	17.48	58	448	14
0.2	0.069	16.27	90	272	44
0.3	0.109	15.01	106	175	76
0.4	0.156	13.69	112	125	103
0.5	0.215	12.33	111	99	123
0.6	0.288	10.95	107	81	146
0.7	0.382	9.57	98	61	184
0.8	0.506	8.18	81	36	260
0.9	0.686	6.73	51	12	401
1.0	1.000	5.11	0	0	643

B_{11}/cm^{3}mol^{-1}	B_{22}/cm^{3}mol^{-1}	B_{12}/cm^{3}mol^{-1}	V_1/cm^{3}mol^{-1}	V_2/cm^{3}mol^{-1}
-2140	-2000	-2070	78	82

$A(1)$ = 0.1435E+00 $A(2)$ = -0.1555E-01 $A(3)$ = 0.7932E-01

---

1. Dimethyl sulfoxide, C_2H_6OS
2. Benzene, C_6H_6

***T*/K = 298.15**
***System Number*: 21**

x_1	y_1	P/kPa	G^E/J mol^{-1}	μ^E_1/J mol^{-1}	μ^E_2/J mol^{-1}
0.0	0.000	12.65	0	4476	0
0.1	0.002	11.74	368	3038	72
0.2	0.003	11.16	606	2087	236
0.3	0.004	10.63	747	1451	446
0.4	0.005	10.03	813	1009	682
0.5	0.006	9.30	816	686	946
0.6	0.007	8.41	762	437	1250
0.7	0.009	7.29	653	245	1606
0.8	0.012	5.76	490	107	2021
0.9	0.021	3.50	271	25	2482
1.0	1.000	0.08	0	0	2950

B_{11}/cm^{3}mol^{-1}	B_{22}/cm^{3}mol^{-1}	B_{12}/cm^{3}mol^{-1}	V_1/cm^{3}mol^{-1}	V_2/cm^{3}mol^{-1}
-5970	-1480	-3520	72	89

$A(1)$ = 0.1316E+01 $A(2)$ = -0.2098E+00 $A(3)$ = 0.1815E+00 $A(4)$ = -0.9802E-01 $A(5)$ = 0.0000E+00

TABLE 3.1.1 (VLEG)
ISOTHERMAL VAPOR-LIQUID EQUILIBRIA AND EXCESS GIBBS ENERGIES OF BINARY MIXTURES (continued)

1. Dimethyl sulfoxide, C_2H_6OS
2. Toluene, C_7H_8

T/K = 323.15
System Number: 332

x_1	y_1	P/kPa	G^E/J mol^{-1}	μ^E_1/J mol^{-1}	μ^E_2/J mol^{-1}
0.0	0.000	12.53	0	5358	0
0.1	0.015	11.77	455	3872	76
0.2	0.020	11.31	768	2754	271
0.3	0.023	10.99	959	1922	546
0.4	0.025	10.66	1048	1311	873
0.5	0.028	10.20	1051	865	1236
0.6	0.032	9.48	977	541	1632
0.7	0.039	8.41	834	306	2068
0.8	0.051	6.83	625	140	2564
0.9	0.085	4.40	349	37	3151
1.0	1.000	0.41	0	0	3874

B_{11}/cm^{3}mol^{-1}	B_{22}/cm^{3}mol^{-1}	B_{12}/cm^{3}mol^{-1}	V_1/cm^{3}mol^{-1}	V_2/cm^{3}mol^{-1}
-4010	-2080	-3140	74	110

$A(1) = 0.1565E+01$ $A(2) = -0.2761E+00$ $A(3) = 0.1534E+00$

1. 1,2-Ethanediol, $C_2H_6O_2$
2. 2-Propanone, C_3H_6O

T/K = 323.15
System Number: 324

x_1	y_1	P/kPa	G^E/J mol^{-1}	μ^E_1/J mol^{-1}	μ^E_2/J mol^{-1}
0.0	0.000	81.83	0	5478	0
0.1	0.001	75.49	468	4018	74
0.2	0.001	71.81	798	2946	261
0.3	0.001	69.27	1012	2144	527
0.4	0.001	67.05	1125	1529	856
0.5	0.001	64.60	1149	1048	1249
0.6	0.001	61.31	1086	668	1713
0.7	0.002	56.14	940	376	2257
0.8	0.002	47.08	710	166	2885
0.9	0.003	30.41	396	41	3595
1.0	1.000	0.11	0	0	4364

B_{11}/cm^{3}mol^{-1}	B_{22}/cm^{3}mol^{-1}	B_{12}/cm^{3}mol^{-1}	V_1/cm^{3}mol^{-1}	V_2/cm^{3}mol^{-1}
-1970	-1430	-1480	57	77

$A(1) = 0.1710E+01$ $A(2) = -0.1499E+00$ $A(3) = 0.1214E+00$ $A(4) = -0.5722E-01$ $A(5) = 0.0000E+00$

1. Ethylamine, C_2H_7N
2. Butane, C_4H_{10}

T/K = 293.15
System Number: 152

x_1	y_1	P/kPa	G^E/J mol^{-1}	μ^E_1/J mol^{-1}	μ^E_2/J mol^{-1}
0.0	0.000	207.0	0	2775	0
0.1	0.135	218.9	245	2174	31
0.2	0.216	222.9	430	1678	118
0.3	0.274	222.8	559	1269	254
0.4	0.322	220.0	634	931	436
0.5	0.367	215.0	658	653	663
0.6	0.414	207.7	632	427	939
0.7	0.471	197.1	555	248	1272
0.8	0.551	181.4	426	115	1673
0.9	0.686	156.8	243	30	2155
1.0	1.000	116.4	0	0	2736

B_{11}/cm^{3}mol^{-1}	B_{22}/cm^{3}mol^{-1}	B_{12}/cm^{3}mol^{-1}	V_1/cm^{3}mol^{-1}	V_2/cm^{3}mol^{-1}
-820	-770	-790	66	100

$A(1) = 0.1080E+01$ $A(2) = -0.8088E-02$ $A(3) = 0.5053E-01$

TABLE 3.1.1 (VLEG) ISOTHERMAL VAPOR-LIQUID EQUILIBRIA AND EXCESS GIBBS ENERGIES OF BINARY MIXTURES (continued)

1. Ethylamine, C_2H_7N
2. Hexane, C_6H_{14}

***T*/K = 293.15**
***System Number*: 153**

x_1	y_1	P/kPa	G^E/J mol^{-1}	μ^E_1/J mol^{-1}	μ^E_2/J mol^{-1}
0.0	0.000	16.1	0	2649	0
0.1	0.643	42.1	237	2124	27
0.2	0.765	59.4	420	1683	105
0.3	0.819	71.8	553	1310	229
0.4	0.851	81.1	637	991	401
0.5	0.875	88.4	671	718	625
0.6	0.894	94.5	655	483	912
0.7	0.912	99.9	584	289	1274
0.8	0.933	105.1	456	137	1731
0.9	0.959	110.6	264	37	2303
1.0	1.000	116.4	0	0	3019

B_{11}/cm^{3}mol^{-1}	B_{22}/cm^{3}mol^{-1}	B_{12}/cm^{3}mol^{-1}	V_1/cm^{3}mol^{-1}	V_2/cm^{3}mol^{-1}
-820	-2080	-1350	66	131

$A(1) = 0.1102E+01$ $A(2) = 0.7592E-01$ $A(3) = 0.6130E-01$

---

1. Propanenitrile, C_3H_5N
2. Ethylbenzene, C_8H_{10}

***T*/K = 353.15**
***System Number*: 169**

x_1	y_1	P/kPa	G^E/J mol^{-1}	μ^E_1/J mol^{-1}	μ^E_2/J mol^{-1}
0.0	0.000	16.91	0	2831	0
0.1	0.428	27.18	243	2091	38
0.2	0.575	33.77	415	1545	133
0.3	0.659	38.65	529	1137	268
0.4	0.718	42.61	591	823	436
0.5	0.766	46.01	607	576	638
0.6	0.809	49.01	579	378	881
0.7	0.850	51.71	506	220	1173
0.8	0.892	54.19	387	102	1529
0.9	0.940	56.48	220	27	1958
1.0	1.000	58.55	0	0	2470

B_{11}/cm^{3}mol^{-1}	B_{22}/cm^{3}mol^{-1}	B_{12}/cm^{3}mol^{-1}	V_1/cm^{3}mol^{-1}	V_2/cm^{3}mol^{-1}
-2270	-2130	-2200	77	131

$A(1) = 0.8270E+00$ $A(2) = -0.4273E-01$ $A(3) = 0.7579E-01$ $A(4) = -0.1878E-01$ $A(5) = 0.0000E+00$

---

1. Propene, C_3H_6
2. Propane, C_3H_8

***T*/K = 293.15**
***System Number*: 81**

x_1	y_1	P/kPa	G^E/J mol^{-1}	μ^E_1/J mol^{-1}	μ^E_2/J mol^{-1}
0.0	0.000	836.0	0	108	0
0.1	0.120	859.2	11	108	0
0.2	0.234	882.1	21	95	3
0.3	0.341	903.7	29	76	9
0.4	0.443	923.8	34	54	21
0.5	0.540	942.2	36	34	37
0.6	0.634	959.0	33	18	57
0.7	0.727	974.5	28	7	77
0.8	0.819	989.0	20	1	93
0.9	0.910	1003.0	10	0	101
1.0	1.000	1017.0	0	0	93

B_{11}/cm^{3}mol^{-1}	B_{22}/cm^{3}mol^{-1}	B_{12}/cm^{3}mol^{-1}	V_1/cm^{3}mol^{-1}	V_2/cm^{3}mol^{-1}
-360	-393	-377	82	88

$A(1) = 0.5840E-01$ $A(2) = -0.3103E-02$ $A(3) = -0.1725E-01$

TABLE 3.1.1 (VLEG) ISOTHERMAL VAPOR-LIQUID EQUILIBRIA AND EXCESS GIBBS ENERGIES OF BINARY MIXTURES (continued)

1. 1,3-Dibromopropane, $C_3H_6Br_2$
2. Nonane, C_9H_{20}

T/K = 348.15
System Number: 112

x_1	y_1	*P*/kPa	G^E/J mol^{-1}	μ^E_1/J mol^{-1}	μ^E_2/J mol^{-1}
0.0	0.000	7.85	0	3333	0
0.1	0.143	8.34	300	2704	33
0.2	0.232	8.55	536	2179	125
0.3	0.297	8.60	711	1730	275
0.4	0.348	8.55	827	1338	486
0.5	0.391	8.42	881	991	771
0.6	0.432	8.23	869	683	1148
0.7	0.477	7.94	785	417	1643
0.8	0.537	7.48	620	203	2291
0.9	0.650	6.60	363	56	3132
1.0	1.000	4.67	0	0	4213

B_{11}/cm^{3}mol^{-1}	B_{22}/cm^{3}mol^{-1}	B_{12}/cm^{3}mol^{-1}	V_1/cm^{3}mol^{-1}	V_2/cm^{3}mol^{-1}
-2740	-3640	-2600	107	190

A(1) = 0.1217E+01 A(2) = 0.1520E+00 A(3) = 0.8680E-01

1. 2-Propanone, C_3H_6O
2. 1-Chlorobutane, C_4H_9Cl

T/K = 298.16
System Number: 273

x_1	y_1	*P*/kPa	G^E/J mol^{-1}	μ^E_1/J mol^{-1}	μ^E_2/J mol^{-1}
0.0	0.000	13.66	0	1285	0
0.1	0.273	17.03	117	1056	12
0.2	0.434	19.75	208	844	50
0.3	0.543	21.98	275	655	112
0.4	0.626	23.85	317	492	200
0.5	0.695	25.47	334	354	314
0.6	0.755	26.88	325	238	455
0.7	0.812	28.12	290	143	631
0.8	0.868	29.20	226	70	854
0.9	0.928	30.13	132	19	1141
1.0	1.000	30.85	0	0	1521

B_{11}/cm^{3}mol^{-1}	B_{22}/cm^{3}mol^{-1}	B_{12}/cm^{3}mol^{-1}	V_1/cm^{3}mol^{-1}	V_2/cm^{3}mol^{-1}
-2040	-1650	-1300	74	105

A(1) = 0.5383E+00 A(2) = 0.3216E-01 A(3) = 0.2751E-01 A(4) = 0.1544E-01 A(5) = 0.0000E+00

1. 2-Propanone, C_3H_6O
2. Diethyl ether, $C_4H_{10}O$

T/K = 298.06
System Number: 320

x_1	y_1	*P*/kPa	G^E/J mol^{-1}	μ^E_1/J mol^{-1}	μ^E_2/J mol^{-1}
0.0	0.000	71.25	0	2092	0
0.1	0.081	70.52	180	1555	27
0.2	0.142	68.95	309	1150	98
0.3	0.194	66.84	393	840	201
0.4	0.244	64.27	438	597	331
0.5	0.298	61.18	447	407	487
0.6	0.358	57.48	422	257	670
0.7	0.433	52.96	364	143	882
0.8	0.536	47.30	274	62	1122
0.9	0.697	40.06	153	15	1389
1.0	1.000	30.67	0	0	1673

B_{11}/cm^{3}mol^{-1}	B_{22}/cm^{3}mol^{-1}	B_{12}/cm^{3}mol^{-1}	V_1/cm^{3}mol^{-1}	V_2/cm^{3}mol^{-1}
-2050	-1240	-870	74	105

A(1) = 0.7211E+00 A(2) = -0.6490E-01 A(3) = 0.3850E-01 A(4) = -0.1960E-01 A(5) = 0.0000E+00

TABLE 3.1.1 (VLEG)
ISOTHERMAL VAPOR-LIQUID EQUILIBRIA AND EXCESS GIBBS ENERGIES OF BINARY MIXTURES (continued)

1. 2-Propanone, C_3H_6O
2. Diethylamine, $C_4H_{11}N$

T/K = 298.03
System Number: 287

x_1	y_1	P/kPa	G^E/J mol^{-1}	μ^E_1/J mol^{-1}	μ^E_2/J mol^{-1}
0.0	0.000	31.37	0	1798	0
0.1	0.157	33.86	156	1354	23
0.2	0.263	35.28	268	1009	83
0.3	0.346	36.09	343	741	172
0.4	0.419	36.47	383	532	284
0.5	0.490	36.52	393	369	417
0.6	0.562	36.26	374	241	573
0.7	0.640	35.65	327	141	759
0.8	0.730	34.65	250	66	984
0.9	0.842	33.08	142	18	1261
1.0	1.000	30.65	0	0	1608

B_{11}/cm^{3}mol^{-1}	B_{22}/cm^{3}mol^{-1}	B_{12}/cm^{3}mol^{-1}	V_1/cm^{3}mol^{-1}	V_2/cm^{3}mol^{-1}
-2040	-1430	-1840	74	104

$A(1) = 0.6345E+00$ $A(2) = -0.3850E-01$ $A(3) = 0.5270E-01$

1. 2-Propanone, C_3H_6O
2. Pyridine, C_5H_5N

T/K = 303.15
System Number: 300

x_1	y_1	P/kPa	G^E/J mol^{-1}	μ^E_1/J mol^{-1}	μ^E_2/J mol^{-1}
0.0	0.000	3.65	0	666	0
0.1	0.577	7.83	57	497	9
0.2	0.744	11.60	99	371	31
0.3	0.826	15.13	126	276	62
0.4	0.876	18.53	142	204	101
0.5	0.910	21.84	147	147	147
0.6	0.936	25.08	142	100	204
0.7	0.956	28.28	126	61	276
0.8	0.973	31.46	98	30	370
0.9	0.987	34.67	57	9	495
1.0	1.000	37.98	0	0	663

B_{11}/cm^{3}mol^{-1}	B_{22}/cm^{3}mol^{-1}	B_{12}/cm^{3}mol^{-1}	V_1/cm^{3}mol^{-1}	V_2/cm^{3}mol^{-1}
-1850	-2300	-1650	74	81

$A(1) = 0.2333E+00$ $A(2) = -0.6901E-03$ $A(3) = 0.3027E-01$

1. Propanal, C_3H_6O
2. Pentane, C_5H_{12}

T/K = 313.15
System Number: 242

x_1	y_1	P/kPa	G^E/J mol^{-1}	μ^E_1/J mol^{-1}	μ^E_2/J mol^{-1}
0.0	0.000	115.4	0	3865	0
0.1	0.179	129.6	334	2894	49
0.2	0.263	134.8	575	2182	173
0.3	0.319	136.5	740	1648	350
0.4	0.364	136.3	837	1232	574
0.5	0.404	134.7	872	894	850
0.6	0.445	131.8	845	610	1198
0.7	0.490	127.3	753	371	1644
0.8	0.552	119.9	588	179	2221
0.9	0.666	106.0	341	49	2967
1.0	1.000	76.1	0	0	3920

B_{11}/cm^{3}mol^{-1}	B_{22}/cm^{3}mol^{-1}	B_{12}/cm^{3}mol^{-1}	V_1/cm^{3}mol^{-1}	V_2/cm^{3}mol^{-1}
-1230	-1100	-780	84	116

$A(1) = 0.1340E+01$ $A(2) = 0.3336E-01$ $A(3) = 0.1551E+00$ $A(4) = -0.2270E-01$ $A(5) = 0.0000E+00$

TABLE 3.1.1 (VLEG)
ISOTHERMAL VAPOR-LIQUID EQUILIBRIA AND EXCESS GIBBS ENERGIES OF BINARY MIXTURES (continued)

1. 2-Propanone, C_3H_6O **T/K = 313.15**
2. Chlorobenzene, C_6H_5Cl ***System Number*: 283**

x_1	y_1	P/kPa	G^E/J mol^{-1}	μ^E_1/J mol^{-1}	μ^E_2/J mol^{-1}
0.0	0.000	3.56	0	1355	0
0.1	0.711	11.24	114	975	19
0.2	0.832	17.51	195	722	63
0.3	0.886	23.13	248	545	121
0.4	0.917	28.35	280	409	195
0.5	0.939	33.24	291	295	288
0.6	0.955	37.87	281	195	410
0.7	0.968	42.34	248	112	565
0.8	0.979	46.83	190	49	753
0.9	0.990	51.53	107	12	966
1.0	1.000	56.60	0	0	1181

B_{11}/cm^{3}mol^{-1}	B_{22}/cm^{3}mol^{-1}	B_{12}/cm^{3}mol^{-1}	V_1/cm^{3}mol^{-1}	V_2/cm^{3}mol^{-1}
-1620	-2970	-1600	76	104

$A(1)$ = 0.4476E+00 $A(2)$ = 0.4927E-02 $A(3)$ = 0.3932E-01 $A(4)$ = -0.3837E-01 $A(5)$ = 0.0000E+00

1. 2-Propanone, C_3H_6O **T/K = 298.15**
2. Benzene, C_6H_6 ***System Number*: 247**

x_1	y_1	P/kPa	G^E/J mol^{-1}	μ^E_1/J mol^{-1}	μ^E_2/J mol^{-1}
0.0	0.000	12.67	0	1366	0
0.1	0.284	16.07	118	1015	18
0.2	0.440	18.64	201	743	65
0.3	0.546	20.74	255	535	135
0.4	0.630	22.54	282	376	220
0.5	0.700	24.16	287	254	319
0.6	0.763	25.66	270	162	431
0.7	0.823	27.05	233	93	560
0.8	0.880	28.35	176	43	710
0.9	0.938	29.57	99	11	889
1.0	1.000	30.72	0	0	1107

B_{11}/cm^{3}mol^{-1}	B_{22}/cm^{3}mol^{-1}	B_{12}/cm^{3}mol^{-1}	V_1/cm^{3}mol^{-1}	V_2/cm^{3}mol^{-1}
-2050	-1480	-1600	74	89

$A(1)$ = 0.4624E+00 $A(2)$ = -0.5217E-01 $A(3)$ = 0.3654E-01

1. 2-Propanone, C_3H_6O **T/K = 298.15**
2. Cyclohexane, C_6H_{12} ***System Number*: 246**

x_1	y_1	P/kPa	G^E/J mol^{-1}	μ^E_1/J mol^{-1}	μ^E_2/J mol^{-1}
0.0	0.000	12.99	0	5400	0
0.1	0.549	26.91	458	3907	75
0.2	0.625	31.00	777	2851	259
0.3	0.655	32.67	984	2088	511
0.4	0.675	33.62	1097	1518	817
0.5	0.692	34.25	1126	1071	1182
0.6	0.709	34.68	1075	710	1624
0.7	0.730	34.93	942	417	2168
0.8	0.764	34.88	723	194	2839
0.9	0.832	34.00	411	51	3657
1.0	1.000	30.72	0	0	4630

B_{11}/cm^{3}mol^{-1}	B_{22}/cm^{3}mol^{-1}	B_{12}/cm^{3}mol^{-1}	V_1/cm^{3}mol^{-1}	V_2/cm^{3}mol^{-1}
-2050	-1740	-1360	74	108

$A(1)$ = 0.1818E+01 $A(2)$ = -0.8975E-01 $A(3)$ = 0.2055E+00 $A(4)$ = -0.6564E-01 $A(5)$ = 0.0000E+00

TABLE 3.1.1 (VLEG) ISOTHERMAL VAPOR-LIQUID EQUILIBRIA AND EXCESS GIBBS ENERGIES OF BINARY MIXTURES (continued)

1. 2-Propanone, C_3H_6O
2. Heptane, C_7H_{16}

***T*/K = 338.15**
***System Number*: 245**

x_1	y_1	P/kPa	G^E/J mol^{-1}	μ^E_1/J mol^{-1}	μ^E_2/J mol^{-1}
0.0	0.000	33.8	0	4600	0
0.1	0.566	73.2	385	3247	67
0.2	0.670	91.0	648	2356	222
0.3	0.726	103.0	821	1773	414
0.4	0.769	113.1	926	1374	627
0.5	0.805	121.9	973	1070	876
0.6	0.834	128.8	963	801	1206
0.7	0.857	133.5	887	539	1697
0.8	0.878	136.5	723	290	2456
0.9	0.909	138.3	441	88	3625
1.0	1.000	135.2	0	0	5376

B_{11}/cm^{3}mol^{-1}	B_{22}/cm^{3}mol^{-1}	B_{12}/cm^{3}mol^{-1}	V_1/cm^{3}mol^{-1}	V_2/cm^{3}mol^{-1}
-1230	-1960	-920	79	155

$A(1) = 0.1384E+01$ $A(2) = 0.1380E+00$ $A(3) = 0.3900E+00$

---

1. Methyl ethanoate, $C_3H_6O_2$
2. Pentane, C_5H_{12}

***T*/K = 298.15**
***System Number*: 265**

x_1	y_1	P/kPa	G^E/J mol^{-1}	μ^E_1/J mol^{-1}	μ^E_2/J mol^{-1}
0.0	0.000	68.31	0	3860	0
0.1	0.135	72.49	340	3011	44
0.2	0.203	73.41	596	2328	164
0.3	0.246	73.01	776	1774	348
0.4	0.277	71.97	883	1320	592
0.5	0.304	70.45	922	944	899
0.6	0.331	68.35	892	632	1282
0.7	0.365	65.25	791	376	1757
0.8	0.418	60.09	614	179	2351
0.9	0.533	50.19	353	48	3097
1.0	1.000	28.81	0	0	4038

B_{11}/cm^{3}mol^{-1}	B_{22}/cm^{3}mol^{-1}	B_{12}/cm^{3}mol^{-1}	V_1/cm^{3}mol^{-1}	V_2/cm^{3}mol^{-1}
-1650	-1260	-982	80	116

$A(1) = 0.1487E+01$ $A(2) = 0.3591E-01$ $A(3) = 0.1056E+00$

---

1. Methyl ethanoate, $C_3H_6O_2$
2. Benzene, C_6H_6

***T*/K = 303.15**
***System Number*: 30**

x_1	y_1	P/kPa	G^E/J mol^{-1}	μ^E_1/J mol^{-1}	μ^E_2/J mol^{-1}
0.0	0.000	15.91	0	825	0
0.1	0.245	19.05	75	684	7
0.2	0.408	21.80	135	554	31
0.3	0.526	24.22	180	434	71
0.4	0.618	26.38	208	326	129
0.5	0.694	28.32	219	231	206
0.6	0.760	30.08	212	151	304
0.7	0.821	31.71	188	87	424
0.8	0.880	33.24	145	39	567
0.9	0.938	34.68	82	10	734
1.0	1.000	36.05	0	0	926

B_{11}/cm^{3}mol^{-1}	B_{22}/cm^{3}mol^{-1}	B_{12}/cm^{3}mol^{-1}	V_1/cm^{3}mol^{-1}	V_2/cm^{3}mol^{-1}
-1600	-1420	-1130	80	90

$A(1) = 0.3472E+00$ $A(2) = 0.2011E-01$ $A(3) = 0.0000E+00$

TABLE 3.1.1 (VLEG)
ISOTHERMAL VAPOR-LIQUID EQUILIBRIA AND EXCESS GIBBS ENERGIES OF BINARY MIXTURES (continued)

1. 1,3-Dioxolane, $C_3H_6O_2$ — ***T*/K = 313.15**
2. Cyclohexane, C_6H_{12} — ***System Number*: 64**

x_1	y_1	*P*/kPa	G^E/J mol^{-1}	μ^E_1/J mol^{-1}	μ^E_2/J mol^{-1}
0.0	0.000	24.62	0	4002	0
0.1	0.281	31.46	354	3135	45
0.2	0.391	34.64	621	2430	169
0.3	0.451	36.20	808	1858	359
0.4	0.493	36.97	922	1392	609
0.5	0.527	37.29	966	1009	922
0.6	0.561	37.26	939	690	1313
0.7	0.597	36.85	839	423	1810
0.8	0.646	35.83	658	209	2457
0.9	0.736	33.35	385	59	3317
1.0	1.000	26.56	0	0	4474

B_{11}/cm^{3}mol^{-1}	B_{22}/cm^{3}mol^{-1}	B_{12}/cm^{3}mol^{-1}	V_1/cm^{3}mol^{-1}	V_2/cm^{3}mol^{-1}
-1910	-1490	-1290	71	111

A(1) = 0.1484E+01 *A*(2) = 0.6660E-01 *A*(3) = 0.1442E+00 *A*(4) = 0.2400E-01 *A*(5) = 0.0000E+00

1. Methyl ethanoate, $C_3H_6O_2$ — ***T*/K = 308.15**
2. Cyclohexane, C_6H_{12} — ***System Number*: 22**

x_1	y_1	*P*/kPa	G^E/J mol^{-1}	μ^E_1/J mol^{-1}	μ^E_2/J mol^{-1}
0.0	0.000	20.07	0	3752	0
0.1	0.443	33.09	341	3088	36
0.2	0.573	40.18	609	2455	147
0.3	0.632	43.96	802	1891	335
0.4	0.667	46.11	921	1411	594
0.5	0.694	47.48	966	1013	919
0.6	0.719	48.42	939	687	1318
0.7	0.748	49.01	839	421	1813
0.8	0.785	49.14	658	210	2451
0.9	0.847	48.30	385	60	3309
1.0	1.000	44.32	0	0	4502

B_{11}/cm^{3}mol^{-1}	B_{22}/cm^{3}mol^{-1}	B_{12}/cm^{3}mol^{-1}	V_1/cm^{3}mol^{-1}	V_2/cm^{3}mol^{-1}
-1550	-1560	-1210	80	110

A(1) = 0.1508E+01 *A*(2) = 0.7312E-01 *A*(3) = 0.1025E+00 *A*(4) = 0.7316E-01 *A*(5) = 0.0000E+00

1. Methyl ethanoate, $C_3H_6O_2$ — ***T*/K = 323.15**
2. 1-Hexene, C_6H_{12} — ***System Number*: 266**

x_1	y_1	*P*/kPa	G^E/J mol^{-1}	μ^E_1/J mol^{-1}	μ^E_2/J mol^{-1}
0.0	0.000	64.69	0	2234	0
0.1	0.214	75.02	207	1910	17
0.2	0.348	82.39	377	1588	75
0.3	0.439	87.39	508	1276	179
0.4	0.507	90.59	595	982	338
0.5	0.563	92.44	636	713	559
0.6	0.614	93.23	625	476	849
0.7	0.667	92.98	560	279	1216
0.8	0.733	91.37	436	129	1667
0.9	0.830	87.47	251	33	2209
1.0	1.000	79.33	0	0	2850

B_{11}/cm^{3}mol^{-1}	B_{22}/cm^{3}mol^{-1}	B_{12}/cm^{3}mol^{-1}	V_1/cm^{3}mol^{-1}	V_2/cm^{3}mol^{-1}
-1300	-1380	-1300	81	131

A(1) = 0.9462E+00 *A*(2) = 0.1147E+00 *A*(3) = 0.0000E+00

TABLE 3.1.1 (VLEG) ISOTHERMAL VAPOR-LIQUID EQUILIBRIA AND EXCESS GIBBS ENERGIES OF BINARY MIXTURES (continued)

1. 1,3-Dioxolane, $C_3H_6O_2$
2. Heptane, C_7H_{16}

***T*/K = 313.15**
***System Number*: 65**

x_1	y_1	P/kPa	G^E/J mol^{-1}	μ^E_1/J mol^{-1}	μ^E_2/J mol^{-1}
0.0	0.000	12.24	0	4389	0
0.1	0.470	21.35	391	3483	48
0.2	0.584	25.56	688	2701	185
0.3	0.634	27.61	896	2054	400
0.4	0.665	28.77	1021	1532	680
0.5	0.690	29.55	1068	1114	1022
0.6	0.714	30.08	1040	774	1438
0.7	0.740	30.38	933	489	1969
0.8	0.772	30.39	739	252	2687
0.9	0.828	29.77	438	75	3709
1.0	1.000	26.56	0	0	5201

B_{11}/cm^{3}mol^{-1}	B_{22}/cm^{3}mol^{-1}	B_{12}/cm^{3}mol^{-1}	V_1/cm^{3}mol^{-1}	V_2/cm^{3}mol^{-1}
-1910	-2450	-1600	71	150

$A(1)$ = 0.1641E+01 $A(2)$ = 0.7100E-01 $A(3)$ = 0.2010E+00 $A(4)$ = 0.8500E-01 $A(5)$ = 0.0000E+00

1. Dimethyl carbonate, $C_3H_6O_3$
2. Cyclohexane, C_6H_{12}

***T*/K = 298.15**
***System Number*: 271**

x_1	y_1	P/kPa	G^E/J mol^{-1}	μ^E_1/J mol^{-1}	μ^E_2/J mol^{-1}
0.0	0.000	13.01	0	5801	0
0.1	0.248	16.11	495	4243	79
0.2	0.303	16.73	843	3111	276
0.3	0.324	16.84	1070	2287	549
0.4	0.338	16.83	1196	1677	875
0.5	0.352	16.71	1233	1208	1258
0.6	0.367	16.50	1187	831	1720
0.7	0.386	16.14	1053	516	2306
0.8	0.415	15.46	824	259	3085
0.9	0.491	13.62	481	74	4147
1.0	1.000	7.19	0	0	5602

B_{11}/cm^{3}mol^{-1}	B_{22}/cm^{3}mol^{-1}	B_{12}/cm^{3}mol^{-1}	V_1/cm^{3}mol^{-1}	V_2/cm^{3}mol^{-1}
-2000	-1740	-1400	85	109

$A(1)$ = 0.1990E+01 $A(2)$ = -0.4000E-01 $A(3)$ = 0.3100E+00

1. 1-Chloropropane, C_3H_7Cl
2. Benzene, C_6H_6

***T*/K = 308.15**
***System Number*: 127**

x_1	y_1	P/kPa	G^E/J mol^{-1}	μ^E_1/J mol^{-1}	μ^E_2/J mol^{-1}
0.0	0.000	19.78	0	-465	0
0.1	0.251	23.72	-36	-270	-10
0.2	0.445	28.26	-54	-137	-33
0.3	0.589	33.14	-58	-53	-60
0.4	0.696	38.16	-54	-5	-86
0.5	0.777	43.18	-43	16	-103
0.6	0.840	48.13	-31	21	-108
0.7	0.890	52.99	-18	17	-100
0.8	0.932	57.79	-8	9	-76
0.9	0.968	62.56	-2	3	-38
1.0	1.000	67.36	0	0	11

B_{11}/cm^{3}mol^{-1}	B_{22}/cm^{3}mol^{-1}	B_{12}/cm^{3}mol^{-1}	V_1/cm^{3}mol^{-1}	V_2/cm^{3}mol^{-1}
-980	-1360	-1170	90	90

$A(1)$ = -0.6760E-01 $A(2)$ = 0.9300E-01 $A(3)$ = -0.2100E-01

TABLE 3.1.1 (VLEG)
ISOTHERMAL VAPOR-LIQUID EQUILIBRIA AND EXCESS GIBBS ENERGIES OF BINARY MIXTURES (continued)

1. 1-Chloropropane, C_3H_7Cl
2. Cyclohexane, C_6H_{12}

***T*/K = 308.15**
***System Number*: 126**

x_1	y_1	P/kPa	G^E/J mol^{-1}	μ^E_1/J mol^{-1}	μ^E_2/J mol^{-1}
0.0	0.000	20.08	0	777	0
0.1	0.329	27.06	77	754	2
0.2	0.514	33.59	148	674	17
0.3	0.631	39.40	206	561	55
0.4	0.711	44.45	248	434	124
0.5	0.771	48.86	268	308	227
0.6	0.821	52.83	263	197	363
0.7	0.866	56.53	234	108	529
0.8	0.909	60.13	179	45	716
0.9	0.954	63.72	100	10	912
1.0	1.000	67.36	0	0	1102

B_{11}/cm^{3}mol^{-1}	B_{22}/cm^{3}mol^{-1}	B_{12}/cm^{3}mol^{-1}	V_1/cm^{3}mol^{-1}	V_2/cm^{3}mol^{-1}
-980	-1560	-1250	90	110

$A(1) = 0.4177E+00$ $A(2) = 0.6350E-01$ $A(3) = -0.5100E-01$

---

1. 2-Fluoropropane, C_3H_7F
2. Butane, C_4H_{10}

***T*/K = 303.15**
***System Number*: 143**

x_1	y_1	P/kPa	G^E/J mol^{-1}	μ^E_1/J mol^{-1}	μ^E_2/J mol^{-1}
0.0	0.000	402.0	0	1307	0
0.1	0.111	410.1	118	1073	12
0.2	0.203	413.0	212	859	50
0.3	0.283	411.8	280	667	114
0.4	0.356	407.0	322	496	206
0.5	0.427	398.9	338	349	327
0.6	0.501	387.2	326	226	477
0.7	0.583	371.4	288	129	658
0.8	0.681	350.4	221	58	871
0.9	0.811	322.4	125	15	1116
1.0	1.000	285.0	0	0	1396

B_{11}/cm^{3}mol^{-1}	B_{22}/cm^{3}mol^{-1}	B_{12}/cm^{3}mol^{-1}	V_1/cm^{3}mol^{-1}	V_2/cm^{3}mol^{-1}
-800	-700	-750	78	102

$A(1) = 0.5362E+00$ $A(2) = 0.1779E-01$ $A(3) = 0.0000E+00$

---

1. N,N-Dimethylmethanamide, C_3H_7NO
2. 2-Propanol, C_3H_8O

***T*/K = 353.15**
***System Number*: 302**

x_1	y_1	P/kPa	G^E/J mol^{-1}	μ^E_1/J mol^{-1}	μ^E_2/J mol^{-1}
0.0	0.000	92.49	0	-229	0
0.1	0.009	83.81	-28	-300	3
0.2	0.021	75.01	-57	-282	-1
0.3	0.036	65.93	-81	-215	-24
0.4	0.058	56.70	-94	-131	-69
0.5	0.088	47.63	-93	-55	-132
0.6	0.132	38.98	-79	0	-198
0.7	0.195	30.93	-55	26	-245
0.8	0.294	23.44	-28	26	-242
0.9	0.475	16.20	-5	11	-150
1.0	1.000	8.49	0	0	79

B_{11}/cm^{3}mol^{-1}	B_{22}/cm^{3}mol^{-1}	B_{12}/cm^{3}mol^{-1}	V_1/cm^{3}mol^{-1}	V_2/cm^{3}mol^{-1}
-2360	-1100	-940	82	84

$A(1) = -0.1271E+00$ $A(2) = 0.5248E-01$ $A(3) = 0.1016E+00$

TABLE 3.1.1 (VLEG)
ISOTHERMAL VAPOR-LIQUID EQUILIBRIA AND EXCESS GIBBS ENERGIES OF BINARY MIXTURES (continued)

1. N,N-Dimethylmethanamide, C_3H_7NO
2. 2-Methylpropanal, C_4H_8O

***T*/K = 333.15**
***System Number*: 303**

x_1	y_1	P/kPa	G^E/J mol^{-1}	μ^E_1/J mol^{-1}	μ^E_2/J mol^{-1}
0.0	0.000	87.79	0	1171	0
0.1	0.006	79.56	104	917	13
0.2	0.013	71.89	181	699	52
0.3	0.020	64.53	234	516	113
0.4	0.028	57.22	262	365	194
0.5	0.038	49.76	268	244	293
0.6	0.052	41.94	253	150	408
0.7	0.073	33.57	217	81	536
0.8	0.113	24.46	162	34	674
0.9	0.211	14.43	90	8	822
1.0	1.000	3.35	0	0	975

B_{11}/cm^{3}mol^{-1}	B_{22}/cm^{3}mol^{-1}	B_{12}/cm^{3}mol^{-1}	V_1/cm^{3}mol^{-1}	V_2/cm^{3}mol^{-1}
-2850	-1280	-1560	80	97

$A(1) = 0.3875E+00$ $A(2) = -0.3545E-01$ $A(3) = 0.0000E+00$

---

1. N,N-Dimethylmethanamide, C_3H_7NO
2. Ethyl ethanoate, $C_4H_8O_2$

***T*/K = 333.15**
***System Number*: 304**

x_1	y_1	P/kPa	G^E/J mol^{-1}	μ^E_1/J mol^{-1}	μ^E_2/J mol^{-1}
0.0	0.000	55.69	0	1680	0
0.1	0.011	50.90	148	1301	20
0.2	0.021	46.54	257	981	76
0.3	0.031	42.40	330	715	165
0.4	0.043	38.27	368	499	281
0.5	0.057	33.94	374	328	420
0.6	0.075	29.25	350	198	578
0.7	0.103	24.00	299	105	751
0.8	0.152	18.03	222	44	934
0.9	0.272	11.19	121	10	1122
1.0	1.000	3.35	0	0	1311

B_{11}/cm^{3}mol^{-1}	B_{22}/cm^{3}mol^{-1}	B_{12}/cm^{3}mol^{-1}	V_1/cm^{3}mol^{-1}	V_2/cm^{3}mol^{-1}
-2850	-1400	-1480	80	103

$A(1) = 0.5400E+00$ $A(2) = -0.6664E-01$ $A(3) = 0.0000E+00$

---

1. N,N-Dimethylmethanamide, C_3H_7NO
2. Benzene, C_6H_6

***T*/K = 303.15**
***System Number*: 188**

x_1	y_1	P/kPa	G^E/J mol^{-1}	μ^E_1/J mol^{-1}	μ^E_2/J mol^{-1}
0.0	0.000	15.9	0	2367	0
0.1	0.009	14.7	188	1483	44
0.2	0.016	13.6	297	932	139
0.3	0.022	12.5	353	596	249
0.4	0.030	11.3	371	391	358
0.5	0.041	9.9	362	260	464
0.6	0.056	8.4	331	168	576
0.7	0.078	6.8	280	98	706
0.8	0.117	5.0	209	45	865
0.9	0.214	3.1	116	12	1057
1.0	1.000	0.7	0	0	1276

B_{11}/cm^{3}mol^{-1}	B_{22}/cm^{3}mol^{-1}	B_{12}/cm^{3}mol^{-1}	V_1/cm^{3}mol^{-1}	V_2/cm^{3}mol^{-1}
-4000	-1420	-2490	78	90

$A(1) = 0.5745E+00$ $A(2) = -0.1625E+00$ $A(3) = 0.1481E+00$ $A(4) = -0.5404E-01$ $A(5) = 0.0000E+00$

TABLE 3.1.1 (VLEG) ISOTHERMAL VAPOR-LIQUID EQUILIBRIA AND EXCESS GIBBS ENERGIES OF BINARY MIXTURES (continued)

1. N,N-Dimethylmethanamide, C_3H_7NO
2. Butyl ethanoate, $C_6H_{12}O_2$

***T*/K = 363.15**
***System Number*: 305**

x_1	y_1	P/kPa	G^E/J mol^{-1}	μ^E_1/J mol^{-1}	μ^E_2/J mol^{-1}
0.0	0.000	31.47	0	2121	0
0.1	0.073	30.78	188	1676	23
0.2	0.134	29.88	332	1309	88
0.3	0.188	28.82	433	1004	189
0.4	0.241	27.61	495	749	326
0.5	0.295	26.23	518	535	501
0.6	0.355	24.63	502	356	720
0.7	0.428	22.73	445	211	991
0.8	0.525	20.37	345	99	1326
0.9	0.680	17.27	198	27	1741
1.0	1.000	12.90	0	0	2256

B_{11}/cm^{3}mol^{-1}	B_{22}/cm^{3}mol^{-1}	B_{12}/cm^{3}mol^{-1}	V_1/cm^{3}mol^{-1}	V_2/cm^{3}mol^{-1}
-2160	-1920	-1430	83	146

$A(1) = 0.6863E+00$ $A(2) = 0.2230E-01$ $A(3) = 0.3850E-01$

---

1. N,N-Dimethylmethanamide, C_3H_7NO
2. Toluene, C_7H_8

***T*/K = 313.15**
***System Number*: 190**

x_1	y_1	P/kPa	G^E/J mol^{-1}	μ^E_1/J mol^{-1}	μ^E_2/J mol^{-1}
0.0	0.000	7.893	0	2580	0
0.1	0.036	7.469	220	1878	36
0.2	0.062	7.071	372	1338	130
0.3	0.084	6.673	465	928	266
0.4	0.107	6.241	507	624	429
0.5	0.133	5.743	506	401	611
0.6	0.167	5.150	467	241	805
0.7	0.216	4.441	394	130	1011
0.8	0.297	3.588	291	56	1230
0.9	0.460	2.559	160	14	1470
1.0	1.000	1.300	0	0	1739

B_{11}/cm^{3}mol^{-1}	B_{22}/cm^{3}mol^{-1}	B_{12}/cm^{3}mol^{-1}	V_1/cm^{3}mol^{-1}	V_2/cm^{3}mol^{-1}
-3500	-2270	-2840	79	108

$A(1) = 0.7769E+00$ $A(2) = -0.1615E+00$ $A(3) = 0.5264E-01$

---

1. 1-Nitropropane, $C_3H_7NO_2$
2. Chlorobenzene, C_6H_5Cl

***T*/K = 348.15**
***System Number*: 201**

x_1	y_1	P/kPa	G^E/J mol^{-1}	μ^E_1/J mol^{-1}	μ^E_2/J mol^{-1}
0.0	0.000	15.80	0	2013	0
0.1	0.140	16.75	164	1328	34
0.2	0.234	17.17	265	869	114
0.3	0.314	17.35	320	572	211
0.4	0.392	17.36	341	385	311
0.5	0.473	17.26	337	266	407
0.6	0.558	17.05	313	184	507
0.7	0.648	16.74	272	120	628
0.8	0.745	16.33	211	64	797
0.9	0.856	15.74	123	20	1053
1.0	1.000	14.85	0	0	1448

B_{11}/cm^{3}mol^{-1}	B_{22}/cm^{3}mol^{-1}	B_{12}/cm^{3}mol^{-1}	V_1/cm^{3}mol^{-1}	V_2/cm^{3}mol^{-1}
-3150	-1560	-2250	94	107

$A(1) = 0.4654E+00$ $A(2) = -0.9766E-01$ $A(3) = 0.1325E+00$

TABLE 3.1.1 (VLEG)
ISOTHERMAL VAPOR-LIQUID EQUILIBRIA AND EXCESS GIBBS ENERGIES OF BINARY MIXTURES (continued)

1. 1-Propanol, C_3H_8O
2. Heptane, C_7H_{16}

***T*/K = 323.00**
***System Number*: 222**

x_1	y_1	P/kPa	G^E/J mol^{-1}	μ^E_1/J mol^{-1}	μ^E_2/J mol^{-1}
0.0	0.000	18.94	0	8247	0
0.1	0.285	25.31	622	4765	161
0.2	0.310	25.78	980	3185	429
0.3	0.339	26.02	1197	2355	701
0.4	0.361	26.04	1316	1754	1025
0.5	0.371	25.94	1341	1223	1460
0.6	0.378	25.81	1267	769	2015
0.7	0.397	25.25	1094	429	2644
0.8	0.448	23.55	827	206	3311
0.9	0.562	20.05	471	66	4119
1.0	1.000	12.22	0	0	5508

B_{11}/cm^{3}mol^{-1}	B_{22}/cm^{3}mol^{-1}	B_{12}/cm^{3}mol^{-1}	V_1/cm^{3}mol^{-1}	V_2/cm^{3}mol^{-1}
-2170	-2240	-1150	77	152

$A(1)$ = 0.1998E+01 $A(2)$ = -0.1764E+00 $A(3)$ = 0.1380E+00 $A(4)$ = -0.3334E+00 $A(5)$ = 0.4249E+00

1. 2,4-Dioxapentane, $C_3H_8O_2$
2. Hexane, C_6H_{14}

***T*/K = 313.15**
***System Number*: 212**

x_1	y_1	P/kPa	G^E/J mol^{-1}	μ^E_1/J mol^{-1}	μ^E_2/J mol^{-1}
0.0	0.000	37.32	0	1858	0
0.1	0.329	50.83	169	1549	16
0.2	0.494	61.39	307	1279	64
0.3	0.596	69.77	412	1037	145
0.4	0.668	76.47	485	815	264
0.5	0.723	81.80	521	611	432
0.6	0.770	86.03	519	425	660
0.7	0.813	89.38	472	261	966
0.8	0.858	91.99	376	127	1371
0.9	0.915	93.78	221	35	1898
1.0	1.000	94.06	0	0	2575

B_{11}/cm^{3}mol^{-1}	B_{22}/cm^{3}mol^{-1}	B_{12}/cm^{3}mol^{-1}	V_1/cm^{3}mol^{-1}	V_2/cm^{3}mol^{-1}
-940	-1670	-1210	92	134

$A(1)$ = 0.8010E+00 $A(2)$ = 0.1378E+00 $A(3)$ = 0.5030E-01

1. Propylamine, C_3H_9N
2. Diethylamine, $C_4H_{11}N$

***T*/K = 297.97**
***System Number*: 183**

x_1	y_1	P/kPa	G^E/J mol^{-1}	μ^E_1/J mol^{-1}	μ^E_2/J mol^{-1}
0.0	0.000	31.40	0	163	0
0.1	0.135	32.73	15	132	2
0.2	0.258	33.99	26	104	7
0.3	0.370	35.18	34	80	15
0.4	0.474	36.32	39	59	26
0.5	0.572	37.41	41	41	41
0.6	0.664	38.44	39	26	59
0.7	0.752	39.44	34	15	80
0.8	0.837	40.39	26	7	104
0.9	0.919	41.30	15	2	132
1.0	1.000	42.18	0	0	163

B_{11}/cm^{3}mol^{-1}	B_{22}/cm^{3}mol^{-1}	B_{12}/cm^{3}mol^{-1}	V_1/cm^{3}mol^{-1}	V_2/cm^{3}mol^{-1}
-1100	-1430	-1240	83	104

$A(1)$ = 0.6581E-01 $A(2)$ = -0.6000E-04 $A(3)$ = 0.0000E+00

TABLE 3.1.1 (VLEG)
ISOTHERMAL VAPOR-LIQUID EQUILIBRIA AND EXCESS GIBBS ENERGIES OF BINARY MIXTURES (continued)

1. Propylamine, C_3H_9N
2. Hexane, C_6H_{14}

***T*/K = 293.15**
***System Number*: 154**

x_1	y_1	*P*/kPa	G^E/J mol^{-1}	μ^E_1/J mol^{-1}	μ^E_2/J mol^{-1}
0.0	0.000	16.23	0	2353	0
0.1	0.327	22.03	209	1867	25
0.2	0.473	25.84	370	1464	96
0.3	0.561	28.48	484	1123	209
0.4	0.623	30.37	552	832	366
0.5	0.673	31.76	576	583	570
0.6	0.719	32.79	555	376	823
0.7	0.766	33.56	487	212	1128
0.8	0.822	34.06	371	93	1484
0.9	0.896	34.17	209	23	1883
1.0	1.000	33.68	0	0	2313

B_{11}/cm^{3}mol^{-1}	B_{22}/cm^{3}mol^{-1}	B_{12}/cm^{3}mol^{-1}	V_1/cm^{3}mol^{-1}	V_2/cm^{3}mol^{-1}
-1160	-2080	-1580	82	131

A(1) = 0.9460E+00 *A*(2) = 0.1097E-01 *A*(3) = 0.1120E-01 *A*(4) = -0.1931E-01 *A*(5) = 0.0000E+00

1. Furan, C_4H_4O
2. Oxolane, C_4H_8O

***T*/K = 303.15**
***System Number*: 327**

x_1	y_1	*P*/kPa	G^E/J mol^{-1}	μ^E_1/J mol^{-1}	μ^E_2/J mol^{-1}
0.0	0.000	27.12	0	-897	0
0.1	0.226	31.49	-80	-711	-10
0.2	0.414	36.67	-141	-554	-37
0.3	0.564	42.59	-183	-421	-82
0.4	0.683	49.17	-209	-310	-142
0.5	0.775	56.32	-217	-217	-217
0.6	0.847	63.94	-209	-141	-310
0.7	0.902	71.93	-183	-81	-421
0.8	0.944	80.15	-140	-37	-554
0.9	0.976	88.42	-80	-10	-710
1.0	1.000	96.52	0	0	-894

B_{11}/cm^{3}mol^{-1}	B_{22}/cm^{3}mol^{-1}	B_{12}/cm^{3}mol^{-1}	V_1/cm^{3}mol^{-1}	V_2/cm^{3}mol^{-1}
-830	-1280	-1120	73	82

A(1) = -0.3445E+00 *A*(2) = 0.6030E-03 *A*(3) = -0.1069E-01

1. Thiophene, C_4H_4S
2. Benzene, C_6H_6

***T*/K = 328.15**
***System Number*: 341**

x_1	y_1	*P*/kPa	G^E/J mol^{-1}	μ^E_1/J mol^{-1}	μ^E_2/J mol^{-1}
0.0	0.000	44.47	0	254	0
0.1	0.092	44.12	23	202	3
0.2	0.183	43.69	40	165	9
0.3	0.275	43.21	54	137	18
0.4	0.368	42.68	64	113	31
0.5	0.463	42.10	70	91	50
0.6	0.559	41.45	72	68	78
0.7	0.658	40.72	68	45	121
0.8	0.762	39.88	56	23	186
0.9	0.873	38.87	34	7	282
1.0	1.000	37.61	0	0	418

B_{11}/cm^{3}mol^{-1}	B_{22}/cm^{3}mol^{-1}	B_{12}/cm^{3}mol^{-1}	V_1/cm^{3}mol^{-1}	V_2/cm^{3}mol^{-1}
-1200	-1160	-1180	82	93

A(1) = 0.1030E+00 *A*(2) = 0.3001E-01 *A*(3) = 0.2008E-01

TABLE 3.1.1 (VLEG)
ISOTHERMAL VAPOR-LIQUID EQUILIBRIA AND EXCESS GIBBS ENERGIES OF BINARY MIXTURES (continued)

1. Thiophene, C_4H_4S
2. Heptane, C_7H_{16}

***T*/K = 328.15**
***System Number*: 340**

x_1	y_1	*P*/kPa	G^E/J mol^{-1}	μ^E_1/J mol^{-1}	μ^E_2/J mol^{-1}
0.0	0.000	23.09	0	1649	0
0.1	0.228	27.15	151	1369	15
0.2	0.370	30.21	270	1095	64
0.3	0.475	32.61	358	869	138
0.4	0.561	34.61	417	699	230
0.5	0.637	36.34	452	568	337
0.6	0.705	37.77	463	453	478
0.7	0.766	38.83	444	331	708
0.8	0.821	39.40	381	195	1125
0.9	0.881	39.35	247	65	1885
1.0	1.000	37.61	0	0	3214

B_{11}/cm^{3}mol^{-1}	B_{22}/cm^{3}mol^{-1}	B_{12}/cm^{3}mol^{-1}	V_1/cm^{3}mol^{-1}	V_2/cm^{3}mol^{-1}
-1200	-2140	-1620	82	154

A(1) = 0.6630E+00 *A*(2) = 0.1696E+00 *A*(3) = 0.2281E+00 *A*(4) = 0.1171E+00 *A*(5) = 0.0000E+00

1. Ethanoic anhydride, $C_4H_6O_3$
2. Ethyl ethanoate, $C_4H_8O_2$

***T*/K = 348.15**
***System Number*: 331**

x_1	y_1	*P*/kPa	G^E/J mol^{-1}	μ^E_1/J mol^{-1}	μ^E_2/J mol^{-1}
0.0	0.000	94.77	0	909	0
0.1	0.016	86.70	77	653	13
0.2	0.032	78.94	130	476	44
0.3	0.051	71.24	164	346	87
0.4	0.074	63.48	183	247	140
0.5	0.102	55.59	186	167	205
0.6	0.140	47.51	175	103	284
0.7	0.194	39.08	150	53	375
0.8	0.282	30.15	111	21	473
0.9	0.460	20.59	60	4	566
1.0	1.000	10.45	0	0	631

B_{11}/cm^{3}mol^{-1}	B_{22}/cm^{3}mol^{-1}	B_{12}/cm^{3}mol^{-1}	V_1/cm^{3}mol^{-1}	V_2/cm^{3}mol^{-1}
-1470	-1230	-1380	101	106

A(1) = 0.2571E+00 *A*(2) = -0.2650E-01 *A*(3) = 0.8942E-02 *A*(4) = -0.2146E-01 *A*(5) = 0.0000E+00

1. cis-2-Butene, C_4H_8
2. Butane, C_4H_{10}

***T*/K = 298.15**
***System Number*: 82**

x_1	y_1	*P*/kPa	G^E/J mol^{-1}	μ^E_1/J mol^{-1}	μ^E_2/J mol^{-1}
0.0	0.000	243.6	0	117	0
0.1	0.094	241.8	12	114	0
0.2	0.188	239.8	22	105	2
0.3	0.282	237.8	32	92	6
0.4	0.377	235.6	39	76	15
0.5	0.473	233.1	44	58	29
0.6	0.570	230.3	45	41	51
0.7	0.669	227.0	42	25	80
0.8	0.772	223.3	34	12	120
0.9	0.882	219.0	20	3	170
1.0	1.000	213.9	0	0	234

B_{11}/cm^{3}mol^{-1}	B_{22}/cm^{3}mol^{-1}	B_{12}/cm^{3}mol^{-1}	V_1/cm^{3}mol^{-1}	V_2/cm^{3}mol^{-1}
-748	-740	-719	91	102

A(1) = 0.7076E-01 *A*(2) = 0.2347E-01 *A*(3) = 0.0000E+00

TABLE 3.1.1 (VLEG)
ISOTHERMAL VAPOR-LIQUID EQUILIBRIA AND EXCESS GIBBS ENERGIES OF BINARY MIXTURES (continued)

1. trans-2-Butene, C_4H_8
2. Butane, C_4H_{10}

***T*/K = 298.15**
***System Number*: 83**

x_1	y_1	*P*/kPa	G^E/J mol^{-1}	μ^E_1/J mol^{-1}	μ^E_2/J mol^{-1}
0.0	0.000	243.6	0	82	0
0.1	0.100	243.5	8	86	0
0.2	0.199	243.4	17	83	0
0.3	0.298	243.3	25	76	3
0.4	0.396	243.0	31	64	9
0.5	0.494	242.4	36	51	20
0.6	0.591	241.6	37	36	38
0.7	0.688	240.4	35	23	64
0.8	0.788	238.8	29	11	99
0.9	0.891	236.6	17	3	144
1.0	1.000	233.8	0	0	203

B_{11}/cm^{3}mol^{-1}	B_{22}/cm^{3}mol^{-1}	B_{12}/cm^{3}mol^{-1}	V_1/cm^{3}mol^{-1}	V_2/cm^{3}mol^{-1}
-767	-740	-728	94	102

A(1) = 0.5735E-01 *A*(2) = 0.2435E-01 *A*(3) = 0.0000E+00

1. Oxolane, C_4H_8O
2. Ethyl ethanoate, $C_4H_8O_2$

***T*/K = 313.15**
***System Number*: 325**

x_1	y_1	*P*/kPa	G^E/J mol^{-1}	μ^E_1/J mol^{-1}	μ^E_2/J mol^{-1}
0.0	0.000	25.05	0	327	0
0.1	0.163	27.01	29	265	3
0.2	0.300	28.83	52	209	13
0.3	0.417	30.53	69	160	29
0.4	0.521	32.13	79	118	52
0.5	0.614	33.63	82	82	82
0.6	0.700	35.06	79	52	118
0.7	0.780	36.42	69	29	160
0.8	0.856	37.73	52	13	209
0.9	0.929	38.99	29	3	265
1.0	1.000	40.20	0	0	327

B_{11}/cm^{3}mol^{-1}	B_{22}/cm^{3}mol^{-1}	B_{12}/cm^{3}mol^{-1}	V_1/cm^{3}mol^{-1}	V_2/cm^{3}mol^{-1}
-1120	-1800	-1160	83	101

A(1) = 0.1257E+00 *A*(2) = 0.0000E+00 *A*(3) = 0.0000E+00

1. 2-Methylpropanal, C_4H_8O
2. Ethyl ethanoate, $C_4H_8O_2$

***T*/K = 313.15**
***System Number*: 326**

x_1	y_1	*P*/kPa	G^E/J mol^{-1}	μ^E_1/J mol^{-1}	μ^E_2/J mol^{-1}
0.0	0.000	25.05	0	400	0
0.1	0.170	27.27	34	283	6
0.2	0.308	29.26	56	200	20
0.3	0.426	31.11	70	141	40
0.4	0.531	32.86	77	100	62
0.5	0.625	34.56	78	69	87
0.6	0.710	36.20	74	47	114
0.7	0.789	37.80	64	29	148
0.8	0.863	39.35	50	14	191
0.9	0.933	40.86	29	4	250
1.0	1.000	42.31	0	0	331

B_{11}/cm^{3}mol^{-1}	B_{22}/cm^{3}mol^{-1}	B_{12}/cm^{3}mol^{-1}	V_1/cm^{3}mol^{-1}	V_2/cm^{3}mol^{-1}
-1540	-1800	-1200	94	101

A(1) = 0.1199E+00 *A*(2) = -0.1318E-01 *A*(3) = 0.2042E-01

TABLE 3.1.1 (VLEG)
ISOTHERMAL VAPOR-LIQUID EQUILIBRIA AND EXCESS GIBBS ENERGIES OF BINARY MIXTURES (continued)

1. Oxolane, C_4H_8O
2. Pyrrolidine, C_4H_9N

***T*/K = 333.35**
***System Number*: 293**

x_1	y_1	*P*/kPa	G^E/J mol^{-1}	μ^E_1/J mol^{-1}	μ^E_2/J mol^{-1}
0.0	0.000	40.23	0	559	0
0.1	0.211	46.06	50	452	6
0.2	0.366	51.38	89	358	22
0.3	0.487	56.26	117	274	50
0.4	0.587	60.78	134	201	89
0.5	0.672	65.03	140	140	140
0.6	0.747	69.04	134	89	201
0.7	0.815	72.88	117	50	274
0.8	0.879	76.58	89	22	358
0.9	0.940	80.17	50	6	452
1.0	1.000	83.69	0	0	559

B_{11}/cm^{3}mol^{-1}	B_{22}/cm^{3}mol^{-1}	B_{12}/cm^{3}mol^{-1}	V_1/cm^{3}mol^{-1}	V_2/cm^{3}mol^{-1}
-960	-1050	-940	84	87

A(1) = 0.2015E+00 *A*(2) = 0.0000E+00 *A*(3) = 0.0000E+00

---

1. 2-Butanone, C_4H_8O
2. Benzene, C_6H_6

***T*/K = 313.15**
***System Number*: 262**

x_1	y_1	*P*/kPa	G^E/J mol^{-1}	μ^E_1/J mol^{-1}	μ^E_2/J mol^{-1}
0.0	0.000	24.38	0	716	0
0.1	0.114	24.89	59	480	12
0.2	0.212	25.16	95	314	41
0.3	0.303	25.26	115	200	78
0.4	0.393	25.24	121	125	119
0.5	0.485	25.14	117	76	158
0.6	0.579	24.96	105	45	195
0.7	0.677	24.72	87	25	232
0.8	0.779	24.43	64	12	272
0.9	0.886	24.08	35	3	321
1.0	1.000	23.65	0	0	387

B_{11}/cm^{3}mol^{-1}	B_{22}/cm^{3}mol^{-1}	B_{12}/cm^{3}mol^{-1}	V_1/cm^{3}mol^{-1}	V_2/cm^{3}mol^{-1}
-2000	-1300	-1550	92	91

A(1) = 0.1797E+00 *A*(2) = -0.6307E-01 *A*(3) = 0.3208E-01

---

1. 2-Butanone, C_4H_8O
2. 1,4-Cyclohexadiene, C_6H_8

***T*/K = 313.15**
***System Number*: 261**

x_1	y_1	*P*/kPa	G^E/J mol^{-1}	μ^E_1/J mol^{-1}	μ^E_2/J mol^{-1}
0.0	0.000	17.40	0	2284	0
0.1	0.218	20.31	194	1648	32
0.2	0.336	21.98	327	1168	116
0.3	0.419	23.03	407	807	236
0.4	0.490	23.74	443	538	380
0.5	0.557	24.22	441	341	541
0.6	0.626	24.50	405	199	713
0.7	0.700	24.60	339	102	893
0.8	0.785	24.50	248	41	1075
0.9	0.884	24.19	133	9	1253
1.0	1.000	23.66	0	0	1419

B_{11}/cm^{3}mol^{-1}	B_{22}/cm^{3}mol^{-1}	B_{12}/cm^{3}mol^{-1}	V_1/cm^{3}mol^{-1}	V_2/cm^{3}mol^{-1}
-2000	-1470	-1520	92	96

A(1) = 0.6771E+00 *A*(2) = -0.1533E+00 *A*(3) = 0.3408E-01 *A*(4) = -0.1290E-01 *A*(5) = 0.0000E+00

TABLE 3.1.1 (VLEG)
ISOTHERMAL VAPOR-LIQUID EQUILIBRIA AND EXCESS GIBBS ENERGIES OF BINARY MIXTURES (continued)

1. 2-Butanone, C_4H_8O
2. Cyclohexene, C_6H_{10}

***T*/K = 313.15**
***System Number*: 260**

x_1	y_1	*P*/kPa	G^E/J mol^{-1}	μ^E_1/J mol^{-1}	μ^E_2/J mol^{-1}
0.0	0.000	22.62	0	3288	0
0.1	0.220	26.63	279	2362	47
0.2	0.318	28.37	468	1675	167
0.3	0.382	29.21	585	1173	333
0.4	0.437	29.62	640	808	528
0.5	0.492	29.72	643	542	745
0.6	0.551	29.50	601	346	983
0.7	0.620	28.95	517	201	1252
0.8	0.704	27.97	390	95	1571
0.9	0.817	26.36	220	26	1967
1.0	1.000	23.65	0	0	2475

B_{11}/cm^{3}mol^{-1}	B_{22}/cm^{3}mol^{-1}	B_{12}/cm^{3}mol^{-1}	V_1/cm^{3}mol^{-1}	V_2/cm^{3}mol^{-1}
-2000	-1470	-1520	92	104

A(1) = 0.9883E+00 *A*(2) = -0.1561E+00 *A*(3) = 0.1183E+00

1. Oxolane, C_4H_8O
2. Cyclohexane, C_6H_{12}

***T*/K = 313.15**
***System Number*: 214**

x_1	y_1	*P*/kPa	G^E/J mol^{-1}	μ^E_1/J mol^{-1}	μ^E_2/J mol^{-1}
0.0	0.000	24.61	0	1604	0
0.1	0.224	28.81	141	1248	18
0.2	0.362	31.85	247	954	70
0.3	0.463	34.15	319	712	150
0.4	0.544	35.94	360	515	257
0.5	0.615	37.37	371	355	387
0.6	0.683	38.50	354	227	543
0.7	0.751	39.38	308	129	725
0.8	0.822	39.99	234	59	937
0.9	0.903	40.30	132	15	1184
1.0	1.000	40.20	0	0	1473

B_{11}/cm^{3}mol^{-1}	B_{22}/cm^{3}mol^{-1}	B_{12}/cm^{3}mol^{-1}	V_1/cm^{3}mol^{-1}	V_2/cm^{3}mol^{-1}
-1120	-1490	-1200	83	111

A(1) = 0.5700E+00 *A*(2) = -0.2520E-01 *A*(3) = 0.2100E-01

1. 2-Butanone, C_4H_8O
2. Cyclohexane, C_6H_{12}

***T*/K = 313.15**
***System Number*: 249**

x_1	y_1	*P*/kPa	G^E/J mol^{-1}	μ^E_1/J mol^{-1}	μ^E_2/J mol^{-1}
0.0	0.000	24.61	0	4116	0
0.1	0.249	30.24	352	3008	57
0.2	0.339	32.32	596	2172	203
0.3	0.390	33.18	751	1545	410
0.4	0.430	33.52	827	1078	660
0.5	0.470	33.52	837	729	945
0.6	0.516	33.15	786	467	1264
0.7	0.572	32.33	678	270	1630
0.8	0.647	30.86	514	127	2061
0.9	0.765	28.30	290	34	2589
1.0	1.000	23.66	0	0	3253

B_{11}/cm^{3}mol^{-1}	B_{22}/cm^{3}mol^{-1}	B_{12}/cm^{3}mol^{-1}	V_1/cm^{3}mol^{-1}	V_2/cm^{3}mol^{-1}
-2000	-1490	-1560	92	111

A(1) = 0.1286E+01 *A*(2) = -0.1657E+00 *A*(3) = 0.1295E+00

TABLE 3.1.1 (VLEG)
ISOTHERMAL VAPOR-LIQUID EQUILIBRIA AND EXCESS GIBBS ENERGIES OF BINARY MIXTURES (continued)

1. 2-Butanone, C_4H_8O
2. 1-Hexene, C_6H_{12}

***T*/K = 333.15**
***System Number*: 250**

x_1	y_1	*P*/kPa	G^E/J mol^{-1}	μ^E_1/J mol^{-1}	μ^E_2/J mol^{-1}
0.0	0.000	90.62	0	2883	0
0.1	0.120	94.05	250	2164	37
0.2	0.196	94.72	429	1603	135
0.3	0.254	93.98	546	1167	280
0.4	0.306	92.27	608	828	461
0.5	0.359	89.66	621	566	675
0.6	0.417	86.03	587	364	923
0.7	0.488	81.14	509	209	1210
0.8	0.583	74.53	387	97	1547
0.9	0.727	65.36	218	26	1952
1.0	1.000	51.98	0	0	2447

B_{11}/cm^{3}mol^{-1}	B_{22}/cm^{3}mol^{-1}	B_{12}/cm^{3}mol^{-1}	V_1/cm^{3}mol^{-1}	V_2/cm^{3}mol^{-1}
-1670	-1270	-880	94	132

A(1) = 0.8964E+00 *A*(2) = -0.7870E-01 *A*(3) = 0.6560E-01

---

1. Oxolane, C_4H_8O
2. Heptane, C_7H_{16}

***T*/K = 298.15**
***System Number*: 66**

x_1	y_1	*P*/kPa	G^E/J mol^{-1}	μ^E_1/J mol^{-1}	μ^E_2/J mol^{-1}
0.0	0.000	6.01	0	1213	0
0.1	0.378	8.76	115	1080	7
0.2	0.560	11.15	213	930	34
0.3	0.667	13.20	293	771	88
0.4	0.738	14.92	349	609	175
0.5	0.790	16.37	378	453	303
0.6	0.832	17.61	377	309	480
0.7	0.870	18.70	343	185	712
0.8	0.907	19.71	271	87	1007
0.9	0.949	20.67	158	23	1371
1.0	1.000	21.60	0	0	1813

B_{11}/cm^{3}mol^{-1}	B_{22}/cm^{3}mol^{-1}	B_{12}/cm^{3}mol^{-1}	V_1/cm^{3}mol^{-1}	V_2/cm^{3}mol^{-1}
-1360	-2840	-2100	82	147

A(1) = 0.6103E+00 *A*(2) = 0.1210E+00 *A*(3) = 0.0000E+00

---

1. 2-Butanone, C_4H_8O
2. Heptane, C_7H_{16}

***T*/K = 323.15**
***System Number*: 74**

x_1	y_1	*P*/kPa	G^E/J mol^{-1}	μ^E_1/J mol^{-1}	μ^E_2/J mol^{-1}
0.0	0.000	18.90	0	4256	0
0.1	0.390	28.73	361	3071	60
0.2	0.497	32.81	610	2204	211
0.3	0.554	35.02	766	1576	419
0.4	0.599	36.52	846	1119	664
0.5	0.641	37.65	860	779	940
0.6	0.683	38.48	815	519	1258
0.7	0.730	38.97	712	314	1639
0.8	0.785	39.04	548	154	2122
0.9	0.862	38.39	315	43	2757
1.0	1.000	36.14	0	0	3611

B_{11}/cm^{3}mol^{-1}	B_{22}/cm^{3}mol^{-1}	B_{12}/cm^{3}mol^{-1}	V_1/cm^{3}mol^{-1}	V_2/cm^{3}mol^{-1}
-1790	-2240	-1370	93	152

A(1) = 0.1280E+01 *A*(2) = -0.1200E+00 *A*(3) = 0.1840E+00

TABLE 3.1.1 (VLEG) ISOTHERMAL VAPOR-LIQUID EQUILIBRIA AND EXCESS GIBBS ENERGIES OF BINARY MIXTURES (continued)

1. Butanal, C_4H_8O
2. Heptane, C_7H_{16}

***T*/K = 318.15**
***System Number*: 243**

x_1	y_1	*P*/kPa	G^E/J mol^{-1}	μ^E_1/J mol^{-1}	μ^E_2/J mol^{-1}
0.0	0.000	15.24	0	2911	0
0.1	0.378	22.44	263	2372	29
0.2	0.521	26.88	467	1854	120
0.3	0.598	29.67	610	1397	273
0.4	0.649	31.56	693	1015	478
0.5	0.692	32.99	719	709	728
0.6	0.734	34.13	690	469	1021
0.7	0.779	35.02	608	282	1369
0.8	0.829	35.62	471	139	1800
0.9	0.895	35.78	273	40	2369
1.0	1.000	35.03	0	0	3161

B_{11}/cm^{3}mol^{-1}	B_{22}/cm^{3}mol^{-1}	B_{12}/cm^{3}mol^{-1}	V_1/cm^{3}mol^{-1}	V_2/cm^{3}mol^{-1}
-1590	-2340	-1480	101	150

A(1) = 0.1087E+01 *A*(2) = -0.1434E-01 *A*(3) = 0.6111E-01 *A*(4) = 0.6147E-01 *A*(5) = 0.0000E+00

1. 2-Methylpropanal, C_4H_8O
2. Heptane, C_7H_{16}

***T*/K = 318.15**
***System Number*: 244**

x_1	y_1	*P*/kPa	G^E/J mol^{-1}	μ^E_1/J mol^{-1}	μ^E_2/J mol^{-1}
0.0	0.000	15.24	0	3129	0
0.1	0.462	26.14	266	2280	43
0.2	0.597	32.40	454	1680	147
0.3	0.672	36.77	577	1243	292
0.4	0.724	40.19	646	911	470
0.5	0.766	43.01	666	645	688
0.6	0.804	45.38	638	426	955
0.7	0.840	47.41	560	248	1286
0.8	0.881	49.22	429	114	1690
0.9	0.931	50.79	243	29	2173
1.0	1.000	51.99	0	0	2725

B_{11}/cm^{3}mol^{-1}	B_{22}/cm^{3}mol^{-1}	B_{12}/cm^{3}mol^{-1}	V_1/cm^{3}mol^{-1}	V_2/cm^{3}mol^{-1}
-1470	-2340	-1400	107	150

A(1) = 0.1007E+01 *A*(2) = -0.3238E-01 *A*(3) = 0.9929E-01 *A*(4) = -0.4401E-01 *A*(5) = 0.0000E+00

1. 2-Butanone, C_4H_8O
2. 2,2,4-Trimethylpentane, C_8H_{18}

***T*/K = 313.15**
***System Number*: 248**

x_1	y_1	*P*/kPa	G^E/J mol^{-1}	μ^E_1/J mol^{-1}	μ^E_2/J mol^{-1}
0.0	0.000	12.95	0	3547	0
0.1	0.362	18.66	312	2748	41
0.2	0.490	21.69	545	2116	152
0.3	0.560	23.50	707	1612	319
0.4	0.609	24.66	805	1204	539
0.5	0.649	25.44	841	867	815
0.6	0.685	25.95	815	585	1159
0.7	0.724	26.21	725	353	1592
0.8	0.773	26.18	565	170	2143
0.9	0.848	25.60	327	47	2849
1.0	1.000	23.65	0	0	3756

B_{11}/cm^{3}mol^{-1}	B_{22}/cm^{3}mol^{-1}	B_{12}/cm^{3}mol^{-1}	V_1/cm^{3}mol^{-1}	V_2/cm^{3}mol^{-1}
-2000	-2610	-1820	92	169

A(1) = 0.1291E+01 *A*(2) = 0.4001E-01 *A*(3) = 0.1112E+00

TABLE 3.1.1 (VLEG)
ISOTHERMAL VAPOR-LIQUID EQUILIBRIA AND EXCESS GIBBS ENERGIES OF BINARY MIXTURES (continued)

1. Ethyl ethanoate, $C_4H_8O_2$
2. 1-Chlorobutane, C_4H_9Cl

***T*/K = 298.15**
***System Number*: 24**

x_1	y_1	P/kPa	G^E/J mol^{-1}	μ^E_1/J mol^{-1}	μ^E_2/J mol^{-1}
0.0	0.000	13.60	0	659	0
0.1	0.113	13.84	59	523	7
0.2	0.213	13.98	104	410	27
0.3	0.306	14.06	136	317	58
0.4	0.396	14.07	155	238	100
0.5	0.483	14.01	163	171	155
0.6	0.572	13.90	158	115	224
0.7	0.663	13.72	141	68	310
0.8	0.761	13.46	110	33	419
0.9	0.870	13.11	63	9	554
1.0	1.000	12.63	0	0	724

B_{11}/cm^{3}mol^{-1}	B_{22}/cm^{3}mol^{-1}	B_{12}/cm^{3}mol^{-1}	V_1/cm^{3}mol^{-1}	V_2/cm^{3}mol^{-1}
-1900	-1660	-1700	98	105

$A(1) = 0.2631E+00$ $A(2) = 0.1296E-01$ $A(3) = 0.1584E-01$

---

1. Ethyl ethanoate, $C_4H_8O_2$
2. Diethylamine, $C_4H_{11}N$

***T*/K = 297.98**
***System Number*: 288**

x_1	y_1	P/kPa	G^E/J mol^{-1}	μ^E_1/J mol^{-1}	μ^E_2/J mol^{-1}
0.0	0.000	31.39	0	750	0
0.1	0.056	29.97	68	613	7
0.2	0.111	28.52	121	489	29
0.3	0.167	27.02	160	377	66
0.4	0.226	25.46	183	280	119
0.5	0.291	23.80	192	196	188
0.6	0.366	22.02	185	126	272
0.7	0.457	20.07	162	72	374
0.8	0.574	17.90	124	32	493
0.9	0.739	15.43	70	8	629
1.0	1.000	12.59	0	0	783

B_{11}/cm^{3}mol^{-1}	B_{22}/cm^{3}mol^{-1}	B_{12}/cm^{3}mol^{-1}	V_1/cm^{3}mol^{-1}	V_2/cm^{3}mol^{-1}
-1900	-1430	-2483	98	104

$A(1) = 0.3094E+00$ $A(2) = 0.6600E-02$ $A(3) = 0.0000E+00$

---

1. 1,4-Dioxane, $C_4H_8O_2$
2. Oxane, $C_5H_{10}O$

***T*/K = 313.15**
***System Number*: 308**

x_1	y_1	P/kPa	G^E/J mol^{-1}	μ^E_1/J mol^{-1}	μ^E_2/J mol^{-1}
0.0	0.000	18.34	0	793	0
0.1	0.073	17.85	72	661	7
0.2	0.143	17.30	130	536	29
0.3	0.213	16.71	174	422	67
0.4	0.283	16.07	201	318	123
0.5	0.357	15.37	212	227	198
0.6	0.438	14.58	207	149	294
0.7	0.531	13.70	183	86	411
0.8	0.644	12.67	141	39	551
0.9	0.790	11.47	81	10	716
1.0	1.000	10.03	0	0	906

B_{11}/cm^{3}mol^{-1}	B_{22}/cm^{3}mol^{-1}	B_{12}/cm^{3}mol^{-1}	V_1/cm^{3}mol^{-1}	V_2/cm^{3}mol^{-1}
-1790	-1480	-1560	87	100

$A(1) = 0.3263E+00$ $A(2) = 0.2175E-01$ $A(3) = 0.0000E+00$

TABLE 3.1.1 (VLEG)
ISOTHERMAL VAPOR-LIQUID EQUILIBRIA AND EXCESS GIBBS ENERGIES OF BINARY MIXTURES (continued)

1. 1,4-Dioxane, $C_4H_8O_2$
2. Piperidine, $C_5H_{11}N$

***T*/K = 313.15**
***System Number*: 289**

x_1	y_1	*P*/kPa	G^E/J mol^{-1}	μ^E_1/J mol^{-1}	μ^E_2/J mol^{-1}
0.0	0.000	8.62	0	1349	0
0.1	0.163	9.32	120	1075	14
0.2	0.282	9.82	212	838	56
0.3	0.378	10.17	277	636	123
0.4	0.460	10.42	315	464	216
0.5	0.536	10.58	327	321	333
0.6	0.611	10.65	313	206	474
0.7	0.689	10.65	273	116	640
0.8	0.775	10.56	208	52	832
0.9	0.875	10.37	117	13	1053
1.0	1.000	10.03	0	0	1304

B_{11}/cm^{3}mol^{-1}	B_{22}/cm^{3}mol^{-1}	B_{12}/cm^{3}mol^{-1}	V_1/cm^{3}mol^{-1}	V_2/cm^{3}mol^{-1}
-1790	-1800	-1800	87	101

A(1) = 0.5022E+00 *A*(2) = -0.8662E-02 *A*(3) = 0.7203E-02

1. 1,4-Dioxane, $C_4H_8O_2$
2. Hexafluorobenzene, C_6F_6

***T*/K = 303.15**
***System Number*: 386**

x_1	y_1	*P*/kPa	G^E/J mol^{-1}	μ^E_1/J mol^{-1}	μ^E_2/J mol^{-1}
0.0	0.000	14.34	0	555	0
0.1	0.057	13.68	52	499	3
0.2	0.118	13.02	100	474	7
0.3	0.185	12.35	146	456	13
0.4	0.257	11.69	188	427	29
0.5	0.333	11.03	224	376	72
0.6	0.412	10.38	247	302	164
0.7	0.496	9.71	248	210	337
0.8	0.591	8.96	217	114	630
0.9	0.724	7.96	140	34	1090
1.0	1.000	6.31	0	0	1771

B_{11}/cm^{3}mol^{-1}	B_{22}/cm^{3}mol^{-1}	B_{12}/cm^{3}mol^{-1}	V_1/cm^{3}mol^{-1}	V_2/cm^{3}mol^{-1}
-2020	-1970	-2000	86	117

A(1) = 0.3551E+00 *A*(2) = 0.2412E+00 *A*(3) = 0.1064E+00

1. Ethyl ethanoate, $C_4H_8O_2$
2. Chlorobenzene, C_6H_5Cl

***T*/K = 313.15**
***System Number*: 284**

x_1	y_1	*P*/kPa	G^E/J mol^{-1}	μ^E_1/J mol^{-1}	μ^E_2/J mol^{-1}
0.0	0.000	3.59	0	215	0
0.1	0.448	5.87	19	163	3
0.2	0.642	8.09	32	124	9
0.3	0.752	10.26	42	94	20
0.4	0.823	12.41	47	70	32
0.5	0.873	14.53	49	51	48
0.6	0.910	16.65	48	35	68
0.7	0.940	18.75	43	21	93
0.8	0.963	20.84	34	11	126
0.9	0.983	22.95	20	3	169
1.0	1.000	25.07	0	0	227

B_{11}/cm^{3}mol^{-1}	B_{22}/cm^{3}mol^{-1}	B_{12}/cm^{3}mol^{-1}	V_1/cm^{3}mol^{-1}	V_2/cm^{3}mol^{-1}
-1800	-2970	-2300	101	104

A(1) = 0.7603E-01 *A*(2) = 0.2232E-02 *A*(3) = 0.8808E-02

TABLE 3.1.1 (VLEG)
ISOTHERMAL VAPOR-LIQUID EQUILIBRIA AND EXCESS GIBBS ENERGIES OF BINARY MIXTURES (continued)

1. 1,4-Dioxane, $C_4H_8O_2$
2. Benzene, C_6H_6

***T*/K = 298.15**
***System Number*: 217**

x_1	y_1	P/kPa	G^E/J mol^{-1}	μ^E_1/J mol^{-1}	μ^E_2/J mol^{-1}
0.0	0.000	12.59	0	429	0
0.1	0.047	11.91	37	314	6
0.2	0.095	11.22	62	221	22
0.3	0.148	10.53	77	149	46
0.4	0.208	9.81	83	94	75
0.5	0.276	9.06	81	55	107
0.6	0.359	8.28	73	29	139
0.7	0.461	7.47	59	12	169
0.8	0.591	6.63	42	4	195
0.9	0.764	5.77	22	1	212
1.0	1.000	4.89	0	0	221

B_{11}/cm^{3}mol^{-1}	B_{22}/cm^{3}mol^{-1}	B_{12}/cm^{3}mol^{-1}	V_1/cm^{3}mol^{-1}	V_2/cm^{3}mol^{-1}
-2160	-1480	-1300	86	89

$A(1)$ = 0.1310E+00 $A(2)$ = -0.4200E-01 $A(3)$ = 0.0000E+00

1. Ethyl ethanoate, $C_4H_8O_2$
2. Aniline, C_6H_7N

***T*/K = 297.49**
***System Number*: 292**

x_1	y_1	P/kPa	G^E/J mol^{-1}	μ^E_1/J mol^{-1}	μ^E_2/J mol^{-1}
0.0	0.000	0.08	0	127	0
0.1	0.945	1.23	-1	-106	11
0.2	0.974	2.32	-18	-184	23
0.3	0.984	3.43	-40	-193	25
0.4	0.990	4.59	-61	-172	14
0.5	0.994	5.80	-76	-136	-16
0.6	0.996	7.08	-83	-92	-70
0.7	0.997	8.40	-80	-49	-151
0.8	0.999	9.73	-63	-15	-252
0.9	0.999	11.01	-34	0	-334
1.0	1.000	12.24	0	0	-306

B_{11}/cm^{3}mol^{-1}	B_{22}/cm^{3}mol^{-1}	B_{12}/cm^{3}mol^{-1}	V_1/cm^{3}mol^{-1}	V_2/cm^{3}mol^{-1}
-1910	-3810	-3250	98	92

$A(1)$ = -0.1230E+00 $A(2)$ = -0.9661E-01 $A(3)$ = 0.4106E-01 $A(4)$ = 0.8890E-02 $A(5)$ = 0.4570E-01

1. 1,4-Dioxane, $C_4H_8O_2$
2. Cyclohexane, C_6H_{12}

***T*/K = 303.15**
***System Number*: 216**

x_1	y_1	P/kPa	G^E/J mol^{-1}	μ^E_1/J mol^{-1}	μ^E_2/J mol^{-1}
0.0	0.000	16.24	0	3074	0
0.1	0.126	16.76	307	3023	6
0.2	0.219	17.07	593	2714	63
0.3	0.274	17.08	829	2254	219
0.4	0.299	16.96	994	1732	502
0.5	0.307	16.89	1071	1218	924
0.6	0.308	16.86	1050	766	1477
0.7	0.316	16.63	928	410	2139
0.8	0.351	15.56	706	167	2865
0.9	0.464	12.56	392	36	3596
1.0	1.000	6.35	0	0	4254

B_{11}/cm^{3}mol^{-1}	B_{22}/cm^{3}mol^{-1}	B_{12}/cm^{3}mol^{-1}	V_1/cm^{3}mol^{-1}	V_2/cm^{3}mol^{-1}
-2020	-1650	-1300	86	109

$A(1)$ = 0.1700E+01 $A(2)$ = 0.2340E+00 $A(3)$ = -0.2460E+00

TABLE 3.1.1 (VLEG)
ISOTHERMAL VAPOR-LIQUID EQUILIBRIA AND EXCESS GIBBS ENERGIES OF BINARY MIXTURES (continued)

1. Ethyl ethanoate, $C_4H_8O_2$
2. Cyclohexane, C_6H_{12}

***T*/K = 313.15**
***System Number*: 269**

x_1	y_1	*P*/kPa	G^E/J mol^{-1}	μ^E_1/J mol^{-1}	μ^E_2/J mol^{-1}
0.0	0.000	24.62	0	3777	0
0.1	0.226	29.39	309	2547	61
0.2	0.318	31.20	510	1777	194
0.3	0.383	32.15	634	1284	356
0.4	0.442	32.66	700	944	538
0.5	0.498	32.81	717	683	751
0.6	0.553	32.63	686	464	1020
0.7	0.610	32.09	604	276	1369
0.8	0.681	31.02	465	128	1817
0.9	0.790	28.96	265	32	2359
1.0	1.000	25.05	0	0	2964

B_{11}/cm^{3}mol^{-1}	B_{22}/cm^{3}mol^{-1}	B_{12}/cm^{3}mol^{-1}	V_1/cm^{3}mol^{-1}	V_2/cm^{3}mol^{-1}
-1800	-1490	-1300	101	111

A(1) = 0.1102E+01 *A*(2) = -0.5200E-01 *A*(3) = 0.1930E+00 *A*(4) = -0.1040E+00 *A*(5) = 0.0000E+00

1. 1,4-Dioxane, $C_4H_8O_2$
2. Heptane, C_7H_{16}

***T*/K = 298.15**
***System Number*: 23**

x_1	y_1	*P*/kPa	G^E/J mol^{-1}	μ^E_1/J mol^{-1}	μ^E_2/J mol^{-1}
0.0	0.000	6.010	0	3444	0
0.1	0.223	7.052	315	2889	30
0.2	0.334	7.603	571	2361	123
0.3	0.397	7.860	762	1867	288
0.4	0.437	7.951	885	1416	532
0.5	0.464	7.953	937	1014	861
0.6	0.488	7.897	914	668	1284
0.7	0.517	7.757	813	387	1807
0.8	0.567	7.431	629	177	2439
0.9	0.674	6.664	359	45	3185
1.0	1.000	4.890	0	0	4054

B_{11}/cm^{3}mol^{-1}	B_{22}/cm^{3}mol^{-1}	B_{12}/cm^{3}mol^{-1}	V_1/cm^{3}mol^{-1}	V_2/cm^{3}mol^{-1}
-2160	-2840	-2500	86	147

A(1) = 0.1513E+01 *A*(2) = 0.1230E+00 *A*(3) = 0.0000E+00

1. Ethyl ethanoate, $C_4H_8O_2$
2. Heptane, C_7H_{16}

***T*/K = 323.15**
***System Number*: 268**

x_1	y_1	*P*/kPa	G^E/J mol^{-1}	μ^E_1/J mol^{-1}	μ^E_2/J mol^{-1}
0.0	0.000	18.85	0	2873	0
0.1	0.335	25.96	255	2264	31
0.2	0.478	30.43	449	1772	118
0.3	0.563	33.46	587	1369	252
0.4	0.623	35.62	673	1034	432
0.5	0.672	37.20	707	750	665
0.6	0.716	38.35	689	509	960
0.7	0.760	39.13	616	307	1336
0.8	0.811	39.53	481	148	1816
0.9	0.881	39.34	279	40	2430
1.0	1.000	37.90	0	0	3214

B_{11}/cm^{3}mol^{-1}	B_{22}/cm^{3}mol^{-1}	B_{12}/cm^{3}mol^{-1}	V_1/cm^{3}mol^{-1}	V_2/cm^{3}mol^{-1}
-1650	-2240	-1480	102	152

A(1) = 0.1053E+01 *A*(2) = 0.6350E-01 *A*(3) = 0.8000E-01

TABLE 3.1.1 (VLEG) ISOTHERMAL VAPOR-LIQUID EQUILIBRIA AND EXCESS GIBBS ENERGIES OF BINARY MIXTURES (continued)

1. 1-Bromobutane, C_4H_9Br
2. 1-Chlorobutane, C_4H_9Cl

***T*/K = 323.15**
***System Number*: 137**

x_1	y_1	P/kPa	G^E/J mol^{-1}	μ^E_1/J mol^{-1}	μ^E_2/J mol^{-1}
0.0	0.000	39.44	0	-259	0
0.1	0.043	36.99	-22	-183	-4
0.2	0.094	34.56	-36	-123	-14
0.3	0.155	32.17	-44	-78	-29
0.4	0.225	29.84	-46	-45	-47
0.5	0.306	27.56	-44	-23	-65
0.6	0.401	25.35	-38	-9	-81
0.7	0.512	23.19	-30	-2	-94
0.8	0.643	21.07	-20	0	-101
0.9	0.802	18.98	-10	0	-101
1.0	1.000	16.89	0	0	-91

B_{11}/cm^{3}mol^{-1}	B_{22}/cm^{3}mol^{-1}	B_{12}/cm^{3}mol^{-1}	V_1/cm^{3}mol^{-1}	V_2/cm^{3}mol^{-1}
-1600	-1300	-1450	109	109

$A(1)$ = -0.6517E-01 $A(2)$ = 0.3140E-01 $A(3)$ = 0.0000E+00

---

1. 1-Bromobutane, C_4H_9Br
2. Benzene, C_6H_6

***T*/K = 343.15**
***System Number*: 105**

x_1	y_1	P/kPa	G^E/J mol^{-1}	μ^E_1/J mol^{-1}	μ^E_2/J mol^{-1}
0.0	0.000	73.42	0	-74	0
0.1	0.053	69.62	-7	-58	-1
0.2	0.112	65.83	-11	-44	-3
0.3	0.178	62.07	-15	-33	-7
0.4	0.253	58.34	-17	-23	-12
0.5	0.337	54.63	-17	-16	-18
0.6	0.434	50.97	-16	-10	-26
0.7	0.545	47.34	-14	-5	-34
0.8	0.673	43.76	-10	-2	-43
0.9	0.823	40.22	-6	-1	-52
1.0	1.000	36.73	0	0	-63

B_{11}/cm^{3}mol^{-1}	B_{22}/cm^{3}mol^{-1}	B_{12}/cm^{3}mol^{-1}	V_1/cm^{3}mol^{-1}	V_2/cm^{3}mol^{-1}
-1400	-1030	-1200	114	94

$A(1)$ = -0.2392E-01 $A(2)$ = 0.2010E-02 $A(3)$ = 0.0000E+00

---

1. 1-Bromobutane, C_4H_9Br
2. Heptane, C_7H_{16}

***T*/K = 323.15**
***System Number*: 110**

x_1	y_1	P/kPa	G^E/J mol^{-1}	μ^E_1/J mol^{-1}	μ^E_2/J mol^{-1}
0.0	0.000	18.67	0	1301	0
0.1	0.126	19.35	112	966	17
0.2	0.224	19.72	192	712	62
0.3	0.310	19.88	243	518	126
0.4	0.392	19.89	271	371	205
0.5	0.472	19.77	277	258	297
0.6	0.555	19.53	264	169	405
0.7	0.643	19.15	230	100	534
0.8	0.741	18.63	176	48	691
0.9	0.855	17.92	101	13	890
1.0	1.000	16.93	0	0	1144

B_{11}/cm^{3}mol^{-1}	B_{22}/cm^{3}mol^{-1}	B_{12}/cm^{3}mol^{-1}	V_1/cm^{3}mol^{-1}	V_2/cm^{3}mol^{-1}
-1600	-2240	-1900	109	153

$A(1)$ = 0.4130E+00 $A(2)$ = -0.2940E-01 $A(3)$ = 0.4200E-01

TABLE 3.1.1 (VLEG) ISOTHERMAL VAPOR-LIQUID EQUILIBRIA AND EXCESS GIBBS ENERGIES OF BINARY MIXTURES (continued)

1. 1-Chlorobutane, C_4H_9Cl
2. Pentane, C_5H_{12}

***T*/K = 298.15**
***System Number*: 128**

x_1	y_1	P/kPa	G^E/J mol^{-1}	μ^E_1/J mol^{-1}	μ^E_2/J mol^{-1}
0.0	0.000	68.28	0	1224	0
0.1	0.032	63.72	108	950	14
0.2	0.063	59.32	188	722	54
0.3	0.095	54.95	242	533	117
0.4	0.129	50.48	272	379	200
0.5	0.169	45.78	278	256	301
0.6	0.218	40.72	263	160	418
0.7	0.285	35.14	227	88	551
0.8	0.387	28.90	171	39	699
0.9	0.568	21.80	95	10	863
1.0	1.000	13.61	0	0	1044

B_{11}/cm^{3}mol^{-1}	B_{22}/cm^{3}mol^{-1}	B_{12}/cm^{3}mol^{-1}	V_1/cm^{3}mol^{-1}	V_2/cm^{3}mol^{-1}
-1660	-1260	-1450	105	116

$A(1)$ = 0.4489E+00 $A(2)$ = -0.3626E-01 $A(3)$ = 0.8424E-02

---

1. 1-Chlorobutane, C_4H_9Cl
2. Chlorobenzene, C_6H_5Cl

***T*/K = 298.17**
***System Number*: 140**

x_1	y_1	P/kPa	G^E/J mol^{-1}	μ^E_1/J mol^{-1}	μ^E_2/J mol^{-1}
0.0	0.000	1.64	0	331	0
0.1	0.499	2.96	27	216	6
0.2	0.684	4.20	43	142	18
0.3	0.783	5.41	52	96	33
0.4	0.847	6.60	56	69	48
0.5	0.891	7.79	57	51	62
0.6	0.924	8.98	54	38	79
0.7	0.949	10.15	49	26	101
0.8	0.969	11.32	39	15	136
0.9	0.986	12.48	24	5	194
1.0	1.000	13.65	0	0	286

B_{11}/cm^{3}mol^{-1}	B_{22}/cm^{3}mol^{-1}	B_{12}/cm^{3}mol^{-1}	V_1/cm^{3}mol^{-1}	V_2/cm^{3}mol^{-1}
-1660	-3130	-2300	105	102

$A(1)$ = 0.9169E-01 $A(2)$ = -0.9130E-02 $A(3)$ = 0.3273E-01

---

1. 1-Chlorobutane, C_4H_9Cl
2. Benzene, C_6H_6

***T*/K = 298.15**
***System Number*: 129**

x_1	y_1	P/kPa	G^E/J mol^{-1}	μ^E_1/J mol^{-1}	μ^E_2/J mol^{-1}
0.0	0.000	12.76	0	188	0
0.1	0.109	12.92	13	83	5
0.2	0.211	13.03	17	25	15
0.3	0.312	13.11	16	-3	24
0.4	0.412	13.17	12	-11	28
0.5	0.513	13.24	9	-9	26
0.6	0.613	13.31	6	-4	19
0.7	0.713	13.39	4	1	11
0.8	0.810	13.48	3	3	6
0.9	0.905	13.56	3	1	15
1.0	1.000	13.63	0	0	46

B_{11}/cm^{3}mol^{-1}	B_{22}/cm^{3}mol^{-1}	B_{12}/cm^{3}mol^{-1}	V_1/cm^{3}mol^{-1}	V_2/cm^{3}mol^{-1}
-1660	-1480	-1570	105	89

$A(1)$ = 0.1373E-01 $A(2)$ = -0.2863E-01 $A(3)$ = 0.3334E-01

TABLE 3.1.1 (VLEG) ISOTHERMAL VAPOR-LIQUID EQUILIBRIA AND EXCESS GIBBS ENERGIES OF BINARY MIXTURES (continued)

1. 1-Chlorobutane, C_4H_9Cl
2. Aniline, C_6H_7N

***T*/K = 298.20**
***System Number*: 178**

x_1	y_1	P/kPa	G^E/J mol^{-1}	μ^E_1/J mol^{-1}	μ^E_2/J mol^{-1}
0.0	0.000	0.08	0	2845	0
0.1	0.980	3.48	255	2289	29
0.2	0.989	5.70	453	1809	114
0.3	0.992	7.22	595	1396	251
0.4	0.993	8.34	682	1040	443
0.5	0.994	9.22	714	738	690
0.6	0.995	9.99	691	485	999
0.7	0.996	10.73	611	282	1377
0.8	0.997	11.53	471	130	1834
0.9	0.998	12.47	269	34	2381
1.0	1.000	13.66	0	0	3034

B_{11}/cm^{3}mol^{-1}	B_{22}/cm^{3}mol^{-1}	B_{12}/cm^{3}mol^{-1}	V_1/cm^{3}mol^{-1}	V_2/cm^{3}mol^{-1}
-1660	-3800	-2500	105	92

$A(1) = 0.1152E+01$ $A(2) = 0.3817E-01$ $A(3) = 0.3357E-01$

---

1. 1-Chlorobutane, C_4H_9Cl
2. 1,4-Cyclohexadiene, C_6H_8

***T*/K = 293.15**
***System Number*: 136**

x_1	y_1	P/kPa	G^E/J mol^{-1}	μ^E_1/J mol^{-1}	μ^E_2/J mol^{-1}
0.0	0.000	6.95	0	493	0
0.1	0.164	7.51	42	350	7
0.2	0.296	7.99	69	243	26
0.3	0.409	8.41	86	165	52
0.4	0.509	8.80	93	109	81
0.5	0.602	9.16	92	70	113
0.6	0.689	9.50	84	43	147
0.7	0.771	9.83	71	24	182
0.8	0.850	10.14	53	11	221
0.9	0.926	10.44	29	3	266
1.0	1.000	10.73	0	0	320

B_{11}/cm^{3}mol^{-1}	B_{22}/cm^{3}mol^{-1}	B_{12}/cm^{3}mol^{-1}	V_1/cm^{3}mol^{-1}	V_2/cm^{3}mol^{-1}
-1750	-1720	-1670	104	94

$A(1) = 0.1506E+00$ $A(2) = -0.3546E-01$ $A(3) = 0.1618E-01$

---

1. 1-Chlorobutane, C_4H_9Cl
2. Cyclohexene, C_6H_{10}

***T*/K = 293.15**
***System Number*: 135**

x_1	y_1	P/kPa	G^E/J mol^{-1}	μ^E_1/J mol^{-1}	μ^E_2/J mol^{-1}
0.0	0.000	9.41	0	738	0
0.1	0.137	9.85	64	561	9
0.2	0.250	10.18	111	415	35
0.3	0.349	10.43	141	297	74
0.4	0.440	10.61	156	204	124
0.5	0.528	10.74	157	132	182
0.6	0.615	10.82	146	79	247
0.7	0.704	10.86	124	41	317
0.8	0.797	10.86	92	17	389
0.9	0.895	10.82	50	4	463
1.0	1.000	10.73	0	0	537

B_{11}/cm^{3}mol^{-1}	B_{22}/cm^{3}mol^{-1}	B_{12}/cm^{3}mol^{-1}	V_1/cm^{3}mol^{-1}	V_2/cm^{3}mol^{-1}
-1750	-1720	-1630	104	101

$A(1) = 0.2581E+00$ $A(2) = -0.4117E-01$ $A(3) = 0.3436E-02$

TABLE 3.1.1 (VLEG)
ISOTHERMAL VAPOR-LIQUID EQUILIBRIA AND EXCESS GIBBS ENERGIES OF BINARY MIXTURES (continued)

1. 1-Chlorobutane, C_4H_9Cl
2. Cyclohexane, C_6H_{12}

***T*/K = 293.15**
***System Number*: 133**

x_1	y_1	P/kPa	G^E/J mol^{-1}	μ^E_1/J mol^{-1}	μ^E_2/J mol^{-1}
0.0	0.000	10.34	0	1292	0
0.1	0.146	10.97	113	985	16
0.2	0.255	11.38	195	734	60
0.3	0.344	11.64	248	533	127
0.4	0.425	11.78	277	373	213
0.5	0.503	11.83	281	248	314
0.6	0.582	11.80	264	153	430
0.7	0.666	11.68	226	84	559
0.8	0.760	11.48	169	36	700
0.9	0.869	11.17	94	9	855
1.0	1.000	10.73	0	0	1025

B_{11}/cm^{3}mol^{-1}	B_{22}/cm^{3}mol^{-1}	B_{12}/cm^{3}mol^{-1}	V_1/cm^{3}mol^{-1}	V_2/cm^{3}mol^{-1}
-1750	-1830	-1680	104	108

$A(1) = 0.4613E+00$ $A(2) = -0.5475E-01$ $A(3) = 0.1396E-01$

1. 1-Chlorobutane, C_4H_9Cl
2. Toluene, C_7H_8

***T*/K = 298.16**
***System Number*: 130**

x_1	y_1	P/kPa	G^E/J mol^{-1}	μ^E_1/J mol^{-1}	μ^E_2/J mol^{-1}
0.0	0.000	3.83	0	244	0
0.1	0.290	4.88	17	109	7
0.2	0.471	5.85	22	34	19
0.3	0.599	6.80	21	0	30
0.4	0.698	7.76	17	-9	35
0.5	0.777	8.74	13	-5	32
0.6	0.840	9.73	10	2	23
0.7	0.891	10.73	9	6	15
0.8	0.934	11.71	8	6	15
0.9	0.969	12.69	6	3	37
1.0	1.000	13.65	0	0	96

B_{11}/cm^{3}mol^{-1}	B_{22}/cm^{3}mol^{-1}	B_{12}/cm^{3}mol^{-1}	V_1/cm^{3}mol^{-1}	V_2/cm^{3}mol^{-1}
-1660	-2620	-2100	105	107

$A(1) = 0.2125E-01$ $A(2) = -0.2983E-01$ $A(3) = 0.4718E-01$

1. 1-Chlorobutane, C_4H_9Cl
2. Heptane, C_7H_{16}

***T*/K = 323.20**
***System Number*: 25**

x_1	y_1	P/kPa	G^E/J mol^{-1}	μ^E_1/J mol^{-1}	μ^E_2/J mol^{-1}
0.0	0.000	18.66	0	1065	0
0.1	0.235	22.14	91	786	14
0.2	0.387	24.95	156	577	51
0.3	0.500	27.35	198	420	103
0.4	0.592	29.48	220	301	166
0.5	0.672	31.44	225	210	240
0.6	0.743	33.25	215	139	327
0.7	0.809	34.92	188	83	432
0.8	0.872	36.47	144	40	561
0.9	0.934	37.91	83	11	728
1.0	1.000	39.22	0	0	945

B_{11}/cm^{3}mol^{-1}	B_{22}/cm^{3}mol^{-1}	B_{12}/cm^{3}mol^{-1}	V_1/cm^{3}mol^{-1}	V_2/cm^{3}mol^{-1}
-1300	-2240	-1760	109	153

$A(1) = 0.3356E+00$ $A(2) = -0.2238E-01$ $A(3) = 0.3835E-01$

TABLE 3.1.1 (VLEG)
ISOTHERMAL VAPOR-LIQUID EQUILIBRIA AND EXCESS GIBBS ENERGIES OF BINARY MIXTURES (continued)

1. 1-Chlorobutane, C_4H_9Cl
2. 2,2,4-Trimethylpentane, C_8H_{18}

***T*/K = 293.15**
***System Number*: 134**

x_1	y_1	P/kPa	G^E/J mol^{-1}	μ^E_1/J mol^{-1}	μ^E_2/J mol^{-1}
0.0	0.000	5.16	0	1004	0
0.1	0.243	6.17	92	834	9
0.2	0.402	7.02	165	676	37
0.3	0.515	7.74	219	530	86
0.4	0.603	8.36	253	399	156
0.5	0.677	8.89	267	284	251
0.6	0.742	9.35	260	186	371
0.7	0.803	9.76	230	107	518
0.8	0.864	10.13	177	49	693
0.9	0.928	10.45	101	12	898
1.0	1.000	10.73	0	0	1135

B_{11}/cm^{3}mol^{-1}	B_{22}/cm^{3}mol^{-1}	B_{12}/cm^{3}mol^{-1}	V_1/cm^{3}mol^{-1}	V_2/cm^{3}mol^{-1}
-1750	-3120	-1975	104	166

$A(1)$ = 0.4387E+00 $A(2)$ = 0.2696E-01 $A(3)$ = 0.0000E+00

---

1. Pyrrolidine, C_4H_9N
2. Cyclohexane, C_6H_{12}

***T*/K = 333.15**
***System Number*: 159**

x_1	y_1	P/kPa	G^E/J mol^{-1}	μ^E_1/J mol^{-1}	μ^E_2/J mol^{-1}
0.0	0.000	51.89	0	1928	0
0.1	0.125	53.90	166	1450	24
0.2	0.221	54.85	290	1141	77
0.3	0.304	55.17	381	916	152
0.4	0.379	55.03	443	726	255
0.5	0.449	54.46	474	545	403
0.6	0.515	53.45	470	373	615
0.7	0.585	51.92	424	219	903
0.8	0.670	49.56	331	97	1267
0.9	0.793	45.81	190	23	1687
1.0	1.000	39.92	0	0	2115

B_{11}/cm^{3}mol^{-1}	B_{22}/cm^{3}mol^{-1}	B_{12}/cm^{3}mol^{-1}	V_1/cm^{3}mol^{-1}	V_2/cm^{3}mol^{-1}
-1050	-1240	-1080	87	114

$A(1)$ = 0.6846E+00 $A(2)$ = 0.1029E+00 $A(3)$ = 0.4522E-01 $A(4)$ = -0.6911E-01 $A(5)$ = 0.0000E+00

---

1. Morpholine, C_4H_9NO
2. 1-Butanol, $C_4H_{10}O$

***T*/K = 363.35**
***System Number*: 301**

x_1	y_1	P/kPa	G^E/J mol^{-1}	μ^E_1/J mol^{-1}	μ^E_2/J mol^{-1}
0.0	0.000	34.63	0	-2645	0
0.1	0.050	32.29	-214	-1727	-46
0.2	0.132	30.28	-345	-1120	-151
0.3	0.236	28.83	-415	-735	-277
0.4	0.351	27.90	-441	-497	-404
0.5	0.471	27.36	-436	-348	-524
0.6	0.590	27.11	-407	-246	-649
0.7	0.708	27.10	-355	-163	-804
0.8	0.822	27.38	-277	-89	-1030
0.9	0.922	28.00	-163	-28	-1385
1.0	1.000	28.98	0	0	-1941

B_{11}/cm^{3}mol^{-1}	B_{22}/cm^{3}mol^{-1}	B_{12}/cm^{3}mol^{-1}	V_1/cm^{3}mol^{-1}	V_2/cm^{3}mol^{-1}
-1200	-1590	-1080	94	100

$A(1)$ = -0.5779E+00 $A(2)$ = 0.1165E+00 $A(3)$ = -0.1812E+00

TABLE 3.1.1 (VLEG)
ISOTHERMAL VAPOR-LIQUID EQUILIBRIA AND EXCESS GIBBS ENERGIES OF BINARY MIXTURES (continued)

1. Morpholine, C_4H_9NO
2. Octane, C_8H_{18}

T/K = 353.35
System Number: 186

x_1	y_1	*P*/kPa	G^E/J mol^{-1}	μ^E_1/J mol^{-1}	μ^E_2/J mol^{-1}
0.0	0.000	23.33	0	3907	0
0.1	0.215	27.18	354	3214	36
0.2	0.327	29.24	636	2615	142
0.3	0.396	30.29	849	2088	318
0.4	0.444	30.74	990	1617	571
0.5	0.480	30.80	1056	1195	918
0.6	0.510	30.59	1043	820	1377
0.7	0.542	30.10	942	498	1979
0.8	0.588	29.03	743	240	2757
0.9	0.683	26.45	434	65	3754
1.0	1.000	19.59	0	0	5018

B_{11}/cm^{3}mol^{-1}	B_{22}/cm^{3}mol^{-1}	B_{12}/cm^{3}mol^{-1}	V_1/cm^{3}mol^{-1}	V_2/cm^{3}mol^{-1}
-1300	-2450	-1630	93	174

A(1) = 0.1438E+01 *A*(2) = 0.1891E+00 *A*(3) = 0.8065E-01

1. 1-Butanol, $C_4H_{10}O$
2. 2-Furaldehyde, $C_5H_4O_2$

T/K = 368.15
System Number: 329

x_1	y_1	*P*/kPa	G^E/J mol^{-1}	μ^E_1/J mol^{-1}	μ^E_2/J mol^{-1}
0.0	0.000	10.97	0	2921	0
0.1	0.478	19.14	265	2400	27
0.2	0.634	24.92	475	1937	109
0.3	0.713	29.11	630	1524	247
0.4	0.763	32.24	730	1158	444
0.5	0.800	34.66	772	835	708
0.6	0.831	36.62	754	558	1048
0.7	0.861	38.31	673	329	1474
0.8	0.894	39.84	523	154	2002
0.9	0.937	41.25	301	41	2647
1.0	1.000	42.41	0	0	3429

B_{11}/cm^{3}mol^{-1}	B_{22}/cm^{3}mol^{-1}	B_{12}/cm^{3}mol^{-1}	V_1/cm^{3}mol^{-1}	V_2/cm^{3}mol^{-1}
-1530	-1990	-860	99	89

A(1) = 0.1009E+01 *A*(2) = 0.8290E-01 *A*(3) = 0.2860E-01

1. 2-Methyl-2-propanol, $C_4H_{10}O$
2. Benzene, C_6H_6

T/K = 313.15
System Number: 238

x_1	y_1	*P*/kPa	G^E/J mol^{-1}	μ^E_1/J mol^{-1}	μ^E_2/J mol^{-1}
0.0	0.000	24.37	0	5101	0
0.1	0.171	27.44	402	3188	92
0.2	0.224	27.93	644	2127	274
0.3	0.263	27.91	785	1489	484
0.4	0.301	27.56	851	1050	718
0.5	0.339	26.94	854	716	991
0.6	0.381	26.00	797	455	1310
0.7	0.436	24.57	683	260	1672
0.8	0.519	22.40	515	124	2081
0.9	0.660	19.14	291	36	2581
1.0	1.000	13.79	0	0	3314

B_{11}/cm^{3}mol^{-1}	B_{22}/cm^{3}mol^{-1}	B_{12}/cm^{3}mol^{-1}	V_1/cm^{3}mol^{-1}	V_2/cm^{3}mol^{-1}
-2530	-1300	-1050	99	91

A(1) = 0.1311E+01 *A*(2) = -0.2118E+00 *A*(3) = 0.1764E+00 *A*(4) = -0.1314E+00 *A*(5) = 0.1281E+00

TABLE 3.1.1 (VLEG)
ISOTHERMAL VAPOR-LIQUID EQUILIBRIA AND EXCESS GIBBS ENERGIES OF BINARY MIXTURES (continued)

1. 1-Butanol, $C_4H_{10}O$
2. Cyclohexane, C_6H_{12}

***T*/K = 318.15**
***System Number*: 226**

x_1	y_1	P/kPa	G^E/J mol^{-1}	μ^E_1/J mol^{-1}	μ^E_2/J mol^{-1}
0.0	0.000	29.97	0	7515	0
0.1	0.058	30.34	564	4270	153
0.2	0.064	30.05	875	2687	422
0.3	0.072	29.41	1041	1842	698
0.4	0.080	28.44	1112	1288	995
0.5	0.088	27.33	1102	854	1349
0.6	0.098	25.86	1014	512	1767
0.7	0.116	23.39	853	270	2215
0.8	0.151	19.17	628	121	2658
0.9	0.244	12.87	347	36	3140
1.0	1.000	3.41	0	0	3907

B_{11}/cm^{3}mol^{-1}	B_{22}/cm^{3}mol^{-1}	B_{12}/cm^{3}mol^{-1}	V_1/cm^{3}mol^{-1}	V_2/cm^{3}mol^{-1}
-3290	-1420	-2000	94	112

$A(1) = 0.1666E+01$ $A(2) = -0.3741E+00$ $A(3) = 0.1972E+00$ $A(4) = -0.3079E+00$ $A(5) = 0.2957E+00$

---

1. 2-Butanol, $C_4H_{10}O$
2. Cyclohexane, C_6H_{12}

***T*/K = 318.15**
***System Number*: 227**

x_1	y_1	P/kPa	G^E/J mol^{-1}	μ^E_1/J mol^{-1}	μ^E_2/J mol^{-1}
0.0	0.000	29.97	0	6708	0
0.1	0.114	32.00	510	3925	130
0.2	0.135	31.87	802	2546	365
0.3	0.154	31.38	965	1764	622
0.4	0.171	30.60	1037	1230	908
0.5	0.190	29.55	1032	825	1239
0.6	0.214	28.06	956	514	1618
0.7	0.248	25.81	812	285	2043
0.8	0.307	22.33	605	126	2520
0.9	0.440	16.86	334	31	3059
1.0	1.000	8.08	0	0	3627

B_{11}/cm^{3}mol^{-1}	B_{22}/cm^{3}mol^{-1}	B_{12}/cm^{3}mol^{-1}	V_1/cm^{3}mol^{-1}	V_2/cm^{3}mol^{-1}
-2430	-1420	-1800	94	112

$A(1) = 0.1560E+01$ $A(2) = -0.3128E+00$ $A(3) = 0.2170E+00$ $A(4) = -0.1735E+00$ $A(5) = 0.1761E+00$

---

1. 2-Butanol, $C_4H_{10}O$
2. 1-Hexene, C_6H_{12}

***T*/K = 333.15**
***System Number*: 228**

x_1	y_1	P/kPa	G^E/J mol^{-1}	μ^E_1/J mol^{-1}	μ^E_2/J mol^{-1}
0.0	0.000	90.62	0	5510	0
0.1	0.076	91.21	449	3671	91
0.2	0.101	89.63	734	2484	296
0.3	0.117	87.46	899	1713	550
0.4	0.135	84.38	974	1192	828
0.5	0.155	80.19	976	817	1134
0.6	0.179	74.81	912	529	1486
0.7	0.212	67.81	784	302	1908
0.8	0.267	57.93	590	134	2411
0.9	0.392	42.62	328	33	2988
1.0	1.000	18.15	0	0	3595

B_{11}/cm^{3}mol^{-1}	B_{22}/cm^{3}mol^{-1}	B_{12}/cm^{3}mol^{-1}	V_1/cm^{3}mol^{-1}	V_2/cm^{3}mol^{-1}
-2220	-1270	-899	96	132

$A(1) = 0.1409E+01$ $A(2) = -0.2286E+00$ $A(3) = 0.2346E+00$ $A(4) = -0.1170E+00$ $A(5) = 0.0000E+00$

TABLE 3.1.1 (VLEG)
ISOTHERMAL VAPOR-LIQUID EQUILIBRIA AND EXCESS GIBBS ENERGIES OF BINARY MIXTURES (continued)

1. 2-Methyl-2-propanol, $C_4H_{10}O$
2. Cyclohexane, C_6H_{12}

***T*/K = 318.15**
***System Number*: 230**

x_1	y_1	*P*/kPa	G^E/J mol^{-1}	μ^E_1/J mol^{-1}	μ^E_2/J mol^{-1}
0.0	0.000	29.97	0	5874	0
0.1	0.203	35.37	458	3601	109
0.2	0.247	36.12	728	2340	325
0.3	0.279	36.21	878	1596	570
0.4	0.311	35.89	941	1103	833
0.5	0.346	35.19	935	742	1127
0.6	0.387	34.04	866	468	1461
0.7	0.441	32.22	737	267	1835
0.8	0.523	29.41	553	127	2254
0.9	0.662	25.15	311	38	2769
1.0	1.000	18.11	0	0	3533

B_{11}/cm^{3}mol^{-1}	B_{22}/cm^{3}mol^{-1}	B_{12}/cm^{3}mol^{-1}	V_1/cm^{3}mol^{-1}	V_2/cm^{3}mol^{-1}
-2480	-1420	-1900	98	112

A(1) = 0.1414E+01 *A*(2) = -0.2912E+00 *A*(3) = 0.2275E+00 *A*(4) = -0.1514E+00 *A*(5) = 0.1370E+00

1. 1-Butanol, $C_4H_{10}O$
2. Hexane, C_6H_{14}

***T*/K = 333.15**
***System Number*: 224**

x_1	y_1	*P*/kPa	G^E/J mol^{-1}	μ^E_1/J mol^{-1}	μ^E_2/J mol^{-1}
0.0	0.000	76.28	0	7475	0
0.1	0.054	76.42	578	4497	142
0.2	0.063	75.46	913	2922	411
0.3	0.069	73.88	1101	2027	704
0.4	0.077	71.52	1187	1433	1022
0.5	0.085	68.67	1186	977	1395
0.6	0.094	65.07	1104	612	1841
0.7	0.109	59.48	941	336	2353
0.8	0.139	49.91	702	149	2912
0.9	0.219	34.10	389	40	3531
1.0	1.000	8.10	0	0	4328

B_{11}/cm^{3}mol^{-1}	B_{22}/cm^{3}mol^{-1}	B_{12}/cm^{3}mol^{-1}	V_1/cm^{3}mol^{-1}	V_2/cm^{3}mol^{-1}
-2360	-1390	-920	96	138

A(1) = 0.1713E+01 *A*(2) = -0.3018E+00 *A*(3) = 0.2379E+00 *A*(4) = -0.2665E+00 *A*(5) = 0.1798E+00

1. 2-Methyl-1-propanol, $C_4H_{10}O$
2. Toluene, C_7H_8

***T*/K = 313.15**
***System Number*: 229**

x_1	y_1	*P*/kPa	G^E/J mol^{-1}	μ^E_1/J mol^{-1}	μ^E_2/J mol^{-1}
0.0	0.000	7.894	0	5642	0
0.1	0.190	9.100	463	3797	93
0.2	0.231	9.260	756	2541	310
0.3	0.254	9.253	923	1729	577
0.4	0.281	9.148	996	1207	856
0.5	0.313	8.933	998	845	1151
0.6	0.349	8.617	936	561	1498
0.7	0.391	8.190	808	322	1944
0.8	0.452	7.525	609	135	2504
0.9	0.586	6.270	336	27	3113
1.0	1.000	4.029	0	0	3557

B_{11}/cm^{3}mol^{-1}	B_{22}/cm^{3}mol^{-1}	B_{12}/cm^{3}mol^{-1}	V_1/cm^{3}mol^{-1}	V_2/cm^{3}mol^{-1}
-4010	-2270	-1520	95	109

A(1) = 0.1533E+01 *A*(2) = -0.2348E+00 *A*(3) = 0.3292E+00 *A*(4) = -0.1655E+00 *A*(5) = -0.9531E-01

TABLE 3.1.1 (VLEG)
ISOTHERMAL VAPOR-LIQUID EQUILIBRIA AND EXCESS GIBBS ENERGIES OF BINARY MIXTURES (continued)

1. 2-Methyl-2-propanol, $C_4H_{10}O$
2. Toluene, C_7H_8

***T*/K = 313.15**
***System Number*: 239**

x_1	y_1	P/kPa	G^E/J mol^{-1}	μ^E_1/J mol^{-1}	μ^E_2/J mol^{-1}
0.0	0.000	7.89	0	4968	0
0.1	0.400	12.25	405	3318	81
0.2	0.485	13.59	664	2279	261
0.3	0.533	14.29	819	1592	487
0.4	0.572	14.75	891	1110	745
0.5	0.610	15.09	895	754	1035
0.6	0.651	15.30	836	485	1363
0.7	0.700	15.38	720	284	1737
0.8	0.762	15.28	545	137	2177
0.9	0.848	14.86	309	40	2738
1.0	1.000	13.79	0	0	3532

B_{11}/cm^{3}mol^{-1}	B_{22}/cm^{3}mol^{-1}	B_{12}/cm^{3}mol^{-1}	V_1/cm^{3}mol^{-1}	V_2/cm^{3}mol^{-1}
-2530	-2270	-1330	99	109

$A(1)$ = 0.1374E+01 $A(2)$ = -0.2164E+00 $A(3)$ = 0.1916E+00 $A(4)$ = -0.5949E-01 $A(5)$ = 0.6643E-01

---

1. 1-Butanol, $C_4H_{10}O$
2. Heptane, C_7H_{16}

***T*/K = 333.15**
***System Number*: 225**

x_1	y_1	P/kPa	G^E/J mol^{-1}	μ^E_1/J mol^{-1}	μ^E_2/J mol^{-1}
0.0	0.000	28.02	0	7166	0
0.1	0.137	30.60	570	4563	127
0.2	0.158	30.64	919	3042	388
0.3	0.171	30.38	1119	2117	692
0.4	0.186	29.86	1213	1498	1023
0.5	0.202	29.07	1218	1035	1401
0.6	0.220	27.98	1140	666	1852
0.7	0.246	26.33	980	375	2393
0.8	0.292	23.43	737	165	3022
0.9	0.409	18.00	409	41	3727
1.0	1.000	8.01	0	0	4494

B_{11}/cm^{3}mol^{-1}	B_{22}/cm^{3}mol^{-1}	B_{12}/cm^{3}mol^{-1}	V_1/cm^{3}mol^{-1}	V_2/cm^{3}mol^{-1}
-2360	-2050	-1100	96	154

$A(1)$ = 0.1759E+01 $A(2)$ = -0.2640E+00 $A(3)$ = 0.2773E+00 $A(4)$ = -0.2183E+00 $A(5)$ = 0.6886E-01

---

1. 2,5-Dioxahexane, $C_4H_{10}O_2$
2. (Trifluoromethyl)benzene, $C_7H_5F_3$

***T*/K = 350.00**
***System Number*: 285**

x_1	y_1	P/kPa	G^E/J mol^{-1}	μ^E_1/J mol^{-1}	μ^E_2/J mol^{-1}
0.0	0.000	45.27	0	-685	0
0.1	0.136	47.09	-59	-512	-9
0.2	0.273	49.39	-101	-370	-34
0.3	0.404	52.13	-127	-258	-71
0.4	0.525	55.24	-139	-170	-118
0.5	0.633	58.64	-138	-105	-171
0.6	0.728	62.25	-126	-59	-227
0.7	0.811	66.00	-105	-28	-284
0.8	0.883	69.83	-76	-10	-337
0.9	0.945	73.70	-40	-2	-383
1.0	1.000	77.59	0	0	-420

B_{11}/cm^{3}mol^{-1}	B_{22}/cm^{3}mol^{-1}	B_{12}/cm^{3}mol^{-1}	V_1/cm^{3}mol^{-1}	V_2/cm^{3}mol^{-1}
-1150	-1900	-1500	110	124

$A(1)$ = -0.1898E+00 $A(2)$ = 0.4543E-01 $A(3)$ = 0.0000E+00

TABLE 3.1.1 (VLEG) ISOTHERMAL VAPOR-LIQUID EQUILIBRIA AND EXCESS GIBBS ENERGIES OF BINARY MIXTURES (continued)

1. 2,5-Dioxahexane, $C_4H_{10}O_2$
2. Toluene, C_7H_8

***T*/K = 350.00**
***System Number*: 5**

x_1	y_1	P/kPa	G^E/J mol^{-1}	μ^E_1/J mol^{-1}	μ^E_2/J mol^{-1}
0.0	0.000	34.81	0	-71	0
0.1	0.197	39.05	-1	34	-5
0.2	0.361	43.57	6	86	-14
0.3	0.494	48.15	17	102	-19
0.4	0.602	52.66	29	94	-14
0.5	0.691	57.03	38	74	3
0.6	0.767	61.26	43	49	33
0.7	0.834	65.39	41	27	74
0.8	0.894	69.46	33	11	122
0.9	0.949	73.51	19	2	172
1.0	1.000	77.59	0	0	213

B_{11}/cm^{3}mol^{-1}	B_{22}/cm^{3}mol^{-1}	B_{12}/cm^{3}mol^{-1}	V_1/cm^{3}mol^{-1}	V_2/cm^{3}mol^{-1}
-1150	-1600	-1250	110	114

$A(1) = 0.5254E-01$ $A(2) = 0.4878E-01$ $A(3) = -0.2822E-01$

1. 2,5-Dioxahexane, $C_4H_{10}O_2$
2. Methylcyclohexane, C_7H_{14}

***T*/K = 350.00**
***System Number*: 4**

x_1	y_1	P/kPa	G^E/J mol^{-1}	μ^E_1/J mol^{-1}	μ^E_2/J mol^{-1}
0.0	0.000	48.70	0	1622	0
0.1	0.222	56.81	154	1450	10
0.2	0.372	63.54	286	1243	47
0.3	0.478	68.83	391	1018	122
0.4	0.558	72.84	464	792	245
0.5	0.623	75.80	499	576	422
0.6	0.682	77.90	493	383	658
0.7	0.740	79.29	443	222	958
0.8	0.805	79.91	345	101	1322
0.9	0.887	79.53	198	26	1750
1.0	1.000	77.59	0	0	2240

B_{11}/cm^{3}mol^{-1}	B_{22}/cm^{3}mol^{-1}	B_{12}/cm^{3}mol^{-1}	V_1/cm^{3}mol^{-1}	V_2/cm^{3}mol^{-1}
-1150	-1550	-1300	110	136

$A(1) = 0.6859E+00$ $A(2) = 0.1062E+00$ $A(3) = -0.2225E-01$

1. Butylamine, $C_4H_{11}N$
2. Chlorobenzene, C_6H_5Cl

***T*/K = 333.15**
***System Number*: 173**

x_1	y_1	P/kPa	G^E/J mol^{-1}	μ^E_1/J mol^{-1}	μ^E_2/J mol^{-1}
0.0	0.000	8.85	0	608	0
0.1	0.449	14.56	54	484	6
0.2	0.637	19.85	96	386	24
0.3	0.743	24.85	127	307	50
0.4	0.812	29.65	147	240	86
0.5	0.862	34.25	158	181	134
0.6	0.899	38.70	156	127	200
0.7	0.930	43.01	142	80	289
0.8	0.955	47.22	114	40	410
0.9	0.978	51.44	67	11	574
1.0	1.000	55.77	0	0	793

B_{11}/cm^{3}mol^{-1}	B_{22}/cm^{3}mol^{-1}	B_{12}/cm^{3}mol^{-1}	V_1/cm^{3}mol^{-1}	V_2/cm^{3}mol^{-1}
-970	-2200	-1500	104	106

$A(1) = 0.2275E+00$ $A(2) = 0.3342E-01$ $A(3) = 0.2552E-01$

TABLE 3.1.1 (VLEG)
ISOTHERMAL VAPOR-LIQUID EQUILIBRIA AND EXCESS GIBBS ENERGIES OF BINARY MIXTURES (continued)

1. Diethylamine, $C_4H_{11}N$
2. Chlorobenzene, C_6H_5Cl

***T*/K = 313.15**
***System Number*: 174**

x_1	y_1	P/kPa	G^E/J mol^{-1}	μ^E_1/J mol^{-1}	μ^E_2/J mol^{-1}
0.0	0.000	3.53	0	-54	0
0.1	0.656	9.26	9	181	-10
0.2	0.814	15.28	32	235	-19
0.3	0.881	21.05	56	211	-10
0.4	0.918	26.58	75	169	13
0.5	0.942	32.04	88	135	41
0.6	0.960	37.50	94	110	70
0.7	0.973	42.91	95	88	113
0.8	0.983	48.11	88	58	205
0.9	0.992	53.06	62	22	421
1.0	1.000	58.14	0	0	884

B_{11}/cm^{3}mol^{-1}	B_{22}/cm^{3}mol^{-1}	B_{12}/cm^{3}mol^{-1}	V_1/cm^{3}mol^{-1}	V_2/cm^{3}mol^{-1}
-1300	-2970	-2020	107	104

$A(1) = 0.1348E+00$ $A(2) = 0.7198E-01$ $A(3) = 0.2466E-01$ $A(4) = 0.1083E+00$ $A(5) = 0.0000E+00$

1. Butylamine, $C_4H_{11}N$
2. Benzene, C_6H_6

***T*/K = 323.15**
***System Number*: 156**

x_1	y_1	P/kPa	G^E/J mol^{-1}	μ^E_1/J mol^{-1}	μ^E_2/J mol^{-1}
0.0	0.000	36.24	0	793	0
0.1	0.124	37.42	66	550	12
0.2	0.228	38.19	109	382	41
0.3	0.324	38.70	136	268	79
0.4	0.417	39.02	148	190	120
0.5	0.508	39.20	150	136	165
0.6	0.600	39.26	143	94	216
0.7	0.692	39.19	126	60	279
0.8	0.787	38.97	98	31	367
0.9	0.887	38.56	58	9	493
1.0	1.000	37.86	0	0	675

B_{11}/cm^{3}mol^{-1}	B_{22}/cm^{3}mol^{-1}	B_{12}/cm^{3}mol^{-1}	V_1/cm^{3}mol^{-1}	V_2/cm^{3}mol^{-1}
-1100	-1200	-1100	103	92

$A(1) = 0.2237E+00$ $A(2) = -0.2186E-01$ $A(3) = 0.4950E-01$

1. Diethylamine, $C_4H_{11}N$
2. Benzene, C_6H_6

***T*/K = 308.15**
***System Number*: 26**

x_1	y_1	P/kPa	G^E/J mol^{-1}	μ^E_1/J mol^{-1}	μ^E_2/J mol^{-1}
0.0	0.000	19.86	0	223	0
0.1	0.221	23.00	21	196	1
0.2	0.387	26.07	39	171	6
0.3	0.516	29.07	54	146	14
0.4	0.621	31.99	65	121	28
0.5	0.707	34.82	72	96	49
0.6	0.780	37.57	74	70	81
0.7	0.842	40.22	69	44	128
0.8	0.899	42.79	57	22	195
0.9	0.950	45.30	34	6	287
1.0	1.000	47.78	0	0	411

B_{11}/cm^{3}mol^{-1}	B_{22}/cm^{3}mol^{-1}	B_{12}/cm^{3}mol^{-1}	V_1/cm^{3}mol^{-1}	V_2/cm^{3}mol^{-1}
-1370	-1360	-1360	106	91

$A(1) = 0.1127E+00$ $A(2) = 0.3656E-01$ $A(3) = 0.1106E-01$

TABLE 3.1.1 (VLEG)
ISOTHERMAL VAPOR-LIQUID EQUILIBRIA AND EXCESS GIBBS ENERGIES OF BINARY MIXTURES (continued)

1. Diethylamine, $C_4H_{11}N$ **T/K = 323.15**
2. Triethylamine, $C_6H_{15}N$ ***System Number*: 182**

x_1	y_1	P/kPa	G^E/J mol^{-1}	μ^E_1/J mol^{-1}	μ^E_2/J mol^{-1}
0.0	0.000	26.00	0	852	0
0.1	0.289	33.33	63	459	19
0.2	0.453	39.19	93	227	59
0.3	0.572	44.62	101	106	98
0.4	0.668	50.06	97	54	125
0.5	0.749	55.65	88	39	137
0.6	0.817	61.37	78	38	138
0.7	0.874	67.08	68	35	144
0.8	0.921	72.65	55	25	177
0.9	0.962	78.03	35	9	270
1.0	1.000	83.33	0	0	463

B_{11}/cm^{3}mol^{-1}	B_{22}/cm^{3}mol^{-1}	B_{12}/cm^{3}mol^{-1}	V_1/cm^{3}mol^{-1}	V_2/cm^{3}mol^{-1}
-1200	-1650	-1400	108	144

$A(1)$ = 0.1311E+00 $A(2)$ = -0.7235E-01 $A(3)$ = 0.1136E+00

1. Diethylamine, $C_4H_{11}N$ **T/K = 308.15**
2. Heptane, C_7H_{16} ***System Number*: 155**

x_1	y_1	P/kPa	G^E/J mol^{-1}	μ^E_1/J mol^{-1}	μ^E_2/J mol^{-1}
0.0	0.000	9.83	0	693	0
0.1	0.402	14.90	66	629	4
0.2	0.592	19.60	124	538	20
0.3	0.702	23.91	169	440	53
0.4	0.776	27.87	201	345	105
0.5	0.830	31.56	217	258	175
0.6	0.872	35.03	217	182	269
0.7	0.908	38.33	199	115	394
0.8	0.939	41.50	160	59	564
0.9	0.969	44.61	96	17	804
1.0	1.000	47.78	0	0	1149

B_{11}/cm^{3}mol^{-1}	B_{22}/cm^{3}mol^{-1}	B_{12}/cm^{3}mol^{-1}	V_1/cm^{3}mol^{-1}	V_2/cm^{3}mol^{-1}
-1370	-2570	-1910	106	148

$A(1)$ = 0.3386E+00 $A(2)$ = 0.6480E-01 $A(3)$ = 0.2085E-01 $A(4)$ = 0.2420E-01 $A(5)$ = 0.0000E+00

1. 2-Furaldehyde, $C_5H_4O_2$ **T/K = 368.15**
2. Dibutyl ether, $C_8H_{18}O$ ***System Number*: 328**

x_1	y_1	P/kPa	G^E/J mol^{-1}	μ^E_1/J mol^{-1}	μ^E_2/J mol^{-1}
0.0	0.000	23.14	0	4695	0
0.1	0.148	24.89	418	3741	49
0.2	0.226	25.45	741	2968	185
0.3	0.276	25.48	977	2329	397
0.4	0.311	25.25	1128	1788	689
0.5	0.339	24.85	1196	1319	1073
0.6	0.364	24.32	1176	909	1576
0.7	0.391	23.57	1059	556	2233
0.8	0.431	22.25	836	271	3094
0.9	0.527	19.23	489	75	4218
1.0	1.000	10.97	0	0	5676

B_{11}/cm^{3}mol^{-1}	B_{22}/cm^{3}mol^{-1}	B_{12}/cm^{3}mol^{-1}	V_1/cm^{3}mol^{-1}	V_2/cm^{3}mol^{-1}
-1990	-3150	-1110	89	184

$A(1)$ = 0.1563E+01 $A(2)$ = 0.1603E+00 $A(3)$ = 0.1312E+00

TABLE 3.1.1 (VLEG)
ISOTHERMAL VAPOR-LIQUID EQUILIBRIA AND EXCESS GIBBS ENERGIES OF BINARY MIXTURES (continued)

1. Pyridine, C_5H_5N
2. Benzene, C_6H_6

***T*/K = 298.15**
***System Number*: 172**

x_1	y_1	*P*/kPa	G^E/J mol^{-1}	μ^E_1/J mol^{-1}	μ^E_2/J mol^{-1}
0.0	0.000	12.70	0	627	0
0.1	0.029	11.79	54	467	8
0.2	0.059	10.91	93	342	30
0.3	0.093	10.02	117	246	62
0.4	0.132	9.11	130	172	102
0.5	0.180	8.17	132	116	147
0.6	0.240	7.20	124	73	200
0.7	0.322	6.19	107	42	259
0.8	0.440	5.13	80	19	326
0.9	0.631	4.01	45	5	406
1.0	1.000	2.80	0	0	501

B_{11}/cm^{3}mol^{-1}	B_{22}/cm^{3}mol^{-1}	B_{12}/cm^{3}mol^{-1}	V_1/cm^{3}mol^{-1}	V_2/cm^{3}mol^{-1}
-2460	-1480	-1200	81	89

A(1) = 0.2126E+00 *A*(2) = -0.2540E-01 *A*(3) = 0.1501E-01

---

1. Pyridine, C_5H_5N
2. Cyclohexane, C_6H_{12}

***T*/K = 298.15**
***System Number*: 171**

x_1	y_1	*P*/kPa	G^E/J mol^{-1}	μ^E_1/J mol^{-1}	μ^E_2/J mol^{-1}
0.0	0.000	13.00	0	4611	0
0.1	0.079	13.06	385	3233	68
0.2	0.112	12.82	645	2320	226
0.3	0.133	12.50	811	1687	435
0.4	0.151	12.12	899	1216	688
0.5	0.169	11.67	918	841	994
0.6	0.188	11.11	869	533	1371
0.7	0.214	10.32	750	289	1824
0.8	0.261	9.00	562	118	2337
0.9	0.383	6.64	308	25	2860
1.0	1.000	2.79	0	0	3293

B_{11}/cm^{3}mol^{-1}	B_{22}/cm^{3}mol^{-1}	B_{12}/cm^{3}mol^{-1}	V_1/cm^{3}mol^{-1}	V_2/cm^{3}mol^{-1}
-2460	-1740	-1480	81	108

A(1) = 0.1481E+01 *A*(2) = -0.1225E+00 *A*(3) = 0.1210E+00 *A*(4) = -0.1290E+00 *A*(5) = 0.0000E+00

---

1. Pyridine, C_5H_5N
2. Heptane, C_7H_{16}

***T*/K = 333.15**
***System Number*: 170**

x_1	y_1	*P*/kPa	G^E/J mol^{-1}	μ^E_1/J mol^{-1}	μ^E_2/J mol^{-1}
0.0	0.000	28.05	0	3611	0
0.1	0.149	30.03	334	3070	30
0.2	0.237	30.87	605	2499	131
0.3	0.291	31.03	806	1963	310
0.4	0.330	30.82	935	1491	564
0.5	0.363	30.36	991	1092	891
0.6	0.395	29.65	975	758	1300
0.7	0.432	28.58	881	476	1825
0.8	0.482	26.90	701	244	2528
0.9	0.580	23.63	416	72	3518
1.0	1.000	14.79	0	0	4952

B_{11}/cm^{3}mol^{-1}	B_{22}/cm^{3}mol^{-1}	B_{12}/cm^{3}mol^{-1}	V_1/cm^{3}mol^{-1}	V_2/cm^{3}mol^{-1}
-1450	-2040	-1200	84	154

A(1) = 0.1432E+01 *A*(2) = 0.1456E+00 *A*(3) = 0.1139E+00 *A*(4) = 0.9645E-01 *A*(5) = 0.0000E+00

TABLE 3.1.1 (VLEG)
ISOTHERMAL VAPOR-LIQUID EQUILIBRIA AND EXCESS GIBBS ENERGIES OF BINARY MIXTURES (continued)

1. Methyl 2-methylpropenoate, $C_5H_8O_2$
2. Hexane, C_6H_{14}

***T*/K = 333.15**
***System Number*: 267**

x_1	y_1	P/kPa	G^E/J mol^{-1}	μ^E_1/J mol^{-1}	μ^E_2/J mol^{-1}
0.0	0.000	76.49	0	2569	0
0.1	0.068	74.58	227	2000	29
0.2	0.119	72.12	396	1533	112
0.3	0.161	69.31	512	1149	239
0.4	0.201	66.16	579	835	408
0.5	0.242	62.54	598	580	616
0.6	0.289	58.24	572	375	867
0.7	0.350	52.96	500	215	1163
0.8	0.438	46.17	382	99	1514
0.9	0.595	37.04	216	26	1929
1.0	1.000	24.09	0	0	2423

B_{11}/cm^{3}mol^{-1}	B_{22}/cm^{3}mol^{-1}	B_{12}/cm^{3}mol^{-1}	V_1/cm^{3}mol^{-1}	V_2/cm^{3}mol^{-1}
-1796	-1385	-1350	112	139

$A(1)$ = 0.8638E+00 $A(2)$ = -0.2648E-01 $A(3)$ = 0.3732E-01

1. 1-Pentene, C_5H_{10}
2. Chlorobenzene, C_6H_5Cl

***T*/K = 320.00**
***System Number*: 132**

x_1	y_1	P/kPa	G^E/J mol^{-1}	μ^E_1/J mol^{-1}	μ^E_2/J mol^{-1}
0.0	0.000	4.9	0	1359	0
0.1	0.847	29.8	123	1118	13
0.2	0.919	51.2	221	902	51
0.3	0.946	69.8	293	708	115
0.4	0.961	86.5	339	536	208
0.5	0.970	101.7	358	385	332
0.6	0.978	116.0	349	255	490
0.7	0.984	129.9	311	149	687
0.8	0.989	144.1	241	69	929
0.9	0.995	159.1	138	18	1220
1.0	1.000	175.7	0	0	1568

B_{11}/cm^{3}mol^{-1}	B_{22}/cm^{3}mol^{-1}	B_{12}/cm^{3}mol^{-1}	V_1/cm^{3}mol^{-1}	V_2/cm^{3}mol^{-1}
-900	-2440	-1360	115	105

$A(1)$ = 0.5388E+00 $A(2)$ = 0.3944E-01 $A(3)$ = 0.1129E-01

1. Cyclopentane, C_5H_{10}
2. 2,3-Dimethylbutane, C_6H_{14}

***T*/K = 298.15**
***System Number*: 79**

x_1	y_1	P/kPa	G^E/J mol^{-1}	μ^E_1/J mol^{-1}	μ^E_2/J mol^{-1}
0.0	0.000	31.29	0	50	0
0.1	0.132	32.47	5	41	1
0.2	0.254	33.64	8	32	2
0.3	0.368	34.78	11	25	5
0.4	0.474	35.90	12	18	8
0.5	0.573	37.01	13	13	13
0.6	0.667	38.10	12	8	18
0.7	0.757	39.17	11	5	25
0.8	0.841	40.24	8	2	32
0.9	0.922	41.29	5	1	41
1.0	1.000	42.33	0	0	50

B_{11}/cm^{3}mol^{-1}	B_{22}/cm^{3}mol^{-1}	B_{12}/cm^{3}mol^{-1}	V_1/cm^{3}mol^{-1}	V_2/cm^{3}mol^{-1}
-1060	-1570	-1320	95	131

$A(1)$ = 0.2020E-01 $A(2)$ = 0.0000E+00 $A(3)$ = 0.0000E+00

TABLE 3.1.1 (VLEG)
ISOTHERMAL VAPOR-LIQUID EQUILIBRIA AND EXCESS GIBBS ENERGIES OF BINARY MIXTURES (continued)

1. Oxane, $C_5H_{10}O$
2. Cyclohexane, C_6H_{12}

***T*/K = 298.15**
***System Number*: 215**

x_1	y_1	P/kPa	G^E/J mol^{-1}	μ^E_1/J mol^{-1}	μ^E_2/J mol^{-1}
0.0	0.000	13.01	0	1133	0
0.1	0.104	13.13	101	907	12
0.2	0.192	13.13	179	708	47
0.3	0.271	13.03	234	536	104
0.4	0.346	12.84	266	389	183
0.5	0.420	12.57	275	267	283
0.6	0.499	12.21	262	169	403
0.7	0.587	11.74	228	94	542
0.8	0.691	11.15	173	41	700
0.9	0.822	10.41	97	10	875
1.0	1.000	9.46	0	0	1067

B_{11}/cm^{3}mol^{-1}	B_{22}/cm^{3}mol^{-1}	B_{12}/cm^{3}mol^{-1}	V_1/cm^{3}mol^{-1}	V_2/cm^{3}mol^{-1}
-1710	-1740	-1720	98	109

$A(1)$ = 0.4436E+00 $A(2)$ = -0.1330E-01 $A(3)$ = 0.0000E+00

---

1. N-Methylpyrrolidine, $C_5H_{11}N$
2. Cyclohexane, C_6H_{12}

***T*/K = 298.15**
***System Number*: 160**

x_1	y_1	P/kPa	G^E/J mol^{-1}	μ^E_1/J mol^{-1}	μ^E_2/J mol^{-1}
0.0	0.000	13.01	0	573	0
0.1	0.120	13.35	49	426	8
0.2	0.225	13.59	84	307	29
0.3	0.321	13.75	106	212	60
0.4	0.413	13.85	115	139	99
0.5	0.504	13.89	114	85	143
0.6	0.596	13.89	104	47	189
0.7	0.690	13.85	86	22	235
0.8	0.789	13.77	62	8	277
0.9	0.892	13.67	33	2	313
1.0	1.000	13.54	0	0	340

B_{11}/cm^{3}mol^{-1}	B_{22}/cm^{3}mol^{-1}	B_{12}/cm^{3}mol^{-1}	V_1/cm^{3}mol^{-1}	V_2/cm^{3}mol^{-1}
-1683	-1720	-1378	104	109

$A(1)$ = 0.1840E+00 $A(2)$ = -0.4703E-01 $A(3)$ = 0.0000E+00

---

1. Piperidine, $C_5H_{11}N$
2. Cyclohexane, C_6H_{12}

***T*/K = 298.15**
***System Number*: 27**

x_1	y_1	P/kPa	G^E/J mol^{-1}	μ^E_1/J mol^{-1}	μ^E_2/J mol^{-1}
0.0	0.000	13.01	0	2130	0
0.1	0.057	12.59	174	1418	36
0.2	0.098	12.08	283	946	117
0.3	0.136	11.49	345	643	217
0.4	0.179	10.79	372	452	319
0.5	0.230	9.99	373	326	421
0.6	0.292	9.09	354	233	534
0.7	0.370	8.12	313	155	681
0.8	0.473	7.05	246	84	897
0.9	0.633	5.79	146	26	1233
1.0	1.000	4.02	0	0	1749

B_{11}/cm^{3}mol^{-1}	B_{22}/cm^{3}mol^{-1}	B_{12}/cm^{3}mol^{-1}	V_1/cm^{3}mol^{-1}	V_2/cm^{3}mol^{-1}
-2410	-1740	-1340	99	109

$A(1)$ = 0.6025E+00 $A(2)$ = -0.7689E-01 $A(3)$ = 0.1798E+00

TABLE 3.1.1 (VLEG) ISOTHERMAL VAPOR-LIQUID EQUILIBRIA AND EXCESS GIBBS ENERGIES OF BINARY MIXTURES (continued)

1. N,N-Diethylmethanamide, $C_5H_{11}NO$
2. Benzene, C_6H_6

***T*/K = 313.15**
***System Number*: 189**

x_1	y_1	*P*/kPa	G^E/J mol^{-1}	μ^E_1/J mol^{-1}	μ^E_2/J mol^{-1}
0.0	0.000	24.39	0	934	0
0.1	0.002	22.15	71	529	20
0.2	0.005	20.03	106	285	62
0.3	0.008	17.87	119	146	107
0.4	0.012	15.59	117	73	146
0.5	0.017	13.19	106	37	175
0.6	0.025	10.71	90	19	196
0.7	0.038	8.18	71	10	213
0.8	0.063	5.63	49	4	230
0.9	0.130	3.05	26	1	248
1.0	1.000	0.44	0	0	265

B_{11}/cm^{3}mol^{-1}	B_{22}/cm^{3}mol^{-1}	B_{12}/cm^{3}mol^{-1}	V_1/cm^{3}mol^{-1}	V_2/cm^{3}mol^{-1}
-3300	-1300	-2150	114	91

A(1) = 0.1626E+00 *A*(2) = -0.1065E+00 *A*(3) = 0.6769E-01 *A*(4) = -0.2203E-01 *A*(5) = 0.0000E+00

1. Butyl methyl ether, $C_5H_{12}O$
2. Benzene, C_6H_6

***T*/K = 343.15**
***System Number*: 206**

x_1	y_1	*P*/kPa	G^E/J mol^{-1}	μ^E_1/J mol^{-1}	μ^E_2/J mol^{-1}
0.0	0.000	73.5	0	159	0
0.1	0.136	76.7	14	129	2
0.2	0.260	79.8	25	102	6
0.3	0.373	82.8	33	78	14
0.4	0.478	85.6	38	57	25
0.5	0.576	88.4	40	40	40
0.6	0.668	91.1	38	25	57
0.7	0.756	93.7	33	14	78
0.8	0.840	96.2	25	6	102
0.9	0.921	98.6	14	2	129
1.0	1.000	101.0	0	0	159

B_{11}/cm^{3}mol^{-1}	B_{22}/cm^{3}mol^{-1}	B_{12}/cm^{3}mol^{-1}	V_1/cm^{3}mol^{-1}	V_2/cm^{3}mol^{-1}
-1190	-1030	-1090	128	95

A(1) = 0.5570E-01 *A*(2) = 0.0000E+00 *A*(3) = 0.0000E+00

1. tert-Butyl methyl ether, $C_5H_{12}O$
2. Benzene, C_6H_6

***T*/K = 343.15**
***System Number*: 208**

x_1	y_1	*P*/kPa	G^E/J mol^{-1}	μ^E_1/J mol^{-1}	μ^E_2/J mol^{-1}
0.0	0.000	73.5	0	316	0
0.1	0.204	83.6	29	263	3
0.2	0.361	93.2	52	209	12
0.3	0.486	102.4	68	157	30
0.4	0.590	111.1	77	110	55
0.5	0.678	119.6	79	71	87
0.6	0.755	128.0	74	40	124
0.7	0.824	136.2	62	19	163
0.8	0.888	144.4	45	7	200
0.9	0.946	152.6	24	1	231
1.0	1.000	160.9	0	0	249

B_{11}/cm^{3}mol^{-1}	B_{22}/cm^{3}mol^{-1}	B_{12}/cm^{3}mol^{-1}	V_1/cm^{3}mol^{-1}	V_2/cm^{3}mol^{-1}
-1050	-1030	-1030	125	94

A(1) = 0.1107E+00 *A*(2) = -0.1172E-01 *A*(3) = -0.1184E-01

TABLE 3.1.1 (VLEG)
ISOTHERMAL VAPOR-LIQUID EQUILIBRIA AND EXCESS GIBBS ENERGIES OF BINARY MIXTURES (continued)

1. tert-Butyl methyl ether, $C_5H_{12}O$
2. Hexane, C_6H_{14}

T/K = 333.15
***System Number*: 207**

x_1	y_1	P/kPa	G^E/J mol^{-1}	μ^E_1/J mol^{-1}	μ^E_2/J mol^{-1}
0.0	0.000	76.4	0	565	0
0.1	0.166	82.9	51	458	6
0.2	0.300	88.6	90	362	23
0.3	0.414	93.8	119	277	51
0.4	0.513	98.4	136	203	90
0.5	0.603	102.6	141	141	141
0.6	0.686	106.4	136	90	203
0.7	0.766	109.9	119	51	277
0.8	0.843	113.0	90	23	362
0.9	0.921	115.9	51	6	458
1.0	1.000	118.5	0	0	565

B_{11}/cm^{3}mol^{-1}	B_{22}/cm^{3}mol^{-1}	B_{12}/cm^{3}mol^{-1}	V_1/cm^{3}mol^{-1}	V_2/cm^{3}mol^{-1}
-1140	-1390	-1240	124	139

$A(1)$ = 0.2041E+00 $A(2)$ = 0.0000E+00 $A(3)$ = 0.0000E+00

---

1. Butyl methyl ether, $C_5H_{12}O$
2. Heptane, C_7H_{16}

T/K = 323.15
***System Number*: 205**

x_1	y_1	P/kPa	G^E/J mol^{-1}	μ^E_1/J mol^{-1}	μ^E_2/J mol^{-1}
0.0	0.000	18.91	0	490	0
0.1	0.253	22.92	44	397	5
0.2	0.424	26.63	78	314	20
0.3	0.549	30.09	103	240	44
0.4	0.646	33.36	118	176	78
0.5	0.725	36.48	123	123	123
0.6	0.792	39.48	118	78	176
0.7	0.851	42.39	103	44	240
0.8	0.904	45.26	78	20	314
0.9	0.953	48.10	44	5	397
1.0	1.000	50.94	0	0	490

B_{11}/cm^{3}mol^{-1}	B_{22}/cm^{3}mol^{-1}	B_{12}/cm^{3}mol^{-1}	V_1/cm^{3}mol^{-1}	V_2/cm^{3}mol^{-1}
-1390	-2240	-1741	125	152

$A(1)$ = 0.1824E+00 $A(2)$ = 0.0000E+00 $A(3)$ = 0.0000E+00

---

1. Hexafluorobenzene, C_6F_6
2. Benzene, C_6H_6

T/K = 303.15
***System Number*: 6**

x_1	y_1	P/kPa	G^E/J mol^{-1}	μ^E_1/J mol^{-1}	μ^E_2/J mol^{-1}
0.0	0.000	15.90	0	451	0
0.1	0.094	15.91	26	103	17
0.2	0.175	15.74	22	-99	52
0.3	0.257	15.48	-1	-194	82
0.4	0.347	15.16	-31	-215	92
0.5	0.449	14.82	-59	-189	70
0.6	0.560	14.51	-80	-140	9
0.7	0.678	14.28	-88	-86	-93
0.8	0.796	14.18	-78	-40	-232
0.9	0.904	14.19	-49	-10	-400
1.0	1.000	14.32	0	0	-585

B_{11}/cm^{3}mol^{-1}	B_{22}/cm^{3}mol^{-1}	B_{12}/cm^{3}mol^{-1}	V_1/cm^{3}mol^{-1}	V_2/cm^{3}mol^{-1}
-1970	-1420	-1700	117	90

$A(1)$ = -0.9432E-01 $A(2)$ = -0.2055E+00 $A(3)$ = 0.6778E-01

TABLE 3.1.1 (VLEG)
ISOTHERMAL VAPOR-LIQUID EQUILIBRIA AND EXCESS GIBBS ENERGIES OF BINARY MIXTURES (continued)

1. Hexafluorobenzene, C_6F_6 — ***T*/K = 323.15**
2. Cyclohexane, C_6H_{12} — ***System Number*: 382**

x_1	y_1	*P*/kPa	G^E/J mol^{-1}	μ^E_1/J mol^{-1}	μ^E_2/J mol^{-1}
0.0	0.000	36.12	0	3711	0
0.1	0.216	42.45	313	2647	54
0.2	0.308	44.96	526	1889	186
0.3	0.369	46.13	660	1353	363
0.4	0.423	46.62	729	971	567
0.5	0.476	46.62	743	690	796
0.6	0.532	46.12	708	472	1063
0.7	0.594	45.08	624	294	1395
0.8	0.668	43.31	486	148	1835
0.9	0.777	40.24	283	43	2442
1.0	1.000	34.04	0	0	3289

B_{11}/cm^{3}mol^{-1}	B_{22}/cm^{3}mol^{-1}	B_{12}/cm^{3}mol^{-1}	V_1/cm^{3}mol^{-1}	V_2/cm^{3}mol^{-1}
-1630	-1350	-1500	120	112

A(1) = 0.1106E+01 *A*(2) = -0.7853E-01 *A*(3) = 0.1964E+00

---

1. Hexafluorobenzene, C_6F_6 — ***T*/K = 283.15**
2. Triethylamine, $C_6H_{15}N$ — ***System Number*: 385**

x_1	y_1	*P*/kPa	G^E/J mol^{-1}	μ^E_1/J mol^{-1}	μ^E_2/J mol^{-1}
0.0	0.000	4.200	0	1794	0
0.1	0.200	4.764	161	1453	18
0.2	0.326	5.143	287	1148	72
0.3	0.416	5.395	377	879	161
0.4	0.487	5.560	431	646	287
0.5	0.550	5.659	449	449	449
0.6	0.612	5.703	431	287	646
0.7	0.678	5.693	377	161	879
0.8	0.756	5.616	287	72	1148
0.9	0.857	5.447	161	18	1453
1.0	1.000	5.144	0	0	1794

B_{11}/cm^{3}mol^{-1}	B_{22}/cm^{3}mol^{-1}	B_{12}/cm^{3}mol^{-1}	V_1/cm^{3}mol^{-1}	V_2/cm^{3}mol^{-1}
-3310	-2130	-2790	113	137

A(1) = 0.7622E+00 *A*(2) = 0.0000E+00 *A*(3) = 0.0000E+00

---

1. Hexafluorobenzene, C_6F_6 — ***T*/K = 303.15**
2. Toluene, C_7H_8 — ***System Number*: 383**

x_1	y_1	*P*/kPa	G^E/J mol^{-1}	μ^E_1/J mol^{-1}	μ^E_2/J mol^{-1}
0.0	0.000	4.89	0	-227	0
0.1	0.213	5.61	-35	-437	10
0.2	0.370	6.27	-85	-512	22
0.3	0.504	6.96	-138	-493	14
0.4	0.624	7.73	-183	-418	-27
0.5	0.728	8.63	-214	-315	-112
0.6	0.815	9.65	-222	-209	-242
0.7	0.884	10.78	-206	-117	-414
0.8	0.936	11.97	-163	-50	-616
0.9	0.973	13.16	-93	-11	-832
1.0	1.000	14.30	0	0	-1039

B_{11}/cm^{3}mol^{-1}	B_{22}/cm^{3}mol^{-1}	B_{12}/cm^{3}mol^{-1}	V_1/cm^{3}mol^{-1}	V_2/cm^{3}mol^{-1}
-1970	-2500	-2240	117	107

A(1) = -0.3390E+00 *A*(2) = -0.1611E+00 *A*(3) = 0.8775E-01

TABLE 3.1.1 (VLEG) ISOTHERMAL VAPOR-LIQUID EQUILIBRIA AND EXCESS GIBBS ENERGIES OF BINARY MIXTURES (continued)

1. Hexafluorobenzene, C_6F_6 **2. 1,4-Dimethylbenzene, C_8H_{10}**

***T*/K = 313.15** ***System Number*: 384**

x_1	y_1	P/kPa	G^E/J mol^{-1}	μ^E_1/J mol^{-1}	μ^E_2/J mol^{-1}
0.0	0.000	2.65	0	-861	0
0.1	0.386	3.90	-95	-999	6
0.2	0.588	5.17	-195	-979	1
0.3	0.722	6.62	-286	-856	-42
0.4	0.818	8.36	-355	-680	-138
0.5	0.885	10.42	-392	-488	-296
0.6	0.930	12.75	-392	-310	-514
0.7	0.961	15.26	-351	-166	-781
0.8	0.980	17.80	-269	-67	-1077
0.9	0.992	20.22	-150	-14	-1373
1.0	1.000	22.45	0	0	-1630

B_{11}/cm^{3}mol^{-1}	B_{22}/cm^{3}mol^{-1}	B_{12}/cm^{3}mol^{-1}	V_1/cm^{3}mol^{-1}	V_2/cm^{3}mol^{-1}
-1800	-3200	-3350	118	125

$A(1)$ = -0.6021E+00 $A(2)$ = -0.1478E+00 $A(3)$ = 0.1237E+00

1. Bromobenzene, C_6H_5Br **2. Benzene, C_6H_6**

***T*/K = 353.15** ***System Number*: 109**

x_1	y_1	P/kPa	G^E/J mol^{-1}	μ^E_1/J mol^{-1}	μ^E_2/J mol^{-1}
0.0	0.000	101.0	0	-94	0
0.1	0.010	91.4	-5	-9	-4
0.2	0.023	81.8	-2	45	-14
0.3	0.040	72.3	6	75	-23
0.4	0.061	63.0	17	84	-28
0.5	0.087	54.0	28	79	-23
0.6	0.124	45.1	37	64	-4
0.7	0.177	36.4	41	43	34
0.8	0.262	27.6	37	22	97
0.9	0.434	18.5	25	6	190
1.0	1.000	8.8	0	0	316

B_{11}/cm^{3}mol^{-1}	B_{22}/cm^{3}mol^{-1}	B_{12}/cm^{3}mol^{-1}	V_1/cm^{3}mol^{-1}	V_2/cm^{3}mol^{-1}
-2900	-960	-1760	111	96

$A(1)$ = 0.3790E-01 $A(2)$ = 0.6976E-01 $A(3)$ = 0.0000E+00

1. Bromobenzene, C_6H_5Br **2. Cyclohexane, C_6H_{12}**

***T*/K = 298.15** ***System Number*: 107**

x_1	y_1	P/kPa	G^E/J mol^{-1}	μ^E_1/J mol^{-1}	μ^E_2/J mol^{-1}
0.0	0.000	13.01	0	2106	0
0.1	0.009	11.94	183	1593	26
0.2	0.017	11.01	315	1182	98
0.3	0.025	10.13	402	856	207
0.4	0.033	9.25	447	601	344
0.5	0.042	8.30	454	402	506
0.6	0.054	7.24	427	251	690
0.7	0.073	6.01	367	139	897
0.8	0.106	4.56	275	62	1129
0.9	0.190	2.79	153	16	1392
1.0	1.000	0.58	0	0	1691

B_{11}/cm^{3}mol^{-1}	B_{22}/cm^{3}mol^{-1}	B_{12}/cm^{3}mol^{-1}	V_1/cm^{3}mol^{-1}	V_2/cm^{3}mol^{-1}
-5000	-1740	-3090	106	108

$A(1)$ = 0.7325E+00 $A(2)$ = -0.8366E-01 $A(3)$ = 0.3339E-01

TABLE 3.1.1 (VLEG)
ISOTHERMAL VAPOR-LIQUID EQUILIBRIA AND EXCESS GIBBS ENERGIES OF BINARY MIXTURES (continued)

1. Chlorobenzene, C_6H_5Cl
2. Benzene, C_6H_6

***T*/K = 298.15**
***System Number*: 120**

x_1	y_1	*P*/kPa	G^E/J mol^{-1}	μ^E_1/J mol^{-1}	μ^E_2/J mol^{-1}
0.0	0.000	12.76	0	144	0
0.1	0.015	11.66	11	88	3
0.2	0.032	10.56	18	56	8
0.3	0.053	9.46	21	38	14
0.4	0.080	8.36	23	29	19
0.5	0.115	7.24	24	24	23
0.6	0.162	6.13	23	20	28
0.7	0.230	5.02	22	15	37
0.8	0.337	3.91	19	9	57
0.9	0.529	2.79	12	3	92
1.0	1.000	1.63	0	0	152

B_{11}/cm^{3}mol^{-1}	B_{22}/cm^{3}mol^{-1}	B_{12}/cm^{3}mol^{-1}	V_1/cm^{3}mol^{-1}	V_2/cm^{3}mol^{-1}
-3130	-1480	-2070	102	89

A(1) = 0.3794E-01 *A*(2) = 0.1500E-02 *A*(3) = 0.2180E-01

---

1. Chlorobenzene, C_6H_5Cl
2. Aniline, C_6H_7N

***T*/K = 343.15**
***System Number*: 179**

x_1	y_1	*P*/kPa	G^E/J mol^{-1}	μ^E_1/J mol^{-1}	μ^E_2/J mol^{-1}
0.0	0.000	1.42	0	2133	0
0.1	0.654	3.72	192	1721	22
0.2	0.785	5.46	340	1358	86
0.3	0.844	6.84	446	1045	190
0.4	0.879	7.97	511	778	333
0.5	0.905	8.97	535	554	517
0.6	0.925	9.87	519	368	744
0.7	0.943	10.72	460	218	1023
0.8	0.960	11.56	357	104	1368
0.9	0.978	12.44	205	28	1800
1.0	1.000	13.40	0	0	2348

B_{11}/cm^{3}mol^{-1}	B_{22}/cm^{3}mol^{-1}	B_{12}/cm^{3}mol^{-1}	V_1/cm^{3}mol^{-1}	V_2/cm^{3}mol^{-1}
-1860	-2580	-2100	107	95

A(1) = 0.7506E+00 *A*(2) = 0.2605E-01 *A*(3) = 0.3476E-01 *A*(4) = 0.1174E-01 *A*(5) = 0.0000E+00

---

1. Chlorobenzene, C_6H_5Cl
2. Triethylamine, $C_6H_{15}N$

***T*/K = 343.15**
***System Number*: 177**

x_1	y_1	*P*/kPa	G^E/J mol^{-1}	μ^E_1/J mol^{-1}	μ^E_2/J mol^{-1}
0.0	0.000	54.93	0	877	0
0.1	0.033	51.27	73	612	14
0.2	0.066	47.67	121	417	47
0.3	0.103	44.01	149	278	93
0.4	0.144	40.20	160	181	145
0.5	0.195	36.22	157	114	199
0.6	0.260	32.05	143	69	255
0.7	0.346	27.69	120	38	312
0.8	0.468	23.14	89	17	374
0.9	0.658	18.39	49	5	447
1.0	1.000	13.37	0	0	538

B_{11}/cm^{3}mol^{-1}	B_{22}/cm^{3}mol^{-1}	B_{12}/cm^{3}mol^{-1}	V_1/cm^{3}mol^{-1}	V_2/cm^{3}mol^{-1}
-1860	-1400	-1600	107	148

A(1) = 0.2200E+00 *A*(2) = -0.5957E-01 *A*(3) = 0.2793E-01

TABLE 3.1.1 (VLEG)
ISOTHERMAL VAPOR-LIQUID EQUILIBRIA AND EXCESS GIBBS ENERGIES OF BINARY MIXTURES (continued)

1. Chlorobenzene, C_6H_5Cl
2. Toluene, C_7H_8

***T*/K = 343.15**
***System Number*: 121**

x_1	y_1	*P*/kPa	G^E/J mol^{-1}	μ^E_1/J mol^{-1}	μ^E_2/J mol^{-1}
0.0	0.000	27.18	0	-48	0
0.1	0.052	25.77	-4	-36	-1
0.2	0.110	24.36	-7	-26	-2
0.3	0.175	22.95	-9	-18	-5
0.4	0.248	21.56	-10	-12	-8
0.5	0.332	20.17	-10	-7	-12
0.6	0.428	18.79	-9	-4	-16
0.7	0.538	17.42	-7	-2	-20
0.8	0.667	16.06	-5	-1	-23
0.9	0.818	14.71	-3	0	-26
1.0	1.000	13.37	0	0	-28

B_{11}/cm^{3}mol^{-1}	B_{22}/cm^{3}mol^{-1}	B_{12}/cm^{3}mol^{-1}	V_1/cm^{3}mol^{-1}	V_2/cm^{3}mol^{-1}
-1860	-1740	-1800	107	113

A(1) = -0.1339E-01 *A*(2) = 0.3583E-02 *A*(3) = 0.0000E+00

1. Chlorobenzene, C_6H_5Cl
2. Heptane, C_7H_{16}

***T*/K = 353.15**
***System Number*: 119**

x_1	y_1	*P*/kPa	G^E/J mol^{-1}	μ^E_1/J mol^{-1}	μ^E_2/J mol^{-1}
0.0	0.000	57.09	0	1450	0
0.1	0.055	54.54	132	1203	13
0.2	0.108	51.93	238	979	53
0.3	0.159	49.23	317	774	121
0.4	0.210	46.42	368	590	220
0.5	0.265	43.44	390	426	355
0.6	0.327	40.18	382	285	528
0.7	0.403	36.49	341	167	746
0.8	0.506	32.13	266	78	1016
0.9	0.669	26.72	153	20	1343
1.0	1.000	19.65	0	0	1737

B_{11}/cm^{3}mol^{-1}	B_{22}/cm^{3}mol^{-1}	B_{12}/cm^{3}mol^{-1}	V_1/cm^{3}mol^{-1}	V_2/cm^{3}mol^{-1}
-1500	-1730	-1610	108	160

A(1) = 0.5319E+00 *A*(2) = 0.4892E-01 *A*(3) = 0.1073E-01

1. Chlorobenzene, C_6H_5Cl
2. Ethylbenzene, C_8H_{10}

***T*/K = 293.15**
***System Number*: 131**

x_1	y_1	*P*/kPa	G^E/J mol^{-1}	μ^E_1/J mol^{-1}	μ^E_2/J mol^{-1}
0.0	0.000	0.97	0	1	0
0.1	0.123	0.99	0	1	0
0.2	0.241	1.02	0	1	0
0.3	0.352	1.04	0	1	0
0.4	0.458	1.07	0	0	0
0.5	0.559	1.10	0	0	0
0.6	0.655	1.12	0	0	0
0.7	0.747	1.15	0	0	1
0.8	0.835	1.17	0	0	1
0.9	0.919	1.20	0	0	1
1.0	1.000	1.23	0	0	1

B_{11}/cm^{3}mol^{-1}	B_{22}/cm^{3}mol^{-1}	B_{12}/cm^{3}mol^{-1}	V_1/cm^{3}mol^{-1}	V_2/cm^{3}mol^{-1}
-3330	-4020	-3560	102	122

A(1) = 0.4576E-03 *A*(2) = 0.0000E+00 *A*(3) = 0.0000E+00

TABLE 3.1.1 (VLEG)
ISOTHERMAL VAPOR-LIQUID EQUILIBRIA AND EXCESS GIBBS ENERGIES OF BINARY MIXTURES (continued)

1. Fluorobenzene, C_6H_5F
2. Benzene, C_6H_6

***T*/K = 348.15**
***System Number*: 28**

x_1	y_1	*P*/kPa	G^E/J mol^{-1}	μ^E_1/J mol^{-1}	μ^E_2/J mol^{-1}
0.0	0.000	86.26	0	117	0
0.1	0.090	85.36	10	88	2
0.2	0.181	84.38	17	63	6
0.3	0.273	83.32	22	47	11
0.4	0.368	82.22	25	40	15
0.5	0.466	81.09	27	38	16
0.6	0.567	79.96	29	37	18
0.7	0.669	78.81	31	32	28
0.8	0.774	77.60	29	21	61
0.9	0.882	76.21	21	8	140
1.0	1.000	74.39	0	0	301

B_{11}/cm^{3}mol^{-1}	B_{22}/cm^{3}mol^{-1}	B_{12}/cm^{3}mol^{-1}	V_1/cm^{3}mol^{-1}	V_2/cm^{3}mol^{-1}
-1180	-1000	-1090	101	95

A(1) = 0.3769E-01 *A*(2) = 0.1499E-01 *A*(3) = 0.3440E-01 *A*(4) = 0.1676E-01 *A*(5) = 0.0000E+00

---

1. Fluorobenzene, C_6H_5F
2. Toluene, C_7H_8

***T*/K = 323.15**
***System Number*: 144**

x_1	y_1	*P*/kPa	G^E/J mol^{-1}	μ^E_1/J mol^{-1}	μ^E_2/J mol^{-1}
0.0	0.000	12.28	0	53	0
0.1	0.216	14.12	5	43	1
0.2	0.381	15.94	8	34	2
0.3	0.513	17.75	11	26	5
0.4	0.620	19.55	13	19	8
0.5	0.709	21.35	13	13	13
0.6	0.784	23.13	13	8	19
0.7	0.849	24.92	11	5	26
0.8	0.906	26.70	8	2	34
0.9	0.956	28.48	5	1	43
1.0	1.000	30.26	0	0	53

B_{11}/cm^{3}mol^{-1}	B_{22}/cm^{3}mol^{-1}	B_{12}/cm^{3}mol^{-1}	V_1/cm^{3}mol^{-1}	V_2/cm^{3}mol^{-1}
-1400	-2080	-1720	96	110

A(1) = 0.1969E-01 *A*(2) = 0.0000E+00 *A*(3) = 0.0000E+00

---

1. Nitrobenzene, $C_6H_5NO_2$
2. Toluene, C_7H_8

***T*/K = 373.15**
***System Number*: 194**

x_1	y_1	*P*/kPa	G^E/J mol^{-1}	μ^E_1/J mol^{-1}	μ^E_2/J mol^{-1}
0.0	0.000	74.75	0	1533	0
0.1	0.007	67.91	137	1210	18
0.2	0.013	61.73	237	882	76
0.3	0.020	55.86	299	605	168
0.4	0.028	49.85	326	401	277
0.5	0.038	43.35	326	266	386
0.6	0.052	36.27	304	181	489
0.7	0.075	28.74	265	124	596
0.8	0.114	20.95	208	73	749
0.9	0.206	12.73	126	26	1029
1.0	1.000	2.84	0	0	1564

B_{11}/cm^{3}mol^{-1}	B_{22}/cm^{3}mol^{-1}	B_{12}/cm^{3}mol^{-1}	V_1/cm^{3}mol^{-1}	V_2/cm^{3}mol^{-1}
-5770	-1360	-3050	110	117

A(1) = 0.4202E+00 *A*(2) = -0.7749E-01 *A*(3) = 0.7889E-01 *A*(4) = 0.8255E-01 *A*(5) = 0.0000E+00

TABLE 3.1.1 (VLEG)
ISOTHERMAL VAPOR-LIQUID EQUILIBRIA AND EXCESS GIBBS ENERGIES OF BINARY MIXTURES (continued)

1. Nitrobenzene, $C_6H_5NO_2$
2. Propylbenzene, C_9H_{12}

***T*/K = 373.15**
***System Number*: 195**

x_1	y_1	*P*/kPa	G^E/J mol^{-1}	μ^E_1/J mol^{-1}	μ^E_2/J mol^{-1}
0.0	0.000	16.72	0	2095	0
0.1	0.031	15.65	183	1599	26
0.2	0.058	14.62	316	1201	95
0.3	0.085	13.61	406	885	200
0.4	0.113	12.56	455	634	335
0.5	0.145	11.43	466	436	497
0.6	0.185	10.19	443	280	687
0.7	0.240	8.80	385	160	909
0.8	0.326	7.18	293	73	1170
0.9	0.492	5.25	165	19	1479
1.0	1.000	2.84	0	0	1849

B_{11}/cm^{3}mol^{-1}	B_{22}/cm^{3}mol^{-1}	B_{12}/cm^{3}mol^{-1}	V_1/cm^{3}mol^{-1}	V_2/cm^{3}mol^{-1}
-5770	-2370	-3820	110	151

A(1) = 0.6013E+00 *A*(2) = -0.3974E-01 *A*(3) = 0.3437E-01

---

1. Nitrobenzene, $C_6H_5NO_2$
2. Butylbenzene, $C_{10}H_{14}$

***T*/K = 373.15**
***System Number*: 196**

x_1	y_1	*P*/kPa	G^E/J mol^{-1}	μ^E_1/J mol^{-1}	μ^E_2/J mol^{-1}
0.0	0.000	7.46	0	2282	0
0.1	0.069	7.27	200	1756	27
0.2	0.125	7.04	348	1329	102
0.3	0.174	6.77	447	983	217
0.4	0.222	6.46	502	705	367
0.5	0.273	6.11	515	483	548
0.6	0.332	5.69	489	308	761
0.7	0.406	5.20	425	175	1009
0.8	0.509	4.59	322	79	1296
0.9	0.672	3.83	181	20	1630
1.0	1.000	2.84	0	0	2021

B_{11}/cm^{3}mol^{-1}	B_{22}/cm^{3}mol^{-1}	B_{12}/cm^{3}mol^{-1}	V_1/cm^{3}mol^{-1}	V_2/cm^{3}mol^{-1}
-5770	-3270	-4400	110	169

A(1) = 0.6644E+00 *A*(2) = -0.4209E-01 *A*(3) = 0.2906E-01

1.Benzene, C_6H_6
2. Phenol, C_6H_6O

***T*/K = 353.15**
***System Number*: 241**

x_1	y_1	*P*/kPa	G^E/J mol^{-1}	μ^E_1/J mol^{-1}	μ^E_2/J mol^{-1}
0.0	0.000	2.1	0	3913	0
0.1	0.917	23.4	295	2289	73
0.2	0.951	36.5	474	1649	181
0.3	0.966	48.7	599	1338	283
0.4	0.975	59.3	686	1087	419
0.5	0.981	67.8	730	832	628
0.6	0.985	74.9	725	593	921
0.7	0.988	81.5	663	393	1295
0.8	0.991	88.0	541	227	1799
0.9	0.995	94.1	338	81	2651
1.0	1.000	101.4	0	0	4397

B_{11}/cm^{3}mol^{-1}	B_{22}/cm^{3}mol^{-1}	B_{12}/cm^{3}mol^{-1}	V_1/cm^{3}mol^{-1}	V_2/cm^{3}mol^{-1}
-960	-3120	-960	96	92

A(1) = 0.9949E+00 *A*(2) = 0.1388E+00 *A*(3) = 0.1364E+00 *A*(4) = -0.5640E-01 *A*(5) = 0.2837E+00

TABLE 3.1.1 (VLEG) ISOTHERMAL VAPOR-LIQUID EQUILIBRIA AND EXCESS GIBBS ENERGIES OF BINARY MIXTURES (continued)

1. Benzene, C_6H_6
2. Aniline, C_6H_7N

T/K = 343.15
System Number: 163

x_1	y_1	P/kPa	G^E/J mol^{-1}	μ^E_1/J mol^{-1}	μ^E_2/J mol^{-1}
0.0	0.000	1.42	0	1817	0
0.1	0.904	13.57	167	1532	15
0.2	0.950	23.64	303	1267	62
0.3	0.967	32.09	407	1022	144
0.4	0.975	39.25	478	794	267
0.5	0.981	45.45	512	586	438
0.6	0.985	50.99	506	399	667
0.7	0.989	56.16	457	240	964
0.8	0.992	61.34	360	114	1343
0.9	0.996	66.92	209	30	1819
1.0	1.000	73.45	0	0	2408

B_{11}/cm^{3}mol^{-1}	B_{22}/cm^{3}mol^{-1}	B_{12}/cm^{3}mol^{-1}	V_1/cm^{3}mol^{-1}	V_2/cm^{3}mol^{-1}
-1030	-2580	-1680	94	95

A(1) = 0.7175E+00 A(2) = 0.1035E+00 A(3) = 0.2299E-01

1. Benzene, C_6H_6
2. Cyclohexane, C_6H_{12}

T/K = 293.15
System Number: 9

x_1	y_1	P/kPa	G^E/J mol^{-1}	μ^E_1/J mol^{-1}	μ^E_2/J mol^{-1}
0.0	0.000	10.33	0	1277	0
0.1	0.141	10.89	115	1038	12
0.2	0.250	11.26	205	830	49
0.3	0.341	11.50	271	648	110
0.4	0.422	11.63	313	489	196
0.5	0.497	11.66	330	351	309
0.6	0.571	11.61	321	233	452
0.7	0.649	11.45	285	137	631
0.8	0.737	11.17	222	64	852
0.9	0.847	10.72	127	17	1120
1.0	1.000	10.02	0	0	1446

B_{11}/cm^{3}mol^{-1}	B_{22}/cm^{3}mol^{-1}	B_{12}/cm^{3}mol^{-1}	V_1/cm^{3}mol^{-1}	V_2/cm^{3}mol^{-1}
-1540	-1830	-1685	89	108

A(1) = 0.5411E+00 A(2) = 0.3467E-01 A(3) = 0.1732E-01

1. Benzene, C_6H_6
2. 1-Hexene, C_6H_{12}

T/K = 298.15
System Number: 94

x_1	y_1	P/kPa	G^E/J mol^{-1}	μ^E_1/J mol^{-1}	μ^E_2/J mol^{-1}
0.0	0.000	24.66	0	917	0
0.1	0.072	23.99	82	745	9
0.2	0.140	23.24	148	604	34
0.3	0.208	22.41	197	484	74
0.4	0.276	21.50	230	378	131
0.5	0.348	20.52	246	284	208
0.6	0.426	19.43	244	198	313
0.7	0.514	18.20	222	123	454
0.8	0.621	16.78	177	60	643
0.9	0.766	15.01	104	17	892
1.0	1.000	12.68	0	0	1219

B_{11}/cm^{3}mol^{-1}	B_{22}/cm^{3}mol^{-1}	B_{12}/cm^{3}mol^{-1}	V_1/cm^{3}mol^{-1}	V_2/cm^{3}mol^{-1}
-1480	-1710	-1595	89	126

A(1) = 0.3969E+00 A(2) = 0.6094E-01 A(3) = 0.3378E-01

TABLE 3.1.1 (VLEG)
ISOTHERMAL VAPOR-LIQUID EQUILIBRIA AND EXCESS GIBBS ENERGIES OF BINARY MIXTURES (continued)

1.Benzene, C_6H_6
2. Hexane, C_6H_{14}

***T*/K = 298.15**
***System Number*: 91**

x_1	y_1	P/kPa	G^E/J mol^{-1}	μ^E_1/J mol^{-1}	μ^E_2/J mol^{-1}
0.0	0.000	20.15	0	1324	0
0.1	0.099	20.22	122	1127	10
0.2	0.185	20.13	223	943	43
0.3	0.262	19.90	302	770	101
0.4	0.333	19.55	356	607	189
0.5	0.401	19.08	384	454	314
0.6	0.470	18.46	383	314	486
0.7	0.544	17.67	349	191	715
0.8	0.636	16.61	277	92	1015
0.9	0.766	15.08	162	25	1399
1.0	1.000	12.68	0	0	1884

B_{11}/cm^{3}mol^{-1}	B_{22}/cm^{3}mol^{-1}	B_{12}/cm^{3}mol^{-1}	V_1/cm^{3}mol^{-1}	V_2/cm^{3}mol^{-1}
-1480	-1970	-1725	89	132

$A(1)$ = 0.6202E+00 $A(2)$ = 0.1130E+00 $A(3)$ = 0.2698E-01

1.Benzene, C_6H_6
2. Dipropyl ether, $C_6H_{14}O$

***T*/K = 343.15**
***System Number*: 209**

x_1	y_1	P/kPa	G^E/J mol^{-1}	μ^E_1/J mol^{-1}	μ^E_2/J mol^{-1}
0.0	0.000	53.01	0	153	0
0.1	0.137	55.38	13	112	2
0.2	0.261	57.65	22	88	6
0.3	0.375	59.86	30	73	11
0.4	0.482	62.03	35	63	16
0.5	0.581	64.16	39	54	24
0.6	0.673	66.23	41	43	38
0.7	0.760	68.22	40	30	62
0.8	0.842	70.10	34	17	103
0.9	0.921	71.85	22	5	169
1.0	1.000	73.44	0	0	271

B_{11}/cm^{3}mol^{-1}	B_{22}/cm^{3}mol^{-1}	B_{12}/cm^{3}mol^{-1}	V_1/cm^{3}mol^{-1}	V_2/cm^{3}mol^{-1}
-1030	-1400	-1200	95	149

$A(1)$ = 0.5454E-01 $A(2)$ = 0.2070E-01 $A(3)$ = 0.1986E-01

1.Benzene, C_6H_6
2. Triethylamine, $C_6H_{15}N$

***T*/K = 333.15**
***System Number*: 32**

x_1	y_1	P/kPa	G^E/J mol^{-1}	μ^E_1/J mol^{-1}	μ^E_2/J mol^{-1}
0.0	0.000	38.70	0	386	0
0.1	0.144	40.78	36	339	3
0.2	0.270	42.71	67	291	11
0.3	0.383	44.50	92	242	27
0.4	0.484	46.12	110	193	54
0.5	0.577	47.59	119	146	93
0.6	0.664	48.89	120	101	147
0.7	0.747	50.01	110	62	221
0.8	0.828	50.96	87	30	318
0.9	0.912	51.71	51	8	442
1.0	1.000	52.24	0	0	599

B_{11}/cm^{3}mol^{-1}	B_{22}/cm^{3}mol^{-1}	B_{12}/cm^{3}mol^{-1}	V_1/cm^{3}mol^{-1}	V_2/cm^{3}mol^{-1}
-1110	-1520	-1300	93	145

$A(1)$ = 0.1723E+00 $A(2)$ = 0.3845E-01 $A(3)$ = 0.5361E-02

TABLE 3.1.1 (VLEG) ISOTHERMAL VAPOR-LIQUID EQUILIBRIA AND EXCESS GIBBS ENERGIES OF BINARY MIXTURES (continued)

1.Benzene, C_6H_6
2. Phosphoric tris(dimethylamide), $C_6H_{18}N_3OP$

***T/K* = 298.15**
***System Number*: 204**

x_1	y_1	P/kPa	G^E/J mol^{-1}	μ^E_1/J mol^{-1}	μ^E_2/J mol^{-1}
0.0	0.000	0.01	0	-1624	0
0.1	0.992	0.75	-146	-1310	-16
0.2	0.997	1.66	-259	-1032	-66
0.3	0.998	2.75	-339	-787	-147
0.4	0.999	4.00	-387	-576	-261
0.5	0.999	5.37	-402	-398	-406
0.6	1.000	6.84	-385	-254	-582
0.7	1.000	8.35	-337	-142	-790
0.8	1.000	9.86	-256	-63	-1028
0.9	1.000	11.32	-144	-16	-1296
1.0	1.000	12.67	0	0	-1594

B_{11}/cm^{3}mol^{-1}	B_{22}/cm^{3}mol^{-1}	B_{12}/cm^{3}mol^{-1}	V_1/cm^{3}mol^{-1}	V_2/cm^{3}mol^{-1}
-1480	-6900	-4180	89	175

$A(1)$ = -0.6490E+00 $A(2)$ = 0.6000E-02 $A(3)$ = 0.0000E+00

1.Benzene, C_6H_6
2. 1-Heptene, C_7H_{14}

***T/K* = 328.15**
***System Number*: 59**

x_1	y_1	P/kPa	G^E/J mol^{-1}	μ^E_1/J mol^{-1}	μ^E_2/J mol^{-1}
0.0	0.000	27.49	0	783	0
0.1	0.181	30.36	71	649	7
0.2	0.321	32.86	129	534	27
0.3	0.435	35.06	173	433	61
0.4	0.532	37.00	203	341	111
0.5	0.617	38.70	218	256	180
0.6	0.694	40.17	217	179	275
0.7	0.767	41.41	198	110	403
0.8	0.838	42.42	158	54	573
0.9	0.914	43.17	93	15	797
1.0	1.000	43.60	0	0	1087

B_{11}/cm^{3}mol^{-1}	B_{22}/cm^{3}mol^{-1}	B_{12}/cm^{3}mol^{-1}	V_1/cm^{3}mol^{-1}	V_2/cm^{3}mol^{-1}
-1160	-1930	-1545	93	147

$A(1)$ = 0.3199E+00 $A(2)$ = 0.5581E-01 $A(3)$ = 0.2279E-01

1.Benzene, C_6H_6
2. Heptane, C_7H_{16}

***T/K* = 333.15**
***System Number*: 33**

x_1	y_1	P/kPa	G^E/J mol^{-1}	μ^E_1/J mol^{-1}	μ^E_2/J mol^{-1}
0.0	0.000	28.05	0	890	0
0.1	0.213	32.25	83	782	6
0.2	0.368	36.03	155	679	24
0.3	0.487	39.43	214	577	58
0.4	0.582	42.42	258	474	114
0.5	0.661	45.03	285	369	201
0.6	0.728	47.24	290	265	328
0.7	0.790	49.08	270	167	511
0.8	0.851	50.56	220	83	765
0.9	0.917	51.66	132	23	1109
1.0	1.000	52.23	0	0	1563

B_{11}/cm^{3}mol^{-1}	B_{22}/cm^{3}mol^{-1}	B_{12}/cm^{3}mol^{-1}	V_1/cm^{3}mol^{-1}	V_2/cm^{3}mol^{-1}
-1110	-2050	-1580	93	154

$A(1)$ = 0.4112E+00 $A(2)$ = 0.1216E+00 $A(3)$ = 0.3160E-01

TABLE 3.1.1 (VLEG) ISOTHERMAL VAPOR-LIQUID EQUILIBRIA AND EXCESS GIBBS ENERGIES OF BINARY MIXTURES (continued)

1.Benzene, C_6H_6
2. 3-Methylhexane, C_7H_{16}

***T*/K = 303.15**
***System Number*: 89**

x_1	y_1	*P*/kPa	G^E/J mol^{-1}	μ^E_1/J mol^{-1}	μ^E_2/J mol^{-1}
0.0	0.000	10.39	0	1069	0
0.1	0.201	11.74	105	1014	4
0.2	0.350	12.94	200	908	23
0.3	0.462	13.95	280	769	70
0.4	0.548	14.73	338	614	154
0.5	0.619	15.33	370	456	284
0.6	0.681	15.76	371	308	465
0.7	0.742	16.06	338	181	702
0.8	0.808	16.20	266	83	997
0.9	0.890	16.16	154	21	1349
1.0	1.000	15.84	0	0	1758

B_{11}/cm^{3}mol^{-1}	B_{22}/cm^{3}mol^{-1}	B_{12}/cm^{3}mol^{-1}	V_1/cm^{3}mol^{-1}	V_2/cm^{3}mol^{-1}
-1420	-2400	-1910	90	149

A(1) = 0.5873E+00 *A*(2) = 0.1366E+00 *A*(3) = -0.2640E-01

1.Benzene, C_6H_6
2. 1,4-Dimethylbenzene, C_8H_{10}

***T*/K = 308.15**
***System Number*: 92**

x_1	y_1	*P*/kPa	G^E/J mol^{-1}	μ^E_1/J mol^{-1}	μ^E_2/J mol^{-1}
0.0	0.000	2.04	0	192	0
0.1	0.529	3.91	16	138	3
0.2	0.712	5.72	27	99	10
0.3	0.807	7.50	34	70	19
0.4	0.865	9.26	38	48	31
0.5	0.905	11.01	38	33	43
0.6	0.934	12.75	36	21	57
0.7	0.956	14.50	31	12	74
0.8	0.973	16.25	23	6	93
0.9	0.988	18.00	13	2	118
1.0	1.000	19.76	0	0	150

B_{11}/cm^{3}mol^{-1}	B_{22}/cm^{3}mol^{-1}	B_{12}/cm^{3}mol^{-1}	V_1/cm^{3}mol^{-1}	V_2/cm^{3}mol^{-1}
-1360	-3380	-2370	90	125

A(1) = 0.5946E-01 *A*(2) = -0.8130E-02 *A*(3) = 0.7300E-02

1.Benzene, C_6H_6
2. 1-Octene, C_8H_{16}

***T*/K = 303.15**
***System Number*: 34**

x_1	y_1	*P*/kPa	G^E/J mol^{-1}	μ^E_1/J mol^{-1}	μ^E_2/J mol^{-1}
0.0	0.000	3.04	0	695	0
0.1	0.418	4.72	62	562	7
0.2	0.607	6.27	112	468	23
0.3	0.718	7.74	152	395	47
0.4	0.792	9.13	182	331	82
0.5	0.845	10.44	201	268	134
0.6	0.885	11.66	207	201	216
0.7	0.917	12.78	196	134	343
0.8	0.945	13.83	163	70	536
0.9	0.971	14.86	101	20	821
1.0	1.000	15.91	0	0	1227

B_{11}/cm^{3}mol^{-1}	B_{22}/cm^{3}mol^{-1}	B_{12}/cm^{3}mol^{-1}	V_1/cm^{3}mol^{-1}	V_2/cm^{3}mol^{-1}
-1420	-3920	-2670	90	159

A(1) = 0.3189E+00 *A*(2) = 0.1057E+00 *A*(3) = 0.6234E-01

TABLE 3.1.1 (VLEG) ISOTHERMAL VAPOR-LIQUID EQUILIBRIA AND EXCESS GIBBS ENERGIES OF BINARY MIXTURES (continued)

1.Benzene, C_6H_6
2. Dibutyl ether, $C_8H_{18}O$

***T*/K = 308.15**
***System Number*: 211**

x_1	y_1	P/kPa	G^E/J mol^{-1}	μ^E_1/J mol^{-1}	μ^E_2/J mol^{-1}
0.0	0.000	1.54	0	9	0
0.1	0.586	3.36	1	14	0
0.2	0.762	5.20	3	29	-3
0.3	0.847	7.06	7	40	-7
0.4	0.896	8.92	12	45	-9
0.5	0.928	10.77	18	44	-8
0.6	0.950	12.60	23	38	-1
0.7	0.967	14.41	25	29	16
0.8	0.980	16.20	24	18	51
0.9	0.991	17.97	17	6	118
1.0	1.000	19.75	0	0	249

B_{11}/cm^{3}mol^{-1}	B_{22}/cm^{3}mol^{-1}	B_{12}/cm^{3}mol^{-1}	V_1/cm^{3}mol^{-1}	V_2/cm^{3}mol^{-1}
-1360	-3600	-2180	90	172

$A(1)$ = 0.2789E-01 $A(2)$ = 0.4074E-01 $A(3)$ = 0.1248E-01 $A(4)$ = 0.6000E-02 $A(5)$ = 0.9900E-02

1.Benzene, C_6H_6
2. Octamethylcyclotetrasiloxane, $C_8H_{24}O_4Si_4$

***T*/K = 298.15**
***System Number*: 338**

x_1	y_1	P/kPa	G^E/J mol^{-1}	μ^E_1/J mol^{-1}	μ^E_2/J mol^{-1}
0.0	0.000	0.13	0	198	0
0.1	0.920	1.50	21	220	-1
0.2	0.963	2.88	44	238	-4
0.3	0.978	4.27	69	243	-6
0.4	0.986	5.64	93	236	-2
0.5	0.990	6.97	116	218	13
0.6	0.993	8.25	133	191	48
0.7	0.996	9.46	142	149	126
0.8	0.997	10.56	134	94	298
0.9	0.999	11.58	95	33	654
1.0	1.000	12.69	0	0	1347

B_{11}/cm^{3}mol^{-1}	B_{22}/cm^{3}mol^{-1}	B_{12}/cm^{3}mol^{-1}	V_1/cm^{3}mol^{-1}	V_2/cm^{3}mol^{-1}
-1480	0	0	89	0

$A(1)$ = 0.1865E+00 $A(2)$ = 0.1660E+00 $A(3)$ = 0.9693E-01 $A(4)$ = 0.6567E-01 $A(5)$ = 0.2829E-01

1.Benzene, C_6H_6
2. Hexadecane, $C_{16}H_{34}$

***T*/K = 308.15**
***System Number*: 53**

x_1	y_1	P/kPa	G^E/J mol^{-1}	μ^E_1/J mol^{-1}	μ^E_2/J mol^{-1}
0.0	0.000	0.00	0	26	0
0.1	1.000	2.07	9	136	-5
0.2	1.000	4.19	26	170	-10
0.3	1.000	6.32	44	180	-13
0.4	1.000	8.46	64	185	-17
0.5	1.000	10.60	85	189	-20
0.6	1.000	12.70	105	182	-10
0.7	1.000	14.67	121	154	45
0.8	1.000	16.44	123	100	211
0.9	1.000	18.05	92	36	592
1.0	1.000	19.79	0	0	1342

B_{11}/cm^{3}mol^{-1}	B_{22}/cm^{3}mol^{-1}	B_{12}/cm^{3}mol^{-1}	V_1/cm^{3}mol^{-1}	V_2/cm^{3}mol^{-1}
-1360	0	0	91	0

$A(1)$ = 0.1325E+00 $A(2)$ = 0.1632E+00 $A(3)$ = 0.1345E+00 $A(4)$ = 0.9380E-01 $A(5)$ = 0.0000E+00

TABLE 3.1.1 (VLEG) ISOTHERMAL VAPOR-LIQUID EQUILIBRIA AND EXCESS GIBBS ENERGIES OF BINARY MIXTURES (continued)

1. Phenol, C_6H_6O
2. Octane, C_8H_{18}

***T*/K = 383.15**
***System Number*: 240**

x_1	y_1	*P*/kPa	G^E/J mol^{-1}	μ^E_1/J mol^{-1}	μ^E_2/J mol^{-1}
0.0	0.000	64.79	0	6038	0
0.1	0.078	64.13	569	5258	48
0.2	0.112	62.92	1027	4161	243
0.3	0.123	62.05	1350	3135	585
0.4	0.128	61.45	1540	2305	1030
0.5	0.132	60.71	1608	1658	1558
0.6	0.137	59.67	1561	1135	2199
0.7	0.142	58.33	1392	691	3027
0.8	0.152	55.44	1087	327	4126
0.9	0.194	44.84	626	83	5517
1.0	1.000	9.18	0	0	7061

B_{11}/cm^{3}mol^{-1}	B_{22}/cm^{3}mol^{-1}	B_{12}/cm^{3}mol^{-1}	V_1/cm^{3}mol^{-1}	V_2/cm^{3}mol^{-1}
-2200	-1900	-2050	96	181

A(1) = 0.2020E+01 *A*(2) = 0.6278E-01 *A*(3) = 0.2137E+00 *A*(4) = 0.9779E-01 *A*(5) = -0.1775E+00

---

1. Aniline, C_6H_7N
2. Toluene, C_7H_8

***T*/K = 353.15**
***System Number*: 164**

x_1	y_1	*P*/kPa	G^E/J mol^{-1}	μ^E_1/J mol^{-1}	μ^E_2/J mol^{-1}
0.0	0.000	38.84	0	2574	0
0.1	0.014	35.73	228	2020	29
0.2	0.026	33.01	399	1555	111
0.3	0.037	30.48	518	1165	240
0.4	0.048	27.99	585	843	414
0.5	0.060	25.39	604	580	629
0.6	0.075	22.49	576	370	885
0.7	0.098	19.11	501	209	1184
0.8	0.137	14.94	381	93	1529
0.9	0.233	9.59	214	24	1925
1.0	1.000	2.45	0	0	2378

B_{11}/cm^{3}mol^{-1}	B_{22}/cm^{3}mol^{-1}	B_{12}/cm^{3}mol^{-1}	V_1/cm^{3}mol^{-1}	V_2/cm^{3}mol^{-1}
-2350	-1600	-1900	96	114

A(1) = 0.8230E+00 *A*(2) = -0.3335E-01 *A*(3) = 0.2032E-01

---

1. Aniline, C_6H_7N
2. Methylcyclohexane, C_7H_{14}

***T*/K = 363.15**
***System Number*: 162**

x_1	y_1	*P*/kPa	G^E/J mol^{-1}	μ^E_1/J mol^{-1}	μ^E_2/J mol^{-1}
0.0	0.000	73.70	0	6060	0
0.1	0.028	69.69	532	4675	71
0.2	0.040	66.76	926	3582	263
0.3	0.046	64.64	1200	2711	552
0.4	0.050	62.97	1361	2010	928
0.5	0.053	61.44	1417	1437	1398
0.6	0.056	59.67	1369	964	1976
0.7	0.060	56.96	1213	577	2696
0.8	0.068	51.57	941	276	3602
0.9	0.095	38.60	543	75	4751
1.0	1.000	3.89	0	0	6216

B_{11}/cm^{3}mol^{-1}	B_{22}/cm^{3}mol^{-1}	B_{12}/cm^{3}mol^{-1}	V_1/cm^{3}mol^{-1}	V_2/cm^{3}mol^{-1}
-2130	-1410	-1740	97	138

A(1) = 0.1877E+01 *A*(2) = 0.2590E-01 *A*(3) = 0.1557E+00

TABLE 3.1.1 (VLEG)
ISOTHERMAL VAPOR-LIQUID EQUILIBRIA AND EXCESS GIBBS ENERGIES OF BINARY MIXTURES (continued)

1. 1-Hexyne, C_6H_{10}
2. Heptane, C_7H_{16}

***T*/K = 343.15**
***System Number*: 87**

x_1	y_1	*P*/kPa	G^E/J mol^{-1}	μ^E_1/J mol^{-1}	μ^E_2/J mol^{-1}
0.0	0.000	40.4	0	782	0
0.1	0.251	48.9	75	713	4
0.2	0.420	56.8	141	626	20
0.3	0.542	64.0	195	527	53
0.4	0.635	70.5	235	422	110
0.5	0.709	76.3	257	318	196
0.6	0.772	81.5	258	219	317
0.7	0.828	86.2	236	132	479
0.8	0.882	90.4	188	63	688
0.9	0.938	94.3	110	17	950
1.0	1.000	97.9	0	0	1271

B_{11}/cm^{3}mol^{-1}	B_{22}/cm^{3}mol^{-1}	B_{12}/cm^{3}mol^{-1}	V_1/cm^{3}mol^{-1}	V_2/cm^{3}mol^{-1}
-1300	-1900	-1600	123	156

A(1) = 0.3598E+00 *A*(2) = 0.8563E-01 *A*(3) = 0.0000E+00

---

1. Cyclohexane, C_6H_{12}
2. N-Methylpiperidine, $C_6H_{13}N$

***T*/K = 298.15**
***System Number*: 161**

x_1	y_1	*P*/kPa	G^E/J mol^{-1}	μ^E_1/J mol^{-1}	μ^E_2/J mol^{-1}
0.0	0.000	4.80	0	179	0
0.1	0.239	5.69	15	122	3
0.2	0.410	6.54	25	91	8
0.3	0.542	7.39	32	76	13
0.4	0.647	8.23	37	67	18
0.5	0.732	9.07	42	59	24
0.6	0.802	9.90	44	49	37
0.7	0.862	10.70	43	35	62
0.8	0.912	11.49	38	20	109
0.9	0.958	12.25	25	6	189
1.0	1.000	13.01	0	0	316

B_{11}/cm^{3}mol^{-1}	B_{22}/cm^{3}mol^{-1}	B_{12}/cm^{3}mol^{-1}	V_1/cm^{3}mol^{-1}	V_2/cm^{3}mol^{-1}
-1740	-3310	-1850	109	122

A(1) = 0.6706E-01 *A*(2) = 0.2771E-01 *A*(3) = -0.3264E-01

---

1. Cyclohexane, C_6H_{12}
2. Hexane, C_6H_{14}

***T*/K = 308.15**
***System Number*: 40**

x_1	y_1	*P*/kPa	G^E/J mol^{-1}	μ^E_1/J mol^{-1}	μ^E_2/J mol^{-1}
0.0	0.000	30.57	0	222	0
0.1	0.073	29.68	20	186	2
0.2	0.149	28.78	37	156	7
0.3	0.228	27.85	50	130	16
0.4	0.312	26.89	60	105	30
0.5	0.401	25.91	65	81	49
0.6	0.496	24.89	66	58	77
0.7	0.599	23.82	61	36	117
0.8	0.713	22.68	49	18	172
0.9	0.844	21.44	29	5	247
1.0	1.000	20.05	0	0	347

B_{11}/cm^{3}mol^{-1}	B_{22}/cm^{3}mol^{-1}	B_{12}/cm^{3}mol^{-1}	V_1/cm^{3}mol^{-1}	V_2/cm^{3}mol^{-1}
-1560	-1760	-1660	110	133

A(1) = 0.1014E+00 *A*(2) = 0.2450E-01 *A*(3) = 0.9700E-02

TABLE 3.1.1 (VLEG)
ISOTHERMAL VAPOR-LIQUID EQUILIBRIA AND EXCESS GIBBS ENERGIES OF BINARY MIXTURES (continued)

1. 1-Hexene, C_6H_{12}
2. Hexane, C_6H_{14}

***T*/K = 328.15**
***System Number*: 57**

x_1	y_1	P/kPa	G^E/J mol^{-1}	μ^E_1/J mol^{-1}	μ^E_2/J mol^{-1}
0.0	0.000	64.43	0	108	0
0.1	0.119	65.94	10	85	1
0.2	0.232	67.38	17	65	5
0.3	0.340	68.74	22	48	10
0.4	0.443	70.03	24	34	18
0.5	0.542	71.27	25	23	27
0.6	0.638	72.46	24	14	38
0.7	0.731	73.60	20	8	50
0.8	0.822	74.70	15	3	63
0.9	0.912	75.76	8	1	78
1.0	1.000	76.78	0	0	93

B_{11}/cm^{3}mol^{-1}	B_{22}/cm^{3}mol^{-1}	B_{12}/cm^{3}mol^{-1}	V_1/cm^{3}mol^{-1}	V_2/cm^{3}mol^{-1}
-1320	-1450	-1380	132	138

$A(1)$ = 0.3680E-01 $A(2)$ = -0.2863E-02 $A(3)$ = 0.0000E+00

---

1. Cyclohexane, C_6H_{12}
2. 2,3-Dimethylbutane, C_6H_{14}

***T*/K = 298.15**
***System Number*: 78**

x_1	y_1	P/kPa	G^E/J mol^{-1}	μ^E_1/J mol^{-1}	μ^E_2/J mol^{-1}
0.0	0.000	31.29	0	314	0
0.1	0.050	29.63	29	265	3
0.2	0.103	27.96	52	219	11
0.3	0.161	26.29	70	174	26
0.4	0.225	24.60	82	133	48
0.5	0.298	22.87	87	96	78
0.6	0.382	21.10	85	64	118
0.7	0.483	19.26	76	37	167
0.8	0.608	17.32	59	17	227
0.9	0.771	15.25	34	4	299
1.0	1.000	13.02	0	0	383

B_{11}/cm^{3}mol^{-1}	B_{22}/cm^{3}mol^{-1}	B_{12}/cm^{3}mol^{-1}	V_1/cm^{3}mol^{-1}	V_2/cm^{3}mol^{-1}
-1740	-1570	-1640	108	111

$A(1)$ = 0.1406E+00 $A(2)$ = 0.1398E-01 $A(3)$ = 0.0000E+00

---

1. Cyclohexane, C_6H_{12}
2. Toluene, C_7H_8

***T*/K = 298.15**
***System Number*: 38**

x_1	y_1	P/kPa	G^E/J mol^{-1}	μ^E_1/J mol^{-1}	μ^E_2/J mol^{-1}
0.0	0.000	3.80	0	1248	0
0.1	0.358	5.36	110	970	14
0.2	0.529	6.62	192	745	54
0.3	0.636	7.68	249	562	115
0.4	0.712	8.61	282	413	195
0.5	0.773	9.46	293	291	294
0.6	0.823	10.24	281	192	415
0.7	0.869	10.97	248	113	563
0.8	0.911	11.68	191	53	743
0.9	0.954	12.36	109	14	965
1.0	1.000	13.03	0	0	1238

B_{11}/cm^{3}mol^{-1}	B_{22}/cm^{3}mol^{-1}	B_{12}/cm^{3}mol^{-1}	V_1/cm^{3}mol^{-1}	V_2/cm^{3}mol^{-1}
-1740	-2620	-2270	109	107

$A(1)$ = 0.4724E+00 $A(2)$ = -0.2100E-02 $A(3)$ = 0.2910E-01

TABLE 3.1.1 (VLEG) ISOTHERMAL VAPOR-LIQUID EQUILIBRIA AND EXCESS GIBBS ENERGIES OF BINARY MIXTURES (continued)

1. Cyclohexane, C_6H_{12}
2. Heptane, C_7H_{16}

***T*/K = 298.15**
***System Number*: 80**

x_1	y_1	P/kPa	G^E/J mol^{-1}	μ^E_1/J mol^{-1}	μ^E_2/J mol^{-1}
0.0	0.000	6.09	0	166	0
0.1	0.200	6.86	15	141	1
0.2	0.357	7.61	28	119	5
0.3	0.485	8.34	38	99	12
0.4	0.591	9.06	45	80	22
0.5	0.682	9.76	49	62	37
0.6	0.760	10.44	50	44	59
0.7	0.828	11.10	46	28	89
0.8	0.890	11.75	37	14	131
0.9	0.946	12.38	22	4	189
1.0	1.000	13.00	0	0	265

B_{11}/cm^{3}mol^{-1}	B_{22}/cm^{3}mol^{-1}	B_{12}/cm^{3}mol^{-1}	V_1/cm^{3}mol^{-1}	V_2/cm^{3}mol^{-1}
-1740	-2840	-2290	109	147

$A(1)$ = 0.7968E-01 $A(2)$ = 0.1993E-01 $A(3)$ = 0.7172E-02

1. 1-Hexene, C_6H_{12}
2. Octane, C_8H_{18}

***T*/K = 328.15**
***System Number*: 84**

x_1	y_1	P/kPa	G^E/J mol^{-1}	μ^E_1/J mol^{-1}	μ^E_2/J mol^{-1}
0.0	0.000	8.43	0	-98	0
0.1	0.485	14.84	-9	-88	-1
0.2	0.680	21.34	-17	-76	-3
0.3	0.785	27.96	-24	-63	-7
0.4	0.850	34.68	-28	-50	-14
0.5	0.895	41.51	-31	-38	-24
0.6	0.928	48.45	-31	-26	-39
0.7	0.953	55.46	-28	-15	-58
0.8	0.972	62.55	-22	-7	-83
0.9	0.988	69.67	-13	-2	-113
1.0	1.000	76.78	0	0	-150

B_{11}/cm^{3}mol^{-1}	B_{22}/cm^{3}mol^{-1}	B_{12}/cm^{3}mol^{-1}	V_1/cm^{3}mol^{-1}	V_2/cm^{3}mol^{-1}
-1320	-3060	-2180	132	169

$A(1)$ = -0.4543E-01 $A(2)$ = -0.9640E-02 $A(3)$ = 0.0000E+00

1. Cyclohexane, C_6H_{12}
2. *cis*-Bicyclo[4.4.0]decane, $C_{10}H_{18}$

***T*/K = 298.15**
***System Number*: 60**

x_1	y_1	P/kPa	G^E/J mol^{-1}	μ^E_1/J mol^{-1}	μ^E_2/J mol^{-1}
0.0	0.000	0.11	0	741	0
0.1	0.938	1.64	57	439	15
0.2	0.969	2.94	88	242	49
0.3	0.980	4.15	98	123	88
0.4	0.987	5.36	97	57	123
0.5	0.991	6.60	87	26	148
0.6	0.994	7.86	73	13	163
0.7	0.996	9.14	57	9	171
0.8	0.998	10.43	41	6	180
0.9	0.999	11.71	22	2	201
1.0	1.000	13.00	0	0	252

B_{11}/cm^{3}mol^{-1}	B_{22}/cm^{3}mol^{-1}	B_{12}/cm^{3}mol^{-1}	V_1/cm^{3}mol^{-1}	V_2/cm^{3}mol^{-1}
-1740	-8300	-3600	108	155

$A(1)$ = 0.1401E+00 $A(2)$ = -0.9858E-01 $A(3)$ = 0.6010E-01

TABLE 3.1.1 (VLEG)
ISOTHERMAL VAPOR-LIQUID EQUILIBRIA AND EXCESS GIBBS ENERGIES OF BINARY MIXTURES (continued)

1. Cyclohexane, C_6H_{12}
2. Hexadecane, $C_{16}H_{34}$

***T*/K = 308.15**
***System Number*: 54**

x_1	y_1	*P*/kPa	G^E/J mol^{-1}	μ^E_1/J mol^{-1}	μ^E_2/J mol^{-1}
0.0	0.000	0.00	0	-594	0
0.1	1.000	1.62	-56	-523	-4
0.2	1.000	3.33	-104	-452	-17
0.3	1.000	5.15	-142	-379	-41
0.4	1.000	7.07	-171	-305	-81
0.5	1.000	9.11	-187	-232	-141
0.6	1.000	11.25	-188	-163	-226
0.7	1.000	13.47	-173	-101	-343
0.8	1.000	15.73	-139	-49	-499
0.9	1.000	17.96	-82	-13	-702
1.0	1.000	20.09	0	0	-961

B_{11}/cm^{3}mol^{-1}	B_{22}/cm^{3}mol^{-1}	B_{12}/cm^{3}mol^{-1}	V_1/cm^{3}mol^{-1}	V_2/cm^{3}mol^{-1}
-1560	0	0	110	0

A(1) = -0.2912E+00 *A*(2) = -0.7162E-01 *A*(3) = -0.1243E-01

1.Butyl ethanoate, $C_6H_{12}O_2$
2. Heptane, C_7H_{16}

***T*/K = 347.85**
***System Number*: 270**

x_1	y_1	*P*/kPa	G^E/J mol^{-1}	μ^E_1/J mol^{-1}	μ^E_2/J mol^{-1}
0.0	0.000	48.22	0	1795	0
0.1	0.071	46.84	172	1630	10
0.2	0.135	45.36	319	1366	58
0.3	0.188	43.79	430	1059	161
0.4	0.235	42.10	497	754	326
0.5	0.280	40.20	515	484	547
0.6	0.332	37.84	485	269	809
0.7	0.401	34.70	409	120	1084
0.8	0.506	30.47	295	35	1333
0.9	0.683	25.06	154	3	1506
1.0	1.000	18.92	0	0	1541

B_{11}/cm^{3}mol^{-1}	B_{22}/cm^{3}mol^{-1}	B_{12}/cm^{3}mol^{-1}	V_1/cm^{3}mol^{-1}	V_2/cm^{3}mol^{-1}
-2470	-1830	-1380	141	158

A(1) = 0.7128E+00 *A*(2) = -0.4400E-01 *A*(3) = -0.1360E+00

1. Hexane, C_6H_{14}
2. Dibutyl ether, $C_8H_{18}O$

***T*/K = 308.15**
***System Number*: 210**

x_1	y_1	*P*/kPa	G^E/J mol^{-1}	μ^E_1/J mol^{-1}	μ^E_2/J mol^{-1}
0.0	0.000	1.54	0	169	0
0.1	0.695	4.56	16	146	1
0.2	0.835	7.54	29	122	6
0.3	0.896	10.48	39	98	13
0.4	0.929	13.38	46	77	25
0.5	0.951	16.26	49	57	41
0.6	0.966	19.11	49	39	63
0.7	0.978	21.95	44	24	91
0.8	0.987	24.79	35	12	128
0.9	0.994	27.66	21	3	176
1.0	1.000	30.57	0	0	240

B_{11}/cm^{3}mol^{-1}	B_{22}/cm^{3}mol^{-1}	B_{12}/cm^{3}mol^{-1}	V_1/cm^{3}mol^{-1}	V_2/cm^{3}mol^{-1}
-1760	-3600	-2340	133	172

A(1) = 0.7659E-01 *A*(2) = 0.1184E-01 *A*(3) = 0.3250E-02 *A*(4) = 0.2000E-02 *A*(5) = 0.0000E+00

TABLE 3.1.1 (VLEG)
ISOTHERMAL VAPOR-LIQUID EQUILIBRIA AND EXCESS GIBBS ENERGIES OF BINARY MIXTURES (continued)

1. Hexane, C_6H_{14}
2. Decane, $C_{10}H_{22}$

T/K = 308.15
***System Number*: 77**

x_1	y_1	P/kPa	G^E/J mol^{-1}	μ^E_1/J mol^{-1}	μ^E_2/J mol^{-1}
0.0	0.000	0.36	0	-41	0
0.1	0.901	3.28	-4	-33	0
0.2	0.954	6.23	-7	-26	-2
0.3	0.972	9.21	-8	-19	-4
0.4	0.982	12.22	-10	-13	-7
0.5	0.988	15.25	-10	-8	-11
0.6	0.992	18.29	-9	-5	-16
0.7	0.995	21.34	-8	-2	-20
0.8	0.997	24.41	-5	-1	-24
0.9	0.999	27.48	-3	0	-28
1.0	1.000	30.56	0	0	-29

B_{11}/cm^{3}mol^{-1}	B_{22}/cm^{3}mol^{-1}	B_{12}/cm^{3}mol^{-1}	V_1/cm^{3}mol^{-1}	V_2/cm^{3}mol^{-1}
-1760	-6540	-4150	133	198

$A(1)$ = -0.1510E-01 $A(2)$ = 0.2200E-02 $A(3)$ = 0.1400E-02

1. Dipropyl ether, $C_6H_{14}O$
2. Heptane, C_7H_{16}

T/K = 343.15
***System Number*: 103**

x_1	y_1	P/kPa	G^E/J mol^{-1}	μ^E_1/J mol^{-1}	μ^E_2/J mol^{-1}
0.0	0.000	40.47	0	257	0
0.1	0.135	42.17	23	208	3
0.2	0.256	43.75	41	165	10
0.3	0.366	45.21	54	126	23
0.4	0.469	46.57	62	93	41
0.5	0.565	47.85	64	64	64
0.6	0.657	49.03	62	41	93
0.7	0.745	50.13	54	23	126
0.8	0.831	51.16	41	10	165
0.9	0.916	52.12	23	3	208
1.0	1.000	53.01	0	0	257

B_{11}/cm^{3}mol^{-1}	B_{22}/cm^{3}mol^{-1}	B_{12}/cm^{3}mol^{-1}	V_1/cm^{3}mol^{-1}	V_2/cm^{3}mol^{-1}
-1400	-1900	-1660	149	156

$A(1)$ = 0.9014E-01 $A(2)$ = 0.0000E+00 $A(3)$ = 0.0000E+00

1. 3,6-Dioxaoctane, $C_6H_{14}O_2$
2. Heptane, C_7H_{16}

T/K = 343.15
***System Number*: 213**

x_1	y_1	P/kPa	G^E/J mol^{-1}	μ^E_1/J mol^{-1}	μ^E_2/J mol^{-1}
0.0	0.000	40.58	0	1561	0
0.1	0.068	39.42	135	1179	20
0.2	0.127	38.05	234	883	71
0.3	0.183	36.50	299	655	147
0.4	0.240	34.75	336	477	243
0.5	0.303	32.80	347	336	358
0.6	0.374	30.60	332	223	496
0.7	0.459	28.12	292	133	663
0.8	0.569	25.25	225	64	872
0.9	0.727	21.81	129	17	1136
1.0	1.000	17.45	0	0	1474

B_{11}/cm^{3}mol^{-1}	B_{22}/cm^{3}mol^{-1}	B_{12}/cm^{3}mol^{-1}	V_1/cm^{3}mol^{-1}	V_2/cm^{3}mol^{-1}
-2000	-1900	-1900	162	156

$A(1)$ = 0.4862E+00 $A(2)$ = -0.1520E-01 $A(3)$ = 0.4580E-01

TABLE 3.1.1 (VLEG)
ISOTHERMAL VAPOR-LIQUID EQUILIBRIA AND EXCESS GIBBS ENERGIES OF BINARY MIXTURES (continued)

1. Triethylamine, $C_6H_{15}N$
2. Heptane, C_7H_{16}

***T*/K = 333.15**
***System Number*: 157**

x_1	y_1	P/kPa	G^E/J mol^{-1}	μ^E_1/J mol^{-1}	μ^E_2/J mol^{-1}
0.0	0.000	28.05	0	29	0
0.1	0.133	29.16	3	33	0
0.2	0.257	30.27	6	33	0
0.3	0.373	31.38	10	31	0
0.4	0.480	32.48	13	28	2
0.5	0.579	33.57	15	23	7
0.6	0.673	34.64	16	17	14
0.7	0.761	35.69	15	11	25
0.8	0.844	36.72	13	5	41
0.9	0.923	37.72	8	2	63
1.0	1.000	38.70	0	0	93

B_{11}/cm^{3}mol^{-1}	B_{22}/cm^{3}mol^{-1}	B_{12}/cm^{3}mol^{-1}	V_1/cm^{3}mol^{-1}	V_2/cm^{3}mol^{-1}
-1520	-2050	-1770	145	154

$A(1) = 0.2106E-01$ $A(2) = 0.1149E-01$ $A(3) = 0.1003E-02$

---

1. Triethylamine, $C_6H_{15}N$
2. 1,3,5-Trimethylbenzene, C_9H_{12}

***T*/K = 303.15**
***System Number*: 158**

x_1	y_1	P/kPa	G^E/J mol^{-1}	μ^E_1/J mol^{-1}	μ^E_2/J mol^{-1}
0.0	0.000	0.55	0	313	0
0.1	0.720	1.77	33	323	0
0.2	0.850	2.94	62	271	10
0.3	0.903	4.04	84	202	33
0.4	0.933	5.09	96	138	67
0.5	0.953	6.12	98	91	106
0.6	0.967	7.15	93	60	144
0.7	0.978	8.19	82	39	182
0.8	0.987	9.23	64	23	231
0.9	0.994	10.25	39	8	318
1.0	1.000	11.30	0	0	487

B_{11}/cm^{3}mol^{-1}	B_{22}/cm^{3}mol^{-1}	B_{12}/cm^{3}mol^{-1}	V_1/cm^{3}mol^{-1}	V_2/cm^{3}mol^{-1}
-1950	-4750	-3150	141	140

$A(1) = 0.1561E+00$ $A(2) = -0.1207E-01$ $A(3) = 0.2637E-02$ $A(4) = 0.4658E-01$ $A(5) = 0.0000E+00$

---

1. Phosphoric tris(dimethylamide), $C_6H_{18}N_3OP$
2. Heptane, C_7H_{16}

***T*/K = 298.15**
***System Number*: 203**

x_1	y_1	P/kPa	G^E/J mol^{-1}	μ^E_1/J mol^{-1}	μ^E_2/J mol^{-1}
0.0	0.000	6.094	0	6128	0
0.1	0.001	5.730	485	3882	108
0.2	0.001	5.556	784	2623	324
0.3	0.001	5.386	960	1851	578
0.4	0.001	5.193	1045	1306	870
0.5	0.001	4.978	1049	881	1218
0.6	0.001	4.698	978	545	1628
0.7	0.001	4.242	834	296	2089
0.8	0.002	3.452	621	130	2584
0.9	0.003	2.147	344	35	3126
1.0	1.000	0.007	0	0	3813

B_{11}/cm^{3}mol^{-1}	B_{22}/cm^{3}mol^{-1}	B_{12}/cm^{3}mol^{-1}	V_1/cm^{3}mol^{-1}	V_2/cm^{3}mol^{-1}
-6900	-2840	-4870	175	147

$A(1) = 0.1693E+01$ $A(2) = -0.2720E+00$ $A(3) = 0.1630E+00$ $A(4) = -0.1950E+00$ $A(5) = 0.1490E+00$

TABLE 3.1.1 (VLEG)
ISOTHERMAL VAPOR-LIQUID EQUILIBRIA AND EXCESS GIBBS ENERGIES OF BINARY MIXTURES (continued)

1. Hexadecafluoroheptane, C_7F_{16}
2. 2,2,4-Trimethylpentane, C_8H_{18}

***T*/K = 303.15**
***System Number*: 381**

x_1	y_1	*P*/kPa	G^E/J mol^{-1}	μ^E_1/J mol^{-1}	μ^E_2/J mol^{-1}
0.0	0.000	8.33	0	6590	0
0.1	0.531	16.72	563	4817	90
0.2	0.579	18.20	956	3500	320
0.3	0.585	18.36	1207	2527	642
0.4	0.586	18.37	1339	1803	1030
0.5	0.588	18.39	1366	1257	1476
0.6	0.596	18.42	1297	833	1992
0.7	0.611	18.35	1134	500	2613
0.8	0.642	18.04	872	243	3388
0.9	0.716	16.94	500	67	4392
1.0	1.000	13.08	0	0	5714

B_{11}/cm^{3}mol^{-1}	B_{22}/cm^{3}mol^{-1}	B_{12}/cm^{3}mol^{-1}	V_1/cm^{3}mol^{-1}	V_2/cm^{3}mol^{-1}
-2300	-2850	-2500	228	167

$A(1) = 0.2168E+01$ $A(2) = -0.1737E+00$ $A(3) = 0.2729E+00$

1. Toluene, C_7H_8
2. 1-Heptene, C_7H_{14}

***T*/K = 328.15**
***System Number*: 101**

x_1	y_1	*P*/kPa	G^E/J mol^{-1}	μ^E_1/J mol^{-1}	μ^E_2/J mol^{-1}
0.0	0.000	27.49	0	1192	0
0.1	0.076	26.95	97	786	20
0.2	0.141	26.21	157	527	65
0.3	0.207	25.29	192	368	117
0.4	0.278	24.24	210	271	168
0.5	0.356	23.07	214	208	220
0.6	0.443	21.82	207	157	282
0.7	0.541	20.48	188	109	373
0.8	0.653	19.04	152	60	520
0.9	0.792	17.36	93	19	762
1.0	1.000	15.15	0	0	1142

B_{11}/cm^{3}mol^{-1}	B_{22}/cm^{3}mol^{-1}	B_{12}/cm^{3}mol^{-1}	V_1/cm^{3}mol^{-1}	V_2/cm^{3}mol^{-1}
-1990	-1930	-1960	110	149

$A(1) = 0.3136E+00$ $A(2) = -0.9194E-02$ $A(3) = 0.1142E+00$

1. Toluene, C_7H_8
2. Methylcyclohexane, C_7H_{14}

***T*/K = 333.15**
***System Number*: 96**

x_1	y_1	*P*/kPa	G^E/J mol^{-1}	μ^E_1/J mol^{-1}	μ^E_2/J mol^{-1}
0.0	0.000	27.08	0	771	0
0.1	0.088	26.78	71	660	6
0.2	0.172	26.39	130	545	27
0.3	0.252	25.91	175	431	65
0.4	0.330	25.32	203	323	123
0.5	0.410	24.62	214	227	202
0.6	0.494	23.80	208	146	301
0.7	0.587	22.81	183	81	421
0.8	0.695	21.63	140	36	558
0.9	0.828	20.21	79	9	710
1.0	1.000	18.51	0	0	873

B_{11}/cm^{3}mol^{-1}	B_{22}/cm^{3}mol^{-1}	B_{12}/cm^{3}mol^{-1}	V_1/cm^{3}mol^{-1}	V_2/cm^{3}mol^{-1}
-1900	-1760	-1830	111	133

$A(1) = 0.3094E+00$ $A(2) = 0.1841E-01$ $A(3) = -0.1272E-01$

TABLE 3.1.1 (VLEG)
ISOTHERMAL VAPOR-LIQUID EQUILIBRIA AND EXCESS GIBBS ENERGIES OF BINARY MIXTURES (continued)

1. Toluene, C_7H_8
2. 2,2,4-Trimethylpentane, C_8H_{18}

***T*/K = 373.15**
***System Number*: 95**

x_1	y_1	P/kPa	G^E/J mol^{-1}	μ^E_1/J mol^{-1}	μ^E_2/J mol^{-1}
0.0	0.000	103.5	0	590	0
0.1	0.090	102.3	62	639	-2
0.2	0.181	101.2	125	605	5
0.3	0.267	99.8	180	519	34
0.4	0.349	98.2	220	407	95
0.5	0.429	96.1	241	290	192
0.6	0.511	93.5	239	183	322
0.7	0.600	90.1	213	98	480
0.8	0.705	85.9	163	40	655
0.9	0.834	80.6	91	9	830
1.0	1.000	74.2	0	0	983

B_{11}/cm^{3}mol^{-1}	B_{22}/cm^{3}mol^{-1}	B_{12}/cm^{3}mol^{-1}	V_1/cm^{3}mol^{-1}	V_2/cm^{3}mol^{-1}
-1360	-1590	-1475	117	182

$A(1)$ = 0.3104E+00 $A(2)$ = 0.6332E-01 $A(3)$ = -0.5691E-01

---

1. 3-Heptene, C_7H_{14}
2. Heptane, C_7H_{16}

***T*/K = 370.40**
***System Number*: 86**

x_1	y_1	P/kPa	G^E/J mol^{-1}	μ^E_1/J mol^{-1}	μ^E_2/J mol^{-1}
0.0	0.000	97.5	0	56	0
0.1	0.106	98.3	5	46	1
0.2	0.211	99.0	9	37	2
0.3	0.313	99.6	12	29	5
0.4	0.414	100.3	14	22	9
0.5	0.513	100.9	15	15	14
0.6	0.612	101.4	14	10	21
0.7	0.710	101.9	13	6	28
0.8	0.807	102.4	10	3	38
0.9	0.903	102.8	5	1	49
1.0	1.000	103.2	0	0	62

B_{11}/cm^{3}mol^{-1}	B_{22}/cm^{3}mol^{-1}	B_{12}/cm^{3}mol^{-1}	V_1/cm^{3}mol^{-1}	V_2/cm^{3}mol^{-1}
-1410	-1500	-1455	150	163

$A(1)$ = 0.1907E-01 $A(2)$ = 0.9455E-03 $A(3)$ = 0.0000E+00

---

1. Methylcyclohexane, C_7H_{14}
2. Ethylbenzene, C_8H_{10}

***T*/K = 313.15**
***System Number*: 98**

x_1	y_1	P/kPa	G^E/J mol^{-1}	μ^E_1/J mol^{-1}	μ^E_2/J mol^{-1}
0.0	0.000	2.86	0	709	0
0.1	0.372	4.12	66	607	5
0.2	0.560	5.27	120	505	24
0.3	0.674	6.32	162	407	57
0.4	0.753	7.29	189	313	107
0.5	0.812	8.19	202	228	177
0.6	0.858	9.03	199	152	270
0.7	0.898	9.84	179	89	387
0.8	0.933	10.62	139	41	531
0.9	0.967	11.41	80	11	705
1.0	1.000	12.22	0	0	911

B_{11}/cm^{3}mol^{-1}	B_{22}/cm^{3}mol^{-1}	B_{12}/cm^{3}mol^{-1}	V_1/cm^{3}mol^{-1}	V_2/cm^{3}mol^{-1}
-2060	-3400	-2700	131	126

$A(1)$ = 0.3110E+00 $A(2)$ = 0.3878E-01 $A(3)$ = 0.0000E+00

TABLE 3.1.1 (VLEG)
ISOTHERMAL VAPOR-LIQUID EQUILIBRIA AND EXCESS GIBBS ENERGIES OF BINARY MIXTURES (continued)

1. 2-Heptanone, $C_7H_{14}O$
2. Dibutyl ether, $C_8H_{18}O$

***T*/K = 363.15**
***System Number*: 323**

x_1	y_1	*P*/kPa	G^E/J mol^{-1}	μ^E_1/J mol^{-1}	μ^E_2/J mol^{-1}
0.0	0.000	19.27	0	1329	0
0.1	0.100	19.37	119	1065	14
0.2	0.187	19.30	210	832	55
0.3	0.265	19.10	275	630	122
0.4	0.340	18.78	312	457	215
0.5	0.415	18.35	323	314	332
0.6	0.495	17.78	308	199	473
0.7	0.584	17.08	268	110	637
0.8	0.689	16.19	203	48	823
0.9	0.822	15.09	114	12	1029
1.0	1.000	13.72	0	0	1256

B_{11}/cm^{3}mol^{-1}	B_{22}/cm^{3}mol^{-1}	B_{12}/cm^{3}mol^{-1}	V_1/cm^{3}mol^{-1}	V_2/cm^{3}mol^{-1}
-1900	-2530	-1490	140	183

A(1) = 0.4281E+00 *A*(2) = -0.1210E-01 *A*(3) = 0.0000E+00

1. Heptane, C_7H_{16}
2. Octane, C_8H_{18}

***T*/K = 328.15**
***System Number*: 76**

x_1	y_1	*P*/kPa	G^E/J mol^{-1}	μ^E_1/J mol^{-1}	μ^E_2/J mol^{-1}
0.0	0.000	8.43	0	360	0
0.1	0.250	10.15	32	291	4
0.2	0.422	11.78	58	230	14
0.3	0.550	13.33	76	176	32
0.4	0.649	14.82	86	130	58
0.5	0.730	16.26	90	90	90
0.6	0.798	17.66	86	58	130
0.7	0.857	19.03	76	32	176
0.8	0.909	20.39	58	14	230
0.9	0.956	21.74	32	4	291
1.0	1.000	23.09	0	0	360

B_{11}/cm^{3}mol^{-1}	B_{22}/cm^{3}mol^{-1}	B_{12}/cm^{3}mol^{-1}	V_1/cm^{3}mol^{-1}	V_2/cm^{3}mol^{-1}
-2140	-3060	-2600	154	170

A(1) = 0.1319E+00 *A*(2) = 0.0000E+00 *A*(3) = 0.0000E+00

1. Heptane, C_7H_{16}
2. Isopropylbenzene, C_9H_{12}

***T*/K = 313.15**
***System Number*: 100**

x_1	y_1	*P*/kPa	G^E/J mol^{-1}	μ^E_1/J mol^{-1}	μ^E_2/J mol^{-1}
0.0	0.000	1.46	0	955	0
0.1	0.557	2.99	88	810	8
0.2	0.725	4.34	159	653	36
0.3	0.807	5.52	211	498	88
0.4	0.856	6.57	241	356	164
0.5	0.892	7.53	249	235	264
0.6	0.919	8.46	235	138	382
0.7	0.942	9.38	201	69	510
0.8	0.963	10.32	148	26	638
0.9	0.982	11.31	80	5	753
1.0	1.000	12.33	0	0	838

B_{11}/cm^{3}mol^{-1}	B_{22}/cm^{3}mol^{-1}	B_{12}/cm^{3}mol^{-1}	V_1/cm^{3}mol^{-1}	V_2/cm^{3}mol^{-1}
-2450	-3650	-3050	149	142

A(1) = 0.3829E+00 *A*(2) = -0.2250E-01 *A*(3) = -0.3858E-01

TABLE 3.1.1 (VLEG) ISOTHERMAL VAPOR-LIQUID EQUILIBRIA AND EXCESS GIBBS ENERGIES OF BINARY MIXTURES (continued)

1. 3-Methylhexane, C_7H_{16}
2. Propylbenzene, C_9H_{12}

***T*/K = 313.15**
***System Number*: 99**

x_1	y_1	P/kPa	G^E/J mol^{-1}	μ^E_1/J mol^{-1}	μ^E_2/J mol^{-1}
0.0	0.000	1.10	0	1321	0
0.1	0.702	3.35	115	1004	16
0.2	0.825	5.18	199	757	60
0.3	0.880	6.78	256	563	124
0.4	0.912	8.23	288	409	206
0.5	0.935	9.59	297	287	306
0.6	0.952	10.90	284	190	425
0.7	0.966	12.18	249	112	570
0.8	0.978	13.45	192	53	747
0.9	0.989	14.76	110	14	968
1.0	1.000	16.15	0	0	1247

B_{11}/cm^{3}mol^{-1}	B_{22}/cm^{3}mol^{-1}	B_{12}/cm^{3}mol^{-1}	V_1/cm^{3}mol^{-1}	V_2/cm^{3}mol^{-1}
-2210	-3920	-3070	150	142

$A(1) = 0.4558E+00$ $A(2) = -0.1423E-01$ $A(3) = 0.3726E-01$

---

1. Ethylbenzene, C_8H_{10}
2. 2,2,4-Trimethylpentane, C_8H_{18}

***T*/K = 313.15**
***System Number*: 97**

x_1	y_1	P/kPa	G^E/J mol^{-1}	μ^E_1/J mol^{-1}	μ^E_2/J mol^{-1}
0.0	0.000	12.96	0	1239	0
0.1	0.035	12.14	112	1010	12
0.2	0.070	11.34	200	805	48
0.3	0.105	10.54	263	624	108
0.4	0.142	9.74	303	466	194
0.5	0.184	8.90	318	330	305
0.6	0.235	8.00	307	215	445
0.7	0.301	7.01	271	124	615
0.8	0.400	5.88	209	57	818
0.9	0.574	4.53	119	15	1057
1.0	1.000	2.87	0	0	1335

B_{11}/cm^{3}mol^{-1}	B_{22}/cm^{3}mol^{-1}	B_{12}/cm^{3}mol^{-1}	V_1/cm^{3}mol^{-1}	V_2/cm^{3}mol^{-1}
-3400	-2610	-2440	126	169

$A(1) = 0.4878E+00$ $A(2) = 0.1854E-01$ $A(3) = 0.6514E-02$

---

1. Ethylbenzene, C_8H_{10}
2. Nonane, C_9H_{20}

***T*/K = 333.15**
***System Number*: 90**

x_1	y_1	P/kPa	G^E/J mol^{-1}	μ^E_1/J mol^{-1}	μ^E_2/J mol^{-1}
0.0	0.000	4.01	0	738	0
0.1	0.205	4.55	69	651	5
0.2	0.358	5.04	129	556	22
0.3	0.477	5.48	175	457	55
0.4	0.573	5.87	208	359	108
0.5	0.654	6.22	225	266	185
0.6	0.726	6.52	224	180	289
0.7	0.793	6.79	203	107	425
0.8	0.858	7.03	160	50	597
0.9	0.925	7.23	93	13	808
1.0	1.000	7.40	0	0	1063

B_{11}/cm^{3}mol^{-1}	B_{22}/cm^{3}mol^{-1}	B_{12}/cm^{3}mol^{-1}	V_1/cm^{3}mol^{-1}	V_2/cm^{3}mol^{-1}
-2720	-4270	-3500	127	187

$A(1) = 0.3251E+00$ $A(2) = 0.5862E-01$ $A(3) = 0.0000E+00$

TABLE 3.1.1 (VLEG)
ISOTHERMAL VAPOR-LIQUID EQUILIBRIA AND EXCESS GIBBS ENERGIES OF BINARY MIXTURES (continued)

1. Octane, C_8H_{18}
2. Propylbenzene, C_9H_{12}

***T*/K = 343.15**
***System Number*: 93**

x_1	y_1	P/kPa	G^E/J mol^{-1}	μ^E_1/J mol^{-1}	μ^E_2/J mol^{-1}
0.0	0.000	4.98	0	951	0
0.1	0.312	6.56	84	746	11
0.2	0.487	7.91	147	571	42
0.3	0.603	9.12	190	423	91
0.4	0.689	10.21	214	300	157
0.5	0.757	11.23	219	201	238
0.6	0.815	12.20	207	124	332
0.7	0.866	13.14	178	67	437
0.8	0.913	14.06	133	29	552
0.9	0.957	14.99	74	7	675
1.0	1.000	15.92	0	0	804

B_{11}/cm^{3}mol^{-1}	B_{22}/cm^{3}mol^{-1}	B_{12}/cm^{3}mol^{-1}	V_1/cm^{3}mol^{-1}	V_2/cm^{3}mol^{-1}
-2760	-3400	-3080	172	147

$A(1)$ = 0.3074E+00 $A(2)$ = -0.2575E-01 $A(3)$ = 0.0000E+00

TABLE 3.1.2
AZEOTROPIC DATA FOR BINARY MIXTURES

Liquid mixtures having an extremum (maximum or minimum) vapor pressure (see Table 3.1.1) at constant temperature, as a function of composition, are called "azeotropic" mixtures or simply "azeotropes". "Non-azeotropic" mixtures are also called "zeotropic". Pressure-maximum azeotropes are usually called "positive" azeotropes, while pressure-minimum azeotropes are usually called "negative azeotropes".

The coordinates of an azeotropic point (maximum or minimum) are the azeotropic temperature T_{Az}, pressure P_{Az}, and liquid-phase composition $x_{1,Az}$ (if the latter is expressed in mole fraction of component 1). At the azeotropic point, the vapor-phase composition $y_{1,Az}$ is the same as the liquid-phase composition.

Most azeotropic points have been determined at atmospheric pressure, P_{Az}/kPa = 101.3. The Table gives the azeotropic points for 305 binary mixtures not only under atmospheric pressure, but also under lower or higher pressures. An asterisk * denotes partial miscibility in the liquid phase.

The components of each system, and the systems themselves, are arranged in a modified Hill order, with substances that do not contain carbon preceding those that do contain carbon.

REFERENCES

1. Malesinski, W., *Azeotropy and Other Theoretical Problems of Vapour-Liquid Equilibrium*, Wiley-Interscience, New York, 1965 (general theory).
2. Horsley, L. H., *Azeotropic Data*, III, American Chemical Society, Washington, D.C., 1973 (a compilation of azeotropic data from the original literature for 16,500 binary and ternary systems).
3. Lide, D. R., Editor, *CRC Handbook of Chemistry and Physics*, 74th Edition, CRC Press, Boca Raton, FL, 1993, 6-158 to 6-190 (general classification and tables for binary and multi-component azeotropes).
4. Ogorodnikov, S. K.; Lesteva, T. M.; Kogan, V. B., *Azeotropic Mixtures* (in Russian), Khimiya, 1971 (azeotropic data for over 21,000 binary and multi-component systems).

Molecular formula					
Component					
1	2	Name	T_{Az}/K	$x_{1,Az}$	P_{Az}/kPa
ClH		Hydrogen chloride			
	H_2O	Water	381.73	0.111	101.3
H_2O		Water			
	$CHCl_3$	Trichloromethane	329.25	0.160	101.3
	CH_2O_2	Methanoic acid	380.35	0.427	101.3
	CH_3NO_2	Nitromethane	356.74	0.511	101.3
	CS_2	Carbon disulfide	315.75	0.109	101.3*
	C_2H_3N	Ethanenitrile	349.65	0.307	101.3*
	$C_2H_5NO_2$	Nitroethane	360.37	0.624	101.3*
	C_2H_6O	Ethanol	351.32	0.096	101.3
	$C_4H_8O_2$	Ethyl ethanoate	343.53	0.312	101.3
	$C_4H_{10}O$	1-Butanol	365.85	0.753	101.3*
	$C_4H_{10}O$	2-Butanol	360.15	0.601	101.3
	C_5H_5N	Pyridine	366.75	0.755	101.3
	$C_5H_{11}N$	Piperidine	365.95	0.718	101.3
	C_5H_{12}	Pentane	307.75	0.054	101.3*
	C_6H_5Cl	Chlorobenzene	363.35	0.712	101.3
	C_6H_6	Benzene	342.40	0.295	101.3*
	C_6H_6O	Phenol	372.67	0.981	101.3
	C_6H_{10}	Cyclohexene	343.95	0.308	101.3*
	C_6H_{12}	Cyclohexane	342.65	0.300	101.3*
	C_6H_{14}	Hexane	334.75	0.221	101.3*
	C_7H_8	Toluene	357.25	0.444	101.3*
	C_7H_{16}	Heptane	352.35	0.452	101.3*
	C_8H_{10}	1,3-Dimethylbenzene	365.15	0.767	101.3*
	C_8H_{10}	Ethylbenzene	365.15	0.744	101.3*

TABLE 3.1.2
AZEOTROPIC DATA FOR BINARY MIXTURES (continued)

Molecular formula					
Component					
1	2	Name	T_{Az}/K	$x_{1, Az}$	P_{Az}/kPa
	C_8H_{18}	Octane	362.75	0.673	101.3*
	$C_8H_{18}O$	Dibutyl ether	366.05	0.781	101.3*
	C_9H_{20}	Nonane	367.95	0.970	101.3*
	$C_{12}H_{27}N$	Tributylamine	372.80	0.976	101.3*
CCl_4		Tetrachloromethane			
	CH_2O_2	Methanoic acid	339.80	0.569	101.3
	CH_3NO_2	Nitromethane	344.45	0.660	101.3
	CH_4O	Methanol	328.85	0.445	101.3
	C_2H_3N	Ethanenitrile	338.25	0.566	101.3
	$C_2H_4C_{12}$	1,2-Dichloroethane	313.15	0.771	29.2
	C_2H_6O	Ethanol	338.19	0.615	101.3
	C_3H_6O	2-Propanone	329.23	0.047	101.3
	$C_3H_6O_3$	Dimethyl carbonate	298.15	0.906	15.45
	C_3H_8O	1-Propanol	346.55	0.820	101.3
	C_4H_9Cl	1-Chlorobutane	313.15	0.864	28.56
	$C_4H_{10}O$	1-Butanol	349.70	0.951	101.3
$CHCl_3$		Trichloromethane			
	CH_4O	Methanol	313.15	0.701	60.59
			323.15	0.670	89.87
	C_2H_6O	Ethanol	323.15	0.863	72.9
	C_3H_6O	2-Propanone	313.15	0.616	41.09
			323.15	0.622	60.62
	C_4H_8O	Oxolane	303.15	0.454	19.5
	$C_6H_{15}N$	Triethylamine	283.15	0.201	4.0
CH_2Cl_2		Dichloromethane			
	CH_4O	Methanol	298.18	0.871	62.5
			398.21	0.650	1222.7
	C_5H_{12}	Pentane	298.19	0.449	82.53
			398.16	0.565	1185.7
CH_2O_2		Methanoic acid			
	CS_2	Carbon disulfide	315.70	0.253	101.3
CH_3NO_2		Nitromethane			
	CH_4O	Methanol	298.15	0.120	17.44
			348.17	0.048	151.89
	CS_2	Carbon disulfide	314.35	0.845	101.3
	C_2H_6O	Ethanol	298.18	0.363	10.23
			348.15	0.232	97.59
			398.17	0.088	504.05
	C_4H_9Cl	1-Chlorobutane	348.16	0.203	96.45
	C_6H_5Cl	Chlorobenzene	348.17	0.881	43.24
CH_4O		Methanol			
	$C_2Cl_3F_3$	1,1,2-Trichlorotrifluoroethane	308.15	0.259	85.25
	$C_2HBrClF_3$	Bromochloro-1,1,1-trifluoroethane	308.15	0.184	65.17
	$C_2H_4Cl_2$	1,2-Dichloroethane	323.15	0.651	65.76
	C_3H_6O	2-Propanone	313.15	0.152	57.49
			323.15	0.198	84.05
			328.65	0.198	101.3
	$C_3H_6O_2$	Methyl ethanoate	326.65	0.352	101.3
	$C_4H_{10}O$	Diethyl ether	298.16	0.030	71.69
			388.15	0.280	998.4
	$C_4H_{11}N$	Diethylamine	297.97	0.740	14.39
	C_5H_{10}	Cyclopentane	311.95	0.263	101.3
	C_5H_{12}	Pentane	304.00	0.145	101.3
	$C_5H_{12}O$	*tert*-Butyl methyl ether	324.42	0.315	101.3
	C_6H_6	Benzene	330.65	0.610	101.3

TABLE 3.1.2
AZEOTROPIC DATA FOR BINARY MIXTURES (continued)

Molecular formula					
Component					
1	2	Name	T_{Az}/K	$x_{1, Az}$	P_{Az}/kPa
	C_6H_{12}	Cyclohexane	327.05	0.601	101.3
	C_7H_8	Toluene	336.65	0.883	101.3
	C_7H_{16}	Heptane	332.25	0.769	101.3*
	C_8H_{18}	Octane	335.90	0.881	101.3
	C_9H_{20}	Nonane	337.25	0.953	101.3
CS_2		Carbon disulfide			
	C_2H_6O	Ethanol	315.75	0.860	101.3
	C_3H_6O	2-Propanone	312.40	0.608	101.3
	C_3H_8O	1-Propanol	318.80	0.931	101.3
	$C_4H_8O_2$	Ethyl ethanoate	319.25	0.974	101.3
$C_2Cl_3F_3$		1,1,2-Trichlorotrifluoroethane			
	C_3H_6O	2-Propanone	308.15	0.698	73.36
	$C_3H_6O_2$	Methyl ethanoate	308.15	0.771	68.11
$C_2HBrClF_3$		Bromochloro-1,1,1-trifluoroethane			
	C_3H_6O	2-Propanone	308.15	0.454	36.13
	$C_3H_6O_2$	Methyl ethanoate	308.15	0.444	36.23
	C_4H_8O	Oxolane	308.15	0.358	26.74
	$C_5H_{12}O$	*tert*-Butyl methyl ether	308.15	0.457	35.28
$C_2H_3Cl_3$		1,1,1-Trichloroethane			
	C_6H_{14}	Hexane	333.15	0.231	77.41
C_2H_3N		Ethanenitrile			
	C_2H_6O	Ethanol	293.15	0.676	10.98
			323.15	0.546	42.42
			345.65	0.469	101.3
			393.15	0.233	449.7
	$C_4H_8O_2$	Ethyl ethanoate	353.15	0.324	116.22
	C_4H_9Cl	1-Chlorobutane	348.18	0.471	116.54
	$C_4H_{10}O$	Diethyl ether	388.17	0.002	903.6
	$C_4H_{11}N$	Diethylamine	298.00	0.116	32.15
			398.33	0.165	703.2
	C_6H_6	Benzene	298.16	0.465	15.95
			313.15	0.472	30.44
			397.86	0.501	415.4
	C_6H_8	1,4-Cyclohexadiene	313.15	0.572	31.0
	C_6H_{10}	Cyclohexene	313.15	0.496	39.83
	C_7H_8	Toluene	343.15	0.900	71.04
			354.55	0.900	101.3
$C_2H_4Cl_2$		1,2-Dichloroethane			
	C_7H_{16}	Heptane	343.15	0.746	71.52
$C_2H_4O_2$		Ethanoic acid			
	$C_4H_8O_2$	1,4-Dioxane	392.65	0.831	101.3
	C_5H_5N	Pyridine	411.25	0.579	101.3
	C_6H_6	Benzene	353.20	0.026	101.3
	C_6H_{12}	Cyclohexane	351.95	0.130	101.3
	C_6H_{14}	Hexane	341.40	0.084	101.3
	$C_6H_{15}N$	Triethylamine	436.15	0.774	101.3
	C_7H_8	Toluene	373.80	0.375	101.3
	C_7H_{16}	Heptane	364.87	0.451	101.3
	C_8H_{10}	Ethylbenzene	387.80	0.774	101.3
	C_8H_{18}	Octane	378.85	0.688	101.3
	C_9H_{20}	Nonane	386.05	0.826	101.3
C_2H_5I		Iodoethane			
	C_6H_{14}	Hexane	337.85	0.420	100.17
C_2H_5N		Aziridine			
	C_6H_{12}	Cyclohexane	298.15	0.734	33.41

TABLE 3.1.2
AZEOTROPIC DATA FOR BINARY MIXTURES (continued)

Molecular formula					
Component					
1	2	Name	T_{Az}/K	$x_{1, Az}$	P_{Az}/kPa
C_2H_6O		Ethanol			
	$C_3H_6O_2$	1,3-Dioxolane	338.15	0.413	82.68
	$C_4H_8O_2$	1,4-Dioxane	323.15	0.835	30.54
	C_4H_9N	Pyrrolidine	333.35	0.510	31.26
	C_5H_{10}	Cyclopentane	317.85	0.110	101.3
	C_5H_{12}	Pentane	307.45	0.076	101.3
	C_6H_6	Benzene	341.05	0.440	101.3
	C_6H_{12}	Cyclohexane	337.95	0.430	101.3
	C_6H_{14}	Hexane	313.15	0.288	49.26
			331.83	0.332	101.3
	C_7H_8	Toluene	303.15	0.696	12.31
			333.15	0.762	51.04
			349.85	0.810	101.3
	C_7H_{16}	Heptane	303.15	0.553	16.25
			313.15	0.581	26.52
			323.15	0.602	42.07
			343.17	0.641	96.47
	C_8H_{18}	Octane	313.15	0.832	19.69
			350.15	0.898	101.3
	C_9H_{20}	Nonane	343.21	0.974	72.63
$C_2H_6O_2$		1,2-Ethanediol			
	C_7H_8	Toluene	383.25	0.034	101.3
	C_7H_{16}	Heptane	371.05	0.048	101.3
	$C_8H_{18}O$	Dibutyl ether	412.65	0.125	101.3
	$C_{10}H_{22}$	Decane	434.15	0.406	101.3
C_2H_6S		2-Thiapropane			
	C_5H_{12}	Pentane	304.95	0.503	101.3
$C_2H_8N_2$		1,2-Ethanediamine			
	C_7H_8	Toluene	377.15	0.406	101.3
C_3H_5N		Propanenitrile			
	C_6H_{14}	Hexane	336.65	0.134	101.3
$C_3H_6Br_2$		1,3-Dibromopropane			
	C_9H_20	Nonane	348.15	0.293	8.6
C_3H_6O		Propanal			
	C_5H_{12}	Pentane	313.15	0.334	136.58
C_3H_6O		2-Propanone			
	$C_3H_6O_2$	Methyl ethanoate	328.95	0.544	101.3
	$C_4H_{10}O$	Diethyl ether	388.30	0.058	912.7
	$C_4H_{11}N$	Diethylamine	298.03	0.466	36.56
			398.10	0.491	765.1
	C_5H_{10}	Cyclopentane	314.15	0.404	101.3
	C_5H_{12}	Pentane	298.15	0.250	77.7
	C_6H_{12}	Cyclohexane	326.15	0.751	101.3
	C_6H_{14}	Hexane	318.15	0.642	86.4
	C_7H_{16}	Heptane	338.15	0.918	138.33
$C_3H_6O_2$		1,3-Dioxolane			
	C_6H_{12}	Cyclohexane	333.15	0.563	77.84
	C_7H_{16}	Heptane	343.15	0.789	93.01
$C_3H_6O_2$		Ethyl methanoate			
	C_5H_{12}	Pentane	305.65	0.294	101.3
$C_3H_6O_2$		Methyl ethanoate			
	C_6H_{12}	Cyclohexane	313.15	0.778	59.49
			328.65	0.801	101.3
	C_6H_{14}	Hexane	324.90	0.642	101.3
$C_3H_6O_2$		Propanoic acid			

TABLE 3.1.2
AZEOTROPIC DATA FOR BINARY MIXTURES (continued)

Molecular formula Component 1	2	Name	T_{Az}/K	$x_{1, Az}$	P_{Az}/kPa
	C_5H_5N	Pyridine	421.75	0.686	101.3
	C_7H_{16}	Heptane	370.97	0.027	101.3
	C_9H_{12}	Propylbenzene	412.65	0.830	101.3
$C_3H_6O_3$		Dimethyl carbonate			
	C_6H_{12}	Cyclohexane	298.15	0.328	16.83
	C_6H_{14}	Hexane	298.15	0.219	22.47
	C_8H_{18}	Octane	298.15	0.842	7.76
C_3H_7N		Azetidine			
	C_6H_{12}	Cyclohexane	298.15	0.813	24.43
C_3H_7NO		*N,N*-Dimethylmethanamide			
	$C_6H_{13}N$	Cyclohexylamine	393.15	0.145	68.77
$C_3H_7NO_2$		1-Nitropropane			
	C_3H_8O	1-Propanol	370.10	0.061	101.3
	C_7H_{16}	Heptane	369.75	0.149	101.3
C_3H_8O		1-Propanol			
	$C_4H_8O_2$	1,4-Dioxane	368.45	0.642	101.3
	C_6H_6	Benzene	350.27	0.209	101.3
	C_6H_{12}	Cyclohexane	347.84	0.241	101.3
	C_7H_{16}	Heptane	303.15	0.282	10.16
			333.15	0.373	39.85
			357.75	0.470	101.3
	C_8H_{18}	Octane	363.15	0.740	87.93
	C_8H_{18}	2,2,4-Trimethylpentane	348.52	0.425	72.83
	C_9H_{20}	Nonane	363.15	0.932	77.82
C_3H_8O		2-Propanol			
	$C_4H_{11}N$	Butylamine	347.85	0.646	101.3
	C_5H_{12}	Pentane	308.65	0.071	101.3
	C_6H_{12}	Cyclohexane	342.55	0.397	101.3
	C_6H_{14}	Hexane	328.21	0.249	78.47
	C_7H_8	Toluene	353.75	0.773	101.3
	C_7H_{16}	Heptane	303.15	0.470	12.97
	C_8H_{18}	Octane	353.15	0.889	95.12
C_3H_8S		2-Thiabutane			
	C_6H_{12}	Methylcyclopentane	338.75	0.664	101.3
	C_6H_{14}	2,2-Dimethylpentane	339.52	0.908	101.3
C_3H_8S		1-Propanethiol			
	C_6H_{12}	Cyclohexane	340.92	0.978	101.3
	C_6H_{14}	Hexane	337.50	0.557	101.3
	$C_6H_{14}O$	Diisopropyl ether	339.00	0.714	101.3
C_4H_4S		Thiophene			
	C_6H_{12}	Cyclohexane	351.05	0.412	101.3
	C_6H_{14}	Hexane	341.61	0.114	101.3
C_4H_8O		Butanal			
	C_6H_{14}	Hexane	333.15	0.296	101.3
	C_7H_{16}	Heptane	318.15	0.888	35.79
C_4H_8O		2-Butanone			
	C_4H_9Cl	1-Chlorobutane	350.15	0.440	101.3
	$C_4H_{11}N$	Butylamine	347.15	0.353	101.3
	C_6H_6	Benzene	313.15	0.333	25.26
			351.48	0.460	101.3
	C_6H_8	1,4-Cyclohexadiene	313.15	0.702	24.6
	C_6H_{10}	Cyclohexene	313.15	0.481	29.72
	C_6H_{12}	Cyclohexane	313.15	0.449	33.56
			344.95	0.438	101.3
	C_6H_{12}	1-Hexene	333.15	0.188	94.7

TABLE 3.1.2
AZEOTROPIC DATA FOR BINARY MIXTURES (continued)

Molecular formula					
Component					
1	2	Name	T_{Az}/K	$x_{1, Az}$	P_{Az}/kPa
	C_6H_{14}	Hexane	333.15	0.324	88.3
	C_7H_{16}	Heptane	323.15	0.764	39.07
			350.15	0.764	101.3
	C_8H_{16}	2,4,4-Trimethyl-1-pentene	313.15	0.827	24.39
	C_8H_{16}	2,4,4-Trimethyl-2-pentene	313.15	0.871	24.07
	C_8H_{18}	2,2,4-Trimethylpentane	313.15	0.744	26.24
C_4H_8O		Oxolane			
	C_6H_{12}	Cyclohexane	298.15	0.855	22.11
			333.15	0.974	83.14
	C_6H_{14}	Hexane	333.15	0.610	91.64
$C_4H_8O_2$		Butanoic acid			
	C_5H_5N	Pyridine	436.35	0.912	101.3
	C_6H_5Cl	Chlorobenzene	404.90	0.035	101.3
	C_8H_{10}	1,2-Dimethylbenzene	416.15	0.118	101.3
$C_4H_8O_2$		1,4-Dioxane			
	C_4H_9Br	1-Bromobutane	371.15	0.580	101.3
	$C_5H_{11}N$	Piperidine	343.15	0.677	36.82
	C_6H_{12}	Cyclohexane	313.15	0.194	26.14
	C_6H_{14}	Hexane	353.15	0.040	142.49
	C_7H_8	Toluene	313.15	0.898	10.29
	C_7H_{16}	Heptane	313.15	0.468	16.06
			353.15	0.489	71.74
	C_8H_{18}	Octane	353.15	0.832	53.41
$C_4H_8O_2$		Ethyl ethanoate			
	C_6H_{12}	Cyclohexane	313.15	0.495	32.81
	C_6H_{14}	Hexane	338.30	0.394	101.3
	C_7H_{16}	Heptane	323.15	0.827	39.56
$C_4H_8O_2$		Methyl propanoate			
	C_4H_9Cl	1-Chlorobutane	349.95	0.392	101.3
$C_4H_8O_2$		Propyl methanoate			
	C_4H_9Cl	1-Chlorobutane	349.25	0.392	101.3
	C_6H_6	Benzene	351.65	0.440	101.3
	C_6H_{12}	Cyclohexane	348.15	0.469	101.3
C_4H_9Br		1-Bromobutane			
	C_7H_{16}	Heptane	323.15	0.355	19.88
C_4H_9Cl		1-Chlorobutane			
	C_6H_{12}	Cyclohexane	313.15	0.538	27.85
			348.31	0.589	96.02
C_4H_9N		Pyrrolidine			
	C_6H_{12}	Cyclohexane	298.15	0.240	13.81
			333.15	0.324	55.24
	C_4H_9NO	Morpholine			
	$C_4H_{10}O$	1-Butanol	383.35	0.790	57.38
	C_8H_{16}	Cyclooctane	393.15	0.794	84.21
	C_8H_{18}	Octane	383.35	0.528	83.15
$C_4H_{10}O$		1-Butanol			
	C_5H_5N	Pyridine	391.75	0.704	101.3
	C_6H_5Cl	Chlorobenzene	388.45	0.659	101.3
	C_6H_{10}	Cyclohexene	355.15	0.055	101.3
	C_6H_{12}	Cyclohexane	318.15	0.043	30.59
	C_6H_{14}	Hexane	333.15	0.036	77.29
	C_7H_8	Toluene	378.65	0.324	101.3
	C_7H_{16}	Heptane	333.15	0.142	30.74
			367.00	0.229	101.3
	C_8H_{10}	1,2-Dimethylbenzene	389.95	0.811	101.3

TABLE 3.1.2
AZEOTROPIC DATA FOR BINARY MIXTURES (continued)

Molecular formula					
Component					
1	2	Name	T_{Az}/K	$x_{1, Az}$	P_{Az}/kPa
	$C_8H_{18}O$	Dibutyl ether	390.80	0.892	101.3
$C_4H_{10}O$		2-Butanol			
	C_6H_6	Benzene	351.65	0.161	101.3
	C_6H_{12}	Cyclohexane	318.15	0.110	32.05
	C_6H_{12}	1-Hexene	333.15	0.052	91.54
	C_6H_{14}	Hexane	333.15	0.098	79.39
	C_7H_{16}	Heptane	361.25	0.439	101.3
$C_4H_{10}O$		Diethyl ether			
	C_5H_{12}	Pentane	306.85	0.553	101.3
$C_4H_{10}O$		2-Methyl-1-propanol			
	C_6H_{14}	Hexane	332.53	0.062	76.66
	C_7H_8	Toluene	313.15	0.239	9.27
	C_7H_{16}	Heptane	333.15	0.225	32.94
	C_8H_{10}	1,4-Dimethylbenzene	313.15	0.610	5.01
	C_8H_{10}	Ethylbenzene	313.15	0.584	5.15
$C_4H_{10}O$		2-Methyl-2-propanol			
	C_6H_6	Benzene	313.15	0.240	27.97
			347.10	0.378	101.3
	C_6H_{12}	Cyclohexane	318.15	0.268	36.24
	C_6H_{14}	Hexane	313.15	0.159	41.02
	C_7H_8	Toluene	313.15	0.700	15.38
	C_7H_{16}	Heptane	313.15	0.500	19.25
			351.15	0.688	101.3
	C_8H_{18}	Octane	313.15	0.888	14.16
$C_4H_{10}O$		Methyl propyl ether			
	C_5H_{12}	Pentane	308.75	0.215	101.3
$C_4H_{10}O_2$		2,5-Dioxahexane			
	C_7H_{16}	2,4-Dimethylpentane	343.15	0.407	82.34
$C_4H_{10}O_2$		2-Ethoxyethanol			
	C_7H_{16}	Heptane	369.65	0.153	101.3
	C_9H_{12}	Propylbenzene	407.75	0.842	101.3
$C_5H_4O_2$		2-Furaldehyde			
	C_7H_{16}	Heptane	371.45	0.055	101.3
	$C_8H_{18}O$	Dibutyl ether	388.15	0.264	50.77
	C_9H_{12}	Propylbenzene	424.55	0.475	101.3
C_5H_5N		Pyridine			
	C_6H_6	Benzene	333.15	0.283	30.99
	C_6H_{12}	Cyclohexane	313.15	0.045	24.82
	C_7H_8	Toluene	313.15	0.145	8.02
			383.25	0.249	101.3
	C_7H_{16}	Heptane	353.15	0.283	62.2
	C_8H_{18}	Octane	369.75	0.666	66.8
	C_8H_{18}	2,2,4-Trimethylpentane	313.15	0.257	14.61
	C_9H_{20}	Nonane	369.75	0.928	58.2
$C_5H_{10}O$		Oxane			
	C_6H_{12}	Cyclohexane	298.15	0.145	13.15
$C_5H_{10}O_3$		Diethyl carbonate			
	C_8H_{18}	Octane	298.65	0.388	2.39
$C_5H_{11}N$		N-Methylpyrrolidine			
	C_6H_{12}	Cyclohexane	298.15	0.546	13.9
$C_5H_{12}O_2$		3,5-Dioxaheptane			
	C_7H_{16}	Heptane	343.15	0.908	56.95
C_6F_6		Hexafluorobenzene			
	$C_6H_{15}N$	Triethylamine	283.15	0.633	5.71
C_6H_6		Benzene			

TABLE 3.1.2
AZEOTROPIC DATA FOR BINARY MIXTURES (continued)

Molecular formula					
Component					
1	2	Name	T_{Az}/K	$x_{1, Az}$	P_{Az}/kPa
	C_6H_{10}	Cyclohexene	352.05	0.635	101.3
	C_6H_{12}	Cyclohexane	350.71	0.538	101.3
	C_6H_{14}	Hexane	298.15	0.095	20.21
C_6H_6O		Phenol			
	C_6H_7N	2-Methylpyridine	458.65	0.752	101.3
	C_7H_9N	2,4-Dimethylpyridine	466.55	0.601	101.3
	C_9H_{12}	1,3,5-Trimethylbenzene	436.65	0.253	101.3
	$C_{10}H_{22}$	Decane	441.15	0.449	101.3
C_6H_7N		Aniline			
	C_9H_{12}	1,3,5-Trimethylbenzene	437.50	0.150	101.3
	$C_{10}H_{22}O$	Dipentyl ether	450.65	0.675	101.3
	$C_{12}H_{26}$	Dodecane	453.52	0.821	101.3
C_6H_7N		2-Methylpyridine			
	C_8H_{18}	Octane	394.27	0.470	101.3
$C_6H_{12}O$		Cyclohexanol			
	C_8H_{10}	1,2-Dimethylbenzene	416.15	0.147	101.3
$C_7H_{14}O$		2-Heptanone			
	$C_8H_{18}O$	Dibutyl ether	393.22	0.161	54.7
$C_8H_{18}O_4$		2,5,8,11-Tetraoxadodecane			
	$C_{12}H_{26}$	Dodecane	435.26	0.409	25.55

TABLE 3.1.3 (TCXX and PCXX)
LIQUID-GAS CRITICAL PROPERTIES OF BINARY MIXTURES

Similarly to pure substances, liquid mixtures in equilibrium with their vapor reach, when heated, a critical point at which all the intensive properties (temperature, pressure, density or molar volume, and composition) of the coexisting phases become identical. The liquid-gas critical properties of mixtures (temperature T_c, pressure P_c, and molar volume V_c) are usually nonlinear functions of the molar composition of the mixture. For completely miscible binary liquid systems, the critical line is a continuous curve joining the critical points of the two pure components.

The following two tables report the T_c and P_c values for some binary completely miscible systems. Only a few V_c data are available for mixtures.

The composition dependence of T_c and P_c can be represented by various empirical equations. A simple equation is obtained by representing the "excess" critical temperature, T_c^E,

$$T_c^E = T_c - (x_1 T_{c1} + x_2 T_{c2}) \tag{1}$$

and pressure P_c^E,

$$P_c^E = P_c - (x_1 P_{c1} + x_2 P_{c2}) \tag{2}$$

by equations of the Redlich-Kister type. Hence

$$T_c / \text{K} = x_1 T_{c1} / \text{K} + x_2 T_{c2} / \text{K} + x_1 x_2 \left[A(1) + A(2)(x_1 - x_2) + A(3)(x_1 - x_2)^2 \right] \tag{3}$$

$$P_c / \text{Pa} = x_1 P_{c1} / \text{Pa} + x_2 P_{c2} / \text{Pa} + x_1 x_2 \left[A(1) + A(2)(x_1 - x_2) + A(3)(x_1 - x_2)^2 \right] \tag{4}$$

where T_{c1}, T_{c2}, P_{c1}, and P_{c2} are the pure component critical parameters.

The tables below give the coefficients *A(i)*, obtained by fitting Equations 3 and 4 to experimental $T_c(x_1)$ and $P_c(x_1)$ data. They also list the calculated values of $T_c(x_1)$ and $P_c(x_1)$ for increments of $x_1 = 0.2$.

The programs, Property Codes TCXX and PCXX, permit the calculation of $T_c(x_1)$, $P_c(x_1)$, $T_c^E(x_1)$, and $P_c^E(x_1)$ over the range $x_1 = 0$ to $x_1 = 1$, with any desired increment, or at any given value of x_1.

The components of each system, and the systems themselves, are arranged in the Hill order, with substances that do not contain carbon preceding those that do contain carbon.

The auxiliary values displayed by the programs have the following meanings:

$V(1) = T_{c1}$ or P_{c1}
$V(2) = T_{c2}$ or P_{c2}

Physical quantity	Symbol	SI unit
Critical temperature	T_c	K
Critical pressure	P_c	Pa
Excess critical temperature	T_c^E	K
Excess critical pressure	P_c^E	Pa
Mole fraction of component *i*	x_i	—

TABLE 3.1.3 (TCXX and PCXX) LIQUID-GAS CRITICAL PROPERTIES OF BINARY MIXTURES (continued)

REFERENCES

1. Rowlinson, J. S., *Liquids and Liquid Mixtures*, Butterworth, London, 1969 (general theory).
2. Reid, R. C.; Prausnitz, J. M.; Poling, B. E., *The Properties of Gases and Liquids*, 4th Ed., McGraw-Hill, New York, 1987(estimation methods).
3. Hicks, C. P.; Young, C. L., The Gas-Liquid Critical Properties of Binary Mixtures, *Chem. Rev.*, 1975, 75, 119 (a compilation of literature data up to 1972).
4. Sadus, R. J., *High Pressure Phase Behaviour of Multicomponent Fluid Mixtures*, Elsevier, Amsterdam, 1991 (theory and data, especially for ternary systems).
5. Knapp, H.; Doering, R.; Oellrich, L. Ploecker, U.; Prausnitz, J. M., *Vapor-Liquid Equilibria for Mixtures of Low Boiling Substances*, DECHEMA, Frankfurt/Main, Germany (data and correlations using equations of state).
6. *Int. DATA Ser., Sel. Data Mixtures, Ser. A* (Ed. H. V. Kehiaian), Thermodynamic Research Center, Texas Engineering Experiment Station, College Station, USA (primary data and correlations).

THE T_c VALUES FOR SOME BINARY COMPLETELY MISCIBLE SYSTEMS

System Number: 1
1. Argon, Ar
2. Nitrogen, N_2

x_1	T_c/K
0.0	126.2
0.2	131.0
0.4	135.9
0.6	141.0
0.8	146.1
1.0	150.9

$A(1)$ = -0.4578E+00
$A(2)$ = 0.1706E+01
$A(3)$ = 0.1062E+01

System Number: 2
1. Argon, Ar
2. Oxygen, O_2

x_1	T_c/K
0.0	154.6
0.2	153.6
0.4	152.7
0.6	151.9
0.8	151.3
1.0	150.9

$A(1)$ = -0.1807E+01
$A(2)$ = -0.6337E+00
$A(3)$ = 0.0000E+00

System Number: 3
1. Argon, Ar
2. Methane, CH_4

x_1	T_c/K
0.0	190.5
0.2	184.3
0.4	177.2
0.6	169.2
0.8	160.4
1.0	150.9

$A(1)$ = 0.1036E+02
$A(2)$ = -0.9526E+00
$A(3)$ = 0.0000E+00

System Number: 4
1. Water, H_2O
2. Ammonia, H_3N

x_1	T_c/K
0.0	405.5
0.2	471.5
0.4	525.3
0.6	570.2
0.8	609.7
1.0	647.1

$A(1)$ = 0.8933E+02
$A(2)$ = -0.3483E+02
$A(3)$ = 0.0000E+00

System Number: 5
1. Water, H_2O
2. Benzene, C_6H_6

x_1	T_c/K
0.0	562.2
0.2	547.3
0.4	538.0
0.6	544.6
0.8	577.5
1.0	647.1

$A(1)$ = -0.2639E+03
$A(2)$ = -0.1082E+03
$A(3)$ = 0.0000E+00

System Number: 6
1. Dihydrogen sulfide, H_2S
2. Carbon dioxide, CO_2

x_1	T_c/K
0.0	304.1
0.2	310.3
0.4	321.1
0.6	335.8
0.8	353.4
1.0	373.2

$A(1)$ = -0.4247E+02
$A(2)$ = 0.8897E+01
$A(3)$ = 0.0000E+00

System Number: 7
1. Dihydrogen sulfide, H_2S
2. Ethane, C_2H_6

x_1	T_c/K
0.0	305.4
0.2	308.8
0.4	317.4
0.6	331.5
0.8	350.5
1.0	373.2

$A(1)$ = -0.6200E+02
$A(2)$ = 0.5220E+01
$A(3)$ = 0.5148E+01

System Number: 8
1. Dihydrogen sulfide, H_2S
2. Propane, C_3H_8

x_1	T_c/K
0.0	369.8
0.2	364.2
0.4	359.5
0.6	357.6
0.8	361.2
1.0	373.2

$A(1)$ = -0.5380E+02
$A(2)$ = -0.2626E+02
$A(3)$ = -0.3296E+01

TABLE 3.1.3 (TCXX and PCXX)
LIQUID-GAS CRITICAL PROPERTIES OF BINARY MIXTURES (continued)

System Number: 9
1. Ammonia, H_3N
2. 2,2,4-Trimethylpentane, C_8H_{18}

x_1	T_c/K
0.0	544.0
0.2	531.5
0.4	518.7
0.6	485.1
0.8	436.6
1.0	405.5

$A(1)$ = 0.1200E+03
$A(2)$ = -0.6149E+02
$A(3)$ = -0.1715E+03

System Number: 10
1. Nitrogen, N_2
2. Oxygen, O_2

x_1	T_c/K
0.0	154.6
0.2	149.4
0.4	143.8
0.6	138.0
0.8	132.0
1.0	126.2

$A(1)$ = 0.2097E+01
$A(2)$ = -0.1985E+01
$A(3)$ = -0.3813E+00

System Number: 11
1. Nitrogen, N_2
2. Methane, CH_4

x_1	T_c/K
0.0	190.5
0.2	180.2
0.4	168.2
0.6	154.5
0.8	140.0
1.0	126.2

$A(1)$ = 0.1266E+02
$A(2)$ = -0.8263E+01
$A(3)$ = -0.5097E+01

System Number: 12
1. Oxygen, O_2
2. Methane, CH_4

x_1	T_c/K
0.0	190.5
0.2	183.1
0.4	176.1
0.6	167.7
0.8	159.2
1.0	154.6

$A(1)$ = -0.1901E+01
$A(2)$ = -0.1239E+02
$A(3)$ = -0.1848E+02

System Number: 13
1. Tetrachloromethane, CCl_4
2. 2-Propanone, C_3H_6O

x_1	T_c/K
0.0	508.1
0.2	515.9
0.4	525.3
0.6	535.3
0.8	545.4
1.0	556.6

$A(1)$ = -0.8360E+01
$A(2)$ = 0.2442E+01
$A(3)$ = -0.6163E+01

System Number: 21
1. Tetrachloromethane, CCl_4
2. Hexafluorobenzene, C_6F_6

x_1	T_c/K
0.0	516.7
0.2	519.6
0.4	525.1
0.6	533.1
0.8	543.6
1.0	556.6

$A(1)$ = -0.3146E+02
$A(2)$ = 0.0000E+00
$A(3)$ = 0.0000E+00

System Number: 14
1. Trichloromethane, $CHCl_3$
2. 2-Propanone, C_3H_6O

x_1	T_c/K
0.0	508.1
0.2	515.1
0.4	521.4
0.6	526.7
0.8	531.4
1.0	536.4

$A(1)$ = 0.7779E+01
$A(2)$ = -0.3649E+01
$A(3)$ = -0.4686E+01

System Number: 22
1. Dichloromethane, CH_2Cl_2
2. Propane, C_3H_8

x_1	T_c/K
0.0	369.8
0.2	388.7
0.4	412.2
0.6	440.2
0.8	472.8
1.0	510.0

$A(1)$ = -0.5704E+02
$A(2)$ = 0.0000E+00
$A(3)$ = 0.0000E+00

System Number: 23
1. Dichloromethane, CH_2Cl_2
2. Pentane, C_5H_{12}

x_1	T_c/K
0.0	469.7
0.2	474.2
0.4	478.5
0.6	484.5
0.8	494.4
1.0	510.0

$A(1)$ = -0.3468E+02
$A(2)$ = -0.2114E+02
$A(3)$ = 0.0000E+00

System Number: 24
1. Dichloromethane, CH_2Cl_2
2. Nonane, C_9H_{20}

x_1	T_c/K
0.0	594.6
0.2	580.8
0.4	567.5
0.6	552.6
0.8	534.2
1.0	510.0

$A(1)$ = 0.3228E+02
$A(2)$ = 0.2175E+02
$A(3)$ = 0.0000E+00

System Number: 15
1. Methane, CH_4
2. Carbon monoxide, CO

x_1	T_c/K
0.0	132.9
0.2	145.7
0.4	158.4
0.6	170.3
0.8	181.2
1.0	190.5

$A(1)$ = 0.1115E+02
$A(2)$ = 0.4641E+01
$A(3)$ = -0.9538E+00

System Number: 16
1. Methane, CH_4
2. Carbon dioxide, CO_2

x_1	T_c/K
0.0	304.1
0.2	279.8
0.4	262.6
0.6	246.2
0.8	224.3
1.0	190.5

$A(1)$ = 0.2970E+02
$A(2)$ = 0.6611E+02
$A(3)$ = 0.0000E+00

TABLE 3.1.3 (TCXX and PCXX) LIQUID-GAS CRITICAL PROPERTIES OF BINARY MIXTURES (continued)

System Number: 17
1. Methane, CH_4
2. Ethane, C_2H_6

x_1	T_c/K
0.0	305.4
0.2	290.9
0.4	272.8
0.6	251.3
0.8	225.0
1.0	190.5

$A(1)$ = 0.5833E+02
$A(2)$ = 0.1575E+02
$A(3)$ = 0.1099E+02

System Number: 27
1. Methylamine, CH_5N
2. Propane, C_3H_8

x_1	T_c/K
0.0	369.8
0.2	371.5
0.4	379.0
0.6	391.9
0.8	409.6
1.0	430.7

$A(1)$ = -0.6185E+02
$A(2)$ = 0.8196E+01
$A(3)$ = 0.2914E+01

System Number: 20
1. Ethene, C_2H_4
2. Heptane, C_7H_{16}

x_1	T_c/K
0.0	540.3
0.2	523.4
0.4	497.0
0.6	454.7
0.8	387.2
1.0	282.3

$A(1)$ = 0.2683E+03
$A(2)$ = 0.9714E+02
$A(3)$ = 0.1886E+02

System Number: 31
1. Ethane, C_2H_6
2. 2-Propanone, C_3H_6O

x_1	T_c/K
0.0	508.1
0.2	486.0
0.4	456.6
0.6	412.3
0.8	356.8
1.0	305.4

$A(1)$ = 0.1185E+03
$A(2)$ = -0.3975E+02
$A(3)$ = -0.7511E+02

System Number: 25
1. Methanol, CH_4O
2. Ethanenitrile, C_2H_3N

x_1	T_c/K
0.0	545.5
0.2	534.0
0.4	526.6
0.6	520.7
0.8	515.6
1.0	512.6

$A(1)$ = -0.2210E+02
$A(2)$ = 0.7382E+01
$A(3)$ = -0.1224E+02

System Number: 28
1. Ethanenitrile, C_2H_3N
2. 2-Butanone, C_4H_8O

x_1	T_c/K
0.0	536.8
0.2	538.1
0.4	539.6
0.6	541.3
0.8	543.1
1.0	545.5

$A(1)$ = -0.2858E+01
$A(2)$ = -0.9818E+00
$A(3)$ = -0.1147E+01

System Number: 29
1. 1,2-Dichloroethane, $C_2H_4Cl_2$
2. Propane, C_3H_8

x_1	T_c/K
0.0	369.8
0.2	397.3
0.4	431.6
0.6	471.2
0.8	514.9
1.0	561.0

$A(1)$ = -0.5836E+02
$A(2)$ = 0.1487E+02
$A(3)$ = 0.0000E+00

System Number: 18
1. Ethane, C_2H_6
2. Benzene, C_6H_6

x_1	T_c/K
0.0	562.2
0.2	498.3
0.4	435.7
0.6	381.6
0.8	339.0
1.0	305.4

$A(1)$ = -0.1061E+03
$A(2)$ = -0.2754E+02
$A(3)$ = 0.3148E+02

System Number: 26
1. Methanol, CH_4O
2. Diethylamine, $C_4H_{11}N$

x_1	T_c/K
0.0	500.0
0.2	497.9
0.4	498.2
0.6	501.4
0.8	506.9
1.0	512.6

$A(1)$ = -0.2722E+02
$A(2)$ = 0.7346E+01
$A(3)$ = 0.7522E+01

System Number: 19
1. Ethene, C_2H_4
2. Propene, C_3H_6

x_1	T_c/K
0.0	364.9
0.2	353.0
0.4	340.0
0.6	324.8
0.8	305.9
1.0	282.3

$A(1)$ = 0.3666E+02
$A(2)$ = 0.1294E+02
$A(3)$ = 0.0000E+00

System Number: 30
1. 1,2-Dichloroethane, $C_2H_4Cl_2$
2. Heptane, C_7H_{14}

x_1	T_c/K
0.0	540.3
0.2	539.2
0.4	540.8
0.6	544.9
0.8	551.7
1.0	561.0

$A(1)$ = -0.3252E+02
$A(2)$ = 0.0000E+00
$A(3)$ = 0.0000E+00

System Number: 32
1. Dimethylamine, C_2H_7N
2. Propane, C_3H_8

x_1	T_c/K
0.0	369.8
0.2	379.4
0.4	391.9
0.6	406.3
0.8	421.8
1.0	437.2

$A(1)$ = -0.1840E+02
$A(2)$ = 0.1003E+02
$A(3)$ = 0.0000E+00

TABLE 3.1.3 (TCXX and PCXX)
LIQUID-GAS CRITICAL PROPERTIES OF BINARY MIXTURES (continued)

System Number: 33
1. Propanenitrile, C_3H_5N
2. Pentane, C_5H_{12}

x_1	T_c/K
0.0	469.7
0.2	474.8
0.4	486.4
0.6	504.8
0.8	529.7
1.0	561.3

$A(1)$ = -0.8288E+02
$A(2)$ = 0.0000E+00
$A(3)$ = 0.0000E+00

System Number: 34
1. Propanenitrile, C_3H_5N
2. Hexane, C_6H_{14}

x_1	T_c/K
0.0	507.7
0.2	505.0
0.4	509.0
0.6	519.7
0.8	537.1
1.0	561.3

$A(1)$ = -0.8401E+02
$A(2)$ = 0.0000E+00
$A(3)$ = 0.0000E+00

System Number: 35
1. Propanenitrile, C_3H_5N
2. Heptane, C_7H_{16}

x_1	T_c/K
0.0	540.3
0.2	533.7
0.4	530.4
0.6	532.6
0.8	542.2
1.0	561.3

$A(1)$ = -0.8032E+02
$A(2)$ = -0.2090E+02
$A(3)$ = 0.0000E+00

System Number: 36
1. 2-Propanone, C_3H_6O
2. Propane, C_3H_8

x_1	T_c/K
0.0	369.8
0.2	392.9
0.4	423.4
0.6	454.9
0.8	483.5
1.0	508.1

$A(1)$ = 0.1470E+01
$A(2)$ = 0.3951E+02
$A(3)$ = -0.1761E+02

System Number: 40
1. 2-Propanone, C_3H_6O
2. Heptane, C_7H_{16}

x_1	T_c/K
0.0	540.3
0.2	529.4
0.4	518.5
0.6	509.9
0.8	505.6
1.0	508.1

$A(1)$ = -0.4173E+02
$A(2)$ = -0.2312E+02
$A(3)$ = 0.0000E+00

System Number: 41
1. Propane, C_3H_8
2. Trimethylamine, C_3H_9N

x_1	T_c/K
0.0	432.9
0.2	421.5
0.4	409.5
0.6	396.9
0.8	383.7
1.0	369.8

$A(1)$ = 0.7866E+01
$A(2)$ = 0.0000E+00
$A(3)$ = 0.0000E+00

System Number: 42
1. Propane, C_3H_8
2. 2-Methylpentane, C_6H_{14}

x_1	T_c/K
0.0	497.7
0.2	482.0
0.4	463.0
0.6	439.0
0.8	408.5
1.0	369.8

$A(1)$ = 0.7197E+02
$A(2)$ = 0.1692E+02
$A(3)$ = 0.0000E+00

System Number: 45
1. 1-Propanol, C_3H_8O
2. Hexane, C_6H_{14}

x_1	T_c/K
0.0	507.7
0.2	503.2
0.4	503.4
0.6	509.2
0.8	520.8
1.0	536.8

$A(1)$ = -0.6667E+02
$A(2)$ = 0.3781E+00
$A(3)$ = 0.7361E+01

System Number: 48
1. 2-Propanol, C_3H_8O
2. Benzene, C_6H_6

x_1	T_c/K
0.0	562.2
0.2	547.6
0.4	534.6
0.6	523.4
0.8	514.5
1.0	508.3

$A(1)$ = -0.2616E+02
$A(2)$ = -0.4153E+01
$A(3)$ = 0.0000E+00

System Number: 49
1. 2-Propanol, C_3H_8O
2. Hexane, C_6H_{14}

x_1	T_c/K
0.0	507.7
0.2	499.6
0.4	494.2
0.6	493.3
0.8	497.9
1.0	508.3

$A(1)$ = -0.5974E+02
$A(2)$ = -0.1058E+02
$A(3)$ = 0.5747E+01

System Number: 52
1. Butanenitrile, C_4H_7N
2. Cyclohexane, C_6H_{12}

x_1	T_c/K
0.0	553.5
0.2	548.8
0.4	550.8
0.6	558.5
0.8	570.4
1.0	585.4

$A(1)$ = -0.6168E+02
$A(2)$ = 0.1305E+02
$A(3)$ = 0.0000E+00

System Number: 54
1. Butanenitrile, C_4H_7N
2. Heptane, C_7H_{16}

x_1	T_c/K
0.0	540.3
0.2	537.5
0.4	542.4
0.6	553.2
0.8	568.2
1.0	585.4

$A(1)$ = -0.6258E+02
$A(2)$ = 0.1886E+02
$A(3)$ = 0.0000E+00

TABLE 3.1.3 (TCXX and PCXX)
LIQUID-GAS CRITICAL PROPERTIES OF BINARY MIXTURES (continued)

System Number: 56
1. 2-Butanone, C_4H_8O
2. Diethylamine, $C_4H_{11}N$

x_1	T_c/K
0.0	500.0
0.2	503.6
0.4	509.3
0.6	516.4
0.8	525.3
1.0	536.8

$A(1)$ = -0.2284E+02
$A(2)$ = -0.2418E+01
$A(3)$ = -0.4982E+01

System Number: 57
1. 2-Butanone, C_4H_8O
2. Hexane, C_6H_{14}

x_1	T_c/K
0.0	507.7
0.2	505.8
0.4	507.8
0.6	513.9
0.8	523.8
1.0	536.8

$A(1)$ = -0.4773E+02
$A(2)$ = 0.3022E+01
$A(3)$ = 0.3254E+01

System Number: 58
1. Oxolane, C_4H_8O
2. Hexafluorobenzene, C_6F_6

x_1	T_c/K
0.0	516.7
0.2	515.4
0.4	516.4
0.6	520.4
0.8	528.1
1.0	540.1

$A(1)$ = -0.4172E+02
$A(2)$ = -0.6786E+01
$A(3)$ = 0.0000E+00

System Number: 59
1. Oxolane, C_4H_8O
2. Benzene, C_6H_6

x_1	T_c/K
0.0	562.2
0.2	558.6
0.4	554.6
0.6	550.2
0.8	545.4
1.0	540.1

$A(1)$ = 0.5220E+01
$A(2)$ = 0.0000E+00
$A(3)$ = 0.0000E+00

System Number: 60
1. Oxolane, C_4H_8O
2. Cyclohexane, C_6H_{12}

x_1	T_c/K
0.0	553.5
0.2	547.6
0.4	543.3
0.6	540.6
0.8	539.6
1.0	540.1

$A(1)$ = -0.2010E+02
$A(2)$ = 0.0000E+00
$A(3)$ = 0.0000E+00

System Number: 61
1. Oxolane, C_4H_8O
2. Hexane, C_6H_{14}

x_1	T_c/K
0.0	507.7
0.2	509.6
0.4	513.5
0.6	519.5
0.8	528.3
1.0	540.1

$A(1)$ = -0.3086E+02
$A(2)$ = -0.4164E+01
$A(3)$ = 0.0000E+00

System Number: 62
1. 1,4-Dioxane, $C_4H_8O_2$
2. Benzene, C_6H_6

x_1	T_c/K
0.0	562.2
0.2	565.8
0.4	570.1
0.6	575.1
0.8	580.7
1.0	587.0

$A(1)$ = -0.8291E+01
$A(2)$ = 0.0000E+00
$A(3)$ = 0.0000E+00

System Number: 63
1. 1,4-Dioxane, $C_4H_8O_2$
2. Cyclohexane, C_6H_{12}

x_1	T_c/K
0.0	553.5
0.2	552.2
0.4	554.5
0.6	560.8
0.8	571.5
1.0	587.0

$A(1)$ = -0.5267E+02
$A(2)$ = -0.4352E+01
$A(3)$ = 0.0000E+00

System Number: 64
1. Diethyl ether, $C_4H_{10}O$
2. Hexane, C_6H_{14}

x_1	T_c/K
0.0	507.7
0.2	499.9
0.4	491.9
0.6	483.7
0.8	475.3
1.0	466.7

$A(1)$ = 0.2359E+01
$A(2)$ = 0.0000E+00
$A(3)$ = 0.0000E+00

System Number: 69
1. 2-Methyl-1-propanol, $C_4H_{10}O$
2. Benzene, C_6H_6

x_1	T_c/K
0.0	562.2
0.2	555.1
0.4	550.1
0.6	547.5
0.8	547.0
1.0	547.8

$A(1)$ = -0.2584E+02
$A(2)$ = 0.2491E+01
$A(3)$ = 0.2915E+01

System Number: 71
1. Diethylamine, $C_4H_{11}N$
2. Hexafluorobenzene, C_6F_6

x_1	T_c/K
0.0	516.7
0.2	514.1
0.4	509.1
0.6	503.7
0.8	500.0
1.0	500.0

$A(1)$ = -0.8145E+01
$A(2)$ = -0.2155E+02
$A(3)$ = 0.0000E+00

System Number: 72
1. Diethylamine, $C_4H_{11}N$
2. Benzene, C_6H_6

x_1	T_c/K
0.0	562.2
0.2	545.8
0.4	532.5
0.6	521.3
0.8	510.8
1.0	500.0

$A(1)$ = -0.1751E+02
$A(2)$ = 0.1204E+02
$A(3)$ = 0.0000E+00

TABLE 3.1.3 (TCXX and PCXX) LIQUID-GAS CRITICAL PROPERTIES OF BINARY MIXTURES (continued)

System Number: 73
1. Butylamine, $C_4H_{11}N$
2. Hexane, C_6H_{14}

x_1	T_c/K
0.0	507.7
0.2	508.2
0.4	510.9
0.6	515.7
0.8	522.7
1.0	531.9

$A(1)$ = -0.2710E+02
$A(2)$ = 0.0000E+00
$A(3)$ = 0.0000E+00

System Number: 75
1. Pentane, C_5H_{12}
2. Benzene, C_6H_6

x_1	T_c/K
0.0	562.2
0.2	548.2
0.4	525.4
0.6	500.3
0.8	479.6
1.0	469.7

$A(1)$ = -0.1294E+02
$A(2)$ = -0.6815E+02
$A(3)$ = 0.0000E+00

System Number: 76
1. Pentane, C_5H_{12}
2. Cyclohexane, C_6H_{12}

x_1	T_c/K
0.0	553.5
0.2	539.7
0.4	524.6
0.6	507.9
0.8	489.7
1.0	469.7

$A(1)$ = 0.1928E+02
$A(2)$ = 0.1147E+01
$A(3)$ = 0.0000E+00

System Number: 77
1. Pentane, C_5H_{12}
2. Hexane, C_6H_{14}

x_1	T_c/K
0.0	507.7
0.2	500.8
0.4	494.0
0.6	486.9
0.8	478.9
1.0	469.7

$A(1)$ = 0.7153E+01
$A(2)$ = 0.4887E+01
$A(3)$ = 0.0000E+00

System Number: 78
1. Pentane, C_5H_{12}
2. Triethylamine, $C_6H_{15}N$

x_1	T_c/K
0.0	535.6
0.2	525.6
0.4	514.0
0.6	500.8
0.8	486.0
1.0	469.7

$A(1)$ = 0.1965E+02
$A(2)$ = 0.0000E+00
$A(3)$ = 0.0000E+00

System Number: 79
1. Pentane, C_5H_{12}
2. Toluene, C_7H_8

x_1	T_c/K
0.0	591.8
0.2	563.6
0.4	535.3
0.6	510.8
0.8	490.1
1.0	469.7

$A(1)$ = -0.3305E+02
$A(2)$ = -0.1316E+01
$A(3)$ = 0.2468E+02

System Number: 81
1. Pentane, C_5H_{12}
2. Decane, $C_{10}H_{22}$

x_1	T_c/K
0.0	617.7
0.2	601.3
0.4	584.0
0.6	560.1
0.8	523.9
1.0	469.7

$A(1)$ = 0.1181E+03
$A(2)$ = 0.5968E+02
$A(3)$ = 0.0000E+00

System Number: 83
1. 2,2-Dimethylpropane, C_5H_{12}
2. Benzene, C_6H_6

x_1	T_c/K
0.0	562.2
0.2	533.6
0.4	503.4
0.6	474.5
0.8	450.3
1.0	433.8

$A(1)$ = -0.3769E+02
$A(2)$ = -0.3266E+02
$A(3)$ = 0.0000E+00

System Number: 84
1. 2,2-Dimethylpropane, C_5H_{12}
2. Cyclohexane, C_6H_{12}

x_1	T_c/K
0.0	553.5
0.2	530.6
0.4	507.1
0.6	483.0
0.8	458.5
1.0	433.8

$A(1)$ = 0.5779E+01
$A(2)$ = -0.1502E+01
$A(3)$ = 0.0000E+00

System Number: 85
1. 2,2-Dimethylpropane, C_5H_{12}
2. Hexane, C_6H_{14}

x_1	T_c/K
0.0	507.7
0.2	496.0
0.4	482.2
0.6	467.0
0.8	450.7
1.0	433.8

$A(1)$ = 0.1617E+02
$A(2)$ = -0.4720E+01
$A(3)$ = 0.0000E+00

System Number: 88
1. Hexafluorobenzene, C_6F_6
2. Triethylamine, $C_6H_{15}N$

x_1	T_c/K
0.0	535.6
0.2	527.7
0.4	521.0
0.6	516.8
0.8	515.6
1.0	516.7

$A(1)$ = -0.3052E+02
$A(2)$ = -0.3907E+01
$A(3)$ = 0.6846E+01

System Number: 89
1. Hexafluorobenzene, C_6F_6
2. Dibutyl ether, $C_8H_{18}O$

x_1	T_c/K
0.0	584.1
0.2	570.0
0.4	554.5
0.6	539.3
0.8	526.2
1.0	516.7

$A(1)$ = -0.1455E+02
$A(2)$ = -0.1762E+02
$A(3)$ = 0.0000E+00

TABLE 3.1.3 (TCXX and PCXX)
LIQUID-GAS CRITICAL PROPERTIES OF BINARY MIXTURES (continued)

System Number: 90
1. Benzene, C_6H_6
2. Cyclohexane, C_6H_{12}

x_1	T_c/K
0.0	553.5
0.2	552.8
0.4	553.2
0.6	554.8
0.8	557.8
1.0	562.2

$A(1)$ = -0.1602E+02
$A(2)$ = -0.1212E+01
$A(3)$ = 0.0000E+00

System Number: 91
1. Benzene, C_6H_6
2. Hexane, C_6H_{14}

x_1	T_c/K
0.0	507.7
0.2	514.7
0.4	522.8
0.6	531.9
0.8	543.9
1.0	562.2

$A(1)$ = -0.3138E+02
$A(2)$ = -0.1864E+02
$A(3)$ = -0.1108E+02

System Number: 94
1. Benzene, C_6H_6
2. Dibutyl ether, $C_8H_{18}O$

x_1	T_c/K
0.0	584.1
0.2	579.8
0.4	575.7
0.6	571.6
0.8	567.2
1.0	562.2

$A(1)$ = 0.2157E+01
$A(2)$ = 0.3179E+01
$A(3)$ = 0.0000E+00

System Number: 97
1. 1-Hexene, C_6H_{12}
2. Decane, $C_{10}H_{22}$

x_1	T_c/K
0.0	617.7
0.2	604.9
0.4	587.6
0.6	567.2
0.8	541.2
1.0	504.1

$A(1)$ = 0.6785E+02
$A(2)$ = 0.2360E+02
$A(3)$ = 0.2224E+02

System Number: 98
1. Hexane, C_6H_{14}
2. Triethylamine, $C_6H_{15}N$

x_1	T_c/K
0.0	535.6
0.2	532.0
0.4	526.4
0.6	520.3
0.8	514.0
1.0	507.7

$A(1)$ = 0.6938E+01
$A(2)$ = -0.6197E+01
$A(3)$ = 0.4080E+01

System Number: 103
1. Hexane, C_6H_{14}
2. Diisopropyl ether, $C_6H_{14}O$

x_1	T_c/K
0.0	500.3
0.2	501.4
0.4	502.7
0.6	504.2
0.8	505.9
1.0	507.7

$A(1)$ = -0.2159E+01
$A(2)$ = 0.0000E+00
$A(3)$ = 0.0000E+00

System Number: 104
1. Hexane, C_6H_{14}
2. Tetradecafluoromethylcyclohexane, C_7F_{14}

x_1	T_c/K
0.0	485.9
0.2	477.7
0.4	472.9
0.6	475.6
0.8	487.4
1.0	507.7

$A(1)$ = -0.9442E+02
$A(2)$ = -0.1761E+02
$A(3)$ = 0.1443E+02

System Number: 106
1. Triethylamine, $C_6H_{15}N$
2. Heptane, C_7H_{16}

x_1	T_c/K
0.0	540.3
0.2	539.8
0.4	539.5
0.6	539.0
0.8	537.8
1.0	535.6

$A(1)$ = 0.5365E+01
$A(2)$ = 0.4058E+01
$A(3)$ = 0.0000E+00

THE P_c VALUES FOR SOME BINARY COMPLETELY MISCIBLE SYSTEMS

System Number: 4
1. Water, H_2O
2. Ammonia, H_3N

x_1	P_c/MPa
0.0	11.350
0.2	16.628
0.4	19.335
0.6	20.749
0.8	21.597
1.0	22.060

$A(1)$ = 0.1376E+08
$A(2)$ = -0.7588E+07
$A(3)$ = 0.3575E+07

System Number: 5
1. Water, H_2O
2. Benzene, C_6H_6

x_1	P_c/MPa
0.0	4.898
0.2	8.029
0.4	8.211
0.6	9.980
0.8	14.999
1.0	22.060

$A(1)$ = -0.1901E+08
$A(2)$ = -0.1733E+08
$A(3)$ = 0.1871E+08

System Number: 6
1. Dihydrogen sulfide, H_2S
2. Carbon dioxide, CO_2

x_1	P_c/MPa
0.0	7.375
0.2	7.525
0.4	7.967
0.6	8.518
0.8	8.941
1.0	8.940

$A(1)$ = 0.3390E+06
$A(2)$ = 0.2486E+07
$A(3)$ = 0.3759E+06

System Number: 7
1. Dihydrogen sulfide, H_2S
2. Ethane, C_2H_6

x_1	P_c/MPa
0.0	4.884
0.2	5.290
0.4	5.902
0.6	6.798
0.8	7.895
1.0	8.940

$A(1)$ = -0.2384E+07
$A(2)$ = 0.8926E+06
$A(3)$ = 0.1074E+07

TABLE 3.1.3 (TCXX and PCXX)
LIQUID-GAS CRITICAL PROPERTIES OF BINARY MIXTURES (continued)

System Number: 8
1. Dihydrogen sulfide, H_2S
2. Propane, C_3H_8

x_1	P_c/MPa
0.0	4.250
0.2	4.916
0.4	5.551
0.6	6.265
0.8	7.282
1.0	8.940

$A(1)$ = -0.2833E+07
$A(2)$ = -0.2332E+07
$A(3)$ = -0.7414E+06

System Number: 9
1. Ammonia, H_3N
2. 2,2,4-Trimethylpentane, C_8H_{18}

x_1	P_c/MPa
0.0	2.568
0.2	4.167
0.4	7.433
0.6	10.078
0.8	11.213
1.0	11.350

$A(1)$ = 0.7851E+07
$A(2)$ = 0.9254E+07
$A(3)$ = -0.9120E+07

System Number: 22
1. Dichloromethane, CH_2Cl_2
2. Propane, C_3H_8

x_1	P_c/MPa
0.0	4.250
0.2	4.644
0.4	5.250
0.6	5.844
0.8	6.202
1.0	6.100

$A(1)$ = 0.1552E+07
$A(2)$ = 0.2336E+07
$A(3)$ = 0.0000E+00

System Number: 23
1. Dichloromethane, CH_2Cl_2
2. Pentane, C_5H_{12}

x_1	P_c/MPa
0.0	3.360
0.2	3.595
0.4	3.818
0.6	4.197
0.8	4.901
1.0	6.100

$A(1)$ = -0.3010E+07
$A(2)$ = -0.1760E+07
$A(3)$ = 0.0000E+00

System Number: 24
1. Dichloromethane, CH_2Cl_2
2. Nonane, C_9H_{20}

x_1	P_c/MPa
0.0	2.280
0.2	3.296
0.4	4.186
0.6	4.950
0.8	5.588
1.0	6.100

$A(1)$ = 0.1575E+07
$A(2)$ = 0.0000E+00
$A(3)$ = 0.0000E+00

System Number: 16
1. Methane, CH_4
2. Carbon dioxide, CO_2

x_1	P_c/MPa
0.0	7.375
0.2	8.606
0.4	8.538
0.6	7.989
0.8	6.954
1.0	4.604

$A(1)$ = 0.9262E+07
$A(2)$ = 0.5421E+05
$A(3)$ = 0.5354E+07

System Number: 17
1. Methane, CH_4
2. Ethane, C_2H_6

x_1	P_c/MPa
0.0	4.884
0.2	5.770
0.4	6.524
0.6	6.849
0.8	6.363
1.0	4.604

$A(1)$ = 0.8073E+07
$A(2)$ = 0.3967E+07
$A(3)$ = 0.5350E+06

System Number: 25
1. Methanol, CH_4O
2. Ethanenitrile, C_2H_3N

x_1	P_c/MPa
0.0	4.850
0.2	5.764
0.4	6.506
0.6	7.115
0.8	7.631
1.0	8.092

$A(1)$ = 0.1414E+07
$A(2)$ = -0.4083E+06
$A(3)$ = 0.0000E+00

System Number: 26
1. Methanol, CH_4O
2. Diethylamine, $C_4H_{11}N$

x_1	P_c/MPa
0.0	3.758
0.2	4.533
0.4	5.389
0.6	6.338
0.8	7.297
1.0	8.092

$A(1)$ = -0.2801E+06
$A(2)$ = 0.8497E+06
$A(3)$ = 0.6055E+06

System Number: 27
1. Methylamine, CH_5N
2. Propane, C_3H_8

x_1	P_c/MPa
0.0	4.250
0.2	4.655
0.4	5.185
0.6	5.916
0.8	6.791
1.0	7.614

$A(1)$ = -0.1626E+07
$A(2)$ = 0.6136E+06
$A(3)$ = 0.8905E+06

System Number: 28
1. Ethanenitrile, C_2H_3N
2. 2-Butanone, C_4H_8O

x_1	P_c/MPa
0.0	4.207
0.2	4.391
0.4	4.554
0.6	4.689
0.8	4.790
1.0	4.850

$A(1)$ = 0.3864E+06
$A(2)$ = 0.6625E+05
$A(3)$ = 0.0000E+00

System Number: 19
1. Ethene, C_2H_4
2. Propene, C_3H_6

x_1	P_c/MPa
0.0	4.601
0.2	4.960
0.4	5.253
0.6	5.443
0.8	5.427
1.0	5.041

$A(1)$ = 0.2180E+07
$A(2)$ = 0.1057E+07
$A(3)$ = 0.4095E+06

TABLE 3.1.3 (TCXX and PCXX)
LIQUID-GAS CRITICAL PROPERTIES OF BINARY MIXTURES (continued)

System Number: 20
1. Ethene, C_2H_4
2. Heptane, C_7H_{16}

x_1	P_c/MPa
0.0	2.756
0.2	4.364
0.4	6.424
0.6	9.164
0.8	10.302
1.0	5.041

$A(1) = 0.1558E+08$
$A(2) = 0.2378E+08$
$A(3) = 0.1636E+08$

System Number: 29
1. 1,2-Dichloroethane, $C_2H_4Cl_2$
2. Propane, C_3H_8

x_1	P_c/MPa
0.0	4.250
0.2	5.108
0.4	6.395
0.6	7.368
0.8	7.284
1.0	5.400

$A(1) = 0.8569E+07$
$A(2) = 0.7742E+07$
$A(3) = 0.0000E+00$

System Number: 30
1. 1,2-Dichloroethane, $C_2H_4Cl_2$
2. Heptane, C_7H_{16}

x_1	P_c/MPa
0.0	2.760
0.2	3.127
0.4	3.486
0.6	3.927
0.8	4.536
1.0	5.400

$A(1) = -0.1555E+07$
$A(2) = -0.9101E+06$
$A(3) = 0.0000E+00$

System Number: 31
1. Ethane, C_2H_6
2. 2-Propanone, C_3H_6O

x_1	P_c/MPa
0.0	4.700
0.2	6.165
0.4	8.368
0.6	9.212
0.8	7.889
1.0	4.880

$A(1) = 0.1700E+08$
$A(2) = 0.8417E+07$
$A(3) = -0.8390E+07$

System Number: 18
1. Ethane, C_2H_6
2. Benzene, C_6H_6

x_1	P_c/MPa
0.0	4.898
0.2	8.543
0.4	9.716
0.6	8.800
0.8	6.707
1.0	4.884

$A(1) = 0.1834E+08$
$A(2) = -0.9517E+07$
$A(3) = -0.3464E+07$

System Number: 32
1. Dimethylamine, C_2H_7N
2. Propane, C_3H_8

x_1	P_c/MPa
0.0	4.250
0.2	4.489
0.4	4.769
0.6	5.038
0.8	5.245
1.0	5.340

$A(1) = 0.4517E+06$
$A(2) = 0.5292E+06$
$A(3) = 0.0000E+00$

System Number: 36
1. 2-Propanone, C_3H_6O
2. Propane, C_3H_8

x_1	P_c/MPa
0.0	4.250
0.2	4.910
0.4	5.632
0.6	5.865
0.8	5.465
1.0	4.700

$A(1) = 0.5413E+07$
$A(2) = 0.1485E+07$
$A(3) = -0.2670E+07$

System Number: 40
1. 2-Propanone, C_3H_6O
2. Heptane, C_7H_{16}

x_1	P_c/MPa
0.0	2.760
0.2	3.321
0.4	3.703
0.6	4.000
0.8	4.301
1.0	4.700

$A(1) = 0.5061E+06$
$A(2) = -0.9571E+06$
$A(3) = 0.0000E+00$

System Number: 41
1. Propane, C_3H_8
2. Trimethylamine, C_3H_9N

x_1	P_c/MPa
0.0	4.087
0.2	4.249
0.4	4.367
0.6	4.408
0.8	4.363
1.0	4.250

$A(1) = 0.9184E+06$
$A(2) = 0.8184E+05$
$A(3) = -0.1670E+06$

System Number: 42
1. Propane, C_3H_8
2. 2-Methylpentane, C_6H_{14}

x_1	P_c/MPa
0.0	3.031
0.2	3.631
0.4	4.241
0.6	4.706
0.8	4.805
1.0	4.250

$A(1) = 0.3452E+07$
$A(2) = 0.2303E+07$
$A(3) = 0.4356E+06$

System Number: 56
1. 2-Butanone, C_4H_8O
2. Diethylamine, $C_4H_{11}N$

x_1	P_c/MPa
0.0	3.758
0.2	3.860
0.4	3.955
0.6	4.045
0.8	4.129
1.0	4.207

$A(1) = 0.7446E+05$
$A(2) = 0.0000E+00$
$A(3) = 0.0000E+00$

System Number: 57
1. 2-Butanone, C_4H_8O
2. Hexane, C_6H_{14}

x_1	P_c/MPa
0.0	3.010
0.2	3.257
0.4	3.504
0.6	3.754
0.8	3.997
1.0	4.207

$A(1) = 0.8029E+05$
$A(2) = 0.1119E+06$
$A(3) = 0.1007E+06$

TABLE 3.1.3 (TCXX and PCXX)
LIQUID-GAS CRITICAL PROPERTIES OF BINARY MIXTURES (continued)

System Number: 97
1. 1-Hexene, C_6H_{12}
2. Decane, $C_{10}H_{22}$

x_1	P_c/MPa
0.0	2.104
0.2	2.446
0.4	2.829
0.6	3.177
0.8	3.363
1.0	3.206

$A(1) = 0.1435E+07$
$A(2) = 0.1334E+07$
$A(3) = 0.3490E+06$

System Number: 104
1. Hexane, C_6H_{14}
2. Tetradecafluoro-methylcyclohexane, C_7F_{14}

x_1	P_c/MPa
0.0	2.019
0.2	2.162
0.4	2.234
0.6	2.396
0.8	2.684
1.0	3.010

$A(1) = -0.8626E+06$
$A(2) = -0.3763E+06$
$A(3) = 0.8104E+06$

TABLE 3.1.4 (LLEX)
LIQUID-LIQUID EQUILIBRIA IN BINARY LIQUID MIXTURES

The *molar Gibbs energy of mixing* G^M is related to the *excess molar Gibbs energy* G^E (Table 3.1.1) through the equation:

$$G^M = RT(x_1 \ln x_1 + x_2 \ln x_2) + G^E \quad (1)$$

where R is the molar gas constant, T is the temperature, and x_1 and $x_2 = 1 - x_1$ are the mole fractions of components 1 and 2, respectively. The function

$$G^M{}_{id} = RT(x_1 \ln x_1 + x_2 \ln x_2) \quad (2)$$

is the *ideal molar Gibbs energy of mixing*. The dependence of $G^M{}_{id}$ on the composition x_1 is represented schematically by curve *a* in Figure 1. It can be seen that the curve has no inflection points. This is also the case of the G^M of real systems when G^E is relatively small. When G^E is large, then inflection points may appear on the G^M curve (curve *b* in Figure 1). In the composition range between the inflection points A and B the system is unstable and splits into two liquid phases. The compositions x_1' and x_1'' of the equilibrium phases are given by the points C and D of tangency of the straight line C-D. The splitting occurs because the Gibbs energy of the two-phase system is lower than the Gibbs energy of the homogenous mixture.

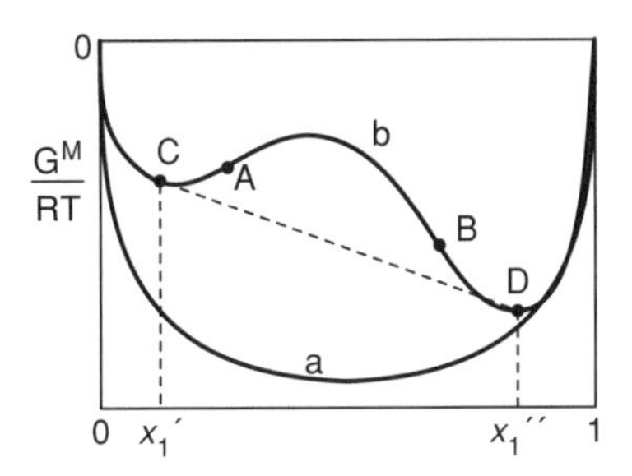

FIGURE 1

The molar Gibbs energy of mixing G^M is a function of T and the pressure P. At given P, the two-phase system may become completely miscible as T increases or decreases. The temperature at which this happens is called a critical solution temperature, T^c, which, together with the corresponding composition x_1^c, characterizes the critical solution point. Most mixtures display an upper critical solution point, i.e., the two-phase mixture becomes completely miscible as T increases. There exist also many cases of mixtures with a lower critical solution point, i.e., the two-phase mixture becomes completely miscible as T decreases. A few mixtures have both types of critical solution points (closed loop miscibility curves).

The liquid-liquid equilibrium curves are difficult to represent accurately by means of correlating equations. A simple equation is the Malesinski equation:

$$T/\mathrm{K} = A(1) + \sum_{i=3}^{n} A(i)\left[\frac{x_1/A(2) - x_2/(1-A(2))}{x_1/A(2) + x_2/(1-A(2))}\right]^{2(i-2)} \quad (3)$$

where $A(1)$ is the critical solution temperature T^c and $A(2)$ is the mole fraction of component 1 at the critical solution point, x_1^c. The table gives the coefficients $A(i)$, obtained by fitting Equation 3 to experimental $T(x_1)$ data. The table lists the calculated values of T for nine mole fractions x_1. All data are at atmospheric pressure or saturation pressure unless otherwise stated. The program, Property Code LLEX, permits the calculation of T at selected increments of x_1, or at a given value of x_1, within a specified range of compositions.

The components of each system, and the systems, are arranged in a modified Hill order, with substances that do not contain carbon preceding those that do contain carbon.

TABLE 3.1.4 (LLEX)
LIQUID-LIQUID EQUILIBRIA IN BINARY LIQUID MIXTURES (continued)

Physical quantity	Symbol	SI unit
Mole fraction of component *i*	x_i	—
Temperature	T	K

REFERENCES

1. Novak, J. P.; Matous, J.; Pick, J., *Liquid-Liquid Equilibria*, Academia, Prague, 1987 (general theory).
2. H. V. Kehiaian; C. S. Kehiaian, Organic Systems (Mixtures), in: *Bulletin of Chemical Thermodynamics*, Freeman, R. D., Editor, Thermochemistry Inc., 1969-1985 (17 volumes) (a compilation of bibliographical liquid-liquid equilibrium data).
3. Wisniak, J.; Tamir, A., *Liquid-Liquid Equilibrium and Extraction. A Literature Source Book*, Elsevier, Amsterdam, Part A, 1980; Part B, 1981; Supplement 1, 1985 (a compilation of bibliographical liquid-liquid equilibrium data).
4. Sorensen, J. M.; Arlt, W. *Liquid-Liquid Equilibrium Data Collection, Binary Systems*, DECHEMA, Frankfurt/Main, Germany, 1979 (an extensive compilation of numerical liquid-liquid equilibrium data and correlation coefficients).
5. Macedo, E. A.; Rasmussen, P., *Liquid-Liquid Equilibrium Data Collection, Binary, Ternary, and Quaternary Systems*, DECHEMA, Frankfurt/Main, Germany, 1987 (an extensive compilation of numerical liquid-liquid equilibrium data, published from 1979 until the end of 1984, and correlation coefficients).
6. *Int. DATA Ser., Sel. Data Mixtures, Ser. A* (Ed. H. V. Kehiaian), Thermodynamic Research Center, Texas Engineering Experiment Station, College Station, USA, 1973-1993 (21 volumes) (primary liquid-liquid equilibrium data and correlations)

System Number: 57
1. Water, H_2O
2. Nitroethane, $C_2H_5NO_2$

x_1	T/K
0.020	298.0
0.140	347.5
0.260	384.8
0.380	409.9
0.500	425.3
0.620	435.4
0.740	442.5
0.860	438.0
0.980	365.6

T^c/K = 0.4431E+03
x^c_1 = 0.7741E+00
$A(1)$ = -0.6860E+02
$A(2)$ = 0.8185E+02
$A(3)$ = -0.1677E+03

System Number: 58
1. Water, H_2O
2. Propenal, C_3H_4O

x_1	T/K
0.140	272.9
0.230	299.8
0.320	322.0
0.410	339.0
0.500	351.0
0.590	358.2
0.680	361.1
0.770	359.3
0.860	346.7

T^c/K = 0.3612E+03
x^c_1 = 0.6972E+00
$A(1)$ = -0.5229E+02
$A(2)$ = -0.8611E+02
$A(3)$ = 0.0000E+00

System Number: 59
1. Water, H_2O
2. Propanenitrile, C_3H_5N

x_1	T/K
0.150	289.3
0.250	318.4
0.350	343.7
0.450	363.8
0.550	377.5
0.650	384.5
0.750	386.5
0.850	384.8
0.950	347.2

T^c/K = 0.3865E+03
x^c_1 = 0.7679E+00
$A(1)$ = -0.1478E+02
$A(2)$ = -0.1307E+03
$A(3)$ = 0.0000E+00

System Number: 3
1. Water, H_2O
2. 2-Furaldehyde, $C_5H_4O_2$

x_1	T/K
0.200	301.4
0.280	319.0
0.360	336.3
0.440	352.8
0.520	367.8
0.600	380.2
0.680	388.9
0.760	393.1
0.840	394.0

T^c/K = 0.3940E+03
x^c_1 = 0.8356E+00
$A(1)$ = -0.9166E+01
$A(2)$ = -0.1261E+03
$A(3)$ = 0.0000E+00

System Number: 73
1. Water, H_2O
2. Phenol, C_6H_6O

x_1	T/K
0.650	285.9
0.690	299.5
0.730	312.3
0.770	323.5
0.810	332.2
0.850	337.3
0.890	338.7
0.930	338.7
0.970	325.1

T^c/K = 0.3387E+03
x^c_1 = 0.9138E+00
$A(1)$ = 0.5613E+01
$A(2)$ = -0.2289E+03
$A(3)$ = 0.0000E+00

System Number: 10
1. Tetrachloromethane, CCl_4
2. Hexadecafluoroheptane, C_7F_{16}

x_1	T/K
0.300	298.5
0.380	309.7
0.460	318.8
0.540	325.5
0.620	329.7
0.700	331.6
0.780	332.0
0.860	330.6
0.940	314.7

T^c/K = 0.3320E+03
x^c_1 = 0.7715E+00
$A(1)$ = -0.9546E+01
$A(2)$ = -0.7696E+02
$A(3)$ = 0.0000E+00

System Number: 11
1. Trichloromethane, $CHCl_3$
2. Hexadecafluoroheptane, C_7F_{16}

x_1	T/K
0.200	293.8
0.290	310.4
0.380	324.8
0.470	336.2
0.560	344.4
0.650	349.3
0.740	351.3
0.830	351.7
0.920	347.9

T^c/K = 0.3517E+03
x^c_1 = 0.8117E+00
$A(1)$ = -0.7193E+01
$A(2)$ = -0.4448E+02
$A(3)$ = -0.4879E+02

System Number: 40
1. Methanoic acid, CH_2O_2
2. Benzene, C_6H_6

x_1	T/K
0.150	293.4
0.250	322.1
0.350	337.3
0.450	343.7
0.550	345.5
0.650	345.3
0.750	341.7
0.850	325.2
0.950	273.0

T^c/K = 0.3456E+03
x^c_1 = 0.5888E+00
$A(1)$ = -0.1752E+02
$A(2)$ = -0.1118E+03
$A(3)$ = 0.0000E+00

TABLE 3.1.4 (LLEX)
LIQUID-LIQUID EQUILIBRIA IN BINARY LIQUID MIXTURES (continued)

System Number: 48
1. Nitromethane, CH_3NO_2
2. Carbon disulfide, CS_2

x_1	T/K
0.200	331.6
0.280	336.0
0.360	336.5
0.440	336.4
0.520	336.4
0.600	336.4
0.680	334.7
0.760	328.3
0.840	311.1

T^c/K = 0.3363E+03
x^c_1 = 0.4511E+00
$A(1)$ = 0.5146E+01
$A(2)$ = -0.5200E+02
$A(3)$ = -0.8791E+02

System Number: 49
1. Nitromethane, CH_3NO_2
2. 1,2-Ethanediol, $C_2H_6O_2$

x_1	T/K
0.350	305.8
0.425	309.9
0.500	312.3
0.575	313.1
0.650	313.2
0.725	313.1
0.800	311.6
0.875	305.1
0.950	294.5

T^c/K = 0.3132E+03
x^c_1 = 0.6542E+00
$A(1)$ = -0.8215E+00
$A(2)$ = -0.1027E+03
$A(3)$ = 0.9295E+02

System Number: 13
1. Nitromethane, CH_3NO_2
2. trans-2-Pentene, C_5H_{10}

x_1	T/K
0.300	322.7
0.360	327.2
0.420	328.6
0.480	328.8
0.540	328.8
0.600	328.7
0.660	327.7
0.720	323.7
0.780	313.8

T^c/K = 0.3288E+03
x^c_1 = 0.5181E+00
$A(1)$ = 0.2291E+01
$A(2)$ = -0.1922E+03
$A(3)$ = 0.0000E+00

System Number: 14
1. Nitromethane, CH_3NO_2
2. *cis*-2-Hexene, C_6H_{12}

x_1	T/K
0.300	325.2
0.360	329.4
0.420	331.4
0.480	332.3
0.540	332.5
0.600	332.5
0.660	332.2
0.720	330.9
0.780	327.3

T^c/K = 0.3325E+03
x^c_1 = 0.5667E+00
$A(1)$ = -0.5626E+01
$A(2)$ = -0.8935E+02
$A(3)$ = 0.0000E+00

System Number: 15
1. Nitromethane, CH_3NO_2
2. *trans*-2-Hexene, C_6H_{12}

x_1	T/K
0.300	333.4
0.360	336.8
0.420	338.9
0.480	340.1
0.540	340.6
0.600	340.6
0.660	340.2
0.720	339.0
0.780	336.2

T^c/K = 0.3406E+03
x^c_1 = 0.5780E+00
$A(1)$ = -0.1364E+02
$A(2)$ = -0.4665E+02
$A(3)$ = 0.0000E+00

System Number: 4
1. Nitromethane, CH_3NO_2
2. Hexane, C_6H_{14}

x_1	T/K
0.350	368.7
0.410	372.4
0.470	374.3
0.530	375.2
0.590	375.5
0.650	375.4
0.710	375.0
0.770	373.3
0.830	368.2

T^c/K = 0.3755E+03
x^c_1 = 0.6152E+00
$A(1)$ = -0.7114E+01
$A(2)$ = -0.8276E+02
$A(3)$ = 0.0000E+00

System Number: 1
1. Nitromethane, CH_3NO_2
2. Heptane, C_7H_{16}

x_1	T/K
0.350	372.0
0.420	377.0
0.490	379.8
0.560	381.0
0.630	381.4
0.700	381.3
0.770	380.4
0.840	376.3
0.910	362.1

T^c/K = 0.3814E+03
x^c_1 = 0.6552E+00
$A(1)$ = -0.7819E+01
$A(2)$ = -0.7164E+02
$A(3)$ = 0.0000E+00

System Number: 20
1. Methanol, CH_4O
2. Carbon disulfide, CS_2

x_1	T/K
0.100	293.1
0.200	307.2
0.300	307.3
0.400	307.5
0.500	305.7
0.600	298.2
0.700	282.2
0.800	256.7
0.900	221.9

T^c/K = 0.3073E+03
x^c_1 = 0.3071E+00
$A(1)$ = 0.1041E+02
$A(2)$ = -0.1393E+03
$A(3)$ = 0.0000E+00

System Number: 79
1. Methanol, CH_4O
2. Methylcyclopentane, C_6H_{12}
P = 5 MPa

x_1	T/K
0.280	300.4
0.330	303.1
0.380	304.2
0.430	304.6
0.480	304.7
0.530	304.6
0.580	304.5
0.630	303.8
0.680	302.0

T^c/K = 0.3047E+03
x^c_1 = 0.4911E+00
$A(1)$ = -0.3457E+01
$A(2)$ = -0.1123E+03
$A(3)$ = 0.0000E+00

System Number: 51
1. Methanol, CH_4O
2. Cyclohexane, C_6H_{12}

x_1	T/K
0.150	298.2
0.230	310.2
0.310	316.1
0.390	318.0
0.470	318.3
0.550	318.2
0.630	317.4
0.710	314.1
0.790	306.2

T^c/K = 0.3183E+03
x^c_1 = 0.4904E+00
$A(1)$ = -0.4003E+01
$A(2)$ = -0.1052E+03
$A(3)$ = 0.5212E+02

System Number: 81
1. Methanol, CH_4O
2. Hexane, C_6H_{14}

x_1	T/K
0.200	291.0
0.280	299.3
0.360	303.7
0.440	305.8
0.520	306.4
0.600	306.3
0.680	305.2
0.760	301.9
0.840	293.6

T^c/K = 0.3065E+03
x^c_1 = 0.5495E+00
$A(1)$ = -0.1345E+02
$A(2)$ = -0.5108E+02
$A(3)$ = 0.0000E+00

System Number: 87
1. Methanol, CH_4O
2. Hexane, C_6H_{14}
P = 1 MPa

x_1	T/K
0.200	284
0.270	299
0.340	305
0.410	307
0.480	307
0.550	307
0.620	307
0.690	305
0.760	297

T^c/K = 0.3075E+03
x^c_1 = 0.5123E+00
$A(1)$ = 0.7496E+01
$A(2)$ = -0.1834E+03
$A(3)$ = 0.0000E+00

TABLE 3.1.4 (LLEX)
LIQUID-LIQUID EQUILIBRIA IN BINARY LIQUID MIXTURES (continued)

System Number: 88
1. Methanol, CH_4O
2. Hexane, C_6H_{14}
P = 120 MPa

x_1	T/K
0.200	315
0.270	326
0.340	331
0.410	334
0.480	335
0.550	334
0.620	330
0.690	324
0.760	314

T^c/K = 0.3353E+03
x^c_1 = 0.4674E+00
$A(1)$ = -0.4506E+02
$A(2)$ = -0.6115E+02
$A(3)$ = 0.0000E+00

System Number: 82
1. Methanol, CH_4O
2. Cycloheptane, C_7H_{14}

x_1	T/K
0.100	305.6
0.190	322.7
0.280	332.7
0.370	337.7
0.460	339.7
0.550	340.3
0.640	340.0
0.730	338.1
0.820	331.7

T^c/K = 0.3403E+03
x^c_1 = 0.5637E+00
$A(1)$ = -0.1091E+02
$A(2)$ = -0.5365E+02
$A(3)$ = 0.0000E+00

System Number: 83
1. Methanol, CH_4O
2. Cyclooctane, C_8H_{16}

x_1	T/K
0.080	314.0
0.170	331.7
0.260	344.1
0.350	351.9
0.440	356.0
0.530	357.8
0.620	358.3
0.710	357.8
0.800	354.7

T^c/K = 0.3583E+03
x^c_1 = 0.6223E+00
$A(1)$ = -0.1094E+02
$A(2)$ = -0.5405E+02
$A(3)$ = 0.0000E+00

System Number: 80
1. Methanol, CH_4O
2. 2,2,4-Trimethylpentane, C_8H_{18}
P = 5 MPa

x_1	T/K
0.280	306.0
0.350	311.7
0.420	315.2
0.490	317.0
0.560	317.8
0.630	318.0
0.700	317.7
0.770	316.3
0.840	311.6

T^c/K = 0.3180E+03
x^c_1 = 0.6264E+00
$A(1)$ = -0.9257E+01
$A(2)$ = -0.5575E+02
$A(3)$ = 0.0000E+00

System Number: 89
1. Methanol, CH_4O
2. 2,2,4-Trimethylpentane, C_8H_{18}
P = 120 MPa

x_1	T/K
0.280	322.4
0.350	332.6
0.420	338.4
0.490	341.2
0.560	342.1
0.630	342.2
0.700	342.0
0.770	340.4
0.840	333.0

T^c/K = 0.3422E+03
x^c_1 = 0.6290E+00
$A(1)$ = -0.4597E+01
$A(2)$ = -0.1164E+03
$A(3)$ = 0.0000E+00

System Number: 84
1. Methanol, CH_4O
2. Hexylbenzene, $C_{12}H_{18}$

x_1	T/K
0.250	276.9
0.320	285.8
0.390	292.7
0.460	297.8
0.530	301.1
0.600	302.9
0.670	303.6
0.740	303.7
0.810	302.7

T^c/K = 0.3037E+03
x^c_1 = 0.7118E+00
$A(1)$ = -0.1035E+02
$A(2)$ = -0.6171E+02
$A(3)$ = 0.0000E+00

System Number: 52
1. Carbon disulfide, CS_2
2. Ethanenitrile, C_2H_3N

x_1	T/K
0.200	302.0
0.280	314.7
0.360	321.2
0.440	323.9
0.520	324.6
0.600	324.7
0.680	324.3
0.760	321.9
0.840	312.9

T^c/K = 0.3247E+03
x^c_1 = 0.5793E+00
$A(1)$ = -0.3293E+01
$A(2)$ = -0.9139E+02
$A(3)$ = 0.0000E+00

System Number: 47
1. Carbon disulfide, CS_2
2. Methyl ethanoate, $C_3H_6O_2$

x_1	T/K
0.100	181.7
0.200	200.3
0.300	211.6
0.400	217.5
0.500	220.0
0.600	220.7
0.700	220.4
0.800	217.6
0.900	204.5

T^c/K = 0.2208E+03
x^c_1 = 0.6219E+00
$A(1)$ = -0.9636E+01
$A(2)$ = -0.4506E+02
$A(3)$ = -0.1234E+02

System Number: 21
1. Carbon disulfide, CS_2
2. 1-Propanol, C_3H_8O

x_1	T/K
0.450	171.7
0.510	185.6
0.570	195.5
0.630	201.9
0.690	205.4
0.750	207.1
0.810	207.9
0.870	207.6
0.930	203.0

T^c/K = 0.2079E+03
x^c_1 = 0.8310E+00
$A(1)$ = -0.1447E+02
$A(2)$ = 0.6063E+01
$A(3)$ = -0.2285E+03

System Number: 22
1. Carbon disulfide, CS_2
2. 2-Propanol, C_3H_8O

x_1	T/K
0.340	217.5
0.420	227.1
0.500	235.8
0.580	242.4
0.660	245.8
0.740	246.4
0.820	246.3
0.900	243.9
0.980	203.4

T^c/K = 0.2462E+03
x^c_1 = 0.7882E+00
$A(1)$ = 0.9308E+01
$A(2)$ = -0.1479E+03
$A(3)$ = 0.7660E+02

System Number: 53
1. Carbon disulfide, CS_2
2. Ethanoic anhydride, $C_4H_6O_3$

x_1	T/K
0.240	274.3
0.330	284.6
0.420	292.7
0.510	298.4
0.600	301.6
0.690	302.9
0.780	302.9
0.870	300.3
0.960	276.7

T^c/K = 0.3030E+03
x^c_1 = 0.7435E+00
$A(1)$ = -0.8596E+01
$A(2)$ = -0.5561E+02
$A(3)$ = 0.0000E+00

System Number: 23
1. Carbon disulfide, CS_2
2. 1-Butanol, $C_4H_{10}O$

x_1	T/K
0.700	177.7
0.730	180.6
0.760	182.7
0.790	184.3
0.820	185.1
0.850	185.4
0.880	185.4
0.910	185.3
0.940	183.4

T^c/K = 0.1854E+03
x^c_1 = 0.8791E+00
$A(1)$ = -0.5690E+00
$A(2)$ = -0.1077E+03
$A(3)$ = 0.0000E+00

TABLE 3.1.4 (LLEX)
LIQUID-LIQUID EQUILIBRIA IN BINARY LIQUID MIXTURES (continued)

System Number: 24
1. Carbon disulfide, CS_2
2. 2-Methyl-1-propanol, $C_4H_{10}O$

x_1	T/K
0.450	186.6
0.510	198.8
0.570	208.3
0.630	214.9
0.690	218.7
0.750	220.2
0.810	220.5
0.870	220.5
0.930	218.6

T^c/K = 0.2205E+03
x^c_1 = 0.8442E+00
$A(1)$ = 0.1784E+01
$A(2)$ = -0.4589E+02
$A(3)$ = -0.1320E+03

System Number: 25
1. Carbon disulfide, CS_2
2. 2-Butanol, $C_4H_{10}O$

x_1	T/K
0.400	187.7
0.470	200.0
0.540	209.6
0.610	216.2
0.680	219.7
0.750	220.8
0.820	220.7
0.890	220.7
0.960	203.2

T^c/K = 0.2207E+03
x^c_1 = 0.8278E+00
$A(1)$ = 0.5294E+01
$A(2)$ = -0.6899E+02
$A(3)$ = -0.7185E+02

System Number: 78
1. Ethanenitrile, C_2H_3N
2. Cyclohexene, C_6H_{10}

x_1	T/K
0.200	293.7
0.265	297.7
0.330	300.0
0.395	301.0
0.460	301.3
0.525	301.2
0.590	300.7
0.655	299.2
0.720	296.5

T^c/K = 0.3013E+03
x^c_1 = 0.4756E+00
$A(1)$ = -0.1015E+02
$A(2)$ = -0.6077E+02
$A(3)$ = 0.5754E+02

System Number: 77
1. Ethanenitrile, C_2H_3N
2. Cyclohexane, C_6H_{12}

x_1	T/K
0.100	316.5
0.180	335.0
0.260	344.1
0.340	347.8
0.420	349.0
0.500	349.2
0.580	348.6
0.660	346.6
0.740	341.4

T^c/K = 0.3492E+03
x^c_1 = 0.4792E+00
$A(1)$ = -0.1260E+02
$A(2)$ = -0.6604E+02
$A(3)$ = 0.0000E+00

System Number: 5
1. Methyl methanoate, $C_2H_4O_2$
2. Hexane, C_6H_{14}

x_1	T/K
0.270	256.4
0.340	261.5
0.410	264.5
0.480	266.1
0.550	266.8
0.620	267.0
0.690	266.8
0.760	265.8
0.830	262.4

T^c/K = 0.2670E+03
x^c_1 = 0.6222E+00
$A(1)$ = -0.6699E+01
$A(2)$ = -0.4881E+02
$A(3)$ = 0.0000E+00

System Number: 30
1. Methyl methanoate, $C_2H_4O_2$
2. 2-Methylpentane, C_6H_{14}

x_1	T/K
0.100	228.3
0.200	245.6
0.300	255.7
0.400	260.8
0.500	263.0
0.600	263.7
0.700	263.5
0.800	261.4
0.900	250.8

T^c/K = 0.2637E+03
x^c_1 = 0.6303E+00
$A(1)$ = -0.8609E+01
$A(2)$ = -0.3052E+02
$A(3)$ = -0.2343E+02

System Number: 31
1. Methyl methanoate, $C_2H_4O_2$
2. 3-Methylpentane, C_6H_{14}

x_1	T/K
0.090	223.2
0.200	243.6
0.310	255.7
0.420	261.2
0.530	262.6
0.640	262.5
0.750	262.5
0.860	257.3
0.970	222.4

T^c/K = 0.2625E+03
x^c_1 = 0.6361E+00
$A(1)$ = 0.3846E+01
$A(2)$ = -0.6673E+02
$A(3)$ = 0.0000E+00

System Number: 65
1. Methyl methanoate, $C_2H_4O_2$
2. Heptane, C_7H_{16}

x_1	T/K
0.250	259.7
0.320	264.9
0.390	269.4
0.460	272.5
0.530	274.0
0.600	274.4
0.670	274.4
0.740	274.3
0.810	272.9

T^c/K = 0.2744E+03
x^c_1 = 0.6612E+00
$A(1)$ = 0.1574E+01
$A(2)$ = -0.1025E+03
$A(3)$ = 0.8194E+02

System Number: 26
1. Methyl methanoate, $C_2H_4O_2$
2. 1-Octene, C_8H_{16}

x_1	T/K
0.250	227.4
0.330	235.2
0.410	240.8
0.490	244.3
0.570	246.0
0.650	246.6
0.730	246.7
0.810	246.0
0.890	240.8

T^c/K = 0.2467E+03
x^c_1 = 0.7035E+00
$A(1)$ = -0.3735E+01
$A(2)$ = -0.5323E+02
$A(3)$ = 0.0000E+00

System Number: 27
1. Methyl methanoate, $C_2H_4O_2$
2. *trans*-2-Octene, C_8H_{16}

x_1	T/K
0.170	221.0
0.270	233.7
0.370	242.6
0.470	248.1
0.570	250.5
0.670	251.2
0.770	251.0
0.870	246.7
0.970	216.4

T^c/K = 0.2512E+03
x^c_1 = 0.6957E+00
$A(1)$ = -0.5867E+01
$A(2)$ = -0.5352E+02
$A(3)$ = 0.0000E+00

System Number: 28
1. Methyl methanoate, $C_2H_4O_2$
2. *trans*-3-Octene, C_8H_{16}

x_1	T/K
0.300	240.7
0.380	246.4
0.460	250.5
0.540	253.0
0.620	254.2
0.700	254.5
0.780	254.0
0.860	250.8
0.940	236.7

T^c/K = 0.2545E+03
x^c_1 = 0.6987E+00
$A(1)$ = -0.1037E+02
$A(2)$ = -0.3995E+02
$A(3)$ = 0.0000E+00

System Number: 29
1. Methyl methanoate, $C_2H_4O_2$
2. *trans*-4-Octene, C_8H_{16}

x_1	T/K
0.150	222.2
0.250	235.9
0.350	245.5
0.450	251.5
0.550	254.4
0.650	255.4
0.750	255.3
0.850	252.7
0.950	231.9

T^c/K = 0.2555E+03
x^c_1 = 0.6967E+00
$A(1)$ = -0.6927E+01
$A(2)$ = -0.4574E+02
$A(3)$ = -0.8833E+01

TABLE 3.1.4 (LLEX)
LIQUID-LIQUID EQUILIBRIA IN BINARY LIQUID MIXTURES (continued)

System Number: 54
1. Nitroethane, $C_2H_5NO_2$
2. Hexane, C_6H_{14}

x_1	T/K
0.250	298.0
0.310	302.5
0.370	304.3
0.430	304.7
0.490	304.8
0.550	304.8
0.610	304.6
0.670	303.5
0.730	300.2

T^c/K = 0.3048E+03
x^c_1 = 0.5020E+00
$A(1)$ = 0.1341E+01
$A(2)$ = -0.1115E+03
$A(3)$ = 0.0000E+00

System Number: 55
1. Nitroethane, $C_2H_5NO_2$
2. Octane, C_8H_{18}

x_1	T/K
0.100	279.1
0.200	298.2
0.300	308.4
0.400	312.8
0.500	314.2
0.600	314.2
0.700	312.7
0.800	306.8
0.900	289.3

T^c/K = 0.3143E+03
x^c_1 = 0.5542E+00
$A(1)$ = -0.1196E+02
$A(2)$ = -0.5508E+02
$A(3)$ = 0.0000E+00

System Number: 56
1. Nitroethane, $C_2H_5NO_2$
2. 2,2,4-Trimethylpentane, C_8H_{18}

x_1	T/K
0.100	272.5
0.200	287.8
0.300	298.2
0.400	301.9
0.500	302.6
0.600	302.6
0.700	301.8
0.800	296.3
0.900	277.4

T^c/K = 0.3027E+03
x^c_1 = 0.5514E+00
$A(1)$ = -0.2503E+01
$A(2)$ = -0.7163E+02
$A(3)$ = 0.0000E+00

System Number: 6
1. Ethane, C_2H_6
2. 2-Propanone, C_3H_6O

x_1	T/K
0.200	173.9
0.280	191.8
0.360	204.9
0.440	213.6
0.520	218.9
0.600	221.8
0.680	223.6
0.760	224.4
0.840	223.5

T^c/K = 0.2244E+03
x^c_1 = 0.7666E+00
$A(1)$ = -0.1859E+02
$A(2)$ = 0.1610E+02
$A(3)$ = -0.1137E+03

System Number: 85
1. Ethanol, C_2H_6O
2. Dodecane, $C_{12}H_{26}$

x_1	T/K
0.300	272.4
0.370	279.0
0.440	283.1
0.510	285.2
0.580	286.0
0.650	286.2
0.720	286.0
0.790	284.7
0.860	279.0

T^c/K = 0.2862E+03
x^c_1 = 0.6549E+00
$A(1)$ = -0.5596E+01
$A(2)$ = -0.7253E+02
$A(3)$ = 0.0000E+00

System Number: 86
1. Ethanol, C_2H_6O
2. Dodecane, $C_{12}H_{26}$
P = 120 MPa

x_1	T/K
0.300	292.0
0.370	299.7
0.440	304.6
0.510	307.2
0.580	308.4
0.650	308.7
0.720	308.5
0.790	306.8
0.860	300.0

T^c/K = 0.3087E+03
x^c_1 = 0.6569E+00
$A(1)$ = -0.9996E+01
$A(2)$ = -0.7854E+02
$A(3)$ = 0.0000E+00

System Number: 90
1. Dimethyl sulfoxide, C_2H_6OS
2. 1,3,5-Trimethylbenzene, C_9H_{12}

x_1	T/K
0.270	284.7
0.320	287.6
0.370	289.3
0.420	290.3
0.470	290.7
0.520	290.9
0.570	290.8
0.620	290.4
0.670	289.6

T^c/K = 0.2909E+03
x^c_1 = 0.5270E+00
$A(1)$ = -0.1134E+02
$A(2)$ = -0.5209E+02
$A(3)$ = 0.0000E+00

System Number: 12
1. 1,2-Ethanediol, $C_2H_6O_2$
2. Triethylamine, $C_6H_{15}N$

x_1	T/K
0.260	343.8
0.320	337.0
0.380	333.3
0.440	331.6
0.500	331.0
0.560	330.9
0.620	331.1
0.680	332.1
0.740	335.0

T^c/K = 0.3309E+03
x^c_1 = 0.5515E+00
$A(1)$ = 0.9811E+01
$A(2)$ = 0.1034E+03
$A(3)$ = 0.0000E+00

System Number: 7
1. 2-Propanone, C_3H_6O
2. Propane, C_3H_8

x_1	T/K
0.100	209.3
0.190	220.5
0.280	222.6
0.370	222.5
0.460	221.0
0.550	217.4
0.640	211.0
0.730	201.2
0.820	188.0

T^c/K = 0.2228E+03
x^c_1 = 0.3164E+00
$A(1)$ = -0.1468E+02
$A(2)$ = -0.5648E+02
$A(3)$ = 0.0000E+00

System Number: 60
1. 2-Propanone, C_3H_6O
2. 1,2,3-Propanetriol, $C_3H_8O_3$

x_1	T/K
0.250	332.8
0.330	350.0
0.410	360.6
0.490	366.1
0.570	368.3
0.650	368.8
0.730	367.6
0.810	361.6
0.890	339.3

T^c/K = 0.3688E+03
x^c_1 = 0.6373E+00
$A(1)$ = -0.2041E+02
$A(2)$ = -0.1231E+03
$A(3)$ = 0.0000E+00

System Number: 8
1. 2-Propanone, C_3H_6O
2. Butane, C_4H_{10}

x_1	T/K
0.100	210.6
0.200	221.6
0.300	222.8
0.400	222.8
0.500	222.6
0.600	220.6
0.700	215.1
0.800	204.2
0.900	187.0

T^c/K = 0.2228E+03
x^c_1 = 0.3649E+00
$A(1)$ = 0.1233E+01
$A(2)$ = -0.6131E+02
$A(3)$ = 0.0000E+00

System Number: 17
1. 2-Propanone, C_3H_6O
2. Cyclopentane, C_5H_{10}

x_1	T/K
0.050	197.8
0.150	220.4
0.250	224.2
0.350	224.3
0.450	224.3
0.550	223.7
0.650	220.7
0.750	213.4
0.850	200.4

T^c/K = 0.2243E+03
x^c_1 = 0.3675E+00
$A(1)$ = 0.2094E+01
$A(2)$ = -0.5463E+02
$A(3)$ = -0.4557E+01

TABLE 3.1.4 (LLEX)
LIQUID-LIQUID EQUILIBRIA IN BINARY LIQUID MIXTURES (continued)

System Number: 9
1. 2-Propanone, C_3H_6O
2. Pentane, C_5H_{12}

x_1	T/K
0.100	217.7
0.200	228.8
0.300	230.1
0.400	230.4
0.500	230.3
0.600	229.8
0.700	228.2
0.800	221.2
0.900	199.0

T^c/K = 0.2304E+03
x^c_1 = 0.4191E+00
$A(1)$ = -0.6733E+01
$A(2)$ = 0.2263E+02
$A(3)$ = -0.1008E+03

System Number: 18
1. 2-Propanone, C_3H_6O
2. Cycloheptane, C_7H_{14}

x_1	T/K
0.120	238.7
0.200	247.8
0.280	251.4
0.360	252.4
0.440	252.6
0.520	252.6
0.600	252.3
0.680	250.9
0.760	247.2

T^c/K = 0.2526E+03
x^c_1 = 0.4648E+00
$A(1)$ = -0.1498E+01
$A(2)$ = -0.4634E+02
$A(3)$ = 0.0000E+00

System Number: 19
1. 2-Propanone, C_3H_6O
2. Cyclooctane, C_8H_{16}

x_1	T/K
0.300	259.9
0.360	261.5
0.420	262.3
0.480	262.6
0.540	262.7
0.600	262.6
0.660	262.4
0.720	261.6
0.780	259.6

T^c/K = 0.2627E+03
x^c_1 = 0.5477E+00
$A(1)$ = -0.3112E+01
$A(2)$ = -0.3963E+02
$A(3)$ = 0.0000E+00

System Number: 32
1. Methyl ethanoate, $C_3H_6O_2$
2. Hexane, C_6H_{14}

x_1	T/K
0.240	221.7
0.310	225.4
0.380	227.2
0.450	227.8
0.520	228.0
0.590	228.0
0.660	227.8
0.730	226.8
0.800	223.8

T^c/K = 0.2280E+03
x^c_1 = 0.5505E+00
$A(1)$ = -0.2151E+01
$A(2)$ = -0.4573E+02
$A(3)$ = 0.0000E+00

System Number: 34
1. Methyl ethanoate, $C_2H_4O_2$
2. 2-Methylpentane, C_6H_{14}

x_1	T/K
0.240	217.3
0.310	221.4
0.380	223.5
0.450	224.3
0.520	224.5
0.590	224.5
0.660	224.2
0.730	223.1
0.800	219.9

T^c/K = 0.2245E+03
x^c_1 = 0.5538E+00
$A(1)$ = -0.3441E+01
$A(2)$ = -0.4792E+02
$A(3)$ = 0.0000E+00

System Number: 35
1. Methyl ethanoate, $C_3H_6O_2$
2. 3-Methylpentane, C_6H_{14}

x_1	T/K
0.160	210.1
0.240	217.0
0.320	220.7
0.400	222.4
0.480	222.9
0.560	223.0
0.640	222.8
0.720	221.6
0.800	218.3

T^c/K = 0.2230E+03
x^c_1 = 0.5428E+00
$A(1)$ = -0.5333E+01
$A(2)$ = -0.3720E+02
$A(3)$ = 0.0000E+00

System Number: 33
1. Methyl ethanoate, $C_3H_6O_2$
2. Heptane, C_7H_{16}

x_1	T/K
0.240	225.3
0.320	230.3
0.400	233.2
0.480	234.6
0.560	235.1
0.640	235.1
0.720	234.6
0.800	232.3
0.880	225.5

T^c/K = 0.2351E+03
x^c_1 = 0.6055E+00
$A(1)$ = -0.5213E+01
$A(2)$ = -0.4791E+02
$A(3)$ = 0.1785E+02

System Number: 36
1. Methyl ethanoate, $C_3H_6O_2$
2. Octane, C_8H_{18}

x_1	T/K
0.150	216.8
0.250	229.1
0.350	236.7
0.450	240.5
0.550	241.7
0.650	241.8
0.750	241.7
0.850	238.7
0.950	218.4

T^c/K = 0.2418E+03
x^c_1 = 0.6565E+00
$A(1)$ = 0.5909E+00
$A(2)$ = -0.5332E+02
$A(3)$ = 0.0000E+00

System Number: 61
1. 1,2-Propanediol, $C_3H_8O_2$
2. Benzene, C_6H_6

x_1	T/K
0.050	332
0.150	351
0.250	352
0.350	352
0.450	351
0.550	350
0.650	343
0.750	323
0.850	284

T^c/K = 0.3530E+03
x^c_1 = 0.2659E+00
$A(1)$ = -0.2245E+02
$A(2)$ = 0.1274E+03
$A(3)$ = -0.2757E+03

System Number: 62
1. 1,2,3-Propanetriol, $C_3H_8O_3$
2. 1-Hexanol, $C_6H_{14}O$

x_1	T/K
0.100	277.4
0.200	327.3
0.300	363.7
0.400	387.4
0.500	400.4
0.600	406.4
0.700	408.6
0.800	408.6
0.900	400.8

T^c/K = 0.4089E+03
x^c_1 = 0.7514E+00
$A(1)$ = -0.1785E+02
$A(2)$ = -0.2290E+02
$A(3)$ = -0.1540E+03

System Number: 50
1. 1,2,3-Propanetriol, $C_3H_8O_3$
2. Methyl phenyl ketone, C_8H_8O

x_1	T/K
0.040	373.4
0.150	417.0
0.260	442.3
0.370	454.2
0.480	458.1
0.590	458.8
0.700	458.0
0.810	450.6
0.920	417.5

T^c/K = 0.4588E+03
x^c_1 = 0.5893E+00
$A(1)$ = -0.8968E+01
$A(2)$ = -0.9770E+02
$A(3)$ = 0.0000E+00

System Number: 63
1. Ethanoic anhydride, $C_4H_6O_3$
2. Cyclohexane, C_6H_{12}

x_1	T/K
0.100	303.2
0.190	320.9
0.280	325.1
0.370	325.6
0.460	325.5
0.550	324.8
0.640	321.6
0.730	313.7
0.820	298.8

T^c/K = 0.3256E+03
x^c_1 = 0.3986E+00
$A(1)$ = -0.1441E+01
$A(2)$ = -0.8371E+02
$A(3)$ = 0.0000E+00

TABLE 3.1.4 (LLEX)
LIQUID-LIQUID EQUILIBRIA IN BINARY LIQUID MIXTURES (continued)

System Number: 64
1. Bis(2-chloroethyl) ether, $C_4H_8Cl_2O$
2. 2,2,4-Trimethylpentane, C_8H_{18}

x_1	T/K
0.200	280.2
0.270	286.4
0.340	289.2
0.410	290.2
0.480	290.5
0.550	290.4
0.620	289.9
0.690	288.2
0.760	284.1

T^c/K = 0.2905E+03
x^c_1 = 0.4987E+00
$A(1)$ = -0.6958E+01
$A(2)$ = -0.6097E+02
$A(3)$ = 0.0000E+00

System Number: 66
1. 2-Furaldehyde, $C_5H_4O_2$
2. Cyclohexane, C_6H_{12}

x_1	T/K
0.050	289
0.150	329
0.250	337
0.350	339
0.450	339
0.550	337
0.650	333
0.750	323
0.850	301

T^c/K = 0.3394E+03
x^c_1 = 0.3897E+00
$A(1)$ = -0.1564E+02
$A(2)$ = -0.2010E+02
$A(3)$ = -0.7575E+02

System Number: 67
1. 2-Furaldehyde, $C_5H_4O_2$
2. Methylcyclopentane, C_6H_{12}

x_1	T/K
0.080	296
0.180	331
0.280	338
0.380	340
0.480	340
0.580	338
0.680	335
0.780	324
0.880	291

T^c/K = 0.3405E+03
x^c_1 = 0.4331E+00
$A(1)$ = -0.2169E+02
$A(2)$ = 0.2740E+02
$A(3)$ = -0.1624E+03

System Number: 69
1. 2-Furaldehyde, $C_5H_4O_2$
2. Methylcyclohexane, C_7H_{14}

x_1	T/K
0.050	281
0.160	335
0.270	341
0.380	345
0.490	346
0.600	343
0.710	339
0.820	329
0.930	280

T^c/K = 0.3467E+03
x^c_1 = 0.4531E+00
$A(1)$ = -0.4857E+02
$A(2)$ = 0.1466E+03
$A(3)$ = -0.2472E+03

System Number: 68
1. 2-Furaldehyde, $C_5H_4O_2$
2. Heptane, C_7H_{16}

x_1	T/K
0.150	334.5
0.250	355.1
0.350	363.1
0.450	366.2
0.550	366.9
0.650	365.4
0.750	360.4
0.850	344.7
0.950	294.6

T^c/K = 0.3669E+03
x^c_1 = 0.5343E+00
$A(1)$ = -0.2454E+02
$A(2)$ = -0.2422E+02
$A(3)$ = -0.7873E+02

System Number: 70
1. Pentane, C_5H_{12}
2. Phenol, C_6H_6O

x_1	T/K
0.300	316.6
0.370	323.3
0.440	327.2
0.510	329.1
0.580	329.8
0.650	329.9
0.720	329.3
0.790	326.8
0.860	318.3

T^c/K = 0.3299E+03
x^c_1 = 0.6269E+00
$A(1)$ = -0.9558E+01
$A(2)$ = -0.8009E+02
$A(3)$ = 0.0000E+00

System Number: 71
1. Tetradecafluorohexane, C_6F_{14}
2. Hexane, C_6H_{14}

x_1	T/K
0.100	282.5
0.180	292.8
0.260	295.3
0.340	295.8
0.420	295.6
0.500	294.9
0.580	292.7
0.660	288.5
0.740	281.5

T^c/K = 0.2958E+03
x^c_1 = 0.3545E+00
$A(1)$ = -0.6276E+01
$A(2)$ = -0.5450E+02
$A(3)$ = 0.0000E+00

System Number: 72
1. Nitrobenzene, $C_6H_5NO_2$
2. Hexane, C_6H_{14}

x_1	T/K
0.150	277.4
0.230	288.2
0.310	292.3
0.390	293.1
0.470	293.1
0.550	292.2
0.630	288.8
0.710	282.0
0.790	272.7

T^c/K = 0.2932E+03
x^c_1 = 0.4218E+00
$A(1)$ = -0.7091E+01
$A(2)$ = -0.1470E+03
$A(3)$ = 0.1404E+03

System Number: 2
1. Aniline, C_6H_7N
2. Hexane, C_6H_{14}

x_1	T/K
0.100	303.6
0.200	321.4
0.300	329.3
0.400	332.0
0.500	332.6
0.600	332.4
0.700	330.6
0.800	324.1
0.900	307.0

T^c/K = 0.3326E+03
x^c_1 = 0.5191E+00
$A(1)$ = -0.7825E+01
$A(2)$ = -0.5369E+02
$A(3)$ = -0.1364E+01

System Number: 74
1. Aniline, C_6H_7N
2. 2,2,4-Trimethylpentane, C_8H_{18}

x_1	T/K
0.150	314.3
0.240	335.4
0.330	346.7
0.420	351.4
0.510	352.6
0.600	352.7
0.690	352.4
0.780	348.9
0.870	333.7

T^c/K = 0.3527E+03
x^c_1 = 0.5861E+00
$A(1)$ = -0.1523E+01
$A(2)$ = -0.1022E+03
$A(3)$ = 0.0000E+00

System Number: 37
1. Dimethyl butanedioate, $C_6H_{10}O_4$
2. Heptane, C_7H_{16}

x_1	T/K
0.170	315.5
0.250	324.5
0.330	327.2
0.410	327.7
0.490	327.7
0.570	327.4
0.650	325.8
0.730	320.8
0.810	309.8

T^c/K = 0.3277E+03
x^c_1 = 0.4565E+00
$A(1)$ = -0.1084E+01
$A(2)$ = -0.8619E+02
$A(3)$ = 0.0000E+00

System Number: 41
1. 1,4-Butanediol dimethanoate, $C_6H_{10}O_4$
2. Heptane, C_7H_{16}

x_1	T/K
0.150	352.4
0.230	363.0
0.310	367.7
0.390	369.5
0.470	369.7
0.550	368.6
0.630	365.8
0.710	360.4
0.790	351.0

T^c/K = 0.3698E+03
x^c_1 = 0.4419E+00
$A(1)$ = -0.2324E+02
$A(2)$ = -0.4943E+02
$A(3)$ = 0.0000E+00

TABLE 3.1.4 (LLEX)
LIQUID-LIQUID EQUILIBRIA IN BINARY LIQUID MIXTURES (continued)

System Number: 44
1. 1,2-Ethanediol diethanoate, $C_6H_{10}O_4$
2. Heptane, C_7H_{16}

x_1	T/K
0.120	317.5
0.200	328.0
0.280	332.1
0.360	333.4
0.440	333.6
0.520	333.3
0.600	332.1
0.680	329.1
0.760	323.3

T^c/K = 0.3336E+03
x^c_1 = 0.4327E+00
$A(1)$ = -0.9338E+01
$A(2)$ = -0.4914E+02
$A(3)$ = 0.0000E+00

System Number: 16
1. Tetradecafluoromethylcyclohexane, C_7F_{14}
2. Heptane, C_7H_{16}

x_1	T/K
0.200	299.0
0.270	299.2
0.340	299.3
0.410	299.4
0.480	299.4
0.550	299.4
0.620	299.3
0.690	299.2
0.760	298.9

T^c/K = 0.2994E+03
x^c_1 = 0.4583E+00
$A(1)$ = -0.9898E+00
$A(2)$ = -0.1251E+01
$A(3)$ = 0.0000E+00

System Number: 75
1. Hexadecafluoroheptane, C_7F_{16}
2. Heptane, C_7H_{16}

x_1	T/K
0.060	294.1
0.150	317.2
0.240	322.4
0.330	323.2
0.420	323.0
0.510	322.0
0.600	318.9
0.690	312.4
0.780	301.4

T^c/K = 0.3232E+03
x^c_1 = 0.3538E+00
$A(1)$ = -0.5435E+01
$A(2)$ = -0.6560E+02
$A(3)$ = 0.0000E+00

System Number: 76
1. Hexadecafluoroheptane, C_7F_{16}
2. 2,2,4-Trimethylpentane, C_8H_{18}

x_1	T/K
0.160	290.6
0.230	295.3
0.300	296.6
0.370	296.9
0.440	296.8
0.510	296.2
0.580	294.8
0.650	291.7
0.720	286.3

T^c/K = 0.2969E+03
x^c_1 = 0.3822E+00
$A(1)$ = -0.5827E+01
$A(2)$ = -0.5945E+02
$A(3)$ = 0.0000E+00

System Number: 42
1. Heptane, C_7H_{16}
2. 1,6-Hexanediol dimethanoate, $C_8H_{14}O_4$

x_1	T/K
0.250	305.1
0.330	320.6
0.410	327.9
0.490	331.1
0.570	332.8
0.650	333.7
0.730	333.4
0.810	331.3
0.890	321.9

T^c/K = 0.3338E+03
x^c_1 = 0.6716E+00
$A(1)$ = -0.2275E+02
$A(2)$ = 0.4313E+02
$A(3)$ = -0.2053E+03

System Number: 38
1. Heptane, C_7H_{16}
2. Dimethyl hexanedioate, $C_8H_{14}O_4$

x_1	T/K
0.320	291.1
0.380	294.1
0.440	296.2
0.500	297.5
0.560	298.2
0.620	298.5
0.680	298.5
0.740	298.1
0.800	296.7

T^c/K = 0.2985E+03
x^c_1 = 0.6488E+00
$A(1)$ = -0.8403E+01
$A(2)$ = -0.3612E+02
$A(3)$ = 0.0000E+00

System Number: 46
1. Heptane, C_7H_{16}
2. Benzyl ethanoate, $C_9H_{10}O_2$

x_1	T/K
0.250	251.5
0.320	257.5
0.390	260.8
0.460	262.4
0.530	263.0
0.600	263.1
0.670	262.8
0.740	261.7
0.810	258.0

T^c/K = 0.2631E+03
x^c_1 = 0.5911E+00
$A(1)$ = -0.6408E+01
$A(2)$ = -0.5915E+02
$A(3)$ = 0.0000E+00

System Number: 45
1. Heptane, C_7H_{16}
2. Methyl phenylethanoate, $C_9H_{10}O_2$

x_1	T/K
0.300	271.5
0.360	275.9
0.420	278.0
0.480	278.7
0.540	278.8
0.600	278.8
0.660	278.8
0.720	278.1
0.780	275.4

T^c/K = 0.2788E+03
x^c_1 = 0.5802E+00
$A(1)$ = 0.1568E+01
$A(2)$ = -0.1007E+03
$A(3)$ = 0.0000E+00

System Number: 39
1. Heptane, C_7H_{16}
2. Dimethyl octanedioate, $C_{10}H_{18}O_4$

x_1	T/K
0.450	261.4
0.500	264.1
0.550	266.0
0.600	267.1
0.650	267.7
0.700	268.0
0.750	268.0
0.800	267.6
0.850	266.2

T^c/K = 0.2680E+03
x^c_1 = 0.7274E+00
$A(1)$ = -0.6292E+01
$A(2)$ = -0.6056E+02
$A(3)$ = 0.0000E+00

System Number: 43
1. Heptane, C_7H_{16}
2. 1,8-Octanediol dimethanoate, $C_{10}H_{18}O_4$

x_1	T/K
0.450	290.1
0.500	292.2
0.550	294.0
0.600	295.5
0.650	296.6
0.700	297.3
0.750	297.7
0.800	297.4
0.850	296.2

T^c/K = 0.2977E+03
x^c_1 = 0.7567E+00
$A(1)$ = -0.1580E+02
$A(2)$ = -0.1872E+02
$A(3)$ = 0.0000E+00

3.2. VOLUMETRIC PROPERTIES

The definitions of the basic volumetric properties of mixtures, i.e., the *density*, the *specific volume*, and the *molar volume*, are the same as for pure substances (see Section 2.2). The *equation of state* of mixtures contains additional composition variables. The derived functions, i.e., the *cubic expansion coefficient*, the *isothermal compressibility*, and the *pressure coefficient* are defined similarly as for pure substances, but with partial derivatives taken at constant composition.

The most characteristic volumetric property of mixtures is the volume change upon mixing the pure components. When fixed amounts of pure substances are mixed at constant pressure, the volume of the mixture generally differs from the sum of the volumes of the original substances. The following table deals with this difference for binary mixtures of representative liquid substances.

TABLE 3.2.1 (VETX)
DENSITIES AND EXCESS VOLUMES OF BINARY LIQUID MIXTURES

The volume of mixing of two liquids at constant temperature and pressure is defined as the change in volume when two initially separated components are mixed. Similar to other excess functions of mixing, the excess volume is defined as the difference between the volume of mixing of a real mixture and the volume of mixing of an ideal mixture. Since the latter is by definition zero, the excess volume equals the volume of mixing. The *molar excess volume* V^E equals the excess volume divided by the total amount of substance (number of moles) of the two components. It is a function of the pressure P, the temperature T, and the composition of the mixture. The latter is usually expressed in terms of the mole fractions x_1 and $x_2 = 1 - x_1$ of components 1 and 2, respectively.

If V is the molar volume of the mixture, i.e., the total volume divided by the total amount of substance, and V_1 and V_2 are the molar volumes of the pure components (Table 2.2.2), then:

$$V^E = V - (x_1 V_1 + x_2 V_2) \tag{1}$$

The specific volume of a mixture, v, is defined similarly to the specific volume of a pure substance, $v = V/M$, where $M = x_1 M_1 + x_2 M_2$ is the average molar mass of the mixture. The *density* ρ is the reciprocal of v.

The molar excess volume is a measure of the pressure dependence of the molar excess Gibbs energy (Table 3.1.1) through the equation:

$$V^E = (\partial G^E / \partial P)_T \tag{2}$$

Most V^E measurements have been reported at atmospheric pressure and room temperature. The molar excess volume may be positive or negative throughout the composition range or may change sign (S-shaped). The molecular interactions, as well as the molecular structure of the components, influence the sign and magnitude of V^E.

The V^E curves can be represented as a function of x_1 by various empirical equations. A simple equation is the so-called Redlich-Kister equation:

$$V^E / \mathrm{m^3\,mol^{-1}} = x_1 x_2 \sum_{i=1}^{n} A(i)(x_1 - x_2)^{i-1} \tag{3}$$

The table gives the coefficients $A(i)$, at the temperature T, obtained by fitting Equation 3 to experimental V^E (x_1) data. The table lists the calculated values of V^E for increments of $x_1 = 0.1$.

The program, Property Code VETX, permits the calculation of V^E, V, ρ, and v at selected increments of x_1 in the full range of composition, from $x_1 = 0$ to $x_1 = 1$, or at any given value of x_1.

The components of each system, and the systems themselves, are arranged in a modified Hill order, with substances that do not contain carbon preceding those that do contain carbon.

Physical quantity	Symbol	SI unit
Mole fraction of component i	x_i	—
Temperature	T	K
Molar excess volume	V^E	$m^3\ mol^{-1}$
Molar volume	V	$m^3\ mol^{-1}$
Density	ρ	$kg\ m^{-3}$
Specific volume	v	$m^3\ kg^{-1}$

REFERENCES

1. Battino, R.; *Chem. Rev.*, 71, 5, 1971 (a review of experimental techniques, theory, and equimolar excess volume data for binary systems published over the period 1955-1968).
2. Handa, Y. P.; Benson, G. C.; *Fluid Phase Equilib.*, 3, 185, 1979 (a review of experimental techniques and excess volume data for about 1000 binary, mainly nonaqueous, systems and correlation coefficients, published over the period 1969-1978).

TABLE 3.2.1 (VETX)
DENSITIES AND EXCESS VOLUMES OF BINARY LIQUID MIXTURES (continued)

3. Edward, J. T.; Farrell, P. G.; Shahidi, F.; *J. Chem. Soc., Faraday Trans.*, 1, 73, 705, 1977; Shahidi, F.; Farrell, P. G.; Edward, J. T.; *J. Chem. Soc., Faraday Trans.*, 1, 73, 715, 1977 (partial molar volumes of 119 organic substances in water).
4. Kehiaian, H. V.; Kehiaian, C. S.; Organic Systems (Mixtures), in: *Bulletin of Chemical Thermodynamics*, Freeman, R. D., Editor, Thermochemistry Inc., 1969-1985 (17 volumes) (a compilation of bibliographical excess volume data).
5. Wisniak, J.; Tamir, A.; *Mixing and Excess Properties. A Literature Source Book*, Elsevier, Amsterdam, 1978; Supplement 1, 1982; Supplement 2, 1986 (a compilation of bibliographical excess volume data).
6. Lacmann, R.; Synowietz, C.; *Landolt-Börnstein Numerical Data and Functional Relationships in Science and Technology*, Vol. IV/1, Densities of Liquid Systems. Springer-Verlag, Berlin-Heidelberg, 1974 (a compilation of numerical density data for binary and multicomponent systems).
7. D'Ans, J.; Surawski, H.; Synowietz, C.; *Landolt-Börnstein Numerical Data and Functional Relationships in Science and Technology*, Vol. IV/1b, Densities of Binary Aqueous Systems and Heat Capacities, Springer-Verlag, Berlin-Heidelberg, 1976 (a compilation of numerical data).
8. *Int. DATA Ser., Sel. Data Mixtures, Ser. A* (Ed. H. V. Kehiaian), Thermodynamic Research Center, Texas Engineering Experiment Station, College Station, USA, 1973-1993 (21 volumes) (primary excess volume data and correlations)

System Number: 456 **T/K = 298.15**
1. Water, H_2O
2. Methanol, CH_4O
$V(1)$/cm^{3}mol^{-1} = 18.07
$V(2)$/cm^{3}mol^{-1} = 40.73

x_1	V^E/cm^{3}mol^{-1}
0.1	-0.342
0.2	-0.603
0.3	-0.804
0.4	-0.945
0.5	-1.010
0.6	-0.982
0.7	-0.849
0.8	-0.616
0.9	-0.312

$A(1)$ = -0.4040E-05 $A(4)$ = 0.9917E-06
$A(2)$ = -0.4271E-06 $A(5)$ = 0.0000E+00
$A(3)$ = 0.6427E-06 $A(6)$ = 0.0000E+00

System Number: 454 **T/K = 298.15**
1. Water, H_2O
2. Oxolane, C_4H_8O
$V(1)$/cm^{3}mol^{-1} = 18.07
$V(2)$/cm^{3}mol^{-1} = 81.70

x_1	V^E/cm^{3}mol^{-1}
0.1	-0.1646
0.2	-0.3638
0.3	-0.5447
0.4	-0.6864
0.5	-0.7854
0.6	-0.8419
0.7	-0.8461
0.8	-0.7638
0.9	-0.5232

$A(1)$ = -0.3142E-05 $A(4)$ = -0.1450E-05
$A(2)$ = -0.1562E-05 $A(5)$ = 0.0000E+00
$A(3)$ = -0.1062E-05 $A(6)$ = 0.0000E+00

System Number: 356 **T/K = 298.15**
1. Tetrachloromethane, CCl_4
2. 1,2-Dichloroethane, $C_2H_4Cl_2$
$V(1)$/cm^{3}mol^{-1} = 97.09
$V(2)$/cm^{3}mol^{-1} = 79.40

x_1	V^E/cm^{3}mol^{-1}
0.1	0.062
0.2	0.137
0.3	0.207
0.4	0.259
0.5	0.289
0.6	0.294
0.7	0.273
0.8	0.223
0.9	0.137

$A(1)$ = 0.1156E-05 $A(4)$ = 0.2542E-06
$A(2)$ = 0.3552E-06 $A(5)$ = 0.0000E+00
$A(3)$ = -0.8000E-07 $A(6)$ = 0.0000E+00

System Number: 355 **T/K = 298.15**
1. Tetrachloromethane, CCl_4
2. 1-Chlorobutane, C_4H_9Cl
$V(1)$/cm^{3}mol^{-1} = 97.09
$V(2)$/cm^{3}mol^{-1} = 105.08

x_1	V^E/cm^{3}mol^{-1}
0.1	-0.072
0.2	-0.112
0.3	-0.132
0.4	-0.139
0.5	-0.136
0.6	-0.122
0.7	-0.099
0.8	-0.068
0.9	-0.032

$A(1)$ = -0.5422E-06 $A(4)$ = 0.1652E-06
$A(2)$ = 0.1703E-06 $A(5)$ = 0.0000E+00
$A(3)$ = -0.5255E-07 $A(6)$ = 0.0000E+00

System Number: 8 **T/K = 298.15**
1. Tetrachloromethane, CCl_4
2. Diethyl ether, $C_4H_{10}O$
$V(1)$/cm^{3}mol^{-1} = 97.09
$V(2)$/cm^{3}mol^{-1} = 104.72

x_1	V^E/cm^{3}mol^{-1}
0.1	-0.2851
0.2	-0.4954
0.3	-0.6340
0.4	-0.7046
0.5	-0.7115
0.6	-0.6601
0.7	-0.5562
0.8	-0.4065
0.9	-0.2184

$A(1)$ = -0.2846E-05
$A(2)$ = 0.4629E-06
$A(3)$ = 0.7637E-07

System Number: 348 **T/K = 298.15**
1. Tetrachloromethane, CCl_4
2. Cyclopentane, C_5H_{10}
$V(1)$/cm^{3}mol^{-1} = 97.09
$V(2)$/cm^{3}mol^{-1} = 94.72

x_1	V^E/cm^{3}mol^{-1}
0.1	-0.0136
0.2	-0.0242
0.3	-0.0314
0.4	-0.0350
0.5	-0.0351
0.6	-0.0320
0.7	-0.0261
0.8	-0.0181
0.9	-0.0090

$A(1)$ = -0.1405E-06
$A(2)$ = 0.3185E-07
$A(3)$ = 0.2290E-07

TABLE 3.2.1 (VETX)
DENSITIES AND EXCESS VOLUMES OF BINARY LIQUID MIXTURES (continued)

System Number: 1 **T/K = 298.15**
1. Tetrachloromethane, CCl_4
2. *tert*-Butyl methyl ether, $C_5H_{12}O$
$V(1)$/cm^{3}mol^{-1} = 97.09
$V(2)$/cm^{3}mol^{-1} = 119.90

x_1	V^E/cm^{3}mol^{-1}
0.1	-0.308
0.2	-0.431
0.3	-0.472
0.4	-0.483
0.5	-0.482
0.6	-0.462
0.7	-0.409
0.8	-0.309
0.9	-0.163

$A(1)$ = -0.1927E-05 $A(4)$ = 0.1314E-05
$A(2)$ = 0.1655E-06 $A(5)$ = 0.0000E+00
$A(3)$ = -0.1070E-05 $A(6)$ = 0.0000E+00

System Number: 75 **T/K = 303.15**
1. Tetrachloromethane, CCl_4
2. Cyclohexane, C_6H_{12}
$V(1)$/cm^{3}mol^{-1} = 97.65
$V(2)$/cm^{3}mol^{-1} = 109.39

x_1	V^E/cm^{3}mol^{-1}
0.1	0.0620
0.2	0.1088
0.3	0.1414
0.4	0.1607
0.5	0.1672
0.6	0.1610
0.7	0.1419
0.8	0.1093
0.9	0.0625

$A(1)$ = 0.6688E-06
$A(2)$ = 0.3030E-08
$A(3)$ = 0.3564E-07

System Number: 12 **T/K = 298.15**
1. Tetrachloromethane, CCl_4
2. Dipropyl ether, $C_6H_{14}O$
$V(1)$/cm^{3}mol^{-1} = 97.09
$V(2)$/cm^{3}mol^{-1} = 137.72

x_1	V^E/cm^{3}mol^{-1}
0.1	-0.1245
0.2	-0.2194
0.3	-0.2855
0.4	-0.3234
0.5	-0.3339
0.6	-0.3177
0.7	-0.2755
0.8	-0.2080
0.9	-0.1159

$A(1)$ = -0.1336E-05
$A(2)$ = 0.5960E-07
$A(3)$ = 0.0000E+00

System Number: 375 **T/K = 298.15**
1. Tetrachloromethane, CCl_4
2. Dibutyl ether, $C_8H_{18}O$
$V(1)$/cm^{3}mol^{-1} = 97.09
$V(2)$/cm^{3}mol^{-1} = 170.44

x_1	V^E/cm^{3}mol^{-1}
0.1	-0.0580
0.2	-0.1035
0.3	-0.1368
0.4	-0.1574
0.5	-0.1645
0.6	-0.1572
0.7	-0.1356
0.8	-0.1003
0.9	-0.0537

$A(1)$ = -0.6579E-06 $A(4)$ = 0.4650E-07
$A(2)$ = 0.0000E+00 $A(5)$ = 0.0000E+00
$A(3)$ = 0.5890E-07 $A(6)$ = 0.0000E+00

System Number: 46 **T/K = 298.15**
1. Trichloromethane, $CHCl_3$
2. Oxolane, C_4H_8O
$V(1)$/cm^{3}mol^{-1} = 80.68
$V(2)$/cm^{3}mol^{-1} = 81.70

x_1	V^E/cm^{3}mol^{-1}
0.1	-0.1015
0.2	-0.1911
0.3	-0.2646
0.4	-0.3169
0.5	-0.3422
0.6	-0.3360
0.7	-0.2957
0.8	-0.2220
0.9	-0.1199

$A(1)$ = -0.1369E-05 $A(4)$ = 0.1190E-06
$A(2)$ = -0.2040E-06 $A(5)$ = 0.0000E+00
$A(3)$ = 0.2170E-06 $A(6)$ = 0.0000E+00

System Number: 52 **T/K = 298.15**
1. Trichloromethane, $CHCl_3$
2. 1,4-Dioxane, $C_4H_8O_2$
$V(1)$/cm^{3}mol^{-1} = 80.68
$V(2)$/cm^{3}mol^{-1} = 85.71

x_1	V^E/cm^{3}mol^{-1}
0.1	0.005
0.2	-0.015
0.3	-0.058
0.4	-0.114
0.5	-0.171
0.6	-0.214
0.7	-0.230
0.8	-0.204
0.9	-0.129

$A(1)$ = -0.6840E-06 $A(4)$ = 0.1870E-06
$A(2)$ = -0.1054E-05 $A(5)$ = 0.0000E+00
$A(3)$ = -0.5000E-08 $A(6)$ = 0.0000E+00

System Number: 7 **T/K = 298.15**
1. Trichloromethane, $CHCl_3$
2. Diethyl ether, $C_4H_{10}O$
$V(1)$/cm^{3}mol^{-1} = 80.68
$V(2)$/cm^{3}mol^{-1} = 104.72

x_1	V^E/cm^{3}mol^{-1}
0.1	-0.479
0.2	-0.869
0.3	-1.150
0.4	-1.308
0.5	-1.335
0.6	-1.234
0.7	-1.026
0.8	-0.735
0.9	-0.388

$A(1)$ = -0.5338E-05 $A(4)$ = -0.2288E-06
$A(2)$ = 0.7783E-06 $A(5)$ = -0.3460E-06
$A(3)$ = 0.1033E-05 $A(6)$ = 0.0000E+00

System Number: 2 **T/K = 298.15**
1. Trichloromethane, $CHCl_3$
2. *tert*-Butyl methyl ether, $C_5H_{12}O$
$V(1)$/cm^{3}mol^{-1} = 80.68
$V(2)$/cm^{3}mol^{-1} = 119.94

x_1	V^E/cm^{3}mol^{-1}
0.1	-0.355
0.2	-0.667
0.3	-0.887
0.4	-1.005
0.5	-1.040
0.6	-1.014
0.7	-0.939
0.8	-0.800
0.9	-0.530

$A(1)$ = -0.4159E-05 $A(4)$ = -0.1878E-05
$A(2)$ = -0.1217E-07 $A(5)$ = -0.4565E-06
$A(3)$ = -0.1184E-05 $A(6)$ = 0.0000E+00

System Number: 14 **T/K = 298.15**
1. Trichloromethane, $CHCl_3$
2. Dipropyl ether, $C_6H_{14}O$
$V(1)$/cm^{3}mol^{-1} = 80.68
$V(2)$/cm^{3}mol^{-1} = 137.72

x_1	V^E/cm^{3}mol^{-1}
0.1	-0.3118
0.2	-0.5497
0.3	-0.7273
0.4	-0.8373
0.5	-0.8680
0.6	-0.8138
0.7	-0.6803
0.8	-0.4850
0.9	-0.2516

$A(1)$ = -0.3472E-05 $A(4)$ = 0.2878E-06
$A(2)$ = 0.2338E-06 $A(5)$ = -0.4565E-06
$A(3)$ = 0.8275E-06 $A(6)$ = 0.0000E+00

TABLE 3.21 (VETX)
DENSITIES AND EXCESS VOLUMES OF BINARY LIQUID MIXTURES (continued)

System Number: 51 **T/K = 298.15**
1. Dichloromethane, CH_2Cl_2
2. 2-Propanone, C_3H_6O
$V(1)$/cm^{3}mol^{-1} = 64.50
$V(2)$/cm^{3}mol^{-1} = 74.04

x_1	V^E/cm^{3}mol^{-1}
0.1	0.0547
0.2	0.0931
0.3	0.1175
0.4	0.1299
0.5	0.1317
0.6	0.1240
0.7	0.1071
0.8	0.0812
0.9	0.0458

$A(1)$ = 0.5269E-06
$A(2)$ = -0.6188E-07
$A(3)$ = 0.4910E-07

System Number: 15 **T/K = 298.15**
1. Dichloromethane, CH_2Cl_2
2. Methyl ethanoate, $C_3H_6O_2$
$V(1)$/cm^{3}mol^{-1} = 64.50
$V(2)$/cm^{3}mol^{-1} = 79.84

x_1	V^E/cm^{3}mol^{-1}
0.1	0.1079
0.2	0.1929
0.3	0.2548
0.4	0.2930
0.5	0.3070
0.6	0.2965
0.7	0.2610
0.8	0.2000
0.9	0.1132

$A(1)$ = 0.1228E-05
$A(2)$ = 0.3686E-07
$A(3)$ = 0.0000E+00

System Number: 3 **T/K = 298.15**
1. Dichloromethane, CH_2Cl_2
2. *tert*-Butyl methyl ether, $C_5H_{12}O$
$V(1)$/cm^{3}mol^{-1} = 64.50
$V(2)$/cm^{3}mol^{-1} = 119.94

x_1	V^E/cm^{3}mol^{-1}
0.1	-0.226
0.2	-0.378
0.3	-0.462
0.4	-0.494
0.5	-0.491
0.6	-0.464
0.7	-0.419
0.8	-0.348
0.9	-0.224

$A(1)$ = -0.1962E-05 $A(4)$ = -0.5053E-06
$A(2)$ = 0.3374E-06 $A(5)$ = 0.0000E+00
$A(3)$ = -0.8448E-06 $A(6)$ = 0.0000E+00

System Number: 17 **T/K = 298.15**
1. Methanol, CH_4O
2. Ethanol, C_2H_6O
$V(1)$/cm^{3}mol^{-1} = 40.73
$V(2)$/cm^{3}mol^{-1} = 58.69

x_1	V^E/cm^{3}mol^{-1}
0.1	0.0030
0.2	0.0052
0.3	0.0068
0.4	0.0079
0.5	0.0085
0.6	0.0087
0.7	0.0082
0.8	0.0069
0.9	0.0043

$A(1)$ = 0.3411E-07
$A(2)$ = 0.8868E-08
$A(3)$ = 0.9955E-08

System Number: 19 **T/K = 298.15**
1. Methanol, CH_4O
2. 1-Propanol, C_3H_8O
$V(1)$/cm^{3}mol^{-1} = 40.73
$V(2)$/cm^{3}mol^{-1} = 75.16

x_1	V^E/cm^{3}mol^{-1}
0.1	0.0141
0.2	0.0267
0.3	0.0373
0.4	0.0453
0.5	0.0503
0.6	0.0515
0.7	0.0483
0.8	0.0396
0.9	0.0240

$A(1)$ = 0.2011E-06 $A(4)$ = 0.7714E-08
$A(2)$ = 0.6400E-07 $A(5)$ = 0.0000E+00
$A(3)$ = 0.1666E-07 $A(6)$ = 0.0000E+00

System Number: 314 **T/K = 298.15**
1. Methanol, CH_4O
2. 2-Butanol, $C_4H_{10}O$
$V(1)$/cm^{3}mol^{-1} = 40.73
$V(2)$/cm^{3}mol^{-1} = 92.38

x_1	V^E/cm^{3}mol^{-1}
0.1	0.0253
0.2	0.0479
0.3	0.0666
0.4	0.0805
0.5	0.0884
0.6	0.0892
0.7	0.0820
0.8	0.0654
0.9	0.0385

$A(1)$ = 0.3535E-06
$A(2)$ = 0.9120E-07
$A(3)$ = 0.1700E-08

System Number: 313 **T/K = 298.15**
1. Methanol, CH_4O
2. 1-Butanol, $C_4H_{10}O$
$V(1)$/cm^{3}mol^{-1} = 40.73
$V(2)$/cm^{3}mol^{-1} = 91.99

x_1	V^E/cm^{3}mol^{-1}
0.1	0.0172
0.2	0.0342
0.3	0.0498
0.4	0.0634
0.5	0.0738
0.6	0.0798
0.7	0.0794
0.8	0.0693
0.9	0.0449

$A(1)$ = 0.2952E-06 $A(4)$ = 0.3580E-07
$A(2)$ = 0.1700E-06 $A(5)$ = 0.0000E+00
$A(3)$ = 0.7780E-07 $A(6)$ = 0.0000E+00

System Number: 318 **T/K = 298.15**
1. Methanol, CH_4O
2. 1-Decanol, $C_{10}H_{22}O$
$V(1)$/cm^{3}mol^{-1} = 40.73
$V(2)$/cm^{3}mol^{-1} = 191.56

x_1	V^E/cm^{3}mol^{-1}
0.1	0.0280
0.2	0.0643
0.3	0.0995
0.4	0.1383
0.5	0.1783
0.6	0.2111
0.7	0.2300
0.8	0.2305
0.9	0.1882

$A(1)$ = 0.7134E-06 $A(4)$ = -0.2929E-07
$A(2)$ = 0.7580E-06 $A(5)$ = 0.6611E-06
$A(3)$ = 0.3395E-06 $A(6)$ = 0.9107E-06

System Number: 353 **T/K = 298.15**
1. Tetrachloroethene, C_2Cl_4
2. Cyclopentane, C_5H_{10}
$V(1)$/cm^{3}mol^{-1} = 102.73
$V(2)$/cm^{3}mol^{-1} = 94.72

x_1	V^E/cm^{3}mol^{-1}
0.1	-0.0047
0.2	-0.0081
0.3	-0.0097
0.4	-0.0093
0.5	-0.0071
0.6	-0.0035
0.7	0.0004
0.8	0.0035
0.9	0.0040

$A(1)$ = -0.2834E-07
$A(2)$ = 0.6050E-07
$A(3)$ = 0.3855E-07

TABLE 3.2.1 (VETX)
DENSITIES AND EXCESS VOLUMES OF BINARY LIQUID MIXTURES (continued)

System Number: 141 T/K = 298.15
1. 1,2-Dichloroethane, $C_2H_4Cl_2$
2. 1-Chlorobutane, C_4H_9Cl
$V(1)$/cm^{3}mol^{-1} = 79.40
$V(2)$/cm^{3}mol^{-1} = 105.08

x_1	V^E/cm^{3}mol^{-1}
0.1	0.066
0.2	0.128
0.3	0.171
0.4	0.194
0.5	0.201
0.6	0.193
0.7	0.172
0.8	0.134
0.9	0.076

$A(1)$ = 0.8024E-06 $A(4)$ = 0.1334E-06
$A(2)$ = -0.1777E-07 $A(5)$ = -0.2130E-06
$A(3)$ = 0.1181E-06 $A(6)$ = 0.0000E+00

System Number: 124 T/K = 298.15
1. 1,2-Dichloroethane, $C_2H_4Cl_2$
2. Heptane, C_7H_{16}
$V(1)$/cm^{3}mol^{-1} = 79.40
$V(2)$/cm^{3}mol^{-1} = 147.48

x_1	V^E/cm^{3}mol^{-1}
0.1	0.3762
0.2	0.6487
0.3	0.8252
0.4	0.9130
0.5	0.9197
0.6	0.8529
0.7	0.7200
0.8	0.5285
0.9	0.2860

$A(1)$ = 0.3679E-05
$A(2)$ = -0.6260E-06
$A(3)$ = 0.0000E+00

System Number: 315 T/K = 298.15
1. Ethanol, C_2H_6O
2. 1-Butanol, $C_4H_{10}O$
$V(1)$/cm^{3}mol^{-1} = 58.69
$V(2)$/cm^{3}mol^{-1} = 91.99

x_1	V^E/cm^{3}mol^{-1}
0.1	0.0023
0.2	0.0047
0.3	0.0073
0.4	0.0098
0.5	0.0121
0.6	0.0136
0.7	0.0137
0.8	0.0119
0.9	0.0076

$A(1)$ = 0.4827E-07 $A(4)$ = -0.3798E-08
$A(2)$ = 0.3900E-07 $A(5)$ = 0.0000E+00
$A(3)$ = 0.1048E-07 $A(6)$ = 0.0000E+00

System Number: 220 T/K = 298.15
1. Ethanol, C_2H_6O
2. Nonane, C_9H_{20}
$V(1)$/cm^{3}mol^{-1} = 58.69
$V(2)$/cm^{3}mol^{-1} = 179.70

x_1	V^E/cm^{3}mol^{-1}
0.1	0.276
0.2	0.387
0.3	0.455
0.4	0.501
0.5	0.518
0.6	0.507
0.7	0.471
0.8	0.404
0.9	0.269

$A(1)$ = 0.2073E-05 $A(4)$ = 0.4867E-06
$A(2)$ = 0.4176E-07 $A(5)$ = 0.1373E-05
$A(3)$ = 0.6087E-06 $A(6)$ = -0.9793E-06

System Number: 20 T/K = 298.15
1. Ethanol, C_2H_6O
2. 1-Decanol, $C_{10}H_{22}O$
$V(1)$/cm^{3}mol^{-1} = 58.69
$V(2)$/cm^{3}mol^{-1} = 191.56

x_1	V^E/cm^{3}mol^{-1}
0.1	0.0209
0.2	0.0415
0.3	0.0590
0.4	0.0748
0.5	0.0885
0.6	0.0984
0.7	0.1025
0.8	0.0980
0.9	0.0750

$A(1)$ = 0.3541E-06 $A(4)$ = 0.5554E-07
$A(2)$ = 0.2441E-06 $A(5)$ = 0.1862E-06
$A(3)$ = 0.1602E-06 $A(6)$ = 0.2337E-06

System Number: 262 T/K = 298.15
1. 2-Butanone, C_4H_8O
2. Benzene, C_6H_6
$V(1)$/cm^{3}mol^{-1} = 90.17
$V(2)$/cm^{3}mol^{-1} = 89.42

x_1	V^E/cm^{3}mol^{-1}
0.1	-0.022
0.2	-0.046
0.3	-0.068
0.4	-0.084
0.5	-0.093
0.6	-0.092
0.7	-0.082
0.8	-0.062
0.9	-0.034

$A(1)$ = -0.3710E-06
$A(2)$ = -0.8712E-07
$A(3)$ = 0.9107E-07

System Number: 249 T/K = 298.15
1. 2-Butanone, C_4H_8O
2. Cyclohexane, C_6H_{12}
$V(1)$/cm^{3}mol^{-1} = 90.17
$V(2)$/cm^{3}mol^{-1} = 108.75

x_1	V^E/cm^{3}mol^{-1}
0.1	0.397
0.2	0.661
0.3	0.815
0.4	0.880
0.5	0.873
0.6	0.805
0.7	0.683
0.8	0.509
0.9	0.284

$A(1)$ = 0.3492E-05
$A(2)$ = -0.7882E-06
$A(3)$ = 0.4570E-06

System Number: 74 T/K = 298.15
1. 2-Butanone, C_4H_8O
2. Heptane, C_7H_{16}
$V(1)$/cm^{3}mol^{-1} = 90.17
$V(2)$/cm^{3}mol^{-1} = 147.48

x_1	V^E/cm^{3}mol^{-1}
0.1	0.367
0.2	0.599
0.3	0.734
0.4	0.797
0.5	0.803
0.6	0.757
0.7	0.658
0.8	0.503
0.9	0.285

$A(1)$ = 0.3212E-05 $A(4)$ = -0.2340E-06
$A(2)$ = -0.4140E-06 $A(5)$ = 0.0000E+00
$A(3)$ = 0.6400E-06 $A(6)$ = 0.0000E+00

System Number: 24 T/K = 298.15
1. Ethyl ethanoate, $C_4H_8O_2$
2. 1-Chlorobutane, C_4H_9Cl
$V(1)$/cm^{3}mol^{-1} = 98.49
$V(2)$/cm^{3}mol^{-1} = 105.08

x_1	V^E/cm^{3}mol^{-1}
0.1	0.1437
0.2	0.2526
0.3	0.3288
0.4	0.3739
0.5	0.3886
0.6	0.3734
0.7	0.3280
0.8	0.2516
0.9	0.1429

$A(1)$ = 0.1554E-05
$A(2)$ = -0.4943E-08
$A(3)$ = 0.5910E-07

TABLE 3.2.1 (VETX)
DENSITIES AND EXCESS VOLUMES OF BINARY LIQUID MIXTURES (continued)

System Number: 269 **T/K = 298.15**
1. Ethyl ethanoate, $C_4H_8O_2$
2. Cyclohexane, C_6H_{12}
$V(1)/cm^3mol^{-1}$ = 98.49
$V(2)/cm^3mol^{-1}$ = 108.75

x_1	V^E/cm^3mol^{-1}
0.1	0.523
0.2	0.870
0.3	1.083
0.4	1.191
0.5	1.206
0.6	1.134
0.7	0.976
0.8	0.731
0.9	0.403

$A(1)$ = 0.4823E-05 $A(4)$ = -0.4100E-06
$A(2)$ = -0.5730E-06 $A(5)$ = 0.0000E+00
$A(3)$ = 0.5010E-06 $A(6)$ = 0.0000E+00

System Number: 268 **T/K = 298.15**
1. Ethyl ethanoate, $C_4H_8O_2$
2. Heptane, C_7H_{16}
$V(1)/cm^3mol^{-1}$ = 98.49
$V(2)/cm^3mol^{-1}$ = 147.48

x_1	V^E/cm^3mol^{-1}
0.1	0.4495
0.2	0.7652
0.3	0.9708
0.4	1.0824
0.5	1.1095
0.6	1.0561
0.7	0.9223
0.8	0.7050
0.9	0.3994

$A(1)$ = 0.4438E-05 $A(4)$ = -0.1230E-06
$A(2)$ = -0.2690E-06 $A(5)$ = 0.0000E+00
$A(3)$ = 0.4350E-06 $A(6)$ = 0.0000E+00

System Number: 217 **T/K = 298.15**
1. 1,4-Dioxane, $C_4H_8O_2$
2. Benzene, C_6H_6
$V(1)/cm^3mol^{-1}$ = 85.71
$V(2)/cm^3mol^{-1}$ = 89.42

x_1	V^E/cm^3mol^{-1}
0.1	-0.0429
0.2	-0.0677
0.3	-0.0771
0.4	-0.0750
0.5	-0.0653
0.6	-0.0514
0.7	-0.0362
0.8	-0.0218
0.9	-0.0094

$A(1)$ = -0.2610E-06 $A(4)$ = -0.2200E-07
$A(2)$ = 0.2470E-06 $A(5)$ = 0.1850E-07
$A(3)$ = -0.5800E-07 $A(6)$ = 0.0000E+00

System Number: 216 **T/K = 298.15**
1. 1,4-Dioxane, $C_4H_8O_2$
2. Cyclohexane, C_6H_{12}
$V(1)/cm^3mol^{-1}$ = 85.71
$V(2)/cm^3mol^{-1}$ = 108.75

x_1	V^E/cm^3mol^{-1}
0.1	0.438
0.2	0.722
0.3	0.887
0.4	0.959
0.5	0.953
0.6	0.881
0.7	0.747
0.8	0.554
0.9	0.304

$A(1)$ = 0.3813E-05 $A(4)$ = -0.1966E-06
$A(2)$ = -0.8043E-06 $A(5)$ = 0.0000E+00
$A(3)$ = 0.4909E-06 $A(6)$ = 0.0000E+00

System Number: 23 **T/K = 298.15**
1. 1,4-Dioxane, $C_4H_8O_2$
2. Heptane, C_7H_{16}
$V(1)/cm^3mol^{-1}$ = 85.71
$V(2)/cm^3mol^{-1}$ = 147.48

x_1	V^E/cm^3mol^{-1}
0.1	0.327
0.2	0.547
0.3	0.677
0.4	0.732
0.5	0.725
0.6	0.664
0.7	0.558
0.8	0.411
0.9	0.225

$A(1)$ = 0.2898E-05
$A(2)$ = -0.7067E-06
$A(3)$ = 0.2662E-06

System Number: 25 **T/K = 298.15**
1. 1-Chlorobutane, C_4H_9Cl
2. Heptane, C_7H_{16}
$V(1)/cm^3mol^{-1}$ = 105.08
$V(2)/cm^3mol^{-1}$ = 147.48

x_1	V^E/cm^3mol^{-1}
0.1	0.1146
0.2	0.1973
0.3	0.2500
0.4	0.2757
0.5	0.2772
0.6	0.2577
0.7	0.2195
0.8	0.1641
0.9	0.0915

$A(1)$ = 0.1109E-05 $A(4)$ = 0.4500E-07
$A(2)$ = -0.1889E-06 $A(5)$ = 0.0000E+00
$A(3)$ = 0.5650E-07 $A(6)$ = 0.0000E+00

System Number: 226 **T/K = 298.15**
1. 1-Butanol, $C_4H_{10}O$
2. Cyclohexane, C_6H_{12}
$V(1)/cm^3mol^{-1}$ = 91.99
$V(2)/cm^3mol^{-1}$ = 108.75

x_1	V^E/cm^3mol^{-1}
0.1	0.235
0.2	0.345
0.3	0.403
0.4	0.423
0.5	0.406
0.6	0.361
0.7	0.298
0.8	0.219
0.9	0.117

$A(1)$ = 0.1622E-05 $A(4)$ = 0.3337E-06
$A(2)$ = -0.6548E-06 $A(5)$ = 0.4335E-06
$A(3)$ = 0.2382E-06 $A(6)$ = -0.9273E-06

System Number: 225 **T/K = 298.15**
1. 1-Butanol, $C_4H_{10}O$
2. Heptane, C_7H_{16}
$V(1)/cm^3mol^{-1}$ = 91.99
$V(2)/cm^3mol^{-1}$ = 147.48

x_1	V^E/cm^3mol^{-1}
0.1	0.171
0.2	0.218
0.3	0.232
0.4	0.223
0.5	0.191
0.6	0.145
0.7	0.100
0.8	0.063
0.9	0.026

$A(1)$ = 0.7635E-06 $A(4)$ = 0.5180E-06
$A(2)$ = -0.8334E-06 $A(5)$ = 0.7373E-06
$A(3)$ = 0.4788E-07 $A(6)$ = -0.1228E-05

System Number: 79 **T/K = 298.15**
1. Cyclopentane, C_5H_{10}
2. 2,3-Dimethylbutane, C_6H_{14}
$V(1)/cm^3mol^{-1}$ = 94.72
$V(2)/cm^3mol^{-1}$ = 131.16

x_1	V^E/cm^3mol^{-1}
0.1	-0.0968
0.2	-0.1722
0.3	-0.2260
0.4	-0.2582
0.5	-0.2690
0.6	-0.2582
0.7	-0.2260
0.8	-0.1722
0.9	-0.0968

$A(1)$ = -0.1076E-05
$A(2)$ = 0.0000E+00
$A(3)$ = 0.0000E+00

TABLE 3.2.1 (VETX)
DENSITIES AND EXCESS VOLUMES OF BINARY LIQUID MIXTURES (continued)

System Number: 6 T/K = 313.15
1. Hexafluorobenzene, C_6F_6
2. Benzene, C_6H_6
$V(1)$/cm^{3}mol^{-1} = 118.27
$V(2)$/cm^{3}mol^{-1} = 90.99

x_1	V^E/cm^{3}mol^{-1}
0.1	0.414
0.2	0.676
0.3	0.812
0.4	0.846
0.5	0.801
0.6	0.697
0.7	0.552
0.8	0.379
0.9	0.192

$A(1)$ = 0.3205E-05
$A(2)$ = -0.1547E-05
$A(3)$ = 0.2535E-06

System Number: 382 T/K = 313.15
1. Hexafluorobenzene, C_6F_6
2. Cyclohexane, C_6H_{12}
$V(1)$/cm^{3}mol^{-1} = 118.27
$V(2)$/cm^{3}mol^{-1} = 111.29

x_1	V^E/cm^{3}mol^{-1}
0.1	1.0143
0.2	1.7538
0.3	2.2452
0.4	2.5105
0.5	2.5671
0.6	2.4278
0.7	2.1004
0.8	1.5883
0.9	0.8902

$A(1)$ = 0.1027E-04
$A(2)$ = -0.8616E-06
$A(3)$ = 0.4874E-06

System Number: 385 T/K = 298.15
1. Hexafluorobenzene, C_6F_6
2. Triethylamine, $C_6H_{15}N$
$V(1)$/cm^{3}mol^{-1} = 115.25
$V(2)$/cm^{3}mol^{-1} = 139.95

x_1	V^E/cm^{3}mol^{-1}
0.1	0.515
0.2	0.910
0.3	1.184
0.4	1.336
0.5	1.370
0.6	1.291
0.7	1.104
0.8	0.819
0.9	0.447

$A(1)$ = 0.5482E-05
$A(2)$ = -0.4740E-06
$A(3)$ = -0.2200E-06

System Number: 107 T/K = 298.15
1. Bromobenzene, C_6H_5Br
2. Cyclohexane, C_6H_{12}
$V(1)$/cm^{3}mol^{-1} = 105.50
$V(2)$/cm^{3}mol^{-1} = 108.75

x_1	V^E/cm^{3}mol^{-1}
0.1	0.146
0.2	0.233
0.3	0.274
0.4	0.281
0.5	0.262
0.6	0.224
0.7	0.176
0.8	0.120
0.9	0.061

$A(1)$ = 0.1046E-05
$A(2)$ = -0.5881E-06
$A(3)$ = 0.1560E-06

System Number: 61 T/K = 298.15
1. Benzene, C_6H_6
2. 1-Hexyne, C_6H_{10}
$V(1)$/cm^{3}mol^{-1} = 89.42
$V(2)$/cm^{3}mol^{-1} = 115.70

x_1	V^E/cm^{3}mol^{-1}
0.1	0.020
0.2	0.030
0.3	0.035
0.4	0.037
0.5	0.039
0.6	0.041
0.7	0.041
0.8	0.037
0.9	0.025

$A(1)$ = 0.1573E-06
$A(2)$ = 0.3397E-07
$A(3)$ = 0.1425E-06

System Number: 9 T/K = 298.15
1. Benzene, C_6H_6
2. Cyclohexane, C_6H_{12}
$V(1)$/cm^{3}mol^{-1} = 89.42
$V(2)$/cm^{3}mol^{-1} = 108.75

x_1	V^E/cm^{3}mol^{-1}
0.1	0.2424
0.2	0.4265
0.3	0.5545
0.4	0.6284
0.5	0.6497
0.6	0.6197
0.7	0.5393
0.8	0.4091
0.9	0.2294

$A(1)$ = 0.2599E-05
$A(2)$ = -0.9043E-07
$A(3)$ = 0.3469E-07

System Number: 59 T/K = 298.15
1. Benzene, C_6H_6
2. 1-Heptene, C_7H_{14}
$V(1)$/cm^{3}mol^{-1} = 89.42
$V(2)$/cm^{3}mol^{-1} = 141.75

x_1	V^E/cm^{3}mol^{-1}
0.1	0.1132
0.2	0.1871
0.3	0.2409
0.4	0.2816
0.5	0.3078
0.6	0.3128
0.7	0.2882
0.8	0.2274
0.9	0.1290

$A(1)$ = 0.1231E-05 $A(4)$ = -0.3573E-06
$A(2)$ = 0.3387E-06 $A(5)$ = 0.0000E+00
$A(3)$ = 0.1785E-06 $A(6)$ = 0.0000E+00

System Number: 33 T/K = 298.15
1. Benzene, C_6H_6
2. Heptane, C_7H_{16}
$V(1)$/cm^{3}mol^{-1} = 89.42
$V(2)$/cm^{3}mol^{-1} = 147.48

x_1	V^E/cm^{3}mol^{-1}
0.1	0.1945
0.2	0.3515
0.3	0.4715
0.4	0.5521
0.5	0.5905
0.6	0.5839
0.7	0.5293
0.8	0.4217
0.9	0.2514

$A(1)$ = 0.2362E-05 $A(4)$ = 0.1070E-06
$A(2)$ = 0.3270E-06 $A(5)$ = 0.1050E-06
$A(3)$ = 0.1130E-06 $A(6)$ = 0.0000E+00

System Number: 92 T/K = 298.15
1. Benzene, C_6H_6
2. 1,4-Dimethylbenzene, C_8H_{10}
$V(1)$/cm^{3}mol^{-1} = 89.42
$V(2)$/cm^{3}mol^{-1} = 123.94

x_1	V^E/cm^{3}mol^{-1}
0.1	0.0668
0.2	0.1214
0.3	0.1635
0.4	0.1926
0.5	0.2075
0.6	0.2067
0.7	0.1882
0.8	0.1496
0.9	0.0880

$A(1)$ = 0.8300E-06
$A(2)$ = 0.1470E-06
$A(3)$ = 0.4700E-07

TABLE 3.2.1 (VETX)
DENSITIES AND EXCESS VOLUMES OF BINARY LIQUID MIXTURES (continued)

System Number: 211 **T/K = 298.15**
1. Benzene, C_6H_6
2. Dibutyl ether, $C_8H_{18}O$
$V(1)/cm^3mol^{-1}$ = 89.42
$V(2)/cm^3mol^{-1}$ = 170.44

x_1	V^E/cm^3mol^{-1}
0.1	0.0282
0.2	0.0580
0.3	0.0863
0.4	0.1108
0.5	0.1298
0.6	0.1409
0.7	0.1407
0.8	0.1234
0.9	0.0806

$A(1)$ = 0.5192E-06 $A(4)$ = 0.8400E-07
$A(2)$ = 0.3103E-06 $A(5)$ = 0.0000E+00
$A(3)$ = 0.1331E-06 $A(6)$ = 0.0000E+00

System Number: 58 **T/K = 298.15**
1. 1-Hexyne, C_6H_{10}
2. Cyclohexane, C_6H_{12}
$V(1)/cm^3mol^{-1}$ = 115.70
$V(2)/cm^3mol^{-1}$ = 108.75

x_1	V^E/cm^3mol^{-1}
0.1	0.3073
0.2	0.4996
0.3	0.6146
0.4	0.6722
0.5	0.6800
0.6	0.6378
0.7	0.5436
0.8	0.3979
0.9	0.2094

$A(1)$ = 0.2720E-05 $A(4)$ = -0.5359E-06
$A(2)$ = -0.3367E-06 $A(5)$ = 0.0000E+00
$A(3)$ = 0.2350E-06 $A(6)$ = 0.0000E+00

System Number: 40 **T/K = 298.15**
1. Cyclohexane, C_6H_{12}
2. Hexane, C_6H_{14}
$V(1)/cm^3mol^{-1}$ = 108.75
$V(2)/cm^3mol^{-1}$ = 131.60

x_1	V^E/cm^3mol^{-1}
0.1	0.0270
0.2	0.0540
0.3	0.0807
0.4	0.1058
0.5	0.1266
0.6	0.1402
0.7	0.1422
0.8	0.1266
0.9	0.0840

$A(1)$ = 0.5065E-06 $A(4)$ = 0.6200E-07
$A(2)$ = 0.3560E-06 $A(5)$ = 0.4000E-07
$A(3)$ = 0.1460E-06 $A(6)$ = 0.0000E+00

System Number: 78 **T/K = 298.15**
1. Cyclohexane, C_6H_{12}
2. 2,3-Dimethylbutane, C_6H_{14}
$V(1)/cm^3mol^{-1}$ = 108.75
$V(2)/cm^3mol^{-1}$ = 131.16

x_1	V^E/cm^3mol^{-1}
0.1	-0.0423
0.2	-0.0744
0.3	-0.0965
0.4	-0.1092
0.5	-0.1133
0.6	-0.1092
0.7	-0.0970
0.8	-0.0762
0.9	-0.0448

$A(1)$ = -0.4530E-06 $A(4)$ = -0.3000E-07
$A(2)$ = 0.1700E-08 $A(5)$ = 0.0000E+00
$A(3)$ = -0.4880E-07 $A(6)$ = 0.0000E+00

System Number: 264 **T/K = 298.15**
1. Cyclohexane, C_6H_{12}
2. Benzaldehyde, C_7H_6O
$V(1)/cm^3mol^{-1}$ = 108.75
$V(2)/cm^3mol^{-1}$ = 102.03

x_1	V^E/cm^3mol^{-1}
0.1	0.105
0.2	0.212
0.3	0.314
0.4	0.403
0.5	0.470
0.6	0.502
0.7	0.488
0.8	0.411
0.9	0.255

$A(1)$ = 0.1879E-05
$A(2)$ = 0.1037E-05
$A(3)$ = 0.1908E-06

System Number: 80 **T/K = 298.15**
1. Cyclohexane, C_6H_{12}
2. Heptane, C_7H_{16}
$V(1)/cm^3mol^{-1}$ = 108.75
$V(2)/cm^3mol^{-1}$ = 147.48

x_1	V^E/cm^3mol^{-1}
0.1	0.0822
0.2	0.1547
0.3	0.2148
0.4	0.2613
0.5	0.2923
0.6	0.3051
0.7	0.2941
0.8	0.2501
0.9	0.1588

$A(1)$ = 0.1169E-05 $A(4)$ = 0.1254E-06
$A(2)$ = 0.4518E-06 $A(5)$ = 0.0000E+00
$A(3)$ = 0.2655E-06 $A(6)$ = 0.0000E+00

System Number: 60 **T/K = 298.15**
1. Cyclohexane, C_6H_{12}
2. *cis*-Bicyclo[4.4.0]decane, $C_{10}H_{18}$
$V(1)/cm^3mol^{-1}$ = 108.75
$V(2)/cm^3mol^{-1}$ = 154.84

x_1	V^E/cm^3mol^{-1}
0.1	-0.035
0.2	-0.071
0.3	-0.110
0.4	-0.149
0.5	-0.182
0.6	-0.203
0.7	-0.203
0.8	-0.175
0.9	-0.109

$A(1)$ = -0.7280E-06 $A(4)$ = 0.8154E-07
$A(2)$ = -0.5687E-06 $A(5)$ = 0.0000E+00
$A(3)$ = -0.1141E-06 $A(6)$ = 0.0000E+00

System Number: 62 **T/K = 298.15**
1. Cyclohexane, C_6H_{12}
2. *trans*-Bicyclo[4.4.0]decane, $C_{10}H_{18}$
$V(1)/cm^3mol^{-1}$ = 108.75
$V(2)/cm^3mol^{-1}$ = 159.65

x_1	V^E/cm^3mol^{-1}
0.1	-0.0139
0.2	-0.0272
0.3	-0.0416
0.4	-0.0577
0.5	-0.0745
0.6	-0.0896
0.7	-0.0981
0.8	-0.0931
0.9	-0.0647

$A(1)$ = -0.2980E-06 $A(4)$ = -0.3439E-07
$A(2)$ = -0.3308E-06 $A(5)$ = 0.0000E+00
$A(3)$ = -0.2170E-06 $A(6)$ = 0.0000E+00

System Number: 57 **T/K = 298.15**
1. 1-Hexene, C_6H_{12}
2. Hexane, C_6H_{14}
$V(1)/cm^3mol^{-1}$ = 125.90
$V(2)/cm^3mol^{-1}$ = 131.60

x_1	V^E/cm^3mol^{-1}
0.1	0.0331
0.2	0.0485
0.3	0.0543
0.4	0.0550
0.5	0.0525
0.6	0.0468
0.7	0.0378
0.8	0.0254
0.9	0.0114

$A(1)$ = 0.2099E-06 $A(4)$ = -0.1083E-06
$A(2)$ = -0.8120E-07 $A(5)$ = 0.0000E+00
$A(3)$ = 0.5820E-07 $A(6)$ = 0.0000E+00

TABLE 3.2.1 (VETX)
DENSITIES AND EXCESS VOLUMES OF BINARY LIQUID MIXTURES (continued)

System Number: 210 **T/K = 298.15**
1. Hexane, C_6H_{14}
2. Dibutyl ether, $C_8H_{18}O$
$V(1)$/cm^{3}mol^{-1} = 131.60
$V(2)$/cm^{3}mol^{-1} = 170.44

x_1	V^E/cm^{3}mol^{-1}
0.1	-0.0140
0.2	-0.0256
0.3	-0.0347
0.4	-0.0411
0.5	-0.0443
0.6	-0.0441
0.7	-0.0401
0.8	-0.0317
0.9	-0.0185

$A(1)$ = -0.1773E-06
$A(2)$ = -0.3170E-07
$A(3)$ = -0.5300E-08

System Number: 77 **T/K = 298.15**
1. Hexane, C_6H_{14}
2. Decane, $C_{10}H_{22}$
$V(1)$/cm^{3}mol^{-1} = 131.60
$V(2)$/cm^{3}mol^{-1} = 195.89

x_1	V^E/cm^{3}mol^{-1}
0.1	-0.0660
0.2	-0.1225
0.3	-0.1689
0.4	-0.2038
0.5	-0.2251
0.6	-0.2299
0.7	-0.2145
0.8	-0.1746
0.9	-0.1051

$A(1)$ = -0.9005E-06
$A(2)$ = -0.2715E-06
$A(3)$ = -0.7800E-07

System Number: 103 **T/K = 298.15**
1. Dipropyl ether, $C_6H_{14}O$
2. Heptane, C_7H_{16}
$V(1)$/cm^{3}mol^{-1} = 137.72
$V(2)$/cm^{3}mol^{-1} = 147.48

x_1	V^E/cm^{3}mol^{-1}
0.1	0.0915
0.2	0.1618
0.3	0.2113
0.4	0.2406
0.5	0.2499
0.6	0.2395
0.7	0.2094
0.8	0.1596
0.9	0.0899

$A(1)$ = 0.9997E-06
$A(2)$ = -0.1108E-07
$A(3)$ = 0.1286E-07

System Number: 213 **T/K = 298.15**
1. 3,6-Dioxaoctane, $C_6H_{14}O_2$
2. Heptane, C_7H_{16}
$V(1)$/cm^{3}mol^{-1} = 141.51
$V(2)$/cm^{3}mol^{-1} = 147.48

x_1	V^E/cm^{3}mol^{-1}
0.1	0.308
0.2	0.523
0.3	0.660
0.4	0.731
0.5	0.742
0.6	0.698
0.7	0.600
0.8	0.450
0.9	0.249

$A(1)$ = 0.2969E-05 $A(4)$ = -0.9600E-07
$A(2)$ = -0.3443E-06 $A(5)$ = 0.0000E+00
$A(3)$ = 0.1980E-06 $A(6)$ = 0.0000E+00

3.3. CALORIMETRIC PROPERTIES

The calorimetric properties of a mixture, such as enthalpy and heat capacity, are most conveniently expressed as excess functions, i.e., as differences between the function in a real mixture and an ideal mixture. The following three tables cover the most important of these calorimetric properties for various types of mixtures.

The definitions of the basic calorimetric properties of mixtures, such as the molar or specific *isobaric* or *isochoric heat capacities* are the same as for pure substances (see Section 2.3), with partial derivatives taken at constant composition.

Characteristic calorimetric properties such as enthalpy and heat capacity change upon mixing the pure components. When fixed amounts of pure substances are mixed at constant pressure, the enthalpy or heat capacity of the mixture generally differs from the sum of the values for the original substances. The following tables deal with this difference for binary mixtures of representative liquid substances. Tables 3.3.1 and 3.3.2 cover enthalpy and heat capacity, respectively, over the full composition range, while Table 3.3.3 gives properties at infinite dilution.

TABLE 3.3.1 (HETX)
EXCESS ENTHALPIES OF BINARY LIQUID MIXTURES

The enthalpy of mixing of two liquids at constant temperature and pressure (often called the "heat of mixing") equals the change in enthalpy between the mixture and the initially separated components. As with other excess functions of mixing, the excess enthalpy is defined as the difference between the enthalpy of mixing of a real mixture and the enthalpy of mixing of an ideal mixture. Since the latter is by definition zero, the excess enthalpy equals the enthalpy, or the heat, of mixing. The *molar excess enthalpy* H^E equals the excess enthalpy divided by the total amount of substance (number of moles) of the two components. It is a function of the pressure P, the temperature T, and the composition of the mixture. The latter is usually expressed in terms of the mole fractions x_1 and $x_2 = 1 - x_1$ of components 1 and 2, respectively. The *partial molar excess enthalpies* of the components, $H^E{}_1$ and $H^E{}_2$, are related to H^E through the equation:

$$H^E = x_1 H^E{}_1 + x_2 H^E{}_2 \tag{1}$$

The Gibbs-Helmholtz equation relates H^E to the molar excess Gibbs energy G^E (Table 3.1.1) through the equation:

$$H^E = \left[\partial\left(G^E / T\right) / \partial(1/T)\right]_P \tag{2}$$

The temperature dependence of H^E is given by the excess molar heat capacity $C_p{}^E$ (see Table 3.3.2).

Most H^E measurements have been reported at atmospheric pressure. The pressure effect, measurable only at high pressures, is related to the excess molar volume V^E (Table 3.2.1) through the equation:

$$\left(\partial H^E / \partial P\right)_T = V^E - T\left(\partial V^E / \partial T\right)_P \tag{3}$$

The molar excess enthalpy may be positive or negative in the whole composition range or may change sign (S-shaped). Mixtures of two non-polar substances or a polar or associated substance with a non-polar substance usually have positive H^E. Mixtures of two strongly interacting (complexing) substances usually have negative H^E (e.g., trichloromethane + 2-propanone, System No. 280). If one of the complexing substances is strongly polar or associated, then S-shaped H^E curves may appear (e.g., water + oxolane, System No. 454).

The H^E curves can be represented as a function of x_1 by various empirical equations. A simple equation is the so-called Redlich-Kister equation:

$$H^E / \mathrm{J\,mol^{-1}} = x_1 x_2 \sum_{i=1}^{n} A(i)\left(x_1 - x_2\right)^{i-1} \tag{4}$$

The Table gives the coefficients $A(i)$, at the temperature T, obtained by fitting Equation 4 to experimental H^E (x_1) data. The Table lists the calculated values of H^E for increments of $x_1 = 0.1$.

The program, Property Code HETX, permits the calculation of H^E at selected increments of x_1 in the whole range of composition, from $x_1 = 0$ to $x_1 = 1$, or at any given value of x_1.

The components of each system, and the systems, are arranged in a modified Hill order, with substances that do not contain carbon preceding those that do contain carbon.

Physical quantity	Symbol	SI unit
Mole fraction of component i	x_i	—
Temperature	T	K
Molar excess enthalpy	H^E	J mol^{-1}

REFERENCES

1. Van Ness, H. C.; Abbott, M. M., *Classical Thermodynamics of Non-electrolyte Solutions*, McGraw-Hill, New York, 1982 (general theory).
2. Kehiaian, H. V.; Kehiaian, C. S., *Organic Systems (Mixtures)*, in: *Bulletin of Chemical Thermodynamics*, Freeman, R. D., Editor, Thermochemistry Inc., 1969-1985 (17 volumes) (a compilation of bibliographical excess enthalpy data).

TABLE 3.3.1 (HETX)
EXCESS ENTHALPIES OF BINARY LIQUID MIXTURES (continued)

3. Wisniak, J.; Tamir, A., *Mixing and Excess Properties. A Literature Source Book*, Elsevier, Amsterdam, 1978; Supplement 1, 1982; Supplement 2, 1986 (a compilation of bibliographical excess enthalpy data).
4. Beggerow, G., Landolt-Börnstein Numerical Data and Functional Relationships in Science and Technology, Vol. 2, *Heats of Mixing and Solution*, Springer-Verlag, Berlin-Heidelberg, 1976 (a compilation of numerical data on heats of mixing and solution).
5. Christensen, J. J.; Rowley, R. L.; Izatt, R. M., *Handbook of Heats of Mixing*, John Wiley & Sons, New York, 1982; Supplementary Volume, 1988 (an extensive compilation of numerical excess enthalpy data and correlation coefficients and a diskette with computer programs for prediction of excess enthalpies).
6. Christensen, C.; Gmehling, J.; Rasmussen, P; Weidlich, U., *Heats of Mixing Data Collection*, DECHEMA, Frankfurt/Main, Germany, 1984 (2 parts); Gmehling, J.; Holderbaum, T., 1988, 1991 (2 supplements) (an extensive compilation of numerical excess enthalpy data and correlation coefficients).
7. *Int. DATA Ser., Sel. Data Mixtures, Ser. A* (Ed. H. V. Kehiaian), Thermodynamic Research Center, Texas Engineering Experiment Station, College Station, USA, 1973-1993 (21 volumes) (primary excess enthalpy data and correlations)

System Number: 456 T/K = 298.15
1. Water, H_2O
2. Methanol, CH_4O

x_1	H^E/J mol^{-1}
0.1	-250
0.2	-450
0.3	-602
0.4	-715
0.5	-802
0.6	-864
0.7	-883
0.8	-812
0.9	-563

A(1) = -0.3208E+04 A(4) = -0.1051E+04
A(2) = -0.1506E+04 A(5) = 0.0000E+00
A(3) = -0.2047E+04 A(6) = 0.0000E+00

System Number: 453 T/K = 323.15
1. Water, H_2O
2. Ethanol, C_2H_6O

x_1	H^E/J mol^{-1}
0.1	-32
0.2	-66
0.3	-73
0.4	-71
0.5	-96
0.6	-177
0.7	-312
0.8	-442
0.9	-422

A(1) = -0.3833E+03 A(4) = -0.2676E+04
A(2) = -0.9958E+03 A(5) = 0.0000E+00
A(3) = -0.3339E+04 A(6) = 0.0000E+00

System Number: 454 T/K = 298.15
1. Water, H_2O
2. Oxolane, C_4H_8O

x_1	H^E/J mol^{-1}
0.1	281
0.2	286
0.3	170
0.4	-13
0.5	-218
0.6	-406
0.7	-569
0.8	-712
0.9	-714

A(1) = -0.8730E+03 A(4) = -0.1644E+04
A(2) = -0.4022E+04 A(5) = -0.4024E+04
A(3) = 0.1752E+03 A(6) = -0.4486E+04

System Number: 457 T/K = 293.15
1. Water, H_2O
2. Pyridine, C_5H_5N

x_1	H^E/J mol^{-1}
0.1	-275
0.2	-645
0.3	-1069
0.4	-1443
0.5	-1668
0.6	-1687
0.7	-1499
0.8	-1142
0.9	-653

A(1) = -0.6670E+04 A(4) = -0.1264E+03
A(2) = -0.2541E+04 A(5) = -0.2318E+04
A(3) = 0.3853E+04 A(6) = 0.0000E+00

System Number: 458 T/K = 298.15
1. Water, H_2O
2. 2-Methylpyridine, C_6H_7N

x_1	H^E/J mol^{-1}
0.1	-435
0.2	-1021
0.3	-1587
0.4	-2021
0.5	-2261
0.6	-2279
0.7	-2068
0.8	-1626
0.9	-946

A(1) = -0.9044E+04 A(4) = -0.1422E+04
A(2) = -0.2637E+04 A(5) = 0.0000E+00
A(3) = 0.2143E+04 A(6) = 0.0000E+00

System Number: 459 T/K = 298.15
1. Water, H_2O
2. 3-Methylpyridine, C_6H_7N

x_1	H^E/J mol^{-1}
0.1	-283
0.2	-700
0.3	-1134
0.4	-1493
0.5	-1709
0.6	-1740
0.7	-1566
0.8	-1193
0.9	-653

A(1) = -0.6836E+04
A(2) = -0.2571E+04
A(3) = 0.2556E+04

TABLE 3.3.1 (HETX)
EXCESS ENTHALPIES OF BINARY LIQUID MIXTURES (continued)

System Number: 460 T/K = 298.15
1. Water, H_2O
2. 4-Methylpyridine, C_6H_7N

x_1	H^E/J mol^{-1}
0.1	-306
0.2	-725
0.3	-1152
0.4	-1502
0.5	-1713
0.6	-1745
0.7	-1577
0.8	-1211
0.9	-671

$A(1)$ = -0.6852E+04
$A(2)$ = -0.2528E+04
$A(3)$ = 0.2226E+04

System Number: 461 T/K = 298.15
1. Water, H_2O
2. 2,6-Dimethylpyridine, C_7H_9N

x_1	H^E/J mol^{-1}
0.1	-593
0.2	-1247
0.3	-1829
0.4	-2263
0.5	-2515
0.6	-2568
0.7	-2403
0.8	-1984
0.9	-1229

$A(1)$ = -0.1006E+05 $A(4)$ = -0.2081E+04
$A(2)$ = -0.3087E+04 $A(5)$ = 0.0000E+00
$A(3)$ = -0.1002E+03 $A(6)$ = 0.0000E+00

System Number: 73 T/K = 298.15
1. Trichlorofluoromethane, $C\ Cl_3F$
2. 2,2,4-Trimethylpentane, C_8H_{18}

x_1	H^E/J mol^{-1}
0.1	48
0.2	93
0.3	132
0.4	165
0.5	186
0.6	194
0.7	184
0.8	152
0.9	92

$A(1)$ = 0.7456E+03
$A(2)$ = 0.3082E+03
$A(3)$ = 0.5041E+02

System Number: 357 T/K = 298.15
1. Tetrachloromethane, CCl_4
2. Trichloromethane, $CHCl_3$

x_1	H^E/J mol^{-1}
0.1	84
0.2	149
0.3	195
0.4	223
0.5	232
0.6	223
0.7	195
0.8	149
0.9	84

$A(1)$ = 0.9281E+03
$A(2)$ = 0.0000E+00
$A(3)$ = 0.0000E+00

System Number: 368 T/K = 298.15
1. Tetrachloromethane, CCl_4
2. Ethanenitrile, C_2H_3N

x_1	H^E/J mol^{-1}
0.1	274.5
0.2	481.7
0.3	628.5
0.4	727.9
0.5	781.8
0.6	785.2
0.7	735.0
0.8	627.9
0.9	429.3

$A(1)$ = 0.3127E+04 $A(4)$ = 0.1136E+03
$A(2)$ = 0.5899E+03 $A(5)$ = 0.9953E+03
$A(3)$ = 0.5864E+03 $A(6)$ = 0.1008E+04

System Number: 356 T/K = 313.15
1. Tetrachloromethane, CCl_4
2. 1,2-Dichloroethane, $C_2H_4Cl_2$

x_1	H^E/J mol^{-1}
0.1	250
0.2	437
0.3	567
0.4	641
0.5	663
0.6	633
0.7	552
0.8	421
0.9	237

$A(1)$ = 0.2651E+04
$A(2)$ = -0.8475E+02
$A(3)$ = 0.8343E+02

System Number: 10 T/K = 298.10
1. Tetrachloromethane, CCl_4
2. Dimethyl sulfoxide, C_2H_6OS

x_1	H^E/J mol^{-1}
0.1	21.6
0.2	60.0
0.3	99.2
0.4	138.0
0.5	180.7
0.6	229.9
0.7	277.5
0.8	296.7
0.9	233.9

$A(1)$ = 0.7228E+03 $A(4)$ = 0.8608E+03
$A(2)$ = 0.9231E+03 $A(5)$ = 0.0000E+00
$A(3)$ = 0.1089E+04 $A(6)$ = 0.0000E+00

System Number: 42 T/K = 303.15
1. Tetrachloromethane, CCl_4
2. Furan, C_4H_4O

x_1	H^E/J mol^{-1}
0.1	118.9
0.2	209.6
0.3	273.5
0.4	311.3
0.5	323.6
0.6	310.6
0.7	272.3
0.8	208.3
0.9	117.8

$A(1)$ = 0.1295E+04
$A(2)$ = -0.7257E+01
$A(3)$ = 0.3173E+02

System Number: 11 T/K = 303.15
1. Tetrachloromethane, CCl_4
2. Oxolane, C_4H_8O

x_1	H^E/J mol^{-1}
0.1	-292.3
0.2	-509.0
0.3	-653.0
0.4	-727.7
0.5	-737.2
0.6	-686.4
0.7	-580.8
0.8	-426.5
0.9	-230.4

$A(1)$ = -0.2949E+04
$A(2)$ = 0.4297E+03
$A(3)$ = 0.7010E+02

TABLE 3.3.1 (HETX)
EXCESS ENTHALPIES OF BINARY LIQUID MIXTURES (continued)

System Number: 49 **T/K = 298.15**
1. Tetrachloromethane, CCl_4
2. 1,4-Dioxane, $C_4H_8O_2$

x_1	H^E/J mol^{-1}
0.1	-87.5
0.2	-158.6
0.3	-210.8
0.4	-242.0
0.5	-250.4
0.6	-235.9
0.7	-199.4
0.8	-143.9
0.9	-74.8

$A(1)$ = -0.1002E+04 $A(4)$ = 0.4128E+02
$A(2)$ = 0.6152E+02 $A(5)$ = 0.0000E+00
$A(3)$ = 0.1566E+03 $A(6)$ = 0.0000E+00

System Number: 354 **T/K = 298.15**
1. Tetrachloromethane, CCl_4
2. 1-Bromobutane, C_4H_9Br

x_1	H^E/J mol^{-1}
0.1	-34
0.2	-56
0.3	-63
0.4	-58
0.5	-43
0.6	-24
0.7	-7
0.8	4
0.9	6

$A(1)$ = -0.1725E+03 $A(4)$ = -0.1280E+03
$A(2)$ = 0.3560E+03 $A(5)$ = 0.0000E+00
$A(3)$ = 0.3000E+02 $A(6)$ = 0.0000E+00

System Number: 355 **T/K = 313.15**
1. Tetrachloromethane, CCl_4
2. 1-Chlorobutane, C_4H_9Cl

x_1	H^E/J mol^{-1}
0.1	-2
0.2	-2
0.3	-2
0.4	-1
0.5	0
0.6	1
0.7	2
0.8	3
0.9	2

$A(1)$ = 0.5538E+00
$A(2)$ = 0.2550E+02
$A(3)$ = 0.0000E+00

System Number: 8 **T/K = 298.15**
1. Tetrachloromethane, CCl_4
2. Diethyl ether, $C_4H_{10}O$

x_1	H^E/J mol^{-1}
0.1	-175
0.2	-311
0.3	-408
0.4	-466
0.5	-485
0.6	-466
0.7	-408
0.8	-311
0.9	-175

$A(1)$ = -0.1942E+04
$A(2)$ = 0.0000E+00
$A(3)$ = 0.0000E+00

System Number: 348 **T/K = 298.15**
1. Tetrachloromethane, CCl_4
2. Cyclopentane, C_5H_{10}

x_1	H^E/J mol^{-1}
0.1	27.7
0.2	49.6
0.3	65.6
0.4	75.7
0.5	79.7
0.6	77.4
0.7	68.6
0.8	53.0
0.9	30.3

$A(1)$ = 0.3187E+03
$A(2)$ = 0.1778E+02
$A(3)$ = 0.5730E+01

System Number: 376 **T/K = 298.15**
1. Tetrachloromethane, CCl_4
2. Diethyl carbonate, $C_5H_{10}O_3$

x_1	H^E/J mol^{-1}
0.1	-29
0.2	-56
0.3	-74
0.4	-77
0.5	-66
0.6	-42
0.7	-12
0.8	15
0.9	24

$A(1)$ = -0.2622E+03
$A(2)$ = 0.3679E+03
$A(3)$ = 0.3706E+03

System Number: 50 **T/K = 298.15**
1. Tetrachloromethane, CCl_4
2. Hexafluorobenzene, C_6F_6

x_1	H^E/J mol^{-1}
0.1	115
0.2	228
0.3	333
0.4	424
0.5	490
0.6	521
0.7	503
0.8	422
0.9	261

$A(1)$ = 0.1959E+04
$A(2)$ = 0.1009E+04
$A(3)$ = 0.2018E+03

System Number: 48 **T/K = 298.15**
1. Tetrachloromethane, CCl_4
2. Benzene, C_6H_6

x_1	H^E/J mol^{-1}
0.1	44.2
0.2	76.5
0.3	98.3
0.4	110.7
0.5	114.3
0.6	109.4
0.7	95.9
0.8	73.6
0.9	41.8

$A(1)$ = 0.4572E+03 $A(4)$ = -0.5126E+01
$A(2)$ = -0.1349E+02 $A(5)$ = 0.0000E+00
$A(3)$ = 0.3282E+02 $A(6)$ = 0.0000E+00

System Number: 75 **T/K = 298.15**
1. Tetrachloromethane, CCl_4
2. Cyclohexane, C_6H_{12}

x_1	H^E/J mol^{-1}
0.1	58.4
0.2	104.2
0.3	137.4
0.4	158.1
0.5	166.0
0.6	160.9
0.7	142.4
0.8	109.8
0.9	62.6

$A(1)$ = 0.6642E+03
$A(2)$ = 0.2938E+02
$A(3)$ = 0.1267E+02

TABLE 3.3.1 (HETX)
EXCESS ENTHALPIES OF BINARY LIQUID MIXTURES (continued)

System Number: 12 **T/K = 298.15**
1. Tetrachloromethane, CCl_4
2. Dipropyl ether, $C_6H_{14}O$

x_1	H^E/J mol^{-1}
0.1	-96.6
0.2	-169.1
0.3	-223.1
0.4	-259.6
0.5	-276.2
0.6	-269.5
0.7	-236.6
0.8	-177.1
0.9	-94.9

A(1) = -0.1105E+04 A(4) = 0.1917E+03
A(2) = -0.1109E+03 A(5) = 0.0000E+00
A(3) = 0.6351E+02 A(6) = 0.0000E+00

System Number: 13 **T/K = 298.15**
1. Tetrachloromethane, CCl_4
2. Heptane, C_7H_{16}

x_1	H^E/J mol^{-1}
0.1	116.3
0.2	207.6
0.3	276.3
0.4	323.3
0.5	347.7
0.6	347.5
0.7	318.7
0.8	256.0
0.9	152.6

A(1) = 0.1391E+04
A(2) = 0.2522E+03
A(3) = 0.1606E+03

System Number: 335 **T/K = 298.15**
1. Tribromomethane, $CHBr_3$
2. Dimethyl sulfoxide, C_2H_6OS

x_1	H^E/J mol^{-1}
0.1	-799
0.2	-1516
0.3	-2102
0.4	-2518
0.5	-2730
0.6	-2710
0.7	-2438
0.8	-1900
0.9	-1087

A(1) = -0.1092E+05
A(2) = -0.2001E+04
A(3) = 0.6860E+03

System Number: 401 **T/K = 298.15**
1. Trichloromethane, $CHCl_3$
2. Ethanenitrile, C_2H_3N

x_1	H^E/J mol^{-1}
0.1	-171
0.2	-351
0.3	-527
0.4	-678
0.5	-783
0.6	-822
0.7	-775
0.8	-626
0.9	-369

A(1) = -0.3133E+04 A(4) = 0.2082E+03
A(2) = -0.1508E+04 A(5) = 0.0000E+00
A(3) = 0.2158E+03 A(6) = 0.0000E+00

System Number: 280 **T/K = 298.15**
1. Trichloromethane, $CHCl_3$
2. 2-Propanone, C_3H_6O

x_1	H^E/J mol^{-1}
0.1	-493
0.2	-957
0.3	-1380
0.4	-1719
0.5	-1925
0.6	-1951
0.7	-1769
0.8	-1372
0.9	-772

A(1) = -0.7699E+04 A(4) = 0.7920E+03
A(2) = -0.2446E+04 A(5) = -0.4400E+03
A(3) = 0.1325E+04 A(6) = 0.0000E+00

System Number: 70 **T/K = 308.15**
1. Trichloromethane, $CHCl_3$
2. Methyl ethanoate, $C_3H_6O_2$

x_1	H^E/J mol^{-1}
0.1	-409
0.2	-793
0.3	-1135
0.4	-1406
0.5	-1574
0.6	-1605
0.7	-1472
0.8	-1157
0.9	-660

A(1) = -0.6295E+04 A(4) = 0.5549E+03
A(2) = -0.2093E+04 A(5) = 0.0000E+00
A(3) = 0.5540E+03 A(6) = 0.0000E+00

System Number: 47 **T/K = 303.15**
1. Trichloromethane, $CHCl_3$
2. Furan, C_4H_4O

x_1	H^E/J mol^{-1}
0.1	-49.3
0.2	-96.5
0.3	-138.2
0.4	-171.0
0.5	-191.4
0.6	-196.4
0.7	-182.7
0.8	-147.3
0.9	-87.3

A(1) = -0.7658E+03
A(2) = -0.2645E+03
A(3) = 0.1072E+02

System Number: 46 **T/K = 303.15**
1. Trichloromethane, $CHCl_3$
2. Oxolane, C_4H_8O

x_1	H^E/J mol^{-1}
0.1	-810
0.2	-1527
0.3	-2118
0.4	-2528
0.5	-2707
0.6	-2624
0.7	-2277
0.8	-1693
0.9	-918

A(1) = -0.1083E+05 A(4) = 0.4179E+03
A(2) = -0.1013E+04 A(5) = -0.7272E+03
A(3) = 0.2384E+04 A(6) = 0.0000E+00

System Number: 69 **T/K = 308.15**
1. Trichloromethane, $CHCl_3$
2. 2-Butanone, C_4H_8O

x_1	H^E/J mol^{-1}
0.1	-597
0.2	-1130
0.3	-1587
0.4	-1928
0.5	-2106
0.6	-2089
0.7	-1862
0.8	-1430
0.9	-808

A(1) = -0.8425E+04 A(4) = 0.3724E+03
A(2) = -0.1700E+04 A(5) = -0.7649E+03
A(3) = 0.1455E+04 A(6) = 0.0000E+00

TABLE 3.3.1 (HETX) EXCESS ENTHALPIES OF BINARY LIQUID MIXTURES (continued)

System Number: 52 T/K = 303.15
1. Trichloromethane, $CHCl_3$
2. 1,4-Dioxane, $C_4H_8O_2$

x_1	H^E/J mol^{-1}
0.1	-490
0.2	-960
0.3	-1387
0.4	-1744
0.5	-1992
0.6	-2081
0.7	-1959
0.8	-1577
0.9	-915

A(1) = -0.7966E+04 A(4) = 0.9335E+03
A(2) = -0.3552E+04 A(5) = 0.5091E+03
A(3) = -0.7937E+02 A(6) = 0.0000E+00

System Number: 71 T/K = 308.15
1. Trichloromethane, $CHCl_3$
2. Ethyl ethanoate, $C_4H_8O_2$

x_1	H^E/J mol^{-1}
0.1	-503
0.2	-981
0.3	-1402
0.4	-1727
0.5	-1915
0.6	-1927
0.7	-1740
0.8	-1342
0.9	-747

A(1) = -0.7658E+04 A(4) = 0.6538E+03
A(2) = -0.2115E+04 A(5) = 0.0000E+00
A(3) = 0.1112E+04 A(6) = 0.0000E+00

System Number: 7 T/K = 298.15
1. Trichloromethane, $CHCl_3$
2. Diethyl ether, $C_4H_{10}O$

x_1	H^E/J mol^{-1}
0.1	-865
0.2	-1527
0.3	-2080
0.4	-2480
0.5	-2645
0.6	-2526
0.7	-2135
0.8	-1539
0.9	-822

A(1) = -0.1058E+05 A(4) = 0.1307E+04
A(2) = -0.5365E+03 A(5) = -0.3160E+04
A(3) = 0.3916E+04 A(6) = 0.0000E+00

System Number: 14 T/K = 298.15
1. Trichloromethane, $CHCl_3$
2. Dipropyl ether, $C_6H_{14}O$

x_1	H^E/J mol^{-1}
0.1	-650
0.2	-1145
0.3	-1549
0.4	-1833
0.5	-1947
0.6	-1860
0.7	-1582
0.8	-1157
0.9	-632

A(1) = -0.7787E+04 A(4) = 0.6610E+03
A(2) = -0.2987E+03 A(5) = -0.2148E+04
A(3) = 0.2419E+04 A(6) = 0.0000E+00

System Number: 410 T/K = 298.15
1. Dichloromethane, CH_2Cl_2
2. Ethanenitrile, C_2H_3N

x_1	H^E/J mol^{-1}
0.1	-75
0.2	-148
0.3	-214
0.4	-267
0.5	-303
0.6	-316
0.7	-300
0.8	-248
0.9	-151

A(1) = -0.1210E+04 A(4) = -0.2947E+02
A(2) = -0.5084E+03 A(5) = 0.0000E+00
A(3) = -0.7424E+02 A(6) = 0.0000E+00

System Number: 336 T/K = 298.15
1. Dichloromethane, CH_2Cl_2
2. Dimethyl sulfoxide, C_2H_6OS

x_1	H^E/J mol^{-1}
0.1	-382
0.2	-694
0.3	-938
0.4	-1127
0.5	-1247
0.6	-1265
0.7	-1152
0.8	-906
0.9	-536

A(1) = -0.4987E+04 A(4) = 0.1645E+04
A(2) = -0.1500E+04 A(5) = -0.4940E+03
A(3) = 0.1400E+03 A(6) = -0.1508E+04

System Number: 51 T/K = 303.15
1. Dichloromethane, CH_2Cl_2
2. 2-Propanone, C_3H_6O

x_1	H^E/J mol^{-1}
0.1	-290.9
0.2	-534.3
0.3	-721.5
0.4	-844.5
0.5	-896.4
0.6	-872.0
0.7	-768.5
0.8	-585.8
0.9	-327.2

A(1) = -0.3586E+04 A(4) = 0.5730E+02
A(2) = -0.2888E+03 A(5) = 0.0000E+00
A(3) = 0.2379E+03 A(6) = 0.0000E+00

System Number: 15 T/K = 298.15
1. Dichloromethane, CH_2Cl_2
2. Methyl ethanoate, $C_3H_6O_2$

x_1	H^E/J mol^{-1}
0.1	-262
0.2	-485
0.3	-654
0.4	-759
0.5	-793
0.6	-756
0.7	-649
0.8	-479
0.9	-258

A(1) = -0.3174E+04
A(2) = 0.3068E+02
A(3) = 0.4491E+03

System Number: 45 T/K = 303.15
1. Dichloromethane, CH_2Cl_2
2. Furan, C_4H_4O

x_1	H^E/J mol^{-1}
0.1	12.2
0.2	18.3
0.3	20.2
0.4	19.3
0.5	16.8
0.6	13.6
0.7	10.2
0.8	6.9
0.9	3.7

A(1) = 0.6725E+02
A(2) = -0.5908E+02
A(3) = 0.3223E+02

TABLE 3.3.1 (HETX)
EXCESS ENTHALPIES OF BINARY LIQUID MIXTURES (continued)

System Number: 44 **T/K = 303.15**
1. Dichloromethane, CH_2Cl_2
2. Oxolane, C_4H_8O

x_1	H^E/J mol^{-1}
0.1	-492
0.2	-882
0.3	-1157
0.4	-1310
0.5	-1339
0.6	-1250
0.7	-1052
0.8	-762
0.9	-402

$A(1)$ = -0.5357E+04
$A(2)$ = 0.6245E+03
$A(3)$ = 0.6076E+03

System Number: 16 **T/K = 298.15**
1. Dichloromethane, CH_2Cl_2
2. Ethyl ethanoate, $C_4H_8O_2$

x_1	H^E/J mol^{-1}
0.1	-316
0.2	-585
0.3	-789
0.4	-915
0.5	-955
0.6	-908
0.7	-776
0.8	-571
0.9	-305

$A(1)$ = -0.3820E+04
$A(2)$ = 0.7376E+02
$A(3)$ = 0.5833E+03

System Number: 43 **T/K = 303.15**
1. Dichloromethane, CH_2Cl_2
2. 1,4-Dioxane, $C_4H_8O_2$

x_1	H^E/J mol^{-1}
0.1	-298
0.2	-571
0.3	-802
0.4	-974
0.5	-1071
0.6	-1077
0.7	-981
0.8	-771
0.9	-444

$A(1)$ = -0.4284E+04 $A(4)$ = 0.1068E+03
$A(2)$ = -0.1080E+04 $A(5)$ = 0.0000E+00
$A(3)$ = 0.2481E+03 $A(6)$ = 0.0000E+00

System Number: 39 **T/K = 298.15**
1. Dichloromethane, CH_2Cl_2
2. Benzene, C_6H_6

x_1	H^E/J mol^{-1}
0.1	-30.2
0.2	-56.2
0.3	-74.5
0.4	-83.4
0.5	-82.2
0.6	-71.8
0.7	-54.2
0.8	-32.9
0.9	-12.8

$A(1)$ = -0.3288E+03
$A(2)$ = 0.1210E+03
$A(3)$ = 0.1400E+03

System Number: 125 **T/K = 298.15**
1. Dichloromethane, CH_2Cl_2
2. Toluene, C_7H_8

x_1	H^E/J mol^{-1}
0.1	-93
0.2	-166
0.3	-216
0.4	-239
0.5	-237
0.6	-211
0.7	-166
0.8	-109
0.9	-50

$A(1)$ = -0.9469E+03
$A(2)$ = 0.2950E+03
$A(3)$ = 0.2385E+03

System Number: 17 **T/K = 298.15**
1. Methanol, CH_4O
2. Ethanol, C_2H_6O

x_1	H^E/J mol^{-1}
0.1	1.2
0.2	2.2
0.3	3.1
0.4	3.9
0.5	4.5
0.6	4.7
0.7	4.4
0.8	3.5
0.9	2.0

$A(1)$ = 0.1792E+02 $A(4)$ = -0.4002E+01
$A(2)$ = 0.8231E+01 $A(5)$ = 0.0000E+00
$A(3)$ = -0.3723E+00 $A(6)$ = 0.0000E+00

System Number: 18 **T/K = 298.15**
1. Methanol, CH_4O
2. Methyl ethanoate, $C_3H_6O_2$

x_1	H^E/J mol^{-1}
0.1	455
0.2	759
0.3	938
0.4	1016
0.5	1009
0.6	932
0.7	791
0.8	591
0.9	329

$A(1)$ = 0.4037E+04
$A(2)$ = -0.8754E+03
$A(3)$ = 0.5024E+03

System Number: 19 **T/K = 298.15**
1. Methanol, CH_4O
2. 1-Propanol, C_3H_8O

x_1	H^E/J mol^{-1}
0.1	23.1
0.2	44.0
0.3	61.3
0.4	74.3
0.5	82.1
0.6	83.8
0.7	78.2
0.8	63.6
0.9	38.2

$A(1)$ = 0.3285E+03 $A(4)$ = 0.8784E+01
$A(2)$ = 0.9912E+02 $A(5)$ = -0.9668E+01
$A(3)$ = 0.2507E+02 $A(6)$ = 0.0000E+00

System Number: 68 **T/K = 298.15**
1. Methanol, CH_4O
2. Ethyl ethanoate, $C_4H_8O_2$

x_1	H^E/J mol^{-1}
0.1	444
0.2	758
0.3	954
0.4	1044
0.5	1039
0.6	950
0.7	791
0.8	571
0.9	304

$A(1)$ = 0.4155E+04
$A(2)$ = -0.9729E+03
$A(3)$ = 0.0000E+00

TABLE 3.3.1 (HETX)
EXCESS ENTHALPIES OF BINARY LIQUID MIXTURES (continued)

System Number: 309 **T/K = 298.15**
1. Methanol, CH_4O
2. Diethyl ether, $C_4H_{10}O$

x_1	H^E/J mol^{-1}
0.1	325.1
0.2	451.1
0.3	485.2
0.4	476.8
0.5	441.0
0.6	377.1
0.7	283.7
0.8	169.4
0.9	59.7

$A(1)$ = 0.1764E+04 $A(4)$ = -0.1340E+04
$A(2)$ = -0.9850E+03 $A(5)$ = 0.3500E+03
$A(3)$ = 0.3600E+03 $A(6)$ = 0.0000E+00

System Number: 314 **T/K = 298.15**
1. Methanol, CH_4O
2. 2-Butanol, $C_4H_{10}O$

x_1	H^E/J mol^{-1}
0.1	-76.7
0.2	-118.2
0.3	-132.3
0.4	-126.1
0.5	-106.3
0.6	-79.0
0.7	-49.9
0.8	-24.0
0.9	-6.0

$A(1)$ = -0.4251E+03
$A(2)$ = 0.4908E+03
$A(3)$ = -0.5400E+02

System Number: 353 **T/K = 298.15**
1. Tetrachloroethene, C_2Cl_4
2. Cyclopentane, C_5H_{10}

x_1	H^E/J mol^{-1}
0.1	87.3
0.2	153.8
0.3	200.4
0.4	227.5
0.5	235.6
0.6	225.2
0.7	196.3
0.8	149.2
0.9	83.8

$A(1)$ = 0.9425E+03
$A(2)$ = -0.2387E+02
$A(3)$ = 0.1239E+02

System Number: 297 **T/K = 323.15**
1. Ethanenitrile, C_2H_3N
2. 1,2-Ethanediol, $C_2H_6O_2$

x_1	H^E/J mol^{-1}
0.1	499
0.2	860
0.3	1143
0.4	1356
0.5	1490
0.6	1531
0.7	1461
0.8	1246
0.9	810

$A(1)$ = 0.5959E+04 $A(4)$ = 0.5550E+03
$A(2)$ = 0.1807E+04 $A(5)$ = 0.1145E+04
$A(3)$ = 0.1317E+04 $A(6)$ = 0.0000E+00

System Number: 167 **T/K = 298.15**
1. Ethanenitrile, C_2H_3N
2. Benzene, C_6H_6

x_1	H^E/J mol^{-1}
0.1	129
0.2	239
0.3	329
0.4	399
0.5	444
0.6	458
0.7	433
0.8	357
0.9	218

$A(1)$ = 0.1777E+04
$A(2)$ = 0.6172E+03
$A(3)$ = 0.2342E+03

System Number: 141 **T/K = 313.15**
1. 1,2-Dichloroethane, $C_2H_4Cl_2$
2. 1-Chlorobutane, C_4H_9Cl

x_1	H^E/J mol^{-1}
0.1	106
0.2	219
0.3	304
0.4	348
0.5	352
0.6	327
0.7	282
0.8	221
0.9	137

$A(1)$ = 0.1409E+04 $A(4)$ = 0.7117E+03
$A(2)$ = -0.2455E+03 $A(5)$ = 0.0000E+00
$A(3)$ = -0.9011E+02 $A(6)$ = 0.0000E+00

System Number: 124 **T/K = 293.15**
1. 1,2-Dichloroethane, $C_2H_4Cl_2$
2. Heptane, C_7H_{16}

x_1	H^E/J mol^{-1}
0.1	467
0.2	953
0.3	1347
0.4	1601
0.5	1705
0.6	1670
0.7	1509
0.8	1217
0.9	748

$A(1)$ = 0.6818E+04 $A(4)$ = 0.2040E+04
$A(2)$ = 0.6410E+03 $A(5)$ = 0.0000E+00
$A(3)$ = -0.1080E+03 $A(6)$ = 0.0000E+00

System Number: 145 **T/K = 303.15**
1. Iodoethane, C_2H_5I
2. Hexane, C_6H_{14}

x_1	H^E/J mol^{-1}
0.1	283
0.2	531
0.3	722
0.4	844
0.5	890
0.6	861
0.7	756
0.8	578
0.9	328

$A(1)$ = 0.3561E+04 $A(4)$ = 0.2200E+03
$A(2)$ = 0.1670E+03 $A(5)$ = 0.0000E+00
$A(3)$ = -0.2580E+03 $A(6)$ = 0.0000E+00

System Number: 102 **T/K = 323.15**
1. Ethanol, C_2H_6O
2. 1,4-Dioxane, $C_4H_8O_2$

x_1	H^E/J mol^{-1}
0.1	602.9
0.2	1055.5
0.3	1368.3
0.4	1550.8
0.5	1610.6
0.6	1551.4
0.7	1372.3
0.8	1065.7
0.9	616.4

$A(1)$ = 0.6442E+04 $A(4)$ = 0.1456E+03
$A(2)$ = 0.4628E+00 $A(5)$ = 0.0000E+00
$A(3)$ = 0.5177E+03 $A(6)$ = 0.0000E+00

TABLE 3.3.1 (HETX)
EXCESS ENTHALPIES OF BINARY LIQUID MIXTURES (continued)

System Number: 67 **T/K = 298.15**
1. Ethanol, C_2H_6O
2. Ethyl ethanoate, $C_4H_8O_2$

x_1	H^E/J mol^{-1}
0.1	596
0.2	943
0.3	1151
0.4	1261
0.5	1279
0.6	1204
0.7	1040
0.8	795
0.9	464

$A(1)$ = 0.5116E+04 $A(4)$ = -0.5311E+03
$A(2)$ = -0.5778E+03 $A(5)$ = 0.1186E+04
$A(3)$ = 0.4478E+03 $A(6)$ = 0.0000E+00

System Number: 220 **T/K = 303.15**
1. Ethanol, C_2H_6O
2. Nonane, C_9H_{20}

x_1	H^E/J mol^{-1}
0.1	530
0.2	686
0.3	726
0.4	739
0.5	732
0.6	689
0.7	605
0.8	484
0.9	312

$A(1)$ = 0.2927E+04 $A(4)$ = -0.1648E+04
$A(2)$ = -0.4561E+03 $A(5)$ = 0.2530E+04
$A(3)$ = 0.1118E+04 $A(6)$ = 0.0000E+00

System Number: 20 **T/K = 298.15**
1. Ethanol, C_2H_6O
2. 1-Decanol, $C_{10}H_{22}O$

x_1	H^E/J mol^{-1}
0.1	85.4
0.2	157.9
0.3	218.8
0.4	267.3
0.5	301.0
0.6	316.5
0.7	309.6
0.8	272.0
0.9	183.7

$A(1)$ = 0.1204E+04 $A(4)$ = 0.2147E+03
$A(2)$ = 0.5035E+03 $A(5)$ = 0.2388E+03
$A(3)$ = 0.3017E+03 $A(6)$ = 0.1016E+03

System Number: 21 **T/K = 298.10**
1. Dimethyl sulfoxide, C_2H_6OS
2. Benzene, C_6H_6

x_1	H^E/J mol^{-1}
0.1	252
0.2	418
0.3	520
0.4	573
0.5	586
0.6	563
0.7	503
0.8	399
0.9	237

$A(1)$ = 0.2343E+04
$A(2)$ = -0.1014E+03
$A(3)$ = 0.5824E+03

System Number: 324 **T/K = 323.15**
1. 1,2-Ethanediol, $C_2H_6O_2$
2. 2-Propanone, C_3H_6O

x_1	H^E/J mol^{-1}
0.1	573.6
0.2	886.2
0.3	1040.1
0.4	1090.5
0.5	1063.3
0.6	969.2
0.7	812.5
0.8	596.9
0.9	326.0

$A(1)$ = 0.4253E+04 $A(4)$ = -0.7590E+03
$A(2)$ = -0.1234E+04 $A(5)$ = 0.3700E+03
$A(3)$ = 0.9260E+03 $A(6)$ = 0.0000E+00

System Number: 273 **T/K = 298.15**
1. 2-Propanone, C_3H_6O
2. 1-Chlorobutane, C_4H_9Cl

x_1	H^E/J mol^{-1}
0.1	173.8
0.2	301.5
0.3	388.6
0.4	439.0
0.5	455.0
0.6	437.7
0.7	386.4
0.8	299.0
0.9	171.9

$A(1)$ = 0.1820E+04
$A(2)$ = -0.1288E+02
$A(3)$ = 0.1572E+03

System Number: 320 **T/K = 298.15**
1. 2-Propanone, C_3H_6O
2. Diethyl ether, $C_4H_{10}O$

x_1	H^E/J mol^{-1}
0.1	433
0.2	623
0.3	701
0.4	719
0.5	691
0.6	613
0.7	490
0.8	334
0.9	165

$A(1)$ = 0.2762E+04 $A(4)$ = -0.1270E+04
$A(2)$ = -0.1051E+04 $A(5)$ = 0.8700E+03
$A(3)$ = 0.3200E+03 $A(6)$ = 0.0000E+00

System Number: 245 **T/K = 323.15**
1. 2-Propanone, C_3H_6O
2. Heptane, C_7H_{16}

x_1	H^E/J mol^{-1}
0.1	799
0.2	1303
0.3	1599
0.4	1748
0.5	1782
0.6	1711
0.7	1525
0.8	1202
0.9	707

$A(1)$ = 0.7130E+04 $A(4)$ = -0.4132E+03
$A(2)$ = -0.3743E+03 $A(5)$ = 0.0000E+00
$A(3)$ = 0.1941E+04 $A(6)$ = 0.0000E+00

System Number: 30 **T/K = 298.15**
1. Methyl ethanoate, $C_3H_6O_2$
2. Benzene, C_6H_6

x_1	H^E/J mol^{-1}
0.1	148
0.2	255
0.3	327
0.4	368
0.5	382
0.6	369
0.7	327
0.8	255
0.9	148

$A(1)$ = 0.1528E+04
$A(2)$ = 0.2866E+01
$A(3)$ = 0.1832E+03

TABLE 3.3.1 (HETX)
EXCESS ENTHALPIES OF BINARY LIQUID MIXTURES (continued)

System Number: 22 T/K = 298.15
1. Methyl ethanoate, $C_3H_6O_2$
2. Cyclohexane, C_6H_{12}

x_1	H^E/J mol^{-1}
0.1	738
0.2	1232
0.3	1536
0.4	1692
0.5	1725
0.6	1649
0.7	1461
0.8	1146
0.9	673

$A(1)$ = 0.6900E+04
$A(2)$ = -0.4474E+03
$A(3)$ = 0.1469E+04

System Number: 64 T/K = 298.15
1. 1,3-Dioxolane, $C_3H_6O_2$
2. Cyclohexane, C_6H_{12}

x_1	H^E/J mol^{-1}
0.1	707
0.2	1154
0.3	1420
0.4	1553
0.5	1581
0.6	1521
0.7	1373
0.8	1118
0.9	699

$A(1)$ = 0.6325E+04 $A(4)$ = 0.4620E+03
$A(2)$ = -0.3512E+03 $A(5)$ = 0.6261E+03
$A(3)$ = 0.1930E+04 $A(6)$ = 0.0000E+00

System Number: 400 T/K = 298.15
1. Methyl ethanoate, $C_3H_6O_2$
2. Heptane, C_7H_{16}

x_1	H^E/J mol^{-1}
0.1	645
0.2	1129
0.3	1471
0.4	1684
0.5	1773
0.6	1735
0.7	1560
0.8	1230
0.9	721

$A(1)$ = 0.7092E+04
$A(2)$ = 0.5243E+03
$A(3)$ = 0.7812E+03

System Number: 65 T/K = 298.15
1. 1,3-Dioxolane, $C_3H_6O_2$
2. Heptane, C_7H_{16}

x_1	H^E/J mol^{-1}
0.1	755
0.2	1274
0.3	1621
0.4	1815
0.5	1866
0.6	1788
0.7	1602
0.8	1305
0.9	836

$A(1)$ = 0.7464E+04 $A(4)$ = 0.1399E+04
$A(2)$ = -0.3380E+03 $A(5)$ = 0.1755E+04
$A(3)$ = 0.1023E+04 $A(6)$ = 0.0000E+00

System Number: 271 T/K = 298.15
1. Dimethyl carbonate, $C_3H_6O_3$
2. Cyclohexane, C_6H_{12}

x_1	H^E/J mol^{-1}
0.1	867
0.2	1403
0.3	1724
0.4	1897
0.5	1947
0.6	1872
0.7	1657
0.8	1278
0.9	723

$A(1)$ = 0.7787E+04 $A(4)$ = -0.1245E+04
$A(2)$ = -0.2030E+03 $A(5)$ = 0.0000E+00
$A(3)$ = 0.1641E+04 $A(6)$ = 0.0000E+00

System Number: 127 T/K = 298.15
1. 1-Chloropropane, C_3H_7Cl
2. Benzene, C_6H_6

x_1	H^E/J mol^{-1}
0.1	2
0.2	2
0.3	0
0.4	-3
0.5	-8
0.6	-12
0.7	-15
0.8	-15
0.9	-11

$A(1)$ = -0.3080E+02
$A(2)$ = -0.9030E+02
$A(3)$ = -0.2900E+02

System Number: 126 T/K = 298.15
1. 1-Chloropropane, C_3H_7Cl
2. Cyclohexane, C_6H_{12}

x_1	H^E/J mol^{-1}
0.1	294
0.2	476
0.3	584
0.4	638
0.5	647
0.6	613
0.7	531
0.8	399
0.9	219

$A(1)$ = 0.2589E+04 $A(4)$ = -0.4350E+03
$A(2)$ = -0.2450E+03 $A(5)$ = 0.0000E+00
$A(3)$ = 0.4050E+03 $A(6)$ = 0.0000E+00

System Number: 212 T/K = 298.15
1. 2,4-Dioxapentane, $C_3H_8O_2$
2. Hexane, C_6H_{14}

x_1	H^E/J mol^{-1}
0.1	383.4
0.2	675.0
0.3	867.9
0.4	992.9
0.5	1064.0
0.6	1067.7
0.7	976.7
0.8	770.9
0.9	447.3

$A(1)$ = 0.4256E+04 $A(4)$ = -0.1320E+04
$A(2)$ = 0.8300E+03 $A(5)$ = -0.6000E+03
$A(3)$ = 0.9450E+03 $A(6)$ = 0.1120E+04

System Number: 262 T/K = 298.15
1. 2-Butanone, C_4H_8O
2. Benzene, C_6H_6

x_1	H^E/J mol^{-1}
0.1	8.3
0.2	-3.9
0.3	-22.1
0.4	-38.5
0.5	-49.5
0.6	-53.8
0.7	-50.7
0.8	-40.1
0.9	-22.6

$A(1)$ = -0.1982E+03 $A(4)$ = -0.9344E+02
$A(2)$ = -0.1553E+03 $A(5)$ = 0.6134E+02
$A(3)$ = 0.1462E+03 $A(6)$ = 0.0000E+00

TABLE 3.3.1 (HETX)
EXCESS ENTHALPIES OF BINARY LIQUID MIXTURES (continued)

System Number: 214 T/K = 298.15
1. Oxolane, C_4H_8O
2. Cyclohexane, C_6H_{12}

x_1	H^E/J mol^{-1}
0.1	277
0.2	488
0.3	630
0.4	706
0.5	720
0.6	680
0.7	591
0.8	453
0.9	262

$A(1)$ = 0.2880E+04 $A(4)$ = 0.2673E+03
$A(2)$ = -0.2779E+03 $A(5)$ = 0.0000E+00
$A(3)$ = 0.1751E+03 $A(6)$ = 0.0000E+00

System Number: 249 T/K = 298.15
1. 2-Butanone, C_4H_8O
2. Cyclohexane, C_6H_{12}

x_1	H^E/J mol^{-1}
0.1	586
0.2	973
0.3	1206
0.4	1310
0.5	1303
0.6	1202
0.7	1020
0.8	767
0.9	438

$A(1)$ = 0.5213E+04 $A(4)$ = 0.1810E+03
$A(2)$ = -0.1139E+04 $A(5)$ = 0.4100E+03
$A(3)$ = 0.4800E+03 $A(6)$ = 0.0000E+00

System Number: 243 T/K = 298.15
1. Butanal, C_4H_8O
2. Heptane, C_7H_{16}

x_1	H^E/J mol^{-1}
0.1	547
0.2	905
0.3	1117
0.4	1215
0.5	1218
0.6	1136
0.7	975
0.8	733
0.9	409

$A(1)$ = 0.4871E+04 $A(4)$ = -0.2368E+03
$A(2)$ = -0.8088E+03 $A(5)$ = 0.0000E+00
$A(3)$ = 0.6924E+03 $A(6)$ = 0.0000E+00

System Number: 66 T/K = 298.15
1. Oxolane, C_4H_8O
2. Heptane, C_7H_{16}

x_1	H^E/J mol^{-1}
0.1	282
0.2	497
0.3	651
0.4	749
0.5	791
0.6	776
0.7	699
0.8	552
0.9	324

$A(1)$ = 0.3164E+04
$A(2)$ = 0.2884E+03
$A(3)$ = 0.3194E+03

System Number: 74 T/K = 298.15
1. 2-Butanone, C_4H_8O
2. Heptane, C_7H_{16}

x_1	H^E/J mol^{-1}
0.1	591.8
0.2	964.6
0.3	1194.6
0.4	1314.2
0.5	1338.5
0.6	1274.3
0.7	1120.0
0.8	864.6
0.9	491.9

$A(1)$ = 0.5354E+04 $A(4)$ = -0.1340E+03
$A(2)$ = -0.4100E+03 $A(5)$ = 0.1270E+03
$A(3)$ = 0.9600E+03 $A(6)$ = -0.4840E+03

System Number: 24 T/K = 298.15
1. Ethyl ethanoate, $C_4H_8O_2$
2. 1-Chlorobutane, C_4H_9Cl

x_1	H^E/J mol^{-1}
0.1	81.4
0.2	150.3
0.3	203.8
0.4	239.2
0.5	254.6
0.6	248.6
0.7	220.2
0.8	169.1
0.9	95.5

$A(1)$ = 0.1019E+04
$A(2)$ = 0.9776E+02
$A(3)$ = -0.5638E+02

System Number: 217 T/K = 298.15
1. 1,4-Dioxane, $C_4H_8O_2$
2. Benzene, C_6H_6

x_1	H^E/J mol^{-1}
0.1	-34
0.2	-48
0.3	-49
0.4	-43
0.5	-32
0.6	-19
0.7	-5
0.8	4
0.9	6

$A(1)$ = -0.1290E+03 $A(4)$ = 0.3980E+02
$A(2)$ = 0.2540E+03 $A(5)$ = -0.7710E+02
$A(3)$ = 0.5600E+01 $A(6)$ = 0.0000E+00

System Number: 216 T/K = 298.15
1. 1,4-Dioxane, $C_4H_8O_2$
2. Cyclohexane, C_6H_{12}

x_1	H^E/J mol^{-1}
0.1	655
0.2	1099
0.3	1393
0.4	1558
0.5	1601
0.6	1524
0.7	1331
0.8	1023
0.9	592

$A(1)$ = 0.6405E+04 $A(4)$ = -0.1330E+03
$A(2)$ = -0.3480E+03 $A(5)$ = 0.6640E+03
$A(3)$ = 0.3920E+03 $A(6)$ = 0.0000E+00

System Number: 269 T/K = 298.15
1. Ethyl ethanoate, $C_4H_8O_2$
2. Cyclohexane, C_6H_{12}

x_1	H^E/J mol^{-1}
0.1	596
0.2	979
0.3	1199
0.4	1296
0.5	1296
0.6	1216
0.7	1060
0.8	820
0.9	477

$A(1)$ = 0.5185E+04
$A(2)$ = -0.8240E+03
$A(3)$ = 0.1212E+04

TABLE 3.3.1 (HETX)
EXCESS ENTHALPIES OF BINARY LIQUID MIXTURES (continued)

System Number: 23 T/K = 298.15
1. 1,4-Dioxane, $C_4H_8O_2$
2. Heptane, C_7H_{16}

x_1	H^E/J mol^{-1}
0.1	686.7
0.2	1149.2
0.3	1451.7
0.4	1612.5
0.5	1641.7
0.6	1557.6
0.7	1383.9
0.8	1125.7
0.9	725.7

$A(1)$ = 0.6567E+04 $A(4)$ = 0.1406E+04
$A(2)$ = -0.6285E+03 $A(5)$ = 0.1763E+04
$A(3)$ = 0.8715E+03 $A(6)$ = 0.0000E+00

System Number: 268 T/K = 298.15
1. Ethyl ethanoate, $C_4H_8O_2$
2. Heptane, C_7H_{16}

x_1	H^E/J mol^{-1}
0.1	570
0.2	979
0.3	1253
0.4	1410
0.5	1462
0.6	1412
0.7	1257
0.8	983
0.9	573

$A(1)$ = 0.5849E+04
$A(2)$ = 0.1992E+02
$A(3)$ = 0.7889E+03

System Number: 105 T/K = 298.15
1. 1-Bromobutane, C_4H_9Br
2. Benzene, C_6H_6

x_1	H^E/J mol^{-1}
0.1	30
0.2	41
0.3	40
0.4	34
0.5	26
0.6	17
0.7	8
0.8	1
0.9	-2

$A(1)$ = 0.1031E+03 $A(4)$ = -0.6900E+02
$A(2)$ = -0.1800E+03 $A(5)$ = 0.0000E+00
$A(3)$ = 0.8000E+02 $A(6)$ = 0.0000E+00

System Number: 111 T/K = 298.15
1. 1-Bromobutane, C_4H_9Br
2. Cyclohexane, C_6H_{12}

x_1	H^E/J mol^{-1}
0.1	226
0.2	386
0.3	487
0.4	533
0.5	532
0.6	487
0.7	406
0.8	294
0.9	157

$A(1)$ = 0.2127E+04
$A(2)$ = -0.4810E+03
$A(3)$ = 0.0000E+00

System Number: 129 T/K = 298.15
1. 1-Chlorobutane, C_4H_9Cl
2. Benzene, C_6H_6

x_1	H^E/J mol^{-1}
0.1	26
0.2	43
0.3	48
0.4	42
0.5	30
0.6	16
0.7	5
0.8	0
0.9	0

$A(1)$ = 0.1187E+03 $A(4)$ = 0.1620E+03
$A(2)$ = -0.2820E+03 $A(5)$ = 0.0000E+00
$A(3)$ = 0.4000E+02 $A(6)$ = 0.0000E+00

System Number: 133 T/K = 298.15
1. 1-Chlorobutane, C_4H_9Cl
2. Cyclohexane, C_6H_{12}

x_1	H^E/J mol^{-1}
0.1	247
0.2	442
0.3	558
0.4	595
0.5	569
0.6	502
0.7	415
0.8	317
0.9	193

$A(1)$ = 0.2275E+04 $A(4)$ = 0.9900E+03
$A(2)$ = -0.1007E+04 $A(5)$ = 0.0000E+00
$A(3)$ = 0.2660E+03 $A(6)$ = 0.0000E+00

System Number: 25 T/K = 298.15
1. 1-Chlorobutane, C_4H_9Cl
2. Heptane, C_7H_{16}

x_1	H^E/J mol^{-1}
0.1	158
0.2	300
0.3	411
0.4	481
0.5	505
0.6	479
0.7	407
0.8	295
0.9	154

$A(1)$ = 0.2018E+04
$A(2)$ = -0.2417E+02
$A(3)$ = -0.4444E+03

System Number: 159 T/K = 298.15
1. Pyrrolidine, C_4H_9N
2. Cyclohexane, C_6H_{12}

x_1	H^E/J mol^{-1}
0.1	559
0.2	905
0.3	1104
0.4	1195
0.5	1197
0.6	1116
0.7	951
0.8	703
0.9	379

$A(1)$ = 0.4787E+04 $A(4)$ = -0.6991E+03
$A(2)$ = -0.7996E+03 $A(5)$ = 0.0000E+00
$A(3)$ = 0.6599E+03 $A(6)$ = 0.0000E+00

System Number: 226 T/K = 288.15
1. 1-Butanol, $C_4H_{10}O$
2. Cyclohexane, C_6H_{12}

x_1	H^E/J mol^{-1}
0.1	334
0.2	479
0.3	531
0.4	533
0.5	498
0.6	435
0.7	347
0.8	244
0.9	130

$A(1)$ = 0.1994E+04 $A(4)$ = -0.6538E+03
$A(2)$ = -0.9916E+03 $A(5)$ = 0.6333E+03
$A(3)$ = 0.5067E+03 $A(6)$ = 0.0000E+00

TABLE 3.3.1 (HETX)
EXCESS ENTHALPIES OF BINARY LIQUID MIXTURES (continued)

System Number: 225 **T/K = 303.15**
1. 1-Butanol, $C_4H_{10}O$
2. Heptane, C_7H_{16}

x_1	H^E/J mol^{-1}
0.1	489
0.2	645
0.3	683
0.4	679
0.5	643
0.6	564
0.7	446
0.8	307
0.9	165

$A(1)$ = 0.2571E+04 $A(4)$ = -0.1742E+04
$A(2)$ = -0.1130E+04 $A(5)$ = 0.1927E+04
$A(3)$ = 0.4262E+03 $A(6)$ = 0.0000E+00

System Number: 26 **T/K = 303.15**
1. Diethylamine, $C_4H_{11}N$
2. Benzene, C_6H_6

x_1	H^E/J mol^{-1}
0.1	123
0.2	214
0.3	273
0.4	304
0.5	308
0.6	288
0.7	245
0.8	181
0.9	99

$A(1)$ = 0.1233E+04
$A(2)$ = -0.1705E+03
$A(3)$ = 0.0000E+00

System Number: 171 **T/K = 298.15**
1. Pyridine, C_5H_5N
2. Cyclohexane, C_6H_{12}

x_1	H^E/J mol^{-1}
0.1	562
0.2	955
0.3	1204
0.4	1330
0.5	1348
0.6	1269
0.7	1098
0.8	833
0.9	471

$A(1)$ = 0.5393E+04
$A(2)$ = -0.6340E+03
$A(3)$ = 0.5450E+03

System Number: 79 **T/K = 298.15**
1. Cyclopentane, C_5H_{10}
2. 2,3-Dimethylbutane, C_6H_{14}

x_1	H^E/J mol^{-1}
0.1	-1.8
0.2	-2.7
0.3	-2.8
0.4	-2.5
0.5	-1.8
0.6	-0.9
0.7	-0.1
0.8	0.4
0.9	0.5

$A(1)$ = -0.7023E+01
$A(2)$ = 0.1613E+02
$A(3)$ = 0.0000E+00

System Number: 215 **T/K = 298.15**
1. Oxane, $C_5H_{10}O$
2. Cyclohexane, C_6H_{12}

x_1	H^E/J mol^{-1}
0.1	177
0.2	311
0.3	403
0.4	455
0.5	468
0.6	444
0.7	384
0.8	289
0.9	160

$A(1)$ = 0.1874E+04
$A(2)$ = -0.1155E+03
$A(3)$ = 0.0000E+00

System Number: 27 **T/K = 298.15**
1. Piperidine, $C_5H_{11}N$
2. Cyclohexane, C_6H_{12}

x_1	H^E/J mol^{-1}
0.1	380
0.2	633
0.3	780
0.4	840
0.5	827
0.6	755
0.7	632
0.8	464
0.9	254

$A(1)$ = 0.3309E+04
$A(2)$ = -0.8792E+03
$A(3)$ = 0.3360E+03

System Number: 160 **T/K = 298.15**
1. *N*-Methylpyrrolidine, $C_5H_{11}N$
2. Cyclohexane, C_6H_{12}

x_1	H^E/J mol^{-1}
0.1	105.3
0.2	178.4
0.3	222.8
0.4	241.5
0.5	238.0
0.6	215.4
0.7	177.1
0.8	126.2
0.9	66.1

$A(1)$ = 0.9520E+03
$A(2)$ = -0.2720E+03
$A(3)$ = 0.0000E+00

System Number: 205 **T/K = 298.15**
1. Butyl methyl ether, $C_5H_{12}O$
2. Heptane, C_7H_{16}

x_1	H^E/J mol^{-1}
0.1	112.1
0.2	199.6
0.3	260.3
0.4	295.5
0.5	307.0
0.6	295.1
0.7	258.5
0.8	195.6
0.9	107.1

$A(1)$ = 0.1228E+04 $A(4)$ = -0.4991E+02
$A(2)$ = -0.2815E+01 $A(5)$ = -0.1246E+03
$A(3)$ = 0.6406E+02 $A(6)$ = 0.0000E+00

System Number: 385 **T/K = 298.15**
1. Hexafluorobenzene, C_6F_6
2. Triethylamine, $C_6H_{15}N$

x_1	H^E/J mol^{-1}
0.1	233
0.2	414
0.3	544
0.4	622
0.5	648
0.6	622
0.7	544
0.8	414
0.9	233

$A(1)$ = 0.2590E+04
$A(2)$ = 0.0000E+00
$A(3)$ = 0.0000E+00

TABLE 3.3.1 (HETX)
EXCESS ENTHALPIES OF BINARY LIQUID MIXTURES (continued)

System Number: 109 **T/K = 298.15**
1. Bromobenzene, C_6H_5Br
2. Benzene, C_6H_6

x_1	H^E/J mol^{-1}
0.1	11.3
0.2	19.8
0.3	25.6
0.4	28.8
0.5	29.7
0.6	28.3
0.7	24.6
0.8	18.7
0.9	10.5

$A(1)$ = 0.1187E+03
$A(2)$ = -0.5850E+01
$A(3)$ = 0.4080E+01

System Number: 107 **T/K = 298.15**
1. Bromobenzene, C_6H_5Br
2. Cyclohexane, C_6H_{12}

x_1	H^E/J mol^{-1}
0.1	312
0.2	535
0.3	677
0.4	748
0.5	757
0.6	707
0.7	605
0.8	452
0.9	250

$A(1)$ = 0.3026E+04
$A(2)$ = -0.4316E+03
$A(3)$ = 0.1576E+03

System Number: 28 **T/K = 298.10**
1. Fluorobenzene, C_6H_5F
2. Benzene, C_6H_6

x_1	H^E/J mol^{-1}
0.1	-3.7
0.2	-4.1
0.3	-2.6
0.4	0.0
0.5	2.8
0.6	5.0
0.7	6.2
0.8	5.9
0.9	3.8

$A(1)$ = 0.1123E+02
$A(2)$ = 0.5211E+02
$A(3)$ = -0.1625E+02

System Number: 29 **T/K = 293.15**
1. Nitrobenzene, $C_6H_5NO_2$
2. Benzene, C_6H_6

x_1	H^E/J mol^{-1}
0.1	102.0
0.2	179.5
0.3	231.8
0.4	258.8
0.5	261.6
0.6	241.8
0.7	201.9
0.8	145.4
0.9	76.4

$A(1)$ = 0.1046E+04
$A(2)$ = -0.1777E+03
$A(3)$ = -0.8604E+02

System Number: 61 **T/K = 298.15**
1. Benzene, C_6H_6
2. 1-Hexyne, C_6H_{10}

x_1	H^E/J mol^{-1}
0.1	37.4
0.2	70.5
0.3	98.2
0.4	119.2
0.5	132.0
0.6	134.8
0.7	125.4
0.8	101.6
0.9	60.8

$A(1)$ = 0.5279E+03
$A(2)$ = 0.1622E+03
$A(3)$ = 0.2724E+02

System Number: 9 **T/K = 298.15**
1. Benzene, C_6H_6
2. Cyclohexane, C_6H_{12}

x_1	H^E/J mol^{-1}
0.1	281.3
0.2	502.2
0.3	662.6
0.4	762.2
0.5	800.8
0.6	777.6
0.7	690.4
0.8	535.7
0.9	308.2

$A(1)$ = 0.3203E+04 $A(4)$ = 0.4519E+02
$A(2)$ = 0.1581E+03 $A(5)$ = 0.0000E+00
$A(3)$ = 0.1119E+03 $A(6)$ = 0.0000E+00

System Number: 94 **T/K = 298.15**
1. Benzene, C_6H_6
2. 1-Hexene, C_6H_{12}

x_1	H^E/J mol^{-1}
0.1	203.6
0.2	360.4
0.3	475.7
0.4	552.2
0.5	589.8
0.6	585.7
0.7	534.3
0.8	427.4
0.9	253.8

$A(1)$ = 0.2359E+04
$A(2)$ = 0.3489E+03
$A(3)$ = 0.2843E+03

System Number: 91 **T/K = 298.15**
1. Benzene, C_6H_6
2. Hexane, C_6H_{14}

x_1	H^E/J mol^{-1}
0.1	284
0.2	522
0.3	708
0.4	835
0.5	897
0.6	887
0.7	799
0.8	627
0.9	362

$A(1)$ = 0.3588E+04
$A(2)$ = 0.5461E+03
$A(3)$ = 0.0000E+00

System Number: 32 **T/K = 303.15**
1. Benzene, C_6H_6
2. Triethylamine, $C_6H_{15}N$

x_1	H^E/J mol^{-1}
0.1	96
0.2	184
0.3	258
0.4	314
0.5	347
0.6	353
0.7	326
0.8	261
0.9	154

$A(1)$ = 0.1389E+04
$A(2)$ = 0.4027E+03
$A(3)$ = 0.0000E+00

TABLE 3.3.1 (HETX)
EXCESS ENTHALPIES OF BINARY LIQUID MIXTURES (continued)

System Number: 59 **T/K = 298.15**
1. Benzene, C_6H_6
2. 1-Heptene, C_7H_{14}

x_1	H^E/J mol^{-1}
0.1	181.9
0.2	339.9
0.3	465.6
0.4	554.4
0.5	603.2
0.6	608.3
0.7	564.1
0.8	460.4
0.9	280.6

$A(1)$ = 0.2413E+04 $A(4)$ = 0.2083E+03
$A(2)$ = 0.5528E+03 $A(5)$ = 0.0000E+00
$A(3)$ = 0.2450E+03 $A(6)$ = 0.0000E+00

System Number: 33 **T/K = 298.15**
1. Benzene, C_6H_6
2. Heptane, C_7H_{16}

x_1	H^E/J mol^{-1}
0.1	276
0.2	523
0.3	722
0.4	861
0.5	933
0.6	935
0.7	858
0.8	692
0.9	417

$A(1)$ = 0.3733E+04 $A(4)$ = 0.3447E+03
$A(2)$ = 0.7590E+03 $A(5)$ = 0.0000E+00
$A(3)$ = 0.1803E+03 $A(6)$ = 0.0000E+00

System Number: 92 **T/K = 298.15**
1. Benzene, C_6H_6
2. 1,4-Dimethylbenzene, C_8H_{10}

x_1	H^E/J mol^{-1}
0.1	54.2
0.2	98.4
0.3	132.3
0.4	155.4
0.5	166.9
0.6	165.7
0.7	150.2
0.8	118.9
0.9	69.6

$A(1)$ = 0.6676E+03
$A(2)$ = 0.1069E+03
$A(3)$ = 0.3148E+02

System Number: 34 **T/K = 298.15**
1. Benzene, C_6H_6
2. 1-Octene, C_8H_{16}

x_1	H^E/J mol^{-1}
0.1	192.0
0.2	351.7
0.3	481.1
0.4	578.8
0.5	640.0
0.6	656.5
0.7	617.2
0.8	507.2
0.9	308.6

$A(1)$ = 0.2560E+04
$A(2)$ = 0.8099E+03
$A(3)$ = 0.3452E+03

System Number: 35 **T/K = 298.15**
1. Benzene, C_6H_6
2. 2,2,4-Trimethylpentane, C_8H_{18}

x_1	H^E/J mol^{-1}
0.1	293
0.2	554
0.3	765
0.4	914
0.5	995
0.6	1002
0.7	927
0.8	753
0.9	457

$A(1)$ = 0.3980E+04 $A(4)$ = 0.3716E+03
$A(2)$ = 0.9026E+03 $A(5)$ = 0.0000E+00
$A(3)$ = 0.2983E+03 $A(6)$ = 0.0000E+00

System Number: 211 **T/K = 298.15**
1. Benzene, C_6H_6
2. Dibutyl ether, $C_8H_{18}O$

x_1	H^E/J mol^{-1}
0.1	83.7
0.2	160.1
0.3	226.7
0.4	281.1
0.5	319.5
0.6	337.0
0.7	326.2
0.8	276.5
0.9	174.0

$A(1)$ = 0.1278E+04 $A(4)$ = 0.7440E+02
$A(2)$ = 0.5798E+03 $A(5)$ = 0.0000E+00
$A(3)$ = 0.2395E+03 $A(6)$ = 0.0000E+00

System Number: 36 **T/K = 303.15**
1. Benzene, C_6H_6
2. Tributyl phosphate, $C_{12}H_{27}O_4P$

x_1	H^E/J mol^{-1}
0.1	-56
0.2	-109
0.3	-159
0.4	-204
0.5	-239
0.6	-258
0.7	-254
0.8	-217
0.9	-137

$A(1)$ = -0.9557E+03
$A(2)$ = -0.5631E+03
$A(3)$ = -0.1775E+03

System Number: 53 **T/K = 298.15**
1. Benzene, C_6H_6
2. Hexadecane, $C_{16}H_{34}$

x_1	H^E/J mol^{-1}
0.1	313
0.2	614
0.3	871
0.4	1073
0.5	1215
0.6	1291
0.7	1277
0.8	1124
0.9	744

$A(1)$ = 0.4861E+04 $A(4)$ = 0.1197E+04
$A(2)$ = 0.2227E+04 $A(5)$ = 0.0000E+00
$A(3)$ = 0.1584E+04 $A(6)$ = 0.0000E+00

System Number: 58 **T/K = 298.15**
1. 1-Hexyne, C_6H_{10}
2. Cyclohexane, C_6H_{12}

x_1	H^E/J mol^{-1}
0.1	310.5
0.2	524.5
0.3	657.2
0.4	721.2
0.5	725.9
0.6	678.3
0.7	582.2
0.8	438.7
0.9	246.2

$A(1)$ = 0.2904E+04
$A(2)$ = -0.4465E+03
$A(3)$ = 0.2952E+03

TABLE 3.3.1 (HETX)
EXCESS ENTHALPIES OF BINARY LIQUID MIXTURES (continued)

System Number: 87 T/K = 298.15
1. 1-Hexyne, C_6H_{10}
2. Heptane, C_7H_{16}

x_1	H^E/J mol^{-1}
0.1	229.5
0.2	409.2
0.3	538.7
0.4	617.6
0.5	645.2
0.6	621.3
0.7	545.2
0.8	416.7
0.9	235.1

$A(1)$ = 0.2581E+04
$A(2)$ = 0.3868E+02
$A(3)$ = 0.0000E+00

System Number: 161 T/K = 298.15
1. Cyclohexane, C_6H_{12}
2. *N*-Methylpiperidine, $C_6H_{13}N$

x_1	H^E/J mol^{-1}
0.1	37
0.2	75
0.3	109
0.4	134
0.5	147
0.6	144
0.7	127
0.8	95
0.9	52

$A(1)$ = 0.5869E+03
$A(2)$ = 0.1048E+03
$A(3)$ = -0.1502E+03

System Number: 40 T/K = 298.15
1. Cyclohexane, C_6H_{12}
2. Hexane, C_6H_{14}

x_1	H^E/J mol^{-1}
0.1	64.4
0.2	119.2
0.3	163.5
0.4	196.2
0.5	215.9
0.6	220.2
0.7	206.2
0.8	169.1
0.9	103.1

$A(1)$ = 0.8635E+03 $A(4)$ = 0.3110E+02
$A(2)$ = 0.2488E+03 $A(5)$ = 0.0000E+00
$A(3)$ = 0.1045E+03 $A(6)$ = 0.0000E+00

System Number: 78 T/K = 298.15
1. Cyclohexane, C_6H_{12}
2. 2,3-Dimethylbutane, C_6H_{14}

x_1	H^E/J mol^{-1}
0.1	52.8
0.2	95.6
0.3	127.6
0.4	148.0
0.5	156.1
0.6	151.5
0.7	133.7
0.8	102.6
0.9	58.0

$A(1)$ = 0.6245E+03
$A(2)$ = 0.3639E+02
$A(3)$ = -0.1432E+02

System Number: 57 T/K = 298.15
1. 1-Hexene, C_6H_{12}
2. Hexane, C_6H_{14}

x_1	H^E/J mol^{-1}
0.1	21.3
0.2	37.7
0.3	49.6
0.4	57.2
0.5	60.5
0.6	59.4
0.7	53.5
0.8	42.2
0.9	24.7

$A(1)$ = 0.2421E+03
$A(2)$ = 0.2333E+02
$A(3)$ = 0.2079E+02

System Number: 264 T/K = 298.15
1. Cyclohexane, C_6H_{12}
2. Benzaldehyde, C_7H_6O

x_1	H^E/J mol^{-1}
0.1	480
0.2	887
0.3	1164
0.4	1312
0.5	1358
0.6	1335
0.7	1256
0.8	1089
0.9	735

$A(1)$ = 0.5433E+04 $A(4)$ = 0.2549E+04
$A(2)$ = 0.1360E+03 $A(5)$ = 0.0000E+00
$A(3)$ = 0.2061E+04 $A(6)$ = 0.0000E+00

System Number: 38 T/K = 298.15
1. Cyclohexane, C_6H_{12}
2. Toluene, C_7H_8

x_1	H^E/J mol^{-1}
0.1	209.9
0.2	389.1
0.3	518.0
0.4	596.7
0.5	629.8
0.6	617.9
0.7	554.8
0.8	430.5
0.9	240.4

$A(1)$ = 0.2519E+04 $A(4)$ = -0.1563E+02
$A(2)$ = 0.2213E+03 $A(5)$ = -0.5138E+03
$A(3)$ = 0.3013E+03 $A(6)$ = 0.0000E+00

System Number: 80 T/K = 298.15
1. Cyclohexane, C_6H_{12}
2. Heptane, C_7H_{16}

x_1	H^E/J mol^{-1}
0.1	74.7
0.2	135.3
0.3	183.4
0.4	219.0
0.5	240.6
0.6	245.6
0.7	230.1
0.8	188.7
0.9	114.7

$A(1)$ = 0.9623E+03
$A(2)$ = 0.2780E+03
$A(3)$ = 0.1400E+03

System Number: 60 T/K = 298.15
1. Cyclohexane, C_6H_{12}
2. *cis*-Bicyclo[4.4.0]decane, $C_{10}H_{18}$

x_1	H^E/J mol^{-1}
0.1	20.7
0.2	33.8
0.3	38.6
0.4	36.0
0.5	28.0
0.6	17.3
0.7	6.7
0.8	-1.0
0.9	-3.6

$A(1)$ = 0.1120E+03 $A(4)$ = 0.4315E+02
$A(2)$ = -0.1967E+03 $A(5)$ = 0.0000E+00
$A(3)$ = -0.2655E+02 $A(6)$ = 0.0000E+00

TABLE 3.3.1 (HETX)
EXCESS ENTHALPIES OF BINARY LIQUID MIXTURES (continued)

System Number: 62 **T/K = 298.15**
1. Cyclohexane, C_6H_{12}
2. *trans*-Bicyclo[4.4.0]decane, $C_{10}H_{18}$

x_1	H^E/J mol^{-1}
0.1	9.9
0.2	17.9
0.3	22.7
0.4	23.3
0.5	19.6
0.6	12.4
0.7	3.5
0.8	-4.3
0.9	-7.2

$A(1)$ = 0.7837E+02 $A(4)$ = -0.1008E+02
$A(2)$ = -0.1124E+03 $A(5)$ = 0.0000E+00
$A(3)$ = -0.9967E+02 $A(6)$ = 0.0000E+00

System Number: 54 **T/K = 298.15**
1. Cyclohexane, C_6H_{12}
2. Hexadecane, $C_{16}H_{34}$

x_1	H^E/J mol^{-1}
0.1	151.9
0.2	274.2
0.3	373.4
0.4	451.4
0.5	505.5
0.6	528.2
0.7	507.8
0.8	427.8
0.9	267.0

$A(1)$ = 0.2022E+04
$A(2)$ = 0.7998E+03
$A(3)$ = 0.4771E+03

System Number: 270 **T/K = 298.15**
1. Butyl ethanoate, $C_6H_{12}O_2$
2. Heptane, C_7H_{16}

x_1	H^E/J mol^{-1}
0.1	436
0.2	715
0.3	882
0.4	966
0.5	980
0.6	928
0.7	806
0.8	610
0.9	339

$A(1)$ = 0.3921E+04 $A(4)$ = -0.4600E+03
$A(2)$ = -0.3810E+03 $A(5)$ = 0.0000E+00
$A(3)$ = 0.6050E+03 $A(6)$ = 0.0000E+00

System Number: 210 **T/K = 298.15**
1. Hexane, C_6H_{14}
2. Dibutyl ether, $C_8H_{18}O$

x_1	H^E/J mol^{-1}
0.1	30.6
0.2	55.6
0.3	74.6
0.4	87.3
0.5	93.2
0.6	91.8
0.7	82.5
0.8	64.7
0.9	37.4

$A(1)$ = 0.3729E+03
$A(2)$ = 0.4710E+02
$A(3)$ = 0.8200E+01

System Number: 77 **T/K = 298.15**
1. Hexane, C_6H_{14}
2. Decane, $C_{10}H_{22}$

x_1	H^E/J mol^{-1}
0.1	5.6
0.2	9.9
0.3	12.8
0.4	14.2
0.5	14.4
0.6	13.2
0.7	11.0
0.8	7.9
0.9	4.1

$A(1)$ = 0.5742E+02
$A(2)$ = -0.1069E+02
$A(3)$ = -0.5020E+01

System Number: 37 **T/K = 303.15**
1. Hexane, C_6H_{14}
2. Tributyl phosphate, $C_{12}H_{27}O_4P$

x_1	H^E/J mol^{-1}
0.1	210
0.2	432
0.3	606
0.4	713
0.5	762
0.6	771
0.7	745
0.8	663
0.9	458

$A(1)$ = 0.3050E+04 $A(4)$ = 0.1867E+04
$A(2)$ = 0.5292E+03 $A(5)$ = 0.0000E+00
$A(3)$ = 0.1030E+04 $A(6)$ = 0.0000E+00

System Number: 63 **T/K = 293.15**
1. Hexane, C_6H_{14}
2. Hexadecane, $C_{16}H_{34}$

x_1	H^E/J mol^{-1}
0.1	44.9
0.2	80.7
0.3	107.2
0.4	123.9
0.5	130.6
0.6	126.8
0.7	112.2
0.8	86.4
0.9	49.2

$A(1)$ = 0.5224E+03
$A(2)$ = 0.2986E+02
$A(3)$ = 0.0000E+00

System Number: 103 **T/K = 298.15**
1. Dipropyl ether, $C_6H_{14}O$
2. Heptane, C_7H_{16}

x_1	H^E/J mol^{-1}
0.1	75.1
0.2	132.5
0.3	172.8
0.4	196.7
0.5	204.3
0.6	195.8
0.7	171.3
0.8	130.7
0.9	73.8

$A(1)$ = 0.8170E+03
$A(2)$ = -0.9000E+01
$A(3)$ = 0.1520E+02

System Number: 213 **T/K = 298.15**
1. 3,6-Dioxaoctane, $C_6H_{14}O_2$
2. Heptane, C_7H_{16}

x_1	H^E/J mol^{-1}
0.1	377.9
0.2	631.7
0.3	790.6
0.4	873.1
0.5	889.1
0.6	842.4
0.7	732.4
0.8	556.6
0.9	312.5

$A(1)$ = 0.3557E+04 $A(4)$ = -0.2230E+03
$A(2)$ = -0.3110E+03 $A(5)$ = 0.0000E+00
$A(3)$ = 0.4360E+03 $A(6)$ = 0.0000E+00

TABLE 3.3.1 (HETX)
EXCESS ENTHALPIES OF BINARY LIQUID MIXTURES (continued)

System Number: 204 **T/K = 298.15**
1. Phosphoric tris(dimethylamide), $C_6H_{18}N_3OP$
2. Toluene, C_7H_8

x_1	H^E/J mol^{-1}
0.1	-185
0.2	-346
0.3	-475
0.4	-563
0.5	-603
0.6	-592
0.7	-525
0.8	-404
0.9	-227

$A(1)$ = -0.2413E+04
$A(2)$ = -0.2980E+03
$A(3)$ = 0.1930E+03

System Number: 203 **T/K = 298.15**
1. Phosphoric tris(dimethylamide), $C_6H_{18}N_3OP$
2. Heptane, C_7H_{16}

x_1	H^E/J mol^{-1}
0.1	644
0.2	974
0.3	1156
0.4	1240
0.5	1238
0.6	1156
0.7	997
0.8	756
0.9	422

$A(1)$ = 0.4953E+04 $A(4)$ = -0.4080E+03
$A(2)$ = -0.8560E+03 $A(5)$ = 0.9080E+03
$A(3)$ = 0.9310E+03 $A(6)$ = -0.1046E+04

System Number: 323 **T/K = 298.15**
1. 2-Heptanone, $C_7H_{14}O$
2. Dibutyl ether, $C_8H_{18}O$

x_1	H^E/J mol^{-1}
0.1	222
0.2	361
0.3	441
0.4	477
0.5	478
0.6	445
0.7	378
0.8	278
0.9	149

$A(1)$ = 0.1912E+04 $A(4)$ = -0.2860E+03
$A(2)$ = -0.3280E+03 $A(5)$ = 0.0000E+00
$A(3)$ = 0.2370E+03 $A(6)$ = 0.0000E+00

TABLE 3.3.2 (CPEX)
ISOBARIC EXCESS HEAT CAPACITIES OF BINARY LIQUID MIXTURES

The *molar isobaric (constant-pressure) heat capacity* of a binary mixture, C_p, equals the heat capacity divided by the total amount of substance (number of moles) of the two components. It is a function of the pressure P, the temperature T, and the composition of the mixture. The latter is usually expressed in terms of the mole fractions x_1 and $x_2 = 1 - x_1$ of components 1 and 2, respectively.

The *excess molar heat capacity* C_p^{E} is defined as

$$C_p^{\mathrm{E}} = C_p - \left(x_1 C_{p1} + x_2 C_{p2}\right) \tag{1}$$

where C_{p1} and C_{p2} are the molar heat capacities of the pure liquid components (Table 2.3.2).

The *specific heat capacity* of a mixture, c_p, is defined similarly to the specific heat capacity of a pure substance (Section 2.3), $c_p = C_p/M$, where $M = x_1 M_1 + x_2 M_2$ is the average molar mass of the mixture.

The excess molar heat capacity is a measure of the temperature dependence of the molar excess enthalpy (Table 3.3.1) through the equation:

$$C_p^{\mathrm{E}} = \left[\left(\partial H^{\mathrm{E}} / \partial T\right)\right]_P \tag{2}$$

Most C_p^{E} measurements have been reported at atmospheric pressure and room temperature. The excess molar heat capacity may be positive or negative in the whole composition range, or may change sign. S-shaped and W-shaped C_p^{E} curves are often observed in polar systems.

The C_p^{E} curves can be represented as a function of x_1 by various empirical equations. A simple equation is the so-called Redlich-Kister equation:

$$C_p^{\mathrm{E}} / \mathrm{J\,K^{-1}\,mol^{-1}} = x_1 x_2 \sum_{i=1}^{n} A(i)\left(x_1 - x_2\right)^{i-1} \tag{3}$$

The table gives the coefficients $A(i)$, at the temperature T, obtained by fitting Equation 3 to experimental $C_p^{\mathrm{E}}(x_1)$ data. The calculated values of C_p^{E} for increments of $x_1 = 0.1$ are also listed.

The program, Property Code CPEX, permits the calculation of C_p^{E}, C_p, and c_p at selected increments of x_1 in the whole range of composition, from $x_1 = 0$ to $x_1 = 1$, or at any given value of x_1. The auxiliary values displayed by the program have the following meaning: $V(1) = C_{p1}$; $V(2) = C_{p2}$.

The components of each system, and the systems themselves, are arranged in a modified Hill order, with substances that do not contain carbon preceding those that do contain carbon.

Physical quantity	**Symbol**	**SI unit**
Mole fraction of component i	x_i	—
Temperature	T	K
Excess molar heat capacity	C_p^{E}	$\mathrm{J\,K^{-1}\,mol^{-1}}$
Molar heat capacity	C_p	$\mathrm{J\,K^{-1}\,mol^{-1}}$
Specific heat capacity	c_p	$\mathrm{J\,K^{-1}\,kg^{-1}}$

REFERENCES

1. D'Ans, J.; Surawski, H.; Synowietz, C., *Landolt-Börnstein Numerical Data and Functional Relationships in Science and Technology*, Vol. IV/1b, Densities of Binary Aqueous Systems and Heat Capacities, Springer-Verlag, Berlin-Heidelberg, 1976 (a compilation of numerical data).
2. *Int. DATA Ser., Sel. Data Mixtures, Ser. A* (Ed. Kehiaian, H. V.), Thermodynamic Research Center, Texas Engineering Experiment Station, College Station, USA, 1973-1993 (21 volumes) (primary excess molar heat capacity data and correlations)

TABLE 3.3.2 (CPEX)
ISOBARIC EXCESS HEAT CAPACITIES OF BINARY LIQUID MIXTURES (continued)

System Number: 1 **T/K = 298.15**
1. Water, H_2O
2. Methanol, CH_4O
$C_p(1)$/J K^{-1}mol^{-1} = 75.30
$C_p(2)$/J K^{-1}mol^{-1} = 81.11

x_1	C_p^E/J K^{-1}mol^{-1}
0.1	1.20
0.2	2.47
0.3	3.47
0.4	4.24
0.5	4.95
0.6	5.75
0.7	6.53
0.8	6.76
0.9	5.25

$A(1)$ = 0.1980E+02 $A(4)$ = 0.2070E+02
$A(2)$ = 0.1490E+02 $A(5)$ = 0.0000E+00
$A(3)$ = 0.2510E+02 $A(6)$ = 0.0000E+00

System Number: 48 **T/K = 298.15**
1. Tetrachloromethane, CCl_4
2. Benzene, C_6H_6
$C_p(1)$/J K^{-1}mol^{-1} = 130.66
$C_p(2)$/J K^{-1}mol^{-1} = 136.30

x_1	C_p^E/J K^{-1}mol^{-1}
0.1	0.40
0.2	0.72
0.3	0.96
0.4	1.11
0.5	1.18
0.6	1.15
0.7	1.03
0.8	0.80
0.9	0.45

$A(1)$ = 0.4721E+01
$A(2)$ = 0.4150E+00
$A(3)$ = 0.0000E+00

System Number: 75 **T/K = 298.15**
1. Tetrachloromethane, CCl_4
2. Cyclohexane, C_6H_{12}
$C_p(1)$/J K^{-1}mol^{-1} = 130.66
$C_p(2)$/J K^{-1}mol^{-1} = 154.85

x_1	C_p^E/J K^{-1}mol^{-1}
0.1	-0.21
0.2	-0.38
0.3	-0.49
0.4	-0.56
0.5	-0.59
0.6	-0.58
0.7	-0.53
0.8	-0.42
0.9	-0.25

$A(1)$ = -0.2377E+01
$A(2)$ = -0.2290E+00
$A(3)$ = -0.2940E+00

System Number: 13 **T/K = 298.15**
1. Tetrachloromethane, CCl_4
2. Heptane, C_7H_{16}
$C_p(1)$/J K^{-1}mol^{-1} = 130.66
$C_p(2)$/J K^{-1}mol^{-1} = 224.75

x_1	C_p^E/J K^{-1}mol^{-1}
0.1	-0.471
0.2	-0.865
0.3	-1.181
0.4	-1.411
0.5	-1.545
0.6	-1.565
0.7	-1.450
0.8	-1.173
0.9	-0.702

$A(1)$ = -0.6179E+01
$A(2)$ = -0.1601E+01
$A(3)$ = -0.5262E+00

System Number: 401 **T/K = 298.15**
1. Trichloromethane, $CHCl_3$
2. Ethanenitrile, C_2H_3N
$C_p(1)$/J K^{-1}mol^{-1} = 114.16
$C_p(2)$/J K^{-1}mol^{-1} = 91.46

x_1	C_p^E/J K^{-1}mol^{-1}
0.1	1.068
0.2	2.082
0.3	3.006
0.4	3.784
0.5	4.343
0.6	4.591
0.7	4.419
0.8	3.697
0.9	2.279

$A(1)$ = 0.1737E+02
$A(2)$ = 0.8411E+01
$A(3)$ = 0.1908E+01

System Number: 71 **T/K = 298.15**
1. Trichloromethane, $CHCl_3$
2. Oxolane, C_4H_8O
$C_p(1)$/J K^{-1}mol^{-1} = 114.16
$C_p(2)$/J K^{-1}mol^{-1} = 124.00

x_1	C_p^E/J K^{-1}mol^{-1}
0.1	1.35
0.2	2.77
0.3	4.62
0.4	6.41
0.5	7.51
0.6	7.53
0.7	6.49
0.8	4.73
0.9	2.62

$A(1)$ = 0.3005E+02 $A(4)$ = -0.4850E+01
$A(2)$ = 0.1191E+02 $A(5)$ = 0.2101E+02
$A(3)$ = -0.2593E+02 $A(6)$ = 0.0000E+00

System Number: 52 **T/K = 298.15**
1. Trichloromethane, $CHCl_3$
2. 1,4-Dioxane, $C_4H_8O_2$
$C_p(1)$/J K^{-1}mol^{-1} = 114.16
$C_p(2)$/J K^{-1}mol^{-1} = 152.14

x_1	C_p^E/J K^{-1}mol^{-1}
0.1	0.502
0.2	1.147
0.3	2.168
0.4	3.485
0.5	4.807
0.6	5.743
0.7	5.901
0.8	4.994
0.9	2.947

$A(1)$ = 0.1923E+02 $A(4)$ = -0.1090E+02
$A(2)$ = 0.2396E+02 $A(5)$ = 0.0000E+00
$A(3)$ = -0.1100E+00 $A(6)$ = 0.0000E+00

System Number: 501 **T/K = 298.15**
1. Ethanenitrile, C_2H_3N
2. Dimethyl sulfoxide, C_2H_6OS
$C_p(1)$/J K^{-1}mol^{-1} = 91.46
$C_p(2)$/J K^{-1}mol^{-1} = 148.25

x_1	C_p^E/J K^{-1}mol^{-1}
0.1	-0.32
0.2	-0.55
0.3	-0.70
0.4	-0.78
0.5	-0.82
0.6	-0.80
0.7	-0.73
0.8	-0.58
0.9	-0.35

$A(1)$ = -0.3266E+01
$A(2)$ = -0.1799E+00
$A(3)$ = -0.7310E+00

System Number: 167 **T/K = 298.15**
1. Ethanenitrile, C_2H_3N
2. Benzene, C_6H_6
$C_p(1)$/J K^{-1}mol^{-1} = 91.46
$C_p(2)$/J K^{-1}mol^{-1} = 136.30

x_1	C_p^E/J K^{-1}mol^{-1}
0.1	0.35
0.2	0.39
0.3	0.27
0.4	0.09
0.5	-0.09
0.6	-0.24
0.7	-0.33
0.8	-0.33
0.9	-0.24

$A(1)$ = -0.3706E+00 $A(4)$ = -0.1133E+01
$A(2)$ = -0.3377E+01 $A(5)$ = 0.0000E+00
$A(3)$ = 0.1529E+01 $A(6)$ = 0.0000E+00

TABLE 3.3.2 (CPEX)
ISOBARIC EXCESS HEAT CAPACITIES OF BINARY LIQUID MIXTURES (continued)

System Number: 502 T/K = 298.15
1. Ethanenitrile, C_2H_3N
2. Triethylamine, $C_6H_{15}N$
$C_p(1)$/J K^{-1}mol^{-1} = 91.46
$C_p(2)$/J K^{-1}mol^{-1} = 217.10

x_1	C_p^E/J K^{-1}mol^{-1}
0.1	2.58
0.2	4.20
0.3	5.28
0.4	5.90
0.5	6.04
0.6	5.64
0.7	4.72
0.8	3.38
0.9	1.77

$A(1)$ = 0.2416E+02 $A(4)$ = -0.4750E+01
$A(2)$ = -0.2556E+01 $A(5)$ = 0.4643E+01
$A(3)$ = -0.2934E+01 $A(6)$ = 0.0000E+00

System Number: 124 T/K = 298.15
1. 1,2-Dichloroethane, $C_2H_4Cl_2$
2. Heptane, C_7H_{16}
$C_p(1)$/J K^{-1}mol^{-1} = 128.81
$C_p(2)$/J K^{-1}mol^{-1} = 224.75

x_1	C_p^E/J K^{-1}mol^{-1}
0.1	-0.83
0.2	-1.33
0.3	-1.51
0.4	-1.45
0.5	-1.25
0.6	-1.04
0.7	-0.90
0.8	-0.82
0.9	-0.66

$A(1)$ = -0.5009E+01 $A(4)$ = -0.5140E+01
$A(2)$ = 0.4460E+01 $A(5)$ = -0.1170E+01
$A(3)$ = -0.4320E+01 $A(6)$ = 0.0000E+00

System Number: 21 T/K = 298.15
1. Dimethyl sulfoxide, C_2H_6OS
2. Benzene, C_6H_6
$C_p(1)$/J K^{-1}mol^{-1} = 148.25
$C_p(2)$/J K^{-1}mol^{-1} = 136.30

x_1	C_p^E/J K^{-1}mol^{-1}
0.1	0.99
0.2	0.97
0.3	0.58
0.4	0.17
0.5	-0.13
0.6	-0.32
0.7	-0.46
0.8	-0.56
0.9	-0.51

$A(1)$ = -0.5221E+00 $A(4)$ = -0.8786E+01
$A(2)$ = -0.4782E+01 $A(5)$ = 0.0000E+00
$A(3)$ = 0.5053E+01 $A(6)$ = 0.0000E+00

System Number: 400 T/K = 298.15
1. Methyl ethanoate, $C_3H_6O_2$
2. Heptane, C_7H_{16}
$C_p(1)$/J K^{-1}mol^{-1} = 141.90
$C_p(2)$/J K^{-1}mol^{-1} = 224.75

x_1	C_p^E/J K^{-1}mol^{-1}
0.1	-0.37
0.2	-0.29
0.3	0.08
0.4	0.54
0.5	0.91
0.6	1.02
0.7	0.83
0.8	0.40
0.9	-0.03

$A(1)$ = 0.3629E+01 $A(4)$ = -0.4420E+01
$A(2)$ = 0.5180E+01 $A(5)$ = 0.0000E+00
$A(3)$ = -0.9200E+01 $A(6)$ = 0.0000E+00

System Number: 74 T/K = 298.15
1. 2-Butanone, C_4H_8O
2. Heptane, C_7H_{16}
$C_p(1)$/J K^{-1}mol^{-1} = 158.66
$C_p(2)$/J K^{-1}mol^{-1} = 224.75

x_1	C_p^E/J K^{-1}mol^{-1}
0.1	0.224
0.2	0.790
0.3	1.350
0.4	1.710
0.5	1.798
0.6	1.627
0.7	1.270
0.8	0.826
0.9	0.384

$A(1)$ = 0.7190E+01 $A(4)$ = 0.3300E+01
$A(2)$ = -0.1000E+01 $A(5)$ = 0.0000E+00
$A(3)$ = -0.5950E+01 $A(6)$ = 0.0000E+00

System Number: 216 T/K = 298.15
1. 1,4-Dioxane, $C_4H_8O_2$
2. Cyclohexane, C_6H_{12}
$C_p(1)$/J K^{-1}mol^{-1} = 152.14
$C_p(2)$/J K^{-1}mol^{-1} = 154.85

x_1	C_p^E/J K^{-1}mol^{-1}
0.1	-0.83
0.2	-0.95
0.3	-0.78
0.4	-0.59
0.5	-0.50
0.6	-0.52
0.7	-0.58
0.8	-0.58
0.9	-0.42

$A(1)$ = -0.2008E+01 $A(4)$ = 0.3445E+01
$A(2)$ = 0.6782E+00 $A(5)$ = 0.0000E+00
$A(3)$ = -0.7695E+01 $A(6)$ = 0.0000E+00

System Number: 269 T/K = 298.15
1. Ethyl ethanoate $C_4H_8O_2$
2. Cyclohexane, C_6H_{12}
$C_p(1)$/J K^{-1}mol^{-1} = 170.69
$C_p(2)$/J K^{-1}mol^{-1} = 154.85

x_1	C_p^E/J K^{-1}mol^{-1}
0.1	-0.921
0.2	-1.213
0.3	-1.235
0.4	-1.184
0.5	-1.138
0.6	-1.088
0.7	-0.986
0.8	-0.774
0.9	-0.429

$A(1)$ = -0.4550E+01 $A(4)$ = 0.4030E+01
$A(2)$ = 0.8400E+00 $A(5)$ = 0.0000E+00
$A(3)$ = -0.4610E+01 $A(6)$ = 0.0000E+00

System Number: 268 T/K = 298.15
1. Ethyl ethanoate, $C_4H_8O_2$
2. Heptane, C_7H_{16}
$C_p(1)$/J K^{-1}mol^{-1} = 170.69
$C_p(2)$/J K^{-1}mol^{-1} = 224.75

x_1	C_p^E/J K^{-1}mol^{-1}
0.1	-0.21
0.2	-0.32
0.3	-0.21
0.4	-0.02
0.5	0.11
0.6	0.07
0.7	-0.08
0.8	-0.21
0.9	-0.18

$A(1)$ = 0.4300E+00 $A(4)$ = -0.1140E+01
$A(2)$ = 0.9700E+00 $A(5)$ = 0.6240E+01
$A(3)$ = -0.8010E+01 $A(6)$ = 0.0000E+00

System Number: 25 T/K = 298.15
1. 1-Chlorobutane, C_4H_9Cl
2. Heptane, C_7H_{16}
$C_p(1)$/J K^{-1}mol^{-1} = 159.64
$C_p(2)$/J K^{-1}mol^{-1} = 224.75

x_1	C_p^E/J K^{-1}mol^{-1}
0.1	-0.320
0.2	-0.489
0.3	-0.564
0.4	-0.580
0.5	-0.557
0.6	-0.502
0.7	-0.418
0.8	-0.304
0.9	-0.161

$A(1)$ = -0.2226E+01 $A(4)$ = 0.5000E+00
$A(2)$ = 0.7850E+00 $A(5)$ = 0.0000E+00
$A(3)$ = -0.7000E+00 $A(6)$ = 0.0000E+00

TABLE 3.3.2 (CPEX)
ISOBARIC EXCESS HEAT CAPACITIES OF BINARY LIQUID MIXTURES (continued)

System Number: 170 **T/K = 298.15**
1. Pyridine, C_5H_5N
2. Heptane, C_7H_{16}
$C_p(1)$/J K^{-1}mol^{-1} = 132.70
$C_p(2)$/J K^{-1}mol^{-1} = 224.75

x_1	C_p^E/J K^{-1}mol^{-1}
0.1	-0.29
0.2	-0.07
0.3	0.43
0.4	0.94
0.5	1.26
0.6	1.28
0.7	1.03
0.8	0.63
0.9	0.25

$A(1)$ = 0.5059E+01 $A(4)$ = 0.3218E+00
$A(2)$ = 0.3537E+01 $A(5)$ = 0.3703E+01
$A(3)$ = -0.1059E+02 $A(6)$ = 0.0000E+00

System Number: 500 **T/K = 298.15**
1. Piperidine, $C_5H_{11}N$
2. Heptane, C_7H_{16}
$C_p(1)$/J K^{-1}mol^{-1} = 179.86
$C_p(2)$/J K^{-1}mol^{-1} = 224.75

x_1	C_p^E/J K^{-1}mol^{-1}
0.1	-1.83
0.2	-2.70
0.3	-2.80
0.4	-2.47
0.5	-1.94
0.6	-1.38
0.7	-0.83
0.8	-0.30
0.9	0.11

$A(1)$ = -0.7755E+01 $A(4)$ = 0.3576E+01
$A(2)$ = 0.1119E+02 $A(5)$ = 0.5691E+01
$A(3)$ = -0.6525E+01 $A(6)$ = 0.0000E+00

System Number: 9 **T/K = 298.15**
1. Benzene, C_6H_6
2. Cyclohexane, C_6H_{12}
$C_p(1)$/J K^{-1}mol^{-1} = 136.30
$C_p(2)$/J K^{-1}mol^{-1} = 154.85

x_1	C_p^E/J K^{-1}mol^{-1}
0.1	-1.00
0.2	-1.80
0.3	-2.39
0.4	-2.77
0.5	-2.93
0.6	-2.87
0.7	-2.58
0.8	-2.04
0.9	-1.20

$A(1)$ = -0.1173E+02 $A(4)$ = -0.5590E+00
$A(2)$ = -0.1037E+01 $A(5)$ = 0.0000E+00
$A(3)$ = -0.7850E+00 $A(6)$ = 0.0000E+00

System Number: 91 **T/K = 298.20**
1. Benzene, C_6H_6
2. Hexane, C_6H_{14}
$C_p(1)$/J K^{-1}mol^{-1} = 136.30
$C_p(2)$/J K^{-1}mol^{-1} = 195.63

x_1	C_p^E/J K^{-1}mol^{-1}
0.1	-0.97
0.2	-1.70
0.3	-2.24
0.4	-2.63
0.5	-2.86
0.6	-2.91
0.7	-2.74
0.8	-2.26
0.9	-1.39

$A(1)$ = -0.1144E+02
$A(2)$ = -0.2940E+01
$A(3)$ = -0.2600E+01

System Number: 33 **T/K = 298.20**
1. Benzene, C_6H_6
2. Heptane, C_7H_{16}
$C_p(1)$/J K^{-1}mol^{-1} = 136.30
$C_p(2)$/J K^{-1}mol^{-1} = 224.75

x_1	C_p^E/J K^{-1}mol^{-1}
0.1	-1.07
0.2	-1.94
0.3	-2.60
0.4	-3.07
0.5	-3.34
0.6	-3.41
0.7	-3.25
0.8	-2.75
0.9	-1.75

$A(1)$ = -0.1335E+02 $A(4)$ = -0.1930E+01
$A(2)$ = -0.3540E+01 $A(5)$ = 0.0000E+00
$A(3)$ = -0.3600E+01 $A(6)$ = 0.0000E+00

System Number: 92 **T/K = 298.15**
1. Benzene, C_6H_6
2. 1,4-Dimethylbenzene, C_8H_{10}
$C_p(1)$/J K^{-1}mol^{-1} = 136.30
$C_p(2)$/J K^{-1}mol^{-1} = 181.50

x_1	C_p^E/J K^{-1}mol^{-1}
0.1	-0.17
0.2	-0.33
0.3	-0.47
0.4	-0.57
0.5	-0.64
0.6	-0.65
0.7	-0.60
0.8	-0.48
0.9	-0.28

$A(1)$ = -0.2542E+01
$A(2)$ = -0.7650E+00
$A(3)$ = 0.0000E+00

System Number: 53 **T/K = 298.15**
1. Benzene, C_6H_6
2. Hexadecane, $C_{16}H_{34}$
$C_p(1)$/J K^{-1}mol^{-1} = 136.30
$C_p(2)$/J K^{-1}mol^{-1} = 501.60

x_1	C_p^E/J K^{-1}mol^{-1}
0.1	-2.17
0.2	-4.01
0.3	-5.44
0.4	-6.45
0.5	-7.06
0.6	-7.24
0.7	-6.91
0.8	-5.88
0.9	-3.76

$A(1)$ = -0.2823E+02 $A(4)$ = -0.4800E+01
$A(2)$ = -0.8010E+01 $A(5)$ = 0.0000E+00
$A(3)$ = -0.7350E+01 $A(6)$ = 0.0000E+00

System Number: 40 **T/K = 298.20**
1. Cyclohexane, C_6H_{12}
2. Hexane, C_6H_{14}
$C_p(1)$/J K^{-1}mol^{-1} = 154.85
$C_p(2)$/J K^{-1}mol^{-1} = 195.63

x_1	C_p^E/J K^{-1}mol^{-1}
0.1	-0.43
0.2	-0.75
0.3	-0.99
0.4	-1.18
0.5	-1.30
0.6	-1.34
0.7	-1.28
0.8	-1.07
0.9	-0.67

$A(1)$ = -0.5185E+01
$A(2)$ = -0.1680E+01
$A(3)$ = -0.1380E+01

System Number: 54 **T/K = 298.15**
1. Cyclohexane, C_6H_{12}
2. Hexadecane, $C_{16}H_{34}$
$C_p(1)$/J K^{-1}mol^{-1} = 154.85
$C_p(2)$/J K^{-1}mol^{-1} = 501.60

x_1	C_p^E/J K^{-1}mol^{-1}
0.1	-2.26
0.2	-3.99
0.3	-5.27
0.4	-6.14
0.5	-6.60
0.6	-6.61
0.7	-6.10
0.8	-4.93
0.9	-2.97

$A(1)$ = -0.2641E+02
$A(2)$ = -0.4930E+01
$A(3)$ = -0.4050E+01

TABLE 3.3.2 (CPEX)
ISOBARIC EXCESS HEAT CAPACITIES OF BINARY LIQUID MIXTURES (continued)

System Number: 503 **T/K = 298.15**

1. Hexane, C_6H_{14}
2. Toluene, C_7H_8

$C_p(1)$/J K^{-1}mol^{-1} = 195.63
$C_p(2)$/J K^{-1}mol^{-1} = 157.30

x_1	C_p^E/J K^{-1}mol^{-1}
0.1	-0.37
0.2	-0.69
0.3	-0.94
0.4	-1.13
0.5	-1.22
0.6	-1.22
0.7	-1.11
0.8	-0.87
0.9	-0.51

$A(1)$ = -0.4880E+01
$A(2)$ = -0.9600E+00
$A(3)$ = 0.0000E+00

System Number: 103 **T/K = 298.15**

1. Dipropyl ether, $C_6H_{14}O$
2. Heptane, C_7H_{16}

$C_p(1)$/J K^{-1}mol^{-1} = 221.45
$C_p(2)$/J K^{-1}mol^{-1} = 224.75

x_1	C_p^E/J K^{-1}mol^{-1}
0.1	-0.22
0.2	-0.39
0.3	-0.51
0.4	-0.58
0.5	-0.60
0.6	-0.57
0.7	-0.49
0.8	-0.37
0.9	-0.20

$A(1)$ = -0.2382E+01
$A(2)$ = 0.1360E+00
$A(3)$ = 0.0000E+00

System Number: 213 **T/K = 298.20**

1. 3,6-Dioxaoctane, $C_6H_{14}O_2$
2. Heptane, C_7H_{16}

$C_p(1)$/J K^{-1}mol^{-1} = 261.05
$C_p(2)$/J K^{-1}mol^{-1} = 224.75

x_1	C_p^E/J K^{-1}mol^{-1}
0.1	-0.274
0.2	-0.436
0.3	-0.534
0.4	-0.598
0.5	-0.645
0.6	-0.672
0.7	-0.663
0.8	-0.583
0.9	-0.384

$A(1)$ = -0.2579E+01
$A(2)$ = -0.7686E+00
$A(3)$ = -0.1684E+01

TABLE 3.3.3
ACTIVITY COEFFICIENTS AND PARTIAL MOLAR EXCESS ENTHALPIES AT INFINITE DILUTION

The partial molar excess Gibbs energies μ_i^E of components i (i = 1 or 2) of a binary liquid mixture are related to the activity coefficients γ_i through the relation:

$$\mu_i^E / RT = \ln \gamma_i \tag{1}$$

In Table 3.1.1 (Program VLEG), the dependence of μ_i^E and γ_i on the mole fraction composition x_1 is given for a large number of binary liquid mixtures, as derived from vapor-liquid equilibrium measurements, including the values at infinite dilution, $\mu_i^{E\infty}$ and γ_i^∞, i.e., for $x_1 = 0$ or $x_1 = 1$. Since the latter are obtained by extrapolation using the correlating equation, they are less accurate than at bulk concentrations. Several experimental techniques, such as gas-liquid chromatography, liquid-liquid chromatography, and differential ebulliometry, permit the accurate measurement of the *activity coefficients at infinite dilution*, γ_i^∞ , as a function of temperature T.

The *partial molar excess enthalpies at infinite dilution*, $h_i^{E\infty}$, are related to $\mu_i^{E\infty}$ or γ_i^∞ through the Gibbs-Helmholtz equation (see Table 3.3.1):

$$h_i^{E\infty} = \left[\partial\left(\mu_i^{E\infty} / T\right) / \partial(1/T)\right]_P \tag{2}$$

or

$$h_i^{E\infty} = -RT^2\left(\partial \ln \gamma_i^\infty / \partial T\right)_P \tag{3}$$

Accurate $h_i^{E\infty}$ values can be obtained only by calorimetric measurements at high dilution. The values derived from correlating equations, e.g., the Redlich-Kister equation (Table 3.3.1, Equation 4) by extrapolation, are less accurate, especially in the case of associating components such as alcohols or water.

The table gives the logarithm of the activity coefficients at infinite dilution $\ln\gamma^\infty$, of several solutes in a number of solvents at two temperatures, T_l and T_u. The values were obtained by correlating experimental data taken in the ($T_l - T_u$)-range with the equation

$$\ln \gamma^\infty = a + b / T$$

The coefficient $b = h^{E\infty}/R$ is an average value of the partial molar excess enthalpy of the solute at infinite dilution in the given ($T_l - T_u$)-range. The calculated $h^{E\infty}$ values are tabulated along with calorimetric data (in parentheses), whenever available. The agreement between calculated and calorimetric $h^{E\infty}$ data is satisfactory in most cases. The discrepancies may be due in part to the extrapolation procedure for the excess enthalpies and their dependence on temperature.

The solvents and, for a given solvent, the solutes are arranged in a modified Hill order, with substances that do not contain carbon preceding those that do contain carbon.

REFERENCES

1. Prausnitz, J. M.; Lichtenthaler, R. N.; Azevedo, E. G., *Molecular Thermodynamics of Fluid-Phase Equilibria*, 2nd Edition, Prentice-Hall, Englewood Cliffs, N. J., 1986 (general theory).
2. Tiegs, D.; Gmehling, J.; Medina, A.; Soares, M.; Bastos, J.; Alessi, P.; Kikic, I., *Activity Coefficients at Infinite Dilution*, DECHEMA, Frankfurt/Main, Germany, 1986 (2 parts) (an extensive compilation of numerical data, mainly for organic components, coverning the literature till February 1986)
3. Thomas, E. R., Newman, B. A., Long, T. C., Wood, D. A. and Eckert, C. A. "Limiting activity coefficients of nonpolar and polar solutes in both volatile and nonvolatile solvents by gas chromatography". *J. Chem. Eng. Data*, 27, 399-405, 1982 (experimental data for 35 solutes in 34 different solvents).
4. Thomas, E. R., Newman, B. A., Nicolaides, G. L. and Eckert, C. A. "Limiting activity coefficients from differential ebulliometry". *J. Chem. Eng. Data*, 27, 233-240, 1982 (experimental data for 147 systems as a function of temperature).
5. Trampe, D. M. and Eckert, C. A. "Limiting activity coefficients from an improved differential boiling point technique". *J. Chem. Eng. Data*, 35, 156-162, 1990 (experimental data for 54 systems as a function of temperature).

TABLE 3.3.3
ACTIVITY COEFFICIENTS AND PARTIAL MOLAR EXCESS ENTHALPIES AT INFINITE DILUTION (continued)

6. Trampe, D. M. and Eckert, C. A. Calorimetric measurement of partial molar excess enthalpies at infinite dilution. *J. Chem. Eng. Data*, 36, 112-118, 1991 (experimental data for ca. 200 systems at 298.15 K).
7. Thomas, E. R. and Eckert, C. A. Prediction of limiting activity coefficients by a modified separation of cohesive energy density model and UNIFAC. *Ind. Eng. Chem., Process Des. Dev.*, 23, 194-209, 1984 (predictive method tested on 3357 γ^∞ values).

Solvent/solute		T_l/K	T_u/K	ln γ^∞_1	ln γ^∞_u	$h^{E\infty}_{calc}$ ($h^{E\infty}_{exp}$)/ kJ mol^{-1}
CCl_4	**Tetrachloromethane**					
CH_3NO_2	Nitromethane	315.15	349.15	2.48	2.14	9.15 (9.68)
C_2H_3N	Ethanenitrile	315.15	346.15	2.35	2.10	7.19 (6.46)
C_3H_6O	2-Propanone	328.15	344.15	1.04	0.95	5.33 (3.65)
$C_3H_7NO_2$	2-Nitropropane	315.15	346.15	1.52	1.38	4.11
C_3H_8O	1-Propanol	314.15	344.15	2.79	2.18	18.23 (17.20)
C_4H_8O	2-Butanone	314.15	346.15	0.75	0.69	1.62 (1.67)
C_6H_{12}	Cyclohexane	318.15	346.15	0.10	0.09	0.41 (0.65)
$C_6H_{15}N$	Triethylamine	321.15	347.15	-0.29	-0.21	-2.84 (-4.40)
CS_2	**Carbon disulfide**					
CH_3NO_2	Nitromethane	298.15	318.15	3.71	3.42	11.69 (13.50)
C_2H_6O	Ethanol	303.15	318.15	4.35	4.02	17.86 (21.50)
C_3H_6O	2-Propanone	298.15	318.15	2.11	1.97	5.80 (7.94)
C_4H_8O	2-Butanone	298.15	318.15	1.60	1.50	3.98 (6.57)
$C_4H_8O_2$	1,4-Dioxane	298.15	318.15	1.33	1.21	4.71 (5.75)
C_7H_8	Toluene	298.15	318.15	0.68	0.55	4.85 (1.90)
$CHCl_3$	**Trichloromethane**					
C_2H_3N	Ethanenitrile	298.15	332.15	0.28	0.29	-0.28 (-4.07)
C_6H_6	Benzene	298.15	332.15	-0.29	-0.15	-3.45 (-1.89)
CH_4O	**Methanol**					
CH_3NO_2	Nitromethane	308.15	337.15	1.64	1.36	8.41 (4.65)
C_4H_8O	2-Butanone	308.15	337.15	0.82	0.73	2.94 (2.80)
$C_4H_8O_2$	1,4-Dioxane	308.15	337.15	1.18	1.06	3.54 (4.59)
C_6H_{12}	Cyclohexane	307.15	337.15	3.03	2.79	7.00 (5.54)
C_7H_8	Toluene	308.15	337.15	2.27	2.19	2.60 (2.86)
C_8H_{18}	Octane	308.15	337.15	3.78	3.49	8.79 (6.63)
C_2H_3N	**Ethanenitrile**					
C_5H_5N	Pyridine	315.15	353.15	0.65	0.56	2.34
$C_2H_4Cl_2$	**1,2-Dichloroethane**					
CCl_4	Tetrachloromethane	307.15	355.15	0.62	0.44	3.40 (2.82)
C_6H_{14}	Hexane	318.15	354.15	1.40	1.12	7.32 (6.53)
C_3H_6O	**2-Propanone**					
CCl_4	Tetrachloromethane	298.15	328.15	0.74	0.71	0.74 (0.45)
CS_2	Carbon disulfide	298.15	318.15	1.47	1.33	5.40 (3.31)
$CHCl_3$	Trichloromethane	308.15	328.15	-0.69	-0.64	-2.01 (-3.53)
C_2H_5I	Iodoethane	308.15	326.15	0.78	0.72	3.02
C_2H_6O	Ethanol	298.15	328.15	0.89	0.65	6.46 (4.75)
C_4H_8O	2-Butanone	298.15	328.15	0.07	0.07	0.00 (0.18)
$C_4H_8O_2$	1,4-Dioxane	298.15	328.15	0.31	0.31	0.00 (0.42)
C_6H_6	Benzene	304.15	329.15	0.46	0.43	1.07 (0.79)
C_6H_{14}	Hexane	300.15	328.15	1.90	1.61	7.83 (7.63)
$C_6H_{15}N$	Triethylamine	304.15	327.15	1.38	1.25	4.55 (4.78)
C_7H_8	Toluene	298.15	328.15	0.75	0.59	4.46 (1.08)
C_8H_{18}	Octane	298.15	328.15	2.41	1.95	12.63 (9.74)

TABLE 3.3.3
ACTIVITY COEFFICIENTS AND PARTIAL MOLAR EXCESS ENTHALPIES AT INFINITE DILUTION (continued)

Solvent/solute		T_l/K	T_u/K	$\ln \gamma^\infty_l$	$\ln \gamma^\infty_u$	$h^{E\infty}_{calc}$ ($h^{E\infty}_{exp}$)/ kJ mol^{-1}
C_3H_8O	**1-Propanol**					
CCl_4	Tetrachloromethane	333.15	369.15	1.16	1.11	1.39
C_3H_8O	**2-Propanol**					
C_6H_{12}	Cyclohexane	313.15	355.15	1.64	1.51	2.76 (4.28)
C_6H_{14}	Hexane	323.15	355.15	1.74	1.62	3.36 (4.58)
C_7H_8	Toluene	313.15	355.15	1.61	1.30	6.68 (6.01)
C_7H_{14}	Methylcyclohexane	323.15	355.15	1.80	1.64	4.83 (4.25)
C_7H_{16}	Heptane	324.15	355.15	1.87	1.74	4.01 (5.25)
C_4H_8O	**2-Butanone**					
$C_2H_4Cl_2$	1,2-Dichloroethane	314.15	350.15	-0.27	-0.20	-1.60 (-1.81)
C_2H_6O	Ethanol	314.15	349.15	0.83	0.56	6.98 (5.47)
CH_3NO_2	Nitromethane	314.15	350.15	0.23	0.18	1.04
$C_6H_{15}N$	Triethylamine	316.15	352.15	0.91	0.75	4.04 (2.92)
C_7H_{14}	Methylcyclohexane	314.15	349.15	1.34	1.12	5.67
$C_4H_8O_2$	**Ethyl ethanoate**					
$CHCl_3$	Trichloromethane	310.15	349.15	-0.85	-0.66	-4.44 (-3.54)
C_2H_6O	Ethanol	328.15	349.15	0.94	0.85	3.96
C_6H_{14}	Hexane	308.15	348.15	1.13	0.88	5.52 (5.50)
C_7H_8	Toluene	328.15	349.15	0.21	0.15	2.70 (0.46)
C_8H_{18}	Octane	328.15	349.15	1.28	1.12	7.36 (6.94)
C_4H_9Cl	**1-Chlorobutane**					
C_3H_6O	2-Propanone	309.15	351.15	0.49	0.33	3.32 (1.97)
$C_4H_{10}O$	**1-Butanol**					
C_6H_{12}	Cyclohexane	349.15	390.15	1.37	1.17	5.53
C_7H_8	Toluene	349.15	390.15	1.03	0.82	5.60
C_6H_6	**Benzene**					
C_3H_6O	2-Propanone	332.15	350.15	0.49	0.48	0.69 (1.05)
C_6H_{12}	Cyclohexane	314.15	350.15	0.48	0.37	2.65 (3.16)
C_7H_{16}	Heptane	318.15	349.15	0.65	0.46	5.57 (5.00)
C_6H_{12}	**Cyclohexane**					
$C_2H_4Cl_2$	1,2-Dichloroethane	334.15	351.15	0.93	0.83	5.55 (5.63)
C_2H_6O	Ethanol	313.15	353.15	3.42	2.48	21.48 (23.20)
$C_3H_7NO_2$	2-Nitropropane	338.15	352.15	2.10	1.92	12.16 (9.90)
C_3H_8O	2-Propanol	312.15	353.15	3.16	2.36	18.08 (21.30)
C_4H_9Cl	1-Chlorobutane	315.15	350.15	0.45	0.36	2.28 (2.52)
$C_4H_{10}O$	1-Butanol	313.15	353.15	3.32	2.38	21.52 (23.50)
C_5H_4O	2-Furaldehyde	337.15	349.15	2.75	2.63	9.77
$C_5H_{12}O$	1-Pentanol	312.15	353.15	3.11	2.10	22.63 (23.80)
C_6H_6	Benzene	310.15	352.15	0.39	0.30	2.02 (2.83)
C_6H_{14}	Hexane	323.15	353.15	0.09	0.05	1.22 (1.17)
C_6H_{14}	**Hexane**					
CCl_4	Tetrachloromethane	301.15	340.15	0.19	0.16	0.63 (1.07)
$CHCl_3$	Trichloromethane	301.15	341.15	0.47	0.36	2.29 (2.37)
C_2H_3N	Ethanenitrile	295.15	341.15	3.31	2.50	14.81 (12.50)
$C_2H_4Cl_2$	1,2-Dichloroethane	298.15	339.15	1.15	0.85	6.27 (5.80)
C_4H_8O	2-Butanone	298.15	340.15	1.48	1.23	5.00 (7.07)
$C_4H_8O_2$	Ethyl ethanoate	298.15	339.15	1.23	0.90	6.65 (6.22)

TABLE 3.3.3
ACTIVITY COEFFICIENTS AND PARTIAL MOLAR EXCESS ENTHALPIES AT INFINITE DILUTION (continued)

Solvent/solute		T_l/K	T_u/K	ln γ^∞_l	ln γ^∞_u	$h^{E\infty}_{calc}$ ($h^{E\infty}_{exp}$)/ kJ mol^{-1}
C_6H_{14}	**Hexane (continued)**					
C_4H_9Cl	1-Chlorobutane	301.15	340.15	0.43	0.34	1.84 (2.06)
$C_4H_{10}O$	1-Butanol	301.15	340.15	3.50	2.51	21.44
C_6H_{12}	Cyclohexane	301.15	340.15	0.09	0.06	0.63 (0.68)
C_3H_8O	2-Propanol	318.15	341.15	2.94	2.43	19.80

Section 4
Transport Properties

4.1. VISCOSITY

Transport properties differ from thermodynamic properties in that they describe the behavior of systems that are not in equilibrium. The three properties of most importance are:

- *Viscosity* - which describes momentum transport
- *Thermal conductivity* - which describes energy transport
- *Diffusion coefficient* - which describes mass transport

The tables in this section cover these three properties for various types of systems as a function of temperature.

Viscosity (symbol η) is a measure of the resistance to flow when a fluid is subjected to a sheer stress. It is defined as the proportionality factor between sheer rate and sheer stress:

$$F = \eta A(\mathrm{d}v / \mathrm{d}x) \tag{1}$$

where F is the tangential force required to move a planar surface of area A at velocity v relative to a parallel surface separated from the first by a distance x. The quantity defined in this way is sometimes called the *dynamic viscosity* or *absolute viscosity*. A related quantity called the *kinematic viscosity* (symbol ν) and defined as η divided by density is sometimes used in engineering applications.

It may be shown that the viscosity of an ideal gas made up of hard (non-interacting) spherical particles is given by

$$\eta = (1/3)mDvL \tag{2}$$

where m is the particle mass, D the number density, v the mean velocity, and L the mean free path. Inserting the kinetic theory expressions for these parameters leads to the relation

$$\eta = 2.67\,(MT)^{1/2} / d^2 \tag{3}$$

where M is the molar mass in g mol^{-1}, T the temperature in K, d the hard-sphere diameter in Å (10^{-10} m), and η is the viscosity in μPa s. In this approximation the viscosity is independent of pressure and increases as the square root of temperature. Even for a real gas, the pressure dependence is very slight at least up to pressures in the neighborhood of atmospheric or somewhat greater. The temperature dependence in a real gas is more complicated than implied by this simple equation, although the viscosity still increases monotonically with temperature. Further details on the theory of gas viscosities may be found in References 1 and 2.

Liquids have much higher viscosities than gases at the same temperature, often by one or two orders of magnitude or greater. Also, there is no simple theory to use as a start in describing liquid viscosities. The viscosity of a liquid always decreases monotonically with increasing temperature, and this decrease is generally quite sharp. The behavior can often be described roughly by an empirical equation of the form $\eta = e^{(a+b/T)}$. Liquid viscosities are sensitive to molecular structure; the presence of functional groups which engage in hydrogen bonding (OH, NH_2, etc.) leads to very high viscosity values (see Reference 1).

The SI unit for viscosity is pascal second (Pa s). A convenient submultiple for expressing liquid viscosity is mPa s, which happens to be identical with the older unit centipoise (cP). Since values of gas viscosity are much smaller, they are most often expressed in μPa s.

The following two tables give the viscosity of representative gases and liquids as a function of temperature. For gases the following equation is used:

$$\eta / \mathrm{Pa\,s} = A(1) + A(2)(T/\mathrm{K}) + A(3)(T/\mathrm{K})^2 + A(4)(T/\mathrm{K})^3 + A(5)(T/\mathrm{K})^4 \tag{4}$$

where $A(1) = 0$ for the substances listed here, while for liquids the equation is:

$$\ln(\eta / \mathrm{Pa\,s}) = A(1) + A(2)/[A(3) - (T/\mathrm{K})] + A(4)\ln(T/\mathrm{K}) \tag{5}$$

In each table the recommended value of η (298.15 K) is also given; this may differ somewhat from the value calculated from the equation because of the nature of the fitting procedure. Substances are arranged in a modified Hill order with compounds not containing carbon preceding those that contain carbon.

The programs VIGT and VILT allow the viscosity of these gases and liquids to be calculated at any desired temperature in the valid range.

REFERENCES

1. Reid, R. C.; Prausnitz, J. M.; Poling, B. E., *The Properties of Gases and Liquids, Fourth Edition*, McGraw-Hill, New York, 1987.
2. Kestin, J.; Knierim, K.; Mason, E. A.; Najafi, B.; Ro, S. T.; Waldman, M., "Equilibrium and Transport Properties of the Rare Gases and Their Mixtures at Low Density", *J. Phys. Chem. Ref. Data*, 13, 229-303, 1984.
3. Boushehri, A.; Bzowski, J.; Kestin, J; Mason, E. A., "Equilibrium and Transport Properties of Eleven Polyatomic Gases at Low Densities", *J. Phys. Chem. Ref. Data*, 16, 445, 1987.
4. Viswananath, D. S.; Natarajan, G., *Data Book on the Viscosity of Liquids*, Hemisphere Publishing Corp., New York, 1989.
5. *DIPPR Data Compilation of Pure Compound Properties*, Design Institute for Physical Property Data, American Institute of Chemical Engineers, New York.
6. Liley, P. E.; Makita, T.; Tanaka, Y., *Properties of Inorganic and Organic Fluids*, Hemisphere Publishing Corp., New York, 1988.
7. Riddick, J. A.; Bunger, W. B.; Sakano, T. K., *Organic Solvents, Fourth Edition*, John Wiley & Sons, New York, 1986.
8. Stephan, K.; Lucas, K., *Viscosity of Dense Fluids*, Plenum Press, New York, 1979.
9. Vargaftik, N. B., *Tables of Thermophysical Properties of Liquids and Gases, Second Edition*, John Wiley, New York, 1975.

Physical quantity	Symbol	SI unit
Viscosity	η	Pa s
Temperature	T	K

TABLE 4.1.1 (VIGT)
VISCOSITY OF GASES

SN	Molecular formula	Name	η (298.15 K)/ μPa s	*A*(2)	*A*(3)	*A*(4)	*A*(5)	(*T*/K)-range
1		Air	18.5	7.72488E-08	-5.95238E-11	2.71368E-14		100 - 600
2	Ar	Argon	22.7	7.91722E-08	2.93448E-11	-1.73227E-13	1.41721E-16	100 - 600
3	BF_3	Boron trifluoride	17.0	7.79955E-08	-1.16797E-10	1.94113E-13	-1.27177E-16	200 - 600
4	ClH	Hydrogen chloride	14.5	3.79933E-08	6.50904E-11	-1.12003E-13	4.84072E-17	280 - 500
5	F_2	Fluorine	23.7	8.31385E-08	3.40305E-11	-2.05927E-13	1.74616E-16	100 - 600
6	F_6S	Sulfur hexafluoride	15.3	5.98603E-08	-4.33220E-11	6.06045E-14	-4.40954E-17	220 - 600
7	H_2	Hydrogen	8.9	5.18909E-08	-1.23594E-10	2.06597E-13	-1.30208E-16	100 - 600
8	D_2	Deuterium	12.6	7.33639E-08	-1.78013E-10	3.03746E-13	-1.94901E-16	100 - 600
9	H_2O	Water	9.9	5.75100E-08	-1.73637E-10	3.90133E-13	-2.69021E-16	273 - 600
10	D_2O	Deuterium oxide	10.2	3.05951E-08	4.60673E-11	-1.61920E-13	1.65034E-16	277 - 600
11	H_2S	Dihydrogen sulfide	12.6	1.11342E-07	-5.53007E-10	1.43992E-12	-1.21786E-15	250 - 500
12	H_3N	Ammonia	10.2	3.11803E-08	1.44546E-11	-1.62410E-14	5.35876E-18	240 - 600
13	He	Helium	19.9	1.23010E-07	-3.30253E-10	5.90174E-13	-3.86787E-16	100 - 600
14	Kr	Krypton	25.4	8.56676E-08	1.91155E-11	-8.50390E-14	4.60526E-17	100 - 600
15	NO	Nitrogen oxide	19.1	7.36899E-08	-1.02540E-11	-1.05643E-13	1.09675E-16	130 - 600
16	N_2	Nitrogen	17.8	7.03206E-08	-2.02187E-11	-7.42361E-14	8.22917E-17	100 - 600
17	N_2O	Dinitrogen oxide	14.8	4.60628E-08	4.00122E-11	-1.15789E-13	8.00602E-17	190 - 600
18	Ne	Neon	32.0	1.73146E-07	-3.45430E-10	5.03710E-13	-2.89748E-16	100 - 600
19	O_2	Oxygen	20.7	7.60255E-08	3.03363E-12	-1.15546E-13	1.03070E-16	100 - 600
20	O_2S	Sulfur dioxide	12.8	5.28546E-08	-1.02879E-10	3.28719E-13	-3.21789E-16	200 - 500
21	Xe	Xenon	23.1	7.33337E-08	3.93839E-11	-1.05562E-13	6.29110E-17	170 - 600
22	CCl_4	Tetrachloromethane	10.1	4.00281E-08	-3.04496E-11	3.90479E-14	-1.89803E-17	270 - 600
25	$CHCl_3$	Trichloromethane	10.1	1.07470E-08	1.58031E-10	-3.42405E-13	2.37042E-16	300 - 600
26	CH_4	Methane	11.1	3.95396E-08	2.21994E-12	-4.05854E-14	2.80976E-17	115- 600
27	CH_4O	Methanol	9.9	4.58513E-08	-9.18193E-11	2.11734E-13	-1.58884E-16	300 - 600
23	CO	Carbon oxide	17.8	7.11110E-08	-2.80674E-11	-5.36367E-14	6.27741E-17	100 - 600
24	CO_2	Carbon dioxide	14.9	4.60862E-08	3.64341E-11	-9.57647E-14	6.10915E-17	200 - 600
28	C_2H_2	Ethyne	10.4	4.11346E-08	-2.91850E-11	2.69522E-14	-2.56596E-18	270 - 500
29	C_2H_4	Ethene	10.3	3.65708E-08	-8.05895E-12	1.05235E-14	-1.71147E-17	170 - 600
30	C_2H_6	Ethane	9.4	3.12934E-08	1.22209E-11	-4.68478E-14	3.27545E-17	185 - 600
31	C_2H_6O	Ethanol		1.11885E-07	-5.24834E-10	1.09564E-12	-7.55010E-16	350 - 500
32	C_3H_8	Propane	8.2	2.89298E-08	-4.41619E-12	1.21654E-15	-1.78331E-18	235 - 600
33	C_4H_{10}	Butane	7.5	2.34936E-08	1.59676E-11	-4.34195E-14	3.18921E-17	273 - 600
34	C_4H_{10}	2-Methylpropane	7.5	2.35668E-08	1.68099E-11	-4.74576E-14	3.59496E-17	262 - 600
35	$C_4H_{10}O$	Diethyl ether	7.5	-5.87473E-09	2.38382E-10	-5.90792E-13	4.74181E-16	250 - 500
36	C_5H_{12}	Pentane	6.7	6.35732E-10	1.44593E-10	-3.00353E-13	1.99410E-16	300 - 600
37	C_6H_{14}	Hexane	6.7	4.23114E-08	-1.35350E-10	2.90544E-13	-2.05388E-16	300 - 600
38	C_8H_{18}	Octane	5.6	1.88184E-08	6.56333E-12	-3.27651E-14	3.34044E-17	250 - 600

TABLE 4.1.2 (VILT)
VISCOSITY OF LIQUIDS

SN	Molecular formula	Name	η (298.15 K)/ mPa s	*A*(1)	*A*(2)	*A*(3)	*A*(4)	(*T*/K)- range
1	Br_2	Bromine	0.944	-9.7584	-759.577	26.153		270 - 330
2	Cl_3HSi	Trichlorosilane	0.326	-10.6511	-782.000			265 - 330
3	Cl_3P	Phosphorus trichloride	0.529	-9.7957	-621.698	21.934		200 - 340
4	H_2O	Water	0.893	-10.4349	-507.881	149.390		273 - 380
5	H_4N_2	Hydrazine	0.876	-10.2251	-683.292	83.603		280 - 450
6	Hg	Mercury	1.526	-7.1622	-117.913	124.040		290 - 380
7	CCl_4	Tetrachloromethane	0.908	-11.8180	-1664.539	-47.672		270 - 460
9	$CHBr_3$	Tribromomethane	1.857	15.4049			-3.80750	282 - 350
10	$CHCl_3$	Trichloromethane	0.537	-10.2633	-750.090	23.789		210 - 360
11	CHN	Hydrogen cyanide	0.183	-10.7165	-500.582	61.204		260 - 300
12	CH_2Br_2	Dibromomethane	0.980	-32.6801	-1935.200		3.38060	230 - 370
13	CH_2Cl_2	Dichloromethane	0.413	-10.3955	-729.067	18.104		200 - 380
14	CH_2O_2	Methanoic acid	1.607	-10.2331	-716.357	109.610		282 - 380
15	CH_3I	Iodomethane	0.469	-10.2612	-773.922			270 - 320
16	CH_3NO	Methanamide	3.343	-11.5735	-1288.388	78.757		270 - 320
17	CH_4O	Methanol	0.549	-7.2881	-1065.300		-0.66570	230 - 337
18	CH_5N	Methylamine	0.180	-10.6444	-456.902	71.960		200 - 300
8	CS_2	Carbon disulfide	0.352	-10.1074	-643.043			270 - 320
19	C_2Cl_4	Tetrachloroethene	0.845	-1.9781	-555.000		-1.22160	251 - 390
20	C_2HCl_5	Pentachloroethane	2.254	-10.3902	-1008.693	63.314		270 - 360
21	$C_2HF_3O_2$	Trifluoroethanoic acid	0.808	-11.6048	-1337.042			290 - 340
22	$C_2H_2Cl_2$	*cis*-1,2-Dichloroethene	0.445	-10.3183	-726.420	18.953		200 - 300
23	$C_2H_2Cl_2$	*trans*-1,2-Dichloroethene	0.317	-10.6013	-774.567	-6.345		224 - 350
24	C_2H_3ClO	Acetyl chloride	0.368	-11.9322	-1701.495	-124.560		290 - 330
25	$C_2H_3Cl_3$	1,1,1-Trichloroethane	0.793	-16.4551	-1460.600		0.77524	245 - 420
26	C_2H_3N	Ethanenitrile	0.344	14.4859	423.700		-3.69260	290 - 350
27	$C_2H_4Br_2$	1,2-Dibromoethane	1.595	-9.7666	-691.604	90.197		284 - 400
28	$C_2H_4Cl_2$	1,1-Dichloroethane	0.464	-11.1353	-1121.267	-25.909		280 - 330
29	$C_2H_4Cl_2$	1,2-Dichloroethane	0.779	-10.5873	-886.864	39.584		270 - 360
30	$C_2H_4O_2$	Ethanoic acid	1.056	14.0066			-3.66120	290 - 390
31	C_2H_5Br	Bromoethane	0.374	-11.0103	-1077.886	-47.339		200 - 300
32	C_2H_5Cl	Chloroethane	0.258	-10.2161	-702.000		-0.07020	150 - 370
33	C_2H_5I	Iodoethane	0.556	-10.6513	-1027.183	-27.294		270 - 350
34	C_2H_5NO	*N*-Methylmethanamide	1.678	9.9222	-466.230		-3.13750	270 - 470
35	C_2H_6O	Ethanol	1.074	-12.8880	-1950.174	-24.124		210 - 350
36	C_2H_6OS	Dimethyl sulfoxide	1.987	-10.0358	-745.922	102.620		292 - 340
37	$C_2H_6O_2$	1,2-Ethanediol	16.1	-10.4648	-960.293	146.530		280 - 420
38	C_2H_6S	2-Thiapropane	0.284	-10.6209	-731.992			270 - 310
39	C_2H_6S	Ethanethiol	0.287	-10.7264	-766.784			270 - 300
40	C_2H_7NO	2-Aminoethanol	21.1	-8.5831	-3152.000		-1.02660	290 - 440
41	C_3H_5Br	3-Bromopropene	0.471	-10.6674	-896.465			270 - 350
42	C_3H_5Cl	3-Chloropropene	0.314	-10.9423	-857.414			270 - 320
43	C_3H_5ClO	(Chloromethyl)oxirane	1.073	-14.6161	-1394.600		0.54437	225 - 420
44	C_3H_6O	2-Propanone	0.306	-10.6207	-687.276	26.203		180 - 320
45	C_3H_6O	2-Propen-1-ol	1.218	-12.8367	-1826.618			280 - 370
46	C_3H_6O	Propanal	0.321	-11.9626	-1407.570	-60.958		280 - 330
47	$C_3H_6O_2$	Ethyl methanoate	0.380	-10.5392	-685.249	40.866		270 - 330
48	$C_3H_6O_2$	Methyl ethanoate	0.364	9.6077			-3.07590	270 - 400
49	$C_3H_6O_2$	Propanoic acid	1.030	-11.1001	-1293.017	-8.122		270 - 420
50	C_3H_7Br	1-Bromopropane	0.489	-10.6522	-902.913			270 - 350
51	C_3H_7Br	2-Bromopropane	0.458	-10.8424	-940.514			270 - 330
52	C_3H_7Cl	1-Chloropropane	0.334	-10.9145	-867.867			270 - 320
53	C_3H_7Cl	2-Chloropropane	0.303	-11.1537	-910.258			270 - 310
54	C_3H_7I	1-Iodopropane	0.703	-10.0503	-673.921	56.586		270 - 380
55	C_3H_7I	2-Iodopropane	0.653	-10.5643	-943.185	6.109		270 - 360

TABLE 4.1.2 (VILT)
VISCOSITY OF LIQUIDS (continued)

SN	Molecular formula	Name	η (298.15 K)/ mPa s	$A(1)$	$A(2)$	$A(3)$	$A(4)$	(T/K)-range
56	C_3H_7NO	*N,N*-Dimethylmethanamide	0.794	-8.3809	-129.053	194.250		270 - 320
57	C_3H_8O	1-Propanol	1.945	-13.1894	-1965.256	15.262		210 - 370
58	C_3H_8O	2-Propanol	2.038	-14.5178	-2323.769	18.917		270 - 360
59	$C_3H_8O_2$	1,2-Propanediol	40.4	-293.0701	-17494.000		40.57600	235 - 420
60	$C_3H_8O_3$	1,2,3-Propanetriol	923	-237.0301	-16739.000		31.73400	293 - 500
61	C_3H_8S	1-Propanethiol	0.385	-10.8017	-876.134			270 - 310
62	C_3H_8S	2-Propanethiol	0.357	-11.0064	-891.400	7.659		270 - 320
63	C_3H_9N	Propylamine	0.376	-11.8873	-1193.015			280 - 320
64	C_4H_4O	Furan	0.361	-12.1491	-945.970		0.18419	220 - 310
65	C_4H_5N	Pyrrole	1.225	-9.7068	-499.085	131.910		250 - 350
66	$C_4H_6O_3$	Ethanoic anhydride	0.843	-10.5468	-862.548	49.411		270 - 410
67	C_4H_7N	Butanenitrile	0.553	-11.1031	-1074.133			280 - 400
68	C_4H_8O	2-Butanone	0.405	10.0400			-3.13320	240 - 360
69	C_4H_8O	Oxolane	0.456	-10.3211	-900.920		-0.06913	165 - 340
70	$C_4H_8O_2$	1,4-Dioxane	1.177	-36.7061	-2691.500		3.67410	290 - 360
71	$C_4H_8O_2$	Ethyl ethanoate	0.423	-11.2184	-1040.930	-3.475		270 - 350
72	$C_4H_8O_2$	Methyl propanoate	0.431	-11.5026	-1274.803	-41.506		270 - 350
73	$C_4H_8O_2$	Butanoic acid	1.426	-11.3660	-1434.948			270 - 430
74	$C_4H_8O_2$	2-Methylpropanoic acid	1.226	-11.2804	-1375.910	-2.482		270 - 430
75	$C_4H_8O_2S$	Thiolane 1,1-dioxide		-55.5121	-4380.000		6.38400	301 - 550
76	C_4H_9N	Pyrrolidine	0.704	-9.8666	-470.188	117.840		216 - 340
77	C_4H_9NO	*N,N*-Dimethylethanamide	1.956	-11.0833	-1261.057	37.944		290 - 450
78	C_4H_9NO	Morpholine	2.021	-50.4341	-3611.700		5.63680	280 - 400
79	$C_4H_{10}O$	1-Butanol	2.544	-13.7508	-2318.703			220 - 390
80	$C_4H_{10}O$	2-Butanol	3.096	-12.1459	-1042.818	134.400		280 - 370
81	$C_4H_{10}O$	2-Methyl-2-propanol	4.31	-10.9785	-550.778	198.590		298 - 360
82	$C_4H_{10}O$	Diethyl ether	0.224	6.8571			-2.67850	270 - 305
83	$C_4H_{10}S$	3-Thiapentane	0.422	-11.6267	-1432.760	-73.352		270 - 370
84	$C_4H_{11}N$	Butylamine	0.574	16.5625			-4.21680	270 - 360
85	$C_4H_{11}N$	(2-Methylpropyl)amine	0.571	-5.6305	-236.867	427.060		270 - 330
86	$C_4H_{11}N$	Diethylamine	0.319	-11.7586	-1105.195			280 - 330
87	$C_4H_{11}NO_2$	Bis(2-hydroxyethyl)amine		-35.7261	-6755.500		2.18220	301 - 420
88	C_5H_5N	Pyridine	0.879	-9.4802	-401.847	133.660		270 - 390
89	C_5H_{10}	1-Pentene	0.195	-10.9016	-702.749			180 - 280
90	C_5H_{10}	2-Methyl-2-butene	0.203	-10.9928	-742.100			270 - 310
91	C_5H_{10}	Cyclopentane	0.413	-11.0331	-966.211			270 - 320
92	$C_5H_{10}O$	2-Pentanone	0.470	-11.0492	-1009.568			270 - 380
93	$C_5H_{10}O$	3-Pentanone	0.444	-11.6707	-1446.484	-68.044		270 - 380
94	$C_5H_{10}O_2$	Butyl methanoate	0.644	-11.0978	-1032.917	22.748		270 - 380
95	$C_5H_{10}O_2$	Propyl ethanoate	0.544	-11.2882	-1124.375			270 - 380
96	$C_5H_{10}O_2$	Ethyl propanoate	0.501	-11.5716	-1329.052	-36.360		270 - 370
97	$C_5H_{10}O_2$	Methyl butanoate	0.541	-11.2269	-1104.573			270 - 380
98	$C_5H_{10}O_2$	Methyl 2-methylpropanoate	0.488	-11.1150	-1040.676			270 - 370
99	$C_5H_{11}N$	Piperidine	1.573	-10.5841	-756.330	115.000		280 - 380
100	C_5H_{12}	Pentane	0.224	-10.3402	-516.101	31.920		150 - 330
101	$C_5H_{12}O$	1-Pentanol	3.619	-12.6149	-1604.441	68.729		220 - 410
102	$C_5H_{12}O$	2-Pentanol	3.470	-11.3824	-791.836	159.690		290 - 390
103	$C_5H_{12}O$	3-Pentanol	4.149	-10.9571	-585.916	191.080		290 - 390
104	$C_5H_{12}O$	2-Methyl-1-butanol	4.453	-12.2854	-1269.415	113.410		290 - 400
105	$C_5H_{12}O$	3-Methyl-1-butanol	3.692	-13.2974	-1936.750	46.485		270 - 410
106	$C_5H_{13}N$	Pentylamine	0.702	17.6911			-4.37950	270 - 360
107	$C_6H_4Cl_2$	1,2-Dichlorobenzene	1.324	-22.2161	-1758.200		1.70110	265 - 450
108	$C_6H_4Cl_2$	1,3-Dichlorobenzene	1.044	-30.6631	-2015.700		2.99030	265 - 445
109	C_6H_5Br	Bromobenzene	1.074	-9.7554	-644.586	77.353		270 - 420
110	C_6H_5Cl	Chlorobenzene	0.755	0.2499	-535.200		-1.62070	250 - 540
111	C_6H_5ClO	2-Chlorophenol	3.589	-9.7565	-531.229	169.420		283 - 450

TABLE 4.1.2 (VILT)
VISCOSITY OF LIQUIDS (continued)

SN	Molecular formula	Name	η (298.15 K)/ mPa s	*A*(1)	*A*(2)	*A*(3)	*A*(4)	(*T*/K)-range
112	C_6H_5ClO	3-Chlorophenol	11.9	-8.0075	-206.233	240.530		306 - 340
113	C_6H_5F	Fluorobenzene	0.550	-10.8908	-1009.062			270 - 360
114	C_6H_5I	Iodobenzene	1.554	-9.6955	-708.690	78.655		270 - 425
115	$C_6H_5NO_2$	Nitrobenzene	1.863	-10.1438	-858.519	75.641		280 - 500
116	C_6H_6	Benzene	0.604	-10.3444	-583.406	99.248		280 - 350
117	C_6H_6ClN	2-Chloroaniline	3.316	-10.0805	-758.771	124.580		290 - 490
118	C_6H_6O	Phenol		-10.0326	-615.504	181.960		315 - 460
119	C_6H_7N	Aniline	3.847	-10.2856	-754.603	138.450		280 - 460
120	C_6H_{10}	Cyclohexene	0.625	-11.1447	-1123.086			270 - 360
121	$C_6H_{10}O$	Cyclohexanone	2.017	-11.6815	-1632.487			280 - 430
122	$C_6H_{10}O$	4-Methyl-3-penten-2-one	0.602	-34.9811	-2107.200		3.59760	220 - 400
123	$C_6H_{11}N$	Hexanenitrile	0.912	-11.0064	-1085.116	27.296		290 - 440
124	C_6H_{12}	Cyclohexane	0.894	-11.8505	-1440.129			280 - 360
125	C_6H_{12}	Methylcyclopentane	0.479	-12.6679	-2164.499	-132.640		240 - 330
126	C_6H_{12}	1-Hexene	0.252	-6.9091	-656.900		-0.62835	220 - 330
127	$C_6H_{12}O$	Cyclohexanol		-11.0948	-893.887	189.650		300 - 440
128	$C_6H_{12}O$	2-Hexanone	0.583	-11.4771	-1188.600		0.00766	220 - 400
129	$C_6H_{12}O$	4-Methyl-2-pentanone	0.545	-11.3264	-1136.418			290 - 340
130	$C_6H_{12}O_2$	Butyl ethanoate	0.685	-10.9884	-993.750	29.726		270 - 380
131	$C_6H_{12}O_2$	Ethyl butanoate	0.639	-6.1735	-131.091	409.090		280 - 330
132	$C_6H_{12}O_3$	2,4,6-Trimethyl-1,3,5-trioxane	1.079	-10.9016	-831.648	93.813		286 - 400
133	$C_6H_{13}N$	Cyclohexylamine	1.944	-10.5956	-822.990	109.060		280 - 410
134	C_6H_{14}	Hexane	0.300	-9.7775	-271.843	134.870		270 - 340
135	C_6H_{14}	2-Methylpentane	0.286	-12.5944	-1987.062	-149.900		270 - 330
136	C_6H_{14}	3-Methylpentane	0.306	-10.8546	-824.118			270 - 310
137	$C_6H_{14}O$	Dipropyl ether	0.396	-11.2590	-1021.058			270 - 360
138	$C_6H_{14}O$	1-Hexanol	4.578	-38.5031	-3810.400		3.56930	280 - 425
139	$C_6H_{15}N$	Dipropylamine	0.517	-11.6458	-1215.995			270 - 390
140	$C_6H_{15}N$	Diisopropylamine	0.393	-11.6433	-1243.741	-29.009		280 - 360
141	$C_6H_{15}NO_3$	Tris(2-hydroxyethyl)amine	609	-198.5801	-14579.000		26.18400	295 - 430
142	C_7H_5N	Benzonitrile	1.267	-9.8776	-632.336	100.950		290 - 470
143	C_7H_7Cl	2-Chlorotoluene	0.964	-10.7204	-1069.712	14.885		270 - 440
144	C_7H_7Cl	3-Chlorotoluene	0.823	-10.5403	-934.780	26.217		270 - 440
145	C_7H_7Cl	4-Chlorotoluene	0.837	-10.4749	-875.972	39.664		280 - 440
146	C_7H_8	Toluene	0.555	-13.3621	-1183.000		0.33300	200 - 400
147	C_7H_8O	2-Methylphenol	7.97	-10.0575	-576.291	187.870		303 - 470
148	C_7H_8O	3-Methylphenol	12.9	-10.3462	-688.174	183.390		285 - 480
149	C_7H_8O	Phenylmethanol	5.474	-10.7257	-973.372	121.750		280 - 480
150	C_7H_8O	Anisole	1.056	-11.5452	-1473.447	-15.876		280 - 430
151	C_7H_9N	*N*-Methylaniline	2.042	-10.0227	-617.853	136.780		270 - 440
152	C_7H_9N	Benzylamine	1.624	-10.8832	-1107.152	49.937		290 - 460
153	C_7H_{14}	Methylcyclohexane	0.679	-10.4418	-734.686	64.712		270 - 380
154	C_7H_{14}	1-Heptene	0.340	-10.8286	-845.993	0.529		270 - 370
155	$C_7H_{14}O$	2-Heptanone	0.714	-9.2481	-128.800	233.860		290 - 360
156	$C_7H_{14}O_2$	Heptanoic acid	3.840	-10.8682	-1219.495	68.313		285 - 395
157	C_7H_{16}	Heptane	0.387	-10.8597	-820.020	24.953		190 - 370
158	C_7H_{16}	3-Methylhexane	0.350	-11.2198	-972.220			280 - 310
159	$C_7H_{16}O$	1-Heptanol	5.815	-88.6231	-6463.800		10.84600	290 - 445
160	$C_7H_{16}O$	(±)-2-Heptanol		-11.3819	-939.593	137.510		305 - 430
161	$C_7H_{16}O$	(±)-3-Heptanol		-10.8991	-665.194	180.490		310 - 420
162	$C_7H_{16}O$	4-Heptanol	4.2	-11.0172	-706.986	170.680		290 - 420
163	$C_7H_{17}N$	Heptylamine	1.314	-12.8595	-2157.637	-48.457		280 - 400
164	C_8H_8	Vinylbenzene	0.701	-22.6751	-1758.000		1.67010	245 - 415
165	C_8H_8O	Methyl phenyl ketone	1.7	-10.6497	-1078.301	45.007		370 - 480
166	$C_8H_8O_2$	Methyl benzoate	1.857	-12.4683	-1842.390			280 - 305
167	$C_8H_8O_3$	Methyl 2-hydroxybenzoate		-10.8010	-1223.386	41.528		350 - 500

TABLE 4.1.2 (VILT)
VISCOSITY OF LIQUIDS (continued)

SN	Molecular formula	Name	η (298.15 K)/ mPa s	A(1)	A(2)	A(3)	A(4)	(T/K)-range
168	C_8H_{10}	Ethylbenzene	0.647	-10.4521	-1048.400		-0.07150	210 - 410
169	C_8H_{10}	1,2-Dimethylbenzene	0.760	-11.2659	-1274.688	-14.003		260 - 420
170	C_8H_{10}	1,3-Dimethylbenzene	0.581	-11.1148	-1163.542	-19.347		270 - 410
171	C_8H_{10}	1,4-Dimethylbenzene	0.603	-5.7751	-826.200		-0.77390	280 - 410
172	$C_8H_{10}O$	Ethyl phenyl ether	1.197	-10.9907	-1082.100	44.325		290 - 450
173	$C_8H_{11}N$	*N,N*-Dimethylaniline	1.300	-10.8083	-1114.359	30.460		276 - 470
174	$C_8H_{11}N$	*N*-Ethylaniline	2.047	-10.4940	-803.003	111.510		270 - 480
175	C_8H_{16}	Ethylcyclohexane	0.784	-10.3833	-780.530	56.684		270 - 320
176	$C_8H_{16}O_2$	Octanoic acid	5.020	-9.6255	-628.329	153.080		290 - 370
177	C_8H_{18}	Octane	0.508	-11.7501	-1456.201	-51.438		270 - 400
178	$C_8H_{18}O$	1-Octanol	7.29	-67.5771	-5535.400		7.73830	290 - 425
179	$C_8H_{18}O$	4-Methyl-3-heptanol	1.085	-10.6626	-750.228	102.600		270 - 380
180	$C_8H_{18}O$	5-Methyl-3-heptanol	1.178	-10.8205	-851.128	89.383		270 - 380
181	$C_8H_{18}O$	2-Ethyl-1-hexanol	6.271	-11.4326	-1005.700	140.040		270 - 380
182	$C_8H_{19}N$	Dibutylamine	0.918	-10.8155	-831.026	80.751		270 - 440
183	$C_8H_{19}N$	Bis(2-methylpropyl)amine	0.723	-10.7342	-795.014	71.137		270 - 400
184	C_9H_{10}	Indan	1.357	-9.7462	-576.498	114.780		270 - 400
185	C_9H_{12}	Isopropylbenzene	0.737	-24.4381	-1785.900		1.97200	200 - 420
186	$C_9H_{18}O$	5-Nonanone	1.199	-10.3193	-797.569	76.180		290 - 380
187	$C_9H_{18}O_2$	Nonanoic acid	7.01	-10.6402	-1126.171	99.881		290 - 370
188	C_9H_{20}	Nonane	0.665	-11.0211	-1014.818	24.243		270 - 380
189	$C_9H_{20}O$	1-Nonanol	9.12	-15.2519	-3146.943			280 - 330
190	$C_{10}H_{10}O_4$	Dimethyl 1,2-benzenedi-carboxylate	14.4	-284.8201	-16163.000		39.73000	280 - 400
191	$C_{10}H_{14}$	Butylbenzene	0.950	-11.2216	-1270.981			280 - 360
192	$C_{10}H_{18}$	*cis*-Bicyclo[4.4.0]decane	3.042	-10.2419	-910.626	93.359		240 - 460
193	$C_{10}H_{18}$	*trans*-Bicyclo[4.4.0]decane	1.948	-10.5560	-1020.506	61.638		243 - 440
194	$C_{10}H_{20}O_2$	Decanoic acid		-12.2677	-2205.462			320 - 355
195	$C_{10}H_{22}$	Decane	0.838	-10.5320	-791.905	68.462		245 - 380
196	$C_{10}H_{22}O$	1-Decanol	10.9	-11.9930	-1425.853	107.410		280 - 330
197	$C_{11}H_{24}$	Undecane	1.098	-10.9585	-1075.031	38.719		260 - 480
198	$C_{12}H_{10}O$	Diphenyl ether		-70.8311	-4767.000		8.64090	300 - 420
199	$C_{12}H_{26}$	Dodecane	1.383	-10.4947	-865.703	76.830		270 - 380
200	$C_{13}H_{12}$	Diphenylmethane		-11.1445	-1507.595	11.031		330 - 540
201	$C_{13}H_{28}$	Tridecane	1.724	-10.7842	-1044.453	61.909		270 - 480
202	$C_{14}H_{30}$	Tetradecane	2.128	-18.9641	-2010.900		1.06480	280 - 525
203	$C_{16}H_{22}O_4$	Dibutyl 1,2-benzenedi-carboxylate	16.6	-250.5301	-14426.000		34.76000	240 - 420
204	$C_{16}H_{34}$	Hexadecane	3.032	-11.0925	-1331.654	46.605		292 - 520
205	$C_{18}H_{38}$	Octadecane		-10.1136	-870.838	111.620		302 - 380

4.2. THERMAL CONDUCTIVITY

Thermal conductivity (symbol λ) is the property that describes the transport of energy (in the form of heat) in a system when there is a temperature gradient. It is defined through the equation

$$-(\mathrm{d}Q / \mathrm{d}t) = \lambda A(\mathrm{d}T / \mathrm{d}x)$$

where $dQ/\mathrm{d}t$ is the rate of heat transfer in the x direction through a surface of area A normal to the x axis, and T is the temperature. In a gas, liquid, or polycrystalline solid λ is a scalar quantity. However, in a non-isotropic crystal, λ will in general depend on the crystal direction, so that two or more values are needed to specify the heat transfer properties completely.

The thermal conductivity of an ideal hard-sphere gas can be derived in a similar way to the viscosity; the resulting value is proportional to the square root of temperature. However, real gases behave quite differently from this idealized model, because molecules may store energy in their vibrational and rotational modes. Various approximation methods for dealing with polyatomic gases are discussed in Reference 1.

As in the case of viscosity, the thermal conductivity of a gas is practically independent of pressure up to pressures of a few atmospheres. At moderate pressures and temperatures, λ increases gradually with increasing temperature but becomes very large as the critical point is approached.

Liquids show thermal conductivity values 10 to 100 times larger than low-pressure gases at the same temperature. The thermal conductivity of a liquid usually decreases with increasing temperature, but the dependence on temperature is much weaker than in the case of viscosity. Pressure has very little effect.

The thermal conductivity of solids has a more complex behavior. Starting from $T = 0$ K, λ rises to a maximum in the range 10 to 50 K in the case of metals (the maximum generally comes at a higher temperature for non-metals), and then decreases rapidly, finally levelling off by about 300 K in typical metals. However, this behavior is highly dependent on the purity of the material, and the maximum thermal conductivity may change by an order of magnitude upon a small change in impurity level. Thus it is difficult to treat the low-temperature thermal conductivity as a true property of the material; values of λ have limited meaning unless the impurity levels are precisely specified. For ambient temperature and above, the sensitivity to impurities is not quite so pronounced.

The following three tables give the thermal conductivity of representative gases, liquids, and solids as a function of temperature, as expressed by the equation

$$\lambda / \mathrm{W\,m^{-1}\,K^{-1}} = A(1) + A(2)(T/\mathrm{K}) + A(3)(T/\mathrm{K})^2 + A(4)(T/\mathrm{K})^3 + A(5)(T/\mathrm{K})^4 + A(6)(T/\mathrm{K})^5$$

For the gases in Table 4.2.1, $A(1) = 0$. Recommended values of λ (298.15 K) are also given, which may differ slightly from the values calculated from the fitting equation. Substances are listed in a modified Hill order, with compounds not containing carbon preceding those that contain carbon.

The programs TCGT, TCLT, and TCST permit calculation of the thermal conductivity of these gases, liquids, and solids at any desired temperature in the valid range.

Physical quantity	Symbol	SI unit
Thermal conductivity	λ	W m^{-1} K^{-1}
Temperature	T	K

REFERENCES

1. Reid, R. C.; Prausnitz, J. M.; Poling, B. E., *The Properties of Gases and Liquids, Fourth Edition*, McGraw-Hill, New York, 1987.
2. *DIPPR Data Compilation of Pure Compound Properties*, Design Institute for Physical Property Data, American Institute of Chemical Engineers, New York.
3. Ho, C. Y.; Powell, R. W.; Liley, P. E., "Thermal Conductivity of the Elements", *J. Phys. Chem. Ref. Data*, 2, 279-421, 1972.

4. Kestin, J.; Knierim, K.; Mason, E. A.; Najafi, B.; Ro, S. T.; Waldman, M., "Equilibrium and Transport Properties of the Rare Gases and Their Mixtures at Low Density", *J. Phys. Chem. Ref. Data*, 13, 229-303, 1984.
5. Uribe, F. J., "Thermal Conductivity of Nine Polyatomic Gases at Low Density", *J. Phys. Chem. Ref. Data*, 19, 1123 (1990).
6. Liley, P. E.; Makita, T.; Tanaka, Y., *Properties of Inorganic and Organic Fluids*, Hemisphere Publishing Corp., New York, 1988.
7. Vargaftik, N. B., *Tables of Thermophysical Properties of Liquids and Gases, Second Edition*, John Wiley, New York, 1975.
8. Touloukian, Y. S.; Buyco, E. H., Editors, *Thermophysical Properties of Matter,* Vol. 2: *Thermal Conductivity - Nonmetallic Solids*; Vol. 3: *Thermal Conductivity - Nonmetallic Liquids and Gases*, IFI/Plenum, New York, 1970.

TABLE 4.2.1 (TCGT)
THERMAL CONDUCTIVITY OF GASES

SN	Molecular formula	Name	λ (298.15 K)/ mW m^{-1} K^{-1}	$A(2)$	$A(3)$	$A(4)$	$A(5)$	(T/K)-range
1		Air	26.1	0.0965	-9.960E-06	-9.310E-08	8.882E-11	100 - 600
2	Ar	Argon	17.8	0.0606	3.151E-05	-1.525E-07	1.223E-10	100 - 600
3	BF_3	Boron trifluoride	18.9	0.0076	4.877E-04	-1.375E-06	1.231E-09	250 - 400
4	ClH	Hydrogen chloride	14.4	0.0313	1.127E-04	-2.286E-07	1.398E-10	190 - 600
5	H_2	Hydrogen	185.8	0.7042	-1.470E-04	-3.652E-07	-1.738E-10	100 - 400
6	H_2O	Water	18.6	0.0349	1.511E-04	-2.576E-07	2.050E-10	273 - 600
7	D_2O	Deuterium oxide	18.2	0.0326	1.367E-04	-1.799E-07	1.361E-10	277 - 600
8	H_2S	Dihydrogen sulfide	14.2	-0.0506	6.582E-04	-1.396E-06	9.824E-10	220 - 600
9	H_3N	Ammonia	24.2	0.0255	2.467E-04	-2.317E-07	9.791E-11	240 - 600
10	He	Helium	154.6	0.9398	-2.467E-03	4.396E-06	-2.888E-09	100 - 600
11	Kr	Krypton	9.4	0.0327	1.632E-06	-2.060E-08	1.042E-11	100 - 600
12	NO	Nitrogen oxide	25.8	0.0875	4.180E-05	-2.064E-07	1.790E-10	130 - 600
13	N_2	Nitrogen	25.9	0.1041	-4.737E-05	-6.240E-08	9.320E-11	100 - 600
14	N_2O	Dinitrogen oxide	17.3	0.0189	1.884E-04	-2.083E-07	5.877E-11	190 - 600
15	Ne	Neon	49.5	0.2686	-5.423E-04	8.100E-07	-4.800E-10	100 - 600
16	O_2	Oxygen	26.2	0.0963	-1.484E-05	-6.958E-08	8.279E-11	100 - 600
17	O_2S	Sulfur dioxide	9.6	0.0475	-1.622E-04	4.816E-07	-3.747E-10	250 - 600
18	Xe	Xenon	5.5	0.0209	-1.629E-05	3.703E-08	-3.322E-11	170 - 600
19	CCl_2F_2	Dichlorodifluoromethane	9.8	-0.0043	2.058E-04	-3.296E-07	1.923E-10	260 - 600
20	CF_4	Tetrafluoromethane	15.9	0.0092	2.239E-04	-2.938E-07	1.333E-10	280 - 600
23	$CHCl_3$	Trichloromethane	7.4	0.0173	1.938E-05	3.279E-08	-3.950E-11	250 - 550
24	CH_4	Methane	34.5	0.1146	-1.448E-04	6.495E-07	-5.113E-10	120 - 600
25	CH_4O	Methanol	14.9	-0.0439	4.942E-04	-7.419E-07	4.761E-10	340 - 600
21	CO	Carbon oxide	24.8	0.0872	1.659E-06	-6.481E-08	5.244E-11	220 - 600
22	CO_2	Carbon dioxide	16.7	0.0255	1.263E-04	-7.156E-08	-2.848E-11	200 - 600
26	C_2H_2	Ethyne	21.1	-0.0204	4.816E-04	-6.977E-07	3.578E-10	250 - 600
27	C_2H_4	Ethene	20.2	-0.0771	8.337E-04	-1.469E-06	1.019E-09	300 - 600
28	C_2H_6	Ethane	20.9	0.0299	1.368E-05	5.885E-07	-6.128E-10	100 - 600
29	C_2H_6O	Ethanol	14.2	-0.0629	5.960E-04	-9.375E-07	6.086E-10	300 - 600
30	C_3H_6O	2-Propanone	11.4	-0.0078	1.872E-04	-1.269E-07	5.704E-11	250 - 600
31	C_3H_8	Propane	17.9	0.0158	1.131E-04	1.788E-07	-2.076E-10	235 - 600
32	C_4H_{10}	Butane	16.2	0.0117	1.017E-04	2.050E-07	-2.224E-10	275 - 600
33	C_4H_{10}	2-Methylpropane	16.0	0.0173	6.069E-05	2.909E-07	-2.892E-10	265 - 600
34	$C_4H_{10}O$	Diethyl ether	14.9	0.0138	1.218E-04	4.567E-09	-1.312E-11	250 - 500
35	C_5H_{12}	Pentane	14.2	-0.0035	1.964E-04	-9.099E-08	2.884E-11	300 - 600
36	C_6H_{14}	Hexane	13.2	-0.0032	1.680E-04	-1.218E-08	-5.592E-11	350 - 600

TABLE 4.2.2 (TCLT)
THERMAL CONDUCTIVITY OF LIQUIDS

SN	Molecular formula	Name	λ (298.15 K)/ W m^{-1} K^{-1}	*A*(1)	*A*(2)	*A*(3)	*A*(4)	(*T*/K)-range
1	Cl_4Si	Silicon tetrachloride	0.0994	0.13697	-1.2600E-04			285 - 330
2	H_2O	Water	0.6068	-0.76760	7.5390E-03	-9.8250E-06		280 - 370
3	Hg	Mercury	8.2479	-1.38000	5.0600E-02	-7.5000E-05	4.5600E-08	235 - 600
4	CCl_4	Tetrachloromethane	0.0985	0.16420	-2.2026E-04			250 - 350
6	$CHCl_3$	Trichloromethane	0.1172	0.17740	-2.0180E-04			210 - 400
7	CH_2Br_2	Dibromomethane	0.1085	0.17558	-2.2499E-04			220 - 370
8	CH_4O	Methanol	0.1999	0.28370	-2.8100E-04			175 - 335
5	CS_2	Carbon disulfide	0.1491	0.20230	-1.7850E-04			273 - 315
9	C_2Cl_4	Tetrachloroethene	0.1103	0.18685	-2.5680E-04			250 - 420
10	C_2HCl_3	Trichloroethene	0.1161	0.21390	-3.2800E-04			215 - 360
11	$C_2H_3Cl_3$	1,1,1-Trichloroethane	0.1012	0.16083	-2.0000E-04			273 - 323
12	C_2H_3N	Ethanenitrile	0.1877	0.30703	-4.0020E-04			230 - 350
13	$C_2H_4O_2$	Ethanoic acid	0.1580	0.21230	-1.8200E-04			290 - 390
14	C_2H_5Cl	Chloroethane	0.1191	0.27470	-5.2200E-04			233 - 350
15	C_2H_5NO	*N*-Methylmethanamide	0.2028	0.22835	-8.5600E-05			288 - 468
16	C_2H_6O	Ethanol	0.1692	0.25300	-2.8100E-04			268 - 335
17	$C_2H_6O_2$	1,2-Ethanediol	0.2560	0.25600				273 - 325
18	C_2H_7NO	2-Aminoethanol	0.2993	0.45275	-5.1480E-04			283 - 445
19	C_3H_5ClO	(Chloromethyl)oxirane	0.1308	0.19902	-2.2888E-04			215 - 389
20	C_3H_6O	2-Propanone	0.1614	0.25020	-2.9800E-04			270 - 315
21	$C_3H_6O_2$	Methyl ethanoate	0.1534	0.27770	-4.1700E-04			175 - 385
22	C_3H_7NO	*N*,*N*-Dimethylmethanamide	0.1840	0.26000	-2.5500E-04			296 - 415
23	C_3H_8O	1-Propanol	0.1536	0.20400	-1.6900E-04			165 - 380
24	C_3H_8O	2-Propanol	0.1350	0.20290	-2.2780E-04			185 - 410
25	$C_3H_8O_2$	1,2-Propanediol	0.2004	0.21550	-5.0500E-05			273 - 405
26	$C_3H_8O_3$	1,2,3-Propanetriol	0.2918	0.25800	1.1340E-04			293 - 550
27	C_3H_9N	Trimethylamine	0.1236	0.23813	-3.8397E-04			155 - 275
28	C_4H_4O	Furan	0.1262	0.21980	-3.1405E-04			187 - 297
29	C_4H_4S	Thiophene	0.1995	0.25256	-1.7798E-04			300 - 450
30	C_4H_6	2-Butyne	0.1209	0.21469	-3.1455E-04			240 - 295
31	C_4H_8O	2-Butanone	0.1451	0.21920	-2.4840E-04			185 - 350
32	C_4H_8O	Oxolane	0.1200	0.19428	-2.4900E-04			165 - 340
33	$C_4H_8O_2$	1,4-Dioxane	0.1588	0.30270	-4.8270E-04			285 - 375
34	$C_4H_8O_2$	Ethyl ethanoate	0.1439	0.25010	-3.5630E-04			190 - 350
35	$C_4H_{10}O$	1-Butanol	0.1536	0.21087	-1.9216E-04			260 - 340
36	$C_4H_{10}O$	Diethyl ether	0.1296	0.24900	-4.0050E-04			160 - 430
37	C_5H_5N	Pyridine	0.1651	0.20832	-1.4500E-04			250 - 350
38	C_5H_8	Cyclopentene	0.1287	0.21578	-2.9220E-04			140 - 320
39	C_5H_{10}	1-Pentene	0.1159	0.20760	-3.0770E-04			116 - 295
40	C_5H_{10}	Cyclopentane	0.1262	0.20660	-2.6960E-04			179 - 322
41	C_5H_{12}	Pentane	0.1125	0.25370	-5.7600E-04	3.4400E-07		143 - 470
42	$C_5H_{12}O$	1-Pentanol	0.1528	0.20060	-1.6030E-04			273 - 353
43	C_6H_5Cl	Chlorobenzene	0.1266	0.18390	-1.9220E-04			227 - 404
44	C_6H_6	Benzene	0.1411	0.23930	-3.2920E-04			295 - 350
45	C_6H_6O	Phenol	0.1585	0.18831	-1.0000E-04			323 - 400
46	C_6H_{10}	Cyclohexene	0.1302	0.20270	-2.4300E-04			170 - 355
47	$C_6H_{10}O$	4-Methyl-3-penten-2-one	0.1558	0.24053	-2.8430E-04			220 - 400
48	C_6H_{12}	Cyclohexane	0.1234	0.19813	-2.5050E-04			279 - 353
49	C_6H_{12}	1-Hexene	0.1207	0.21820	-3.2700E-04			143 - 333
50	$C_6H_{12}O$	Cyclohexanol	0.1342	0.17347	-1.3173E-04			298 - 338
51	$C_6H_{12}O$	2-Hexanone	0.1392	0.21076	-2.4000E-04			217 - 400
52	C_6H_{14}	Hexane	0.1196	0.22492	-3.5330E-04			183 - 400
53	$C_6H_{14}O$	1-Hexanol	0.1499	0.20200	-1.7490E-04			228 - 415
54	C_7H_6O	Benzaldehyde	0.1508	0.26990	-3.9930E-04			280 - 400
55	C_7H_8	Toluene	0.1311	0.22050	-3.0000E-04			230 - 360

TABLE 4.2.2 (TCLT)
THERMAL CONDUCTIVITY OF LIQUIDS (continued)

SN	Molecular formula	Name	λ (298.15 K)/ W m^{-1} K^{-1}	*A*(1)	*A*(2)	*A*(3)	*A*(4)	(*T*/K)-range
56	C_7H_8O	Anisole	0.1563	0.23590	-2.6702E-04			235 - 510
57	C_7H_{16}	Heptane	0.1228	0.21250	-3.0100E-04			190 - 370
58	$C_7H_{16}O$	1-Heptanol	0.1594	0.23237	-2.4480E-04			250 - 375
59	C_8H_8	Vinylbenzene	0.1365	0.20215	-2.2010E-04			242 - 412
60	C_8H_{10}	Ethylbenzene	0.1300	0.20149	-2.3988E-04			275 - 425
61	C_8H_{10}	1,2-Dimethylbenzene	0.1313	0.19989	-2.2990E-04			275 - 420
62	C_8H_{10}	1,3-Dimethylbenzene	0.1302	0.20044	-2.3544E-04			275 - 425
63	C_8H_{10}	1,4-Dimethylbenzene	0.1297	0.20003	-2.3573E-04			286 - 413
64	C_8H_{18}	Octane	0.1278	0.21590	-2.9560E-04			223 - 373
65	$C_8H_{18}O$	1-Octanol	0.1611	0.24314	-2.7500E-04			257 - 400
66	C_9H_{12}	Isopropylbenzene	0.1279	0.37307	-1.4070E-03	2.3736E-06	-1.3847E-09	293 - 625
67	C_9H_{12}	1,3,5-Trimethylbenzene	0.1355	0.20330	-2.2740E-04			228 - 635
68	C_9H_{20}	Nonane	0.1310	0.20960	-2.6360E-04			223 - 373
69	$C_9H_{20}O$	1-Nonanol	0.1606	0.22920	-2.3000E-04			268 - 578
70	$C_{10}H_{14}$	4-Isopropyltoluene	0.1219	0.17977	-1.9410E-04			205 - 450
71	$C_{10}H_{22}$	Decane	0.1318	0.20630	-2.5000E-04			243 - 433
72	$C_{10}H_{22}O$	1-Decanol	0.1615	0.22800	-2.2300E-04			280 - 503
73	$C_{11}H_{24}$	Undecane	0.1401	0.20664	-2.2315E-04			275 - 455
74	$C_{12}H_{10}O$	Diphenyl ether	0.1423	0.18720	-1.5060E-04			300 - 530
75	$C_{12}H_{26}$	Dodecane	0.1515	0.21880	-2.2560E-04			263 - 420
76	$C_{12}H_{26}O$	1-Dodecanol	0.1461	0.18929	-1.4482E-04			296 - 535
77	$C_{13}H_{28}$	Tridecane	0.1371	0.19810	-2.0460E-04			293 - 500
78	$C_{14}H_{30}$	Tetradecane	0.1363	0.19570	-1.9930E-04			293 - 510
79	$C_{14}H_{30}O$	1-Tetradecanol	0.1723	0.23150	-1.9870E-04			310 - 560
80	$C_{16}H_{22}O_4$	Dibutyl 1,2-benzenedi-carboxylate	0.1364	0.18090	-1.4940E-04			238 - 450
81	$C_{16}H_{34}$	Hexadecane	0.1397	0.19630	-1.9000E-04			305 - 535
82	$C_{18}H_{38}$	Octadecane	0.1505	0.20310	-1.7650E-04			300 - 590

TABLE 4.2.3 (TCST)
THERMAL CONDUCTIVITY OF SOLIDS

SN	Molecular formula	Name	λ (298.15 K)/ W cm^{-1} K^{-1}	$A(1)$	$A(2)$	$A(3)$	$A(4)$	$A(5)$	$A(6)$	(T/K)-range
1	Ag	Silver	4.29	469.422	-0.240308	5.04491E-04	-5.63673E-07	2.08208E-10		100 - 1100
2	Al	Aluminum	2.37	200.625	0.222807	-3.82422E-04	1.62974E-07			250 - 900
3	Al_2O_3	Dialuminum trioxide	0.362	83.710	-0.224898	2.55842E-04	-1.32727E-07	2.59695E-11		300 - 1700
4	Au	Gold	3.18	332.440	-0.049893	-1.29464E-05				100 - 1100
5	B	Boron	0.274	237.729	-1.618272	4.72040E-03	-6.93422E-06	5.05919E-09	-1.45502E-12	200 - 1000
6	Be	Beryllium	2.01	449.938	-1.326593	2.05321E-03	-1.48356E-06	3.98438E-10		300 - 1300
7	BeO	Beryllium oxide	2.74	718.227	-2.287398	3.34651E-03	-2.58726E-06	1.01634E-09	-1.59142E-13	300 - 1800
8	C	Diamond (Type I)	9.0	6454.312	-46.742095	1.34806E-01	-1.35453E-04			100 - 400
9	Cu	Copper	4.01	571.298	-1.125096	2.54110E-03	-2.52800E-06	8.89684E-10		100 - 1100
10	Fe	Iron	0.804	197.749	-0.822405	2.08161E-03	-2.52201E-06	1.12958E-09		100 - 850
11	Ge	Germanium	0.602	292.742	-1.629895	4.23618E-03	-5.65207E-06	3.72975E-09	-9.59210E-13	200 - 1200
12	Mg	Magnesium	1.56	182.957	-0.176418	3.64937E-04	-2.76266E-07			100 - 600
13	MgO	Magnesium oxide	0.487	109.662	-0.284047	3.07675E-04	-1.52191E-07	2.84787E-11		300 - 1800
14	Na	Sodium	1.42	112.454	0.422991	-2.58811E-03	8.20562E-06	-1.06813E-08		100 - 350
15	Ni	Nickel	0.909	304.842	-2.033412	7.37763E-03	-1.21585E-05	7.33981E-09		100 - 600
16	O_2Si	Silicon dioxide (fused quartz)	0.0133	1.363	-0.000960	2.57365E-06				200 - 1200
17	O_2Th	Thorium dioxide	0.133	23.963	-0.045075	3.22159E-05	-7.67187E-09			300 - 1800
18	O_2Ti	Titanium dioxide	0.084	12.446	-0.015882	6.72378E-06				300 - 1400
19	Pb	Lead	0.353	43.878	-0.051475	9.89562E-05	-7.98866E-08			100 - 600
20	Pt	Platinum	0.716	84.889	-0.094317	2.22059E-04	-2.03976E-07	6.98576E-11		100 - 1100
21	Si	Silicon	1.49	1063.103	-7.302616	2.27572E-02	-3.66949E-05	2.98557E-08	-9.71843E-12	200 - 800
22	Sn	Tin	0.668	155.869	-1.02982	4.70299E-03	-9.72792E-06	7.34291E-09		50 - 500
23	Ti	Titanium	0.219	37.553	-0.083524	1.30450E-04	-8.29088E-08	1.93876E-11		100 - 1600
24	V	Vanadium	0.307	44.143	-0.111828	3.13912E-04	-3.44638E-07	1.37817E-10		100 - 850
25	W	Tungsten	1.73	231.514	-0.253766	2.06280E-04	-7.76072E-08	1.05947E-11		100 - 3500
26	Zn	Zinc	1.16	110.796	0.092165	-3.28527E-04	2.55746E-07			100 - 600
27		95% Cu, 5% Zn	2.49	79.175	1.031716	-1.96675E-03	1.38425E-06			100 - 600
28		90% Cu, 10% Zn	1.93	48.001	0.789340	-1.26208E-03	8.05811E-07			100 - 600
29		80% Cu, 20% Zn	1.43	29.702	0.610792	-9.69002E-04	6.41799E-07			100 - 600
30		65% Fe, 35% Ni	0.0971	1.790	0.045634	-8.39624E-05	6.94141E-08			100 - 600
31		45% Fe, 55% Ni	0.254	9.073	0.134658	-4.05830E-04	5.48611E-07	-2.89893E-10		100 - 600
32		60% Pt, 40% Rh	0.476	33.882	0.050495	-1.50609E-05				300 - 1500

4.3. DIFFUSION COEFFICIENTS

Transport of mass occurs in a system when a concentration gradient is present. In the simplest one-dimensional case, this transport can be described by the generic equation

$$-\mathrm{d}X / \mathrm{d}t = DA(\mathrm{d}c / \mathrm{d}z)$$

where X measures what is being transported (in terms of mass, amount of substance, number of entities, etc.); A is the area of a surface normal to z, the direction of the gradient; c is the concentration (X divided by volume); and D is called the *diffusion coefficient* (or diffusivity). The dimensions of D are $[\mathrm{length}]^2 \cdot [\mathrm{time}]^{-1}$ and the SI unit is $\mathrm{m^2\ s^{-1}}$. Other types of diffusion result from temperature, pressure, or other types of gradients; those are not considered here.

The following tables relate to three types of systems in which diffusion driven by a concentration gradient occurs. The first is a binary gas mixture at low pressures. Here it may be shown that the coefficient describing diffusion of component 1 into component 2, D_{12}, is equal to the coefficient describing diffusion of component 2 into component 1, D_{21}. Thus the system can be described by a single parameter, D_{12}. As long as the pressure is not too high (specifically, as long as binary collisions dominate), D_{12} is inversely proportional to pressure. The temperature dependence of D_{12} is more complex, but can be described by using kinetic theory (Reference 1). There is also a small composition dependence of D_{12}, which is discussed in Reference 1. Self-diffusion in a pure low-pressure gas is a special case (see References 2 and 3).

Table 4.3.1 gives binary diffusion coefficients for a number of gas mixtures at atmospheric pressure (101325 Pa) as a function of temperature. Values of D_{12} at other pressures can be calculated by using the inverse dependence on pressure. The first part of the table gives data for several gases in the presence of a large excess of air. The remainder applies to equimolar mixtures of gases. Each gas pair is ordered alphabetically according to the most common way of writing the formula. The listing of pairs then follows alphabetical order by the first constituent.

The second process considered is diffusion of one liquid, present at very low concentration, into a large excess of another liquid. Values of D_{12} for liquid diffusion are much smaller than those for gases, because of the stronger intermolecular forces in liquids. Typical values for liquids are of order of magnitude $10^{-9}\ \mathrm{m^2\ s^{-1}}$, compared to $10^{-4}\ \mathrm{m^2\ s^{-1}}$ in gases. Although there is no sound theory for liquid diffusion, various approximation methods for treating the temperature and composition dependence are discussed in Reference 5.

Table 4.3.2 gives values of D_{12} for diffusion in some liquid systems at infinite dilution. Although values are given here to two decimal places, most values in the table cannot be relied on to better than 10%. Solvents are listed in alphabetical order, as are the solutes within each solvent group.

Table 4.3.3 give representative values of D_{12} for diffusion of ions in aqueous solution at infinite dilution. All values refer to 298.15 K and were derived from electrical conductivity data. In typical cases the diffusion coefficient increases by 2 to 3% per kelvin as the temperature increases from 298.15 K. The diffusion coefficient of a salt, D_{salt}, may be calculated from the D_{12} values of the constituent ions, D_{12}^{+} and D_{12}^{-}, by the equation

$$D_{\mathrm{salt}} = \frac{(z^{+} + z^{-}) D_{12}^{+} D_{12}^{-}}{z^{+} D_{12}^{+} + z^{-} D_{12}^{-}}$$

where z^{+} and z^{-} are the absolute values of the charges on the positive and negative ions, respectively.

The ions in Table 4.3.3 are arranged in four groups, Inorganic Cations, Inorganic Anions, Organic Cations, and Organic Anions. Within each group entries are listed in alphabetical order, either by formula (for inorganic ions) or name (for organic ions).

Physical quantity	Symbol	SI unit
Diffusion coefficient	D_{12}	$\mathrm{m^2\ s^{-1}}$
Temperature	T	K

REFERENCES

1. Marrero, T. R.; Mason, E. A., "Gaseous Diffusion Coefficients", *J. Phys. Chem. Ref. Data*, 1, 1-118, 1972.
2. Kestin, J.; Knierim, K.; Mason, E. A.; Najafi, B.; Ro, S. T.; Waldman, M., "Equilibrium and Transport Properties of the Rare Gases and Their Mixtures at Low Density", *J. Phys. Chem. Ref. Data*, 13, 229-303, 1984.

3. Boushehri, A.; Bzowski, J.; Kestin, J; Mason, E. A., "Equilibrium and Transport Properties of Eleven Polyatomic Gases at Low Densities", *J. Phys. Chem. Ref. Data*, 16, 445, 1987.
4. Bzowski, J.; Kestin, J.; Mason, E. A.; Uribe, F. J., "Equilibrium and Transport Properties of Gas Mixtures at Low Density: Eleven Polyatomic Gases and Five Noble Gases", J. Phys. Chem. Ref. Data, 19, 1179, 1990.
5. Reid, R. C.; Prausnitz, J. M.; Poling, B. E., *The Properties of Gases and Liquids, Fourth Edition*, McGraw-Hill, New York, 1987.
6. Vargaftik, N. B., *Tables of Thermophysical Properties of Liquids and Gases, Second Edition*, John Wiley, New York, 1975.
7. *Landolt-Börnstein, Numerical Data and Functional Relationships in Science and Technology, Sixth Edition,* II/5a, Transport Phenomena I (Viscosity and Diffusion), Springer-Verlag, Heidelberg, 1969.
8. Lide, D. R., Editor, *CRC Handbook of Chemistry and Physics, 74th Edition*, CRC Press, Boca Raton, FL, 1993.
9. Lobo, V. M. M.; Quaresma, J. L., *Handbook of Electrolyte Solutions*, Physical Science Data Series 41, Elsevier, Amsterdam, 1989.

TABLE 4.3.1
DIFFUSION COEFFICIENTS IN GASES AT A PRESSURE OF 101325 Pa (1 atm)

System	D_{12}/cm² s⁻¹						
	T/K = 200	273.15	293.15	373.15	473.15	573.15	673.15
Large Excess of Air							
Ar-air		0.167	0.148	0.289	0.437	0.612	0.810
CH_4-air			0.106	0.321	0.485	0.678	0.899
CO-air			0.208	0.315	0.475	0.662	0.875
CO_2-air			0.160	0.252	0.390	0.549	0.728
H_2-air		0.668	0.627	1.153	1.747	2.444	3.238
H_2O-air			0.242	0.399	0.638	0.873	1.135
He-air		0.617	0.580	1.057	1.594	2.221	2.933
SF_6-air				0.150	0.233	0.329	0.438
Equimolar Mixture							
Ar-CH_4				0.306	0.467	0.657	0.876
Ar-CO		0.168	0.187	0.290	0.439	0.615	0.815
Ar-CO_2		0.129	0.078	0.235	0.365	0.517	0.689
Ar-H_2		0.698	0.794	1.228	1.876	2.634	3.496
Ar-He	0.381	0.645	0.726	1.088	1.617	2.226	2.911
Ar-Kr	0.064	0.117	0.134	0.210	0.323	0.456	0.605
Ar-N_2		0.168	0.190	0.290	0.439	0.615	0.815
Ar-Ne	0.160	0.277	0.313	0.475	0.710	0.979	1.283
Ar-O_2		0.166	0.189	0.285	0.430	0.600	0.793
Ar-SF_6				0.128	0.202	0.290	0.389
Ar-Xe	0.052	0.095	0.108	0.171	0.264	0.374	0.498
CH_4-H_2			0.782	1.084	1.648	2.311	3.070
CH_4-He			0.723	0.992	1.502	2.101	2.784
CH_4-N_2			0.220	0.317	0.480	0.671	0.890
CH_4-O_2			0.210	0.341	0.523	0.736	0.978
CH_4-SF_6				0.167	0.257	0.363	0.482
CO-CO_2			0.162	0.250	0.384		
CO-H_2	0.408	0.686	0.772	1.162	1.743	2.423	3.196
CO-He	0.365	0.619	0.698	1.052	1.577	2.188	2.882
CO-Kr		0.131	0.581	0.227	0.346	0.485	0.645
CO-N_2	0.133	0.208	0.231	0.336	0.491	0.673	0.878
CO-O_2			0.202	0.307	0.462	0.643	0.849
CO-SF_6				0.144	0.226	0.323	0.432
CO_2-C_3H_8			0.084	0.133	0.209		
CO_2-H_2	0.315	0.552	0.412	0.964	1.470	2.066	2.745
CO_2-H_2O			0.162	0.292	0.496	0.741	1.021
CO_2-He	0.300	0.513	0.400	0.878	1.321		
CO_2-N_2			0.160	0.253	0.392	0.553	0.733
CO_2-N_2O	0.055	0.099	0.113	0.177	0.276		
CO_2-Ne	0.131	0.227	0.199	0.395	0.603	0.847	
CO_2-O_2			0.159	0.248	0.380	0.535	0.710
CO_2-SF_6				0.099	0.155		
D_2-H_2	0.631	1.079	1.219	1.846	2.778	3.866	5.103
H_2-He	0.775	1.320	1.490	2.255	3.394	4.726	6.242
H_2-Kr	0.340	0.601	0.682	1.053	1.607	2.258	2.999
H_2-N_2	0.408	0.686	0.772	1.162	1.743	2.423	3.196
H_2-Ne	0.572	0.982	0.317	1.684	2.541	3.541	4.677
H_2-O_2		0.692	0.756	1.188	1.792	2.497	3.299
H_2-SF_6			0.208	0.649	0.998	1.400	1.851
H_2-Xe		0.513	0.122	0.890	1.349	1.885	2.493
H_2O-N_2			0.242	0.399			

TABLE 4.3.1
DIFFUSION COEFFICIENTS IN GASES AT A PRESSURE OF 101325 Pa (1 atm) (continued)

System	T/K = 200	273.15	293.15	373.15	473.15	573.15	673.15
H_2O-O_2			0.244	0.403	0.645	0.882	1.147
He-Kr	0.330	0.559	0.629	0.942	1.404	1.942	2.550
He-N_2	0.365	0.619	0.698	1.052	1.577	2.188	2.882
He-Ne	0.563	0.948	1.066	1.592	2.362	3.254	4.262
He-O_2		0.641	0.697	1.092	1.640	2.276	2.996
He-SF_6			1.109	0.592	0.871	1.190	1.545
He-Xe	0.282	0.478	0.538	0.807	1.201	1.655	2.168
Kr-N_2		0.131	0.149	0.227	0.346	0.485	0.645
Kr-Ne	0.131	0.228	0.258	0.392	0.587	0.812	1.063
Kr-Xe	0.035	0.064	0.073	0.116	0.181	0.257	0.344
N_2-Ne			0.258	0.483	0.731	1.021	1.351
N_2-O_2			0.202	0.307	0.462	0.643	0.849
N_2-SF_6				0.148	0.231	0.328	0.436
N_2-Xe		0.107	0.708	0.188	0.287	0.404	0.539
Ne-Xe	0.111	0.193	0.219	0.332	0.498	0.688	0.901
O_2-SF_6			0.097	0.154	0.238	0.334	0.441

TABLE 4.3.2
DIFFUSION COEFFICIENTS IN LIQUIDS AT INFINITE DILUTION

Solute	Solvent	T/K	$D_{12}/10^{-5}$ cm^2 s^{-1}
Aniline	Benzene	298	1.96
Benzoic acid	Benzene	298	1.38
Bromobenzene	Benzene	281	1.45
2-Butanone	Benzene	303	2.09
Chloroethene	Benzene	281	1.77
Cyclohexane	Benzene	298	2.09
Ethanoic acid	Benzene	298	2.09
Ethanol	Benzene	288	2.25
Heptane	Benzene	298	1.78
Methanoic acid	Benzene	298	2.28
Toluene	Benzene	298	1.85
1,2,4-Trichlorobenzene	Benzene	281	1.34
Trichloromethane	Benzene	298	2.26
Benzene	1-Butanol	298	1.00
Biphenyl	1-Butanol	298	0.63
Butyric acid	1-Butanol	303	0.51
1,4-Dichlorobenzene	1-Butanol	298	0.82
Hexanedioc acid	1-Butanol	303	0.40
Methanol	1-Butanol	303	0.59
cis-9-Octadecenoic acid	1-Butanol	303	0.25
Propane	1-Butanol	298	1.57
Water	1-Butanol	298	0.56
Benzene	Cyclohexane	298	1.90
Tetrachloromethane	Cyclohexane	298	1.49
Toluene	Cyclohexane	298	1.57
Benzene	Ethanol	298	1.81
Iodine	Ethanol	298	1.32
Iodobenzene	Ethanol	293	1.00
3-Methyl-1-butanol	Ethanol	293	0.81
2-Propen-1-ol	Ethanol	293	0.98
Pyridine	Ethanol	293	1.10
Tetrachloromethane	Ethanol	298	1.50
Water	Ethanol	298	1.24
2-Butanone	Ethyl ethanoate	303	2.93
Ethanoic acid	Ethyl ethanoate	293	2.18
Ethyl benzoate	Ethyl ethanoate	293	1.85
Nitrobenzene	Ethyl ethanoate	293	2.25
2-Propanone	Ethyl ethanoate	293	3.18
Water	Ethyl ethanoate	298	3.20
Benzene	Heptane	298	3.91
Toluene	Heptane	298	3.72
Bromobenzene	Hexane	281	2.60
2-Butanone	Hexane	303	3.74
Dodecane	Hexane	298	2.73
Iodine	Hexane	298	4.45
Propane	Hexane	298	4.87
Tetrachloromethane	Hexane	298	3.70
Toluene	Hexane	298	4.21
Benzoic acid	2-Propanone	298	2.62
Ethanoic acid	2-Propanone	298	3.31
Methanoic acid	2-Propanone	298	3.77
Nitrobenzene	2-Propanone	293	2.94
Water	2-Propanone	298	4.56
Cyclohexane	Tetrachloromethane	298	1.30
Iodine	Tetrachloromethane	303	1.63
Benzene	Toluene	298	2.54
Benzoic acid	Toluene	298	1.49
Cyclohexane	Toluene	298	2.42
Ethanoic acid	Toluene	298	2.26
Methanoic acid	Toluene	298	2.65
Water	Toluene	298	6.19
Benzene	Trichloromethane	298	2.89
2-Butanone	Trichloromethane	298	2.13
Butyl ethanoate	Trichloromethane	298	1.71
Diethyl ether	Trichloromethane	298	2.13
Ethanol	Trichloromethane	288	2.20
Ethyl ethanoate	Trichloromethane	298	2.02
2-Propanone	Trichloromethane	298	2.35
Alanine	Water	298	0.91
Aniline	Water	293	0.92
Arabinose	Water	293	0.69
Benzene	Water	293	1.02
1-Butanol	Water	288	0.77
Chloroethene	Water	298	1.34
Cyclohexane	Water	293	0.84
Diethylamine	Water	293	0.97
1,2-Ethanediol	Water	298	1.16
Ethanenitrile	Water	288	1.26
Ethanoic acid	Water	293	1.19
Ethanol	Water	288	1.00
Ethylbenzene	Water	293	0.81
Ethyl ethanoate	Water	293	1.00
Glucose	Water	288	0.52
Lactose	Water	288	0.38
Maltose	Water	288	0.38
Mannitol	Water	288	0.50
Methane	Water	298	1.49
Methanol	Water	288	1.28
3-Methyl-1-butanol	Water	283	0.69
Methylcyclopentane	Water	293	0.85
Phenol	Water	293	0.89
1,2,3-Propanetriol	Water	288	0.72
1-Propanol	Water	288	0.87
2-Propanone	Water	298	1.28
Propene	Water	298	1.44
2-Propen-1-ol	Water	288	0.90
Pyridine	Water	288	0.58
Raffinose	Water	288	0.33
Sacharose	Water	288	0.38
Toluene	Water	293	0.85
Urea	Water	293	1.18
Urethane	Water	288	0.80

TABLE 4.3.3
DIFFUSION COEFFICIENTS OF IONS IN WATER AT INFINITE DILUTION

Ion	D_{12}(298.15 K)/ 10^{-5} cm^2 s^{-1}
Inorganic Cations	
Ag^+	1.648
Al^{3+}	0.541
Ba^{2+}	0.847
Be^{2+}	0.599
Ca^{2+}	0.792
Cd^{2+}	0.719
Ce^{3+}	0.620
Co^{2+}	0.732
$[Co(NH_3)_6]^{3+}$	0.904
$[Co(en)_3]^{3+}$	0.663
$[Co_2(trien)_3]^{6+}$	0.306
Cr^{3+}	0.595
Cs^+	2.056
Cu^{2+}	0.714
D^+	6.655
Dy^{3+}	0.582
Er^{3+}	0.585
Eu^{3+}	0.602
Fe^{2+}	0.719
Fe^{3+}	0.604
Gd^{3+}	0.597
H^+	9.311
Hg^{2+}	0.913
Hg^{2+}	0.847
Ho^{3+}	0.589
K^+	1.957
La^{3+}	0.619
Li^+	1.029
Mg^{2+}	0.706
Mn^{2+}	0.712
NH_4^+	1.957
$N_2H_5^+$	1.571
Na^+	1.334
Nd^{3+}	0.616
Ni^{2+}	0.666
$[Ni_2(trien)_3]^{4+}$	0.346
Pb^{2+}	0.945
Pr^{3+}	0.617
Ra^{2+}	0.889
Rb^+	2.072
Sc^{3+}	0.574
Sm^{3+}	0.608
Sr^{2+}	0.791
Tl^+	1.989
Tm^{3+}	0.581
UO_2^{2+}	0.426
Y^{3+}	0.550
Yb^{3+}	0.582
Zn^{2+}	0.703
Inorganic Anions	
$Au(CN)_2^-$	1.331
$Au(CN)_4^-$	0.959
$B(C_6H_5)_4^-$	0.559
Br^-	2.080
Br_3^-	1.145
BrO_3^-	1.483
CN^-	2.077
CNO^-	1.720
CO_3^{2-}	0.923
Cl^-	2.032
ClO_2^-	1.385
ClO_3^-	1.720
ClO_4^-	1.792
$[Co(CN)_6]^{3-}$	0.878
CrO_4^{2-}	1.132
F^-	1.475
$[Fe(CN)_6]^{4-}$	0.735
$[Fe(CN)_6]^{3-}$	0.896
$H_2AsO_4^-$	0.905
HCO_3^-	1.185
HF_2^-	1.997
HPO_4^{2-}	0.439
$H_2PO_4^-$	0.879
$H_2PO_2^-$	1.225
HS^-	1.731
HSO_3^-	1.331
HSO_4^-	1.331
$H_2SbO_4^-$	0.825
I^-	2.045
IO_3^-	1.078
IO_4^-	1.451
MnO_4^-	1.632
MoO_4^-	1.984
$N(CN)_2^-$	1.451
NO_2^-	1.912
NO_3^-	1.902
$NH_2SO_3^-$	1.294
N_3^-	1.837
OCN^-	1.720
OH^-	5.273
PF_6^-	1.515
PO_3F^{2-}	0.843
PO_4^{3-}	0.612
$P_2O_7^{4-}$	0.639
$P_3O_9^{3-}$	0.742
$P_3O_{10}^{5-}$	0.581
ReO_4^-	1.462
SCN^-	1.758
SO_3^{2-}	1.064
SO_4^{2-}	1.065
$S_2O_3^{2-}$	1.132
$S_4O_4^{2-}$	0.885
$S_2O_6^{2-}$	1.238
$S_2O_8^{2-}$	1.145
$Sb(OH)_6^-$	0.849
$SeCN^-$	1.723
SeO_4^{2-}	1.008
WO_4^{2-}	0.919

Ion	D_{12}(298.15 K)/ 10^{-5} cm^2 s^{-1}
Organic Cations	
Benzyltrimethylammonium$^+$	0.921
Isobutylammonium$^+$	1.012
Butyltrimethylammonium$^+$	0.895
Decylpyridinium$^+$	0.786
Decyltrimethylammonium$^+$	0.650
Diethylammonium$^+$	1.118
Dimethylammonium$^+$	1.379
Dipropylammonium$^+$	0.802
Dodecylammonium$^+$	0.634
Dodecyltrimethylammonium$^+$	0.602
Ethanolammonium$^+$	1.124
Ethylammonium$^+$	1.257
Ethyltrimethylammonium$^+$	1.078
Hexadecyltrimethylammonium$^+$	0.557
Hexyltrimethylammonium$^+$	0.788
Histidyl$^+$	0.612
Hydroxyethyltrimethylarsonium$^+$	1.049
Methylammonium$^+$	1.563
Octadecylpyridinium$^+$	0.533
Octadecyltributylammonium$^+$	0.442
Octadecyltriethylammonium$^+$	0.477
Octadecyltrimethylammonium$^+$	0.530

TABLE 4.3.3
DIFFUSION COEFFICIENTS OF IONS IN WATER AT INFINITE DILUTION (continued)

Ion	D_{12}(298.15 K)/ 10^{-5} cm^2 s^{-1}	Ion	D_{12}(298.15 K)/ 10^{-5} cm^2 s^{-1}
Octadecyltripropylammonium$^+$	0.458	Decylsulfate$^-$	0.692
Octyltrimethylammonium$^+$	0.706	Dichloroacetate$^-$	1.020
Pentylammonium$^+$	0.985	Diethylbarbiturate^{2-}	0.350
Piperidinium$^+$	0.991	Dihydrogencitrate$^-$	0.799
Propylammonium$^+$	1.086	Dimethylmalonate^{2-}	0.658
Pyrilammonium$^+$	0.647	3,5-Dinitrobenzoate$^-$	0.754
Tetrabutylammonium$^+$	0.519	Dodecylsulfate$^-$	0.639
Tetradecyltrimethylammonium$^+$	0.573	Ethylmalonate$^-$	1.313
Tetraethylammonium$^+$	0.868	Ethylsulfate$^-$	1.055
Tetramethylammonium$^+$	1.196	Fluoroacetate$^-$	1.182
Tetraisopentylammonium$^+$	0.477	Fluorobenzoate$^-$	0.879
Tetrapentylammmonium$^+$	0.466	Formate$^-$	1.454
Tetrapropylammonium$^+$	0.623	Fumarate^{2-}	0.823
Triethylammonium$^+$	0.913	Glutarate^{2-}	0.700
Triethylsulfonium$^+$	0.961	Hydrogenoxalate$^-$	1.070
Trimethylammonium$^+$	1.258	Isovalerate$^-$	0.871
Trimethylhexylammonium$^+$	0.921	Iodoacetate$^-$	1.081
Trimethylsulfonium$^+$	1.369	Lactate$^-$	1.033
Tripropylammonium$^+$	0.695	Malate^{2-}	0.783
		Maleate^{2-}	0.824
Organic Anions		Malonate^{2-}	0.845
		Methylsulfate$^-$	1.299
Acetate$^-$	1.089	Naphthylacetate$^-$	0.756
p-Anisate$^-$	0.772	Oxalate^{2-}	0.987
Azelate^{2-}	0.541	Octylsulfate$^-$	0.772
Benzoate$^-$	0.863	Phenylacetate$^-$	0.815
Bromoacetate$^-$	1.044	*o*-Phthalate^{2-}	0.696
Bromobenzoate$^-$	0.799	*m*-Phthalate^{2-}	0.728
Butyrate$^-$	0.868	Picrate$^-$	0.809
Chloroacetate$^-$	1.124	Pivalate$^-$	0.849
m-Chlorobenzoate$^-$	0.825	Propionate$^-$	0.953
o-Chlorobenzoate$^-$	0.804	Propylsulfate$^-$	0.988
Citrate^{3-}	0.623	Salicylate$^-$	0.959
Crotonate$^-$	0.884	Suberate^{2-}	0.479
Cyanoacetate$^-$	1.156	Succinate^{2-}	0.783
Cyclohexane carboxylate$^-$	0.764	Tartarate^{2-}	0.794
1,1-Cyclopropanedicarboxylate^{2-}	0.711	Trichloroacetate$^-$	0.975

Section 5
Tables of Properties of Individual Substances

TABLE 5.1
THERMOPHYSICAL PROPERTIES OF SOME COMMON FLUIDS

The following tables give various thermodynamic and transport properties in the liquid and gaseous states for nitrogen, oxygen, hydrogen, helium, argon, methane, ethane, and propane at a selection of pressures and temperatures. Please consult the references for information on the uncertainties and the reference states for U, H, and S. More extensive tables may also be found in the references.

Physical quantity	Symbol	SI unit
Amount of substance density	ρ	mol m^{-3}
Molar internal energy	U	J mol^{-1}
Molar enthalpy	H	J mol^{-1}
Molar entropy	S	J K^{-1} mol^{-1}
Molar isochoric heat capacity	C_v	J K^{-1} mol^{-1}
Molar isobaric heat capacity	C_p	J K^{-1} mol^{-1}
Speed of sound	v_s	m s^{-1}
Viscosity	η	Pa s
Thermal conductivity	λ	W m^{-1} K^{-1}
Relative permittivity (dielectric constant)	ε	1

REFERENCES

1. Younglove, B. A., "Thermophysical Properties of Fluids. Part I", *J. Phys. Chem. Ref. Data*, 11, Suppl. 1, 1982.
2. Younglove, B. A.; Ely, J. F., "Thermophysical Properties of Fluids. Part II", *J. Phys. Chem. Ref. Data*, 16, 577, 1987.
3. McCarty, R. D., "Thermodynamic Properties of Helium", *J. Phys. Chem. Ref. Data*, 2, 923, 1973.

Nitrogen (N_2)

T/ K	ρ/ mol L^{-1}	U/ J mol^{-1}	H/ J mol^{-1}	S/ J K^{-1} mol^{-1}	C_v/ J K^{-1} mol^{-1}	C_p/ J K^{-1} mol^{-1}	η/ μPa s	λ/ mW m^{-1} K^{-1}	ε
P = 0.1 MPa (1 bar)									
70	30.017	–3828	–3824	73.8	28.5	57.2	203.9	143.5	1.45269
77.25	28.881	–3411	–3407	79.5	27.8	57.8	152.2	133.8	1.43386
77.25	0.163	1546	2161	151.6	21.6	31.4	5.3	7.6	1.00215
100	0.123	2041	2856	159.5	21.1	30.0	6.8	9.6	1.00162
200	0.060	4140	5800	179.9	20.8	29.2	12.9	18.4	1.00079
300	0.040	6223	8717	191.8	20.8	29.2	18.0	25.8	1.00053
400	0.030	8308	11635	200.2	20.9	29.2	22.2	32.3	1.00040
500	0.024	10414	14573	206.7	21.2	29.6	26.1	38.5	1.00032
600	0.020	12563	17554	212.2	21.8	30.1	29.5	44.5	1.00026
700	0.017	14770	20593	216.8	22.4	30.7	32.8	50.5	1.00023
800	0.015	17044	23698	221.0	23.1	31.4	35.8	56.3	1.00020
900	0.013	19383	26869	224.7	23.7	32.0	38.7	62.0	1.00017
1000	0.012	21786	30103	228.1	24.3	32.6	41.5	67.7	1.00016
1500	0.008	34530	47004	241.8	26.4	34.7	54.0	93.3	1.00010
P = 1 MPa									
70	30.070	–3838	–3805	73.6	28.9	56.9	205.9	144.1	1.45355
80	28.504	–3267	–3232	81.3	27.8	57.7	139.5	130.7	1.42760
90	26.721	–2685	–2648	88.2	26.7	59.4	100.1	115.3	1.39824
100	24.634	–2073	–2032	94.6	26.2	64.4	73.1	98.5	1.36417
103.75	23.727	–1828	–1786	97.1	26.2	67.8	64.8	91.8	1.34947
103.75	1.472	1788	2467	138.1	24.1	45.0	7.6	12.5	1.01954
200	0.614	4048	5675	160.3	21.0	30.4	13.2	19.3	1.00812
300	0.402	6171	8661	172.5	20.9	29.6	18.1	26.3	1.00529
400	0.300	8273	11609	180.9	20.9	29.5	22.4	32.7	1.00395

TABLE 5.1
THERMOPHYSICAL PROPERTIES OF SOME COMMON FLUIDS (continued)

T/ K	ρ/ mol L⁻¹	U/ J mol⁻¹	H/ J mol⁻¹	S/ J K⁻¹ mol⁻¹	C_v/ J K⁻¹ mol⁻¹	C_p/ J K⁻¹ mol⁻¹	η/ μPa s	λ/ mW m⁻¹ K⁻¹	ε
500	0.240	10389	14563	187.5	21.3	29.7	26.1	38.8	1.00315
600	0.200	12544	17554	193.0	21.8	30.2	29.6	44.8	1.00262
700	0.171	14756	20600	197.7	22.4	30.8	32.8	50.7	1.00224
800	0.150	17032	23709	201.8	23.1	31.4	35.9	56.5	1.00196
900	0.133	19374	26884	205.6	23.7	32.1	38.8	62.2	1.00174
1000	0.120	21778	30121	209.0	24.3	32.7	41.5	67.8	1.00157
1500	0.080	34527	47029	222.7	26.4	34.8	54.0	93.4	1.00104
P = 10 MPa									
65.32	31.120	–4176	–3855	68.6	31.8	53.8	275.7	153.8	1.47067
100	26.201	–2328	–1946	92.0	27.4	56.3	90.2	112.3	1.38942
200	7.117	3037	4442	136.4	22.7	45.5	17.6	30.4	1.09698
300	3.989	5667	8174	151.7	21.4	33.4	20.1	31.9	1.05347
400	2.898	7941	11392	161.0	21.3	31.3	23.7	36.7	1.03860
500	2.302	10148	14492	167.9	21.5	30.8	27.1	42.0	1.03055
600	1.918	12361	17575	173.5	21.9	30.9	30.4	47.4	1.02538
700	1.647	14613	20683	178.3	22.5	31.3	33.5	53.0	1.02175
800	1.445	16919	23837	182.5	23.2	31.8	36.4	58.6	1.01904
900	1.288	19283	27046	186.3	23.8	32.4	39.3	64.1	1.01694
1000	1.162	21705	30308	189.8	24.4	32.9	42.0	69.6	1.01526
1500	0.783	34504	47283	203.5	26.5	34.8	54.3	94.7	1.01020

Oxygen (O_2)

T/ K	ρ/ mol L⁻¹	U/ J mol⁻¹	H/ J mol⁻¹	S/ J K⁻¹ mol⁻¹	C_v/ J K⁻¹ mol⁻¹	C_p/ J K⁻¹ mol⁻¹	η/ μPa s	λ/ mW m⁻¹ K⁻¹	ε
P = 0.1 MPa (1 bar)									
60	40.049	–5883	–5880	72.4	34.9	53.4	425.2	188.2	1.55619
80	37.204	–4814	–4812	87.7	31.0	53.6	251.7	166.1	1.51114
100	0.123	2029	2840	172.9	21.4	30.5	7.5	9.3	1.00146
120	0.102	2458	3442	178.4	21.0	29.8	9.0	11.2	1.00121
140	0.087	2881	4035	182.9	20.9	29.5	10.5	13.1	1.00103
160	0.076	3301	4624	186.9	20.9	29.4	11.9	15.0	1.00090
180	0.067	3720	5210	190.3	20.8	29.3	13.3	16.7	1.00080
200	0.060	4138	5796	193.4	20.8	29.3	14.6	18.4	1.00072
220	0.055	4556	6381	196.2	20.8	29.3	15.9	20.1	1.00065
240	0.050	4974	6966	198.8	20.9	29.3	17.2	21.7	1.00060
260	0.046	5393	7552	201.1	20.9	29.3	18.4	23.2	1.00055
280	0.043	5812	8138	203.3	21.0	29.4	19.5	24.8	1.00051
300	0.040	6234	8726	205.3	21.1	29.4	20.6	26.3	1.00048
320	0.038	6657	9316	207.2	21.2	29.5	21.7	27.8	1.00045
340	0.035	7082	9908	209.0	21.3	29.7	22.8	29.3	1.00042
360	0.033	7510	10503	210.7	21.5	29.8	23.8	30.8	1.00040
380	0.032	7941	11100	212.3	21.6	30.0	24.8	32.2	1.00038
P = 1 MPa									
60	40.084	–5887	–5863	72.3	34.9	53.3	428.5	188.4	1.55674
80	37.254	–4822	–4795	87.6	31.0	53.5	253.8	166.4	1.51192
100	34.153	–3741	–3712	99.7	28.5	55.2	155.6	137.9	1.46381
120	1.198	2163	2997	156.7	24.0	40.6	9.4	13.9	1.01429
140	0.950	2683	3735	162.4	22.2	34.4	10.8	14.9	1.01133
160	0.802	3151	4398	166.8	21.5	32.2	12.2	16.3	1.00955
180	0.698	3598	5030	170.5	21.2	31.2	13.5	17.7	1.00831
200	0.620	4035	5647	173.8	21.1	30.6	14.8	19.3	1.00738
220	0.559	4466	6255	176.7	21.0	30.3	16.1	20.8	1.00665

TABLE 5.1
THERMOPHYSICAL PROPERTIES OF SOME COMMON FLUIDS (continued)

T/ K	ρ/ mol L⁻¹	U/ J mol⁻¹	H/ J mol⁻¹	S/ J K⁻¹ mol⁻¹	C_v/ J K⁻¹ mol⁻¹	C_p/ J K⁻¹ mol⁻¹	η/ μPa s	λ/ mW m⁻¹ K⁻¹	ε
240	0.509	4894	6858	179.3	21.0	30.1	17.3	22.3	1.00606
260	0.468	5321	7458	181.7	21.0	29.9	18.5	23.8	1.00556
280	0.433	5748	8056	183.9	21.1	29.9	19.6	25.2	1.00515
300	0.403	6174	8654	186.0	21.1	29.9	20.7	26.7	1.00479
320	0.377	6602	9252	187.9	21.2	29.9	21.8	28.2	1.00448
340	0.355	7032	9851	189.7	21.4	30.0	22.8	29.6	1.00421
360	0.335	7463	10452	191.4	21.5	30.1	23.9	31.1	1.00397
380	0.317	7898	11056	193.1	21.7	30.2	24.9	32.6	1.00376
P = 10 MPa									
60	40.419	–5931	–5684	71.5	35.1	53.0	461.8	189.9	1.56210
80	37.727	–4893	–4628	86.7	31.6	52.7	274.4	168.6	1.51936
100	34.881	–3856	–3570	98.5	29.1	53.4	171.0	141.2	1.47500
120	31.721	–2796	–2481	108.4	27.3	55.9	113.0	115.1	1.42677
140	27.890	–1662	–1304	117.5	26.2	62.9	76.3	91.8	1.36972
160	22.379	-322	125	127.0	26.1	84.8	48.6	71.2	1.29037
180	13.232	1489	2245	139.5	26.6	105.9	26.2	46.8	1.16560
200	8.666	2681	3835	147.9	24.0	60.6	21.2	34.0	1.10650
220	6.868	3424	4880	152.9	22.6	46.4	20.5	30.8	1.08380
240	5.836	4029	5742	156.6	22.0	40.6	20.8	30.1	1.07090
260	5.134	4573	6521	159.7	21.8	37.6	21.4	30.2	1.06219
280	4.613	5086	7254	162.5	21.6	35.8	22.1	30.8	1.05575
300	4.205	5581	7959	164.9	21.6	34.7	22.9	31.6	1.05073
320	3.874	6063	8645	167.1	21.7	33.9	23.7	32.6	1.04667
340	3.598	6538	9318	169.1	21.8	33.4	24.6	33.7	1.04329
360	3.363	7009	9982	171.0	21.9	33.0	25.4	34.9	1.04043
380	3.161	7477	10641	172.8	22.0	32.8	26.3	36.1	1.03796

Hydrogen (H_2)

T/ K	ρ/ mol L⁻¹	U/ J mol⁻¹	H/ J mol⁻¹	S/ J K⁻¹ mol⁻¹	C_v/ J K⁻¹ mol⁻¹	C_p/ J K⁻¹ mol⁻¹	v_s/ m s⁻¹	ε
P = 0.1 MPa (1 bar)								
15	37.738	–605	–603	11.2	9.7	14.4	1319	1.24827
20	35.278	–524	–521	15.8	11.3	19.1	1111	1.23093
40	0.305	491	818	75.6	12.5	21.3	521	1.00186
60	0.201	748	1244	84.3	13.1	21.6	636	1.00122
80	0.151	1030	1694	90.7	15.3	23.7	714	1.00091
100	0.120	1370	2202	96.4	18.7	27.1	773	1.00073
120	0.100	1777	2776	101.6	21.8	30.2	827	1.00061
140	0.086	2237	3401	106.4	23.8	32.2	883	1.00052
160	0.075	2723	4054	110.8	24.6	33.0	940	1.00046
180	0.067	3216	4714	114.7	24.6	32.9	998	1.00041
200	0.060	3703	5367	118.1	24.1	32.4	1054	1.00037
220	0.055	4179	6009	121.2	23.4	31.8	1110	1.00033
240	0.050	4641	6638	123.9	22.8	31.2	1163	1.00030
260	0.046	5093	7256	126.4	22.3	30.6	1214	1.00028
280	0.043	5535	7865	128.6	21.9	30.2	1263	1.00026
300	0.040	5970	8466	130.7	21.6	29.9	1310	1.00024
400	0.030	8093	11421	139.2	21.0	29.3	1518	1.00018
P = 1 MPa								
15	38.109	–609	–583	10.9	10.1	14.1	1315	1.25089
20	35.852	–532	–504	15.5	11.4	18.4	1155	1.23496
40	3.608	399	676	54.1	12.9	28.4	498	1.02209

TABLE 5.1
THERMOPHYSICAL PROPERTIES OF SOME COMMON FLUIDS (continued)

T/ K	ρ/ mol L^{-1}	U/ J mol^{-1}	H/ J mol^{-1}	S/ J K^{-1} mol^{-1}	C_v/ J K^{-1} mol^{-1}	C_p/ J K^{-1} mol^{-1}	v_s/ m s^{-1}	ε
60	2.098	697	1173	64.3	13.2	23.5	635	1.01280
80	1.523	994	1651	71.1	15.4	24.7	719	1.00928
100	1.204	1343	2174	77.0	18.8	27.7	779	1.00733
120	0.999	1756	2758	82.3	21.9	30.6	835	1.00608
140	0.854	2219	3390	87.1	23.9	32.5	891	1.00520
160	0.747	2709	4048	91.5	24.7	33.2	949	1.00454
180	0.663	3204	4712	95.4	24.6	33.1	1006	1.00404
200	0.597	3693	5368	98.9	24.1	32.5	1063	1.00363
220	0.543	4170	6012	102.0	23.5	31.9	1118	1.00330
240	0.498	4634	6643	104.7	22.9	31.2	1171	1.00303
260	0.460	5087	7263	107.2	22.3	30.7	1222	1.00279
280	0.427	5530	7873	109.5	21.9	30.3	1271	1.00259
300	0.399	5966	8475	111.5	21.6	30.0	1317	1.00242
400	0.299	8091	11433	120.1	21.0	29.4	1525	1.00182
P = 10 MPa								
20	39.669	–568	–316	13.0	10.9	15.0	1458	1.26198
40	31.344	–209	110	27.3	13.2	27.0	1171	1.20354
60	21.273	255	725	39.7	13.8	32.5	931	1.13527
80	14.830	686	1360	48.8	15.9	31.1	886	1.09303
100	11.417	1110	1986	55.8	19.3	31.9	904	1.07109
120	9.357	1571	2640	61.8	22.4	33.5	941	1.05801
140	7.969	2068	3323	67.0	24.3	34.6	989	1.04925
160	6.963	2583	4020	71.7	25.0	34.9	1042	1.04294
180	6.195	3099	4713	75.7	24.9	34.4	1096	1.03814
200	5.588	3604	5393	79.3	24.4	33.6	1150	1.03436
220	5.094	4094	6057	82.5	23.7	32.8	1203	1.03129
240	4.683	4569	6704	85.3	23.1	32.0	1254	1.02874
260	4.336	5030	7336	87.8	22.6	31.3	1302	1.02659
280	4.038	5481	7958	90.1	22.1	30.8	1349	1.02475
300	3.780	5924	8570	92.3	21.8	30.4	1394	1.02315
400	2.869	8073	11559	100.9	21.2	29.6	1592	1.01753

Helium (^{4}He)

T/ K	ρ/ mol L^{-1}	U/ J mol^{-1}	H/ J mol^{-1}	S/ J K^{-1} mol^{-1}	C_v/ J K^{-1} mol^{-1}	C_p/ J K^{-1} mol^{-1}	v_s/ m s^{-1}	η/ μPa s	ε
P = 0.1 MPa (1 bar)									
3	35.794	–39	–36	9.8	7.6	9.4	222	3.85	1.05646
4	32.477	–27	–24	13.3	9.1	16.3	185	3.33	1.05114
5	2.935	52	86	39.1	12.7	27.1	120	1.39	1.00456
10	1.238	120	201	55.2	12.5	21.7	185	2.26	1.00192
20	0.602	247	413	69.9	12.5	21.0	264	3.58	1.00093
50	0.240	623	1039	89.0	12.5	20.8	417	6.36	1.00037
100	0.120	1247	2079	103.4	12.5	20.8	589	9.78	1.00019
200	0.060	2494	4158	117.8	12.5	20.8	833	15.14	1.00009
300	0.040	3741	6237	126.3	12.5	20.8	1020	19.93	1.00006
400	0.030	4988	8315	132.3	12.5	20.8	1177	24.29	1.00005
500	0.024	6236	10394	136.9	12.5	20.8	1316	28.36	1.00004
600	0.020	7483	12472	140.7	12.5	20.8	1441	32.22	1.00003
700	0.017	8730	14551	143.9	12.5	20.8	1557	35.89	1.00003
800	0.015	9977	16630	146.7	12.5	20.8	1664	39.43	1.00002
900	0.013	11224	18708	149.1	12.5	20.8	1765	42.85	1.00002
1000	0.012	12471	20787	151.3	12.5	20.8	1861	46.16	1.00002
1500	0.008	18707	31179	159.7	12.5	20.8	2279	61.55	1.00001

TABLE 5.1
THERMOPHYSICAL PROPERTIES OF SOME COMMON FLUIDS (continued)

T/ K	ρ/ mol L^{-1}	U/ J mol^{-1}	H/ J mol^{-1}	S/ J K^{-1} mol^{-1}	C_v/ J K^{-1} mol^{-1}	C_p/ J K^{-1} mol^{-1}	v_s/ m s^{-1}	η/ μPa s	ε
P = 1 MPa									
3	39.703	–42	–16	8.6	7.1	7.8	300	5.63	1.06274
4	38.210	–34	–7	11.2	8.3	10.9	290	5.01	1.06034
5	35.818	–22	6	14.0	9.7	15.1	269	4.38	1.05650
10	15.378	78	143	32.2	12.3	30.5	198	3.07	1.02402
20	6.067	228	393	49.8	12.6	22.9	274	3.94	1.00943
50	2.353	617	1042	69.8	12.5	21.1	428	6.53	1.00365
100	1.186	1245	2089	84.3	12.5	20.9	597	9.89	1.00184
200	0.597	2495	4170	98.7	12.5	20.8	838	15.21	1.00093
300	0.399	3742	6249	107.1	12.5	20.8	1024	19.96	1.00062
400	0.300	4990	8327	113.1	12.5	20.8	1180	24.32	1.00046
500	0.240	6237	10406	117.8	12.5	20.8	1319	28.38	1.00037
600	0.200	7485	12484	121.5	12.5	20.8	1444	32.23	1.00031
700	0.172	8732	14562	124.7	12.5	20.8	1559	35.91	1.00027
800	0.150	9979	16641	127.5	12.5	20.8	1666	39.44	1.00023
900	0.133	11227	18719	130.0	12.5	20.8	1767	42.86	1.00021
1000	0.120	12474	20798	132.2	12.5	20.8	1862	46.17	1.00019
1500	0.080	18710	31190	140.6	12.5	20.8	2280	61.55	1.00012
P = 10 MPa									
4	51.978	–24	169	6.7	6.0	7.3	586	24.27	1.08262
5	51.118	–18	177	8.5	7.9	9.3	576	18.16	1.08122
10	46.872	23	236	16.6	11.0	14.5	546	9.31	1.07432
20	37.092	154	423	29.5	12.6	20.7	498	6.99	1.05854
50	19.192	572	1093	49.9	12.9	22.4	541	8.07	1.03003
100	10.525	1231	2181	65.0	12.8	21.3	674	10.93	1.01640
200	5.605	2500	4284	79.6	12.6	20.9	889	15.82	1.00871
300	3.829	3755	6367	88.0	12.6	20.8	1063	20.25	1.00595
400	2.908	5006	8445	94.0	12.6	20.8	1212	24.54	1.00452
500	2.344	6256	10522	98.6	12.5	20.8	1346	28.56	1.00364
600	1.963	7505	12599	102.4	12.5	20.8	1467	32.38	1.00305
700	1.689	8754	14676	105.6	12.5	20.8	1580	36.04	1.00262
800	1.481	10003	16753	108.4	12.5	20.8	1685	39.56	1.00230
900	1.320	11252	18830	110.9	12.5	20.8	1784	42.96	1.00205
1000	1.189	12500	20907	113.0	12.5	20.8	1877	46.26	1.00185
1500	0.797	18742	31294	121.5	12.5	20.8	2289	61.62	1.00124

Argon (Ar)

T/ K	ρ/ mol L^{-1}	U/ J mol^{-1}	H/ J mol^{-1}	S/ J K^{-1} mol^{-1}	C_v/ J K^{-1} mol^{-1}	C_p/ J K^{-1} mol^{-1}	v_s/ m s^{-1}	η/ μPa s	λ/ mW m^{-1} K^{-1}
P = 0.1 MPa (1 bar)									
85	35.243	–4811	–4808	53.6	23.1	44.7	820	278.8	132.4
90	0.138	1077	1802	129.4	13.1	22.5	174	7.5	6.0
100	0.123	1211	2024	131.8	12.9	21.9	184	8.2	6.6
120	0.102	1471	2456	135.7	12.6	21.4	203	9.8	7.8
140	0.087	1727	2881	139.0	12.6	21.1	220	11.4	9.0
160	0.076	1980	3302	141.8	12.5	21.0	235	13.0	10.2
180	0.067	2232	3722	144.3	12.5	21.0	250	14.5	11.4
200	0.060	2483	4141	146.5	12.5	20.9	263	16.0	12.5
220	0.055	2734	4559	148.5	12.5	20.9	276	17.5	13.7
240	0.050	2984	4976	150.3	12.5	20.9	289	18.9	14.8
260	0.046	3234	5394	152.0	12.5	20.9	300	20.3	15.8
280	0.043	3484	5811	153.5	12.5	20.8	312	21.6	16.9

TABLE 5.1
THERMOPHYSICAL PROPERTIES OF SOME COMMON FLUIDS (continued)

T/ K	ρ/ mol L⁻¹	U/ J mol⁻¹	H/ J mol⁻¹	S/ J K⁻¹ mol⁻¹	C_v/ J K⁻¹ mol⁻¹	C_p/ J K⁻¹ mol⁻¹	v_s/ m s⁻¹	η/ μPa s	λ/ mW m⁻¹ K⁻¹
300	0.040	3734	6227	155.0	12.5	20.8	323	22.9	17.9
320	0.038	3984	6644	156.3	12.5	20.8	333	24.2	18.9
340	0.035	4234	7060	157.6	12.5	20.8	344	25.4	19.9
360	0.033	4484	7477	158.7	12.5	20.8	354	26.6	20.8
380	0.032	4734	7893	159.9	12.5	20.8	363	27.8	21.7
P = 1 MPa									
85	35.307	–4820	–4792	53.5	23.1	44.6	823	281.3	133.0
90	34.542	–4598	–4569	56.1	21.6	44.7	808	242.7	124.2
100	32.909	–4145	–4115	60.9	19.9	46.2	753	185.0	109.2
120	1.181	1210	2057	114.3	14.7	30.1	189	10.3	9.3
140	0.945	1544	2603	118.5	13.5	25.4	212	11.8	10.1
160	0.799	1838	3089	121.8	13.0	23.6	231	13.3	11.1
180	0.697	2116	3551	124.5	12.8	22.7	247	14.8	12.1
200	0.619	2384	3999	126.9	12.7	22.2	262	16.3	13.2
220	0.559	2648	4438	128.9	12.6	21.8	275	17.7	14.2
240	0.509	2908	4873	130.8	12.6	21.6	288	19.1	15.3
260	0.468	3167	5304	132.6	12.6	21.5	301	20.4	16.3
280	0.433	3423	5732	134.2	12.6	21.4	312	21.8	17.3
300	0.403	3679	6159	135.6	12.5	21.3	324	23.1	18.3
320	0.377	3934	6583	137.0	12.5	21.2	334	24.3	19.2
340	0.355	4188	7007	138.3	12.5	21.2	345	25.5	20.2
360	0.335	4441	7429	139.5	12.5	21.1	355	26.7	21.1
380	0.317	4694	7851	140.6	12.5	21.1	365	27.9	22.0
P = 10 MPa									
90	35.208	–4694	–4410	55.0	21.9	43.2	846	265.2	129.5
100	33.744	–4271	–3974	59.6	20.4	44.0	800	205.0	115.1
120	30.525	–3396	–3069	67.8	18.8	46.9	672	131.2	92.1
140	26.609	–2447	–2072	75.5	17.6	54.1	526	85.9	71.7
160	20.816	–1279	–799	83.9	17.4	78.6	357	51.3	52.8
180	12.296	228	1042	94.8	17.3	83.6	257	27.8	32.0
200	8.442	1118	2302	101.4	15.3	48.6	268	23.3	23.6
220	6.776	1661	3137	105.4	14.2	36.8	284	22.8	21.6
240	5.787	2087	3815	108.4	13.7	31.6	300	23.2	21.3
260	5.105	2458	4416	110.8	13.4	28.8	314	23.9	21.4
280	4.596	2798	4974	112.9	13.2	27.1	327	24.8	21.8
300	4.195	3119	5503	114.7	13.1	25.9	339	25.7	22.3
320	3.869	3427	6012	116.3	13.0	25.0	350	26.7	22.9
340	3.596	3726	6506	117.8	13.0	24.4	361	27.7	23.5
360	3.364	4017	6989	119.2	12.9	23.9	372	28.7	24.2
380	3.164	4303	7464	120.5	12.9	23.5	381	29.7	24.9

Methane (CH_4)

T/ K	ρ/ mol L⁻¹	U/ J mol⁻¹	H/ J mol⁻¹	S/ J K⁻¹ mol⁻¹	C_v/ J K⁻¹ mol⁻¹	C_p/ J K⁻¹ mol⁻¹	η/ μPa s	λ/ mW m⁻¹ K⁻¹	ε
P = 0.1 MPa (1 bar)									
100	27.370	–5258	–5254	73.0	33.4	54.1	156.3	208.1	1.65504
125	0.099	3026	4039	156.5	25.4	34.6	5.0	13.4	1.00193
150	0.081	3667	4896	162.7	25.2	34.0	5.9	16.2	1.00159
175	0.069	4301	5743	168.0	25.2	33.8	6.9	19.1	1.00136
200	0.061	4935	6587	172.5	25.3	33.8	7.8	21.9	1.00119
225	0.054	5571	7434	176.5	25.5	34.0	8.7	24.8	1.00105
250	0.048	6216	8288	180.1	26.0	34.4	9.6	27.8	1.00095

TABLE 5.1
THERMOPHYSICAL PROPERTIES OF SOME COMMON FLUIDS (continued)

T/ K	ρ/ mol L^{-1}	*U*/ J mol^{-1}	*H*/ J mol^{-1}	*S*/ J K^{-1} mol^{-1}	C_v/ J K^{-1} mol^{-1}	C_p/ J K^{-1} mol^{-1}	η/ μPa s	λ/ mW m^{-1} K^{-1}	ε
275	0.044	6875	9156	183.4	26.6	35.0	10.4	30.9	1.00086
300	0.040	7552	10042	186.4	27.5	35.9	11.2	34.1	1.00079
325	0.037	8252	10951	189.4	28.5	36.9	12.0	37.6	1.00073
350	0.034	8979	11887	192.1	29.7	38.0	12.8	41.2	1.00068
375	0.032	9737	12853	194.8	30.9	39.3	13.5	45.1	1.00063
400	0.030	10528	13852	197.4	32.3	40.7	14.3	49.1	1.00059
425	0.028	11354	14886	199.9	33.7	42.1	15.0	53.3	1.00056
450	0.027	12215	15956	202.3	35.2	43.5	15.7	57.6	1.00053
500	0.024	14047	18204	207.1	38.0	46.4	17.0	66.5	1.00047
600	0.020	18111	23101	216.0	42.9	51.3	19.4	84.1	1.00039
P = 1 MPa									
100	27.413	–5268	–5231	72.9	33.4	54.0	158.1	208.9	1.65617
125	25.137	–3882	–3842	85.3	32.4	57.4	89.2	168.2	1.59261
150	0.969	3282	4315	140.9	27.9	45.2	6.2	18.4	1.01911
175	0.765	4041	5348	147.3	26.4	38.9	7.1	20.6	1.01507
200	0.644	4736	6289	152.3	25.9	36.8	8.0	23.1	1.01268
225	0.560	5410	7197	156.6	25.9	36.0	8.9	25.8	1.01102
250	0.497	6081	8093	160.4	26.2	35.8	9.7	28.7	1.00979
275	0.448	6758	8991	163.8	26.8	36.1	10.6	31.7	1.00882
300	0.408	7449	9901	167.0	27.6	36.7	11.4	34.9	1.00803
325	0.375	8160	10829	169.9	28.6	37.6	12.1	38.3	1.00738
350	0.347	8897	11781	172.8	29.7	38.6	12.9	41.9	1.00683
375	0.323	9662	12760	175.5	31.0	39.8	13.6	45.7	1.00636
400	0.302	10460	13770	178.1	32.4	41.1	14.4	49.6	1.00595
425	0.284	11291	14814	180.6	33.8	42.4	15.1	53.8	1.00559
450	0.268	12157	15892	183.1	35.2	43.8	15.7	58.1	1.00527
500	0.241	13997	18153	187.8	38.1	46.6	17.0	66.9	1.00474
600	0.200	18073	23070	196.8	43.0	51.4	19.5	84.5	1.00394
P = 10 MPa									
100	27.815	–5362	–5003	72.0	33.8	53.2	175.4	217	1.66668
125	25.754	–4036	–3648	84.1	32.7	55.3	100.4	178.8	1.60895
150	23.441	–2655	–2229	94.4	31.4	58.6	65.7	144.6	1.54553
175	20.613	–1175	–689	103.9	30.3	65.5	44.9	113.4	1.47021
200	16.602	542	1144	113.6	30.1	84.7	29.4	85.8	1.36789
225	10.547	2680	3628	125.3	30.8	102.2	17.6	61.0	1.22352
250	7.013	4289	5714	134.1	29.3	67.4	14.3	47.6	1.14481
275	5.530	5387	7195	139.8	28.7	53.4	13.8	44.1	1.11297
300	4.685	6320	8454	144.2	28.9	48.0	13.9	44.6	1.09513
325	4.115	7192	9622	147.9	29.6	45.8	14.3	46.6	1.08322
350	3.695	8047	10753	151.3	30.5	44.9	14.7	49.2	1.07450
375	3.366	8903	11874	154.4	31.7	44.8	15.2	52.3	1.06773
400	3.101	9774	12999	157.3	32.9	45.2	15.8	55.7	1.06227
425	2.880	10666	14138	160.0	34.3	46.0	16.3	59.4	1.05775
450	2.692	11584	15298	162.7	35.7	46.9	16.9	63.3	1.05392
500	2.389	13507	17692	167.7	38.5	48.9	18.0	71.6	1.04775
600	1.963	17700	22795	177.0	43.3	52.9	20.2	88.3	1.03911

Ethane (C_2H_6)

T/ K	ρ/ mol L^{-1}	*U*/ J mol^{-1}	*H*/ J mol^{-1}	*S*/ J K^{-1} mol^{-1}	C_v/ J K^{-1} mol^{-1}	C_p/ J K^{-1} mol^{-1}	v_s/ m s^{-1}	ε
P = 0.1 MPa (1 bar)								
95	21.50	–14555	–14550	80.2	47.2	68.7	1970	1.93480

TABLE 5.1
THERMOPHYSICAL PROPERTIES OF SOME COMMON FLUIDS (continued)

T/ K	ρ/ mol L^{-1}	U/ J mol^{-1}	H/ J mol^{-1}	S/ J K^{-1} mol^{-1}	C_v/ J K^{-1} mol^{-1}	C_p/ J K^{-1} mol^{-1}	v_s/ m s^{-1}	ε
100	21.32	–14210	–14205	83.8	47.1	69.3	1943	1.92500
125	20.41	–12468	–12463	99.3	45.0	69.8	1775	1.87634
150	19.47	–10717	–10712	112.1	43.4	70.4	1587	1.82726
175	18.49	–8938	-8933	123.1	42.7	72.1	1396	1.77671
200	0.062	5503	7123	210.1	34.5	43.8	258	1.00208
225	0.054	6401	8238	215.4	36.5	45.5	273	1.00183
250	0.049	7349	9401	220.3	38.9	47.7	287	1.00164
275	0.044	8360	10624	224.9	41.6	50.2	300	1.00148
300	0.040	9439	11914	229.4	44.5	53.1	312	1.00136
325	0.037	10592	13278	233.8	47.6	56.1	324	1.00125
350	0.035	11823	14719	238.1	50.7	59.2	335	1.00116
375	0.032	13133	16240	242.3	54.0	62.4	345	1.00108
400	0.030	14525	17841	246.4	57.2	65.6	355	1.00101
450	0.027	17548	21282	254.5	63.6	72.0	375	1.00090
500	0.024	20883	25035	262.4	69.7	78.1	393	1.00081
600	0.020	28429	33415	277.6	80.9	89.3	428	1.00067
P = 1 MPa								
95	21.514	–14562	–14515	80.2	47.3	68.7	1972	1.93537
100	21.334	–14217	–14170	83.7	47.2	69.3	1946	1.92560
125	20.427	–12478	–12429	99.2	45.0	69.8	1778	1.87709
150	19.494	–10731	–10679	112.0	43.4	70.3	1592	1.82823
175	18.515	–8957	–8903	123.0	42.7	72.0	1402	1.77800
200	17.464	–7127	–7070	132.7	42.9	74.9	1209	1.72513
225	16.288	–5199	–5137	141.8	43.8	80.2	1008	1.66733
250	0.564	6762	8534	198.7	41.6	57.5	260	1.01909
275	0.489	7902	9949	204.1	43.2	56.2	280	1.01650
300	0.435	9063	11363	209.0	45.5	57.2	297	1.01467
325	0.393	10273	12815	213.7	48.3	59.1	311	1.01327
350	0.360	11546	14321	218.1	51.3	61.5	325	1.01214
375	0.333	12889	15893	222.5	54.4	64.2	337	1.01121
400	0.310	14306	17534	226.7	57.5	67.1	349	1.01043
450	0.272	17367	21038	234.9	63.8	73.0	370	1.00917
500	0.244	20730	24836	242.9	69.9	78.9	390	1.00819
600	0.201	28313	33278	258.3	81.0	89.8	427	1.00677
P = 10 MPa								
95	21.624	–14626	–14163	79.5	47.4	68.5	2000	1.94104
100	21.448	–14286	–13819	83.0	47.4	69.1	1974	1.93146
125	20.570	–12572	–12086	98.5	45.5	69.3	1814	1.88436
150	19.678	–10858	–10350	111.1	43.9	69.6	1637	1.83753
175	18.758	–9130	–8596	121.9	43.3	70.8	1459	1.79010
200	17.793	–7363	–6801	131.5	43.5	73.0	1284	1.74134
225	16.760	–5535	–4938	140.3	44.3	76.4	1110	1.69017
250	15.620	–3609	–2969	148.6	45.8	81.5	935	1.63488
275	14.301	–1539	–839	156.7	47.9	89.4	758	1.57249
300	12.666	757	1547	165.0	50.8	102.7	577	1.49740
325	10.398	3443	4404	174.1	54.7	129.1	399	1.39745
350	7.292	6643	8015	184.8	58.8	150.1	290	1.26832
375	5.182	9419	11349	194.1	60.0	115.7	289	1.18570
400	4.182	11577	13968	200.8	61.4	96.9	310	1.14797
450	3.204	15379	18500	211.5	65.8	87.5	347	1.11193
500	2.677	19135	22870	220.7	71.2	88.0	378	1.09288
600	2.076	27160	31978	237.3	81.8	94.7	427	1.07142

TABLE 5.1
THERMOPHYSICAL PROPERTIES OF SOME COMMON FLUIDS (continued)

Propane (C_3H_8)

T/ K	ρ/ mol L^{-1}	U/ J mol^{-1}	H/ J mol^{-1}	S/ J K^{-1} mol^{-1}	C_v/ J K^{-1} mol^{-1}	C_p/ J K^{-1} mol^{-1}	v_s/ m s^{-1}	ε
P = 0.1 MPa (1 bar)								
90	16.526	–21486	–21426	87.3	59.2	84.5	2126	2.07988
100	16.295	–20639	–20577	96.2	59.6	85.2	2041	2.05806
125	15.726	–18495	–18432	115.4	59.2	86.5	1856	2.00674
150	15.156	–16319	–16253	131.3	58.9	88.0	1685	1.95796
175	14.577	–14096	–14028	145.0	59.5	90.3	1521	1.91036
200	13.982	–11806	–11735	157.3	61.0	93.5	1359	1.86300
225	13.339	–9395	–9387	168.5	63.4	97.9	1197	1.81487
250	0.050	9194	11213	257.6	57.2	66.8	228	1.00238
275	0.045	10691	12930	264.1	61.6	70.7	239	1.00215
300	0.041	12297	14752	270.5	66.2	75.1	249	1.00195
325	0.037	14019	16689	276.7	71.1	79.8	259	1.00179
350	0.035	15862	18744	282.8	76.0	84.6	269	1.00166
375	0.032	17827	20921	288.8	80.9	89.5	278	1.00154
400	0.030	19912	23217	294.7	85.7	94.3	286	1.00144
450	0.027	24441	28166	306.4	95.2	103.6	303	1.00128
500	0.024	29428	33573	317.7	104.1	112.6	318	1.00115
600	0.020	40677	45658	339.7	120.4	128.8	347	1.00095
P = 1 MPa								
90	16.526	–21486	–21426	87.2	59.3	84.5	2128	2.08034
100	16.295	–20639	–20577	96.2	59.7	85.2	2043	2.05856
125	15.726	–18495	–18432	115.3	59.2	86.4	1859	2.00736
150	15.156	–16319	–16253	131.2	59.0	88.0	1690	1.95873
175	14.577	–14096	–14028	144.9	59.6	90.2	1526	1.91132
200	13.982	–11806	–11735	157.2	61.1	93.4	1365	1.86421
225	13.361	–9424	–9349	168.4	63.4	97.7	1205	1.81642
250	12.696	–6919	–6840	179.0	66.4	103.3	1045	1.76672
275	11.962	–4252	–4169	189.1	70.0	110.8	881	1.71316
300	11.102	–1360	–1270	199.2	74.1	121.9	708	1.65216
325	0.428	13278	15614	255.2	74.1	89.6	233	1.02067
350	0.383	15259	17869	261.9	78.0	91.2	248	1.01846
375	0.349	17318	20183	268.3	82.2	94.2	261	1.01678
400	0.322	19472	22582	274.4	86.7	97.8	272	1.01544
450	0.279	24092	27672	286.4	95.7	105.9	293	1.01337
500	0.248	29137	33172	298.0	104.4	114.1	312	1.01184
600	0.203	40455	45374	320.2	120.5	129.7	344	1.00968
P = 10 MPa								
90	16.590	–21553	–20951	86.5	59.9	84.4	2146	2.08489
100	16.364	–20714	–20103	95.4	60.1	85.1	2068	2.06350
125	15.810	–18595	–17962	114.5	59.6	86.1	1895	2.01342
150	15.259	–16448	–15793	130.3	59.3	87.5	1733	1.96617
175	14.705	–14261	–13581	144.0	59.9	89.5	1577	1.92048
200	14.141	–12016	–11309	156.1	61.4	92.4	1425	1.87557
225	13.562	–9692	–8955	167.2	63.7	96.1	1277	1.83076
250	12.960	–7268	–6496	177.5	66.7	100.7	1133	1.78529
275	12.322	–4721	–3909	187.4	70.2	106.4	991	1.73826
300	11.631	–2027	–1167	196.9	74.1	113.2	851	1.68849
325	10.860	843	1764	206.3	78.4	121.5	715	1.63437
350	9.973	3924	4927	215.7	82.9	132.0	582	1.57361
375	8.905	7270	8393	225.2	87.7	146.1	455	1.50271

TABLE 5.1
THERMOPHYSICAL PROPERTIES OF SOME COMMON FLUIDS (continued)

T/ K	ρ/ mol L^{-1}	U/ J mol^{-1}	H/ J mol^{-1}	S/ J K^{-1} mol^{-1}	C_v/ J K^{-1} mol^{-1}	C_p/ J K^{-1} mol^{-1}	v_S/ m s^{-1}	ε
400	7.561	10957	12279	235.3	93.0	165.7	339	1.41671
450	4.614	18845	21013	255.8	101.8	167.8	249	1.24060
500	3.241	25567	28652	272.0	107.8	142.7	276	1.16439
600	2.242	38131	42591	297.4	121.7	140.5	332	1.11122

TABLE 5.2
THERMODYNAMIC PROPERTIES OF AIR

The following tables give a selection of thermodynamic properties of air in the saturation state (liquid and gas in equilibrium), the liquid state, and the gaseous state. In the table for the saturation state, *P*(boil) is the pressure at which boiling begins at the specified temperature, and *P*(con) is the pressure at which condensation begins. A point *T*, *P*(boil) on the phase diagram of a mixture is often called a bubble point, and *T*, *P*(con) is referred to as a dew point.

Physical quantity	Symbol	SI unit
Density	ρ	kg m^{-3}
Specific enthalpy	h	J kg^{-1}
Specific entropy	s	J K^{-1} kg^{-1}
Isobaric specific heat capacity	c_p	J K^{-1} kg^{-1}

REFERENCES

1. Vasserman, A. A.; Rabinovich, V. A., *Thermophysical Properties of Liquid Air and its Components*, Izdatel'stvo Komiteta, Standartov, Moscow, 1968.
2. Vasserman, A. A., *Thermophysical Properties of Air and Air Components*, Izdatel'stvo Nauka, Moscow, 1966.
3. Sytchev, V. V.; Vasserman, A. A.; Kozlov, A. D.; Spiridonov, G. A.; Tsymarny, V. A., *Thermodynamic Properties of Air*, Hemisphere Publishing Corp., New York, 1987.

Properties in the saturation state:

T/ K	*P*(boil)/ 10^5 Pa	*P*(con)/ 10^5 Pa	ρ (liq)/ g cm^{-3}	ρ (gas)/ kg m^{-3}
65	0.1468	0.0861	0.939	0.464
70	0.3234	0.2052	0.917	1.033
75	0.6366	0.4321	0.894	2.048
80	1.146	0.8245	0.871	3.709
85	1.921	1.453	0.845	6.258
90	3.036	2.397	0.819	9.980
95	4.574	3.748	0.792	15.21
100	6.621	5.599	0.763	22.39
110	12.59	11.22	0.699	45.15
120	21.61	20.14	0.622	87.34
130	34.16	33.32	0.487	184.33
132.55	37.69	37.69	0.313	312.89

Properties of liquid air:

P/ 10^5 Pa	*T*/ K	ρ/ g cm^{-3}	*h*/ J g^{-1}	*s*/ J K^{-1} g^{-1}	c_p/ J K^{-1} g^{-1}
1	75	0.8935	–131.7	2.918	1.843
5	75	0.8942	–131.4	2.916	1.840
5	80	0.8718	–122.3	3.031	1.868
5	85	0.8482	–112.9	3.143	1.901
5	90	0.8230	–103.3	3.250	1.941
5	95	0.7962	–93.5	3.356	1.991
10	75	0.8952	–131.1	2.913	1.836
10	80	0.8729	–122.0	3.028	1.863
10	90	0.8245	–103.1	3.246	1.932
10	100	0.7695	–83.2	3.452	2.041
50	75	0.9025	–128.2	2.892	1.806
50	100	0.7859	–81.8	3.415	1.939

TABLE 5.2
THERMODYNAMIC PROPERTIES OF AIR (continued)

P/ 10^5 Pa	T/ K	ρ/ g cm^{-3}	h/ J g^{-1}	s/ J K^{-1} g^{-1}	c_p/ J K^{-1} g^{-1}
50	125	0.6222	–28.3	3.889	2.614
50	150	0.1879	91.9	4.764	2.721
100	75	0.9111	–124.5	2.867	1.774
100	100	0.8033	–79.4	3.376	1.852
100	125	0.6746	–31.4	3.805	2.062
100	150	0.4871	32.8	4.271	2.832

Properties of air in the gaseous state:

P/ 10^5 Pa	T/ K	ρ/ kg m^{-3}	h/ J g^{-1}	s/ J K^{-1} g^{-1}	c_p/ J K^{-1} g^{-1}
1	100	3.556	98.3	5.759	1.032
1	200	1.746	199.7	6.463	1.007
1	300	1.161	300.3	6.871	1.007
1	500	0.696	503.4	7.389	1.030
1	1000	0.348	1046.6	8.138	1.141
10	200	17.835	195.2	5.766	1.049
10	300	11.643	298.3	6.204	1.021
10	500	6.944	502.9	6.727	1.034
10	1000	3.471	1047.2	7.477	1.142
100	200	213.950	148.8	4.949	1.650
100	300	116.945	279.9	5.486	1.158
100	500	66.934	499.0	6.048	1.073
100	1000	33.613	1052.4	6.812	1.151

TABLE 5.3
PROPERTIES OF LIQUID WATER

This table summarizes properties of liquid water in the range 0 to 100°C. All values except vapor pressure refer to a pressure of 100 kPa (1 bar).

Physical quantity	Symbol	SI unit
Density	ρ	kg m^{-3}
Isobaric specific heat capacity	c_p	J K^{-1} kg^{-1}
Vapor pressure	P_{vap}	Pa
Viscosity	η	Pa s
Thermal conductivity	λ	W m^{-1} K^{-1}
Relative permittivity (dielectric constant)	ε	1
Surface tension	σ	N m^{-1}

REFERENCES

1. Haar, L.; Gallagher, J. S.; Kell, G. S., *NBS/NRC Steam Tables*, Hemisphere Publishing Corp., 1984.
2. Marsh, K. N., Editor, *Recommended Reference Materials for the Realization of Physicochemical Properties*, Blackwell Scientific Publications, Oxford, 1987.
3. Sengers, J. V.; Watson, J. T. R., *J. Phys. Chem. Ref. Data*, 15, 1291, 1986.
4. Archer, D. G.; Wang, P., *J.Phys. Chem. Ref. Data*, 19, 371, 1990.
5. Vargaftik, N. B.; Volkov, B. N.; Voljak, L. D., *J.Phys. Chem. Ref. Data*, 12, 817, 1983.

t/ °C	ρ/ g cm^{-3}	c_p/ J K^{-1} g^{-1}	P_{vap}/ kPa	η/ μPa s	λ/ mW m^{-1} K^{-1}	ε	σ/ mN m^{-1}
0	0.99984	4.2176	0.6113	1793	561.0	87.90	75.64
10	0.99970	4.1921	1.2281	1307	580.0	83.96	74.23
20	0.99821	4.1818	2.3388	1002	598.4	80.20	72.75
30	0.99565	4.1784	4.2455	797.7	615.4	76.60	71.20
40	0.99222	4.1785	7.3814	653.2	630.5	73.17	69.60
50	0.98803	4.1806	12.344	547.0	643.5	69.88	67.94
60	0.98320	4.1843	19.932	466.5	654.3	66.73	66.24
70	0.97778	4.1895	31.176	404.0	663.1	63.73	64.47
80	0.97182	4.1963	47.373	354.4	670.0	60.86	62.67
90	0.96535	4.2050	70.117	314.5	675.3	58.12	60.82
100	0.95840	4.2159	101.325	281.8	679.1	55.51	58.91
Ref.	1—3	2	1, 3	3	3	4	5

TABLE 5.4
DENSITY OF STANDARD MEAN OCEAN WATER

Since the density of water varies slightly with isotopic composition, very accurate measurements are referenced to a standard sample of distilled ocean water maintained by the International Atomic Energy Agency in Vienna. This is called Standard Mean Ocean Water (SMOW). The density of SMOW between 0 and 40°C is given by the following equation:

$$\rho(\mathrm{SMOW})/\mathrm{kg\,m^{-3}} = a_0 + a_1x + a_2x^2 + a_3x^3 + a_4x^4 + a_5x^5$$

where $x = t$/°C and

$a_0 = 999.842\,594$ $\quad a_3 = 1.001\,685 \times 10^{-4}$
$a_1 = 6.793\,952 \times 10^{-2}$ $\quad a_4 = -1.120\,083 \times 10^{-6}$
$a_2 = -9.095\,290 \times 10^{-3}$ $\quad a_5 = 6.536\,332 \times 10^{-9}$

The table below lists values of the density at 0.1°C intervals as calculated from the equation. Temperature values refer to the IPTS-68 scale.

The variation in isotopic composition of most natural waters is sufficiently small that the density is unlikely to differ by more than 0.02 kg m^{-3} from the values in the table. If the isotopic composition of a sample is known, corrections to the table can be applied (see Reference 1).

Physical quantity	**Symbol**	**SI unit**
Density	ρ	kg m^{-3}

REFERENCES

1. Marsh, K. N., Ed., *Recommended Reference Materials for the Realization of Physicochemical Properties*, pp. 13-18, 25-27, Blackwell Scientific Publications, Oxford, 1987.
2. Bigg, P. H., *Br. J. Appl. Phys.*, 8, 521, 1967.

ρ/kg m^{-3}

t_{68}/°C	0.0	0.1	0.2	0.3	0.4	0.5	0.6	0.7	0.8	0.9
0	999.8426	8493	8558	8622	8683	8743	8801	8857	8912	8964
1	999.9015	9065	9112	9158	9202	9244	9284	9323	9360	9395
2	999.9429	9461	9491	9519	9546	9571	9595	9616	9636	9655
3	999.9672	9687	9700	9712	9722	9731	9738	9743	9747	9749
4	999.9750	9748	9746	9742	9736	9728	9719	9709	9696	9683
5	999.9668	9651	9632	9612	9591	9568	9544	9518	9490	9461
6	999.9430	9398	9365	9330	9293	9255	9216	9175	9132	9088
7	999.9043	8996	8948	8898	8847	8794	8740	8684	8627	8569
8	999.8509	8448	8385	8321	8256	8189	8121	8051	7980	7908
9	999.7834	7759	7682	7604	7525	7444	7362	7279	7194	7108
10	999.7021	6932	6842	6751	6658	6564	6468	6372	6274	6174
11	999.6074	5972	5869	5764	5658	5551	5443	5333	5222	5110
12	999.4996	4882	4766	4648	4530	4410	4289	4167	4043	3918
13	999.3792	3665	3536	3407	3276	3143	3010	2875	2740	2602
14	999.2464	2325	2184	2042	1899	1755	1609	1463	1315	1166
15	999.1016	0864	0712	0558	0403	0247	0090	9932*	9772*	9612*
16	998.9450	9287	9123	8957	8791	8623	8455	8285	8114	7942
17	998.7769	7595	7419	7243	7065	6886	6706	6525	6343	6160
18	998.5976	5790	5604	5416	5228	5038	4847	4655	4462	4268
19	998.4073	3877	3680	3481	3282	3081	2880	2677	2474	2269
20	998.2063	1856	1649	1440	1230	1019	0807	0594	0380	0164

TABLE 5.4
DENSITY OF STANDARD MEAN OCEAN WATER (continued)

t_{68}/°C	0.0	0.1	0.2	0.3	0.4	0.5	0.6	0.7	0.8	0.9
21	997.9948	9731	9513	9294	9073	8852	8630	8406	8182	7957
22	997.7730	7503	7275	7045	6815	6584	6351	6118	5883	5648
23	997.5412	5174	4936	4697	4456	4215	3973	3730	3485	3240
24	997.2994	2747	2499	2250	2000	1749	1497	1244	0990	0735
25	997.0480	0223	9965*	9707*	9447*	9186*	8925*	8663*	8399*	8135*
26	996.7870	7604	7337	7069	6800	6530	6259	5987	5714	5441
27	996.5166	4891	4615	4337	4059	3780	3500	3219	2938	2655
28	996.2371	2087	1801	1515	1228	0940	0651	0361	0070	9778*
29	995.9486	9192	8898	8603	8306	8009	7712	7413	7113	6813
30	995.6511	6209	5906	5602	5297	4991	4685	4377	4069	3760
31	995.3450	3139	2827	2514	2201	1887	1572	1255	0939	0621
32	995.0302	9983*	9663*	9342*	9020*	8697*	8373*	8049*	7724*	7397*
33	994.7071	6743	6414	6085	5755	5423	5092	4759	4425	4091
34	994.3756	3420	3083	2745	2407	2068	1728	1387	1045	0703
35	994.0359	0015	9671*	9325*	8978*	8631*	8283*	7934*	7585*	7234*
36	993.6883	6531	6178	5825	5470	5115	4759	4403	4045	3687
37	993.3328	2968	2607	2246	1884	1521	1157	0793	0428	0062
38	992.9695	9328	8960	8591	8221	7850	7479	7107	6735	6361
39	992.5987	5612	5236	4860	4483	4105	3726	3347	2966	2586
40	992.2204	1822	1439	1055	670	285	9899	9513	9125	8737

* The leading figure decreases by 1.0.

TABLE 5.5
STEAM TABLES

This table gives thermodynamic properties of compressed water and superheated steam at selected pressures and temperatures. It was generated from the formulation approved by the International Association for the Properties of Steam in 1984. The reference state is the liquid at the triple point, at which the internal energy and entropy are taken as zero. A duplicate entry in the temperature column indicates a liquid-vapor phase transition; property values are then given for both phases. More extensive tables are given in the reference.

Physical quantity	Symbol	SI unit
Density	ρ	kg m^{-3}
Specific enthalpy	h	J kg^{-1}
Specific entropy	s	J K^{-1} kg^{-1}
Isobaric specific heat capacity	c_p	J K^{-1} kg^{-1}

REFERENCE

1. Haar, L.; Gallagher, J. S.; Kell, G. S., *NBS/NRC Steam Tables*, Hemisphere Publishing Corp., New York, 1984.

P/ MPa	T/ K	ρ/ kg m^{-3}	h/ J g^{-1}	s/ J K^{-1} g^{-1}	c_p/ J K^{-1} g^{-1}
0.1	273.15	999.83	0.06	–0.00015	4.2282
0.1	300	996.57	112.58	0.3928	4.1831
0.1	372.78	958.66	417.51	1.3027	4.2166
0.1	372.78	0.59021	2675.1	7.3589	2.0427
0.1	373.15	0.58958	2675.9	7.3609	2.0418
0.1	400	0.54765	2730.0	7.5010	1.9973
0.1	500	0.43517	2927.9	7.9427	1.9816
0.1	600	0.36186	3128.2	8.3076	2.0272
0.1	700	0.30988	3333.8	8.6245	2.0869
0.1	800	0.27102	3545.8	8.9074	2.1526
0.1	900	0.24085	3764.4	9.1649	2.2216
0.1	1000	0.21673	3990.1	9.4026	2.2921
0.1	1100	0.19701	4222.8	9.6244	2.3621
0.1	1200	0.18058	4462.5	9.8329	2.4302
0.1	1300	0.16668	4708.8	10.030	2.4954
0.1	1400	0.15477	4961.4	10.217	2.5568
0.1	1500	0.14445	5220.0	10.396	2.6143
0.1	1600	0.13542	5484.2	10.566	2.6676
0.1	1700	0.12745	5753.4	10.729	2.7166
0.1	1800	0.12037	6027.3	10.886	2.7617
0.1	1900	0.11403	6305.6	11.036	2.8031
0.1	2000	0.10833	6587.9	11.181	2.8412
1	273.15	1000.30	0.98	–0.0001	4.2233
1	300	996.97	113.41	0.3926	4.1806
1	373.15	958.81	419.74	1.3062	4.2150
1	400	937.92	533.47	1.6005	4.2593
1	453.07	887.15	762.88	2.1388	4.4030
1	453.07	5.1445	2777.7	6.5859	2.5569
1	500	4.5348	2890.0	6.8220	2.2760
1	600	3.6877	3108.0	7.2199	2.1332
1	700	3.1307	3321.1	7.5483	2.1386
1	800	2.7266	3537.0	7.8365	2.1821
1	900	2.4175	3758.0	8.0968	2.2402
1	1000	2.1724	3985.2	8.3361	2.3046

TABLE 5.5
STEAM TABLES (continued)

P/ MPa	*T*/ K	ρ/ kg m^{-3}	*h*/ J g^{-1}	*s*/ J K^{-1} g^{-1}	c_p/ J K^{-1} g^{-1}
1	1100	1.9730	4219.0	8.5588	2.3710
1	1200	1.8074	4459.4	8.7680	2.4368
1	1300	1.6677	4706.3	8.9656	2.5004
1	1400	1.5481	4959.3	9.1531	2.5608
1	1500	1.4445	5218.3	9.3317	2.6175
1	1600	1.3540	5482.7	9.5024	2.6702
1	1700	1.2742	5752.2	9.6657	2.7188
1	1800	1.2034	6026.3	9.8224	2.7636
1	1900	1.1399	6304.8	9.9729	2.8047
1	2000	1.0829	6587.2	10.1180	2.8426
10	273.15	1004.80	10.10	0.0004	4.1765
10	300	1001.00	121.66	0.3900	4.1564
10	373.15	962.98	426.52	1.2992	4.1948
10	400	942.47	539.67	1.5921	4.2360
10	500	838.17	977.04	2.5666	4.5903
10	584.18	688.63	1407.3	3.3591	6.1244
10	584.18	55.477	2724.5	5.6139	6.8973
10	600	49.830	2817.8	5.7716	5.1447
10	700	35.392	3176.1	6.3279	2.8769
10	800	29.128	3442.6	6.6844	2.5324
10	900	25.132	3690.6	6.9764	2.4463
10	1000	22.243	3934.5	7.2334	2.4395
10	1100	20.015	4179.5	7.4670	2.4654
10	1200	18.227	4428.0	7.6832	2.5059
10	1300	16.751	4680.9	7.8856	2.5529
10	1400	15.508	4938.6	8.0766	2.6018
10	1500	14.443	5201.3	8.2577	2.6503
10	1600	13.520	5468.6	8.4303	2.6969
10	1700	12.710	5740.6	8.5951	2.7410
10	1800	11.994	6016.8	8.7530	2.7823
10	1900	11.356	6296.9	8.9044	2.8206
10	2000	10.784	6580.8	9.0500	2.8563
100	273.15	1045.30	95.40	–0.0085	3.9092
100	300	1037.20	201.35	0.3614	3.9846
100	373.15	999.70	495.00	1.2371	4.0392
100	400	981.74	603.77	1.5186	4.0633
100	500	899.23	1015.3	2.4365	4.1854
100	600	791.46	1447.6	3.2237	4.5025
100	700	651.37	1925.2	3.9588	5.0954
100	800	482.17	2466.3	4.6806	5.6041
100	900	343.48	2997.0	5.3065	4.8280
100	1000	265.84	3431.3	5.7649	3.9368
100	1100	220.96	3798.6	6.1154	3.4632
100	1200	191.59	4131.4	6.4051	3.2174
100	1300	170.56	4445.8	6.6568	3.0848
100	1400	154.56	4750.3	6.8825	3.0129
100	1500	141.85	5049.5	7.0890	2.9759
100	1600	131.43	5346.2	7.2804	2.9600
100	1700	122.68	5641.9	7.4597	2.9570
100	1800	115.18	5937.8	7.6289	2.9620
100	1900	108.67	6234.5	7.7893	2.9720
100	2000	102.94	6532.4	7.9420	2.9852

Indexes

SUBSTANCE LIST

In order to conserve space in the tables of this book, chemical substances are identified only by their molecular formulas and primary (systematic) names. The Substance List that follows gives other information about individual substances which is often needed. As in the tables, substances are listed by molecular formula in a modified Hill order, in which all substances not containing carbon are listed before those that do contain carbon.

The following information is given, when available, in the Substance List:

- **CASRN:** Chemical Abstracts Service Registry Number (a unique identifying number assigned to each substance, which is useful for searching many chemical databases, both printed and electronic).
- **M_r:** Molecular weight (also called relative molar mass). Values have been calculated with the 1989 IUPAC Standard Atomic Weights. In this calculation all the significant figures in each atomic weight, up to a maximum of seven, have been kept, but trailing zeros in the sum have been dropped. For radioactive elements for which IUPAC does not specify a standard atomic weight, the atomic mass of the isotope of longest lifetime was used.
- **Name:** The systematic name used in the tables is given first. Below this, in smaller type, the most frequently encountered synonyms are listed.

Formula	CAS RN	M_r	Name
Ac	7440-34-8	227.0278	Actinium
Ag	7440-22-4	107.8682	Silver
AgBr	7785-23-1	187.7722	Silver bromide Silver(I) bromide
$AgBrO_3$	7783-89-3	235.7704	Silver bromate Bromic acid silver(I) salt Silver(I) bromate
AgCl	7783-90-6	143.3209	Silver chloride Silver(I) chloride
$AgClO_3$	7783-92-8	191.3191	Silver chlorate Chloric acid silver(I) salt Silver(I) chlorate
$AgClO_4$	7783-93-9	207.3185	Silver perchlorate Perchloric acid silver(I) salt Silver(I) perchlorate
AgF	7775-41-9	126.8666032	Silver fluoride Silver(I) fluoride
AgF_2	7783-95-1	145.8650064	Silver difluoride Silver(II) fluoride
AgI	7783-96-2	234.77267	Silver iodide Silver(I) iodide
$AgIO_3$	7783-97-3	282.77087	Silver iodate Iodic acid silver(I) salt Silver(I) iodate
$AgNO_3$	7761-88-8	169.87314	Silver nitrate Nitric acid silver(I) salt Silver(I) nitrate
AgO	1301-96-8	123.8676	Silver oxide Silver(II) oxide
Ag_2	12187-06-3	215.7364	Disilver
Ag_2CrO_4	7784-01-2	331.7301	Disilver chromate Chromic acid silver(I) salt Silver(I) chromate
Ag_2O	20667-12-3	231.7358	Disilver oxide Silver(I) oxide
Ag_2O_2	25455-73-6	247.7352	Disilver peroxide Silver(I) peroxide
Ag_2O_3	12002-97-0	263.7346	Disilver trioxide Silver(III) oxide
Ag_2O_4S	10294-26-5	311.8	Disilver sulfate Silver(I) sulfate Sulfuric acid silver(I) salt
Ag_2S	21548-73-2	247.8024	Disilver sulfide Silver(I) sulfide
Ag_2Te	12002-99-2	343.3364	Disilver telluride Silver(I) telluride
Al	7429-90-5	26.981539	Aluminum
AlB_3H_{12}	16962-07-5	71.509819	Aluminum trihydride-tris(borane) Aluminum borohydride
AlBr	22359-97-3	106.885539	Aluminum bromide
$AlBr_3$	7727-15-3	266.693539	Aluminum tribromide Aluminum bromide
AlCl	13595-81-8	62.434239	Aluminum chloride
$AlCl_2$	16603-84-2	97.886939	Aluminum dichloride
$AlCl_3$	7446-70-0	133.339639	Aluminum trichloride Aluminum chloride
$AlCl_3H_3N$		150.370199	Aluminum chloride-ammonia
$AlCl_4Cu$		232.338339	Copper tetrachloroaluminate Copper(I) aluminum tetrachloride Cuprous aluminum tetrachloride
$AlCl_4K$		207.890639	Potassium tetrachloroaluminate Aluminum potassium tetrachloride Aluminum potassium chloride
$AlCl_4Li$		175.733339	Lithium tetrachloroaluminate Aluminum lithium chloride Aluminum lithium tetrachloride
$AlCl_4Na$		191.782107	Sodium tetrachloroaluminate Aluminum sodium tetrachloride Aluminum sodium chloride
AlF	13595-82-9	45.9799422	Aluminum fluoride
AlF_3	7784-18-1	83.9767486	Aluminum trifluoride Aluminum fluoride
AlF_4Na	13821-15-3	125.9649198	Sodium tetrafluoroaluminate
AlF_6Na_3	13775-53-6	209.9412622	Trisodium hexafluoroaluminate Sodium hexafluoroaluminate
AlH	13967-22-1	27.989479	Aluminum hydride
AlH_3	7784-21-6	30.005359	Aluminum trihydride Aluminum hydride
AlH_4K	16903-34-7	70.111599	Potassium tetrahydroaluminate
AlH_4Li	16853-85-3	37.954299	Lithium tetrahydroaluminate Lithium aluminum hydride Lithium aluminum tetrahydride
AlI	29977-41-1	153.886009	Aluminum iodide
AlI_3	7784-23-8	407.694949	Aluminum triiodide Aluminum iodide
$AlKO_8S_2$		258.207039	Aluminum potassium disulfate Sulfuric acid aluminum potassium salt
AlN	24304-00-5	40.988279	Aluminum nitride
AlN_3O_9		212.996359	Aluminum trinitrate Nitric acid aluminum salt Aluminum nitrate
AlO	14457-64-8	42.980939	Aluminum oxide
AlO_4P	7784-30-7	121.952901	Aluminum phosphate Phosphoric acid aluminum salt
AlO_9P_3		263.897425	Aluminum trimetaphosphate Aluminum metaphosphate Metaphosphoric acid aluminum salt
AlP	20859-73-8	57.955301	Aluminum phosphide
AlS	12251-90-0	59.047539	Aluminum sulfide
Al_2	32752-94-6	53.963078	Dialuminum
Al_2Br_6	18898-34-5	533.387078	Dialuminum hexabromide Aluminum hexabromide
Al_2Cl_6	13845-12-0	266.679278	Dialuminum hexachloride Aluminum hexachloride
Al_2F_6	17949-86-9	167.9534972	Dialuminum hexafluoride Aluminum hexafluoride
Al_2I_6	18898-35-6	815.389898	Dialuminum hexaiodide Aluminum hexaiodide
Al_2O	12004-36-3	69.962478	Dialuminum oxide
Al_2O_3	1344-28-1	101.961278	Dialuminum trioxide Aluminum oxide

Formula	CAS RN	M_r	Name
$Al_2O_{12}S_3$	10043-01-3	342.153878	Dialuminum trisulfate
			Aluminum sulfate
			Sulfuric acid aluminum salt
Al_2S_3	1302-81-4	150.161078	Dialuminum trisulfide
			Aluminum sulfide
Am	7440-35-9	243.0614	Americium
Ar	7440-37-1	39.948	Argon
^{36}Ar		35.9675455	(^{36}Ar)Argon
^{40}Ar		39.9623837	(^{40}Ar)Argon
As	7440-38-2	74.92159	Arsenic
$AsBr_3$	7784-33-0	314.63359	Arsenic tribromide
			Arsenic(III) bromide
$AsCl_3$	7784-34-1	181.27969	Arsenic trichloride
			Arsenic(III) chloride
AsF_3	7784-35-2	131.9167996	Arsenic trifluoride
			Arsenic(III) fluoride
AsF_5	7784-36-3	169.913606	Arsenic pentafluoride
			Arsenic(V) fluoride
AsGa	1303-00-0	144.64459	Gallium arsenide
AsH_3	7784-42-1	77.94541	Arsane
			Arsine
AsH_3O_4	7778-39-4	141.94301	Arsenic acid
AsI_3	7784-45-4	455.635	Arsenic triiodide
			Arsenic(III) iodide
AsIn	1303-11-3	189.74159	Indium arsenide
AsO	12005-99-1	90.92099	Arsenic oxide
			Arsenic(II) oxide
As_2	23878-46-8	149.84318	Diarsenic
As_2O_3	1327-53-3	197.84138	Diarsenic trioxide
			Arsenic(III) oxide
			Arsenic trioxide
As_2O_5	1303-28-2	229.84018	Diarsenic pentaoxide
			Arsenic(V) oxide
			Arsenic pentaoxide
			Arsenic pentoxide
As_2S_3	1303-33-9	246.04118	Diarsenic trisulfide
			Arsenic trisulfide
			Arsenic(III) sulfide
At	7440-68-8	209.9871	Astatine
Au	7440-57-5	196.96654	Gold
AuBr	10294-27-6	276.87054	Gold bromide
$AuBr_3$	10294-28-7	436.67854	Gold tribromide
			Gold(III) bromide
AuCl	10294-29-8	232.41924	Gold chloride
$AuCl_3$	13453-07-1	303.32464	Gold trichloride
			Gold(III) chloride
AuF_3	14720-21-9	253.9617496	Gold trifluoride
			Gold(III) fluoride
AuH	13464-75-0	197.97448	Gold hydride
AuI	10294-31-2	323.87101	Gold iodide
Au_2	12187-09-6	393.93308	Digold
B	7440-42-8	10.811	Boron
BBr	19961-29-6	90.715	Boron bromide
			Bromoborane
BBr_3	10294-33-4	250.523	Boron tribromide
			Boron bromide
BCl	20583-55-5	46.2637	Boron chloride
			Chloroborane
BCl_3	10294-34-5	117.1691	Boron trichloride
			Boron chloride
$BCsO_2$	92141-86-1	175.71523	Cesium metaborate
BF	13768-60-0	29.8094032	Boron fluoride
BF (Cont.)			Fluoroborane
BF_3	7637-07-2	67.8062096	Boron trifluoride
			Boron fluoride
$^{10}BF_3$		67.0081465	(^{10}B)Boron trifluoride
			(^{10}B)Boron fluoride
$^{11}BF_3$		68.004515	(^{11}B)Boron trifluoride
			(^{11}B)Boron fluoride
BF_3H_3N	13709-86-9	84.8367696	Ammonia-boron trifluoride
			Azane-boron trifluoride
BF_3H_3P	14931-39-6	101.8037916	Phosphine-boron trifluoride
			Phosphane-boron trifluoride
BF_4Na	13755-29-8	109.7943808	Sodium tetrafluoroborate
BH	13766-26-2	11.81894	Boron hydride
			Borane(1)
BHO_2	13460-50-9	43.81774	Metaboric acid
BH_3	13283-31-3	13.83482	Borane(3)
			Boron trihydride
			Borane
BH_3O_3	10043-35-3	61.83302	Boric acid
BH_4K	13762-51-1	53.94106	Potassium hydride-borane
			Potassium tetrahydroborate
			Potassium tetrahydridoborate
			Potassium borohydride
BH_4Li	16949-15-8	21.78376	Lithium hydride-borane
			Lithium tetrahydroborate
			Lithium borohydride
			Lithium tetrahydridoborate
BH_4Na	16940-66-2	37.832528	Sodium hydride-borane
			Sodium tetrahydridoborate
			Sodium borohydride
			Sodium tetrahydroborate
BI_3	13517-10-7	391.52441	Boron triiodide
			Boron iodide
BKO_2	13709-94-9	81.9081	Potassium metaborate
			Metaboric acid potassium salt
$BLiO_2$	13453-69-5	49.7508	Lithium metaborate
			Metaboric acid lithium salt
BN	10043-11-5	24.81774	Boron nitride
$BNaO_2$	7775-19-1	65.799568	Sodium metaborate
			Metaboric acid sodium salt
BO	12505-77-0	26.8104	Boron oxide
BO_2	13840-88-5	42.8098	Boron dioxide
BO_2Rb	13709-66-5	128.2776	Rubidium metaborate
			Metaboric acid rubidium salt
BS	12228-39-6	42.877	Boron sulfide
B_2	14452-61-0	21.622	Diboron
B_2Cl_4	13701-67-2	163.4328	Tetrachlorodiborane(4)
			Diboron tetrachloride
B_2F_4	13965-73-6	97.6156128	Tetrafluorodiborane(4)
			Diboron tetrafluoride
B_2H_6	19287-45-7	27.66964	Diborane(6)
			Diborane
B_2O_2	13766-28-4	53.6208	Diboron dioxide
B_2O_3	1303-86-2	69.6202	Diboron trioxide
			Boron oxide
B_2S_3	12007-33-9	117.82	Diboron trisulfide
			Boron sulfide
$B_3H_6N_3$	6569-51-3	80.50086	Cyclotriborazane
			Borazine
B_4H_{10}	18283-93-7	53.3234	Tetraborane(10)
$B_4Na_2O_7$	1330-43-4	201.219336	Disodium tetraborate
			Tetraboric acid sodium salt

Formula	CAS RN	M_r	Name
$B_4Na_2O_7$ (Cont.)			Sodium tetraborate
B_5H_9	19624-22-7	63.12646	Pentaborane(9)
B_5H_{11}	18433-84-6	65.14234	Pentaborane(11)
B_6H_{10}	23777-80-2	74.9454	Hexaborane(10)
$B_{10}H_{14}$	17702-41-9	122.22116	Decaborane(14)
Ba	7440-39-3	137.327	Barium
$BaBr_2$	10553-31-8	297.135	Barium dibromide
			Barium bromide
$BaCl_2$	10361-37-2	208.2324	Barium dichloride
			Barium chloride
$BaCl_2O_8$	13465-95-7	336.2276	Barium diperchlorate
			Barium perchlorate
			Perchloric acid barium salt
BaF_2	7787-32-8	175.3238064	Barium difluoride
			Barium fluoride
BaH_2	13477-09-3	139.34288	Barium dihydride
			Barium hydride
BaH_2O_2	17194-00-2	171.34168	Barium dihydroxide
			Barium hydroxide
BaI_2	13718-50-8	391.13594	Barium diiodide
			Barium iodide
BaN_2O_4	13465-94-6	229.33808	Barium dinitrite
			Barium nitrite
			Nitrous acid barium salt
BaN_2O_6	10022-31-8	261.33688	Barium dinitrate
			Barium nitrate
			Nitric acid barium salt
BaO	1304-28-5	153.3264	Barium oxide
BaO_4S	7727-43-7	233.3906	Barium sulfate
			Sulfuric acid barium salt
BaS	21109-95-5	169.393	Barium sulfide
Be	7440-41-7	9.012182	Beryllium
$BeBr_2$	7787-46-4	168.820182	Beryllium dibromide
			Beryllium bromide
$BeCl_2$	7787-47-5	79.917582	Beryllium dichloride
			Beryllium chloride
$BeCl_2O_8$		207.912782	Beryllium diperchlorate
			Beryllium perchlorate
			Perchloric acid beryllium salt
BeF_2	7787-49-7	47.0089884	Beryllium difluoride
			Beryllium fluoride
BeF_4Li_2		98.8877948	Beryllium difluoride-bis(lithium fluoride)
			Beryllium lithium fluoride
BeH_2O_2	13327-32-7	43.026862	Beryllium dihydroxide
			Beryllium hydroxide
BeI_2	7787-53-3	262.821122	Beryllium diiodide
			Beryllium iodide
BeN_2O_6	13597-99-4	133.022062	Beryllium dinitrate
			Beryllium nitrate
			Nitric acid beryllium salt
BeO	1304-56-9	25.011582	Beryllium oxide
BeO_4S	13510-49-1	105.075782	Beryllium sulfate
			Sulfuric acid beryllium salt
BeS	13598-22-6	41.078182	Beryllium sulfide
Bi	7440-69-9	208.98037	Bismuth
$BiBr_3$	7787-58-8	448.69237	Bismuth tribromide
			Bismuth bromide
BiClO	7787-59-9	260.43247	Bismuth oxychloride
$BiCl_3$	7787-60-2	315.33847	Bismuth trichloride
			Bismuth chloride
BiH_3O_3	10361-43-0	260.00239	Bismuth trihydroxide

Formula	CAS RN	M_r	Name
BiH_3O_3 (Cont.)			Bismuth hydroxide
BiI_3	7787-64-6	589.69378	Bismuth triiodide
			Bismuth iodide
Bi_2	12187-12-1	417.96074	Dibismuth
Bi_2O_3	1304-76-3	465.95894	Dibismuth trioxide
			Bismuth oxide
$Bi_2O_{12}S_3$	7787-68-0	706.15154	Dibismuth trisulfate
			Bismuth sulfate
			Sulfuric acid bismuth salt
Bi_2S_3	1345-07-9	514.15874	Dibismuth trisulfide
			Bismuth sulfide
Bk	7440-40-6	247.0703	Berkelium
Br	10097-32-2	79.904	Bromine
BrCl	13863-41-7	115.3567	Bromine chloride
$BrCl_3Si$	13465-74-2	214.3476	Bromotrichlorosilane
BrCs	7787-69-1	212.80943	Cesium bromide
BrCu	7787-70-4	143.45	Copper bromide
			Copper(I) bromide
BrF	13863-59-7	98.9024032	Bromine fluoride
BrF_3	7787-71-5	136.8992096	Bromine trifluoride
			Bromine(III) fluoride
BrF_5	7789-30-2	174.896016	Bromine pentafluoride
			Bromine(V) fluoride
BrGe	25884-11-1	152.514	Germanium bromide
$BrGeH_3$	13569-43-2	155.53782	Bromogermane
BrH	10035-10-6	80.91194	Hydrogen bromide
			Bromane
			Hydrobromic acid
Br^2H	13536-59-9	81.9181017	(^{2}H)Hydrogen bromide
			Deuterium bromide
			Hydrobromic acid-d
			(^{2}H)Bromane
BrHSi	13569-45-4	108.99744	Bromosilanediyl
BrH_3Si	13465-73-1	111.01332	Bromosilane
BrH_4N	12124-97-9	97.9425	Ammonium bromide
BrI	7789-33-5	206.80847	Iodine bromide
BrIn	14280-53-6	194.724	Indium bromide
BrK	7758-02-3	119.0023	Potassium bromide
$BrKO_3$	7758-01-2	167.0005	Potassium bromate
			Bromic acid potassium salt
$BrKO_4$	22207-96-1	182.9999	Potassium perbromate
			Perbromic acid potassium salt
BrLi	7550-35-8	86.845	Lithium bromide
BrNO	13444-87-6	109.91014	Nitrosyl bromide
BrNa	7647-15-6	102.893768	Sodium bromide
$BrNaO_3$	7789-38-0	150.891968	Sodium bromate
			Bromic acid sodium salt
BrO	15656-19-6	95.9034	Bromine oxide
BrO_2	21255-83-4	111.9028	Bromine superoxide
BrRb	7789-39-1	165.3718	Rubidium bromide
BrSi	14791-57-2	107.9895	Bromosilanetriyl
BrTl	7789-40-4	284.2873	Thallium bromide
			Thallium(I) bromide
			Thallous bromide
Br_2	7726-95-6	159.808	Bromine
			Dibromine
Br_2Ca	7789-41-5	199.886	Calcium dibromide
			Calcium bromide
Br_2Cd	7789-42-6	272.219	Cadmium dibromide
			Cadmium bromide
Br_2CdO_6		368.2154	Cadmium dibromate
			Cadmium bromate

Formula	CAS RN	M_r	Name
Br_2CdO_6 (Cont.)			Bromic acid cadmium salt
Br_2Co	7789-43-7	218.7412	Cobalt dibromide
			Cobalt(II) bromide
Br_2Cr	10049-25-9	211.8041	Chromium dibromide
			Chromium(II) bromide
Br_2Cu	7789-45-9	223.354	Copper dibromide
			Copper(II) bromide
Br_2Fe	7789-46-0	215.655	Iron dibromide
			Iron(II) bromide
			Ferrous bromide
Br_2H_2Si	13768-94-0	189.90938	Dibromosilane
Br_2Hg	7789-47-1	360.398	Mercury dibromide
			Mercuric bromide
			Mercury(II) bromide
Br_2Hg_2	15385-58-7	560.988	Dimercury dibromide
			Mercury(I) bromide
			Mercurous bromide
Br_2Mg	7789-48-2	184.113	Magnesium dibromide
			Magnesium bromide
Br_2Mn	13446-03-2	214.74605	Manganese dibromide
			Manganese(II) bromide
Br_2Ni	13462-88-9	218.5014	Nickel dibromide
			Nickel(II) bromide
Br_2Pb	10031-22-8	367.008	Lead dibromide
			Lead(II) bromide
Br_2Pt	13455-12-4	354.888	Platinum dibromide
			Platinum(II) bromide
Br_2S_2	13172-31-1	223.94	Disulfur dibromide
			Sulfur bromide
Br_2Se	22987-45-7	238.768	Selenium dibromide
			Selenium(II) bromide
Br_2Sn	10031-24-0	278.518	Tin dibromide
			Tin(II) bromide
Br_2Sr	10476-81-0	247.428	Strontium dibromide
			Strontium bromide
Br_2Ti	13783-04-5	207.688	Titanium dibromide
			Titanium(II) bromide
Br_2Zn	7699-45-8	225.198	Zinc dibromide
			Zinc bromide
Br_3ClSi	13465-76-4	303.2502	Tribromochlorosilane
Br_3Fe	10031-26-2	295.559	Iron tribromide
			Iron(III) bromide
			Ferric bromide
Br_3Ga	13450-88-9	309.435	Gallium tribromide
			Gallium(III) bromide
Br_3HSi	7789-57-3	268.80544	Tribromosilane
Br_3In	13465-09-3	354.532	Indium tribromide
			Indium(III) bromide
Br_3OP	7789-59-5	286.685162	Phosphoryl tribromide
			Phosphoryl bromide
Br_3P	7789-60-8	270.685762	Phosphorus tribromide
			Phosphorous bromide
			Phosphorus(III) bromide
Br_3PS		302.751762	Thiophosphoryl tribromide
			Phosphorus sulfobromide
Br_3Pt	25985-07-3	434.792	Platinum tribromide
			Platinum(III) bromide
Br_3Re	13569-49-8	425.919	Rhenium tribromide
			Rhenium(III) bromide
Br_3Ru	14014-88-1	340.782	Ruthenium tribromide
			Ruthenium(III) bromide
Br_3Sb	7789-61-9	361.469	Antimony tribromide

Formula	CAS RN	M_r	Name
Br_3Sb (Cont.)			Antimony(III) bromide
Br_3Sc	13465-59-3	284.66791	Scandium tribromide
			Scandium(III) bromide
Br_3Ti	13135-31-4	287.592	Titanium tribromide
			Titanium(III) bromide
Br_3U		477.7409	Uranium tribromide
			Uranium(III) bromide
Br_4Ge	13450-92-5	392.226	Germanium tetrabromide
			Germanium bromide
Br_4Hf	13777-22-5	498.106	Hafnium tetrabromide
			Hafnium(IV) bromide
Br_4Pa	13867-42-0	550.65188	Protactinium tetrabromide
			Protactinium(IV) bromide
Br_4Pt	68938-92-1	514.696	Platinum tetrabromide
			Platinum(IV) bromide
Br_4Si	7789-66-4	347.7015	Silicon tetrabromide
			Silicon(IV) bromide
Br_4Sn	7789-67-5	438.326	Tin tetrabromide
			Tin(IV) bromide
			Tetrabromostannane
Br_4Te	10031-27-3	447.216	Tellurium tetrabromide
			Tellurium(IV) bromide
Br_4Ti	7789-68-6	367.496	Titanium tetrabromide
			Titanium(IV) bromide
Br_4V	13595-30-7	370.5575	Vanadium tetrabromide
			Vanadium(IV) bromide
Br_4Zr	13777-25-8	410.84	Zirconium tetrabromide
			Zirconium(IV) bromide
Br_5P	7789-69-7	430.493762	Phosphorus pentabromide
			Phosphorus(V) bromide
			Phosphoric bromide
Br_5Ta	13451-11-1	580.4679	Tantalum pentabromide
			Tantalum(V) bromide
Br_6W	13701-86-5	663.274	Tungsten hexabromide
			Tungsten(VI) bromide
Ca	7440-70-2	40.078	Calcium
$CaCl_2$	10043-52-4	110.9834	Calcium dichloride
			Calcium chloride
$CaCl_2O_8$		238.9786	Calcium diperchlorate
			Calcium perchlorate
			Perchloric acid calcium salt
CaF_2	7789-75-5	78.0748064	Calcium difluoride
			Calcium fluoride
CaH_2	7789-78-8	42.09388	Calcium dihydride
			Calcium hydride
CaH_2O_2	1305-62-0	74.09268	Calcium dihydroxide
			Calcium hydroxide
$CaH_4O_8P_2$		234.052484	Calcium bis(dihydrogen phosphate)
			Calcium phosphate monobasic
			Phosphoric acid calcium salt
CaI_2	10102-68-8	293.88694	Calcium diiodide
			Calcium iodide
CaN_2O_4		132.08908	Calcium dinitrite
			Calcium nitrite
			Nitrous acid calcium salt
CaN_2O_6	10124-37-5	164.08788	Calcium dinitrate
			Calcium nitrate
			Nitric acid calcium salt
CaO	1305-78-8	56.0774	Calcium oxide
CaO_3S		120.1422	Calcium sulfite
			Sulfurous acid calcium salt

Formula	CAS RN	M_r	Name
CaO_4S	7778-18-9	136.1416	Calcium sulfate
			Sulfuric acid calcium salt
CaS	20548-54-3	72.144	Calcium sulfide
$Ca_3O_8P_2$	7758-87-4	310.176724	Tricalcium bisphosphate
			Calcium phosphate
			Phosphoric acid calcium salt
Cd	7440-43-9	112.411	Cadmium
$CdCl_2$	10108-64-2	183.3164	Cadmium dichloride
			Cadmium chloride
$CdCl_2O_8$		311.3116	Cadmium diperchlorate
			Cadmium perchlorate
			Perchloric acid cadmium salt
CdF_2	7790-79-6	150.4078064	Cadmium difluoride
			Cadmium fluoride
CdH_2O_2	21041-95-2	146.42568	Cadmium dihydroxide
			Cadmium hydroxide
CdI_2	7790-80-9	366.21994	Cadmium diiodide
			Cadmium iodide
CdN_2O_4		204.42208	Cadmium dinitrite
			Cadmium nitrite
			Nitrous acid cadmium salt
CdN_2O_6	10325-94-7	236.42088	Cadmium dinitrate
			Cadmium nitrate
			Nitric acid cadmium salt
CdO	1306-19-0	128.4104	Cadmium oxide
CdO_4S	10124-36-4	208.4746	Cadmium sulfate
			Sulfuric acid cadmium salt
CdS	1306-23-6	144.477	Cadmium sulfide
$CdSe$	1306-24-7	191.371	Cadmium selenide
$CdTe$	1306-25-8	240.011	Cadmium telluride
Ce	7440-45-1	140.115	Cerium
$CeCl_3$	7790-86-5	246.4731	Cerium trichloride
			Cerium(III) chloride
CeN_3O_9		326.12982	Cerium trinitrate
			Cerium(III) nitrate
			Nitric acid cerium(III) salt
CeO_2	1306-38-3	172.1138	Cerium dioxide
			Cerium(IV) oxide
CeS	12014-82-3	172.181	Cerium sulfide
			Cerium(II) sulfide
Ce_2O_3	1345-13-7	328.2282	Dicerium trioxide
			Cerium(III) oxide
$Ce_2O_{12}S_3$	13454-94-9	568.4208	Dicerium trisulfate
			Cerium(III) sulfate
			Sulfuric acid cerium(III) salt
Cf	7440-71-3	251.0796	Californium
Cl	22537-15-1	35.4527	Chlorine
$ClCs$	7647-17-8	168.35813	Cesium chloride
$ClCsO_4$	13454-84-7	232.35573	Cesium perchlorate
			Perchloric acid cesium salt
$ClCu$	7758-89-6	98.9987	Copper chloride
			Cuprous chloride
			Copper(I) chloride
ClF	7790-89-8	54.4511032	Chlorine fluoride
$ClFO_3$	7616-94-6	102.4493032	Perchloryl fluoride
ClF_2N	13637-87-1	87.4562464	Nitrogen chloride difluoride
ClF_2P	14335-40-1	104.4232684	Phosphorus chloride difluoride
ClF_2PS	2524-02-9	136.4892684	Thiophosphoryl chloride difluoride
ClF_3	7790-91-2	92.4479096	Chlorine trifluoride
			Chlorine(III) fluoride
ClF_3Si	14049-36-6	120.5334096	Chlorotrifluorosilane
ClF_5	13637-63-3	130.444716	Chlorine pentafluoride
			Chlorine(V) fluoride
ClF_5S	13780-57-9	162.510716	Sulfur chloride pentafluoride
$ClGe$	21110-21-4	108.0627	Germanium chloride
$ClGeH_3$	13637-65-5	111.08652	Chlorogermane
ClH	7647-01-0	36.46064	Hydrogen chloride
			Chlorane
			Hydrochloric acid
Cl^2H	7698-05-7	37.4668017	(^{2}H)Hydrogen chloride
			Deuterium chloride
			(^{2}H)Chlorane
			Hydrochloric acid-d
$ClHO$	7790-92-3	52.46004	Hypochlorous acid
$ClHO_3S$	7790-94-5	116.52484	Chlorosulfonic acid
$ClHO_4$	7601-90-3	100.45824	Perchloric acid
ClH_3Si	13465-78-6	66.56202	Chlorosilane
ClH_4N	12125-02-9	53.4912	Ammonium chloride
ClH_4NO_4	7790-98-9	117.4888	Ammonium perchlorate
			Perchloric acid ammonium salt
ClH_4P	24567-53-1	70.458222	Phosphonium chloride
$ClH_5N_2O_4$	13762-80-6	132.50348	Hydrazinium perchlorate
ClI	7790-99-0	162.35717	Iodine chloride
$ClIn$	13465-10-6	150.2727	Indium chloride
ClK	7447-40-7	74.551	Potassium chloride
ClK		74.551	Potassium chloride (c,I)
ClK		74.551	Potassium chloride (c,II)
$ClKO_3$	3811-04-9	122.5492	Potassium chlorate
			Chloric acid potassium salt
$ClKO_4$	7778-74-7	138.5486	Potassium perchlorate
			Perchloric acid potassium salt
$ClLi$	7447-41-8	42.3937	Lithium chloride
$ClLiO_3$		90.3919	Lithium chlorate
			Chloric acid lithium salt
$ClLiO_4$	7791-03-9	106.3913	Lithium perchlorate
			Perchloric acid lithium salt
$ClNO$	2696-92-6	65.45884	Nitrosyl chloride
$ClNO_2$	13444-90-1	81.45824	Nitryl chloride
$ClNa$	7647-14-5	58.442468	Sodium chloride
$ClNaO_2$	7758-19-2	90.441268	Sodium chlorite
$ClNaO_3$	7775-09-9	106.440668	Sodium chlorate
			Chloric acid sodium salt
$ClNaO_4$	7601-89-0	122.440068	Sodium perchlorate
			Perchloric acid sodium salt
ClO	14989-30-1	51.4521	Chlorine oxide
$ClOV$	13520-87-1	102.3936	Vanadium oxychloride
ClO_2	10049-04-4	67.4515	Chlorine dioxide
ClO_4Rb	13510-42-4	184.9181	Rubidium perchlorate
			Perchloric acid rubidium salt
$ClRb$	7791-11-9	120.9205	Rubidium chloride
$ClRb$		120.9205	Rubidium chloride (c,I)
$ClRb$		120.9205	Rubidium chloride (c,II)
$ClTl$	7791-12-0	239.836	Thallium chloride
			Thallium(I) chloride
			Thallous chloride
Cl_2	7782-50-5	70.9054	Chlorine
			Dichlorine
Cl_2Co	7646-79-9	129.8386	Cobalt dichloride
			Cobalt(II) chloride
Cl_2CoO_8		257.8338	Cobalt diperchlorate
			Cobalt(II) perchlorate

Formula	CAS RN	M_r	Name
Cl_2CoO_8 (Cont.)			Perchloric acid cobalt(II) salt
Cl_2Cr	10049-05-5	122.9015	Chromium dichloride
			Chromium(II) chloride
Cl_2CrO_2	14977-61-8	154.9003	Chromyl dichloride
Cl_2Cu	7447-39-4	134.4514	Copper dichloride
			Cupric chloride
			Copper(II) chloride
Cl_2CuO_8		262.4466	Copper diperchlorate
			Cupric perchlorate
			Copper(II) perchlorate
			Perchloric acid copper(II) salt
Cl_2FP	15597-63-4	120.8775652	Phosphorus dichloride fluoride
Cl_2F_2Si	18356-71-3	136.9877064	Dichlorodifluorosilane
Cl_2Fe	7758-94-3	126.7524	Iron dichloride
			Iron(II) chloride
			Ferrous chloride
Cl_2H_2Si	4109-96-0	101.00678	Dichlorosilane
Cl_2Hg	7487-94-7	271.4954	Mercury dichloride
			Mercury(I) chloride
			Mercurous chloride
Cl_2HgO_8		399.4906	Mercury diperchlorate
			Mercuric perchlorate
			Mercury(II) perchlorate
			Perchloric acid mercury(II) salt
Cl_2Hg_2	10112-91-1	472.0854	Dimercury dichloride
Cl_2Mg	7786-30-3	95.2104	Magnesium dichloride
			Magnesium chloride
Cl_2Mn	7773-01-5	125.84345	Manganese dichloride
			Manganese(II) chloride
Cl_2MnO_8		253.83865	Manganese diperchlorate
			Manganese(II) perchlorate
			Perchloric acid manganese(II) salt
Cl_2Ni	7718-54-9	129.5988	Nickel dichloride
			Nickel(II) chloride
Cl_2NiO_8		257.594	Nickel diperchlorate
			Nickel(II) perchlorate
			Perchloric acid nickel(II) salt
Cl_2O	7791-21-1	86.9048	Dichlorine oxide
			Chlorinous oxide
			Chlorine oxide
Cl_2OS	7719-09-7	118.9708	Sulfinyl dichloride
			Thionyl chloride
Cl_2OSe	7791-23-3	165.8648	Selenosyl dichloride
			Selenium oxychloride
Cl_2OTi		134.7848	Titanyl dichloride
			Titanyl chloride
Cl_2OZr		178.1288	Zirconosyl dichloride
			Zirconium oxychloride
Cl_2O_2S	7791-25-5	134.9702	Sulfonyl dichloride
			Sulfuryl chloride
			Sulfuryl dichloride
Cl_2O_2U	7791-26-6	340.9331	Uranyl dichloride
			Dichlorodioxouranium
$Cl_2O_5S_2$	7791-27-7	215.0344	Pyrosulfuryl dichloride
			Pyrosulfuryl chloride
			Disulfuryl chloride
Cl_2Pb	7758-95-4	278.1054	Lead dichloride
			Lead(II) chloride
Cl_2Pt	10025-65-7	265.9854	Platinum dichloride
Cl_2Pt (Cont.)			Platinum(II) chloride
Cl_2S	10545-99-0	102.9714	Sulfur dichloride
			Sulfur chloride
Cl_2S_2	10025-67-9	135.0374	Disulfur dichloride
			Sulfur chloride
Cl_2Sm		221.2654	Samarium dichloride
			Samarium(II) chloride
Cl_2Sn	7772-99-8	189.6154	Tin dichloride
			Tin(II) chloride
Cl_2Sr	10476-85-4	158.5254	Strontium dichloride
			Strontium chloride
Cl_2Ti	10049-06-6	118.7854	Titanium dichloride
			Titanium(II) chloride
Cl_2Yb		243.9454	Ytterbium dichloride
			Ytterbium(II) chloride
Cl_2Zn	7646-85-7	136.2954	Zinc dichloride
			Zinc chloride
Cl_2Zr	13762-26-0	162.1294	Zirconium dichloride
$Cl_3CoH_{15}N_5$	13859-51-3	250.4441	Pentaamminechlorocobalt dichloride
			Pentamminechlorocobalt(III) chloride
Cl_3Cr	10025-73-7	158.3542	Chromium trichloride
			Chromium(III) chloride
Cl_3Dy	10025-74-8	268.8581	Dysprosium trichloride
			Dysprosium(III) chloride
Cl_3Er	10138-41-7	273.6181	Erbium trichloride
			Erbium(III) chloride
Cl_3Eu	10025-76-0	258.3231	Europium trichloride
			Europium(III) chloride
Cl_3FSi	14965-52-7	153.4420032	Trichlorofluorosilane
Cl_3Fe	7705-08-0	162.2051	Iron trichloride
			Iron(III) chloride
			Ferric chloride
Cl_3Ga	13450-90-3	176.0811	Gallium trichloride
			Gallium(III) chloride
Cl_3Gd	10138-52-0	263.6081	Gadolinium trichloride
			Gadolinium(III) chloride
Cl_3HSi	10025-78-2	135.45154	Trichlorosilane
Cl_3Ho	10138-62-2	271.28842	Holmium trichloride
			Holmium(III) chloride
Cl_3In	10025-82-8	221.1781	Indium trichloride
			Indium(III) chloride
Cl_3Ir	10025-83-9	298.5781	Iridium trichloride
			Iridium(III) chloride
Cl_3La	10099-58-8	245.2636	Lanthanum trichloride
			Lanthanum(III) chloride
Cl_3Lu	10099-66-8	281.3251	Lutetium trichloride
			Lutetium(III) chloride
Cl_3N	10025-85-1	120.36484	Nitrogen trichloride
Cl_3Nd	10024-93-8	250.5981	Neodymium trichloride
			Neodymium(III) chloride
Cl_3OP	10025-87-3	153.331262	Phosphoryl trichloride
			Phosphorus oxychloride
Cl_3OV	7727-18-6	173.299	Vanadyl trichloride
			Vanadium oxytrichloride
Cl_3Os	13444-93-4	296.5581	Osmium trichloride
			Osmium(III) chloride
Cl_3P	7719-12-2	137.331862	Phosphorus trichloride
			Phosphorus(III) chloride
Cl_3PS	3982-91-0	169.397862	Thiophosphoryl chloride
			Phosphorous thiochloride

Formula	CAS RN	M_r	Name
Cl_3PS (Cont.)			Phosphorous sulfochloride
Cl_3Pr	10361-79-2	247.26575	Praseodymium trichloride
			Praseodymium(III) chloride
Cl_3Pt	25909-39-1	301.4381	Platinum trichloride
			Platinum(III) chloride
Cl_3Re	13569-63-6	292.5651	Rhenium trichloride
Cl_3Rh	10049-07-7	209.2636	Rhodium trichloride
			Rhodium(III) chloride
Cl_3Ru	10049-08-8	207.4281	Ruthenium trichloride
			Ruthenium(III) chloride
Cl_3Sb	10025-91-9	228.1151	Antimony trichloride
			Antimony(III) chloride
Cl_3Sc	10361-84-9	151.31401	Scandium trichloride
Cl_3Sm	10361-82-7	256.7181	Samarium trichloride
			Samarium(III) chloride
Cl_3Ta		287.306	Tantalum trichloride
			Tantalum(III) chloride
Cl_3Tb	10042-88-3	265.28344	Terbium trichloride
			Terbium(III) chloride
Cl_3Ti	7705-07-9	154.2381	Titanium trichloride
			Titanium(III) chloride
Cl_3Tl	13453-32-2	310.7414	Thallium trichloride
			Thallium(III) chloride
			Thallic chloride
Cl_3Tm	13537-18-3	275.29231	Thullium trichloride
			Thullium(III) chloride
Cl_3U	10025-93-1	344.387	Uranium trichloride
			Uranium(III) chloride
Cl_3V	7718-98-1	157.2996	Vanadium trichloride
			Vanadium(III) chloride
Cl_3Y	10361-92-9	195.26395	Yttrium trichloride
			Yttrium(III) chloride
Cl_3Yb	10361-91-8	279.3981	Ytterbium trichloride
			Ytterbium(III) chloride
Cl_4Ge	10038-98-9	214.4208	Tetrachlorogerman
			Germanium tetrachloride
Cl_4Hf	13499-05-3	320.3008	Hafnium tetrachloride
			Hafnium(IV) chloride
Cl_4ORe	13814-76-1	344.0172	Rhenosyl tetrachloride
			Rhenium oxychloride
Cl_4OW	13520-78-0	341.6602	Tungstenosyl tetrachloride
			Tungsten oxychloride
Cl_4Pa	13867-41-9	372.84668	Protactinium tetrachloride
Cl_4Pb	13463-30-4	349.0108	Lead tetrachloride
			Lead(IV) chloride
Cl_4Pt	13454-96-1	336.8908	Platinum tetrachloride
			Platinum(IV) chloride
Cl_4Se	10026-03-6	220.7708	Selenium tetrachloride
			Selenium(IV) chloride
Cl_4Si	10026-04-7	169.8963	Silicon tetrachloride
			Silicon(IV) chloride
Cl_4Sn	7646-78-8	260.5208	Tin tetrachloride
			Tin(IV) chloride
			Tetrachlorostannane
Cl_4Te	10026-07-0	269.4108	Tellurium tetrachloride
			Tellurium(IV) chloride
Cl_4Th	10026-08-1	373.8489	Thorium tetrachloride
			Thorium(IV) chloride
Cl_4Ti	7550-45-0	189.6908	Titanium tetrachloride
			Titanium(IV) chloride
Cl_4U	10026-10-5	379.8397	Uranium tetrachloride
			Uranium(IV) chloride
Cl_4V	7632-51-1	192.7523	Vanadium tetrachloride
			Vanadium(IV) chloride
Cl_4Zr	10026-11-6	233.0348	Zirconium tetrachloride
			Zirconium(IV) chloride
Cl_5Mo	10241-05-1	273.2035	Molybdenum pentachloride
			Molybdenum(V) chloride
Cl_5Nb	10026-12-7	270.16988	Niobium pentachloride
			Niobium(V) chloride
Cl_5P	10026-13-8	208.237262	Phosphorus pentachloride
Cl_5Pa	13760-41-3	408.29938	Protactinium pentachloride
Cl_5Ta	7721-01-9	358.2114	Tantalum pentachloride
			Tantalum(V) chloride
Cl_5W		361.1135	Tungsten pentachloride
			Tungsten(V) chloride
Cl_6HSb		335.48114	Hydrogen hexachloroantimo-nate
Cl_6OSi_2	14986-21-1	284.8866	Hexachlorodisiloxane
Cl_6U	13763-23-0	450.7451	Uranium hexachloride
			Uranium(VI) chloride
Cl_6W	13283-01-7	396.5662	Tungsten hexachloride
			Tungsten(VI) chloride
$Cl_8O_2Si_3$		399.8769	Octachlorotrisiloxane
Cm	7440-51-9	247.0703	Curium
Co	7440-48-4	58.9332	Cobalt
CoF_2	10026-17-2	96.9300064	Cobalt difluoride
			Cobalt(II) fluoride
CoH_2O_2	21041-93-0	92.94788	Cobalt dihydroxide
			Cobalt(II) hydroxide
CoI_2	15238-00-3	312.74214	Cobalt diiodide
			Cobalt(II) iodide
CoN_2O_6	10141-05-6	182.94308	Cobalt dinitrate
			Cobalt(II) nitrate
			Nitric acid cobalt(II) salt
CoO	1307-96-6	74.9326	Cobalt oxide
			Cobalt(II) oxide
CoO_4S	10124-43-3	154.9968	Cobalt sulfate
			Cobalt(II) sulfate
			Sulfuric acid cobalt(II) salt
CoS	1317-42-6	90.9992	Cobalt sulfide
			Cobalt(II) sulfide
Co_2S_3	1332-71-4	214.0644	Dicobalt trisulfide
			Cobalt(III) sulfide
Co_3O_4	1308-06-1	240.7972	Tricobalt tetraoxide
			Cobalt(II,III) oxide
Cr	7440-47-3	51.9961	Chromium
CrF_2	10049-10-2	89.9929064	Chromium difluoride
			Chromium(II) fluoride
CrF_3	7788-97-8	108.9913096	Chromium trifluoride
			Chromium(III) fluoride
CrI_2	13478-28-9	305.80504	Chromium diiodide
			Chromium(II) iodide
CrI_3	13569-75-0	432.70951	Chromium triiodide
			Chromium(III) iodide
CrK_2O_4	7789-00-6	194.1903	Dipotassium chromate
			Potassium chromate
			Chromic acid potassium salt
CrN_3O_9		238.01092	Chromium trinitrate
			Chromium(III) nitrate
			Nitric acid chromium(III) salt
$CrNa_2O_4$	7775-11-3	161.973236	Disodium chromate
			Sodium chromate
			Chromic acid sodium salt

SUBSTANCE LIST (continued)

Formula	CAS RN	M_r	Name
CrO_2	12018-01-8	83.9949	Chromium dioxide
			Chromium(IV) oxide
CrO_4Pb	7758-97-6	323.1937	Lead chromate
			Lead(II) chromate
Cr_2FeO_4	12068-77-8	223.8368	Dichromium iron tetraoxide
			Chromium(III) iron(II) oxide
$Cr_2K_2O_7$	7778-50-9	294.1846	Dipotassium dichromate
			Potassium dichromate
			Dichromic acid potassium salt
$Cr_2Na_2O_7$		261.967536	Disodium dichromate
			Sodium dichromate
			Dichromic acid sodium salt
Cr_2O_3	1308-38-9	151.9904	Dichromium trioxide
			Chromium(III) oxide
$Cr_2O_{12}S_3$		392.183	Dichromium trisulfate
			Chromium(III) sulfate
			Sulfuric acid chromium(III) salt
Cr_3O_4	12018-34-7	219.9859	Trichromium tetraoxide
			Chromium(II,III) oxide
Cs	7440-46-2	132.90543	Cesium
CsF	13400-13-0	151.9038332	Cesium fluoride
CsF_2H	12280-52-3	171.9101764	Cesium hydrogen difluoride
			Cesium bifluoride
CsH	58724-12-2	133.91337	Cesium hydride
CsHO	21351-79-1	149.91277	Cesium hydroxide
$CsHO_4S$	7789-16-4	229.97697	Cesium hydrogen sulfate
			Cesium bisulfate
CsH_2N	22205-57-8	148.92805	Cesium dihydronitride
			Cesium amide
CsI	7789-17-5	259.8099	Cesium iodide
$CsNO_2$		178.91097	Cesium nitrite
			Nitrous acid cesium salt
$CsNO_3$	7789-18-6	194.91037	Cesium nitrate
			Nitric acid cesium salt
CsO_2	12018-61-0	164.90423	Cesium hyperoxide
			Cesium superoxide
Cs_2O	20281-00-9	281.81026	Dicesium oxide
			Cesium oxide
Cs_2O_3S	18832-76-3	345.87506	Dicesium sulfite
			Sulfurous acid dicesium salt
			Cesium sulfite
Cs_2O_4S	10294-54-9	361.87446	Dicesium sulfate
			Cesium sulfate
			Sulfuric acid cesium salt
$Cs_2O_7S_2$		441.93866	Dicesium disulfate
			Cesium pyrosulfate
Cs_2S	12214-16-3	297.87686	Dicesium sulfide
			Cesium sulfide
Cu	7440-50-8	63.546	Copper
CuF	13478-41-6	82.5444032	Copper fluoride
			Copper(I) fluoride
			Cuprous fluoride
CuF_2	7789-19-7	101.5428064	Copper difluoride
			Cupric fluoride
			Copper(II) fluoride
CuH_2O_2	20427-59-2	97.56068	Copper dihydroxide
			Cupric hydroxide
			Copper(II) hydroxide
CuI	7681-65-4	190.45047	Copper iodide
			Copper(I) iodide
			Cuprous iodide
CuN_2O_6	3251-23-8	187.55588	Copper dinitrate
CuN_2O_6 (Cont.)			Cupric nitrate
			Copper(II) nitrate
			Nitric acid copper(II) salt
			Nitric acid cupric salt
CuO	1317-38-0	79.5454	Copper oxide
			Cupric oxide
			Copper(II) oxide
CuO_4S	7758-98-7	159.6096	Copper sulfate
			Cupric sulfate
			Copper(II) sulfate
			Sulfuric acid copper(II) salt
			Sulfuric acid cupric salt
CuO_4W	13587-35-4	311.3936	Copper tungstate
			Copper(II) tungstate
			Tungstic acid copper(II) salt
			Cupric tungstate
			Tungstic acid cupric salt
CuS	1317-40-4	95.612	Copper sulfide
			Copper(II) sulfide
			Cupric sulfide
CuSe	1317-41-5	142.506	Copper selenide
			Copper(II) selenide
			Cupric selenide
Cu_2	12190-70-4	127.092	Dicopper
Cu_2O	1317-39-1	143.0914	Dicopper oxide
			Cuprous oxide
			Copper(I) oxide
Cu_2S	22205-45-4	159.158	Dicopper sulfide
			Cuprous sulfide
			Copper(I) sulfide
Cu_2Te	12019-52-2	254.692	Dicopper telluride
			Cuprous telluride
			Copper(I) telluride
Dy	7429-91-6	162.5	Dysprosium
Dy_2O_3	1308-87-8	372.9982	Didysprosium trioxide
			Dysprosium(III) oxide
Er	7440-52-0	167.26	Erbium
ErF_3	13760-83-3	224.2552096	Erbium trifluoride
			Erbium(III) fluoride
Er_2O_3	12061-16-4	382.5182	Dierbium trioxide
			Erbium(III) oxide
Es	7429-92-7	252.083	Einsteinium
Eu	7440-53-1	151.965	Europium
Eu_2O_3	1308-96-9	351.9282	Dieuropium trioxide
			Europium(III) oxide
Eu_3O_4	12061-63-1	519.8926	Trieuropium tetraoxide
			Europium(II,III) oxide
F	14762-94-8	18.9984032	Fluorine
FGa	13966-78-4	88.7214032	Gallium fluoride
FGe	14929-46-5	91.6084032	Germanium fluoride
$FGeH_3$	13537-30-9	94.6322232	Fluorogermane
FH	7664-39-3	20.0063432	Hydrogen fluoride
			Fluorane
			Hydrofluoric acid
F^2H	14333-26-7	21.0125049	(^{2}H)Hydrogen fluoride
			Deuterium fluoride
			Hydrofluoric acid-d
			(^{2}H)Fluorane
FHO_3S	7789-21-1	100.0705432	Fluorosulfonic acid
FH_3Si	13537-33-2	50.1077232	Fluorosilane
FH_4N	12125-01-8	37.0369032	Ammonium fluoride
FI	13873-84-2	145.9028732	Iodine fluoride

Formula	CAS RN	M_r	Name
FIn	13966-95-5	133.8184032	Indium fluoride
FK	7789-23-3	58.0967032	Potassium fluoride
FLi	7789-24-4	25.9394032	Lithium fluoride
FNO	7789-25-5	49.0045432	Nitrosyl fluoride
FNO_2	10022-50-1	65.0039432	Nitryl fluoride
FNS	13537-38-7	65.0711432	Thionitrosyl fluoride
FNa	7681-49-4	41.9881712	Sodium fluoride
FO	12061-70-0	34.9978032	Fluorine oxide
FRb	13446-74-7	104.4662032	Rubidium fluoride
FTl	7789-27-7	223.3817032	Thallium fluoride Thallium(I) fluoride Thallous fluoride
F_2	7782-41-4	37.9968064	Fluorine Difluorine
F_2Fe	7789-28-8	93.8438064	Iron difluoride Iron(II) fluoride Ferric fluoride
F_2HK	7789-29-9	78.1030464	Potassium hydrogen difluoride Potassium bifluoride
F_2HN	10405-27-3	53.0114864	Difluoroazane Difluoramine
F_2HNa	1333-83-1	61.9945144	Sodium hydrogen difluoride Sodium bifluoride
F_2HRb	12280-64-7	124.4725464	Rubidium hydrogen difluoride Rubidium bifluoride
F_2H_2Si	13824-36-7	68.0981864	Difluorosilane
F_2H_5N	1341-49-7	57.0432464	Ammonium hydrogen difluoride Ammonium bifluoride
F_2Mg	7783-40-6	62.3018064	Magnesium difluoride Magnesium fluoride
F_2N_2	13812-43-6	66.0102864	cis-Difluorodiazene
F_2N_2	13776-62-0	66.0102864	trans-Difluorodiazene
F_2Ni	10028-18-9	96.6902064	Nickel difluoride Nickel(II) fluoride
F_2O	7783-41-7	53.9962064	Oxygen difluoride
F_2OS	7783-42-8	86.0622064	Sulfinyl difluoride Thionyl difluoride Thionyl fluoride
F_2O_2	7783-44-0	69.9956064	Dioxygen difluoride Fluorine peroxide
F_2O_2S	2699-79-8	102.0616064	Sulfonyl difluoride Sulfuryl difluoride Sulfuryl fluoride
F_2O_2U	13536-84-0	308.0245064	Uranyl difluoride Uranyl fluoride
F_2Pb	7783-46-2	245.1968064	Lead difluoride Lead(II) fluoride
F_2Sn	7783-47-3	156.7068064	Tin difluoride Tin(II) fluoride
F_2Sr	7783-48-4	125.6168064	Strontium difluoride Strontium fluoride
F_2Xe	13709-36-9	169.2868064	Xenon difluoride
F_2Zn	7783-49-5	103.3868064	Zinc difluoride Zinc fluoride
F_3Fe	7783-50-8	112.8422096	Iron trifluoride Iron(III) fluoride Ferric fluoride
F_3Ga	7783-51-9	126.7182096	Gallium trifluoride Gallium(III) fluoride
F_3Gd	13765-26-9	214.2452096	Gadolinium trifluoride Gadolinium(III) fluoride
F_3HSi	13465-71-9	86.0886496	Trifluorosilane
F_3Ho	13760-78-6	221.9255296	Holmium trifluoride Holmium(III) fluoride
F_3N	7783-54-2	71.0019496	Nitrogen trifluoride Nitrogen(III) fluoride
F_3NO	13847-65-9	87.0013496	Trifluoroazane oxide Trifluoramine oxide
F_3Nd	13709-42-7	201.2352096	Neodymium trifluoride Neodymium(III) fluoride
F_3OP	13478-20-1	103.9683716	Phosphoryl trifluoride Phosphoryl fluoride
F_3P	7783-55-3	87.9689716	Phosphorus trifluoride Phosphorus(III) fluoride
F_3PS	2404-52-6	120.0349716	Thiophosphoryl trifluoride Thiophosphoryl fluoride
F_3Sb	7783-56-4	178.7522096	Antimony trifluoride Antimony(III) fluoride
F_3Sc	13709-47-2	101.9511196	Scandium trifluoride Scandium(III) fluoride
F_3Sm	13765-24-7	207.3552096	Samarium trifluoride Samarium(III) fluoride
F_3Th	13842-84-7	289.0333096	Thorium trifluoride Thorium(III) fluoride
F_3U	13775-06-9	295.0241096	Uranium trifluoride Uranium(III) fluoride
F_3V	10049-12-4	107.9367096	Vanadium trifluoride Vanadium(III) fluoride
F_3Y	13709-49-4	145.9010596	Yttrium trifluoride Yttrium(III) fluoride
F_4Ge	7783-58-6	148.6036128	Tetrafluorogerman Germanium tetrafluoride
F_4Hf	13709-52-9	254.4836128	Hafnium tetrafluoride Hafnium(IV) fluoride
F_4N_2	10036-47-2	104.0070928	Tetrafluorohydrazine Dinitrogen tetrafluoride
F_4Pb	94217-84-2	283.1936128	Lead tetrafluoride Lead(IV) fluoride
F_4S	7783-60-0	108.0596128	Sulfur tetrafluoride Sulfur(IV) fluoride
F_4Se	13465-66-2	154.9536128	Selenium tetrafluoride Selenium(IV) fluoride
F_4Si	7783-61-1	104.0791128	Silicon tetrafluoride Silicon(IV) fluoride
F_4Th	13709-59-6	308.0317128	Thorium tetrafluoride Thorium(IV) fluoride
F_4U	10049-14-6	314.0225128	Uranium tetrafluoride Uranium(IV) fluoride
F_4V	10049-16-8	126.9351128	Vanadium tetrafluoride Vanadium(IV) fluoride
F_4Xe	13709-61-0	207.2836128	Xenon tetrafluoride
F_4Zr	7783-64-4	167.2176128	Zirconium tetrafluoride Zirconium(IV) fluoride
$F_5H_4NO_4U_2$		653.085916	Ammonium pentafluorodiuranylate
F_5I	7783-66-6	221.896486	Iodine pentafluoride Iodine(V) fluoride
F_5Nb	7783-68-8	187.898396	Niobium pentafluoride Niobium(V) fluoride
F_5P	7647-19-0	125.965778	Phosphorus pentafluoride Phosphorus(V) fluoride

Formula	CAS RN	M_r	Name
F_5Ru		196.062016	Ruthenium pentafluoride
			Ruthenium(V) fluoride
F_5S	10546-01-7	127.058016	Sulfur pentafluoride
			Sulfur(V) fluoride
F_5Sb	7783-70-2	216.749016	Antimony pentafluoride
			Antimony(V) fluoride
F_5Ta	7783-71-3	275.939916	Tantalum pentafluoride
			Tantalum(V) fluoride
F_5V	7783-72-4	145.933516	Vanadium pentafluoride
			Vanadium(V) fluoride
F_6H_2Si	16961-83-4	144.0917992	Hexafluorosilicic acid
			Dihydrogen hexafluorosilicate
$F_6H_8N_2Si$	16919-19-0	178.1529192	Diammonium hexafluorosilicate
			Ammonium hexafluorosilicate
F_6Ir	7783-75-7	306.2104192	Iridium hexafluoride
			Iridium(VI) fluoride
F_6K_2Si	16871-90-2	220.2725192	Dipotassium hexafluorosilicate
			Potassium hexafluorosilicate
F_6Mo	7783-77-9	209.9304192	Molybdenum hexafluoride
			Molybdenum(VI) fluoride
F_6Na_2Si	16893-85-9	188.0554552	Disodium hexafluorosilicate
			Sodium hexafluorosilicate
F_6Os	13768-38-2	304.1904192	Osmium hexafluoride
			Osmium(VI) fluoride
F_6Pt	13693-05-5	309.0704192	Platinum hexafluoride
			Platinum(VI) fluoride
F_6Pu		358.0546192	Plutonium hexafluoride
			Plutonium(VI) fluoride
F_6S	2551-62-4	146.0564192	Sulfur hexafluoride
			Sulfur(VI) fluoride
F_6Se	7783-79-1	192.9504192	Selenium hexafluoride
			Selenium(VI) fluoride
F_6Te	7783-80-4	241.5904192	Tellurium hexafluoride
			Tellurium(VI) fluoride
F_6U	7783-81-5	352.0193192	Uranium hexafluoride
			Uranium(VI) fluoride
F_6W	7783-82-6	297.8404192	Tungsten hexafluoride
			Tungsten(VI) fluoride
Fe	7439-89-6	55.847	Iron
FeI_2	7783-86-0	309.65594	Iron diiodide
			Iron(II) iodide
			Ferrous iodide
FeI_3	15600-49-4	436.56041	Iron triiodide
			Iron(III) iodide
			Ferric iodide
$FeMoO_4$	13718-70-2	215.7846	Iron molybdate
			Iron(II) molybdate
			Ferrous molybdate
FeN_3O_9		241.86182	Iron trinitrate
			Ferric nitrate
			Iron(III) nitrate
			Nitric acid iron(III) salt
			Nitric acid ferric salt
FeO	1345-25-1	71.8464	Iron oxide
			Iron(II) oxide
			Ferrous oxide
FeO_4S	7720-78-7	151.9106	Iron sulfate
			Ferrous sulfate
			Iron(II) sulfate
			Sulfuric acid iron(II) salt
FeO_4S (Cont.)			Sulfuric acid ferrous salt
FeO_4W	13870-24-1	303.6946	Iron tungstate
			Iron(II) tungstate
FeS	1317-37-9	87.913	Iron sulfide
			Ferrous sulfide
			Iron(II) sulfide
FeS_2	1317-66-4	119.979	Iron disulfide
Fe_2O_3	1309-37-1	159.6922	Diiron trioxide
			Ferric oxide
			Iron(III) oxide
Fe_2O_4Si	10179-73-4	203.7771	Diiron silicate
			Iron(II) silicate
			Silicic acid iron(III) salt
$Fe_2O_{12}S_3$		399.8848	Diiron trisulfate
			Ferric sulfate
			Iron(III) sulfate
			Sulfuric acid ferric salt
			Sulfuric acid iron(III) salt
Fe_3O_4	1317-61-9	231.5386	Triiron tetraoxide
			Iron(II,III) oxide
Fm	7440-72-4	257.0951	Fermium
Fr	7440-73-5	223.0197	Francium
Ga	7440-55-3	69.723	Gallium
GaH_3O_3	12023-99-3	120.74502	Gallium trihydroxide
			Gallium(III) hydroxide
GaI_3	13450-91-4	450.43641	Gallium triiodide
			Gallium(III) iodide
GaN	25617-97-4	83.72974	Gallium nitride
			Gallium(III) nitride
GaO	12024-08-7	85.7224	Gallium oxide
			Gallium(II) oxide
GaP	12063-98-8	100.696762	Gallium phosphide
			Gallium(III) phosphide
GaSb	12064-03-8	191.48	Gallium antimonide
Ga_2	74508-24-0	139.446	Digallium
Ga_2O	12024-20-3	155.4454	Digallium oxide
			Gallium(I) oxide
Ga_2O_3	12024-21-4	187.4442	Digallium trioxide
			Gallium(III) oxide
$Ga_2O_{12}S_3$	13494-91-2	427.6368	Digallium trisulfate
			Gallium(III) sulfate
Gd	7440-54-2	157.25	Gadolinium
Gd_2O_3	12064-62-9	362.4982	Digadolinium trioxide
			Gadolinium(III) oxide
Ge	7440-56-4	72.61	Germanium
GeH_3I	13573-02-9	202.53829	Iodogermane
GeH_4	7782-65-2	76.64176	Germane
GeI_4	13450-95-8	580.22788	Germanium tetraiodide
GeO	20619-16-3	88.6094	Germanium oxide
GeO_2	1310-53-8	104.6088	Germanium dioxide
			Germanium oxide
GeP	25324-55-4	103.583762	Germanium phosphide
GeS	12025-32-0	104.676	Germanium sulfide
			Germanium(II) sulfide
Ge_2	12596-05-3	145.22	Digermanium
Ge_2H_6	13818-89-8	151.26764	Digermane
Ge_3H_8	14691-44-2	225.89352	Trigermane
H	12385-13-6	1.00794	Hydrogen
$^1H\,^2H$	13983-20-5	3.0219268	$(^1H,^2H)$Hydrogen
			Hydrogen deuteride
$^1H\,^3H$	14885-60-0	4.0238743	$(^1H,^3H)$Hydrogen
$^2H\,^3H$	14885-61-1	5.030151	$(^2H,^3H)$Hydrogen

Formula	CAS RN	M_r	Name
H^2HO	14940-63-7	19.0214417	(²H)Water Water-d (²H)Oxidane
HI	10034-85-2	127.91241	Hydrogen iodide Iodane Hydroiodic acid
HIO_3	7782-68-5	175.91061	Iodic acid
HK	7693-26-7	40.10624	Potassium hydride
HKO	1310-58-3	56.10564	Potassium hydroxide
HKO_4S	7646-93-7	136.16984	Potassium hydrogen sulfate Potassium bisulfate
HLi	7580-67-8	7.94894	Lithium hydride
HLiO	1310-65-2	23.94834	Lithium hydroxide
HNO_2	7782-77-6	47.01348	Nitrous acid
HNO_3	7697-37-2	63.01288	Nitric acid
$^2HNO_3$	13587-52-5	64.0190417	(²H)Nitric acid Nitric acid-d
HNO_5S	7782-78-7	127.07768	Nitrosylsulfuric acid
HN_3	7782-79-8	43.02816	Hydrogen azide Hydrazoic acid
HNa	7646-69-7	23.997708	Sodium hydride
HNaO	1310-73-2	39.997108	Sodium hydroxide
$HNaO_3S$	7631-90-5	104.061908	Sodium hydrogen sulfite Sodium bisulfite Sulfurous acid sodium salt
$HNaO_4S$	7681-38-1	120.061308	Sodium hydrogen sulfate Sodium bisulfate Sulfuric acid sodium salt
HNaS	16721-80-5	56.063708	Sodium hydrogen sulfide Sodium sulfide Sodium bisulfide
HNa_2O_4P	7558-79-4	141.958838	Disodium hydrogen phos- phate Sodium phosphate dibasic Phosphoric acid disodium salt
HO	3352-57-6	17.00734	Hydroxyl
HORb	1310-82-3	102.47514	Rubidium hydroxide
HOTl	12026-06-1	221.39064	Thallium hydroxide Thallium(I) hydroxide Thallous hydroxide
HO_2	3170-83-0	33.00674	Hydroperoxyl
HO_3P	37267-86-0	79.979902	Metaphosphoric acid
HO_4RbS	15587-72-1	182.53934	Rubidium hydrogen sulfate Rubidium bisulfate
HO_4Re	13768-11-1	251.21254	Perrhenic acid
HO_4Tc		162.91274	Pertechnetic acid
HRb	13446-75-8	86.47574	Rubidium hydride
HS	13940-21-1	33.07394	Mercaptyl
HTa_2	12026-09-4	362.90374	Ditantalum hydride Tantalum hydride
H_2	1333-74-0	2.01588	Hydrogen Dihydrogen
H_2		2.01588	Parahydrogen
$^2H_2$	7782-39-0	4.0282035	(²H₂)Hydrogen Deuterium
$^3H_2$	10028-17-8	6.0320985	(³H₂)Hydrogen Tritium
H_2KN	17242-52-3	55.12092	Potassium dihydronitride Potassium amide
H_2KO_4P	7778-77-0	136.085542	Potassium dihydrogen phos- phate Potassium phosphate mono- basic
H_2KO_4P (Cont.)			Phosphoric acid potassium salt
H_2LiN	7782-89-0	22.96362	Lithium dihydronitride Lithium amide
H_2Mg	7693-27-8	26.32088	Magnesium dihydride Magnesium hydride
H_2MgO_2	1309-42-8	58.31968	Magnesium dihydroxide Magnesium hydroxide
H_2N	17655-31-1	16.02262	Aminyl
H_2NNa	7782-92-5	39.012388	Sodium dihydronitride Sodium amide
H_2NRb	12141-27-4	101.49042	Rubidium dihydronitride Rubidium amide
$H_2N_2O_2$	7782-94-7	62.02816	Nitroazane Nitramide
H_2NaO_4P	7558-80-7	119.97701	Sodium dihydrogen phos- phate Phosphoric acid sodium salt Sodium phosphate monobasic
H_2NiO_2	12054-48-7	92.70808	Nickel dihydroxide Nickel(II) hydroxide
H_2O	7732-18-5	18.01528	Water Oxidane
^2H_2O	7789-20-0	20.0276035	(²H₂)Dihydrogen oxide Water-d2 Deuterium oxide Heavy water (²H₂)Oxidane
H_2O_2	7722-84-1	34.01468	Dihydrogen peroxide Hydrogen peroxide
H_2O_2Sn	12026-24-3	152.72468	Tin dihydroxide Tin(II) hydroxide
H_2O_2Sr	18480-07-4	121.63468	Strontium dihydroxide Strontium hydroxide
H_2O_2Zn	20427-58-1	99.40468	Zinc dihydroxide Zinc hydroxide
H_2O_3Si	7699-41-4	78.09958	Metasilicic acid
H_2O_4S	7664-93-9	98.07948	Sulfuric acid
$^2H_2O_4S$	13813-19-9	100.0918035	(²H₂)Sulfuric acid Sulfuric acid-d2
H_2O_4Se	7783-08-6	144.97348	Selenic acid
H_2S	7783-06-4	34.08188	Dihydrogen sulfide Hydrogen sulfide Sulfane
H_2S_2	13465-07-1	66.14788	Dihydrogen disulfide
H_2Se	7783-07-5	80.97588	Dihydrogen selenide Hydrogen selenide
H_2Sr	13598-33-9	89.63588	Strontium dihydride Strontium hydride
H_2Te	7783-09-7	129.61588	Dihydrogen telluride Hydrogen telluride
H_2Th	16689-88-6	234.05398	Thorium dihydride Thorium(II) hydride
H_2Zr	7704-99-6	93.23988	Zirconium dihydride Zirconium(II) hydride
H_3ISi	13598-42-0	158.01379	Iodosilane
H_3N	7664-41-7	17.03056	Ammonia Azane
^2H_3N	13550-49-7	20.0490453	(²H₃)Ammonia Ammonia-d3 (²H₃)Azane
H_3NO	7803-49-8	33.02996	Hydroxylamine

Formula	CAS RN	M_r	Name
H_3O_2P	6303-21-5	65.996382	Phosphonous acid
			Hypophosphorous acid
H_3O_3P	13598-36-2	81.995782	Phosphorous acid
H_3O_4P	7664-38-2	97.995182	Phosphoric acid
H_3P	7803-51-2	33.997582	Phosphine
			Phosphane
H_3Sb	7803-52-3	124.78082	Stibine
			Stibane
H_3U	13598-56-6	241.05272	Uranium trihydride
			Uranium(III) hydride
H_4IN	12027-06-4	144.94297	Ammonium iodide
H_4N_2	302-01-2	32.04524	Hydrazine
$H_4N_2O_2$	13446-48-5	64.04404	Ammonium nitrite
$H_4N_2O_2S$	7803-58-9	96.11004	Sulfamide
$H_4N_2O_3$	6484-52-2	80.04344	Ammonium nitrate
			Nitric acid ammonium salt
H_4N_4	12164-94-2	60.05872	Ammonium azide
H_4O_4Si	10193-36-9	96.11486	Silicic acid
$H_4O_7P_2$	2466-09-3	177.975084	Diphosphoric acid
H_4P_2	13445-50-6	65.979284	Diphosphane
			Diphosphine
H_4Si	7803-62-5	32.11726	Silane
			Silicon tetrahydride
			Silicon(IV) hydride
H_4Sn	2406-52-2	122.74176	Stannane
H_5NO	1336-21-6	35.04584	Ammonium hydroxide
H_5NO_3S	10192-30-0	99.11064	Ammonium hydrogen sulfite
			Ammonium bisulfite
			Sulfurous acid ammonium salt
H_5NO_4S	7803-63-6	115.11004	Ammonium hydrogen sulfate
			Ammonium bisulfate
			Sulfuric acid ammonium salt
H_6NO_4P	7722-76-1	115.025742	Ammonium dihydrogen phosphate
			Ammonium phosphate mono-basic
			Phosphoric acid ammonium salt
H_6OSi_2	13597-73-4	78.21804	Disiloxane
H_6Si_2	1590-87-0	62.21864	Disilane
$H_8N_2O_3S$		116.1412	Diammonium sulfite
			Ammonium sulfite
			Sulfurous acid diammonium salt
$H_8N_2O_4S$	7783-20-2	132.1406	Diammonium sulfate
			Ammonium sulfate
			Sulfuric acid ammonium salt
H_8Si_3	7783-26-8	92.32002	Trisilane
$H_9N_2O_4P$	7783-28-0	132.056302	Diammonium hydrogen phosphate
			Ammonium phosphate dibasic
			Phosphoric acid diammonium salt
$H_{10}N_2O_7P_2$		212.036204	Diammonium dihydrogen diphosphate
$H_{12}N_3O_4P$	10361-65-6	149.086862	Triammonium phosphate
Ha	53850-35-4		Hahnium
He	7440-59-7	4.002602	Helium
^3He	14762-55-1	3.0160293	(^3He)Helium
^4He		4.0026032	(^4He)Helium
Hf	7440-58-6	178.49	Hafnium
HfI_4	13777-23-6	686.10788	Hafnium tetraiodide
			Hafnium(IV) iodide
HfO_2	12055-23-1	210.4888	Hafnium dioxide
HfO_2 (Cont.)			Hafnium(IV) oxide
Hg	7439-97-6	200.59	Mercury
HgI_2	7774-29-0	454.39894	Mercury diiodide
			Mercuric iodide
			Mercury(II) iodide
HgO	21908-53-2	216.5894	Mercury oxide
			Mercury(II) oxide
			Mercuric oxide
HgO_4S	7783-35-9	296.6536	Mercury sulfate
			Sulfuric acid mercury(II) salt
			Mercury(II) sulfate
HgS	1344-48-5	232.656	Mercury sulfide
			Mercury(II) sulfide
			Mercuric sulfide
HgTe	12068-90-5	328.19	Mercury telluride
			Mercury(II) telluride
Hg_2	12596-25-7	401.18	Dimercury
Hg_2I_2	15385-57-6	654.98894	Dimercury diiodide
			Mercury(I) iodide
			Mercurous iodide
Hg_2O_4S	7783-36-0	497.2436	Dimercury sulfate
			Mercury(I) sulfate
			Mercurous sulfate
Ho	7440-60-0	164.93032	Holmium
Ho_2O_3	12055-62-8	377.85884	Diholmium trioxide
			Holmium(III) oxide
I	14362-44-8	126.90447	Iodine
IIn	13966-94-4	241.72447	Indium iodide
			Indium(I) iodide
IK	7681-11-0	166.00277	Potassium iodide
IKO_3	7758-05-6	214.00097	Potassium iodate
			Iodic acid potassium salt
IKO_4	7790-21-8	230.00037	Potassium periodate
ILi	10377-51-2	133.84547	Lithium iodide
INa	7681-82-5	149.894238	Sodium iodide
$INaO_3$	7681-55-2	197.892438	Sodium iodate
$INaO_4$	7790-28-5	213.891838	Sodium periodate
IRb	7790-29-6	212.37227	Rubidium iodide
ITl	7790-30-9	331.28777	Thallium iodide
			Thallium(I) iodide
			Thallous iodide
I_2	7553-56-2	253.80894	Iodine
			Diiodine
I_2Mg	10377-58-9	278.11394	Magnesium diiodide
			Magnesium iodide
I_2Ni	13462-90-3	312.50234	Nickel diiodide
I_2Pb	10101-63-0	461.00894	Lead diiodide
			Lead(II) iodide
I_2Sn	10294-70-9	372.51894	Tin diiodide
			Tin(II) iodide
I_2Sr	10476-86-5	341.42894	Strontium diiodide
			Strontium iodide
I_2Zn	10139-47-6	319.19894	Zinc diiodide
			Zinc iodide
I_3In	13510-35-5	495.53341	Indium triiodide
I_3Lu	13813-45-1	555.68041	Lutetium triiodide
			Lutetium(III) iodide
I_3Nd	13813-24-6	524.95341	Neodymium triiodide
			Neodymium(III) iodide
I_3P	13455-01-1	411.687172	Phosphorus triiodide
I_3Ru	13896-65-6	481.78341	Ruthenium triiodide
I_3Sb	7790-44-5	502.47041	Antimony triiodide

Formula	CAS RN	M_r	Name
I_3Sb (Cont.)			Antimony(III) iodide
I_4Pt	7790-46-7	702.69788	Platinum tetraiodide
I_4Sb		629.37488	Antimony tetraiodide
			Antimony(IV) iodide
I_4Si	13465-84-4	535.70338	Silicon tetraiodide
			Silicon(IV) iodide
I_4Sn	7790-47-8	626.32788	Tin tetraiodide
			Tin(IV) iodide
			Tetraiodostannane
I_4Ti	7720-83-4	555.49788	Titanium tetraiodide
			Titanium(IV) iodide
I_4V	15831-18-2	558.55938	Vanadium tetraiodide
I_4Zr	13986-26-0	598.84188	Zirconium tetraiodide
In	7440-74-6	114.82	Indium
InO	12136-26-4	130.8194	Indium oxide
			Indium(II) oxide
InP	22398-80-7	145.793762	Indium phosphide
			Indium(III) phosphide
InS	12030-14-7	146.886	Indium sulfide
			Indium(II) sulfide
InSb	1312-41-0	236.577	Indium antimonide
In_2O_3	1312-43-2	277.6382	Diindium trioxide
			Indium(III) oxide
$In_2O_{12}S_3$	13464-82-9	517.8308	Diindium trisulfate
			Indium(III) sulfate
			Sulfuric acid indium(III) salt
In_2S_3	12030-24-9	325.838	Diindium trisulfide
			Indium(III) sulfide
Ir	7439-88-5	192.22	Iridium
IrO_2	12030-49-8	224.2188	Iridium dioxide
			Iridium(IV) oxide
IrS_2	12030-51-2	256.352	Iridium disulfide
			Iridium(IV) sulfide
Ir_2S_3	12136-42-4	480.638	Diiridium trisulfide
			Iridium(III) sulfide
K	7440-09-7	39.0983	Potassium
$KLiO_4S$		142.1029	Lithium potassium sulfate
			Sulfuric acid lithium potassium salt
$KMnO_4$	7722-64-7	158.03395	Potassium permanganate
KNO_2	7758-09-0	85.10384	Potassium nitrite
			Nitrous acid potassium salt
KNO_3	7757-79-1	101.10324	Potassium nitrate
			Nitric acid potassium salt
KO_2	12030-88-5	71.0971	Potassium hyperoxide
			Potassium superoxide
KO_3		87.0965	Potassium ozonide
K_2O	12136-45-7	94.196	Dipotassium oxide
			Potassium oxide
K_2O_2	17014-71-0	110.1954	Dipotassium peroxide
			Potassium peroxide
K_2O_4S	7778-80-5	174.2602	Dipotassium sulfate
			Potassium sulfate
			Sulfuric acid potassium salt
$K_2O_7S_2$		254.3244	Dipotassium disulfate
			Potassium pyrosulfate
K_2S	1312-73-8	110.2626	Dipotassium sulfide
			Potassium sulfide
K_3O_4P	7778-53-2	212.266262	Tripotassium phosphate
			Potassium phosphate
			Phosphoric acid potassium salt
Kr	7439-90-9	83.8	Krypton

Formula	CAS RN	M_r	Name
^{80}Kr		79.91638	(^{80}Kr)Krypton
^{84}Kr		83.911507	(^{84}Kr)Krypton
La	7439-91-0	138.9055	Lanthanum
LaN_3O_9		324.92032	Lanthanum trinitrate
			Lanthanum(III) nitrate
			Nitric acid lanthanum(III) salt
LaS	12031-30-0	170.9715	Lanthanum sulfide
La_2O_3	1312-81-8	325.8092	Dilanthanum trioxide
			Lanthanum(III) oxide
$La_2O_{12}S_3$		566.0018	Dilanthanum trisulfate
			Lanthanum(III) sulfate
			Sulfuric acid lanthanum(III) salt
Li	7439-93-2	6.941	Lithium
$LiNO_2$	13568-33-7	52.94654	Lithium nitrite
			Nitrous acid lithium salt
$LiNO_3$	7790-69-4	68.94594	Lithium nitrate
			Nitric acid lithium salt
Li_2O	12057-24-8	29.8814	Dilithium oxide
			Lithium oxide
Li_2O_2	12031-80-0	45.8808	Dilithium peroxide
			Lithium peroxide
Li_2O_3Si	10102-24-6	89.9657	Dilithium metasilicate
			Metasilicic acid lithium salt
Li_2O_4S	10377-48-7	109.9456	Dilithium sulfate
			Lithium sulfate
			Sulfuric acid lithium salt
Li_2S	12136-58-2	45.948	Dilithium sulfide
			Lithium sulfide
Li_3O_4P	10377-52-3	115.794362	Trilithium phosphate
Lr	22537-19-5	262.11	Lawrencium
Lu	7439-94-3	174.967	Lutetium
Lu_2O_3	12032-20-1	397.9322	Dilutetium trioxide
			Lutetium(III) oxide
Md	7440-11-1	258.1	Mendelevium
Mg	7439-95-4	24.305	Magnesium
MgN_2O_4		116.31608	Magnesium dinitrite
			Magnesium nitrite
			Nitrous acid magnesium salt
MgN_2O_6	10377-60-3	148.31488	Magnesium dinitrate
			Magnesium nitrate
			Nitric acid magnesium salt
MgO	1309-48-4	40.3044	Magnesium oxide
MgO_3S		104.3692	Magnesium sulfite
			Sulfurous acid magnesium salt
MgO_4S	7487-88-9	120.3686	Magnesium sulfate
			Sulfuric acid magnesium salt
MgO_4Se	14986-91-5	167.2626	Magnesium selenate
MgS	12032-36-9	56.371	Magnesium sulfide
Mg_2O_4Si	26686-77-1	140.6931	Dimagnesium silicate
			Silicic acid magnesium salt
Mn	7439-96-5	54.93805	Manganese
MnN_2O_6	10377-66-9	178.94793	Manganese dinitrate
			Manganese(II) nitrate
			Nitric acid manganese(II) salt
$MnNa_2O_4$	15702-33-7	164.915186	Disodium manganate
			Manganic acid sodium salt
MnO	1344-43-0	70.93745	Manganese oxide
			Manganese(II) oxide
MnO_2	1313-13-9	86.93685	Manganese dioxide
			Manganese(IV) oxide
MnO_3Si	7759-00-4	131.02175	Manganese metasilicate
			Metasilicic acid manganese salt

Formula	CAS RN	M_r	Name
MnO_4S		151.00165	Manganese sulfate Sulfuric acid manganese(II) salt
MnS	18820-29-6	87.00405	Manganese sulfide Manganese(II) sulfide
$MnSe$	1313-22-0	133.89805	Manganese selenide
Mn_2O_3	1317-34-6	157.8743	Dimanganese trioxide Manganese(III) oxide
Mn_2O_4Si	13568-32-6	201.9592	Dimanganese silicate Silicic acid manganese salt
Mn_3O_4	1317-35-7	228.81175	Trimanganese tetraoxide Manganese(II,III) oxide
Mo	7439-98-7	95.94	Molybdenum
$MoNa_2O_4$	7631-95-0	205.917136	Disodium molybdate Molybdic acid sodium salt Sodium molybdate
MoO_2	18868-43-4	127.9388	Molybdenum dioxide Molybdenum(IV) oxide
MoO_3	1313-27-5	143.9382	Molybdenum trioxide Molybdenum(VI) oxide
MoO_4Pb	10190-55-3	367.1376	Lead molybdate
MoS_2	1317-33-5	160.072	Molybdenum disulfide Molybdenum(IV) sulfide
$NNaO_2$	7632-00-0	68.995308	Sodium nitrite Nitrous acid sodium salt
$NNaO_3$	7631-99-4	84.994708	Sodium nitrate Nitric acid sodium salt
NO	10102-43-9	30.00614	Nitrogen oxide Nitrogen monoxide
^{14}NO		30.002474	(^{14}N)Nitrogen oxide (^{14}N)Nitrogen monoxide
^{15}NO	15917-77-8	30.9995089	(^{15}N)Nitrogen oxide (^{15}N)Nitrogen monoxide
NO_2	10102-44-0	46.00554	Nitrogen dioxide
NO_2Rb	13825-25-7	131.47334	Rubidium nitrite Nitrous acid rubidium salt
NO_3Rb	13126-12-0	147.47274	Rubidium nitrate Nitric acid rubidium salt
NO_3Tl	10102-45-1	266.38824	Thallium nitrate Nitric acid thallium(I) salt Thallium(I) nitrate Thallous nitrate
N_2	7727-37-9	28.01348	Nitrogen Dinitrogen
$^{14}N_2$		28.006148	($^{14}N_2$)Nitrogen
$^{15}N_2$	29817-79-6	30.0002179	($^{15}N_2$)Nitrogen
N_2NiO_6		182.70328	Nickel dinitrate Nickel(II) nitrate Nitric acid nickel(II) salt
N_2O	10024-97-2	44.01288	Dinitrogen oxide Nitrous oxide
N_2O_3	10544-73-7	76.01168	Dinitrogen trioxide Nitrogen trioxide Nitrogen(III) oxide
N_2O_4	10544-72-6	92.01108	Dinitrogen tetraoxide Nitrogen(I) oxide Nitrogen peroxide
N_2O_4Sr	13470-06-9	179.63108	Strontium dinitrite Strontium nitrite Nitrous acid strontium salt
N_2O_5	10102-03-1	108.01048	Dinitrogen pentaoxide Nitrogen pentoxide Nitrogen(V) oxide
N_2O_6Pb	10099-74-8	331.20988	Lead dinitrate Lead(II) nitrate Nitric acid lead(II) salt
N_2O_6Ra	10213-12-4	350.03528	Radium dinitrate Nitric acid radium salt
N_2O_6Sr	10042-76-9	211.62988	Strontium dinitrate Strontium nitrate Nitric acid strontium salt
N_2O_6Zn	7779-88-6	189.39988	Zinc dinitrate Zinc nitrate Nitric acid zinc salt
N_2O_8U		394.03758	Uranyl dinitrate Uranyl nitrate Nitric acid uranyl salt
N_3Na	26628-22-8	65.009988	Sodium azide
N_3NdO_9		330.25482	Neodymium trinitrate Neodymium(III) nitrate Nitric acid neodymium(III) salt
N_3O_9Pr		326.92247	Praseodymium trinitrate Nitric acid praseodymium(III) salt Praseodymium(III) nitrate
$N_4O_{12}Th$		480.05786	Thorium tetranitrate Thorium(IV) nitrate
$N_4O_{12}Zr$	12372-57-5	339.24376	Zirconium tetranitrate Nitric acid zirconium(IV) salt Zirconium(IV) nitrate
N_4Si_3	12033-89-5	140.28346	Trisilicon tetranitride
Na	7440-23-5	22.989768	Sodium
NaO_2	12034-12-7	54.988568	Sodium hyperoxide Sodium superoxide
NaO_3P		101.96173	Sodium metaphosphate
Na_2O	1313-59-3	61.978936	Disodium oxide Sodium oxide
Na_2O_2	1313-60-6	77.978336	Disodium peroxide Sodium peroxide
Na_2O_3S	7757-83-7	126.043736	Disodium sulfite Sodium sulfite Sulfurous acid sodium salt
Na_2O_3Si	6834-92-0	122.063236	Disodium metasilicate Metasilicic acid sodium salt Sodium metasilicate
Na_2O_4S	7757-82-6	142.043136	Disodium sulfate Sodium sulfate Sulfuric acid sodium salt
$Na_2O_4S_2$	7775-14-6	174.109136	Disodium dithionite Sodium dithionite Sodium hydrosulfite
Na_2O_4W	13472-45-2	293.827136	Disodium tungstate Sodium tungstate Tungstic acid sodium salt
$Na_2O_7S_2$		222.107336	Disodium disulfate Sodium pyrosulfate
Na_2S	1313-82-2	78.045536	Disodium sulfide Sodium sulfide
Na_3O_4P		163.940666	Trisodium phosphate Sodium phosphate Phosphoric acid sodium salt
Nb	7440-03-1	92.90638	Niobium
NbO	12034-57-0	108.90578	Niobium oxide Niobium(II) oxide
NbO_2	12034-59-2	124.90518	Niobium dioxide

Formula	CAS RN	M_r	Name
NbO_2 (Cont.)			Niobium(IV) oxide
Nb_2O_5	1313-96-8	265.80976	Diniobium pentaoxide Niobium(V) oxide Diniobium pentoxide
Nd	7440-00-8	144.24	Neodymium
Nd_2O_3	1313-97-9	336.4782	Dineodymium trioxide Neodymium(III) oxide
Ne	7440-01-9	20.1797	Neon
^{20}Ne		19.9924356	(^{20}Ne)Neon
^{22}Ne		21.9913831	(^{22}Ne)Neon
Ni	7440-02-0	58.6934	Nickel
NiO_4S	7786-81-4	154.757	Nickel sulfate Sulfuric acid nickel(II) salt Nickel(II) sulfate
NiS	16812-54-7	90.7594	Nickel sulfide Nickel(II) sulfide
Ni_2O_3	1314-06-3	165.385	Dinickel trioxide Nickel(III) oxide
No	10028-14-5	259.1009	Nobelium
Np	7439-99-8	237.0482	Neptunium
$^{16}O\,^{18}O$		33.9940749	(^{16}O,^{18}O)Oxygen (^{16}O,^{18}O)Dioxygen
OPb	1317-36-8	223.1994	Lead oxide Lead(II) oxide
OPd	1314-08-5	122.4194	Palladium oxide Palladium(II) oxide
ORa	12143-02-1	242.0248	Radium oxide
ORb_2	18088-11-4	186.935	Dirubidium oxide Rubidium oxide
OS_2	20901-21-7	80.1314	Sulfinyl sulfide Thionyl sulfide
OSn	21651-19-4	134.7094	Tin oxide Tin(II) oxide
OSr	1314-11-0	103.6194	Strontium oxide
OTi	12137-20-1	63.8794	Titanium oxide Titanium(II) oxide
OTl_2	1314-12-1	424.766	Dithallium oxide Thallium(I) oxide Thallous oxide
OV	12035-98-2	66.9409	Vanadium oxide Vanadium(II) oxide
OZn	1314-13-2	81.3894	Zinc oxide
O_2	7782-44-7	31.9988	Oxygen Dioxygen
$^{16}O_2$		31.9898292	($^{16}O_2$)Oxygen ($^{16}O_2$)Dioxygen
O_2Pb	1309-60-0	239.1988	Lead dioxide Lead(IV) oxide
O_2Rb	12137-25-6	117.4666	Rubidium hyperoxide Rubidium superoxide
O_2Rb_2	23611-30-5	202.9344	Dirubidium peroxide Rubidium peroxide
O_2Ru	12036-10-1	133.0688	Ruthenium dioxide Ruthenium(IV) oxide
O_2S	7446-09-5	64.0648	Sulfur dioxide Sulfur(IV) oxide
O_2Se	7446-08-4	110.9588	Selenium dioxide Selenium(IV) oxide
O_2Si	14808-60-7	60.0843	Silicon dioxide Silicon(IV) oxide
O_2Si		60.0843	Silicon dioxide (cr,l) Quartz (fused)
O_2Sn	18282-10-5	150.7088	Tin dioxide Tin(IV) oxide
O_2Te	7446-07-3	159.5988	Tellurium dioxide Tellurium(IV) oxide
O_2Th	1314-20-1	264.0369	Thorium dioxide Thorium(IV) oxide
O_2Ti	13463-67-7	79.8788	Titanium dioxide Titanium(IV) oxide
O_2U	1344-57-6	270.0277	Uranium dioxide Uranium(IV) oxide
O_2W	12036-22-5	215.8488	Tungsten dioxide Tungsten(IV) oxide
O_2Zr	1314-23-4	123.2228	Zirconium dioxide Zirconium(IV) oxide
O_3	10028-15-6	47.9982	Ozone
O_3PbS	7446-10-8	287.2642	Lead sulfite Sulfurous acid lead salt
O_3PbSi	10099-76-0	283.2837	Lead metasilicate Metasilicic acid lead salt
O_3Pr_2	12036-32-7	329.8135	Dipraseodymium trioxide Praseodymium(III) oxide
O_3Rh_2	12036-35-0	253.8092	Dirhodium trioxide Rhodium(III) oxide
O_3S	7446-11-9	80.0642	Sulfur trioxide Sulfur(VI) oxide
O_3Sb_2	1309-64-4	291.5122	Diantimony trioxide Antimony trioxide Antimony(III) oxide
O_3Sc_2	12060-08-1	137.91002	Discandium trioxide Scandium(III) oxide
O_3SiSr	13451-00-8	163.7037	Strontium metasilicate Metasilicic acid strontium salt
O_3Sm_2	12060-58-1	348.7182	Disamarium trioxide Samarium(III) oxide
O_3Tb_2	12036-41-8	365.84888	Diterbium trioxide Terbium(III) oxide
O_3Ti_2	1344-54-3	143.7582	Dititanium trioxide Titanium(III) oxide
O_3Tm_2	12036-44-1	385.86662	Dithullium trioxide Thullium(III) oxide
O_3U	1344-58-7	286.0271	Uranium trioxide Uranium(VI) oxide
O_3V_2	1314-34-7	149.8812	Divanadium trioxide Vanadium(III) oxide
O_3W	1314-35-8	231.8482	Tungsten trioxide Tungsten(VI) oxide
O_3Y_2	1314-36-9	225.8099	Diyttrium trioxide Yttrium(III) oxide
O_3Yb_2	1314-37-0	394.0782	Diytterbium trioxide Ytterbium(III) oxide
O_4Os	20816-12-0	254.1976	Osmium tetraoxide Osmium(VIII) oxide
O_4PbS	7446-14-2	303.2636	Lead sulfate Lead(II) sulfate Sulfuric acid lead(II) salt
O_4PbSe	7446-15-3	350.1576	Lead selenate
O_4Pb_2Si	13566-17-1	506.4831	Dilead silicate Silicic acid lead salt Lead silicate
O_4Pb_3	1314-41-6	685.5976	Trilead tetraoxide Lead(II,III) oxide
O_4RaS	7446-16-4	322.089	Radium sulfate

Formula	CAS RN	M_r	Name
O_4RaS (Cont.)			Sulfuric acid radium salt
O_4Rb_2S	7488-54-2	266.9992	Dirubidium sulfate
			Rubidium sulfate
			Sulfuric acid rubidium salt
O_4Ru	20427-56-9	165.0676	Ruthenium tetraoxide
			Ruthenium(VIII) oxide
O_4SSr	7759-02-6	183.6836	Strontium sulfate
			Sulfuric acid strontium salt
O_4STl_2	7446-18-6	504.8302	Dithallium sulfate
			Thallium(I) sulfate
			Thallous sulfate
			Sulfuric acid thallium(I) salt
			Sulfuric acid thallous salt
O_4SZn	7733-02-0	161.4536	Zinc sulfate
			Sulfuric acid zinc salt
O_4SiSr_2	13597-55-2	267.3231	Distrontium silicate
			Silicic acid strontium salt
O_4SiZn_2	13597-65-4	222.8631	Dizinc silicate
			Silicic acid zinc salt
			Zinc silicate
O_4SiZr	10101-52-7	183.3071	Zirconium silicate
			Silicic acid zirconium salt
O_5P_2	1314-56-3	141.944524	Diphosphorus pentaoxide
			Phosphorus(V) oxide
			Phosphorus pentoxide
O_5Sb_2	1314-60-9	323.511	Diantimony pentaoxide
			Antimony(V) oxide
			Antimony pentoxide
O_5Ta_2	1314-61-0	441.8928	Ditantalum pentaoxide
			Tantalum(V) oxide
O_5Ti_3	12065-65-5	223.637	Trititanium pentaoxide
O_5V_2	1314-62-1	181.88	Divanadium pentaoxide
			Vanadium(V) oxide
			Vanadium pentoxide
O_5V_3	12036-83-8	232.8215	Trivanadium pentaoxide
O_6P_2Sr		245.563924	Strontium dimetaphosphate
			Strontium metaphosphate
O_6SU		366.0913	Uranyl sulfate
			Sulfuric acid uranyl salt
$O_7Rb_2S_2$		347.0634	Dirubidium disulfate
			Rubidium pyrosulfate
O_7Re_2	1314-68-7	484.4098	Dirhenium heptaoxide
			Rhenium(VII) oxide
O_7U_3	12037-04-6	826.0825	Triuranium heptaoxide
			Uranium(IV,IV,VI) oxide
O_8S_2Zr	14644-61-2	283.3512	Zirconium disulfate
O_8U_3	1344-59-8	842.0819	Triuranium octaoxide
			Uranium(IV,VI,VI) oxide
O_9U_4	12037-15-9	1096.1102	Tetrauranium nonaoxide
			Uranium(IV,IV,IV,VI) oxide
$O_{12}N_4Pu$		492.08396	Plutonium tetranitrate
			Plutonium(IV) nitrate
			Nitric acid plutonium(IV) salt
$O_{12}S_3Sc_2$		378.10262	Discandium trisulfate
			Scandium(III) sulfate
			Sulfuric acid scandium(III) salt
$O_{12}S_3Ti_2$		383.9508	Dititanium trisulfate
			Titanium(III) sulfate
			Sulfuric acid titanium(III) salt
$O_{12}S_3Yb_2$	13469-97-1	634.2708	Diytterbium trisulfate
			Ytterbium(III) sulfate
			Sulfuric acid ytterbium(III) salt
Os	7440-04-2	190.2	Osmium
P	7723-14-0	30.973762	Phosphorus
Pa	7440-13-3	231.03588	Protactinium
Pb	7439-92-1	207.2	Lead
PbS	1314-87-0	239.266	Lead sulfide
			Lead(II) sulfide
PbSe	12069-00-0	286.16	Lead selenide
			Lead(II) selenide
PbTe	1314-91-6	334.8	Lead telluride
			Lead(II) telluride
Pd	7440-05-3	106.42	Palladium
PdS	12125-22-3	138.486	Palladium sulfide
			Palladium(II) sulfide
Pm	7440-12-2	144.9127	Promethium
Po	7440-08-6	208.9824	Polonium
Pr	7440-10-0	140.90765	Praseodymium
Pt	7440-06-4	195.08	Platinum
PtS	12038-20-9	227.146	Platinum sulfide
			Platinum(II) sulfide
PtS_2	12038-21-0	259.212	Platinum disulfide
			Platinum(IV) sulfide
Pu	7440-07-5	244.0642	Plutonium
Ra	7440-14-4	226.0254	Radium
Rb	7440-17-7	85.4678	Rubidium
Re	7440-15-5	186.207	Rhenium
Rf	53850-36-5		Rutherfordium
Rh	7440-16-6	102.9055	Rhodium
Rn	10043-92-2	222.0176	Radon
Ru	7440-18-8	101.07	Ruthenium
S	7704-34-9	32.066	Sulfur
S		32.066	Sulfur (c,I)
			Sulfur monoclinic
S		32.066	Sulfur (c,II)
			Sulfur rhombic
SSn	1314-95-0	150.776	Tin sulfide
			Tin(II) sulfide
SSr	1314-96-1	119.686	Strontium sulfide
STl_2	1314-97-2	440.8326	Dithallium sulfide
			Thallium(I) sulfide
			Thallous sulfide
SZn	1314-98-3	97.456	Zinc sulfide
S_3Sb_2	1345-04-6	339.712	Diantimony trisulfide
			Antimony trisulfide
			Antimony(III) sulfide
Sb	7440-36-0	121.757	Antimony
Sc	7440-20-2	44.95591	Scandium
Se	7782-49-2	78.96	Selenium
SeSr	1315-07-7	166.58	Strontium selenide
$SeTl_2$	15572-25-5	487.7266	Dithallium selenide
			Thallium(I) selenide
			Thallous selenide
SeZn	1315-09-9	144.35	Zinc selenide
Si	7440-21-3	28.0855	Silicon
Sm	7440-19-9	150.36	Samarium
Sn	7440-31-5	118.71	Tin
Sn		118.71	Tin (c,I)
Sn		118.71	Tin (c,II)
Sr	7440-24-6	87.62	Strontium
Ta	7440-25-7	180.9479	Tantalum
Tb	7440-27-9	158.92534	Terbium
Tc	7440-26-8	97.9072	Technetium
Te	13494-80-9	127.6	Tellurium

Formula	CAS RN	M_r	Name
Th	7440-29-1	232.0381	Thorium
Ti	7440-32-6	47.88	Titanium
Tl	7440-28-0	204.3833	Thallium
Tm	7440-30-4	168.93421	Thulium
U	7440-61-1	238.0289	Uranium
V	7440-62-2	50.9415	Vanadium
W	7440-33-7	183.85	Tungsten
Xe	7440-63-3	131.29	Xenon
^{133}Xe		132.905888	(^{133}Xe)Xenon
Y	7440-65-5	88.90585	Yttrium
Yb	7440-64-4	173.04	Ytterbium
Zn	7440-66-6	65.39	Zinc
Zr	7440-67-7	91.224	Zirconium
C	7440-44-0	12.011	Carbon
C		12.011	Carbon (cr,I)
			Dymond (Type I)
CAgN	506-64-9	133.88594	Silver cyanide
			Silver(I) cyanide
CAg_2O_3	16920-45-9	275.7456	Disilver carbonate
			Silver(I) carbonate
			Carbonic acid silver(I) salt
$CBaO_3$	513-77-9	197.3362	Barium carbonate
			Carbonic acid barium salt
$CBeO_3$	13106-47-3	69.021382	Beryllium carbonate
			Carbonic acid beryllium salt
$CBrClF_2$	353-59-3	165.3645064	Bromochlorodifluoromethane
$CBrCl_2F$	353-58-2	181.8188032	Bromodichlorofluoromethane
$CBrCl_3$	75-62-7	198.2731	Bromotrichloromethane
$CBrF_3$	75-63-8	148.9102096	Bromotrifluoromethane
			Freon 13B1
			R 13b1
CBrN	506-68-3	105.92174	Cyanogen bromide
CBr_2ClF	353-55-9	226.2701032	Dibromochlorofluoromethane
CBr_2Cl_2	594-18-3	242.7244	Dibromodichloromethane
CBr_2F_2	75-61-6	209.8158064	Dibromodifluoromethane
CBr_2O	593-95-3	187.8184	Carbonyl dibromide
			Carbonyl bromide
CBr_3Cl	594-15-0	287.1757	Tribromochloromethane
CBr_3F	353-54-8	270.7214032	Tribromofluoromethane
CBr_4	558-13-4	331.627	Tetrabromomethane
			Carbon tetrabromide
$CCaO_3$	471-34-1	100.0872	Calcium carbonate
			Carbonic acid calcium salt
$CCdO_3$	513-78-0	172.4202	Cadmium carbonate
			Carbonic acid cadmium salt
CClFO	353-49-1	82.4615032	Carbonyl chloride fluoride
$CClF_3$	75-72-9	104.4589096	Chlorotrifluoromethane
			Freon 13
			R 13
CClN	506-77-4	61.47044	Cyanogen chloride
$CClN_3O_6$	1943-16-4	185.48032	Chlorotrinitromethane
CCl_2F_2	75-71-8	120.9132064	Dichlorodifluoromethane
			Freon 12
			R 12
			CFC 12
CCl_2O	75-44-5	98.9158	Carbonyl dichloride
			Phosgene
			Carbonyl chloride
CCl_3F	75-69-4	137.3675032	Trichlorofluoromethane
CCl_3F (Cont.)			Freon 11
			R 11
			CFC 11
CCl_4	56-23-5	153.8218	Tetrachloromethane
			Carbon tetrachloride
CCl_4S	594-42-3	185.8878	Trichloromethylsulfur chloride
			Perchloro(methyl mercaptan)
			Trichloromethanesulfenyl chloride
			PCM
$CCoO_3$	76868-90-1	118.9424	Cobalt carbonate
			Carbonic acid cobalt(II) salt
$CCsHO_3$	15519-28-5	193.92257	Cesium hydrogen carbonate
			Carbonic acid hydrogen cesium salt
			Cesium bicarbonate
CCs_2O_3	534-17-8	325.82006	Dicesium carbonate
			Carbonic acid cesium salt
CCuN	544-92-3	89.56374	Copper cyanide
			Copper(I) cyanide
			Cuprous cyanide
CCu_2O_3		187.1012	Dicopper carbonate
			Cuprous carbonate
			Copper(I) carbonate
			Carbonic acid copper(I) salt
CFN	1495-50-7	45.0161432	Cyanogen fluoride
CF_2O	353-50-4	66.0072064	Carbonyl difluoride
			Carbonyl fluoride
CF_3I	2314-97-8	195.9106796	Trifluoroiodomethane
CF_4	75-73-0	88.0046128	Tetrafluoromethane
			Perfluoromethane
			Freon 14
			Carbon tetrafluoride
$CFeO_3$	67328-72-7	115.8562	Iron carbonate
			Ferrous carbonate
			Carbonic acid iron(II) salt
CFe_3	12011-67-5	179.552	Triiron carbide
CHBrClF	593-98-6	147.3740432	Bromochlorofluoromethane
$CHBrCl_2$	75-27-4	163.82834	Bromodichloromethane
$CHBrF_2$	1511-62-2	130.9197464	Bromodifluoromethane
$CHBr_2Cl$	124-48-1	208.27964	Chlorodibromomethane
$CHBr_2F$	1868-53-7	191.8253432	Dibromofluoromethane
$CHBr_3$	75-25-2	252.73094	Tribromomethane
			Bromoform
$CHClF_2$	75-45-6	86.4684464	Chlorodifluoromethane
			Freon 22
			R 22
			HCFC 22
$CHCl_2F$	75-43-4	102.9227432	Dichlorofluoromethane
			Freon 21
$CHCl_3$	67-66-3	119.37704	Trichloromethane
			Chloroform
C^2HCl_3	865-49-6	120.3832017	Trichloro(^{2}H)methane
			Deuterochloroform
			Deuterotrichloromethane
			Chloroform-d
CHFO	1493-02-3	48.0167432	Formyl fluoride
CHF_3	75-46-7	70.0141496	Trifluoromethane
			Fluoroform
			Freon 23
CHF_3O_3S	1493-13-6	150.0783496	Trifluoromethanesulfonic acid

Formula	CAS RN	M_r	Name
CH^2H_3O	1849-29-2	35.0606453	($^2H_3$)Methanol Methyl-d3 alcohol
CH^2H_3S		51.1272453	($^2H_3$)Methanethiol Methane-d3-thiol
CHI_3	75-47-8	393.73235	Triiodomethane
$CHKO_2$	590-29-4	84.11604	Potassium methanoate Potassium formate Methanoic acid potassium salt Formic acid potassium salt
$CHKO_3$	298-14-6	100.11544	Potassium hydrogen carbonate Carbonic acid hydrogen potassium salt Potassium bicarbonate
CHN	74-90-8	27.02568	Hydrogen cyanide Hydrocyanic acid Prussic acid
$CHNO$	506-85-4	43.02508	Fulminic acid
$CHNS$	3129-90-6	59.09168	Isothiocyanic acid
$CHNS$	463-56-9	59.09168	Thiocyanic acid
$CHNaO_2$	141-53-7	68.007508	Sodium methanoate Methanoic acid sodium salt Sodium formate Formic acid sodium salt
$CHNaO_3$	144-55-8	84.006908	Sodium hydrogen carbonate Carbonic acid hydrogen sodium salt Sodium bicarbonate
CH_2BrCl	74-97-5	129.38358	Bromochloromethane
CH_2BrF	373-52-4	112.9292832	Bromofluoromethane
CH_2Br_2	74-95-3	173.83488	Dibromomethane Methylene bromide
CH_2ClF	593-70-4	68.4779832	Chlorofluoromethane
CH_2Cl_2	75-09-2	84.93228	Dichloromethane Methylene chloride
CH_2F_2	75-10-5	52.0236864	Difluoromethane Methylene fluoride
$CH_2{}^2H_3N$	5581-55-5	34.0759253	[($^2H_3$)Methyl]amine Methyl-d3-amine
CH_2I_2	75-11-6	267.83582	Diiodomethane
CH_2N_2	420-04-2	42.04036	Cyanamide
CH_2N_2	334-88-3	42.04036	Diazomethane
CH_2O	50-00-0	30.02628	Methanal Formaldehyde
CH_2O_2	64-18-6	46.02568	Methanoic acid Formic acid
CH_3		15.03482	Methyl
CH_3Br	74-83-9	94.93882	Bromomethane Methyl bromide
CH_3Cl	74-87-3	50.48752	Chloromethane Methyl chloride
CH_3Cl_3Si	75-79-6	149.47842	Trichloromethylsilane
CH_3F	593-53-3	34.0332232	Fluoromethane Methyl fluoride
$CH_3{}^2HO$	1455-13-6	33.0483217	Methan(^2H)ol Methyl alcohol-d
$CH_3{}^2HS$	16978-68-0	49.1149217	Methane(^2H)thiol Methanethiol-d
$CH_3{}^2H_2N$	2614-35-9	33.0697635	Methyl($^2H_2$)amine Methylamine-d2
CH_3I	74-88-4	141.93929	Iodomethane Methyl iodide
C^2H_3I	865-50-9	144.9577753	($^2H_3$)Iodomethane Methyl iodide-d3
CH_3NO	75-12-7	45.04096	Methanamide Formamide
CH_3NO_2	75-52-5	61.04036	Nitromethane
$C^2H_3NO_2$	13031-32-8	64.0588453	Nitro($^2H_3$)methane Deuteronitromethane
CH_3NO_3	598-58-3	77.03976	Methyl nitrate Nitric acid methyl ester
CH_3N_3	624-90-8	57.05504	Methyl azide Azidomethane
CH_3NaO_4S		134.088188	Methyl sodium sulfate
CH_4	74-82-8	16.04276	Methane Carbane
$^{12}CH_4$		16.03176	(^{12}C)Methane (^{12}C)Carbane
$^{13}CH_4$		17.0351148	(^{13}C)Methane (^{13}C)Carbane
C^2H_4	558-20-3	20.0674071	($^2H_4$)Methane Methane-d4 Deuteromethane
CH_4Cl_2Si	75-54-7	115.03366	Dichloromethylsilane
CH_4N_2	12211-52-8	44.05624	Ammonium cyanide
CH_4N_2O	57-13-6	60.05564	Urea
CH_4N_2S	1762-95-4	76.12224	Ammonium thiocyanate Thiocyanic acid ammonium salt
CH_4N_2S	62-56-6	76.12224	Thiourea
CH_4O	67-56-1	32.04216	Methanol Methyl alcohol
C^2H_4O	811-98-3	36.0668071	($^2H_4$)Methanol Methyl alcohol-d4
CH_4O_3S	75-75-2	96.10696	Methanesulfonic acid
CH_4S	74-93-1	48.10876	Methanethiol Methyl mercaptan
C^2H_4S		52.1334071	($^2H_4$)Methanethiol Methanethiol-d4
CH_5ClSi	993-00-0	80.5889	Chloromethylsilane
CH_5N	74-89-5	31.05744	Methylamine Aminomethane
C^2H_5N	3767-37-1	36.0882488	($^2H_5$)Methylamine Methylamine-d5
CH_5NO_3	1066-33-7	79.05564	Ammonium hydrogen carbonate Ammonium bicarbonate Carbonic acid hydrogen ammonium salt
CH_5O_3P		96.022662	Dihydrogen methyl phosphite Phosphorous acid methyl ester
CH_6Ge	1449-65-6	90.66864	Methylgermane Trihydromethylgermanium
CH_6N_2	60-34-4	46.07212	Methylhydrazine
$CH_6N_2O_2$	1111-78-0	78.07092	Ammonium carbamate Carbamic acid ammonium salt
CH_6OSi	2171-96-2	62.14354	Methoxysilane
CH_6Si	992-94-9	46.14414	Methylsilane
$CH_8N_2O_3$	506-87-6	96.0862	Diammonium carbonate Ammonium carbonate Carbonic acid ammonium salt
$CH_{12}Cu_2N_4O_3$		255.22344	Dicopper carbonate tetraammoniate Cuprous ammonium carbonate

Formula	CAS RN	M_r	Name
$CH_{12}Cu_2N_4O_3$ (Cont.)			Copper(I) ammonium carbonate
CHg_2O_3	6824-78-8	461.1892	Dimercury carbonate
			Mercurous carbonate
			Carbonic acid mercury(I) salt
CIN	506-78-5	152.92221	Cyanogen iodide
CI_4	507-25-5	519.62888	Tetraiodomethane
CKN	151-50-8	65.11604	Potassium cyanide
$CKNS$		97.18204	Potassium isothiocyanate
			Isothiocyanic acid potassium salt
$CKNS$	333-20-0	97.18204	Potassium thiocyanate
			Thiocyanic acid potassium salt
CK_2O_3	584-08-7	138.2058	Dipotassium carbonate
			Potassium carbonate
			Carbonic acid potassium salt
$CLiNS$		65.02474	Lithium thiocyanate
			Thiocyanic acid lithium salt
CLi_2O_3	554-13-2	73.8912	Dilithium carbonate
			Lithium carbonate
			Carbonic acid lithium salt
$CMgO_3$	82597-01-1	84.3142	Magnesium carbonate
			Carbonic acid magnesium salt
$CMnO_3$	68013-64-9	114.94725	Manganese carbonate
			Carbonic acid manganese(II) salt
$CNNa$	143-33-9	49.007508	Sodium cyanide
$CNNaO$	917-61-3	65.006908	Sodium cyanate
$CNNaS$	540-72-7	81.073508	Sodium thiocyanate
			Thiocyanic acid sodium salt
CN_3NaO_6		173.017388	Trinitrosodiomethane
			Sodium trinitromethane
			Trinitromethylsodium
CN_4O_8	509-14-8	196.03316	Tetranitromethane
CNa_2O_3	497-19-8	105.988736	Disodium carbonate
			Sodium carbonate
			Carbonic acid sodium salt
CO	630-08-0	28.0104	Carbon oxide
			Carbon monoxide
$^{12}C^{16}O$		27.9949146	(^{12}C)Carbon (^{16}O)oxide
			(^{12}C)Carbon (^{16}O)monoxide
$^{13}C^{16}O$		28.9982694	(^{13}C)Carbon (^{16}O)oxide
			(^{13}C)Carbon (^{16}O)monoxide
COS	463-58-1	60.0764	Carbonyl sulfide
			Carbon oxysulfide
$COSe$	1603-84-5	106.9704	Carbonyl selenide
			Carbon oxyselenide
CO_2	124-38-9	44.0098	Carbon dioxide
CO_3Pb	598-63-0	267.2092	Lead carbonate
			Carbonic acid lead(II) salt
CO_3Rb_2	584-09-8	230.9448	Dirubidium carbonate
			Carbonic acid rubidium salt
			Rubidium carbonate
CO_3Sr	1633-05-2	147.6292	Strontium carbonate
			Carbonic acid strontium salt
CO_3Tl_2	6533-73-9	468.7758	Dithallium carbonate
			Thallium(I) carbonate
			Thallous carbonate
			Carbonic acid thallium(I) salt
			Carbonic acid thallous salt
CO_3Zn	3486-35-9	125.3992	Zinc carbonate
			Carbonic acid zinc salt

Formula	CAS RN	M_r	Name
CS_2	75-15-0	76.143	Carbon disulfide
CSe_2	506-80-9	169.931	Carbon diselenide
CSi	409-21-2	40.0965	Silicon carbide
$C_2Br_2ClF_3$	354-51-8	276.2779096	1,2-Dibromo-1-chloro-1,2,2-trifluoroethane
$C_2Br_2F_4$	124-73-2	259.8236128	1,2-Dibromotetrafluoroethane
C_2Br_4	79-28-7	343.638	Tetrabromoethene
			Tetrabromoethylene
C_2Br_6	594-73-0	503.446	Hexabromoethane
C_2Ca	75-20-7	64.1	Calcium acetylide
			Calcium carbide
C_2CaN_2	592-01-8	92.11348	Calcium dicyanide
			Calcium cyanide
$C_2CaN_2S_2$		156.24548	Calcium dithiocyanate
			Calcium thiocyanate
			Thiocyanic acid calcium salt
C_2CaO_4	563-72-4	128.0976	Calcium ethanedioate
			Calcium oxalate
			Oxalic acid calcium salt
			Ethanedioic acid calcium salt
C_2ClF_3	79-38-9	116.4699096	Chlorotrifluoroethene
			Chlorotrifluoroethylene
			CTFE
C_2ClF_5	76-15-3	154.466716	Chloropentafluoroethane
			R 115
C_2Cl_2	7572-29-4	94.9274	Dichloroethyne
			Dichloroacethylene
$C_2Cl_2F_4$	374-07-2	170.9210128	1,1-Dichlorotetrafluoroethane
$C_2Cl_2F_4$	76-14-2	170.9210128	1,2-Dichlorotetrafluoroethane
			Freon 114
			R 114
			CFC 114
$C_2Cl_3F_3$	354-58-5	187.3753096	1,1,1-Trichlorotrifluoroethane
$C_2Cl_3F_3$	76-13-1	187.3753096	1,1,2-Trichlorotrifluoroethane
			Freon 113
			CFC 113
			R 113
C_2Cl_3N	545-06-2	144.38684	Trichloroethanenitrile
			Trichloroacetonitrile
C_2Cl_4	127-18-4	165.8328	Tetrachloroethene
			Tetrachloroethylene
$C_2Cl_4F_2$	76-11-9	203.8296064	Tetrachloro-1,1-difluoroethane
$C_2Cl_4F_2$	76-12-0	203.8296064	Tetrachloro-1,2-difluoroethane
C_2Cl_4O	76-02-8	181.8322	Trichloroacetyl chloride
C_2Cl_6	67-72-1	236.7382	Hexachloroethane
			Perchloroethane
C_2CoO_2		114.954	Cobalt dicarbonyl
			Cobalt carbonyl
C_2F_3N	353-85-5	95.0239496	Trifluoroethanenitrile
			Trifluoroacetonitrile
C_2F_4	116-14-3	100.0156128	Tetrafluoroethene
			Tetrafluoroethylene
C_2F_5I	354-64-3	245.918486	Pentafluoroiodoethane
C_2F_6	76-16-4	138.0124192	Hexafluoroethane
			Perfluoroethane

Formula	CAS RN	M_r	Name
C_2F_6 (Cont.)			Freon 116
C_2F_6O		154.0118192	Bis(trifluoromethyl) ether
			Perfluoro(dimethyl ether)
C_2HBr	593-61-3	104.93394	Bromoethyne
$C_2HBrClF_3$	151-67-7	197.3818496	Bromochloro-1,1,1-trifluoroethane
			Halothane
$C_2HBrClF_3$	354-06-3	197.3818496	1-Bromo-2-chloro-1,1,2-trifluoroethane
C_2HCl	593-63-5	60.48264	Chloroethyne
C_2HClF_2	359-10-4	98.4794464	1-Chloro-2,2-difluoroethene
C_2HClF_4		136.4762528	Chlorotetrafluoroethane
			<undefined positional isomer>
C_2HClF_4	2837-89-0	136.4762528	2-Chloro-1,1,1,2-tetrafluoroethane
			Freon 124
			R 124
C_2HClF_4	354-25-6	136.4762528	1-Chloro-1,1,2,2-tetrafluoroethane
C_2HCl_2F	27156-05-4	114.9337432	1,1-Dichloro-2-fluoroethene
$C_2HCl_2F_3$	306-83-2	152.9305496	1,1-Dichloro-2,2,2-trifluoroethane
			HCFC 123
			R 123
C_2HCl_3	79-01-6	131.38804	Trichloroethene
			Trichloroethylene
$C_2HCl_3F_2$	354-21-2	169.3848464	1,2,2-Trichloro-1,1-difluoroethane
			Freon 122
C_2HCl_3O	79-36-7	147.38744	Dichloroacetyl chloride
C_2HCl_3O	75-87-6	147.38744	Trichloroethanal
			Chloral
			Trichloroacetaldehyde
$C_2HCl_3O_2$	76-03-9	163.38684	Trichloroethanoic acid
			Trichloroacetic acid
C_2HCl_5	76-01-7	202.29344	Pentachloroethane
C_2HF	2713-09-9	44.0283432	Fluoroethyne
C_2HF_3	359-11-5	82.0251496	Trifluoroethene
			Trifluoroethylene
$C_2HF_3O_2$	76-05-1	114.0239496	Trifluoroethanoic acid
			Trifluoroacetic acid
C_2HF_5	354-33-6	120.021956	Pentafluoroethane
			R 125
			HFC 125
$C_2H\,^2H_5O$	1859-08-1	51.0998488	($^2H_5$)Ethanol
			Ethyl-d5 alcohol
$C_2H\,^2H_6N$		51.1212906	Di[($^2H_3$)methyl]amine
			Dimethylamine-d6
C_2H_2	74-86-2	26.03788	Ethyne
			Acetylene
$C_2H_2Br_2$	590-11-4	185.84588	cis-1,2-Dibromoethene
$C_2H_2Br_2$	590-12-5	185.84588	trans-1,2-Dibromoethene
$C_2H_2Br_4$	79-27-6	345.65388	1,1,2,2-Tetrabromoethane
			Acetylene tetrabromide
$C_2H_2CaO_4$	544-17-2	130.11348	Calcium dimethanoate
			Calcium formate
			Methanoic acid calcium salt
			Formic acid calcium salt
$C_2H_2ClF_3$	75-88-7	118.4857896	2-Chloro-1,1,1-trifluoroethane
$C_2H_2Cl_2$	75-35-4	96.94328	1,1-Dichloroethene
			1,1-Dichloroethylene
$C_2H_2Cl_2$ (Cont.)			Vinylidene chloride
$C_2H_2Cl_2$	156-59-2	96.94328	cis-1,2-Dichloroethene
			cis-1,2-Dichloroethylene
$C_2H_2Cl_2$	156-60-5	96.94328	trans-1,2-Dichloroethene
			trans-1,2-Dichloroethylene
$C_2H_2Cl_2F_2$		134.9400864	1,1-Dichloro-1,2-difluoroethane
			Freon 132
$C_2H_2Cl_2F_2$	471-43-2	134.9400864	1,1-Dichloro-2,2-difluoroethane
$C_2H_2Cl_2O$	79-04-9	112.94268	Chloroacetyl chloride
$C_2H_2Cl_2O_2$	79-43-6	128.94208	Dichloroethanoic acid
			Dichloroacetic acid
$C_2H_2Cl_4$	25322-20-7	167.84868	Tetrachloroethane
			<undefined positional isomer>
$C_2H_2Cl_4$	630-20-6	167.84868	1,1,1,2-Tetrachloroethane
$C_2H_2Cl_4$	79-34-5	167.84868	1,1,2,2-Tetrachloroethane
			Acetylene tetrachloride
$C_2H_2F_2$	75-38-7	64.0346864	1,1-Difluoroethene
			1,1-Difluoroethylene
			Vinylidene fluoride
$C_2H_2F_2$	1630-77-9	64.0346864	cis-1,2-Difluoroethene
$C_2H_2F_4$	811-97-2	102.0314928	1,1,1,2-Tetrafluoroethane
			HFC 134a
			R 134a
$C_2H_2F_4$		102.0314928	1,1,2,2-Tetrafluoroethane
			HFC 134
$C_2H_2MgO_4$		114.34048	Magnesium dimethanoate
			Magnesium formate
			Methanoic acid magnesium salt
			Formic acid magnesium salt
C_2H_2O	463-51-4	42.03728	Ketene
			Ethenone
$C_2H_2O_2$	107-22-2	58.03668	Ethanedial
			Glyoxal
$C_2H_2O_3$	298-12-4	74.03608	Formylmethanoic acid
			Glyoxylic acid
$C_2H_2O_4$	144-62-7	90.03548	Ethanedioic acid
			Oxalic acid
$C_2H_2O_4Sr$	592-89-2	177.65548	Strontium dimethanoate
			Strontium formate
			Methanoic acid strontium salt
			Formic acid strontium salt
C_2H_3Br	593-60-2	106.94982	Bromoethene
C_2H_3BrO	506-96-7	122.94922	Acetyl bromide
C_2H_3Cl	75-01-4	62.49852	Chloroethene
			Vinyl chloride
$C_2H_3ClF_2$	75-68-3	100.4953264	1-Chloro-1,1-difluoroethane
			HCFC 142b
			R 142b
C_2H_3ClO	75-36-5	78.49792	Acetyl chloride
C_2H_3ClO	107-20-0	78.49792	Chloroethanal
			Chloroacetaldehyde
$C_2H_3ClO_2$	79-11-8	94.49732	Chloroethanoic acid
			Chloroacetic acid
$C_2H_3Cl_2F$	1717-00-6	116.9496232	1,1-Dichloro-1-fluoroethane
			HCFC 141b
			R 141b
$C_2H_3Cl_3$	25323-89-1	133.40392	Trichloroethane
			<undefined positional isomer>
$C_2H_3Cl_3$	71-55-6	133.40392	1,1,1-Trichloroethane
			Methylchloroform

Formula	CAS RN	M_r	Name
$C_2H_3Cl_3$	79-00-5	133.40392	1,1,2-Trichloroethane
$C_2H_3Cl_3O$	115-20-8	149.40332	2,2,2-Trichloroethanol
$C_2H_3Cl_3O_2Ti$		213.28272	Trichloro(ethanoato)titanium
			Trichloro(acetato)titanium
$C_2H_3Cl_3Si$	75-94-5	161.48942	Trichlorovinylsilane
$C_2H_3CuO_2$	598-54-9	122.59062	Copper ethanoate
			Cuprous acetate
			Copper(I) acetate
			Ethanoic acid copper(I) salt
			Acetic acid copper(I) salt
C_2H_3F	75-02-5	46.0442232	Fluoroethene
			Vinyl fluoride
C_2H_3FO	557-99-3	62.0436232	Acetyl fluoride
$C_2H_3F_3$	420-46-2	84.0410296	1,1,1-Trifluoroethane
			Methylfluoroform
$C_2H_3F_3$	430-66-0	84.0410296	1,1,2-Trifluoroethane
$C_2H_3F_3O$	75-89-8	100.0404296	2,2,2-Trifluoroethanol
C_2H_3I	593-66-8	153.95029	Iodoethene
C_2H_3IO	507-02-8	169.94969	Acetyl iodide
$C_2H_3KO_2$	127-08-2	98.14292	Potassium ethanoate
			Potassium acetate
			Ethanoic acid potassium salt
			Acetic acid potassium salt
C_2H_3N	75-05-8	41.05256	Ethanenitrile
			Acetonitrile
			Cyanomethane
			Ethanonitrile
			Methanecarbonitrile
			Methyl cyanide
C_2H_3N	593-75-9	41.05256	Methyl isocyanide
			Isocyanomethane
C_2H_3NO	624-83-9	57.05196	Methyl isocyanate
			Isocyanic acid methyl ester
$C_2H_3N_2NaO_4$		142.046668	1,1-Dinitro-1-sodioethane
			Sodium 1,1-dinitroethane
$C_2H_3NaO_2$	127-09-3	82.034388	Sodium ethanoate
			Sodium acetate
			Ethanoic acid sodium salt
			Acetic acid sodium salt
C_2H_4	74-85-1	28.05376	Ethene
			Ethylene
C_2H_4BrCl	107-04-0	143.41046	1-Bromo-2-chloroethane
			Ethylene chlorobromide
$C_2H_4Br_2$	557-91-5	187.86176	1,1-Dibromoethane
			Ethylidene bromide
$C_2H_4Br_2$	106-93-4	187.86176	1,2-Dibromoethane
			Ethylene dibromide
			Ethylene bromide
C_2H_4ClF	1615-75-4	82.5048632	1-Chloro-1-fluoroethane
$C_2H_4Cl_2$	75-34-3	98.95916	1,1-Dichloroethane
			Ethylidene chloride
$C_2H_4Cl_2$	107-06-2	98.95916	1,2-Dichloroethane
			Ethylene chloride
			Ethylene dichloride
$C_2H_4Cl_2O$	542-88-1	114.95856	Bis(chloromethyl) ether
$C_2H_4Cl_2O$	598-38-9	114.95856	2,2-Dichloroethanol
$C_2H_4Cl_4Si$	1558-31-2	197.95006	Dichloro(dichloromethyl)-methylsilane
$C_2H_4F_2$	75-37-6	66.0505664	1,1-Difluoroethane
			R 152a
			Ethylidene fluoride
$C_2H_4F_2$ (Cont.)			HFC 152a
$C_2H_4F_2$	624-72-6	66.0505664	1,2-Difluoroethane
$C_2H_4NNaO_2$		97.049068	1-Nitro-1-sodioethane
			Sodium nitroethane
$C_2H_4N_2O_6$	628-96-6	152.06364	1,2-Ethanediol dinitrate
			Ethylene glycol dinitrate
C_2H_4O	75-07-0	44.05316	Ethanal
			Acetaldehyde
			Methanecarbaldehyde
C_2H_4O	75-21-8	44.05316	Oxirane
			Ethylene oxide
$C_2H_4O_2$	64-19-7	60.05256	Ethanoic acid
			Acetic acid
			Methanecarboxylic acid
$C_2H_4O_2$	141-46-8	60.05256	Hydroxyethanal
			Glycolaldehyde
$C_2H_4O_2$	107-31-3	60.05256	Methyl methanoate
			Methanoic acid methyl ester
			Methyl formate
			Formic acid methyl ester
$C_2H_4O_3$	79-14-1	76.05196	Hydroxyethanoic acid
			Glycolic acid
$C_2H_4O_3$	79-21-0	76.05196	Peroxyethanoic acid
			Peroxyacetic acid
C_2H_4Si	1066-27-9	56.13926	Ethynylsilane
C_2H_5Br	74-96-4	108.9657	Bromoethane
			Ethyl bromide
C_2H_5Cl	75-00-3	64.5144	Chloroethane
			Ethyl chloride
C_2H_5ClO	107-07-3	80.5138	2-Chloroethanol
			Ethylene chlorohydrin
C_2H_5ClO	107-30-2	80.5138	Chloromethyl methyl ether
$C_2H_5Cl_3Si$	1558-33-4	163.5053	Dichloro(chloromethyl)-methylsilane
$C_2H_5Cl_3Si$	115-21-9	163.5053	Trichloroethylsilane
C_2H_5F	353-36-6	48.0601032	Fluoroethane
			Ethyl fluoride
$C_2H_5{}^2HO$	925-93-9	47.0752017	(O-2H)Ethanol
			Ethyl alcohol-d
$C_2H_5{}^2H_2N$	5852-45-9	47.0966435	Ethyl($^2H_2$)amine
			Ethylamine-d2
C_2H_5I	75-03-6	155.96617	Iodoethane
			Ethyl iodide
C_2H_5N	151-56-4	43.06844	Aziridine
			Azacyclopropane
			Azirane
			Dihydroazirine
			Ethyleneimine
			Perhydroazirine
C_2H_5NO	60-35-5	59.06784	Ethanamide
			Acetamide
C_2H_5NO	123-39-7	59.06784	N-Methylmethanamide
			N-Methylformamide
			NMF
$C_2H_5NO_2$	56-40-6	75.06724	Aminoethanoic acid
			Glycine
$C_2H_5NO_2$	79-24-3	75.06724	Nitroethane
$C_2H_5NO_3$	625-58-1	91.06664	Ethyl nitrate
			Nitric acid ethyl ester
$C_2H_5NO_3$	625-48-9	91.06664	2-Nitroethanol
$C_2H_5NO_4$	16051-48-2	107.06604	1,2-Ethanediol nitrate
			Ethylene glycol nitrate

Formula	CAS RN	M_r	Name
C_2H_6	74-84-0	30.06964	Ethane
C_2H_6AlCl		92.503879	Chlorodimethylaluminum
C_2H_6Cd	506-82-1	142.48064	Dimethylcadmium
$C_2H_6Cl_2Si$	75-78-5	129.06054	Dichlorodimethylsilane
$C_2H_6Cl_2Si$	1789-58-8	129.06054	Dichloroethylsilane
$C_2H_6{}^2HN$	917-72-6	46.0904817	Dimethyl(^{2}H)amine
			Dimethylamine-d
C_2H_6Hg	593-74-8	230.65964	Dimethylmercury
$C_2H_6N_2O$	598-50-5	74.08252	Methylurea
C_2H_6O	115-10-6	46.06904	Dimethyl ether
			Methyl ether
C_2H_6O	64-17-5	46.06904	Ethanol
			Ethyl alcohol
$C_2{}^2H_6O$	1516-08-1	52.1060106	($^2H_6$)Ethanol
			Ethyl alcohol-d6
C_2H_6OS	67-68-5	78.13504	Dimethyl sulfoxide
			DMSO
			Sulfinylbismethane
			Methyl sulfoxide
$C_2H_6O_2$	107-21-1	62.06844	1,2-Ethanediol
			Ethylene glycol
			Glycol
			1,2-Dihydroxyethane
$C_2H_6O_2S$	67-71-0	94.13444	Dimethyl sulfone
			Methyl sulfone
C_2H_6S	75-08-1	62.13564	Ethanethiol
			Ethyl mercaptan
C_2H_6S	75-18-3	62.13564	2-Thiapropane
			Dimethyl sulfide
			Dimethyl thioether
$C_2H_6S_2$	624-92-0	94.20164	2,3-Dithiabutane
			Dimethyl disulfide
$C_2H_6S_2$	26914-40-9	94.20164	1,2-Ethanedithiol
C_2H_6Zn	544-97-8	95.45964	Dimethylzinc
C_2H_7ClSi	1066-35-9	94.61578	Chlorodimethylsilane
C_2H_7N	124-40-3	45.08432	Dimethylamine
C_2H_7N	75-04-7	45.08432	Ethylamine
			Aminoethane
$C_2{}^2H_7N$		52.1274524	($^2H_7$)Dimethylamine
			Dimethylamine-d7
C_2H_7NO	141-43-5	61.08372	2-Aminoethanol
			Ethanolamine
$C_2H_7O_3P$	868-85-9	110.049542	Hydrogen dimethyl phosphite
			Dimethyl phosphite
			Phosphorous acid dimethyl ester
C_2H_8ClN	506-59-2	81.54496	Dimethylamine hydrochloride
$C_2H_8N_2$	57-14-7	60.099	1,1-Dimethylhydrazine
$C_2H_8N_2$	540-73-8	60.099	1,2-Dimethylhydrazine
$C_2H_8N_2$	107-15-3	60.099	1,2-Ethanediamine
			Ethylenediamine
$C_2H_8N_2O_3$		108.0972	Ethylammonium nitrate
$C_2H_8N_2O_4$	1113-38-8	124.0966	Diammonium ethanedioate
$C_2H_9CuN_2O_2$		156.65174	Copper ethanoate diammoniate
			Cuprous ammonium acetate
			Copper(I) ammonium acetate
C_2H_5+		29.0617	Ethylium
			Ethyl cation
C_2HgN_2	592-04-1	252.62548	Mercury dicyanide
C_2HgN_2 (Cont.)			Mercury(II) cyanide
			Mercuric cyanide
C_2HgO_4	3444-13-1	288.6096	Mercury ethanedioate
C_2I_2	624-74-8	277.83094	Diiodoethyne
C_2I_4	513-92-8	531.63988	Tetraiodoethene
$C_2K_2O_4$	583-52-8	166.2162	Dipotassium ethanedioate
			Potassium oxalate
			Ethanedioic acid potassium salt
			Oxalic acid potassium salt
C_2MgO_4	547-66-0	112.3246	Magnesium ethanedioate
			Magnesium oxalate
			Oxalic acid magnesium salt
			Ethanedioic acid magnesium salt
C_2N_2	460-19-5	52.03548	Ethanedinitrile
			Oxalonitrile
			Cyanogen
$C_2N_2S_2Zn$		181.55748	Zinc dithiocyanate
			Zinc thiocyanate
			Thiocyanic acid zinc salt
$C_2Na_2O_4$	62-76-0	133.999136	Disodium ethanedioate
C_2O_4Pb	814-93-7	295.2196	Lead ethanedioate
C_3BrF_7		248.9258224	1-Bromoheptafluoropropane
$C_3Br_2F_6$	661-95-0	309.8314192	1,2-Dibromohexafluoropropane
			<undefined optical isomer>
C_3ClF_5O	79-53-8	182.477116	Chloropentafluoro-2-propanone
			Chloropentafluoroacetone
$C_3Cl_2F_6$	661-97-2	220.9288192	1,2-Dichlorohexafluoropropane
$C_3Cl_3F_5$	1599-41-3	237.383116	1,2,2-Trichloropentafluoropropane
$C_3Cl_4F_4$	2268-46-4	253.8374128	1,1,1,3-Tetrachlorotetrafluoropropane
C_3CoO_3		142.9644	Cobalt tricarbonyl
C_3F_6	116-15-4	150.0234192	Hexafluoropropene
			Hexafluoropropylene
			Perfluoropropene
			Perfluoropropylene
C_3F_6O	428-59-1	166.0228192	Hexafluoromethyloxirane
			Perfluoropropylene oxide
			Hexafluoro-1,2-epoxypropane
C_3F_6O	684-16-2	166.0228192	Hexafluoro-2-propanone
			Hexafluoroacetone
			Perfluoroacetone
C_3F_8	76-19-7	188.0202256	Octafluoropropane
			Perfluoropropane
			Freon 218
$C_3HCl_2F_5$		202.938356	1,3-Dichloro-1,2,2,3,3-pentafluoropropane
			HCFC 225cb
$C_3HCl_2F_5$		202.938356	1,1-Dichloro-2,2,3,3,3-pentafluoropropane
			HCFC 225ca
$C_3HF_5O_2$	422-64-0	164.031756	Pentafluoropropanoic acid
			Pentafluoropropionic acid
			Perfluoropropionic acid
			Perfluoropropanoic acid
C_3HN	1070-71-9	51.04768	Propynenitrile
			Cyanoacetylene
			Ethynecarbonitrile

Formula	CAS RN	M_r	Name
C_3HN (Cont.)			Ethynyl cyanide
C_3H_2ClN	920-37-6	87.50832	2-Chloropropenenitrile
			2-Chloroacrylonitrile
C_3H_2ClN		87.50832	cis-3-Chloropropenenitrile
			cis-3-Chloroacrylonitrile
C_3H_2ClN		87.50832	trans-3-Chloropropenenitrile
			trans-3-Chloroacrylonitrile
$C_3H_2F_6O$	920-66-1	168.0386992	1,1,1,3,3,3-Hexafluoro-2-propanol
C_3H_2O	624-67-9	54.04828	Propynal
			Ethynecarbaldehyde
$C_3H_3Cl_2F$	430-95-5	128.9606232	1,1-Dichloro-2-fluoropropene
$C_3H_3Cl_3O_2$	598-99-2	177.41372	Methyl trichloroethanoate
			Methyl trichloroacetate
			Trichloroacetic acid methyl ester
			Trichloroethanoic acid methyl ester
$C_3H_3F_3$		96.0520296	Trifluoropropene
			<undefined positional isomer>
$C_3H_3F_3$	677-21-4	96.0520296	3,3,3-Trifluoropropene
$C_3H_3F_3O_2$	431-47-0	128.0508296	Methyl trifluoroethanoate
			Methyl trifluoroacetate
			Trifluoroethanoic acid methyl ester
			Trifluoroacetic acid methyl ester
$C_3H_3F_5$	1814-88-6	134.048836	1,1,1,2,2-Pentafluoropropane
$C_3H_3F_5O$	422-05-9	150.048236	2,2,3,3,3-Pentafluoro-1-propanol
C_3H_3N	107-13-1	53.06356	Propenenitrile
			Acrylonitrile
C_3H_3NO	288-14-2	69.06296	Isoxazole
C_3H_3NO	288-42-6	69.06296	Oxazole
C_3H_3NS	288-47-1	85.12956	Thiazole
C_3H_4	2781-85-3	40.06476	Cyclopropene
C_3H_4	463-49-0	40.06476	Propadiene
			Allene
C_3H_4	74-99-7	40.06476	Propyne
			Methylacetylene
$C_3H_4Cl_2$	542-75-6	110.97016	1,3-Dichloropropene
			1,3-Dichloropropylene
$C_3H_4Cl_2$	78-88-6	110.97016	2,3-Dichloropropene
			2,3-Dichloropropylene
$C_3H_4Cl_2F_2O$	76-38-0	164.9663664	2,2-Dichloro-1,1-difluoro-1-methoxyethane
$C_3H_4Cl_2O_2$	116-54-1	142.96896	Methyl dichloroethanoate
			Methyl dichloroacetate
			Dichloroacetic acid methyl ester
			Dichloroethanoic acid methyl ester
$C_3H_4Cl_3F_3Si$	592-09-6	231.5035696	Trichloro(3,3,3-trifluoropropyl)silane
$C_3H_4Cl_4$		181.87556	Tetrachloropropane
			<undefined positional isomer>
$C_3H_4Cl_4$	812-03-3	181.87556	1,1,1,2-Tetrachloropropane
$C_3H_4Cl_4$	1070-78-6	181.87556	1,1,1,3-Tetrachloropropane
$C_3H_4Cl_4$	13116-60-4	181.87556	1,1,2,2-Tetrachloropropane
$C_3H_4Cl_4$	18495-30-2	181.87556	1,1,2,3-Tetrachloropropane
$C_3H_4Cl_4$	13116-53-5	181.87556	1,2,2,3-Tetrachloropropane
$C_3H_4F_2$	430-63-7	78.0615664	1,1-Difluoropropene
$C_3H_4F_4O$	76-37-9	132.0577728	2,2,3,3-Tetrafluoro-1-propanol
$C_3H_4N_2$	288-32-4	68.07824	Imidazole
$C_3H_4N_2$	288-13-1	68.07824	Pyrazole
C_3H_4O	5009-27-8	56.06416	Cyclopropanone
C_3H_4O	107-02-8	56.06416	Propenal
			Acrolein
			Acrylaldehyde
C_3H_4O	107-19-7	56.06416	2-Propyn-1-ol
			Propargyl alcohol
$C_3H_4O_2$	57-57-8	72.06356	2-Oxetanone
$C_3H_4O_2$	6704-31-0	72.06356	3-Oxetanone
$C_3H_4O_2$	79-10-7	72.06356	Propenoic acid
			Acrylic acid
$C_3H_4O_2$	692-45-5	72.06356	Vinyl methanoate
			Vinyl formate
			Methanoic acid vinyl ester
			Formic acid vinyl ester
$C_3H_4O_3$	96-49-1	88.06296	1,3-Dioxolan-2-one
			Ethylene carbonate
$C_3H_4O_4$	141-82-2	104.06236	Propanedioic acid
			Malonic acid
C_3H_5Br	106-95-6	120.9767	3-Bromopropene
			Allyl bromide
			2-Propenyl bromide
			3-Bromopropylene
C_3H_5BrO	3132-64-7	136.9761	(Bromomethyl)oxirane
			1-Bromo-2,3-epoxypropane
			Epibromohydrin
			<undefined optical isomer>
C_3H_5Cl	16136-84-8	76.5254	cis-1-Chloropropene
C_3H_5Cl	16136-85-9	76.5254	trans-1-Chloropropene
C_3H_5Cl	557-98-2	76.5254	2-Chloropropene
			2-Chloropropylene
C_3H_5Cl	107-05-1	76.5254	3-Chloropropene
			Allyl chloride
			3-Chloropropylene
			2-Propenyl chloride
C_3H_5ClO	106-89-8	92.5248	(Chloromethyl)oxirane
			1-Chloro-2,3-epoxypropane
			Epichlorohydrin
			<undefined optical isomer>
$C_3H_5ClO_2$	541-41-3	108.5242	Ethyl chloromethanoate
			Ethyl chloroformate
			Chloromethanoic acid ethyl ester
			Chloroformic acid ethyl ester
$C_3H_5ClO_2$	96-34-4	108.5242	Methyl chloroethanoate
			Chloroethanoic acid methyl ester
			Methyl chloroacetate
$C_3H_5Cl_3$	25735-29-9	147.4308	Trichloropropane
			<undefined positional isomer>
$C_3H_5Cl_3$	7789-89-1	147.4308	1,1,1-Trichloropropane
$C_3H_5Cl_3$	598-77-6	147.4308	1,1,2-Trichloropropane
$C_3H_5Cl_3$	3175-23-3	147.4308	1,2,2-Trichloropropane
$C_3H_5Cl_3$	96-18-4	147.4308	1,2,3-Trichloropropane
$C_3H_5Cl_3O_2Ti$		227.3096	Trichloro(propanoato)titanium
			Trichloro(propionato)titanium
C_3H_5F	19184-10-2	60.0711032	cis-1-Fluoropropene

Formula	CAS RN	M_r	Name
C_3H_5F	20327-65-5	60.0711032	trans-1-Fluoropropene
C_3H_5F	1184-60-7	60.0711032	2-Fluoropropene
C_3H_5F	818-92-8	60.0711032	3-Fluoropropene
$C_3H_5F_3O$	374-01-6	114.0673096	1,1,1-Trifluoro-2-propanol
$C_3H_5KO_2$		112.1698	Potassium propanoate
			Potassium propionate
			Propanoic acid potassium salt
			Propionic acid potassium salt
C_3H_5N	107-12-0	55.07944	Propanenitrile
			Cyanoethane
			Ethanecarbonitrile
			Ethyl cyanide
			Propanonitrile
			Propionitrile
C_3H_5NO	109-78-4	71.07884	3-Hydroxypropanenitrile
			Ethylene cyanohydrin
			3-Hydroxypropionitrile
			Hydracrylonitrile
C_3H_5NO	79-06-1	71.07884	Propenamide
			Acrylamide
$C_3H_5N_3O_9$	55-63-0	227.08752	1,2,3-Propanetriol trinitrate
			Glycerine trinitrate
			Trinitroglycerol
$C_3H_5NaO_2$	137-40-6	96.061268	Sodium propanoate
			Sodium propionate
			Propanoic acid sodium salt
			Propionic acid sodium salt
C_3H_6	75-19-4	42.08064	Cyclopropane
C_3H_6	115-07-1	42.08064	Propene
			Propylene
C_3H_6		42.08064	Propene (c,I)
			Propylene (c,I)
C_3H_6		42.08064	Propene (c,II)
			Propylene (c,II)
C_3H_6BrCl	34652-54-5	157.43734	Bromochloropropane
			<undefined positional isomer>
C_3H_6BrCl	3017-95-6	157.43734	2-Bromo-1-chloropropane
C_3H_6BrCl	3017-96-7	157.43734	1-Bromo-2-chloropropane
C_3H_6BrCl	109-70-6	157.43734	1-Bromo-3-chloropropane
			Trimethylene bromochloride
$C_3H_6Br_2$	78-75-1	201.88864	1,2-Dibromopropane
$C_3H_6Br_2$	109-64-8	201.88864	1,3-Dibromopropane
$C_3H_6Cl_2$	78-87-5	112.98604	1,2-Dichloropropane
			Propylene dichloride
$C_3H_6Cl_2$	142-28-9	112.98604	1,3-Dichloropropane
$C_3H_6Cl_2O$	616-23-9	128.98544	2,3-Dichloro-1-propanol
$C_3H_6Cl_2O$	96-23-1	128.98544	1,3-Dichloro-2-propanol
$C_3H_6Cl_2Si$	124-70-9	141.07154	Dichloromethylvinylsilane
$C_3H_6NNaO_2$		111.075948	2-Nitro-2-sodiopropane
			Sodium 2-nitropropane
$C_3H_6N_2$	1467-79-4	70.09412	Dimethylcyanamide
$C_3H_6N_6$	108-78-1	126.12108	Melamine
C_3H_6O	75-56-9	58.08004	Methyloxirane
			1,2-Epoxypropane
			Propylene oxide
C_3H_6O	107-25-5	58.08004	Methyl vinyl ether
C_3H_6O	503-30-0	58.08004	Oxetane
			Trimethylene oxide
C_3H_6O	123-38-6	58.08004	Propanal
			Ethanecarbaldehyde
			Propional
			Propionaldehyde

Formula	CAS RN	M_r	Name
C_3H_6O	67-64-1	58.08004	2-Propanone
			Acetone
			Dimethyl ketone
C_3H_6O	107-18-6	58.08004	2-Propen-1-ol
			Allyl alcohol
			2-Propenyl alcohol
$C_3{}^2H_6O$	666-52-4	64.1170106	($^2H_6$)-2-Propanone
			Hexadeuteroacetone
			Hexadeutero-2-propanone
$C_3H_6O_2$	646-06-0	74.07944	1,3-Dioxolane
$C_3H_6O_2$	109-94-4	74.07944	Ethyl methanoate
			Ethyl formate
			Methanoic acid ethyl ester
			Formic acid ethyl ester
$C_3H_6O_2$	556-52-5	74.07944	(Hydroxymethyl)oxirane
			2,3-Epoxy-1-propanol
$C_3H_6O_2$	79-20-9	74.07944	Methyl ethanoate
			Ethanoic acid methyl ester
			Methyl acetate
			Acetic acid methyl ester
$C_3H_6O_2$	79-09-4	74.07944	Propanoic acid
			Ethanecarboxylic acid
			Propionic acid
$C_3H_6O_3$	616-38-6	90.07884	Dimethyl carbonate
			Carbonic acid dimethyl ester
$C_3H_6O_3$	50-21-5	90.07884	2-Hydroxypropanoic acid
			Lactic acid
$C_3H_6O_3$	598-82-3	90.07884	DL-2-Hydroxypropanoic acid
			DL-Lactic acid
$C_3H_6O_3$	110-88-3	90.07884	1,3,5-Trioxane
			Metaformaldehyde
C_3H_6S	287-27-4	74.14664	Thiacyclobutane
C_3H_7Br	106-94-5	122.99258	1-Bromopropane
			Propyl bromide
C_3H_7Br	75-26-3	122.99258	2-Bromopropane
			Isopropyl bromide
C_3H_7Cl	540-54-5	78.54128	1-Chloropropane
			Propyl chloride
C_3H_7Cl	75-29-6	78.54128	2-Chloropropane
			Isopropyl chloride
$C_3H_7Cl_3Si$		177.53218	Chloro(dichloromethyl)-dimethylsilane
C_3H_7F	460-13-9	62.0869832	1-Fluoropropane
C_3H_7F	420-26-8	62.0869832	2-Fluoropropane
			Isopropyl fluoride
$C_3H_7{}^2H_2N$		61.1235235	Isopropyl($^2H_2$)amine
$C_3H_7{}^2H_2N$		61.1235235	(2,2-2H2)Propylamine
			Propyl-2,2-d2-amine
C_3H_7I	107-08-4	169.99305	1-Iodopropane
			Propyl iodide
C_3H_7I	75-30-9	169.99305	2-Iodopropane
C_3H_7N	107-11-9	57.09532	Allylamine
			2-Propenylamine
			1-Amino-2-propene
C_3H_7N	503-29-7	57.09532	Azetidine
			Azacyclobutane
			Trimethyleneimine
C_3H_7N	765-30-0	57.09532	Cyclopropylamine
C_3H_7N	75-55-8	57.09532	Propyleneimine
C_3H_7NO	68-12-2	73.09472	N,N-Dimethylmethanamide
			N,N-Dimethylformamide

Formula	CAS RN	M_r	Name
C_3H_7NO (Cont.)			DMF
C_3H_7NO	627-45-2	73.09472	N-Ethylmethanamide
			N-Ethylformamide
C_3H_7NO	79-16-3	73.09472	N-Methylethanamide
			N-Methylacetamide
C_3H_7NO		73.09472	Propanal oxime
			Propionaldehyde oxime
			Propional oxime
			Ethanecarbaldehyde oxime
C_3H_7NO	127-06-0	73.09472	2-Propanone oxime
			Acetone oxime
$C_3H_7NO_2$	56-41-7	89.09412	(S)-(+)-2-Aminopropanoic acid
			L-Alanine
			(S)-(+)-Alanine
$C_3H_7NO_2$	302-72-7	89.09412	DL-2-Aminopropanoic acid
			DL-Alanine
$C_3H_7NO_2$	338-69-2	89.09412	(R)-(−)-2-Aminopropanoic acid
			(R)-(−)-Alanine
			D-Alanine
$C_3H_7NO_2$	51-79-6	89.09412	Ethyl carbamate
			Carbamic acid ethyl ester
			Urethane
$C_3H_7NO_2$	108-03-2	89.09412	1-Nitropropane
$C_3H_7NO_2$	79-46-9	89.09412	2-Nitropropane
$C_3H_7NO_2S$	52-90-4	121.16012	(R)-(+)-Cysteine
$C_3H_7NO_3$	2902-96-7	105.09352	2-Nitro-1-propanol
$C_3H_7NO_3$	3156-73-8	105.09352	1-Nitro-2-propanol
$C_3H_7NO_3$	56-45-1	105.09352	(S)-(+)-Serine
C_3H_8	74-98-6	44.09652	Propane
$C_3H_8Cl_2Si$	1719-57-9	143.08742	Chloro(chloromethyl)dimethylsilane
$C_3H_8{}^2HN$	65363-33-9	60.1173617	N-Methylethyl(^{2}H)amine
			N-Methylethylamine-d
$C_3H_8N_2O$	598-94-7	88.1094	1,1-Dimethylurea
			N,N-Dimethylurea
$C_3H_8N_2O$	96-31-1	88.1094	1,3-Dimethylurea
			N,N'-Dimethylurea
C_3H_8O	540-67-0	60.09592	Ethyl methyl ether
C_3H_8O	71-23-8	60.09592	1-Propanol
			Propyl alcohol
C_3H_8O	67-63-0	60.09592	2-Propanol
			Isopropanol
			Isopropyl alcohol
$C_3H_8O_2$	109-87-5	76.09532	2,4-Dioxapentane
			Dimethoxymethane
			Dimethyl formal
			Formaldehyde dimethyl acetal
			Methylal
$C_3H_8O_2$	109-86-4	76.09532	2-Methoxyethanol
			1,2-Ethanediol methyl ether
			Methyl Cellosolve
			3-Oxa-1-butanol
			Ethylene glycol methyl ether
$C_3H_8O_2$	57-55-6	76.09532	1,2-Propanediol
			Propylene glycol
$C_3H_8O_2$	504-63-2	76.09532	1,3-Propanediol
			Trimethylene glycol
$C_3H_8O_2S$	594-43-4	108.16132	Ethyl methyl sulfone
$C_3H_8O_3$	56-81-5	92.09472	1,2,3-Propanetriol
			Glycerol
$C_3H_8O_3$ (Cont.)			Glycerine
C_3H_8S	107-03-9	76.16252	1-Propanethiol
			Propyl mercaptan
C_3H_8S	75-33-2	76.16252	2-Propanethiol
			Isopropyl mercaptan
C_3H_8S	624-89-5	76.16252	2-Thiabutane
			Ethyl methyl sulfide
$C_3H_8S_2$	109-80-8	108.22852	1,3-Propanedithiol
C_3H_9Al	75-24-1	72.085999	Trimethylaluminum
$C_3H_9AsO_3$	6596-95-8	168.02425	Trimethoxyarsane
			Trimethyl arsenite
			Trimethoxyarsine
C_3H_9B	593-90-8	55.91546	Trimethylborane
$C_3H_9BO_3$	121-43-7	103.91366	Trimethoxyborane
			Methyl borate
			Trimethyl borate
			Boric acid trimethyl ester
$C_3H_9B_3O_3$	823-96-1	125.53566	Trimethylcycloboroxane
			Methylboric anhydride
			Trimethylboroxine
C_3H_9ClSi	75-77-4	108.64266	Chlorotrimethylsilane
C_3H_9Ga	1445-79-0	114.82746	Trimethylgallium
C_3H_9N	75-31-0	59.1112	Isopropylamine
			2-Aminopropane
C_3H_9N	624-78-2	59.1112	N-Methylethylamine
C_3H_9N	107-10-8	59.1112	Propylamine
			1-Aminopropane
C_3H_9N	75-50-3	59.1112	Trimethylamine
C_3H_9NO	156-87-6	75.1106	3-Amino-1-propanol
			Propanolamine
C_3H_9NO	2799-17-9	75.1106	(S)-(+)-1-Amino-2-propanol
			(S)-(+)-Isopropanolamine
C_3H_9NO	78-96-6	75.1106	DL-1-Amino-2-propanol
			DL-Isopropanolamine
C_3H_9NO	109-83-1	75.1106	2-(Methylamino)ethanol
			N-Methylethanolamine
$C_3H_9O_3P$	121-45-9	124.076422	Trimethyl phosphite
			Methyl phosphite
			Phosphorous acid trimethyl ester
$C_3H_{10}N_2$	78-90-0	74.12588	1,2-Propanediamine
			Propylenediamine
			1,2-Diaminopropane
$C_3H_{10}N_2$	109-76-2	74.12588	1,3-Propanediamine
			1,3-Diaminopropane
$C_3H_{10}Si$	993-07-7	74.1979	Trimethylsilane
$C_3H_{11}CuN_2O_3$		186.67802	Copper 2-hydroxypropanoate diammoniate
			Cuprous ammonium lactate
			Copper(I) ammonium lactate
$C_3H_{12}BN$	1830-95-1	72.94602	Ammonia-trimethylborane
$C_3H_{12}BN$	75-22-9	72.94602	Trimethylamine-borane
$C_4Br_2F_8$	335-48-8	359.8392256	1,4-Dibromooctafluorobutane
$C_4Cl_2F_6$	356-18-3	232.9398192	1,2-Dichlorohexafluorocyclobutane
$C_4Cl_3F_7$	335-44-4	287.3909224	2,2,3-Trichloroheptafluorobutane
			Perfluoro(2,2,3-trichlorobutane)
$C_4Cl_4F_4$		265.8484128	1,1,2,2-Tetrachlorotetrafluorocyclobutane
$C_4Cl_4F_6$	423-38-1	303.8452192	1,1,3,4-Tetrachlorohexafluorobutane

Formula	CAS RN	M_r	Name
$C_4Cl_4F_6$ (Cont.)			Perfluoro(1,1,3,4-tetrachloro-butane)
C_4Cl_6	87-68-3	260.7602	Hexachloro-1,3-butadiene
$C_4F_6O_3$	407-25-0	210.0326192	Trifluoroethanoic anhydride
			Trifluoroacetic anhydride
C_4F_8	115-25-3	200.0312256	Octafluorocyclobutane
			Perfluorocyclobutane
C_4F_{10}	355-25-9	238.028032	Decafluorobutane
			Perfluorobutane
C_4F_{10}	354-92-7	238.028032	Decafluoro(2-methylpropane)
			Perfluoroisobutane
			Deacafluoroisobutane
C_4H_2	460-12-8	50.05988	Butadiyne
			Diacetylene
$C_4H_2Cl_2S$	3172-52-9	153.03128	2,5-Dichlorothiophene
$C_4H_2O_3$	108-31-6	98.05808	1-Oxa-3-cyclopenten-2,5-dione
			Maleic anhydride
			cis-Butenedioic anhydride
C_4H_3ClS	96-43-5	118.58652	2-Chlorothiophene
C_4H_4	689-97-4	52.07576	3-Buten-1-yne
			Vinylacetylene
C_4H_4ClN		101.5352	cis-3-Chloro-2-butenenitrile
			cis-3-Chlorocrotonitrile
C_4H_4ClN		101.5352	trans-3-Chloro-2-butenenitrile
			trans-3-Chlorocrotonitrile
$C_4H_4Cl_2$	1653-19-6	122.98116	2,3-Dichloro-1,3-butadiene
$C_4H_4N_2$	110-61-2	80.08924	Butanedinitrile
			1,2-Dicyanoethane
			Succinonitrile
$C_4H_4N_2$	290-37-9	80.08924	Pyrazine
$C_4H_4N_2$	289-80-5	80.08924	Pyridazine
$C_4H_4N_2$	289-95-2	80.08924	Pyrimidine
$C_4H_4N_2O_2$	66-22-8	112.08804	Uracil
			2,4(1-H,3-H)-Pyrimidinedione
C_4H_4O	110-00-9	68.07516	Furan
$C_4H_4O_2$	674-82-8	84.07456	4-Methyleneoxetan-2-one
			Diketene
$C_4H_4O_3$	108-30-5	100.07396	Oxolan-2,5-dione
			Succinic anhydride
$C_4H_4O_4$	110-16-7	116.07336	cis-Butenedioic acid
			Maleic acid
$C_4H_4O_4$	110-17-8	116.07336	trans-Butenedioic acid
			Fumaric acid
C_4H_4S	110-02-1	84.14176	Thiophene
C_4H_5Cl	126-99-8	88.5364	2-Chloro-1,3-butadiene
$C_4H_5Cl_3$	2431-50-7	159.4418	2,3,4-Trichloro-1-butene
$C_4H_5Cl_3O_2$	515-84-4	191.4406	Ethyl trichloroethanoate
			Ethyl trichloroacetate
			Trichloroethanoic acid ethyl ester
			Trichloroacetic acid ethyl ester
$C_4H_5Cl_3O_2$		191.4406	2,2,2-Trichloroethyl ethanoate
			2,2,2-Trichloroethyl acetate
			Ethanoic acid 2,2,2-trichloroethyl ester
			Acetic acid 2,2,2-trichloroethyl ester
C_4H_5N	1190-76-7	67.09044	cis-2-Butenenitrile
C_4H_5N (Cont.)			cis-Crotononitrile
C_4H_5N	627-26-9	67.09044	trans-2-Butenenitrile
			trans-Crotononitrile
C_4H_5N	109-75-1	67.09044	3-Butenenitrile
			Allyl cyanide
			Vinyl acetonitrile
			2-Propenyl cyanide
C_4H_5N	5500-21-0	67.09044	Cyclopropanecarbonitrile
C_4H_5N	126-98-7	67.09044	2-Methylpropenenitrile
			Methacrylonitrile
			2-Propenecarbonitrile
			Isopropenyl cyanide
C_4H_5N	109-97-7	67.09044	Pyrrole
C_4H_5NO	1476-23-9	83.08984	Allyl isocyanate
			Isocyanic acid allyl ester
			2-Propenyl isocyanate
			Isocyanic acid 2-propenyl ester
$C_4H_5NO_2$	105-34-0	99.08924	Methyl cyanoethanoate
			Methyl cyanoacetate
			Cyanoacetic acid methyl ester
			Cyanoethanoic acid methyl ester
C_4H_5NS	57-06-7	99.15644	Allyl isothiocyanate
			Isothiocyanic acid allyl ester
			2-Propenyl isothiocyanate
			Isothiocyanic acid 2-propenyl ester
C_4H_5NS	693-95-8	99.15644	4-Methylthiazole
$C_4H_5N_3O$	71-30-7	111.10332	Cytosine
C_4H_6	590-19-2	54.09164	1,2-Butadiene
C_4H_6	106-99-0	54.09164	1,3-Butadiene
C_4H_6	107-00-6	54.09164	1-Butyne
C_4H_6	503-17-3	54.09164	2-Butyne
			Crotonylene
C_4H_6	822-35-5	54.09164	Cyclobutene
$C_4H_6BaO_4$	543-80-6	255.41624	Barium diethanoate
			Barium acetate
			Ethanoic acid barium salt
			Acetic acid barium salt
$C_4H_6CaO_4$		158.16724	Calcium diethanoate
			Calcium acetate
			Acetic acid calcium salt
			Ethanoic acid calcium salt
$C_4H_6CdO_4$	543-90-8	230.50024	Cadmium diethanoate
			Cadmium acetate
			Acetic acid cadmium salt
			Ethanoic acid cadmium salt
$C_4H_6Cl_2$	926-57-8	124.99704	1,3-Dichloro-2-butene
$C_4H_6Cl_2$	110-57-6	124.99704	trans-1,4-Dichloro-2-butene
$C_4H_6Cl_2O_2$	535-15-9	156.99584	Ethyl dichloroethanoate
			Ethyl dichloroacetate
			Dichloroethanoic acid ethyl ester
			Dichloroacetic acid ethyl ester
$C_4H_6CoO_4$		177.02244	Cobalt diethanoate
			Cobalt(II) acetate
			Ethanoic acid cobalt(II) salt
			Acetic acid cobalt(II) salt
$C_4H_6CuO_4$		181.63524	Copper diethanoate
			Cupric acetate
			Ethanoic acid copper(II) salt
			Acetic acid copper(II) salt

Formula	CAS RN	M_r	Name
$C_4H_6CuO_4$ (Cont.)			Copper(II) acetate
$C_4H_6MgO_4$	142-72-3	142.39424	Magnesium diethanoate
			Magnesium acetate
			Ethanoic acid magnesium salt
			Acetic acid magnesium salt
$C_4H_6MnO_4$	638-38-0	173.02729	Manganese diethanoate
			Manganese(II) acetate
			Acetic acid manganese(II) salt
			Ethanoic acid manganese(II) salt
$C_4H_6N_2$	1453-58-3	82.10512	3-Methylpyrazole
			5-Methylpyrazole
$C_4H_6N_2$		82.10512	5-Methylpyrazole
			3-Methylpyrazole
C_4H_6O	123-73-9	70.09104	trans-2-Butenal
			trans-Crotonaldehyde
C_4H_6O	78-94-4	70.09104	3-Buten-2-one
			Methyl vinyl ketone
C_4H_6O	1191-95-3	70.09104	Cyclobutanone
C_4H_6O	1191-99-7	70.09104	2,3-Dihydrofuran
C_4H_6O	109-93-3	70.09104	Divinyl ether
C_4H_6O	78-85-3	70.09104	2-Methylpropenal
$C_4H_6O_2$	431-03-8	86.09044	2,3-Butanedione
			Diacetyl
$C_4H_6O_2$	503-64-0	86.09044	cis-2-Butenoic acid
			Isocrotonic acid
$C_4H_6O_2$	107-93-7	86.09044	trans-2-Butenoic acid
			Crotonic acid
$C_4H_6O_2$	110-65-6	86.09044	2-Butyn-1,4-diol
$C_4H_6O_2$	96-33-3	86.09044	Methyl propenoate
			Methyl acrylate
			Propenoic acid methyl ester
			Acrylic acid methyl ester
$C_4H_6O_2$	79-41-4	86.09044	2-Methylpropenoic acid
			Methacrylic acid
$C_4H_6O_2$	96-48-0	86.09044	Oxolan-2-one
			4-Butanolide
			η-Butyrolactone
			4-Hydroxybutyric acid η-lactone
			1-Oxacyclopentan-2-one
			Tetrahydro-2-furanone
$C_4H_6O_2$	108-05-4	86.09044	Vinyl ethanoate
			Vinyl acetate
			Ethanoic acid vinyl ester
			Acetic acid vinyl ester
$C_4H_6O_2S$	77-79-2	118.15644	Thiacyclopent-3-ene 1,1-dioxide
			3-Sulfolene
			Butadiene sulfone
			2,5-Dihydrothiophene-1,1-dioxide
$C_4H_6O_3$	108-24-7	102.08984	Ethanoic anhydride
			Acetic anhydride
$C_4H_6O_3$	108-32-7	102.08984	4-Methyl-1,3-dioxolan-2-one
			Propylene carbonate
$C_4H_6O_4$	110-15-6	118.08924	Butanedioic acid
			Succinic acid
$C_4H_6O_4$	553-90-2	118.08924	Dimethyl oxalate
$C_4H_6O_4Zn$	557-34-6	183.47924	Zinc diethanoate
			Zinc acetate
			Ethanoic acid zinc salt
			Acetic acid zinc salt

Formula	CAS RN	M_r	Name
$C_4H_6O_5$	636-61-3	134.08864	(R)-(+)-Hydroxybutanedioic acid
			(R)-(+)-Malic acid
			(R)-(+)-Hydroxysuccinic acid
			D-Malic acid
$C_4H_6O_5$	617-48-1	134.08864	DL-Hydroxybutanedioic acid
			DL-Malic acid
			DL-Hydroxysuccinic acid
C_4H_6S	1120-59-8	86.15764	2,3-Dihydrothiophene
C_4H_6S	1708-32-3	86.15764	2,5-Dihydrothiophene
$C_4H_7BrO_2$	105-36-2	167.00238	Ethyl bromoethanoate
			Ethyl bromoacetate
			Bromoethanoic acid ethyl ester
			Bromoacetic acid ethyl ester
C_4H_7Cl	513-37-1	90.55228	1-Chloro-2-methylpropene
C_4H_7Cl	563-47-3	90.55228	3-Chloro-2-methylpropene
			Methallyl chloride
$C_4H_7ClO_2$	542-58-5	122.55108	2-Chloroethyl ethanoate
			2-Chloroethyl acetate
			Ethanoic acid 2-chloroethyl ester
			Acetic acid 2-chloroethyl ester
$C_4H_7ClO_2$	105-39-5	122.55108	Ethyl chloroethanoate
			Ethyl chloroacetate
			Chloroethanoic acid ethyl ester
			Chloroacetic acid ethyl ester
$C_4H_7ClO_2$	108-23-6	122.55108	Isopropyl chloromethanoate
			Isopropyl chloroformate
			Chloromethanoic acid isopropyl ester
			Chloroformic acid isopropyl ester
$C_4H_7ClO_2$	109-61-5	122.55108	Propyl chloromethanoate
			Propyl chloroformate
			Chloromethanoic acid propyl ester
			Chloroformic acid propyl ester
$C_4H_7Cl_2F_3Si$	675-62-7	211.0856896	Dichloro(3,3,3-trifluoropropyl)methylsilane
$C_4H_7Cl_3O_2$		193.45648	2,2,2-Trichloro-1-ethoxyethanol
$C_4H_7Cl_3O_2Ti$		241.33648	(Butanoato)trichlorotitanium
			(Butyrato)trichlorotitanium
$C_4H_7Cl_3O_2Ti$		241.33648	Trichloro(2-methylpropanoato)titanium
			Trichloro(isobutyrato)titanium
C_4H_7N	109-74-0	69.10632	Butanenitrile
			1-Cyanopropane
			Butanonitrile
			Butyronitrile
			1-Propanecarbonitrile
			Propyl cyanide
C_4H_7N	78-82-0	69.10632	2-Methylpropanenitrile
			Isobutyronitrile
			Isopropyl cyanide
			2-Methylpropionitrile
C_4H_7NO	33695-59-9	85.10572	2-Methoxypropanenitrile
			2-Methoxypropionitrile
C_4H_7NO	79-39-0	85.10572	2-Methylpropenamide
			Methacrylamide

Formula	CAS RN	M_r	Name
C_4H_7NO	616-45-5	85.10572	2-Pyrrolidinone 2-Pyrrolidone
$C_4H_7NO_4$	56-84-8	133.10392	(S)-(+)-Aminobutanedioic acid (S)-(+)-Aspartic acid (S)-(+)-Aminosuccinic acid L-Aspartic acid
$C_4H_7NO_4$		133.10392	2-Vinyloxyethyl nitrate Ethylene glycol vinyl ether nitrate 3-Oxa-4-pentenyl nitrate
C_4H_8	106-98-9	56.10752	1-Butene
C_4H_8	107-01-7	56.10752	2-Butene <undefined cis/trans isomer>
C_4H_8	590-18-1	56.10752	cis-2-Butene
C_4H_8	624-64-6	56.10752	trans-2-Butene
C_4H_8	287-23-0	56.10752	Cyclobutane
C_4H_8	594-11-6	56.10752	Methylcyclopropane
C_4H_8	115-11-7	56.10752	2-Methylpropene Isobutylene
$C_4H_8Br_2$	110-52-1	215.91552	1,4-Dibromobutane
$C_4H_8Cl_2$	616-21-7	127.01292	1,2-Dichlorobutane
$C_4H_8Cl_2$	1190-22-3	127.01292	1,3-Dichlorobutane
$C_4H_8Cl_2$	110-56-5	127.01292	1,4-Dichlorobutane
$C_4H_8Cl_2O$	111-44-4	143.01232	Bis(2-chloroethyl) ether Chlorex 2-Chloroethyl ether
$C_4H_8Cl_2S$	505-60-2	159.07892	1,5-Dichloro-3-thiapentane Bis(2-chloroethyl) sulfide Bis(2-chloroethyl) thioether
$C_4H_8N_2O_2$	95-45-4	116.1198	Dimethylglyoxime
$C_4H_8N_2O_3$	70-47-3	132.1192	L-Asparagine
C_4H_8O	123-72-8	72.10692	Butanal Butyraldehyde 1-Propanecarbaldehyde
C_4H_8O	78-93-3	72.10692	2-Butanone Ethyl methyl ketone MEK
C_4H_8O	598-32-3	72.10692	(±)-3-Buten-2-ol (±)-Methylvinylcarbinol
C_4H_8O	3266-23-7	72.10692	2,3-Dimethyloxirane 2-Butene oxide 2-Butylene oxide 2,3-Epoxybutane 2,3-Epoxybutylene <undefined cis/trans isomer>
C_4H_8O	1758-33-4	72.10692	cis-2,3-Dimethyloxirane cis-2,3-Epoxybutane cis-2-Butene oxide cis-2-Butylene oxide cis-2,3-Epoxybutylene
C_4H_8O	21490-63-1	72.10692	trans-2,3-Dimethyloxirane trans-2-Butylene oxide trans-2-Butene oxide trans-2,3-Epoxybutene trans-2,3-Epoxybutylene
C_4H_8O	106-88-7	72.10692	Ethyloxirane 1,2-Butylene oxide 1,2-Epoxybutane 1,2-Butene oxide 1,2-Epoxybutylene
C_4H_8O	109-92-2	72.10692	Ethyl vinyl ether
C_4H_8O	116-11-0	72.10692	Isopropenyl methyl ether 2-Methoxypropene
C_4H_8O	78-84-2	72.10692	2-Methylpropanal Isobutyraldehyde 2-Methylpropionaldehyde
C_4H_8O	513-42-8	72.10692	2-Methyl-2-propen-1-ol Methallyl alcohol
C_4H_8O	109-99-9	72.10692	Oxolane Tetrahydrofuran THF
C_4H_8OS	1600-44-8	104.17292	Thiolane-1-oxide Tetrahydrothiophene-1-oxide
$C_4H_8O_2$	107-92-6	88.10632	Butanoic acid Butyric acid 1-Propanecarboxylic acid
$C_4H_8O_2$	497-06-3	88.10632	3-Butene-1,2-diol
$C_4H_8O_2$	110-64-5	88.10632	cis-2-Butene-1,4-diol
$C_4H_8O_2$	821-11-4	88.10632	trans-2-Butene-1,4-diol
$C_4H_8O_2$	505-22-6	88.10632	1,3-Dioxane
$C_4H_8O_2$	123-91-1	88.10632	1,4-Dioxane
$C_4H_8O_2$	141-78-6	88.10632	Ethyl ethanoate Ethanoic acid ethyl ester Ethyl acetate Acetic acid ethyl ester
$C_4H_8O_2$	513-86-0	88.10632	3-Hydroxy-2-butanone Acetoin
$C_4H_8O_2$	625-55-8	88.10632	Isopropyl methanoate Isopropyl formate Methanoic acid isopropyl ester Formic acid isopropyl ester
$C_4H_8O_2$	5878-19-3	88.10632	1-Methoxy-2-propanone Methoxyacetone
$C_4H_8O_2$	497-26-7	88.10632	2-Methyl-1,3-dioxolane
$C_4H_8O_2$	554-12-1	88.10632	Methyl propanoate Methyl propionate Propanoic acid methyl ester Propionic acid methyl ester
$C_4H_8O_2$	79-31-2	88.10632	2-Methylpropanoic acid Isobutyric acid 2-Methylpropionic acid Isobutanoic acid
$C_4H_8O_2$	764-48-7	88.10632	3-Oxa-4-penten-1-ol Ethylene glycol vinyl ether
$C_4H_8O_2$	110-74-7	88.10632	Propyl methanoate Methanoic acid propyl ester Propyl formate Formic acid propyl ester
$C_4H_8O_2S$	126-33-0	120.17232	Thiolane 1,1-dioxide Sulfolane Tetramethylene sulfone Tetrahydrothiophene 1,1-dioxide
$C_4H_8O_2S$	109-03-5	120.17232	1,4-Thioxane-1-oxide 1,4-Oxathiane-4-oxide
$C_4H_8O_3$	594-61-6	104.10572	2-Hydroxy-2-methylpropanoic acid 2-Hydroxyisobutyric acid 2-Methyllactic acid 2-Hydroxyisobutanoic acid
$C_4H_8O_3$	6290-49-9	104.10572	Methyl methoxyethanoate Methyl methoxyacetate Methoxyethanoic acid methyl ester

Formula	CAS RN	M_r	Name
$C_4H_8O_3$ (Cont.)			Methoxyacetic acid methyl ester
C_4H_8S	110-01-0	88.17352	Thiacyclopentane
			Tetrahydrothiophene
$C_4H_9BCl_2O$		154.83126	Butoxydichloroborane
			Butyl dichloroboronate
C_4H_9Br	109-65-9	137.01946	1-Bromobutane
			Butyl bromide
C_4H_9Br	78-76-2	137.01946	2-Bromobutane
C_4H_9Br	78-77-3	137.01946	1-Bromo-2-methylpropane
C_4H_9Br	507-19-7	137.01946	2-Bromo-2-methylpropane
			tert-Butyl bromide
C_4H_9Cl	109-69-3	92.56816	1-Chlorobutane
			Butyl chloride
C_4H_9Cl	78-86-4	92.56816	2-Chlorobutane
			sec-Butyl chloride
C_4H_9Cl	513-36-0	92.56816	1-Chloro-2-methylpropane
			Isobutyl chloride
C_4H_9Cl	507-20-0	92.56816	2-Chloro-2-methylpropane
			tert-Butyl chloride
$C_4H_9ClO_2$	628-89-7	124.56696	5-Chloro-3-oxa-1-pentanol
			2-(2-Chloroethoxy)ethanol
C_4H_9I	542-69-8	184.01993	1-Iodobutane
			Butyl iodide
C_4H_9I	513-48-4	184.01993	2-Iodobutane
C_4H_9I	513-38-2	184.01993	1-Iodo-2-methylpropane
C_4H_9I	558-17-8	184.01993	2-Iodo-2-methylpropane
C_4H_9N	83-34-1	71.1222	3-Methylindole
			Skatole
C_4H_9N	123-75-1	71.1222	Pyrrolidine
			Azacyclopentane
			Azolidine
			Perhydroazole
			Perhydropyrrole
			Tetrahydroazole
			Tetrahydropyrrole
			Tetramethyleneimine
C_4H_9NO	110-69-0	87.1216	Butanal oxime
			Butyraldehyde oxime
C_4H_9NO	96-29-7	87.1216	2-Butanone oxime
			Ethyl methyl ketone oxime
C_4H_9NO	127-19-5	87.1216	N,N-Dimethylethanamide
			N,N-Dimethylacetamide
C_4H_9NO	625-50-3	87.1216	N-Ethylethanamide
			N-Ethylacetamide
C_4H_9NO	110-91-8	87.1216	Morpholine
			Tetrahydro-1,4-oxazine
$C_4H_9NO_2$	627-05-4	103.121	1-Nitrobutane
$C_4H_9NO_3$	609-31-4	119.1204	2-Nitro-1-butanol
$C_4H_9NO_3$	72-19-5	119.1204	L-Threonine
$C_4H_9NO_4$	77-49-6	135.1198	2-Methyl-2-nitro-1,3-propanediol
			1,1-Bis(hydroxymethyl)nitroethane
C_4H_{10}	106-97-8	58.1234	Butane
C_4H_{10}	75-28-5	58.1234	2-Methylpropane
			Isobutane
C_4H_{10}		58.1234	2-Methylpropane (c,I)
			Isobutane (c,I)
C_4H_{10}		58.1234	2-Methylpropane (c,II)
			Isobutane (c,II)
$C_4H_{10}{}^2HN$	997-11-5	74.1442417	Diethyl(^{2}H)amine
$C_4H_{10}{}^2HN$ (Cont.)			Diethylamine-d
$C_4H_{10}Hg$	627-44-1	258.7134	Diethyl mercury
$C_4H_{10}O$	71-36-3	74.1228	1-Butanol
			Butyl alcohol
$C_4H_{10}O$	78-92-2	74.1228	2-Butanol
			sec-Butanol
			sec-Butyl alcohol
$C_4H_{10}O$	60-29-7	74.1228	Diethyl ether
			Ethoxyethane
			Ethyl ether
			Ether
$C_4H_{10}O$	598-53-8	74.1228	Isopropyl methyl ether
$C_4H_{10}O$	78-83-1	74.1228	2-Methyl-1-propanol
			Isobutanol
			Isobutyl alcohol
$C_4H_{10}O$	75-65-0	74.1228	2-Methyl-2-propanol
			tert-Butanol
			tert-Butyl alcohol
$C_4H_{10}O$	557-17-5	74.1228	Methyl propyl ether
$C_4H_{10}OS$	70-29-1	106.1888	Diethyl sulfoxide
$C_4H_{10}O_2$		90.1222	1,3-Butanediol
			<undefined optical isomer>
$C_4H_{10}O_2$	110-63-4	90.1222	1,4-Butanediol
$C_4H_{10}O_2$	513-85-9	90.1222	2,3-Butanediol
			2,3-Butylene glycol
			<undefined optical isomer>
$C_4H_{10}O_2$	24347-58-8	90.1222	(^{2}R,^{3}R)-(−)-2,3-Butanediol
			(−)-2,3-Butylene glycol
			L-2,3-Butylene glycol
$C_4H_{10}O_2$	5341-95-7	90.1222	meso-2,3-Butanediol
			meso-2,3-Butylene glycol
			(R*,S*)-2,3-Butanediol
$C_4H_{10}O_2$	75-91-2	90.1222	tert-Butyl hydroperoxide
$C_4H_{10}O_2$	534-15-6	90.1222	1,1-Dimethoxyethane
			Acetaldehyde dimethyl acetal
			Dimethyl acetal
			3-Methyl-2,4-dioxapentane
$C_4H_{10}O_2$	110-71-4	90.1222	2,5-Dioxahexane
			1,2-Dimethoxyethane
			1,2-Ethanediol dimethyl ether
			Ethylene glycol dimethyl ether
$C_4H_{10}O_2$	110-80-5	90.1222	2-Ethoxyethanol
			Cellosolve
			1,2-Ethanediol ethyl ether
			Ethyl Cellosolve
			3-Oxa-1-pentanol
			Ethylene glycol ethyl ether
$C_4H_{10}O_2S$	597-35-3	122.1882	Diethyl sulfone
			Ethyl sulfone
$C_4H_{10}O_2S$	111-48-8	122.1882	3-Thiapentane-1,5-diol
			Thiodiethylene glycol
			2,2'-Thiodiethanol
			2-Hydroxyethyl sulfide
$C_4H_{10}O_3$	111-46-6	106.1216	3-Oxa-1,5-pentanediol
			Bis(2-hydroxyethyl) ether
			Diethylene glycol
			2-Hydroxyethyl ether
			2,2'-Oxydiethanol
$C_4H_{10}S$	109-79-5	90.1894	1-Butanethiol
			Butyl mercaptan
$C_4H_{10}S$	513-53-1	90.1894	2-Butanethiol
$C_4H_{10}S$	1551-21-9	90.1894	Isopropyl methyl sulfide

Formula	CAS RN	M_r	Name
$C_4H_{10}S$	513-44-0	90.1894	2-Methyl-1-propanethiol Isobutyl mercaptan
$C_4H_{10}S$	75-66-1	90.1894	2-Methyl-2-propanethiol
$C_4H_{10}S$	3877-15-4	90.1894	Methyl propyl sulfide
$C_4H_{10}S$	352-93-2	90.1894	3-Thiapentane Diethyl sulfide Diethyl thioether
$C_4H_{10}S_2$	4532-64-3	122.2554	1,4-Butanedithiol
$C_4H_{10}S_2$	110-81-6	122.2554	3,4-Dithiahexane Diethyl disulfide
$C_4H_{10}S_3$	5418-86-0	154.3214	Tris(methylthio)methane Trimethyl orthothioformate Trithiomethoxymethane
$C_4H_{10}Te$		185.7234	Diethyltellurium
$C_4H_{10}Zn$	557-20-0	123.5134	Diethylzinc
$C_4H_{11}BrSi$	18243-41-9	167.12084	Bromomethyltrimethylsilane
$C_4H_{11}N$	13952-84-6	73.13808	2-Aminobutane sec-Butylamine
$C_4H_{11}N$	33966-50-6	73.13808	(±)-2-Aminobutane (±)-sec-Butylamine
$C_4H_{11}N$	513-49-5	73.13808	(S)-(+)-2-Aminobutane (S)-(+)-sec-Butylamine
$C_4H_{11}N$	13250-12-9	73.13808	(R)-(−)-2-Aminobutane (R)-(−)-sec-Butylamine
$C_4H_{11}N$	109-73-9	73.13808	Butylamine 1-Aminobutane
$C_4H_{11}N$	75-64-9	73.13808	tert-Butylamine 2-Amino-2-methylpropane
$C_4H_{11}N$	109-89-7	73.13808	Diethylamine
$C_4H_{11}N$	598-56-1	73.13808	N,N-Dimethylethylamine
$C_4H_{11}N$	4747-21-1	73.13808	Isopropylmethylamine
$C_4H_{11}N$	78-81-9	73.13808	(2-Methylpropyl)amine 1-Amino-2-methylpropane Isobutylamine
$C_4H_{11}NO$	124-68-5	89.13748	2-Amino-2-methyl-1-propanol
$C_4H_{11}NO$	108-01-0	89.13748	2-(Dimethylamino)ethanol N,N-Dimethylethanolamine
$C_4H_{11}NO_2$	929-06-6	105.13688	2-(2-Aminoethoxy)ethanol
$C_4H_{11}NO_2$	111-42-2	105.13688	Bis(2-hydroxyethyl)amine Diethanolamine 2,2'-Iminodiethanol
$C_4H_{11}NO_2S$		137.20288	N,N-Dimethylethanesulfonamide
$C_4H_{11}O_4P$		154.102702	Butyl dihydrogen phosphate Phosphoric acid butyl ester Butyl phosphate
$C_4H_{12}BrN$	64-20-0	154.05002	Tetramethylammonium bromide
$C_4H_{12}ClN$	3858-78-4	109.59872	Butylamine hydrochloride
$C_4H_{12}ClN$	660-68-4	109.59872	Diethylamine hydrochloride
$C_4H_{12}ClN$	75-57-0	109.59872	Tetramethylammonium chloride
$C_4H_{12}IN$	75-58-1	201.05049	Tetramethylammonium iodide
$C_4H_{12}N_2$	6415-12-9	88.15276	Tetramethylhydrazine
$C_4H_{12}O_4Si$	681-84-5	152.22238	Tetramethoxysilane Tetramethyl orthosilicate
$C_4H_{12}Pb$	75-74-1	267.33928	Tetramethyllead Tetramethylplumbane
$C_4H_{12}Si$	1600-29-9	88.22478	Butylsilane
$C_4H_{12}Si$	75-76-3	88.22478	Tetramethylsilane
$C_4H_{12}Sn$	594-27-4	178.84928	Tetramethylstannane Tetramethyltin
$C_4H_{13}NO$	75-59-2	91.15336	Tetramethylammonium hydroxide
$C_4H_{13}N_3$	111-40-0	103.16744	Bis(2-aminoethyl)amine Diethylenetriamine
C_4NiO_4	13463-39-3	170.735	Nickel tetracarbonyl Nickel carbonyl
$C_5Cl_2F_6$	706-79-6	244.9508192	1,2-Dichlorohexafluorocyclopentene
C_5Cl_6	77-47-4	272.7712	Hexachlorocyclopentadiene
C_5F_{10}	376-77-2	250.039032	Decafluorocyclopentane Perfluorocyclopentane
$C_5F_{10}O$	1623-05-8	266.038432	Decafluoro(propyl vinyl ether) Perfluoro(propyl vinyl ether)
C_5F_{12}	678-26-2	288.0358384	Dodecafluoropentane Perfluoropentane
C_5FeO_5	13463-40-6	195.899	Iron pentacarbonyl Iron carbonyl
$C_5H_2F_6O_2$	1522-22-1	208.0600992	1,1,1,5,5,5-Hexafluoroacetylacetone 1,1,1,5,5,5-Hexafluoro-2,4-pentadione
C_5H_4		64.08676	Bicyclo[1.1.1]penta-1,3-diene
$C_5H_4N_2O_4$	65-86-1	156.09784	2,6-Dioxo-1,2,3,6-tetrahydro-4-pyrimidinecarboxylic acid Orotic acid
$C_5H_4N_4O$	68-94-0	136.11312	Hypoxanthine
$C_5H_4N_4O_2$	69-89-6	152.11252	Xanthine
$C_5H_4N_4O_3$	69-93-2	168.11192	Uric acid
$C_5H_4O_2$	98-01-1	96.08556	2-Furaldehyde Furfural Furfuraldehyde
$C_5H_5MnO_3$		168.03095	Ethylmanganese tricarbonyl
C_5H_5N	110-86-1	79.10144	Pyridine Azine
$C_5H_5N_5$	73-24-5	135.1284	6-Aminopurine Adenine
$C_5H_5N_5O$	73-40-5	151.1278	Guanine
$C_5H_5N_5O$	3373-53-3	151.1278	Isoguanine
C_5H_6	542-92-7	66.10264	1,3-Cyclopentadiene
C_5H_6	78-80-8	66.10264	2-Methyl-1-buten-3-yne Isopropenylacetylene
C_5H_6	1574-40-9	66.10264	cis-3-Penten-1-yne
C_5H_6	2004-69-5	66.10264	trans-3-Penten-1-yne
C_5H_6	646-05-9	66.10264	1-Penten-3-yne
C_5H_6	871-28-3	66.10264	1-Penten-4-yne
$C_5H_6N_2$	109-08-0	94.11612	2-Methylpyrazine
$C_5H_6N_2$	544-13-8	94.11612	Pentanedinitrile Glutaronitrile 1,3-Dicyanopropane
$C_5H_6N_2O_2$	65-71-4	126.11492	Thymine 5-Methyluracil 5-Methyl-2,4(1-H,3-H)-pyrimidinedione
C_5H_6O	534-22-5	82.10204	2-Methylfuran
C_5H_6O	930-27-8	82.10204	3-Methylfuran
$C_5H_6O_2$	98-00-0	98.10144	2-Furylmethanol Furfuryl alcohol
C_5H_6S	554-14-3	98.16864	2-Methylthiophene

Formula	CAS RN	M_r	Name
C_5H_6S	616-44-4	98.16864	3-Methylthiophene
C_5H_7N	4426-11-3	81.11732	Cyclobutanecarbonitrile
C_5H_7N	16545-78-1	81.11732	trans-3-Pentenenitrile
$C_5H_7N_3O$	554-01-8	125.1302	5-Methylcytosine
$C_5H_7N_3O_2$	1123-95-1	141.1296	5-Hydroxymethylcytosine
C_5H_8	142-29-0	68.11852	Cyclopentene
C_5H_8	78-79-5	68.11852	2-Methyl-1,3-butadiene
			Isoprene
C_5H_8	598-23-2	68.11852	3-Methyl-1-butyne
C_5H_8	1489-60-7	68.11852	1-Methylcyclobutene
C_5H_8	1120-56-5	68.11852	Methylenecyclobutane
C_5H_8	504-60-9	68.11852	1,3-Pentadiene
			Piperylene
			<mixture of cis/trans isomers>
C_5H_8	1574-41-0	68.11852	cis-1,3-Pentadiene
			cis-Piperylene
C_5H_8	2004-70-8	68.11852	trans-1,3-Pentadiene
			trans-Piperylene
C_5H_8	591-93-5	68.11852	1,4-Pentadiene
C_5H_8	627-19-0	68.11852	1-Pentyne
C_5H_8	627-21-4	68.11852	2-Pentyne
C_5H_8	157-40-4	68.11852	Spiropentane
$C_5H_8Cl_4$		209.92932	Tetrachloropentane
			<undefined positional isomer>
$C_5H_8Cl_4$	2467-10-9	209.92932	1,1,1,5-Tetrachloropentane
$C_5H_8N_2$		96.132	1,3-Dimethylpyrazole
$C_5H_8N_2$		96.132	1,5-Dimethylpyrazole
C_5H_8O	120-92-3	84.11792	Cyclopentanone
C_5H_8O	765-43-5	84.11792	Cyclopropyl methyl ketone
			Acetylcyclopropane
C_5H_8O	110-87-2	84.11792	3,4-Dihydro-2-H-pyran
C_5H_8O	115-19-5	84.11792	2-Methyl-3-butyn-2-ol
$C_5H_8O_2$	591-87-7	100.11732	Allyl ethanoate
			Allyl acetate
			Acetic acid allyl ester
			Ethanoic acid allyl ester
			2-Propenyl ethanoate
			Ethanoic acid 2-propenyl ester
$C_5H_8O_2$	140-88-5	100.11732	Ethyl propenoate
			Ethyl acrylate
			Propenoic acid ethyl ester
			Acrylic acid ethyl ester
$C_5H_8O_2$	108-22-5	100.11732	Isopropenyl ethanoate
			Isopropenyl acetate
			Ethanoic acid isopropenyl ester
			Acetic acid isopropenyl ester
$C_5H_8O_2$	2868-37-3	100.11732	Methyl cyclopropanecar-boxylate
$C_5H_8O_2$	80-62-6	100.11732	Methyl 2-methylpropenoate
			Methyl methacrylate
			2-Methylpropenoic acid methyl ester
			Methacrylic acid methyl ester
$C_5H_8O_2$	123-54-6	100.11732	2,4-Pentanedione
			Acetylacetone
$C_5H_8O_2$	105-38-4	100.11732	Vinyl propanoate
			Vinyl propionate
			Propanoic acid vinyl ester
			Propionic acid vinyl ester
$C_5H_8O_3$	818-61-1	116.11672	2-Hydroxyethyl propenoate
			Ethylene glycol acrylate
$C_5H_8O_3$	123-76-2	116.11672	4-Oxopentanoic acid
$C_5H_8O_3$ (Cont.)			Levulinic acid
			4-Oxovaleric acid
$C_5H_8O_4$	108-59-8	132.11612	Dimethyl propanedioate
			Dimethyl malonate
			Propanedioic acid dimethyl ester
			Malonic acid dimethyl ester
$C_5H_8O_4$	498-21-5	132.11612	Methylbutanedioic acid
			Methylsuccinic acid
$C_5H_8O_4$	628-51-3	132.11612	Methylene diethanoate
			Methylene diacetate
$C_5H_8O_4$	110-94-1	132.11612	Pentanedioic acid
			Glutaric acid
$C_5H_9ClO_2$	592-34-7	136.57796	Butyl chloromethanoate
			Chloroformic acid butyl ester
			Butyl chloroformate
			Chloromethanoic acid butyl ester
$C_5H_9ClO_3$		152.57736	2-Chloroethyl ethyl carbo-nate
			Carbonic acid 2-chloroethyl ethyl ester
$C_5H_9Cl_3O$	813-99-0	191.48396	2,2,2-Tris(chloromethyl)-ethanol
			Pentaerythritol trichlorohydrin
$C_5H_9Cl_3O_2Ti$		255.36336	Trichloro(2,2-dimethylpro-panoato)titanium
			Trichloro(trimethylacetato)tita-nium
C_5H_9N	630-18-2	83.1332	2,2-Dimethylpropanenitrile
			Pivalonitrile
			Trimethylacetonitrile
C_5H_9N	18936-17-9	83.1332	2-Methylbutanenitrile
C_5H_9N	625-28-5	83.1332	3-Methylbutanenitrile
C_5H_9N	110-59-8	83.1332	Pentanenitrile
C_5H_9NO	111-36-4	99.1326	Butyl isocyanate
			Isocyanic acid butyl ester
C_5H_9NO	872-50-4	99.1326	1-Methyl-2-pyrrolidinone
			1-Methyl-2-pyrrolidone
			N-Methylpyrrolidone
			N-Methyl-2-pyrrolidinone
$C_5H_9NO_2$	4394-85-8	115.132	N-Formylmorpholine
$C_5H_9NO_2$	609-36-9	115.132	DL-2-Pyrrolidinecarboxylic acid
			DL-Proline
$C_5H_9NO_2$	147-85-3	115.132	(S)-(−)-2-Pyrrolidinecar-boxylic acid
			L-Proline
$C_5H_9NO_3$	51-35-4	131.1314	(^{2}S,^{4}R)-(−)-4-Hydroxy-2-pyrrolidinecarboxylic acid
			trans-4-Hydroxy-L-proline
$C_5H_9NO_4$	56-86-0	147.1308	(S)-(+)-2-Aminopentane-dioic acid
			(S)-(+)-Glutamic acid
			L-Glutamic acid
$C_5H_9N_3$	51-45-6	111.14668	2-(4-Imidazolyl)ethylamine
			Histamine
C_5H_{10}	287-92-3	70.1344	Cyclopentane
C_5H_{10}	563-46-2	70.1344	2-Methyl-1-butene
C_5H_{10}	563-45-1	70.1344	3-Methyl-1-butene
C_5H_{10}	513-35-9	70.1344	2-Methyl-2-butene
			β-Isoamylene

Formula	CAS RN	M_r	Name
C_5H_{10}	598-61-8	70.1344	Methylcyclobutane
C_5H_{10}	109-67-1	70.1344	1-Pentene
C_5H_{10}	109-68-2	70.1344	2-Pentene <mixture of cis/trans isomers>
C_5H_{10}	627-20-3	70.1344	cis-2-Pentene
C_5H_{10}	646-04-8	70.1344	trans-2-Pentene
$C_5H_{10}Br_2$	111-24-0	229.9424	1,5-Dibromopentane
$C_5H_{10}ClNO$	88-10-8	135.59324	Diethylcarbamyl chloride Diethylcarbamoyl chloride
$C_5H_{10}Cl_2$	1674-33-5	141.0398	1,2-Dichloropentane
$C_5H_{10}Cl_2$	628-76-2	141.0398	1,5-Dichloropentane
$C_5H_{10}I_2$	628-77-3	323.94334	1,5-Diiodopentane
$C_5H_{10}N_2$	1738-25-6	98.14788	3-(Dimethylamino)propane-nitrile 3-(Dimethylamino)propionitrile
$C_5H_{10}N_2O$		114.14728	Dimethylimidazolidinone <undefined positional isomer>
$C_5H_{10}N_2O$	80-73-9	114.14728	1,3-Dimethyl-2-imidazolidi-none
$C_5H_{10}N_2O_3$	56-85-9	146.14608	L-Glutamine
$C_5H_{10}O$	557-31-3	86.1338	Allyl ethyl ether Ethyl 2-propenyl ether
$C_5H_{10}O$	96-41-3	86.1338	Cyclopentanol Cyclopentyl alcohol
$C_5H_{10}O$	6921-35-3	86.1338	3,3-Dimethyloxetane
$C_5H_{10}O$	630-19-3	86.1338	2,2-Dimethylpropanal 2,2-Dimethylpropionaldehyde Pivalaldehyde Trimethylacetaldehyde
$C_5H_{10}O$	96-17-3	86.1338	2-Methylbutanal 2-Methylbutyraldehyde
$C_5H_{10}O$	590-86-3	86.1338	3-Methylbutanal Isovaleraldehyde 3-Methylbutyraldehyde
$C_5H_{10}O$	563-80-4	86.1338	3-Methyl-2-butanone Isopropyl methyl ketone
$C_5H_{10}O$	556-82-1	86.1338	3-Methyl-2-buten-1-ol
$C_5H_{10}O$	763-32-6	86.1338	3-Methyl-3-buten-1-ol
$C_5H_{10}O$	115-18-4	86.1338	2-Methyl-3-buten-2-ol
$C_5H_{10}O$	10473-14-0	86.1338	3-Methyl-3-buten-2-ol
$C_5H_{10}O$	96-47-9	86.1338	2-Methyloxolane 2-Methyltetrahydrofuran
$C_5H_{10}O$	142-68-7	86.1338	Oxane Tetrahydropyran
$C_5H_{10}O$	110-62-3	86.1338	Pentanal 1-Butanecarbaldehyde Valeraldehyde
$C_5H_{10}O$	107-87-9	86.1338	2-Pentanone Methyl propyl ketone
$C_5H_{10}O$	96-22-0	86.1338	3-Pentanone Diethyl ketone
$C_5H_{10}O_2$	592-84-7	102.1332	Butyl methanoate Butyl formate Methanoic acid butyl ester Formic acid butyl ester
$C_5H_{10}O_2$	762-75-4	102.1332	tert-Butyl methanoate tert-Butyl formate Formic acid tert-butyl ester Methanoic acid tert-butyl ester
$C_5H_{10}O_2$	75-98-9	102.1332	2,2-Dimethylpropanoic acid Trimethylacetic acid Pivalic acid
$C_5H_{10}O_2$	105-37-3	102.1332	Ethyl propanoate Ethyl propionate Propanoic acid ethyl ester Propionic acid ethyl ester
$C_5H_{10}O_2$	115-22-0	102.1332	3-Hydroxy-3-methyl-2-buta-none
$C_5H_{10}O_2$	1071-73-4	102.1332	5-Hydroxy-2-pentanone 3-Acetyl-1-propanol
$C_5H_{10}O_2$	108-21-4	102.1332	Isopropyl ethanoate Isopropyl acetate Ethanoic acid isopropyl ester Acetic acid isopropyl ester
$C_5H_{10}O_2$	623-42-7	102.1332	Methyl butanoate Butanoic acid methyl ester Methyl butyrate Butyric acid methyl ester
$C_5H_{10}O_2$	116-53-0	102.1332	2-Methylbutanoic acid
$C_5H_{10}O_2$	503-74-2	102.1332	3-Methylbutanoic acid Isovaleric acid
$C_5H_{10}O_2$	547-63-7	102.1332	Methyl 2-methylpropanoate Methyl isobutyrate 2-Methylpropanoic acid methyl ester Isobutyric acid methyl ester Methyl isobutanoate Isobutanoic acid methyl ester
$C_5H_{10}O_2$	542-55-2	102.1332	2-Methylpropyl methanoate Isobutyl formate Methanoic acid isobutyl ester Formic acid isobutyl ester Isobutyl methanoate
$C_5H_{10}O_2$	109-52-4	102.1332	Pentanoic acid Valeric acid 1-Butanecarboxylic acid
$C_5H_{10}O_2$	109-60-4	102.1332	Propyl ethanoate Ethanoic acid propyl ester Propyl acetate Acetic acid propyl ester
$C_5H_{10}O_2$	97-99-4	102.1332	2-Tetrahydrofurylmethanol Tetrahydrofurfuryl alcohol
$C_5H_{10}O_2S$	872-93-5	134.1992	3-Methylsulfolane
$C_5H_{10}O_3$	105-58-8	118.1326	Diethyl carbonate Carbonic acid diethyl ester
$C_5H_{10}O_3$	110-49-6	118.1326	2-Methoxyethyl ethanoate 2-Methoxyethyl acetate Ethanoic acid 2-methoxyethyl ester Acetic acid 2-methoxyethyl ester
$C_5H_{10}O_3$	2110-78-3	118.1326	Methyl 2-hydroxy-2-methyl-propanoate Methyl 2-hydroxyisobutyrate Methyl 2-methyllactate 2-Methyllactic acid methyl ester 2-Hydroxy-2-methylpropanoic acid methyl ester 2-Hydroxyisobutyric acid methyl ester
$C_5H_{10}O_3$		118.1326	Methyl 2-methoxypropa-noate Methyl 2-methoxypropionate 2-Methoxypropanoic acid methyl ester

Formula	CAS RN	M_r	Name
$C_5H_{10}O_3$ (Cont.)			2-Methoxypropionic acid methyl ester
$C_5H_{10}O_5$		150.1314	Xylose <undefined optical isomer>
$C_5H_{10}O_5$	58-86-6	150.1314	D-Xylose
$C_5H_{10}O_5$	25990-60-7	150.1314	DL-Xylose
$C_5H_{10}O_5$	609-06-3	150.1314	L-Xylose
$C_5H_{10}S$	1679-07-8	102.2004	Cyclopentanethiol
$C_5H_{10}S$	1613-51-0	102.2004	Thiacyclohexane
$C_5H_{11}Br$	630-17-1	151.04634	1-Bromo-2,2-dimethylpropane Neopentyl bromide 2,2-Dimethylpropyl bromide
$C_5H_{11}Br$	110-53-2	151.04634	1-Bromopentane Amyl bromide Pentyl bromide
$C_5H_{11}Br$	1809-10-5	151.04634	3-Bromopentane
$C_5H_{11}Cl$	107-84-6	106.59504	1-Chloro-3-methylbutane
$C_5H_{11}Cl$	543-59-9	106.59504	1-Chloropentane Amyl chloride Pentyl chloride
$C_5H_{11}Cl$	625-29-6	106.59504	2-Chloropentane
$C_5H_{11}F$	592-50-7	90.1407432	1-Fluoropentane Amyl fluoride Pentyl fluoride
$C_5H_{11}I$	628-17-1	198.04681	1-Iodopentane Pentyl iodide Amyl iodide
$C_5H_{11}N$	1003-03-8	85.14908	Cyclopentylamine
$C_5H_{11}N$	120-94-5	85.14908	N-Methylpyrrolidine
$C_5H_{11}N$	110-89-4	85.14908	Piperidine Azacyclohexane Hexahydroazine Hexahydropyridine Pentamethyleneimine Perhydroazine Perhydropyridine
$C_5H_{11}NO$	617-84-5	101.14848	N,N-Diethylmethanamide N,N-Diethylformamide
$C_5H_{11}NO$	15364-56-4	101.14848	1-Dimethylamino-2-propanone Dimethylaminoacetone
$C_5H_{11}NO$	109-02-4	101.14848	N-Methylmorpholine 4-Methylmorpholine
$C_5H_{11}NO$		101.14848	3-Pentanone oxime Diethyl ketone oxime
$C_5H_{11}NO_2$	72-18-4	117.14788	(S)-(+)-2-Amino-3-methylbutanoic acid L-Valine (S)-(+)-Valine
$C_5H_{11}NO_2$	640-68-6	117.14788	(R)-(−)-2-Amino-3-methylbutanoic acid (R)-(−)-Valine D-Valine
$C_5H_{11}NO_2$	623-78-9	117.14788	Ethyl N-ethylcarbamate N-Ethylurethane
$C_5H_{11}NO_2$	7529-22-8	117.14788	4-Methylmorpholine N-oxide
$C_5H_{11}NO_2$	628-05-7	117.14788	1-Nitropentane
$C_5H_{11}NO_2S$	63-68-3	149.21388	L-Methionine
$C_5H_{11}NO_3$	543-87-3	133.14728	3-Methylbutyl nitrate Isoamyl nitrate
$C_5H_{11}NO_3$ (Cont.)			Isopentyl nitrate
C_5H_{12}	463-82-1	72.15028	2,2-Dimethylpropane Neopentane
C_5H_{12}	78-78-4	72.15028	2-Methylbutane Isopentane
C_5H_{12}	109-66-0	72.15028	Pentane
$C_5H_{12}N_2O$	632-22-4	116.16316	Tetramethylurea
$C_5H_{12}O$	628-28-4	88.14968	Butyl methyl ether
$C_5H_{12}O$	1634-04-4	88.14968	tert-Butyl methyl ether MTBE
$C_5H_{12}O$	75-84-3	88.14968	2,2-Dimethyl-1-propanol
$C_5H_{12}O$	625-54-7	88.14968	Ethyl isopropyl ether
$C_5H_{12}O$	628-32-0	88.14968	Ethyl propyl ether
$C_5H_{12}O$	137-32-6	88.14968	2-Methyl-1-butanol
$C_5H_{12}O$	123-51-3	88.14968	3-Methyl-1-butanol Isoamyl alcohol Isopentanol Isopentyl alcohol
$C_5H_{12}O$	75-85-4	88.14968	2-Methyl-2-butanol tert-Pentyl alcohol tert-Amyl alcohol
$C_5H_{12}O$	598-75-4	88.14968	3-Methyl-2-butanol
$C_5H_{12}O$	6795-87-5	88.14968	Methyl 1-methylpropyl ether sec-Butyl methyl ether
$C_5H_{12}O$	625-44-5	88.14968	Methyl 2-methylpropyl ether Isobutyl methyl ether
$C_5H_{12}O$	71-41-0	88.14968	1-Pentanol Amyl alcohol Pentyl alcohol
$C_5H_{12}O$	6032-29-7	88.14968	2-Pentanol
$C_5H_{12}O$	584-02-1	88.14968	3-Pentanol
$C_5H_{12}O_2$	77-76-9	104.14908	2,2-Dimethoxypropane Acetone dimethyl acetal 3,3-Dimethyl-2,4-dioxapentane
$C_5H_{12}O_2$		104.14908	1,1-Dimethylpropyl hydroperoxide tert-Amyl hydroperoxide tert-Pentyl hydroperoxide
$C_5H_{12}O_2$		104.14908	2,5-Dioxaheptane Ethylene glycol ethyl methyl ether
$C_5H_{12}O_2$	462-95-3	104.14908	3,5-Dioxaheptane Diethoxymethane Diethyl formal Ethylal Formaldehyde diethyl acetal
$C_5H_{12}O_2$	5137-45-1	104.14908	1-Ethoxy-2-methoxyethane
$C_5H_{12}O_2$	2568-33-4	104.14908	3-Methyl-1,3-butanediol
$C_5H_{12}O_2$	7778-85-0	104.14908	3-Methyl-2,5-dioxahexane Propylene glycol dimethyl ether 1,2-Dimethoxypropane
$C_5H_{12}O_2$	109-59-1	104.14908	4-Methyl-3-oxa-1-pentanol Ethylene glycol isopropyl ether 2-Isopropoxyethanol
$C_5H_{12}O_2$	111-29-5	104.14908	1,5-Pentanediol
$C_5H_{12}O_2$	2807-30-9	104.14908	2-Propoxyethanol
$C_5H_{12}O_3$	111-77-3	120.14848	3,6-Dioxa-1-heptanol 2-(2-Methoxyethoxy)ethanol Diethylene glycol methyl ether
$C_5H_{12}O_4$	115-77-5	136.14788	Pentaerythritol
$C_5H_{12}O_5$	87-99-0	152.14728	Xylitol
$C_5H_{12}S$	628-29-5	104.21628	Butyl methyl sulfide

SUBSTANCE LIST (continued)

Formula	CAS RN	M_r	Name
$C_5H_{12}S$	6163-64-0	104.21628	tert-Butyl methyl sulfide
$C_5H_{12}S$	5145-99-3	104.21628	Ethyl isopropyl sulfide
$C_5H_{12}S$	4110-50-3	104.21628	Ethyl propyl sulfide
$C_5H_{12}S$	1878-18-8	104.21628	2-Methyl-1-butanethiol
$C_5H_{12}S$	541-31-1	104.21628	3-Methyl-1-butanethiol
$C_5H_{12}S$	1679-09-0	104.21628	2-Methyl-2-butanethiol
$C_5H_{12}S$	110-66-7	104.21628	1-Pentanethiol
$C_5H_{12}Si$	754-05-2	100.23578	Trimethylvinylsilane
$C_5H_{13}N$	616-39-7	87.16496	N,N-Diethylmethylamine N-Methyldiethylamine
$C_5H_{13}N$	996-35-0	87.16496	N,N-Dimethylisopropyl-amine
$C_5H_{13}N$	19961-27-4	87.16496	N-Ethylisopropylamine
$C_5H_{13}N$	110-68-9	87.16496	N-Methylbutylamine
$C_5H_{13}N$	110-58-7	87.16496	Pentylamine 1-Aminopentane Amylamine
$C_5H_{13}NO$	108-16-7	103.16436	1-Dimethylamino-2-pro-panol
$C_5H_{13}NO_2$	105-59-9	119.16376	N-Methyl-bis(2-hydroxyethyl)-amine N-Methyldiethanolamine
$C_5H_{13}NO_2S$	2374-61-0	151.22976	N,N-Diethylmethanesulfon-amide
$C_5H_{13}NO_3$		135.16316	2-Aminomethyl-2-hydroxy-methyl-1,3-propanediol 2,2,2-Tris(hydroxymethyl)amino-ethane
$C_5H_{14}ClNO_3$		171.6238	2-Aminomethyl-2-hydroxym-ethyl-1,3-propanediol hydrochloride 2,2,2-Tris(hydroxymethyl)amino-ethane hydrochloride
$C_5H_{14}N_2$	109-55-7	102.17964	3-(Dimethylamino)propyl-amine N,N-Dimethyl-1,3-propanedi-amine
$C_5H_{14}N_2$	51-80-9	102.17964	N,N,N',N'-Tetramethyl-methanediamine N,N,N',N'-Tetramethyldiamino-methane
$C_5H_{14}OSi$	1825-62-3	118.25106	Ethoxytrimethylsilane
C_6BrF_5	344-04-7	246.962016	Bromopentafluorobenzene
C_6ClF_5	344-07-0	202.510716	Chloropentafluorobenzene
$C_6Cl_2F_4$	1198-59-0	218.9650128	1,2-Dichlorotetrafluoroben-zene
$C_6Cl_3F_3$	319-88-0	235.4193096	1,3,5-Trichloro-2,4,6-trifluo-robenzene
C_6Cl_6	118-74-1	284.7822	Hexachlorobenzene
C_6CrO_6	13007-92-6	220.0585	Chromium hexacarbonyl Chromium carbonyl
C_6F_5I	827-15-6	293.962486	Pentafluoroiodobenzene
C_6F_6	392-56-3	186.0564192	Hexafluorobenzene Perfluorobenzene
C_6F_{10}	355-75-9	262.050032	Decafluorocyclohexene Perfluorocyclohexene
C_6F_{12}	355-68-0	300.0468384	Dodecafluorocyclohexane Perfluorocyclohexane
C_6F_{12}	755-25-9	300.0468384	Dodecafluoro-1-hexene Perfluoro-1-hexene
$C_6F_{12}O_2$		332.0456384	Dodecafluoro(methoxypro-pyl vinyl ether) Perfluoro(methoxypropyl vinyl ether) <undefined positional isomer>
C_6F_{14}	354-96-1	338.0436448	Tetradecafluoro(2,3-di-methylbutane) Perfluoro(2,3-dimethylbutane)
C_6F_{14}	355-42-0	338.0436448	Tetradecafluorohexane Perfluorohexane
C_6F_{14}	355-04-4	338.0436448	Tetradecafluoro(2-methyl-pentane) Perfluoro(2-methylpentane)
C_6F_{14}	865-71-4	338.0436448	Tetradecafluoro(3-methyl-pentane) Perfluoro(3-methylpentane)
$C_6F_{15}N$	359-70-6	371.048788	Pentadecafluorotriethyl-amine Perfluorotriethylamine
$C_6FeK_3N_6$	13746-66-2	329.24834	Tripotassium ferricyanide Ferricyanic acid potassium salt Potassium ferricyanide
C_6HCl_5O	87-86-5	266.33684	Pentachlorophenol
C_6HCl_6N	1201-30-5	299.79688	3,4,5-Trichloro-2-(trichloro-methyl)pyridine
C_6HF_5	363-72-4	168.065956	Pentafluorobenzene
C_6HF_5O	771-61-9	184.065356	Pentafluorophenol
C_6HF_{11}	308-24-7	282.0563752	Undecafluorocyclohexane
$C_6H_2Cl_4$	95-94-3	215.89268	1,2,4,5-Tetrachlorobenzene
$C_6H_2Cl_4O$	58-90-2	231.89208	2,3,4,6-Tetrachlorophenol
$C_6H_2Cl_5N$	1128-16-1	265.35212	3,5-Dichloro-2-(trichloro-methyl)pyridine
$C_6H_2Cl_5N$		265.35212	3,4,5-Trichloro-2-(dichloro-methyl)pyridine
$C_6H_2F_4$	551-62-2	150.0754928	1,2,3,4-Tetrafluorobenzene
$C_6H_2F_4$	2367-82-0	150.0754928	1,2,3,5-Tetrafluorobenzene
$C_6H_2F_4$	327-54-8	150.0754928	1,2,4,5-Tetrafluorobenzene
$C_6H_2N_3NaO_7$		251.087668	Sodium 2,4,6-trinitropheno-late Sodium picrate Picric acid sodium salt
$C_6H_3Cl_2NO_2$	89-61-2	192.00076	1,4-Dichloro-2-nitrobenzene 2,5-Dichloronitrobenzene
$C_6H_3Cl_2NO_2$	99-54-7	192.00076	1,2-Dichloro-4-nitrobenzene 3,4-Dichloronitrobenzene
$C_6H_3Cl_3$	87-61-6	181.44792	1,2,3-Trichlorobenzene
$C_6H_3Cl_3$	120-82-1	181.44792	1,2,4-Trichlorobenzene
$C_6H_3Cl_3$	108-70-3	181.44792	1,3,5-Trichlorobenzene
$C_6H_3Cl_3O$	88-06-2	197.44732	2,4,6-Trichlorophenol
$C_6H_3Cl_4N$	1197-03-1	230.90736	5-Chloro-2-(trichloromethyl)-pyridine
$C_6H_3Cl_4N$	1929-82-4	230.90736	6-Chloro-2-(trichloromethyl)-pyridine
$C_6H_3Cl_4N$	7041-25-0	230.90736	3,5-Dichloro-2-(dichloro-methyl)pyridine
$C_6H_3F_3$	372-38-3	132.0850296	1,3,5-Trifluorobenzene
$C_6H_3N_3O_7$	88-89-1	229.10584	2,4,6-Trinitrophenol Picric acid
$C_6H_4BrNO_2$	577-19-5	202.0073	1-Bromo-2-nitrobenzene
$C_6H_4BrNO_2$	585-79-5	202.0073	1-Bromo-3-nitrobenzene
$C_6H_4BrNO_2$	586-78-7	202.0073	1-Bromo-4-nitrobenzene
$C_6H_4ClNO_2$	88-73-3	157.556	1-Chloro-2-nitrobenzene
$C_6H_4ClNO_2$	121-73-3	157.556	1-Chloro-3-nitrobenzene
$C_6H_4ClNO_2$	100-00-5	157.556	1-Chloro-4-nitrobenzene

Formula	CAS RN	M_r	Name
$C_6H_4Cl_2$	95-50-1	147.00316	1,2-Dichlorobenzene
$C_6H_4Cl_2$	541-73-1	147.00316	1,3-Dichlorobenzene
$C_6H_4Cl_2$	106-46-7	147.00316	1,4-Dichlorobenzene
$C_6H_4Cl_2O$	120-83-2	163.00256	2,4-Dichlorophenol
$C_6H_4Cl_2O$	87-65-0	163.00256	2,6-Dichlorophenol
$C_6H_4Cl_3N$	4377-37-1	196.4626	2-(Trichloromethyl)pyridine
$C_6H_4FNO_2$	350-46-9	141.1017032	1-Fluoro-4-nitrobenzene
$C_6H_4F_2$	367-11-3	114.0945664	1,2-Difluorobenzene
$C_6H_4F_2$	372-18-9	114.0945664	1,3-Difluorobenzene
$C_6H_4F_2$	540-36-3	114.0945664	1,4-Difluorobenzene
$C_6H_4N_2$	100-54-9	104.11124	3-Cyanopyridine Nicotinonitrile
$C_6H_4N_2$	100-48-1	104.11124	4-Cyanopyridine Isonicotinonitrile
$C_6H_4O_2$	106-51-4	108.09656	1,4-Benzoquinone
C_6H_5Br	108-86-1	157.0097	Bromobenzene Phenyl bromide
C_6H_5Cl	108-90-7	112.5584	Chlorobenzene Phenyl chloride
C_6H_5ClO	95-57-8	128.5578	2-Chlorophenol
C_6H_5ClO	108-43-0	128.5578	3-Chlorophenol
C_6H_5ClO	106-48-9	128.5578	4-Chlorophenol
$C_6H_5Cl_2N$	95-76-1	162.01784	3,4-Dichloroaniline
$C_6H_5Cl_3Si$	98-13-5	211.5493	Trichlorophenylsilane
C_6H_5F	462-06-6	96.1041032	Fluorobenzene Phenyl fluoride
$C_6H_5F_7O_2$	356-27-4	242.0933224	Ethyl heptafluorobutanoate Ethyl heptafluorobutyrate Heptafluorobutanoic acid ethyl ester Heptafluorobutyric acid ethyl ester
$C_6H_5{}^2HO$		95.1192017	(O-2H)Phenol Phenol-d
C_6H_5I	591-50-4	204.01017	Iodobenzene Phenyl iodide
C_6H_5IO	533-58-4	220.00957	2-Iodophenol
$C_6H_5NO_2$	98-95-3	123.11124	Nitrobenzene
$C_6H_5NO_3$	88-75-5	139.11064	2-Nitrophenol
$C_6H_5NO_3$	554-84-7	139.11064	3-Nitrophenol
$C_6H_5NO_3$	100-02-7	139.11064	4-Nitrophenol
C_6H_5NaO		116.094868	Sodium phenolate Sodium phenoxide Sodium phenate
C_6H_6	71-43-2	78.11364	Benzene
C_6H_6	497-20-1	78.11364	Methylenecyclopentadiene Fulvene
$C_6{}^2H_6$	1076-43-3	84.1506106	($^2H_6$)Benzene Hexadeuterobenzene Benzene-d6
C_6H_6ClN	95-51-2	127.57308	2-Chloroaniline
C_6H_6ClN	108-42-9	127.57308	3-Chloroaniline
C_6H_6ClN	106-47-8	127.57308	4-Chloroaniline
$C_6H_6ClO_3P$		192.538302	Hydrogen phenyl chlorophosphonate Phenyl phosphorochloridate Phenyl chlorophosphate
$C_6H_6Cl_2Si$	1631-84-1	177.10454	Dichlorophenylsilane
$C_6H_6Cl_9O_3P$	140-08-9	476.159902	Tris(2,2,2-trichloroethyl) phosphite Phosphorous acid tris(2,2,2-trichloroethyl) ester
$C_6H_6N_2O_2$	88-74-4	138.12592	2-Nitroaniline
$C_6H_6N_2O_2$	99-09-2	138.12592	3-Nitroaniline
$C_6H_6N_2O_2$	100-01-6	138.12592	4-Nitroaniline
C_6H_6O	108-95-2	94.11304	Phenol Hydroxybenzene
C_6H_6O		94.11304	Phenol (c,I) Hydroxybenzene (c,I)
C_6H_6O		94.11304	Phenol (c,II) Hydroxybenzene (c,II)
$C_6H_6O_2$	120-80-9	110.11244	1,2-Benzenediol Pyrocatechol Catechol 1,2-Dihydroxybenzene
$C_6H_6O_2$	108-46-3	110.11244	1,3-Benzenediol Resorcinol 1,3-Dihydroxybenzene
$C_6H_6O_2$	123-31-9	110.11244	1,4-Benzenediol Hydroquinone Quinol 1,4-Dihydroxybenzene
$C_6H_6O_2$	620-02-0	110.11244	5-Methylfuraldehyde 5-Methylfurfural
$C_6H_6O_3$	87-66-1	126.11184	1,2,3-Benzenetriol
C_6H_6S	108-98-5	110.17964	Benzenethiol
C_6H_7N	62-53-3	93.12832	Aniline Benzenamine Phenylamine
C_6H_7N	3047-38-9	93.12832	1-Cyclopentenecarbonitrile
C_6H_7N	109-06-8	93.12832	2-Methylpyridine 2-Picoline
C_6H_7N	108-99-6	93.12832	3-Methylpyridine 3-Picoline
C_6H_7N	108-89-4	93.12832	4-Methylpyridine 4-Picoline
C_6H_7NO	95-55-6	109.12772	2-Aminophenol
$C_6H_7N_5O$	578-76-7	165.15468	7-Methylguanine
C_6H_8	592-57-4	80.12952	1,3-Cyclohexadiene
C_6H_8	628-41-1	80.12952	1,4-Cyclohexadiene
C_6H_8		80.12952	Methylcyclopentadiene <undefined positional isomer>
C_6H_8ClN	142-04-1	129.58896	Aniline hydrochloride
C_6H_8ClN		129.58896	1-Methylpyridinium chloride
$C_6H_8N_2$	111-69-3	108.143	Hexanedinitrile Adiponitrile 1,4-Dicyanobutane
$C_6H_8N_2$	95-54-5	108.143	1,2-Phenylenediamine
$C_6H_8N_2$	108-45-2	108.143	1,3-Phenylenediamine
$C_6H_8N_2$	106-50-3	108.143	1,4-Phenylenediamine
$C_6H_8N_2$	100-63-0	108.143	Phenylhydrazine
$C_6H_8N_2O$	1656-48-0	124.1424	Bis(2-cyanoethyl) ether 2-Cyanoethyl ether 3,3'-Oxydipropionitrile
$C_6H_8N_2O_2$	874-14-6	140.1418	1,3-Dimethyluracil 1,3-Dimethyl-2,4(1-H,3-H)-pyrimidinedione
$C_6H_8O_7$	77-92-9	192.12532	2-Hydroxy-1,2,3-propanetricarboxylic acid Citric acid
C_6H_9N	4254-02-8	95.1442	Cyclopentanecarbonitrile
C_6H_9NO	88-12-0	111.1436	1-Vinyl-2-pyrrolidinone
$C_6H_9NO_3$	641-06-5	143.1424	Tris(acetyl)amine Triacetamide

Formula	CAS RN	M_r	Name
$C_6H_9N_3O_2$		155.15648	2-Amino-3-(4-imidazolyl)-propanoic acid Histidine <undefined optical isomer>
$C_6H_9N_3O_2$	71-00-1	155.15648	(S)-(+)-2-Amino-3-(4-imidazolyl)propanoic acid L-Histidine
$C_6H_9N_3O_2$	4998-57-6	155.15648	DL-2-Amino-3-(4-imidazolyl)-propanoic acid DL-Histidine
C_6H_{10}	110-83-8	82.1454	Cyclohexene Tetrahydrobenzene
C_6H_{10}	513-81-5	82.1454	2,3-Dimethyl-1,3-butadiene
C_6H_{10}	917-92-0	82.1454	3,3-Dimethyl-1-butyne
C_6H_{10}	592-42-7	82.1454	1,5-Hexadiene
C_6H_{10}	693-02-7	82.1454	1-Hexyne
C_6H_{10}	928-49-4	82.1454	3-Hexyne
$C_6H_{10}Cl_2O_4Sn$		335.7584	Dichloro(dipropanoato)tin Dichloro(dipropionato)tin
$C_6H_{10}Cl_2O_4Ti$		264.9284	Dichloro(dipropanoato)titanium Dichloro(dipropionato)titanium
$C_6H_{10}O$	108-94-1	98.1448	Cyclohexanone
$C_6H_{10}O$	35656-02-1	98.1448	4-Methyleneoxane 4-Methylenetetrahydropyran
$C_6H_{10}O$	141-79-7	98.1448	4-Methyl-3-penten-2-one Mesityl oxide
$C_6H_{10}O$	16015-11-5	98.1448	1-Oxa-2-methyl-2-cyclohexene 6-Methyl-3,4-dihydro-2-H-pyran
$C_6H_{10}O$	55230-25-6	98.1448	1-Oxa-2-methyl-3-cyclohexene 2-Methyl-5,6-dihydro-2-H-pyran
$C_6H_{10}O$		98.1448	1-Oxa-4-methyl-3-cyclohexene 4-Methyl-5,6-dihydro-2-H-pyran
$C_6H_{10}O_2$	623-70-1	114.1442	Ethyl trans-2-butenoate trans-2-Butenoic acid ethyl ester Ethyl crotonate Crotonic acid ethyl ester
$C_6H_{10}O_2$	97-63-2	114.1442	Ethyl 2-methylpropenoate Methacrylic acid ethyl ester 2-Methylpropenoic acid methyl ester Ethyl methacrylate
$C_6H_{10}O_2$	110-13-4	114.1442	2,5-Hexanedione Acetonylacetone
$C_6H_{10}O_2$	765-85-5	114.1442	Methyl cyclobutanecarboxylate Cyclobutanecarboxylic acid methyl ester
$C_6H_{10}O_2$	502-44-3	114.1442	2-Oxepanone 6-Caprolactone 6-Hexanolactone
$C_6H_{10}O_2$	123-20-6	114.1442	Vinyl butanoate Vinyl butyrate Butanoic acid vinyl ester Butyric acid vinyl ester
$C_6H_{10}O_2$	1072-96-4	114.1442	4-Vinyl-1,3-dioxane
$C_6H_{10}O_3$	141-97-9	130.1436	Ethyl 3-oxobutanoate Ethyl acetoacetate
$C_6H_{10}O_3$ (Cont.)			Acetylethanoic acid ethyl ester Acetoacetic acid ethyl ester 3-Oxobutanoic acid ethyl ester Ethyl acetylethanoate
$C_6H_{10}O_3$	868-77-9	130.1436	2-Hydroxyethyl 2-methylpropenoate Ethylene glycol methacrylate 1,2-Ethanediol 2-methylpropenoate Methacrylic acid 2-hydroxyethyl ester 2-Methylpropenoic acid 2-hydroxyethyl ester
$C_6H_{10}O_3$	123-62-6	130.1436	Propanoic anhydride Propionic anhydride
$C_6H_{10}O_4$	61836-76-8	146.143	1,4-Butanediol dimethanoate 1,4-Butanediol diformate Tetramethylene diformate Tetramethylene dimethanoate
$C_6H_{10}O_4$	95-92-1	146.143	Diethyl ethanedioate Diethyl oxalate Ethanedioic acid diethyl ester Oxalic acid diethyl ester
$C_6H_{10}O_4$	106-65-0	146.143	Dimethyl butanedioate Butanedioic acid dimethyl ester Dimethyl succinate Succinic acid dimethyl ester
$C_6H_{10}O_4$	111-55-7	146.143	1,2-Ethanediol diethanoate 1,2-Ethanediol diacetate Ethylene diacetate Ethylene diethanoate Ethylene glycol diacetate
$C_6H_{10}O_4$	542-10-9	146.143	Ethylidene diethanoate 1,1-Ethanediol diethanoate Ethylidene diacetate
$C_6H_{10}O_4$	124-04-9	146.143	Hexanedioic acid Adipic acid
$C_6H_{10}O_6$	5057-96-5	178.1418	Dimethyl D-2,3-dihydroxybutanedioate Dimethyl D-tartrate D-Tartaric acid dimethyl ester D-2,3-Dihydroxybutanedioic acid dimethyl ester
$C_6H_{10}O_6$		178.1418	Dimethyl 2,3-dihydroxybutanedioate Dimethyl tartrate Tartaric acid dimethyl ester 2,3-Dihydroxybutanedioic acid dimethyl ester <undefined optical isomer>
$C_6H_{10}O_6$	608-68-4	178.1418	Dimethyl L-2,3-dihydroxybutanedioate Dimethyl L-tartrate L-Tartaric acid dimethyl ester L-2,3-Dihydroxybutanedioic acid dimethyl ester
$C_6H_{10}S$	592-88-1	114.2114	4-Thia-1,6-heptadiene Diallyl sulfide Bis(2-propenyl) sulfide
$C_6H_{11}Br$	108-85-0	163.05734	Bromocyclohexane Cyclohexyl bromide

Formula	CAS RN	M_r	Name
$C_6H_{11}Cl$	542-18-7	118.60604	Chlorocyclohexane
			Cyclohexyl chloride
$C_6H_{11}F$	372-46-3	102.1517432	Fluorocyclohexane
$C_6H_{11}N$	628-73-9	97.16008	Hexanenitrile
			1-Cyanopentane
			Hexanonitrile
			1-Pentanecarbonitrile
			Amyl cyanide
			Pentyl cyanide
$C_6H_{11}NO$	100-64-1	113.15948	Cyclohexanone oxime
$C_6H_{11}NO$	5075-92-3	113.15948	1,5-Dimethyl-2-pyrrolidinone
$C_6H_{11}NO$	2591-86-8	113.15948	N-Formylpiperidine
			1-Formylpiperidine
$C_6H_{11}NO$	105-60-2	113.15948	Hexahydro-2-azepinone
			6-Caprolactam
			6-Hexanelactam
			2-Oxohexamethyleneimine
			ϵ-Caprolactam
$C_6H_{11}NO_2$	1122-60-7	129.15888	Nitrocyclohexane
C_6H_{12}	110-82-7	84.16128	Cyclohexane
C_6H_{12}	563-78-0	84.16128	2,3-Dimethyl-1-butene
C_6H_{12}	563-79-1	84.16128	2,3-Dimethyl-2-butene
C_6H_{12}	760-21-4	84.16128	2-Ethyl-1-butene
C_6H_{12}	4806-61-5	84.16128	Ethylcyclobutane
C_6H_{12}	592-41-6	84.16128	1-Hexene
C_6H_{12}	7688-21-3	84.16128	cis-2-Hexene
C_6H_{12}	4050-45-7	84.16128	trans-2-Hexene
C_6H_{12}	13269-52-8	84.16128	trans-3-Hexene
C_6H_{12}	96-37-7	84.16128	Methylcyclopentane
C_6H_{12}	763-29-1	84.16128	2-Methyl-1-pentene
C_6H_{12}	760-20-3	84.16128	3-Methyl-1-pentene
C_6H_{12}	691-37-2	84.16128	4-Methyl-1-pentene
C_6H_{12}	625-27-4	84.16128	2-Methyl-2-pentene
C_6H_{12}	691-38-3	84.16128	cis-4-Methyl-2-pentene
C_6H_{12}	674-76-0	84.16128	trans-4-Methyl-2-pentene
$C_6{}^2H_{12}$	1735-17-7	96.2352213	($^2H_{12}$)Cyclohexane
			Cyclohexane-d12
$C_6H_{12}Br_2$	629-03-8	243.96928	1,6-Dibromohexane
$C_6H_{12}Cl_2$	2162-92-7	155.06668	1,2-Dichlorohexane
$C_6H_{12}Cl_2$	2163-00-0	155.06668	1,6-Dichlorohexane
$C_6H_{12}Cl_2O_2$	112-26-5	187.06548	1,8-Dichloro-3,6-dioxaoctane
			1,2-Bis(2-chloroethoxy)ethane
			Triethylene glycol dichloride
			Triglycol dichloride
$C_6H_{12}I_2$	629-09-4	337.97022	1,6-Diiodohexane
$C_6H_{12}N_4$	100-97-0	140.18824	Hexamethylenetetramine
			Methenamine
$C_6H_{12}O$	111-34-2	100.16068	Butyl vinyl ether
$C_6H_{12}O$	108-93-0	100.16068	Cyclohexanol
			Cyclohexyl alcohol
$C_6H_{12}O$	75-97-8	100.16068	3,3-Dimethyl-2-butanone
			tert-Butyl methyl ketone
$C_6H_{12}O$	97-96-1	100.16068	2-Ethylbutanal
			2-Ethylbutyraldehyde
$C_6H_{12}O$	1192-22-9	100.16068	3-Ethyl-2,2-dimethyloxirane
			2-Methyl-2-pentene oxide
			2-Methyl-2,3-epoxypentane
$C_6H_{12}O$	66-25-1	100.16068	Hexanal
			1-Pentanecarbaldehyde
			Caproaldehyde

Formula	CAS RN	M_r	Name
$C_6H_{12}O$ (Cont.)			Hexaldehyde
$C_6H_{12}O$	591-78-6	100.16068	2-Hexanone
			Butyl methyl ketone
$C_6H_{12}O$	589-38-8	100.16068	3-Hexanone
			Ethyl propyl ketone
$C_6H_{12}O$	123-15-9	100.16068	2-Methylpentanal
			2-Methylvaleraldehyde
$C_6H_{12}O$	15877-57-3	100.16068	3-Methylpentanal
			3-Methylvaleraldehyde
$C_6H_{12}O$	565-61-7	100.16068	3-Methyl-2-pentanone
			sec-Butyl methyl ketone
$C_6H_{12}O$	108-10-1	100.16068	4-Methyl-2-pentanone
			Isobutyl methyl ketone
			Isopropyl acetone
			MIBK
$C_6H_{12}O$	565-69-5	100.16068	2-Methyl-3-pentanone
			Ethyl isopropyl ketone
$C_6H_{12}O$	592-90-5	100.16068	Oxepane
			Hexamethylene oxide
$C_6H_{12}O_2$	123-86-4	116.16008	Butyl ethanoate
			Butyl acetate
			Ethanoic acid butyl ester
			Acetic acid butyl ester
$C_6H_{12}O_2$	540-88-5	116.16008	tert-Butyl ethanoate
			tert-Butyl acetate
			Ethanoic acid tert-butyl ester
			Acetic acid tert-butyl ester
$C_6H_{12}O_2$	766-15-4	116.16008	4,4-Dimethyl-1,3-dioxane
$C_6H_{12}O_2$	2391-24-4	116.16008	cis-4,5-Dimethyl-1,3-dioxane
$C_6H_{12}O_2$	1121-20-6	116.16008	trans-4,5-Dimethyl-1,3-dioxane
$C_6H_{12}O_2$	105-54-4	116.16008	Ethyl butanoate
			Butanoic acid ethyl ester
			Ethyl butyrate
			Butyric acid ethyl ester
$C_6H_{12}O_2$	1121-61-5	116.16008	4-Ethyl-1,3-dioxane
$C_6H_{12}O_2$	97-62-1	116.16008	Ethyl 2-methylpropanoate
			Ethyl isobutyrate
			2-Methylpropanoic acid ethyl ester
			Isobutyric acid ethyl ester
			Ethyl isobutanoate
			Isobutanoic acid ethyl ester
$C_6H_{12}O_2$	142-62-1	116.16008	Hexanoic acid
			Caproic acid
			1-Pentanecarboxylic acid
$C_6H_{12}O_2$	123-42-2	116.16008	4-Hydroxy-4-methyl-2-pentanone
			Diacetone alcohol
$C_6H_{12}O_2$	19354-27-9	116.16008	2-(Methoxymethyl)oxolane
			Methyl tetrahydrofurfuryl ether
$C_6H_{12}O_2$	110-45-2	116.16008	3-Methylbutyl methanoate
			Isoamyl formate
			Methanoic acid 3-methylbutyl ester
			Formic acid isoamyl ester
$C_6H_{12}O_2$	598-98-1	116.16008	Methyl 2,2-dimethylpropanoate
			Methyl pivalate
			Methyl trimethylacetate
			Pivalic acid methyl ester
			2,2-Dimethylpropanoic acid methyl ester

Formula	CAS RN	M_r	Name
$C_6H_{12}O_2$ (Cont.)			Trimethylacetic acid methyl ester
$C_6H_{12}O_2$	624-24-8	116.16008	Methyl pentanoate
			Methyl valerate
			Pentanoic acid methyl ester
			Valeric acid methyl ester
$C_6H_{12}O_2$		116.16008	1-Methylpropyl ethanoate
			Ethanoic acid 1-methylpropyl ester
			sec-Butyl acetate
			Acetic acid sec-butyl ester
			<undefined optical isomer>
$C_6H_{12}O_2$	105-46-4	116.16008	DL-1-Methylpropyl ethanoate
			DL-sec-Butyl acetate
			DL-Acetic acid sec-butyl ester
			DL-Ethanoic acid 1-methylpropyl ester
$C_6H_{12}O_2$	110-19-0	116.16008	2-Methylpropyl ethanoate
			Isobutyl acetate
			Ethanoic acid isobutyl ester
			Acetic acid isobutyl ester
			Isobutyl ethanoate
$C_6H_{12}O_2$	638-49-3	116.16008	Pentyl methanoate
			Amyl formate
			Methanoic acid pentyl ester
			Formic acid amyl ester
$C_6H_{12}O_2$	106-36-5	116.16008	Propyl propanoate
			Propanoic acid propyl ester
			Propyl propionate
			Propionic acid propyl ester
$C_6H_{12}O_2S$		148.22608	Dimethylthiolane 1,1-dioxide
			Dimethylsulfolane
			Dimethyltetrahydrothiophene 1,-1-dioxide
			<undefined positional isomer>
$C_6H_{12}O_2S$	1003-78-7	148.22608	2,4-Dimethylthiolane 1,1-dioxide
			2,4-Dimethylsulfolane
			2,4-Dimethyltetrahydrothiophene 1,1-dioxide
$C_6H_{12}O_3$		132.15948	3-Methylbutyl hydrogen carbonate
			Isoamyl hydrogen carbonate
			Carbonic acid hydrogen isoamyl ester
			Carbonic acid hydrogen 3-methylbutyl ester
			Carbonic acid hydrogen isopentyl ester
			Isopentyl hydrogen carbonate
$C_6H_{12}O_3$	111-15-9	132.15948	3-Oxapentyl ethanoate
			Ethylene glycol ethyl ether acetate
			Ethanoic acid 3-oxapentyl ester
			Acetic acid 2-ethoxyethyl ester
			2-Ethoxyethyl acetate
$C_6H_{12}O_3$	123-63-7	132.15948	2,4,6-Trimethyl-1,3,5-trioxane
			Paraldehyde
$C_6H_{12}O_4$		148.15888	4,6-Dioxa-7-octene-1,2-diol
$C_6H_{12}O_4$ (Cont.)			Glycerol 1-vinyloxymethyl ether
$C_6H_{12}O_6$	57-48-7	180.15768	D-Fructose
$C_6H_{12}O_6$		180.15768	Glucose
			<undefined optical isomer>
$C_6H_{12}O_6$	50-99-7	180.15768	D-Glucose
			<undefined stereoisomer>
$C_6H_{12}O_6$	492-62-6	180.15768	α-D-Glucose
			Dextrose
$C_6H_{12}O_6$	87-89-8	180.15768	Inositol
$C_6H_{12}S$	1569-69-3	116.22728	Cyclohexanethiol
$C_6H_{13}BO_3$		143.97842	1-Butoxy-2,5-dioxa-1-boracyclopentane
			Butyl ethylene borate
$C_6H_{13}Br$	111-25-1	165.07322	1-Bromohexane
			Hexyl bromide
$C_6H_{13}Cl$	544-10-5	120.62192	1-Chlorohexane
			Hexyl chloride
$C_6H_{13}ClN_2$		148.6354	1,4-Diazabicyclo[2.2.2]octane hydrochloride
			Triethylenediamine hydrochloride
$C_6H_{13}F$	373-14-8	104.1676232	1-Fluorohexane
			Hexyl fluoride
$C_6H_{13}I$	638-45-9	212.07369	1-Iodohexane
			Hexyl iodide
$C_6H_{13}N$	108-91-8	99.17596	Cyclohexylamine
			Aminocyclohexane
$C_6H_{13}N$	109-05-7	99.17596	2-Methylpiperidine
			2-Pipecoline
$C_6H_{13}N$	626-56-2	99.17596	3-Methylpiperidine
			3-Pipecoline
$C_6H_{13}N$	626-58-4	99.17596	4-Methylpiperidine
			4-Pipecoline
$C_6H_{13}N$	626-67-5	99.17596	N-Methylpiperidine
			1-Pipecoline
			1-Methylpiperidine
$C_6H_{13}N$	111-49-9	99.17596	Perhydroazepine
			Azacycloheptane
			Hexahydroazepine
			Hexamethyleneimine
			Homopiperidine
$C_6H_{13}NO$	1119-49-9	115.17536	N-Butylethanamide
			N-Butylacetamide
$C_6H_{13}NO$	685-91-6	115.17536	N,N-Diethylethanamide
			N,N-Diethylacetamide
$C_6H_{13}NO_2$	73-32-5	131.17476	(^{2}S,^{3}S)-(+)-2-Amino-3-methylpentanoic acid
			L-Isoleucine
$C_6H_{13}NO_2$	61-90-5	131.17476	(S)-(+)-2-Amino-4-methylpentanoic acid
			L-Leucine
			(S)-(+)-Leucine
$C_6H_{13}NO_2$	328-38-1	131.17476	(R)-(−)-2-Amino-4-methylpentanoic acid
			D-Leucine
			(R)-(−)-Leucine
$C_6H_{13}NaO_4S$		204.222588	Hexyl sodium sulfate
C_6H_{14}	75-83-2	86.17716	2,2-Dimethylbutane
			Neohexane
C_6H_{14}	79-29-8	86.17716	2,3-Dimethylbutane
			Diisopropyl

Formula	CAS RN	M_r	Name
C_6H_{14}	110-54-3	86.17716	Hexane
C_6H_{14}	107-83-5	86.17716	2-Methylpentane
			Isohexane
C_6H_{14}	96-14-0	86.17716	3-Methylpentane
$C_6H_{14}ClN$		135.6366	Perhydroazepine hydrochloride
			Hexamethylenimine hydrochloride
$C_6H_{14}N_2$	821-67-0	114.19064	Azopropane
$C_6H_{14}N_2$	3114-70-3	114.19064	1,4-Cyclohexanediamine
			1,4-Diaminocyclohexane
			<undefined cis/trans isomer>
$C_6H_{14}N_2$	2615-25-0	114.19064	trans-1,4-Cyclohexanediamine
			trans-1,4-Diaminocyclohexane
$C_6H_{14}N_2O_2$	56-87-1	146.18944	(S)-(+)-2,6-Diaminohexanoic acid
			L-Lysine
			(S)-(+)-Lysine
			L-2,6-Diaminohexanoic acid
$C_6H_{14}N_2O_2$	70-54-2	146.18944	DL-2,6-Diaminohexanoic acid
			DL-Lysine
$C_6H_{14}N_4O_2$	7004-12-8	174.20292	6-Aza-2,7-diamino-7-iminoheptanoic acid
			Arginine
			<undefined optical isomer>
$C_6H_{14}N_4O_2$	74-79-3	174.20292	(S)-(+)-6-Aza-2,7-diamino-7-iminoheptanoic acid
			L-Arginine
$C_6H_{14}N_4O_2$	157-06-2	174.20292	(R)-(−)-6-Aza-2,7-diamino-7-iminoheptanoic acid
			D-Arginine
$C_6H_{14}O$	628-81-9	102.17656	Butyl ethyl ether
$C_6H_{14}O$	108-20-3	102.17656	Diisopropyl ether
$C_6H_{14}O$	1185-33-7	102.17656	2,2-Dimethyl-1-butanol
			2,2-Dimethylbutyl alcohol
$C_6H_{14}O$	19550-30-2	102.17656	2,3-Dimethyl-1-butanol
			2,3-Dimethylbutyl alcohol
$C_6H_{14}O$	624-95-3	102.17656	3,3-Dimethyl-1-butanol
			3,3-Dimethylbutyl alcohol
			Neopentyl carbinol
$C_6H_{14}O$	594-60-5	102.17656	2,3-Dimethyl-2-butanol
$C_6H_{14}O$	464-07-3	102.17656	3,3-Dimethyl-2-butanol
$C_6H_{14}O$	111-43-3	102.17656	Dipropyl ether
			Propyl ether
$C_6H_{14}O$	97-95-0	102.17656	2-Ethyl-1-butanol
			2-Ethylbutyl alcohol
$C_6H_{14}O$	2679-87-0	102.17656	Ethyl 1-methylpropyl ether
			sec-Butyl ethyl ether
$C_6H_{14}O$	627-02-1	102.17656	Ethyl 2-methylpropyl ether
			Ethyl isobutyl ether
$C_6H_{14}O$	111-27-3	102.17656	1-Hexanol
			Hexyl alcohol
$C_6H_{14}O$	626-93-7	102.17656	2-Hexanol
$C_6H_{14}O$	623-37-0	102.17656	3-Hexanol
$C_6H_{14}O$	17015-11-1	102.17656	(±)-3-Hexanol
$C_6H_{14}O$	6210-51-1	102.17656	(S)-(+)-3-Hexanol
$C_6H_{14}O$	13471-42-6	102.17656	(R)-(−)-3-Hexanol
$C_6H_{14}O$	627-08-7	102.17656	Isopropyl propyl ether
$C_6H_{14}O$	994-05-8	102.17656	Methyl 1,1-dimethylpropyl ether
$C_6H_{14}O$ (Cont.)			tert-Amyl methyl ether
			Methyl tert-pentyl ether
$C_6H_{14}O$	105-30-6	102.17656	2-Methyl-1-pentanol
			2-Methylpentyl alcohol
$C_6H_{14}O$	589-35-5	102.17656	3-Methyl-1-pentanol
			3-Methylpentyl alcohol
$C_6H_{14}O$	626-89-1	102.17656	4-Methyl-1-pentanol
			4-Methylpentyl alcohol
$C_6H_{14}O$	590-36-3	102.17656	2-Methyl-2-pentanol
$C_6H_{14}O$	565-60-6	102.17656	3-Methyl-2-pentanol
$C_6H_{14}O$	108-11-2	102.17656	4-Methyl-2-pentanol
$C_6H_{14}O$	565-67-3	102.17656	2-Methyl-3-pentanol
$C_6H_{14}O$	77-74-7	102.17656	3-Methyl-3-pentanol
$C_6H_{14}O$	628-80-8	102.17656	Methyl pentyl ether
			Amyl methyl ether
$C_6H_{14}OS$	4253-91-2	134.24256	Dipropyl sulfoxide
$C_6H_{14}O_2$	111-76-2	118.17596	2-Butoxyethanol
			Butyl Cellosolve
			1,2-Ethanediol butyl ether
			3-Oxa-1-heptanol
			Ethylene glycol butyl ether
$C_6H_{14}O_2$	105-57-7	118.17596	1,1-Diethoxyethane
			Acetal
			Acetaldehyde diethyl acetal
			Diethyl acetal
			4-Methyl-3,5-dioxaheptane
$C_6H_{14}O_2$	629-14-1	118.17596	3,6-Dioxaoctane
			1,2-Diethoxyethane
			1,2-Ethanediol diethyl ether
			Ethylene glycol diethyl ether
$C_6H_{14}O_2$	629-11-8	118.17596	1,6-Hexanediol
			Hexamethylene glycol
$C_6H_{14}O_2$	4439-24-1	118.17596	5-Methyl-3-oxa-1-hexanol
			1,2-Ethanediol isobutyl ether
			Ethylene glycol isobutyl ether
$C_6H_{14}O_2$	107-41-5	118.17596	2-Methyl-2,4-pentanediol
			Hexylene glycol
$C_6H_{14}O_3$	111-90-0	134.17536	3,6-Dioxa-1-octanol
			2-(2-Ethoxyethoxy)ethanol
			3-Oxa-1,5-pentanediol ethyl ether
			Diethylene glycol ethyl ether
$C_6H_{14}O_3$	77-99-6	134.17536	2-Ethyl-2-(hydroxymethyl)-1,3-propanediol
			Trimethylolpropane
$C_6H_{14}O_3$	110-98-5	134.17536	4-Oxa-1,7-heptanediol
			Dipropylene glycol
$C_6H_{14}O_3$	111-96-6	134.17536	2,5,8-Trioxanonane
			Bis(2-methoxyethyl) ether
			Diglyme
			2-Methoxyethyl ether
			3-Oxa-1,5-pentanediol dimethyl ether
			Diethylene glycol dimethyl ether
$C_6H_{14}O_3$	5648-29-3	134.17536	3,5,7-Trioxanonane
			Bis(ethoxymethyl)ether
$C_6H_{14}O_4$	112-27-6	150.17476	3,6-Dioxa-1,8-octanediol
			Triethylene glycol
$C_6H_{14}O_6$	69-65-8	182.17356	D-Mannitol
$C_6H_{14}O_6$	50-70-4	182.17356	D-Sorbitol
			D-Glucitol

Formula	CAS RN	M_r	Name
$C_6H_{14}S$	625-80-9	118.24316	2,4-Dimethyl-3-thiapentane Diisopropyl sulfide
$C_6H_{14}S$	5008-73-1	118.24316	2-Methyl-3-thiahexane Isopropyl propyl sulfide
$C_6H_{14}S$	1741-83-9	118.24316	2-Thiaheptane Methyl pentyl sulfide Amyl methyl sulfide
$C_6H_{14}S$	638-46-0	118.24316	3-Thiaheptane Butyl ethyl sulfide
$C_6H_{14}S$	111-47-7	118.24316	4-Thiaheptane Dipropyl sulfide Dipropyl thioether
$C_6H_{14}S_2$	4253-89-8	150.30916	2,5-Dimethyl-3,4-dithiahexane Diisopropyl disulfide
$C_6H_{15}As$	617-75-4	162.10669	Triethylarsine
$C_6H_{15}AsO_3$	3141-12-6	210.10489	Triethoxyarsane Triethyl arsenite Triethoxyarsine
$C_6H_{15}B$	97-94-9	97.9961	Triethylborane
$C_6H_{15}BO_3$	150-46-9	145.9943	Triethoxyborane Triethyl borate Boric acid triethyl ester
$C_6H_{15}B_3O_3$	3043-60-5	167.6163	Triethylboroxin Triethylcyclotriboroxane Ethylboric anhydride
$C_6H_{15}N$	108-18-9	101.19184	Diisopropylamine
$C_6H_{15}N$	927-62-8	101.19184	N,N-Dimethylbutylamine
$C_6H_{15}N$	918-02-5	101.19184	N,N-Dimethyl-tert-butylamine
$C_6H_{15}N$	142-84-7	101.19184	Dipropylamine
$C_6H_{15}N$	21035-44-9	101.19184	N-Ethyl(1-methylpropyl)amine N-Ethyl-sec-butylamine
$C_6H_{15}N$	13360-63-9	101.19184	N-Ethylbutylamine
$C_6H_{15}N$	111-26-2	101.19184	Hexylamine 1-Aminohexane
$C_6H_{15}N$	21968-17-2	101.19184	N-Isopropylpropylamine
$C_6H_{15}N$	121-44-8	101.19184	Triethylamine
$C_6H_{15}NO$	100-37-8	117.19124	2-(Diethylamino)ethanol N,N-Diethylethanolamine
$C_6H_{15}NO_2$	110-97-4	133.19064	Bis(2-hydroxypropyl)amine Diisopropanolamine
$C_6H_{15}NO_2$	1704-62-7	133.19064	2-(2-Dimethylaminoethoxy)ethanol
$C_6H_{15}NO_2S$		165.25664	N,N-Diethylethanesulfonamide
$C_6H_{15}NO_3$	102-71-6	149.19004	Tris(2-hydroxyethyl)amine Triethanolamine 2,2',2''-Nitrilotriethanol
$C_6H_{15}O_3P$	122-52-1	166.157062	Triethyl phosphite Triethoxyphosphine Triethoxophosphorus Phosphorous acid triethyl ester
$C_6H_{15}O_3Sb$	10433-06-4	256.9403	Triethoxystibine Triethyl antimonate Antimony(III) ethoxide
$C_6H_{15}O_4P$	78-40-0	182.156462	Triethyl phosphate Ethyl phosphate Phosphoric acid triethyl ester
$C_6H_{15}Sb$	617-85-6	208.9421	Triethylstibine
$C_6H_{16}N_2$	124-09-4	116.20652	1,6-Hexanediamine Hexamethylenediamine
$C_6H_{16}N_2$	110-18-9	116.20652	N,N,N',N'-Tetramethyl-1,2-ethanediamine N,N,N',N'-Tetramethyl-1,2-diaminoethane N,N,N',N'-Tetramethylethylenediamine
$C_6H_{16}O_2Si$	78-62-6	148.27734	Diethoxydimethylsilane
$C_6H_{16}O_3Si$	998-30-1	164.27674	Triethoxysilane
$C_6H_{17}N_3$	56-18-8	131.2212	Bis(3-aminopropyl)amine
$C_6H_{18}N_3OP$	680-31-9	179.202302	Phosphoric tris(dimethylamide) Hexamethylphosphotriamide HMPT Hexamethylphosphoric acid triamide HMPA Tris(dimethylamino)phosphine oxide
$C_6H_{18}N_4$	112-24-3	146.23588	1,4,7,10-Tetraazadecane Triethylenetetramine
$C_6H_{18}OSi_2$	107-46-0	162.37932	Hexamethyldisiloxane
C_6MoO_6	13939-06-5	264.0024	Molybdenum hexacarbonyl Molybdenum carbonyl Hexacarbonylmolybdenum
C_6O_6W	14040-11-0	351.9124	Tungsten hexacarbonyl Tungsten carbonyl
C_7F_8	434-64-0	236.0642256	Octafluorotoluene Perfluorotoluene
C_7F_{14}	355-63-5	350.0546448	Tetradecafluoro-1-heptene Perfluoro-1-heptene
C_7F_{14}	355-02-2	350.0546448	Tetradecafluoromethylcyclohexane Perfluoromethylcyclohexane
C_7F_{16}	335-57-9	388.0514512	Hexadecafluoroheptane Perfluoroheptane
C_7HF_{15}	375-83-7	370.060988	1-Hydropentadecafluoroheptane 1-Hydroperfluoroheptane
$C_7H_3Cl_2F_3$		250.4541296	2,6-Dichloro-3,4,5-trifluorotoluene
$C_7H_3Cl_3O_2$	50-31-7	225.45772	2,3,6-Trichlorobenzoic acid
$C_7H_3F_5$	771-56-2	182.092836	2,3,4,5,6-Pentafluorotoluene
$C_7H_4ClF_3$		180.5566696	Chlorotrifluorotoluene <mixture of positional isomers>
$C_7H_4ClF_3$		180.5566696	2-Chloro-3,4,5-trifluorotoluene
C_7H_4ClNO	2909-38-8	153.5676	3-Chlorophenyl isocyanate Isocyanic acid 3-chlorophenyl ester
$C_7H_4Cl_2O$	618-46-2	175.01356	3-Chlorobenzoyl chloride
$C_7H_4F_{13}NO_2$		381.0935416	Ammonium tridecafluoroheptanoate Ammonium perfluoroheptanoate Tridecafluoroheptanoic acid ammonium salt Perfluoroheptanoic acid ammonium salt
C_7H_5ClO	98-88-4	140.5688	Benzoyl chloride Benzenecarbonyl chloride

Formula	CAS RN	M_r	Name
$C_7H_5ClO_2$	118-91-2	156.5682	2-Chlorobenzoic acid
$C_7H_5ClO_2$	74-11-3	156.5682	4-Chlorobenzoic acid
$C_7H_5Cl_3$	98-07-7	195.4748	(Trichloromethyl)benzene
			α,α,α-Trichlorotoluene
			Benzotrichloride
$C_7H_5Cl_3$	2077-46-5	195.4748	2,3,6-Trichlorotoluene
$C_7H_5Cl_3$	6639-30-1	195.4748	2,4,5-Trichlorotoluene
$C_7H_5F_3$	98-08-8	146.1119096	(Trifluoromethyl)benzene
			α,α,α-Trifluorotoluene
			Benzotrifluoride
$C_7H_5KO_2$	582-25-2	160.2138	Potassium benzoate
			Benzoic acid potassium salt
C_7H_5N	100-47-0	103.12344	Benzonitrile
			Benzenecarbonitrile
			Phenyl cyanide
C_7H_5NO	273-53-0	119.12284	Benzoxazole
C_7H_5NO	103-71-9	119.12284	Phenyl isocyanate
			Isocyanic acid phenyl ester
C_7H_5NS	95-16-9	135.18944	Benzothiazole
C_7H_6ClNO		155.58348	Phenylcarbamoyl chloride
$C_7H_6Cl_2$	98-87-3	161.03004	(Dichloromethyl)benzene
			Benzal chloride
			α,α-Dichlorotoluene
$C_7H_6Cl_2$	95-73-8	161.03004	2,4-Dichlorotoluene
$C_7H_6Cl_4Si$		260.02094	Dichloro(dichlorophenyl)-methylsilane
			<undefined positional isomer>
C_7H_6O	100-52-7	106.12404	Benzaldehyde
$C_7H_6O_2$	274-09-9	122.12344	4,5-Benzo-1,3-dioxolane
			1,3-Benzodioxole
			1,2-(Methylenedioxy)benzene
$C_7H_6O_2$	65-85-0	122.12344	Benzoic acid
$C_7H_6O_2$	90-02-8	122.12344	2-Hydroxybenzaldehyde
			Salicylaldehyde
$C_7H_6O_3$	69-72-7	138.12284	2-Hydroxybenzoic acid
			Salicylic acid
$C_7H_7BO_3$		149.94178	2-Methoxy-1,3-dioxa-2-bora-indan
			Methyl 1,2-phenylene borate
C_7H_7Br	591-17-3	171.03658	3-Bromotoluene
C_7H_7Br	106-38-7	171.03658	4-Bromotoluene
C_7H_7Cl	100-44-7	126.58528	Benzyl chloride
			α-Chlorotoluene
			(Chloromethyl)benzene
C_7H_7Cl	95-49-8	126.58528	2-Chlorotoluene
C_7H_7Cl	108-41-8	126.58528	3-Chlorotoluene
C_7H_7Cl	106-43-4	126.58528	4-Chlorotoluene
C_7H_7ClO	59-50-7	142.58468	4-Chloro-3-methylphenol
			4-Chloro-3-cresol
$C_7H_7Cl_3Si$		225.57618	Dichloro(chlorophenyl)-methylsilane
			<undefined positional isomer>
C_7H_7F	95-52-3	110.1309832	2-Fluorotoluene
C_7H_7F	352-70-5	110.1309832	3-Fluorotoluene
C_7H_7F	352-32-9	110.1309832	4-Fluorotoluene
C_7H_7N	100-69-6	105.13932	2-Vinylpyridine
C_7H_7N	100-43-6	105.13932	4-Vinylpyridine
C_7H_7NO	55-21-0	121.13872	Benzamide
C_7H_7NO	103-70-8	121.13872	N-Phenylmethanamide
			Formanilide
$C_7H_7NO_2$	88-72-2	137.13812	2-Nitrotoluene
$C_7H_7NO_2$	99-08-1	137.13812	3-Nitrotoluene
$C_7H_7NO_2$	99-99-0	137.13812	4-Nitrotoluene
$C_7H_7NO_3$	700-38-9	153.13752	5-Methyl-2-nitrophenol
			6-Nitro-3-cresol
$C_7H_7NO_3$	91-23-6	153.13752	2-Nitroanisole
C_7H_8	108-88-3	92.14052	Toluene
			Methylbenzene
			Phenylmethane
$C_7H_8Cl_2Si$	149-74-6	191.13142	Dichloromethylphenylsilane
C_7H_8O	1319-77-3	108.13992	Methylphenol
			Cresol
			<mixture of positional isomers>
C_7H_8O	95-48-7	108.13992	2-Methylphenol
			2-Cresol
C_7H_8O	108-39-4	108.13992	3-Methylphenol
			3-Cresol
C_7H_8O	106-44-5	108.13992	4-Methylphenol
			4-Cresol
C_7H_8O	100-66-3	108.13992	Methyl phenyl ether
			Methoxybenzene
			Anisole
C_7H_8O	100-51-6	108.13992	Phenylmethanol
			Benzyl alcohol
$C_7H_8O_2$	90-05-1	124.13932	2-Methoxyphenol
			Guaiacol
$C_7H_8O_3$		140.13872	2-Furyl-1,3-dioxolane
$C_7H_8O_3S$	104-15-4	172.20472	4-Toluenesulfonic acid
C_7H_9N	100-46-9	107.1552	Benzylamine
C_7H_9N	1855-63-6	107.1552	1-Cyclohexenecarbonitrile
C_7H_9N	583-61-9	107.1552	2,3-Dimethylpyridine
			2,3-Lutidine
C_7H_9N	108-47-4	107.1552	2,4-Dimethylpyridine
			2,4-Lutidine
C_7H_9N	589-93-5	107.1552	2,5-Dimethylpyridine
			2,5-Lutidine
C_7H_9N	108-48-5	107.1552	2,6-Dimethylpyridine
			2,6-Lutidine
C_7H_9N	583-58-4	107.1552	3,4-Dimethylpyridine
C_7H_9N	591-22-0	107.1552	3,5-Dimethylpyridine
			3,5-Lutidine
C_7H_9N	536-78-7	107.1552	3-Ethylpyridine
C_7H_9N	95-53-4	107.1552	2-Methylaniline
			2-Toluidine
C_7H_9N	108-44-1	107.1552	3-Methylaniline
			3-Toluidine
C_7H_9N	106-49-0	107.1552	4-Methylaniline
			4-Toluidine
C_7H_9N	100-61-8	107.1552	N-Methylaniline
C_7H_9NO	90-04-0	123.1546	2-Methoxyaniline
			2-Anisidine
$C_7H_9NO_2$		139.154	2-Cyanoethyl 2-methylpro-penoate
			2-Cyanoethyl methacrylate
			Methacrylic acid 2-cyanoethyl ester
			2-Methylpropenoic acid 2-cyanoethyl ester
$C_7H_{10}BrN$	1906-79-2	188.06714	1-Ethylpyridinium bromide
$C_7H_{10}N_2$	95-80-7	122.16988	2,4-Diaminotoluene
$C_7H_{10}N_2O_2$	4401-71-2	154.16868	1,3,5-Trimethyluracil
			1,3,5-Trimethyl-2,4(1-H,3-H)-pyrimidinedione

Formula	CAS RN	M_r	Name
$C_7H_{10}N_2O_2$ (Cont.)			1,3-Dimethylthymine
$C_7H_{10}N_2O_2$	13509-52-9	154.16868	1,3,6-Trimethyluracil
			1,3,6-Trimethyl-2,4(1-H,3-H)-pyrimidinedione
$C_7H_{10}O$	497-38-1	110.1558	Bicyclo[2.2.1]heptan-2-one
			2-Norbornanone
			Norcamphor
$C_7H_{10}O$	1121-37-5	110.1558	Dicyclopropyl ketone
$C_7H_{10}O_4$		158.154	Methyl vinyl butanedioate
			Methyl vinyl succinate
			Butanedioic acid methyl vinyl ester
			Succinic acid methyl vinyl ester
$C_7H_{11}N$	766-05-2	109.17108	Cyclohexanecarbonitrile
C_7H_{12}	279-23-2	96.17228	Bicyclo[2.2.1]heptane
			Norbornane
C_7H_{12}	2146-38-5	96.17228	1-Ethylcyclopentene
C_7H_{12}	694-35-9	96.17228	3-Ethylcyclopentene
C_7H_{12}	628-71-7	96.17228	1-Heptyne
C_7H_{12}	1119-65-9	96.17228	2-Heptyne
C_7H_{12}	4625-24-5	96.17228	1-Methylbicyclo[3.1.0]hexane
C_7H_{12}		96.17228	Methylcyclohexene
			<undefined positional isomer>
C_7H_{12}	591-49-1	96.17228	1-Methylcyclohexene
C_7H_{12}	591-48-0	96.17228	3-Methylcyclohexene
$C_7H_{12}ClN$		145.63172	7-Chloroheptanenitrile
			7-Chloroenanthonitrile
$C_7H_{12}ClNO$		161.63112	1-(Chloroformyl)homopiperidine
			Hexamethylenimine-1-carbonyl chloride
			N-(Chloroformyl)homopiperidine
$C_7H_{12}ClNO$	13654-91-6	161.63112	6-Chlorohexyl isocyanate
$C_7H_{12}Cl_4$		237.98308	Tetrachloroheptane
			<undefined positional isomer>
$C_7H_{12}Cl_4$	3922-36-9	237.98308	1,1,1,7-Tetrachloroheptane
$C_7H_{12}O$	502-42-1	112.17168	Cycloheptanone
$C_7H_{12}O$	1331-22-2	112.17168	Methylcyclohexanone
			<undefined positional isomer>
$C_7H_{12}O$	583-60-8	112.17168	2-Methylcyclohexanone
$C_7H_{12}O$	591-24-2	112.17168	(±)-3-Methylcyclohexanone
$C_7H_{12}O$	13368-65-5	112.17168	(R)-(+)-3-Methylcyclohexanone
$C_7H_{12}O$	589-92-4	112.17168	4-Methylcyclohexanone
$C_7H_{12}O_2$	141-32-2	128.17108	Butyl propenoate
			Butyl acrylate
			Propenoic acid butyl ester
			Acrylic acid butyl ester
$C_7H_{12}O_2$	4351-54-6	128.17108	Cyclohexyl methanoate
			Cyclohexyl formate
			Methanoic acid cyclohexyl ester
			Formic acid cyclohexyl ester
$C_7H_{12}O_2$	2210-28-8	128.17108	Propyl 2-methylpropenaoate
			Propyl methacrylate
			2-Methylpropenoic acid propyl ester
			Methacrylic acid propyl ester
$C_7H_{12}O_3$		144.17048	2,5-Dioxa-6-heptenyloxirane
			Ethylene glycol glycidyl vinyl ether
$C_7H_{12}O_3$	5185-97-7	144.17048	4-Oxopentyl ethanoate
			Ethanoic acid 4-oxopentyl ester
			Acetic acid 4-oxopentyl ester
$C_7H_{12}O_4$	105-53-3	160.16988	Diethyl propanedioate
			Diethyl malonate
			Malonic acid diethyl ester
			Propanedioic acid diethyl ester
$C_7H_{12}O_4$	1119-40-0	160.16988	Dimethyl pentanedioate
			Dimethyl glutarate
			Pentanedioic acid dimethyl ester
			Glutaric acid dimethyl ester
$C_7H_{12}O_4$	111-16-0	160.16988	Heptanedioic acid
			Pimelic acid
$C_7H_{12}O_4$	627-91-8	160.16988	Methyl hexanedioate
			Methyl adipate
			Hexanedioic acid methyl ester
			Adipic acid methyl ester
$C_7H_{12}O_4$	623-84-7	160.16988	1,2-Propanediol diethanoate
			Propylene glycol diacetate
$C_7H_{13}ClO_2$	6092-54-2	164.63172	Hexyl chloromethanoate
			Hexyl chloroformate
			Chloromethanoic acid hexyl ester
			Chloroformic acid hexyl ester
$C_7H_{13}NO$	2556-73-2	127.18636	1-Methylhexahydro-2-azepinone
			N-Methylcaprolactam
			N-Methyl-ϵ-caprolactam
			1-Methyl-6-caprolactam
			1-Methyl-6-hexanelactam
$C_7H_{13}NO$		127.18636	2-Perhydroazocinone
			Hexamethyleneformamide
C_7H_{14}	291-64-5	98.18816	Cycloheptane
C_7H_{14}	1638-26-2	98.18816	1,1-Dimethylcyclopentane
C_7H_{14}	1192-18-3	98.18816	cis-1,2-Dimethylcyclopentane
C_7H_{14}	822-50-4	98.18816	trans-1,2-Dimethylcyclopentane
C_7H_{14}	2532-58-3	98.18816	cis-1,3-Dimethylcyclopentane
C_7H_{14}	625-65-0	98.18816	2,4-Dimethyl-2-pentene
C_7H_{14}	1640-89-7	98.18816	Ethylcyclopentane
C_7H_{14}	592-76-7	98.18816	1-Heptene
C_7H_{14}	6443-92-1	98.18816	cis-2-Heptene
C_7H_{14}	14686-13-6	98.18816	trans-2-Heptene
C_7H_{14}	592-78-9	98.18816	3-Heptene
			<undefined cis/trans isomer>
C_7H_{14}	7642-10-6	98.18816	cis-3-Heptene
C_7H_{14}	14686-14-7	98.18816	trans-3-Heptene
C_7H_{14}	108-87-2	98.18816	Methylcyclohexane
C_7H_{14}	2738-19-4	98.18816	2-Methyl-2-hexene
$C_7H_{14}O$	502-41-0	114.18756	Cycloheptanol
			Cycloheptyl alcohol
$C_7H_{14}O$	564-04-5	114.18756	2,2-Dimethyl-3-pentanone
$C_7H_{14}O$	565-80-0	114.18756	2,4-Dimethyl-3-pentanone
			Isobutyron
			Diisopropyl ketone
$C_7H_{14}O$	111-71-7	114.18756	Heptanal
			1-Hexanecarbaldehyde
			Heptyl aldehyde
$C_7H_{14}O$	110-43-0	114.18756	2-Heptanone
			Methyl pentyl ketone

Formula	CAS RN	M_r	Name
$C_7H_{14}O$ (Cont.)			Amyl methyl ketone
$C_7H_{14}O$	106-35-4	114.18756	3-Heptanone
			Butyl ethyl ketone
$C_7H_{14}O$	123-19-3	114.18756	4-Heptanone
			Dipropyl ketone
$C_7H_{14}O$	590-67-0	114.18756	1-Methylcyclohexanol
$C_7H_{14}O$	583-59-5	114.18756	2-Methylcyclohexanol
$C_7H_{14}O$	7443-70-1	114.18756	(±)-cis-2-Methylcyclohexanol
$C_7H_{14}O$	7443-52-9	114.18756	(±)-trans-2-Methylcyclohexanol
$C_7H_{14}O$	591-23-1	114.18756	3-Methylcyclohexanol
$C_7H_{14}O$	5454-79-5	114.18756	cis-3-Methylcyclohexanol
$C_7H_{14}O$	7443-55-2	114.18756	trans-3-Methylcyclohexanol
$C_7H_{14}O$	589-91-3	114.18756	4-Methylcyclohexanol
$C_7H_{14}O$	7731-28-4	114.18756	cis-4-Methylcyclohexanol
$C_7H_{14}O$	7731-29-5	114.18756	trans-4-Methylcyclohexanol
$C_7H_{14}O$	110-12-3	114.18756	5-Methyl-2-hexanone
$C_7H_{14}O$	5063-65-0	114.18756	Pentyloxirane
			1,2-Epoxyheptane
			Amyloxirane
$C_7H_{14}O_2$	590-01-2	130.18696	Butyl propanoate
			Butyl propionate
			Propanoic acid butyl ester
			Propionic acid butyl ester
$C_7H_{14}O_2$	3938-95-2	130.18696	Ethyl 2,2-dimethylpropanoate
$C_7H_{14}O_2$	108-64-5	130.18696	Ethyl 3-methylbutanoate
$C_7H_{14}O_2$	539-82-2	130.18696	Ethyl pentanoate
$C_7H_{14}O_2$	111-14-8	130.18696	Heptanoic acid
			1-Hexanecarboxylic acid
$C_7H_{14}O_2$	123-92-2	130.18696	3-Methylbutyl ethanoate
			Ethanoic acid 3-methylbutyl ester
			Isoamyl acetate
			Acetic acid isoamyl ester
$C_7H_{14}O_2$	106-70-7	130.18696	Methyl hexanoate
			Hexanoic acid methyl ester
			Methyl caproate
			Caproic acid methyl ester
$C_7H_{14}O_2$	540-42-1	130.18696	2-Methylpropyl propanoate
			Isobutyl propanoate
			Isobutyl propionate
			Propanoic acid 2-methylpropyl ester
			Propionic acid isobutyl ester
$C_7H_{14}O_2$	628-63-7	130.18696	Pentyl ethanoate
			Amyl acetate
			Ethanoic acid pentyl ester
			Acetic acid amyl ester
$C_7H_{14}O_2$	105-66-8	130.18696	Propyl butanoate
			Propyl butyrate
			Butanoic acid propyl ester
			Butyric acid propyl ester
$C_7H_{14}O_2$	644-49-5	130.18696	Propyl 2-methylpropanoate
			Propyl isobutyrate
			Propyl isobutanoate
			Isobutyric acid propyl ester
			Isobutanoic acid propyl ester
$C_7H_{14}O_3$		146.18636	4-(2-Hydroxyethyl)-4-methyl-1,3-dioxane
$C_7H_{15}Br$	629-04-9	179.1001	1-Bromoheptane
$C_7H_{15}Br$ (Cont.)			Heptyl bromide
$C_7H_{15}Cl$	629-06-1	134.6488	1-Chloroheptane
			Heptyl chloride
$C_7H_{15}I$	4282-40-0	226.10057	1-Iodoheptane
			Heptyl iodide
$C_7H_{15}N$	100-60-7	113.20284	N-Methylcyclohexylamine
$C_7H_{15}NO$	1620-14-0	129.20224	1-Diethylamino-2-propanone
			Diethylaminoacetone
$C_7H_{15}NO$	43018-61-7	129.20224	5-Dimethylamino-2-pentanone
			(2-Dimethylaminoethyl)acetone
$C_7H_{15}NO_2$		145.20164	Ethyl N,N-diethylcarbamate
$C_7H_{15}NaO_3S$		202.250068	Sodium heptanesulfonate
C_7H_{16}	590-35-2	100.20404	2,2-Dimethylpentane
C_7H_{16}	565-59-3	100.20404	2,3-Dimethylpentane
C_7H_{16}	108-08-7	100.20404	2,4-Dimethylpentane
C_7H_{16}	562-49-2	100.20404	3,3-Dimethylpentane
C_7H_{16}	617-78-7	100.20404	3-Ethylpentane
C_7H_{16}	142-82-5	100.20404	Heptane
C_7H_{16}	591-76-4	100.20404	2-Methylhexane
C_7H_{16}	589-34-4	100.20404	3-Methylhexane
C_7H_{16}	464-06-2	100.20404	2,2,3-Trimethylbutane
$C_7H_{16}O$	3073-92-5	116.20344	Butyl propyl ether
$C_7H_{16}O$	595-41-5	116.20344	2,3-Dimethyl-3-pentanol
$C_7H_{16}O$	14602-88-1	116.20344	2-Ethylbutyl methyl ether
$C_7H_{16}O$	597-49-9	116.20344	3-Ethyl-3-pentanol
			Triethylcarbinol
$C_7H_{16}O$	17952-11-3	116.20344	Ethyl pentyl ether
			Amyl ethyl ether
$C_7H_{16}O$	111-70-6	116.20344	1-Heptanol
			Heptyl alcohol
$C_7H_{16}O$	543-49-7	116.20344	(±)-2-Heptanol
$C_7H_{16}O$	589-82-2	116.20344	(±)-3-Heptanol
$C_7H_{16}O$	589-55-9	116.20344	4-Heptanol
$C_7H_{16}O$	4747-07-3	116.20344	Hexyl methyl ether
$C_7H_{16}O$	624-22-6	116.20344	2-Methyl-1-hexanol
$C_7H_{16}O$	14476-10-9	116.20344	Methyl 2-methylpentyl ether
$C_7H_{16}O$	594-83-2	116.20344	2,3,3-Trimethyl-2-butanol
$C_7H_{16}O_2$	4744-08-5	132.20284	1,1-Diethoxypropane
			4-Ethyl-3,5-dioxaheptane
			Propionaldehyde diethyl acetal
$C_7H_{16}O_2$	126-84-1	132.20284	2,2-Diethoxypropane
			Acetone diethyl acetal
			4,4-Dimethyl-3,5-dioxaheptane
$C_7H_{16}O_3$	6881-94-3	148.20224	3,6-Dioxa-1-nonanol
			Diethylene glycol propyl ether
$C_7H_{16}O_3$	101750-15-6	148.20224	4,8-Dioxa-1-nonanol
			Dipropylene glycol methyl ether
$C_7H_{16}O_3$	122-51-0	148.20224	Triethoxymethane
			Orthomethanoic acid triethyl ester
			Triethyl orthoformate
			Triethyl orthomethanoate
			Orthoformic acid triethyl ester
$C_7H_{16}O_4$	102-52-3	164.20164	1,1,3,3-Tetramethoxypropane
			Malonaldehyde bis(dimethyl acetal)
$C_7H_{16}O_4$	112-35-6	164.20164	3,6,9-Trioxa-1-decanol
			Triethylene glycol methyl ether
$C_7H_{17}N$	3405-42-3	115.21872	N,N-Dipropylmethylamine

Formula	CAS RN	M_r	Name
$C_7H_{17}N$	111-68-2	115.21872	Heptylamine 1-Aminoheptane
$C_7H_{18}N_2$	110-95-2	130.2334	N,N,N',N'-Tetramethyl-1,3-propanediamine N,N,N',N'-Tetramethyl-1,3-diaminopropane
$C_7H_{18}N_2O$	5966-51-8	146.2328	1,3-Bis(dimethylamino)-2-propanol N,N,N',N'-Tetramethyl-1,3-diamino-2-propanol
$C_7H_{22}O_2Si_3$	2895-07-0	222.50698	Heptamethyltrisiloxane
$C_8Co_2O_8$	10210-68-1	341.9496	Dicobalt octacarbonyl Cobalt carbonyl
$C_8F_{15}LiO_2$		420.003848	Lithium pentadecafluorooctanoate Lithium perfluorooctanoate Perfluorooctanoic acid lithium salt Pentadecafluorooctanoic acid lithium salt
$C_8F_{15}NaO_2$	18017-22-6	436.052616	Sodium pentadecafluorooctanoate Sodium perfluorooctanoate Pentadecafluorooctanoic acid sodium salt Perfluorooctanoic acid sodium salt
$C_8F_{16}O$	335-36-4	416.0618512	Hexadecafluoro(2-butyloxolane) Perfluoro(2-butyltetrahydrofuran)
$C_8F_{17}I$	507-63-1	545.9653244	Heptadecafluoro-1-iodooctane Perfluorooctyl iodide
C_8F_{18}	307-34-6	438.0592576	Octadecafluorooctane Perfluorooctane
$C_8F_{18}O$	308-48-5	454.0586576	Bis(nonafluorobutyl) ether Perfluoro(dibutyl ether)
$C_8F_{18}O_2$		470.0580576	Octadecafluoro(2,9-dioxadecane) Perfluoro(1,6-dimethoxyhexane)-
$C_8H_4Cl_2O_2$	100-20-9	203.02396	1,4-Benzenedicarbonyl dichloride Terephthaloyl dichloride
$C_8H_4Cl_4O$		257.92996	4-(Trichloromethyl)benzoyl chloride
$C_8H_4F_6$	402-31-3	214.1101792	1,3-Bis(trifluoromethyl)-benzene
$C_8H_4O_3$	85-44-9	148.11796	1,2-Benzenedicarboxylic anhydride Phthalic anhydride
$C_8H_5Cl_3O$		223.4852	4-(Dichloromethyl)benzoyl chloride
$C_8H_5F_{13}O$		364.1063416	3,3,4,4,5,5,6,6,7,7,8,8,8-Tridecafluoro-1-octanol 2-(Perfluorohexyl)ethanol
$C_8H_5MnO_3$	12079-65-1	204.06395	Cyclopentadienylmanganese tricarbonyl
$C_8H_6Cl_2O$	876-08-4	189.04044	4-(Chloromethyl)benzoyl chloride
$C_8H_6N_2$	91-19-0	130.14912	Quinoxaline
$C_8H_6N_2$ (Cont.)			1,4-Benzodiazine 1,4-Diazanaphthalene
$C_8H_6O_4$	121-91-5	166.13324	Isophthalic acid
$C_8H_6O_4$	88-99-3	166.13324	Phthalic acid
$C_8H_6O_4$	100-21-0	166.13324	Terephthalic acid
C_8H_6S	95-15-8	134.20164	2,3-Benzothiophene Thianaphthene Thionaphthene Benzo[b]thiophene
C_8H_7ClO	874-60-2	154.59568	4-Toluoyl chloride
$C_8H_7ClO_2$	620-73-5	170.59508	Phenyl chloroethanoate Phenyl chloroacetate Chloroethanoic acid phenyl ester Chloroacetic acid phenyl ester
C_8H_7N	120-72-9	117.15032	Indole
C_8H_7N	140-29-4	117.15032	Phenylethanenitrile Benzyl cyanide Phenylacetonitrile
C_8H_7N	529-19-1	117.15032	2-Tolunitrile
C_8H_7N	620-22-4	117.15032	3-Tolunitrile
C_8H_7N	104-85-8	117.15032	4-Tolunitrile
C_8H_8	100-42-5	104.15152	Vinylbenzene Styrene
C_8H_8O	98-86-2	120.15092	Methyl phenyl ketone Acetophenone
C_8H_8O	96-09-3	120.15092	Phenyloxirane (Epoxyethyl)benzene Epoxystyrene Styrene oxide
C_8H_8O	104-87-0	120.15092	4-Tolualdehyde
$C_8H_8O_2$	493-09-4	136.15032	2,3-Benzo-1,4-dioxane 1,4-Benzodioxan 1,2-(Ethylenedioxy)benzene
$C_8H_8O_2$	135-02-4	136.15032	2-Methoxybenzaldehyde 2-Anisaldehyde
$C_8H_8O_2$	591-31-1	136.15032	3-Methoxybenzaldehyde 3-Anisaldehyde
$C_8H_8O_2$	123-11-5	136.15032	4-Methoxybenzaldehyde 4-Anisaldehyde
$C_8H_8O_2$	93-58-3	136.15032	Methyl benzoate Benzoic acid methyl ester
$C_8H_8O_2$	122-79-2	136.15032	Phenyl ethanoate Phenyl acetate Ethanoic acid phenyl ester Acetic acid phenyl ester
$C_8H_8O_2$	103-82-2	136.15032	Phenylethanoic acid Phenylacetic acid
$C_8H_8O_2$	118-90-1	136.15032	2-Toluic acid
$C_8H_8O_2$	99-04-7	136.15032	3-Toluic acid
$C_8H_8O_2$	99-94-5	136.15032	4-Toluic acid
$C_8H_8O_3$	119-36-8	152.14972	Methyl 2-hydroxybenzoate Methyl salicylate 2-Hydroxybenzoic acid methyl ester Salicylic acid methyl ester
$C_8H_9BO_3$		163.96866	2-Ethoxy-1,3-dioxa-2-bora-indan Ethyl 1,2-phenylene borate
C_8H_9Cl	672-65-1	140.61216	(1-Chloroethyl)benzene
C_8H_9N	496-15-1	119.1662	Indoline
C_8H_9N	140-76-1	119.1662	2-Methyl-5-vinylpyridine

Formula	CAS RN	M_r	Name
C_8H_9NO	613-91-2	135.1656	Methyl phenyl ketone oxime Acetophenone oxime
C_8H_9NO	103-84-4	135.1656	N-Phenylethanamide Acetanilide
$C_8H_9NO_2$	612-22-6	151.165	2-Ethylnitrobenzene
$C_8H_9NO_2$	100-12-9	151.165	4-Ethylnitrobenzene
C_8H_{10}	95-47-6	106.1674	1,2-Dimethylbenzene 1,2-Xylene
C_8H_{10}	108-38-3	106.1674	1,3-Dimethylbenzene 1,3-Xylene
C_8H_{10}	106-42-3	106.1674	1,4-Dimethylbenzene 1,4-Xylene
C_8H_{10}	100-41-4	106.1674	Ethylbenzene
$C_8H_{10}O$	526-75-0	122.1668	2,3-Dimethylphenol 2,3-Xylenol
$C_8H_{10}O$	105-67-9	122.1668	2,4-Dimethylphenol 2,4-Xylenol
$C_8H_{10}O$	95-87-4	122.1668	2,5-Dimethylphenol 2,5-Xylenol
$C_8H_{10}O$	576-26-1	122.1668	2,6-Dimethylphenol 2,6-Xylenol
$C_8H_{10}O$	95-65-8	122.1668	3,4-Dimethylphenol 3,4-Xylenol
$C_8H_{10}O$	108-68-9	122.1668	3,5-Dimethylphenol 3,5-Xylenol
$C_8H_{10}O$	90-00-6	122.1668	2-Ethylphenol
$C_8H_{10}O$	620-17-7	122.1668	3-Ethylphenol
$C_8H_{10}O$	123-07-9	122.1668	4-Ethylphenol
$C_8H_{10}O$	103-73-1	122.1668	Ethyl phenyl ether Phenetole Ethoxybenzene
$C_8H_{10}O$	578-58-5	122.1668	2-Methylanisole 2-Methoxytoluene
$C_8H_{10}O$	100-84-5	122.1668	3-Methylanisole 3-Methoxytoluene
$C_8H_{10}O$	104-93-8	122.1668	4-Methylanisole 4-Methoxytoluene
$C_8H_{10}O$	98-85-1	122.1668	1-Phenylethanol sec-Phenethyl alcohol α-Methylbenzyl alcohol
$C_8H_{10}O$	60-12-8	122.1668	2-Phenylethanol 2-Phenethyl alcohol Phenethyl alcohol
$C_8H_{10}O_2$	91-16-7	138.1662	1,2-Dimethoxybenzene Veratrole
$C_8H_{10}O_2$	612-16-8	138.1662	2-Methoxybenzyl alcohol 2-Anisyl alcohol
$C_8H_{10}O_2$	6971-51-3	138.1662	3-Methoxybenzyl alcohol 3-Anisyl alcohol
$C_8H_{10}O_2$	105-13-5	138.1662	4-Methoxybenzyl alcohol 4-Anisyl alcohol
$C_8H_{10}O_3S$	57352-34-8	186.2316	Ethylbenzenesulfonic acid <undefined positional isomer>
$C_8H_{10}O_4$	2274-11-5	170.165	1,2-Ethanediol dipropenoate Ethylene glycol diacrylate Ethylene dipropenoate Ethylene diacrylate
$C_8H_{11}ClSi$	768-33-2	170.71354	Chlorodimethylphenylsilane
$C_8H_{11}N$	95-68-1	121.18208	2,4-Dimethylaniline 2,4-Xylidine
$C_8H_{11}N$	87-62-7	121.18208	2,6-Dimethylaniline 2,6-Xylidine
$C_8H_{11}N$	121-69-7	121.18208	N,N-Dimethylaniline
$C_8H_{11}N$	103-69-5	121.18208	N-Ethylaniline
$C_8H_{11}N$	104-90-5	121.18208	5-Ethyl-2-methylpyridine
$C_8H_{11}N$	529-21-5	121.18208	3-Ethyl-4-methylpyridine
$C_8H_{11}N$	64-04-0	121.18208	(2-Phenylethyl)amine Phenethylamine
$C_8H_{11}N$	1462-84-6	121.18208	2,3,6-Trimethylpyridine
$C_8H_{11}N$	108-75-8	121.18208	2,4,6-Trimethylpyridine 2,4,6-Collidine
$C_8H_{11}NO$	94-70-2	137.18148	2-Ethoxyaniline 2-Phenetidine
$C_8H_{11}NO$	156-43-4	137.18148	4-Ethoxyaniline 4-Phenetidine
C_8H_{12}	111-78-4	108.18328	1,5-Cyclooctadiene
C_8H_{12}	100-40-3	108.18328	4-Vinylcyclohexene
$C_8H_{12}N_2$	629-40-3	136.19676	Octanedinitrile Suberodinitrile 1,6-Dicyanohexane
$C_8H_{12}N_2O_2$	822-06-0	168.19556	1,6-Diisocyanatohexane Hexamethylene diisocyanate
$C_8H_{12}O$		124.18268	cis-9-Oxabicyclo[6.1.0]non-4-ene 5,6-Epoxy-cis-cyclooctene cis-Cyclooctene-5,6-epoxide
$C_8H_{12}O_3$		156.18148	2-Vinyloxyethyl 2-methylpropenoate 2-Vinyloxyethyl methacrylate 2-Methylpropenoic acid 2-vinyloxyethyl ester Methacrylic acid 2-vinyloxyethyl ester
$C_8H_{12}O_4$		172.18088	Methyl vinyl pentanedioate Methyl vinyl glutarate Pentanedioic acid methyl vinyl ester Glutaric acid methyl vinyl ester
C_8H_{14}	627-58-7	110.19916	2,5-Dimethyl-1,5-hexadiene
C_8H_{14}	764-13-6	110.19916	2,5-Dimethyl-2,4-hexadiene
C_8H_{14}	3710-30-3	110.19916	1,7-Octadiene
C_8H_{14}	629-05-0	110.19916	1-Octyne
C_8H_{14}	2809-67-8	110.19916	2-Octyne
C_8H_{14}	15232-76-5	110.19916	3-Octyne
C_8H_{14}	1942-45-6	110.19916	4-Octyne
C_8H_{14}	695-12-5	110.19916	Vinylcyclohexane
$C_8H_{14}O$	2816-57-1	126.19856	2,6-Dimethylcyclohexanone <mixture of isomers>
$C_8H_{14}O$		126.19856	2-Ethylhexenal <undefined positional isomer>
$C_8H_{14}O$	64344-45-2	126.19856	trans-2-Ethyl-2-hexenal
$C_8H_{14}O$	19600-63-6	126.19856	5-Hexenyloxirane 1,2-Epoxy-7-octene 1-Octene-7,8-epoxide
$C_8H_{14}O$	110-93-0	126.19856	6-Methyl-5-hepten-2-one
$C_8H_{14}O$	286-62-4	126.19856	9-Oxabicyclo[6.1.0]nonane 1,2-Epoxycyclooctane Cyclooctene oxide Cyclooctane 1,2-epoxide
$C_8H_{14}O_2$	97-88-1	142.19796	Butyl 2-methylpropenoate Butyl methacrylate 2-Methylpropenoic acid butyl ester Methacrylic acid butyl ester

Formula	CAS RN	M_r	Name
$C_8H_{14}O_2$	622-45-7	142.19796	Cyclohexyl ethanoate
			Cyclohexyl acetate
			Ethanoic acid cyclohexyl ester
			Acetic acid cyclohexyl ester
$C_8H_{14}O_3$	106-31-0	158.19736	Butanoic anhydride
			Butyric anhydride
$C_8H_{14}O_3$	97-72-3	158.19736	2-Methylpropanoic anhydride
			Isobutyric anhydride
$C_8H_{14}O_3$	764-99-8	158.19736	3,5,6-Trioxa-1,10-undecadiene
			Diethylene glycol divinyl ether
$C_8H_{14}O_4$	628-67-1	174.19676	1,4-Butanediol diethanoate
			1,4-Butanediol diacetate
			Tetramethylene diacetate
			Tetramethylene diethanoate
			1,4-Butylene glycol diacetate
$C_8H_{14}O_4$		174.19676	2,3-Butanediol diethanoate
			2,3-Butylene glycol diacetate
$C_8H_{14}O_4$		174.19676	meso-2,3-Butanediol diethanoate
			meso-2,3-Butylene glycol diacetate
$C_8H_{14}O_4$	627-93-0	174.19676	Dimethyl hexanedioate
			Dimethyl adipate
			Hexanedioic acid dimethyl ester
			Adipic acid dimethyl ester
$C_8H_{14}O_4$	626-86-8	174.19676	Ethyl hexanedioate
			Ethyl adipate
			Hexanedioic acid ethyl ester
			Adipic acid ethyl ester
$C_8H_{14}O_4$	123-25-1	174.19676	Ethyl succinate
$C_8H_{14}O_4$	61836-77-9	174.19676	1,6-Hexanediol dimethanoate
			Hexamethylene diformate
			Hexamethylene dimethanoate
			1,6-Hexanediol diformate
$C_8H_{14}O_4$	505-48-6	174.19676	Octanedioic acid
			Suberic acid
$C_8H_{15}N$	124-12-9	125.21384	Octanenitrile
$C_8H_{15}NO_2$	2867-47-2	157.21264	2-(Dimethylamino)ethyl 2-methylpropenoate
			2-(Dimethylamino)ethyl methacrylate
			2-Methylpropenoic acid 2--(dimethylamino)ethyl ester
			Methacrylic acid 2-(dimethylamino)ethyl ester
$C_8H_{15}NO_2$		157.21264	6-Dimethylamino-2,4-hexanedione
			(Dimethylaminomethyl)acetylacetone
$C_8H_{15}NaO_2$	1984-06-1	166.195668	Sodium octanoate
			Sodium caprylate
			Caprylic acid sodium salt
			Octanoic acid sodium salt
C_8H_{16}	292-64-8	112.21504	Cyclooctane
C_8H_{16}	590-66-9	112.21504	1,1-Dimethylcyclohexane
C_8H_{16}	2207-01-4	112.21504	cis-1,2-Dimethylcyclohexane
C_8H_{16}	6876-23-9	112.21504	trans-1,2-Dimethylcyclohexane
C_8H_{16}	591-21-9	112.21504	1,3-Dimethylcyclohexane
C_8H_{16} (Cont.)			<mixture of cis/trans isomers>
C_8H_{16}	638-04-0	112.21504	cis-1,3-Dimethylcyclohexane
C_8H_{16}	2207-03-6	112.21504	trans-1,3-Dimethylcyclohexane
C_8H_{16}	589-90-2	112.21504	1,4-Dimethylcyclohexane
			<mixture of cis/trans isomers>
C_8H_{16}	624-29-3	112.21504	cis-1,4-Dimethylcyclohexane
C_8H_{16}	2207-04-7	112.21504	trans-1,4-Dimethylcyclohexane
C_8H_{16}	1678-91-7	112.21504	Ethylcyclohexane
C_8H_{16}	16747-50-5	112.21504	1-Ethyl-1-methylcyclopentane
C_8H_{16}	3875-51-2	112.21504	Isopropylcyclopentane
C_8H_{16}	627-97-4	112.21504	2-Methyl-2-heptene
C_8H_{16}	111-66-0	112.21504	1-Octene
C_8H_{16}	111-67-1	112.21504	2-Octene
			<mixture of cis/trans isomers>
C_8H_{16}	7642-04-8	112.21504	cis-2-Octene
C_8H_{16}	13389-42-9	112.21504	trans-2-Octene
C_8H_{16}	14850-22-7	112.21504	cis-3-Octene
C_8H_{16}	14919-01-8	112.21504	trans-3-Octene
C_8H_{16}	7642-15-1	112.21504	cis-4-Octene
C_8H_{16}	14850-23-8	112.21504	trans-4-Octene
C_8H_{16}	2040-96-2	112.21504	Propylcyclopentane
C_8H_{16}	107-39-1	112.21504	2,4,4-Trimethyl-1-pentene
			α-Diisobutylene
C_8H_{16}	107-40-4	112.21504	2,4,4-Trimethyl-2-pentene
			β-Diisobutylene
$C_8H_{16}Br_2$	4549-32-0	272.02304	1,8-Dibromooctane
$C_8H_{16}Cl_4O_4Si$	18290-84-1	346.10894	Tetrakis(2-chloroethoxy)silane
			Tetrakis(2-chloroethyl) orthosilicate
$C_8H_{16}O$	4442-79-9	128.21444	2-Cyclohexylethanol
			2-Cyclohexylethyl alcohol
			Cyclohexaneethanol
$C_8H_{16}O$	123-05-7	128.21444	2-Ethylhexanal
$C_8H_{16}O$	124-13-0	128.21444	Octanal
			1-Heptanecarbaldehyde
$C_8H_{16}O$	111-13-7	128.21444	2-Octanone
			Hexyl methyl ketone
$C_8H_{16}O$	5857-36-3	128.21444	2,2,4-Trimethyl-3-pentanone
$C_8H_{16}O_2$	109-21-7	144.21384	Butyl butanoate
			Butyl butyrate
			Butanoic acid butyl ester
			Butyric acid butyl ester
$C_8H_{16}O_2$	123-66-0	144.21384	Ethyl hexanoate
$C_8H_{16}O_2$	149-57-5	144.21384	(±)-2-Ethylhexanoic acid
$C_8H_{16}O_2$	142-92-7	144.21384	Hexyl ethanoate
			Hexyl acetate
			Ethanoic acid hexyl ester
			Acetic acid hexyl ester
$C_8H_{16}O_2$	105-68-0	144.21384	3-Methylbutyl propanoate
			Isopentyl propanoate
			Propanoic acid 3-methylbutyl ester
			Propanoic acid isopentyl ester
			Isoamyl propionate
			Propionic acid isoamyl ester
$C_8H_{16}O_2$	106-73-0	144.21384	Methyl heptanoate
			Methyl enanthate

Formula	CAS RN	M_r	Name
$C_8H_{16}O_2$ (Cont.)			Heptanoic acid methyl ester
			Enanthic acid methyl ester
			Oenanthic acid methyl ester
			Methyl oenanthate
$C_8H_{16}O_2$		144.21384	Methyl 2-methylhexanoate
			Methyl isoheptanoate
			2-Methylhexanoic acid methyl ester
			Isoheptanoic acid methyl ester
$C_8H_{16}O_2$	539-90-2	144.21384	2-Methylpropyl butanoate
			Isobutyl butanoate
			Butanoic acid 2-methylpropyl ester
			Butyric acid isobutyl ester
			Isobutyl butyrate
$C_8H_{16}O_2$	97-85-8	144.21384	2-Methylpropyl 2-methylpropanoate
			Isobutyl isobutanoate
			Isobutyl isobutyrate
			Isobutanoic acid isobutyl ester
			Isobutyric acid isobutyl ester
			2-Methylpropanoic acid 2-methylpropyl ester
$C_8H_{16}O_2$	124-07-2	144.21384	Octanoic acid
			Caprylic acid
			1-Heptanecarboxylic acid
$C_8H_{16}O_2$	557-00-6	144.21384	Propyl 3-methylbutanoate
$C_8H_{16}O_3$	112-07-2	160.21324	2-Butoxyethyl ethanoate
			2-Butoxyethyl acetate
			Ethanoic acid 2-butoxyethyl ester
			Butyl Cellosolve acetate
			Ethylene glycol butyl ether acetate
			Acetic acid 2-butoxyethyl ester
$C_8H_{16}O_3$	816-50-2	160.21324	Butyl 2-hydroxy-2-methylpropanoate
			2-Hydroxy-2-methylpropanoic acid butyl ester
			Butyl 2-hydroxy-2-methylpropionate
			Butyl 2-methyllactate
			2-Hydroxy-2-methylpropionic acid butyl ester
			2-Methyllactic acid butyl ester
$C_8H_{16}O_4$	294-93-9	176.21264	1,4,7,10-Tetraoxacyclododecane
			12-Crown-4
$C_8H_{16}O_4$		176.21264	3,6,9-Trioxa-10-undecen-1-ol
			Triethylene glycol vinyl ether
$C_8H_{17}Br$	111-83-1	193.12698	1-Bromooctane
			Octyl bromide
$C_8H_{17}Cl$	123-04-6	148.67568	1-Chloro-2-ethylhexane
			2-Ethylhexyl chloride
			3-(Chloromethyl)heptane
$C_8H_{17}Cl$	111-85-3	148.67568	1-Chlorooctane
			Octyl chloride
$C_8H_{17}F$	463-11-6	132.2213832	1-Fluorooctane
$C_8H_{17}I$	629-27-6	240.12745	1-Iodooctane
			Octyl iodide
$C_8H_{17}I$	557-36-8	240.12745	2-Iodooctane
$C_8H_{17}N$	98-94-2	127.22972	N,N-Dimethylcyclohexylamine

Formula	CAS RN	M_r	Name
$C_8H_{17}NO$		143.22912	Octanal oxime
			Octyl aldehyde oxime
			Caprylic aldehyde oxime
$C_8H_{17}NO$	7207-49-0	143.22912	2-Octanone oxime
$C_8H_{17}NO$	7207-50-3	143.22912	3-Octanone oxime
$C_8H_{17}NO$	7207-51-4	143.22912	4-Octanone oxime
$C_8H_{17}NaO_4S$	142-31-4	232.276348	Octyl sodium sulfate
C_8H_{18}	590-73-8	114.23092	2,2-Dimethylhexane
C_8H_{18}	584-94-1	114.23092	2,3-Dimethylhexane
C_8H_{18}	589-43-5	114.23092	2,4-Dimethylhexane
C_8H_{18}	592-13-2	114.23092	2,5-Dimethylhexane
C_8H_{18}	563-16-6	114.23092	3,3-Dimethylhexane
C_8H_{18}	583-48-2	114.23092	3,4-Dimethylhexane
C_8H_{18}	619-99-8	114.23092	3-Ethylhexane
C_8H_{18}	609-26-7	114.23092	3-Ethyl-2-methylpentane
C_8H_{18}	1067-08-9	114.23092	3-Ethyl-3-methylpentane
C_8H_{18}	592-27-8	114.23092	2-Methylheptane
C_8H_{18}	589-81-1	114.23092	3-Methylheptane
C_8H_{18}	589-53-7	114.23092	4-Methylheptane
C_8H_{18}	111-65-9	114.23092	Octane
C_8H_{18}	594-82-1	114.23092	2,2,3,3-Tetramethylbutane
C_8H_{18}	564-02-3	114.23092	2,2,3-Trimethylpentane
C_8H_{18}	540-84-1	114.23092	2,2,4-Trimethylpentane
			Isooctane
C_8H_{18}	560-21-4	114.23092	2,3,3-Trimethylpentane
C_8H_{18}	565-75-3	114.23092	2,3,4-Trimethylpentane
$C_8H_{18}N_2$	2159-75-3	142.2444	Azobutane
$C_8H_{18}N_2$	150-77-6	142.2444	N,N,N',N',-Tetraethyl-1,2-ethanediamine
			N,N,N',N'-Tetraethylethylenediamine
$C_8H_{18}N_2O_2$	122-96-3	174.2432	1,4-Bis(2-hydroxyethyl)piperazine
$C_8H_{18}O$	6863-58-7	130.23032	Bis(1-methylpropyl) ether
			Di-sec-butyl ether
$C_8H_{18}O$	628-55-7	130.23032	Bis(2-methylpropyl) ether
			Isobutyl ether
			Diisobutyl ether
$C_8H_{18}O$	17071-47-5	130.23032	Butyl 2-methylpropyl ether
			Butyl isobutyl ether
$C_8H_{18}O$	142-96-1	130.23032	Dibutyl ether
			Butyl ether
$C_8H_{18}O$	6163-66-2	130.23032	Di-tert-butyl ether
$C_8H_{18}O$	104-76-7	130.23032	2-Ethyl-1-hexanol
$C_8H_{18}O$	629-32-3	130.23032	Heptyl methyl ether
$C_8H_{18}O$	5582-82-1	130.23032	3-Methyl-3-heptanol
$C_8H_{18}O$	14979-39-6	130.23032	4-Methyl-3-heptanol
$C_8H_{18}O$	18720-65-5	130.23032	5-Methyl-3-heptanol
$C_8H_{18}O$	111-87-5	130.23032	1-Octanol
			Octyl alcohol
			Capryl alcohol
$C_8H_{18}O$	4128-31-8	130.23032	2-Octanol
$C_8H_{18}O$	123-96-6	130.23032	(±)-2-Octanol
$C_8H_{18}O$	20296-29-1	130.23032	3-Octanol
$C_8H_{18}O$	589-62-8	130.23032	4-Octanol
$C_8H_{18}OS$	2168-93-6	162.29632	Dibutyl sulfoxide
$C_8H_{18}O_2$	110-05-4	146.22972	Di-tert-butyl peroxide
$C_8H_{18}O_2$	18854-56-3	146.22972	1,2-Dipropoxyethane
$C_8H_{18}O_2$	629-41-4	146.22972	1,8-Octanediol
			Octamethylene glycol
$C_8H_{18}O_2S$	598-04-9	178.29572	Dibutyl sulfone
			Butyl sulfone

Formula	CAS RN	M_r	Name
$C_8H_{18}O_3$	112-34-5	162.22912	3,6-Dioxa-1-decanol 2-(2-Butoxyethoxy)ethanol Butyl carbitol Diethylene glycol butyl ether
$C_8H_{18}O_3$		162.22912	4,8-Dioxa-1-decanol Dipropylene glycol ethyl ether
$C_8H_{18}O_3$	78-39-7	162.22912	1,1,1-Triethoxyethane Orthoethanoic acid triethyl ester Triethyl orthoacetate Triethyl orthoethanoate Orthoacetic acid triethyl ester
$C_8H_{18}O_3$	112-36-7	162.22912	3,6,9-Trioxaundecane Bis(2-ethoxyethyl) ether 2-Ethoxyethyl ether Diethyl carbitol 3-Oxa-1,5-pentanediol diethyl ether Diethylene glycol diethyl ether
$C_8H_{18}O_4$	112-49-2	178.22852	2,5,8,11-Tetraoxadodecane 1,2-Bis(2-methoxyethoxy)ethane Triethylene glycol dimethyl ether Triglyme
$C_8H_{18}O_5$	112-60-7	194.22792	3,6,9-Trioxa-1,11-undecane-diol Tetraethylene glycol
$C_8H_{18}S$	592-65-4	146.29692	Bis(2-methylpropyl) sulfide Diisobutyl sulfide
$C_8H_{18}S$	107-47-1	146.29692	Di-tert-butyl sulfide tert-Butyl sulfide
$C_8H_{18}S$	544-40-1	146.29692	5-Thianonane Dibutyl sulfide Dibutyl thioether
$C_8H_{19}N$	110-96-3	129.2456	Bis(2-methylpropyl)amine Diisobutylamine
$C_8H_{19}N$	111-92-2	129.2456	Dibutylamine
$C_8H_{19}N$	104-75-6	129.2456	2-Ethylhexylamine
$C_8H_{19}N$	111-86-4	129.2456	Octylamine 1-Aminooctane
$C_8H_{19}O_4P$	107-66-4	210.210222	Dibutyl hydrogen phosphate Phosphoric acid dibutyl ester Dibutyl phosphate
$C_8H_{20}B_2O$	7318-84-5	153.8682	Tetraethyldiboroxane Oxybis(diethylborane) Diethylborinic anhydride
$C_8H_{20}BrN$	71-91-0	210.15754	Tetraethylammonium bromide
$C_8H_{20}BrNO_4$	4328-04-5	274.15514	Tetrakis(2-hydroxyethyl)-ammonium bromide Tetraethanolammonium bromide
$C_8H_{20}ClN$		165.70624	Tetraethylammonium chloride
$C_8H_{20}GeO_4$	14165-55-0	252.8544	Tetraethoxygermane Germanium(IV) ethoxide
$C_8H_{20}IN$	68-05-3	257.15801	Tetraethylammonium iodide
$C_8H_{20}N_2$	111-51-3	144.26028	N,N,N',N'-Tetramethyl-1,4-butanediamine N,N,N',N'-Tetramethyl-1,4-diaminobutane
$C_8H_{20}N_2O_2S$	2832-49-7	208.32508	Tetraethylsulfamide
$C_8H_{20}O_4Si$	78-10-4	208.3299	Tetraethoxysilane Tetraethyl orthosilicate
$C_8H_{20}O_4Sn$		298.9544	Tetraethoxytin Tetraethoxystannane
$C_8H_{20}Pb$	78-00-2	323.4468	Tetraethyllead Tetraethylplumbane
$C_8H_{20}Si$	631-36-7	144.3323	Tetraethylsilane
$C_8H_{22}N_4$	10563-26-5	174.28964	N,N'-Bis(3-aminopropyl)-1,2-ethanediamine N,N'-Bis(3-aminopropyl)ethylenediamine
$C_8H_{24}O_2Si_3$	107-51-7	236.53386	Octamethyltrisiloxane
$C_8H_{24}O_4Si_4$	556-67-2	296.61816	Octamethylcyclotetrasiloxane
C_9F_{20}	375-96-2	488.067064	Icosafluorononane Perfluorononane
$C_9F_{21}N$	338-83-0	521.0722072	Henicosafluorotripropylamine Perfluorotripropylamine
$C_9H_4F_{17}NO_2$		481.1091544	Ammonium heptadecafluorononanoate Ammonium perfluorononanoate Perfluorononanoic acid ammonium salt Heptadecafluorononanoic acid ammonium salt
$C_9H_4O_5$	552-30-7	192.12776	1,2,4-Benzenetricarboxylic anhydride Trimellitic anhydride
$C_9H_5NbO_4$		270.04268	Cyclopentadienylniobium tetracarbonyl
$C_9H_6N_2O_2$	584-84-9	174.15892	2,4-Diisocyanatotoluene Tolylene 2,4-diisocyanate 4-Methyl-1,3-phenylene diisocyanate
$C_9H_6N_2O_2$	91-08-7	174.15892	2,6-Diisocyanatotoluene Tolylene 2,6-diisocyanate 2-Methyl-1,3-phenylene diisocyanate
C_9H_7N	119-65-3	129.16132	Isoquinoline 2-Azanaphthalene
C_9H_7N	91-22-5	129.16132	Quinoline 1-Azanaphthalene
C_9H_7NO	59-31-4	145.16072	2-Hydroxyquinoline 2-Quinolinol
C_9H_7NO	611-36-9	145.16072	4-Hydroxyquinoline 4-Quinolinol
C_9H_8	95-13-6	116.16252	Indene
$C_9H_8O_2$	140-10-3	148.16132	trans-3-Phenylpropenoic acid Cinnamic acid
C_9H_{10}	873-49-4	118.1784	Cyclopropylbenzene
C_9H_{10}	496-11-7	118.1784	Indan
C_9H_{10}	98-83-9	118.1784	Isopropenylbenzene α-Methylstyrene
C_9H_{10}	611-15-4	118.1784	2-Methylvinylbenzene 2-Methylstyrene 2-Vinyltoluene
C_9H_{10}	100-80-1	118.1784	3-Methylvinylbenzene 3-Methylstyrene 3-Vinyltoluene

Formula	CAS RN	M_r	Name
C_9H_{10}	622-97-9	118.1784	4-Methylvinylbenzene
			4-Methylstyrene
			4-Vinyltoluene
$C_9H_{10}O$	122-00-9	134.1778	Methyl 4-tolyl ketone
			4-Acetyltoluene
$C_9H_{10}O_2$		150.1772	2,3-Benzo-1,4-dioxepane
			1,2-(Trimethylenedioxy)benzene
$C_9H_{10}O_2$	33632-34-7	150.1772	2,3-Benzo-5-methyl-1,4-dioxane
			5-Methyl-1,4-benzodioxan
			1,2-(Propylenedioxy)benzene
$C_9H_{10}O_2$	140-11-4	150.1772	Benzyl ethanoate
			Benzyl acetate
			Ethanoic acid benzyl ester
			Acetic acid benzyl ester
$C_9H_{10}O_2$	93-89-0	150.1772	Ethyl benzoate
			Benzoic acid ethyl ester
$C_9H_{10}O_2$	99-36-5	150.1772	Methyl 3-methylbenzoate
			Methyl 3-toluate
			3-Methylbenzoic acid methyl ester
			3-Toluic acid methyl ester
$C_9H_{10}O_2$	99-75-2	150.1772	Methyl 4-methylbenzoate
			Methyl 4-toluate
			4-Toluic acid methyl ester
			4-Methylbenzoic acid methyl ester
$C_9H_{10}O_2$	101-41-7	150.1772	Methyl phenylethanoate
			Methyl phenylacetate
			Phenylethanoic acid methyl ester
			Phenylacetic acid methyl ester
$C_9H_{10}O_3$	118-61-6	166.1766	Ethyl 2-hydroxybenzoate
			Ethyl salicylate
			2-Hydroxybenzoic acid ethyl ester
			Salicylic acid ethyl ester
$C_9H_{10}O_3$	120-47-8	166.1766	Ethyl 4-hydroxybenzoate
			4-Hydroxybenzoic acid ethyl ester
$C_9H_{10}O_3$	606-45-1	166.1766	Methyl 2-methoxybenzoate
			Methyl 2-anisate
			2-Methoxybenzoic acid methyl ester
			2-Anisic acid methyl ester
$C_9H_{10}O_3$	121-98-2	166.1766	Methyl 4-methoxybenzoate
			Methyl 4-anisate
			4-Methoxybenzoic acid methyl ester
			4-Anisic acid methyl ester
$C_9H_{11}BO_3$		177.99554	2-Propoxy-1,3-dioxa-2-bora-indan
			Propyl 1,2-phenylene borate
$C_9H_{11}N$	635-46-1	133.19308	1,2,3,4-Tetrahydroquinoline
$C_9H_{11}NO_2$	673-06-3	165.19188	(R)-(+)-2-Amino-3-phenyl-propanoic acid
			(R)-(+)-Phenylalanine
			D-Phenylalanine
$C_9H_{11}NO_2$	63-91-2	165.19188	(S)-(−)-2-Amino-3-phenylpro-panoic acid
			L-Phenylalanine
			(S)-(−)-Phenylalanine
$C_9H_{11}NO_3$	60-18-4	181.19128	L-Tyrosine
C_9H_{12}	3048-65-5	120.19428	Bicyclo[4.3.0]nona-3,7-diene
			4,7,8,9-Tetrahydroindene
C_9H_{12}	16219-75-3	120.19428	2-Ethylidenebicyclo[2.1.1]-hept-5-ene
			5-Ethylidene-2-norbornene
C_9H_{12}	611-14-3	120.19428	2-Ethyltoluene
C_9H_{12}	620-14-4	120.19428	3-Ethyltoluene
C_9H_{12}	622-96-8	120.19428	4-Ethyltoluene
C_9H_{12}	98-82-8	120.19428	Isopropylbenzene
			Cumene
C_9H_{12}	103-65-1	120.19428	Propylbenzene
C_9H_{12}	526-73-8	120.19428	1,2,3-Trimethylbenzene
C_9H_{12}	95-63-6	120.19428	1,2,4-Trimethylbenzene
			Pseudocumene
C_9H_{12}	108-67-8	120.19428	1,3,5-Trimethylbenzene
			Mesitylene
C_9H_{12}	3048-64-4	120.19428	2-Vinylbicyclo[2.2.1]hept-5-ene
			5-Vinyl-2-norbornene
$C_9H_{12}O$	539-30-0	136.19368	Benzyl ethyl ether
$C_9H_{12}O$	99-89-8	136.19368	4-Isopropylphenol
$C_9H_{12}O$	2741-16-4	136.19368	Isopropyl phenyl ether
$C_9H_{12}O$	122-97-4	136.19368	3-Phenyl-1-propanol
			3-Phenylpropyl alcohol
$C_9H_{12}O$	617-94-7	136.19368	2-Phenyl-2-propanol
$C_9H_{12}O$	527-60-6	136.19368	2,4,6-Trimethylphenol
			Mesitol
$C_9H_{13}F_7O$		270.1904424	1-(Heptafluoroisopropoxy)-hexane
			1-(Perfluoroisopropoxy)hexane
$C_9H_{13}N$		135.20896	Butylpyridine
			<undefined positional isomer>
$C_9H_{13}N$	539-32-2	135.20896	3-Butylpyridine
$C_9H_{13}N$	609-72-3	135.20896	N,N-Dimethyl-2-toluidine
$C_9H_{13}N$	2038-57-5	135.20896	(3-Phenylpropyl)amine
C_9H_{14}	2146-39-6	122.21016	2-Vinylbicyclo[2.2.1]heptane
			2-Vinylnorbornane
$C_9H_{14}O$	78-59-1	138.20956	3,5,5-Trimethyl-2-cyclohexen-1-one
			Isophorone
$C_9H_{14}O_4$		186.20776	Methyl vinyl hexanedioate
			Methyl vinyl adipate
			Hexanedioic acid methyl vinyl ester
			Adipic acid methyl vinyl ester
$C_9H_{14}O_6$	102-76-1	218.20656	1,2,3-Propanetriol tris(-ethanoate)
			Glycerine triacetate
			Triacetin
$C_9H_{15}Cl_3O_2$		261.575	2,2,2-Tris(chloromethyl)-ethyl butanoate
			Pentaerythritetrichlorohydrin butyrate
			Butanoic acid 2,2,2-tris(chloro-methyl)ethyl ester
C_9H_{16}	2423-01-0	124.22604	1-Butylcyclopentene
C_9H_{16}	22531-00-6	124.22604	3-Butylcyclopentene
C_9H_{16}	2146-41-0	124.22604	2-Ethylbicyclo[2.2.1]heptane
			2-Ethylnorbornane
C_9H_{16}	3452-09-3	124.22604	1-Nonyne
C_9H_{16}	20184-89-8	124.22604	3-Nonyne

Formula	CAS RN	M_r	Name
$C_9H_{16}Cl_4$	1561-48-4	266.03684	1,1,1,9-Tetrachlorononane
$C_9H_{16}O$	873-94-9	140.22544	3,3,5-Trimethylcyclohexanone Dihydroisophorone
$C_9H_{16}O_2$	5726-19-2	156.22484	2-Methylcyclohexyl ethanoate 2-Methylcyclohexyl acetate Ethanoic acid 2-methylcyclohexyl ester Acetic acid 2-methylcyclohexyl ester
$C_9H_{16}O_4$	133-13-1	188.22364	Diethyl ethylpropanedioate Diethyl ethylmalonate Ethylmalonic acid diethyl ester Ethylpropanedioic acid diethyl ester
$C_9H_{16}O_4$	1732-08-7	188.22364	Dimethyl heptanedioate Dimethyl pimelate Heptanedioic acid dimethyl ester Pimelic acid dimethyl ester
$C_9H_{16}O_4$	123-99-9	188.22364	Nonanedioic acid Azelaic acid
$C_9H_{16}O_4$		188.22364	Propyl hexanedioate Propyl adipate Hexanedioic acid propyl ester Adipic acid propyl ester
C_9H_{18}	2040-95-1	126.24192	Butylcyclopentane
C_9H_{18}	696-29-7	126.24192	Isopropylcyclohexane
C_9H_{18}	124-11-8	126.24192	1-Nonene
C_9H_{18}	6434-78-2	126.24192	trans-2-Nonene
C_9H_{18}	20063-92-7	126.24192	trans-3-Nonene
C_9H_{18}	10405-85-3	126.24192	trans-4-Nonene
C_9H_{18}	1678-92-8	126.24192	Propylcyclohexane
C_9H_{18}	1795-27-3	126.24192	1,cis-3,cis-5-Trimethylcyclohexane
C_9H_{18}	1795-26-2	126.24192	1,cis-3,trans-5-Trimethylcyclohexane
$C_9H_{18}O$	108-83-8	142.24132	2,6-Dimethyl-4-heptanone Diisobutyl ketone
$C_9H_{18}O$	124-19-6	142.24132	Nonanal 1-Octanecarbaldehyde Nonyl aldehyde Pelargonaldehyde
$C_9H_{18}O$	821-55-6	142.24132	2-Nonanone Heptyl methyl ketone
$C_9H_{18}O$	502-56-7	142.24132	5-Nonanone Dibutyl ketone
$C_9H_{18}O_2$	112-06-1	158.24072	Heptyl ethanoate Heptyl acetate Ethanoic acid heptyl ester Acetic acid heptyl ester
$C_9H_{18}O_2$	106-27-4	158.24072	3-Methylbutyl butanoate Butanoic acid 3-methylbutyl ester Butanoic acid isopentyl ester Isopentyl butanoate Isoamyl butyrate Butyric acid isoamyl ester
$C_9H_{18}O_2$	111-11-5	158.24072	Methyl octanoate Methyl caprylate Octanoic acid methyl ester Caprylic acid methyl ester
$C_9H_{18}O_2$	589-59-3	158.24072	2-Methylpropyl 3-methylbutanoate Isobutyl 3-methylbutanoate 3-Methylbutanoic acid 2-methylpropyl ester Isobutyl isovalerate Isovaleric acid isobutyl ester
$C_9H_{18}O_2$	112-05-0	158.24072	Nonanoic acid Pelargonic acid 1-Octanecarboxylic acid
$C_9H_{18}O_3$	539-92-4	174.24012	Bis(2-methylpropyl) carbonate Diisobutyl carbonate Carbonic acid diisobutyl ester Carbonic acid bis(2-methylpropyl) ester
$C_9H_{18}O_3$	542-52-9	174.24012	Dibutyl carbonate Carbonic acid dibutyl ester
$C_9H_{18}O_5$	3610-27-3	206.23892	3,6,9-Trioxadecyl ethanoate Triethylene glycol methyl ether acetate
$C_9H_{19}F$	463-18-3	146.2482632	1-Fluorononane Nonyl fluoride
$C_9H_{19}N$	4945-48-6	141.2566	N-Butylpiperidine 1-Butylpiperidine
C_9H_{20}	3074-77-9	128.2578	2,3-Diethylpentane
C_9H_{20}	1067-20-5	128.2578	3,3-Diethylpentane
C_9H_{20}	1071-26-7	128.2578	2,2-Dimethylheptane
C_9H_{20}	3221-61-2	128.2578	2-Methyloctane
C_9H_{20}	2216-33-3	128.2578	3-Methyloctane
C_9H_{20}	2216-34-4	128.2578	4-Methyloctane
C_9H_{20}	111-84-2	128.2578	Nonane
C_9H_{20}	7154-79-2	128.2578	2,2,3,3-Tetramethylpentane
C_9H_{20}	1186-53-4	128.2578	2,2,3,4-Tetramethylpentane
C_9H_{20}	1070-87-7	128.2578	2,2,4,4-Tetramethylpentane
C_9H_{20}	16747-38-9	128.2578	2,3,3,4-Tetramethylpentane
C_9H_{20}	16747-26-5	128.2578	2,2,4-Trimethylhexane
C_9H_{20}	3522-94-9	128.2578	2,2,5-Trimethylhexane
C_9H_{20}	1069-53-0	128.2578	2,3,5-Trimethylhexane
$C_9H_{20}O$	929-56-6	144.2572	Methyl octyl ether
$C_9H_{20}O$	143-08-8	144.2572	1-Nonanol Nonyl alcohol
$C_9H_{20}O$	3452-97-9	144.2572	3,5,5-Trimethyl-1-hexanol
$C_9H_{20}O_3$		176.256	4,8-Dioxa-1-undecanol Dipropylene glycol propyl ether
$C_9H_{21}AsO_3$		252.18553	Triisopropoxyarsane Triisopropyl arsenite Triisopropoxyarsine
$C_9H_{21}AsO_3$	15606-91-4	252.18553	Tripropoxyarsane Tripropyl arsenite Tripropoxyarsine Arsenous acid tripropyl ester
$C_9H_{21}BO_3$	5419-55-6	188.07494	Triisopropoxyborane Triisopropyl borate Boric acid triisopropyl ester
$C_9H_{21}BO_3$	688-71-1	188.07494	Tripropoxyborane Tripropyl borate Boric acid tripropyl ester
$C_9H_{21}N$	112-20-9	143.27248	Nonylamine 1-Aminononane
$C_9H_{21}N$	102-69-2	143.27248	Tripropylamine

Formula	CAS RN	M_r	Name
$C_9H_{21}O_3Sb$		299.02094	Triisopropoxystibine Triisopropyl antimonate
$C_9H_{21}O_3Sb$		299.02094	Tripropoxystibine Tripropyl antimonate
$C_9H_{21}O_4P$		224.237102	Triisopropyl phosphate Phosphoric acid triisopropyl ester
$C_{10}F_8$	313-72-4	272.0972256	Octafluoronaphthalene Perfluoronaphthalene
$C_{10}F_{18}$	306-94-5	462.0812576	Octadecafluorobicyclo[4.4.0]-decane Perfluorodecalin Perfluorodecahydronaphthalene Octadecafluorodecahydronaphthalene
$C_{10}F_{19}LiO_2$		520.0194608	Lithium nonadecafluorodecanoate Lithium perfluorodecanoate Perfluorodecanoic acid lithium salt Nonadecafluorodecanoic acid lithium salt
$C_{10}F_{22}$	307-45-9	538.0748704	Docosafluorodecane Perfluorodecane
$C_{10}F_{22}$	3021-63-4	538.0748704	Docosafluoro(2,7-dimethyloctane) Perfluoro(2,7-dimethyloctane)
$C_{10}F_{22}O$	464-36-8	554.0742704	Bis(undecafluoropentyl) ether Perfluoro(dipentyl ether) Perfluoro(diamyl ether)
$C_{10}H_5F_{15}O_2$		442.124548	Ethyl pentadecafluorooctanoate Ethyl perfluorooctanoate Pentadecafluorooctanoic acid ethyl ester Perfluorooctanoic acid ethyl ester
$C_{10}H_7Br$	90-11-9	207.06958	1-Bromonaphthalene
$C_{10}H_7Cl$	90-13-1	162.61828	1-Chloronaphthalene 1-Naphthyl chloride
$C_{10}H_7Cl$		162.61828	2-Chloronaphthalene
$C_{10}H_8$	275-51-4	128.17352	Azulene
$C_{10}H_8$	91-20-3	128.17352	Naphthalene
$C_{10}H_8F_{14}O_2$		426.1499648	1,4-Bis(heptafluoroisopropoxy)butane 1,4-Bis(perfluoroisopropoxy)-butane
$C_{10}H_8N_2$	366-18-7	156.187	2,2'-Dipyridyl 2,2'-Bipyridine
$C_{10}H_8O$	90-15-3	144.17292	1-Naphthol 1-Hydroxynaphthalene
$C_{10}H_8O$	135-19-3	144.17292	2-Naphthol 2-Hydroxynaphthalene
$C_{10}H_9$		129.18146	Methylvinylbenzene Methylstyrene Vinyltoluene <mixture of positional isomers>
$C_{10}H_9N$	91-63-4	143.1882	2-Methylquinoline Quinaldine
$C_{10}H_9N$	491-35-0	143.1882	4-Methylquinoline Lepidine
$C_{10}H_9N$	612-60-2	143.1882	7-Methylquinoline
$C_{10}H_{10}$	108-57-6	130.1894	1,3-Divinylbenzene
$C_{10}H_{10}$	77-73-6	130.1894	Tricyclo[5,2,1,0|2,6]deca-3,8-diene Dicyclopentadiene Cyclopentadiene dimer
$C_{10}H_{10}O$	529-34-0	146.1888	3,4-Dihydro-1(2-H)-naphthalenone α-Tetralon
$C_{10}H_{10}O_4$	131-11-3	194.187	Dimethyl 1,2-benzenedicarboxylate Dimethyl phthalate 1,2-Benzenedicarboxylic acid dimethyl ester Phthalic acid dimethyl ester
$C_{10}H_{10}O_4$	1459-93-4	194.187	Dimethyl 1,3-benzenedicarboxylate Dimethyl isophthalate 1,3-Benzenedicarboxylic acid dimethyl ester Isophthalic acid dimethyl ester
$C_{10}H_{10}O_4$	120-61-6	194.187	Dimethyl 1,4-benzenedicarboxylate Dimethyl terephthalate Terephthalic acid dimethyl ester 1,4-Benzenedicarboxylic acid dimethyl ester
$C_{10}H_{11}Cl$		166.65004	5-Chloro-1,2,3,4-tetrahydronaphthalene
$C_{10}H_{11}N$	769-68-6	145.20408	DL-2-Phenylbutanenitrile DL-2-Phenylbutyronitrile
$C_{10}H_{12}$	28106-30-1	132.20528	Ethylvinylbenzene Ethylstyrene <undefined positional isomer>
$C_{10}H_{12}$	3454-07-7	132.20528	1-Ethyl-4-vinylbenzene 4-Ethylstyrene
$C_{10}H_{12}$	1195-32-0	132.20528	4-Isopropenyltoluene
$C_{10}H_{12}$	119-64-2	132.20528	1,2,3,4-Tetrahydronaphthalene Tetralin
$C_{10}H_{12}FeN_2O_8$		344.06096	Ethylenediaminetetraacetic acid iron(III)
$C_{10}H_{12}O$	104-46-1	148.20468	1-Methoxy-4-(1-propenyl)-benzene Anethole 4-(1-Propenyl)anisole <undefined cis/trans isomer>
$C_{10}H_{12}O$	4180-23-8	148.20468	trans-1-Methoxy-4-(1-propenyl)benzene trans-Anethole trans-4-(1-Propenyl)anisole
$C_{10}H_{12}O$	529-33-9	148.20468	1,2,3,4-Tetrahydro-1-naphthol α-Tetralol <undefined optical isomer>
$C_{10}H_{12}O$	23357-45-1	148.20468	(R)-(−)-1,2,3,4-Tetrahydro-1-naphthol (R)-(−)-α-Tetralol
$C_{10}H_{12}O$	1125-78-6	148.20468	5,6,7,8-Tetrahydro-2-naphthol

Formula	CAS RN	M_r	Name
$C_{10}H_{12}O_2$	101-97-3	164.20408	Ethyl phenylethanoate
			Ethyl phenylacetate
			Phenylethanoic acid ethyl ester
			Phenylacetic acid ethyl ester
$C_{10}H_{12}O_2$	2315-68-6	164.20408	Propyl benzoate
			Benzoic acid propyl ester
$C_{10}H_{12}O_2S$		196.27008	Allyl tolyl sulfone
			2-Propenyl tolyl sulfone
			<undefined positional isomer>
$C_{10}H_{12}O_2S$		196.27008	Allyl 4-tolyl sulfone
			2-Propenyl 4-tolyl sulfone
$C_{10}H_{13}BO_3$		192.02242	2-Butoxy-1,3-dioxa-2-bora-indan
			Butyl 1,2-phenylene borate
$C_{10}H_{14}$	104-51-8	134.22116	Butylbenzene
$C_{10}H_{14}$	98-06-6	134.22116	tert-Butylbenzene
$C_{10}H_{14}$	25340-17-4	134.22116	Diethylbenzene
			<mixture of positional iso-mers>
$C_{10}H_{14}$	135-01-3	134.22116	1,2-Diethylbenzene
$C_{10}H_{14}$	141-93-5	134.22116	1,3-Diethylbenzene
$C_{10}H_{14}$	105-05-5	134.22116	1,4-Diethylbenzene
$C_{10}H_{14}$	933-98-2	134.22116	1-Ethyl-2,3-dimethylbenzene
			3-Ethyl-1,2-xylene
$C_{10}H_{14}$	527-84-4	134.22116	2-Isopropyltoluene
			2-Cymene
$C_{10}H_{14}$	535-77-3	134.22116	3-Isopropyltoluene
			3-Cymene
$C_{10}H_{14}$	99-87-6	134.22116	4-Isopropyltoluene
			4-Cymene
$C_{10}H_{14}$	135-98-8	134.22116	(1-Methylpropyl)benzene
			sec-Butylbenzene
			2-Phenylbutane
$C_{10}H_{14}$	538-93-2	134.22116	(2-Methylpropyl)benzene
			Isobutylbenzene
$C_{10}H_{14}$	95-93-2	134.22116	1,2,4,5-Tetramethylbenzene
			Durene
$C_{10}H_{14}N_2$		162.23464	3-(N-Methyl-2-pyrrolidinyl)-pyridine
			Nicotine
			<undefined optical isomer>
$C_{10}H_{14}N_2$	54-11-5	162.23464	(S)-(−)-3-(N-Methyl-2-pyrroli-dinyl)pyridine
			(S)-(−)-Nicotine
$C_{10}H_{14}O$	88-18-6	150.22056	2-tert-Butylphenol
$C_{10}H_{14}O$	98-54-4	150.22056	4-tert-Butylphenol
$C_{10}H_{14}O$	89-83-8	150.22056	2-Isopropyl-5-methylphenol
			Thymol
$C_{10}H_{14}O_4$	97-90-5	198.21876	1,2-Ethanediol bis(2-methyl-propenoate)
			Ethylene glycol dimethacrylate
$C_{10}H_{15}N$	91-66-7	149.23584	N,N-Diethylaniline
$C_{10}H_{16}$	79-92-5	136.23704	2,2-Dimethyl-3-methylenebi-cyclo[2.2.1]heptane
			Camphene
			<undefined optical isomer>
$C_{10}H_{16}$	5794-03-6	136.23704	(+)-2,2-Dimethyl-3-methyl-enebicyclo[2.2.1]heptane
			(+)-Camphene
$C_{10}H_{16}$	127-91-3	136.23704	7,7-Dimethyl-2-methylenebi-cyclo[3.1.1]heptane
			β-Pinene
$C_{10}H_{16}$ (Cont.)			2(10)-Pinene
$C_{10}H_{16}$	18172-67-3	136.23704	(^{1}S)-(−)-7,7-Dimethyl-2-methylenebicyclo[3.1.1]-heptane
			(^{1}S)-(−)-β-Pinene
			(^{1}S)-(−)-2(10)-Pinene
$C_{10}H_{16}$		136.23704	1,4-Dimethyl-4-vinylcyclohex-ene
$C_{10}H_{16}$	138-86-3	136.23704	4-Isopropenyl-1-methylcyclo-hexene
			Dipentene
			p-Mentha-1,8-diene
			Limonene (inactive)
$C_{10}H_{16}$	5989-27-5	136.23704	(R)-(+)-4-Isopropenyl-1-methylcyclohexene
			(R)-(+)-Limonene
			(R)-(+)-p-Mentha-1,8-diene
$C_{10}H_{16}$	5989-54-8	136.23704	(S)-(−)-4-Isopropenyl-1-methylcyclohexene
			(S)-(−)-Limonene
$C_{10}H_{16}$	586-62-9	136.23704	4-Isopropylidene-1-methylcy-clohexene
			1,4(8)-p-Menthadiene
			Terpinolene
			Isoterpinene
$C_{10}H_{16}$	99-86-5	136.23704	1-Isopropyl-4-methyl-1,3-cyclohexadiene
			α-Terpinene
$C_{10}H_{16}$	99-85-4	136.23704	1-Isopropyl-4-methyl-1,4-cyclohexadiene
			η-Terpinene
$C_{10}H_{16}$	1461-27-4	136.23704	(R)-(+)-1-Methyl-5-isopro-penylcyclohexene
			(R)-(+)-Sylvestrene
			(R)-(+)-Carvestrene
$C_{10}H_{16}$	7785-26-4	136.23704	(^{1}S)-(−)-2,7,7-Trimethylbicy-clo[3.1.1]heptane
			(^{1}S)-(−)-α-Pinene
			(^{1}S)-(−)-2-Pinene
$C_{10}H_{16}$	80-56-8	136.23704	2,7,7-Trimethylbicyclo[3.1.1]-hept-2-ene
			α-Pinene
			2-Pinene
			<undefined optical isomer>
$C_{10}H_{16}$	7785-70-8	136.23704	(^{1}R)-(+)-2,7,7-Trimethylbicy-clo[3.1.1]hept-2-ene
			(^{1}R)-(+)-α-Pinene
			(^{1}R)-(+)-2-Pinene
$C_{10}H_{16}$	554-61-0	136.23704	3,7,7-Trimethylbicyclo[4.1.0]-hept-2-ene
			2-Carene
$C_{10}H_{16}$	13466-78-9	136.23704	3,7,7-Trimethylbicyclo[4.1.0]-hept-3-ene
			3-Carene
$C_{10}H_{16}N_2O_8$	60-00-4	292.24572	Ethylenedinitrilotetraetha-noic acid
			Ethylenediaminetetraacetic acid
			EDTA
			Ethylenedinitrilotetraacetic acid
$C_{10}H_{16}O$	5392-40-5	152.23644	3,7-Dimethyl-2,6-octadienal
			Citral
			<mixture of cis/trans isomers>

Formula	CAS RN	M_r	Name
$C_{10}H_{16}O$		152.23644	1,3,3-Trimethylbicyclo[2.2.1]-heptane-2-one 1,3,3-Trimethyl-2-norbornanone Fenchone <undefined optical isomer>
$C_{10}H_{16}O$	1195-79-5	152.23644	(^{1}R)-(−)-1,3,3-Trimethylbicyclo[2.2.1]heptan-2-one (^{1}R)-(−)-1,3,3-Trimethyl-2-norbornanone (^{1}R)-(−)-Fenchone
$C_{10}H_{16}O$	76-22-2	152.23644	1,7,7-Trimethylbicyclo[2.2.1]-heptan-2-one Camphor 1,7,7-Trimethyl-2-norbornanone <undefined optical isomer>
$C_{10}H_{16}O$	21368-68-3	152.23644	(±)-1,7,7-Trimethylbicyclo-[2.2.1]heptan-2-one (±)-Camphor (±)-1,7,7-Trimethyl-2-norbornanone
$C_{10}H_{16}O$	464-49-3	152.23644	1,7,7-Trimethyl-(^{1}R)-bicyclo-[2.2.1]heptan-2-one (^{1}R)-(+)-Camphor d-Camphor 1,7,7-Trimethyl-(^{1}R)-2-norbornanone
$C_{10}H_{16}O$		152.23644	1,3,3-Trimethylbicyclo[2.2.1]-heptan-6-one 1,5,5-Trimethyl-2-bornanone Isofenchone
$C_{10}H_{16}O_4$		200.23464	Ethyl vinyl hexanedioate Ethyl vinyl adipate Hexanedioic acid ethyl vinyl ester Adipic acid ethyl vinyl ester
$C_{10}H_{16}O_4$		200.23464	Methyl vinyl heptanedioate Methyl vinyl pimelate Heptanedioic acid methyl vinyl ester Pimelic acid methyl vinyl ester
$C_{10}H_{17}NO$	6837-24-7	167.25112	1-Cyclohexyl-2-pyrrolidinone
$C_{10}H_{17}NO$	2792-42-9	167.25112	(^{1}R)-(+)-1,7,7-Trimethylbicyclo[2.2.1]heptan-2-one oxime (^{1}R)-(+)-Camphoroxime (^{1}R)-(+)-1,7,7-Trimethyl-2-bornanone oxime
$C_{10}H_{18}$	91-17-8	138.25292	Bicyclo[4.4.0]decane Decahydronaphthalene Decalin <mixture of cis/trans isomers>
$C_{10}H_{18}$	493-01-6	138.25292	cis-Bicyclo[4.4.0]decane cis-Decahydronaphthalene cis-Decalin
$C_{10}H_{18}$	493-02-7	138.25292	trans-Bicyclo[4.4.0]decane trans-Decahydronaphthalene trans-Decalin
$C_{10}H_{18}$	2051-25-4	138.25292	1,3-Decadiene
$C_{10}H_{18}$	764-93-2	138.25292	1-Decyne
$C_{10}H_{18}$		138.25292	2,2,3-Trimethylbicyclo[2.2.1]-heptane

Formula	CAS RN	M_r	Name
$C_{10}H_{18}$ (Cont.)			Isocamphane
$C_{10}H_{18}O$	26902-25-0	154.25232	Bis(3-methyl-2-butenyl) ether
$C_{10}H_{18}O$	106-24-1	154.25232	3,7-Dimethyl-2,6-octadien-1-ol Geraniol
$C_{10}H_{18}O$	106-25-2	154.25232	cis-3,7-Dimethyl-2,6-octadien-1-ol Nerol
$C_{10}H_{18}O$	106-23-0	154.25232	3,7-Dimethyl-6-octenal Citronellal
$C_{10}H_{18}O$	470-82-6	154.25232	1,8-Epoxy-p-menthane Cineole Eucalyptol
$C_{10}H_{18}O$	14073-97-3	154.25232	(−)-2-Isopropyl-5-methylcyclohexanone (−)-Menthone
$C_{10}H_{18}O$	10458-14-7	154.25232	2-Isopropyl-5-methylcyclohexanone Menthone <undefined optical isomer>
$C_{10}H_{18}O_4$		202.25052	Butyl hexanedioate Butyl adipate Hexanedioic acid butyl ester Adipic acid butyl ester
$C_{10}H_{18}O_4$	111-20-6	202.25052	Decanedioic acid Sebacic acid
$C_{10}H_{18}O_4$	141-28-6	202.25052	Diethyl hexanedioate Diethyl adipate Hexanedioic acid diethyl ester Adipic acid diethyl ester
$C_{10}H_{18}O_4$	1732-09-8	202.25052	Dimethyl octanedioate Dimethyl suberate Octanedioic acid dimethyl ester Suberic acid dimethyl ester
$C_{10}H_{18}O_4$		202.25052	1,2-Ethanediol dibutanoate Ethylene glycol dibutyrate
$C_{10}H_{18}O_4$		202.25052	Ethyl pentyl propanedioate Amyl ethyl malonate Malonic acid amyl ethyl ester Propanedioic acid ethyl pentyl ester
$C_{10}H_{18}O_4$	29803-25-6	202.25052	1,8-Octanediol dimethanoate 1,8-Octanediol diformate Octamethylene diformate Octamethylene dimethanoate
$C_{10}H_{18}O_4$		202.25052	3,6,9,12-Tetraoxa-1,13-tetradecadiene Triethylene glycol divinyl ether
$C_{10}H_{19}N$	1975-78-6	153.2676	Decanenitrile
$C_{10}H_{19}NO_2$	105-16-8	185.2664	2-(Diethylamino)ethyl 2-methylpropenoate 2-(Diethylamino)ethyl methacrylate Methacrylic acid 2-(diethylamino)ethyl ester 2-Methylpropenoic acid 2--(diethylamino)ethyl ester
$C_{10}H_{20}$	1678-93-9	140.2688	Butylcyclohexane
$C_{10}H_{20}$	293-96-9	140.2688	Cyclodecane
$C_{10}H_{20}$	872-05-9	140.2688	1-Decene

Formula	CAS RN	M_r	Name
$C_{10}H_{20}$	19398-89-1	140.2688	trans-4-Decene
$C_{10}H_{20}$	4984-01-4	140.2688	3,7-Dimethyl-1-octene
			<undefined optical isomer>
$C_{10}H_{20}$	13827-59-3	140.2688	(±)-3,7-Dimethyl-1-octene
$C_{10}H_{20}$	1117-83-5	140.2688	(R)-(−)-3,7-Dimethyl-1-octene
$C_{10}H_{20}$	7058-01-7	140.2688	(1-Methylpropyl)cyclohexane
			sec-Butylcyclohexane
$C_{10}H_{20}O$	112-31-2	156.2682	Decanal
			1-Nonanecarbaldehyde
$C_{10}H_{20}O$	106-22-9	156.2682	3,7-Dimethyl-6-octen-1-ol
			β-Citronellol
$C_{10}H_{20}O$	1490-04-6	156.2682	2-Isopropyl-5-methylcyclohexanol
			Menthol
			<undefined optical isomer>
$C_{10}H_{20}O$	89-78-1	156.2682	(±)-2-Isopropyl-5-methylcyclohexanol
			(±)-Menthol
$C_{10}H_{20}OS$		188.3342	2-Hexylthiolane-1-oxide
			2-Hexyltetrahydrothiophene-1-oxide
$C_{10}H_{20}OS$	62290-30-6	188.3342	3-Pentylthiane 1-oxide
			3-Amylthiane 1-oxide
$C_{10}H_{20}O_2$	334-48-5	172.2676	Decanoic acid
			Capric acid
			1-Nonanecarboxylic acid
$C_{10}H_{20}O_2$	103-09-3	172.2676	2-Ethylhexyl acetate
$C_{10}H_{20}O_2$	106-32-1	172.2676	Ethyl octanoate
$C_{10}H_{20}O_2$	659-70-1	172.2676	3-Methylbutyl 3-methylbutanoate
			Isopentyl isopentanoate
			Isopentyl 3-methylbutanoate
			Isoamyl isovalerate
			Isovaleric acid isoamyl ester
			3-Methylbutanoic acid 3-methylbutyl ester
			Isopentanoic acid isopentyl ester
$C_{10}H_{20}O_2$	1731-84-6	172.2676	Methyl nonanoate
			Methyl pelargonate
			Nonanoic acid methyl ester
			Pelargonic acid methyl ester
$C_{10}H_{20}O_2$	112-14-1	172.2676	Octyl ethanoate
			Octyl acetate
			Ethanoic acid octyl ester
			Acetic acid octyl ester
$C_{10}H_{20}O_2$	2173-56-0	172.2676	Pentyl pentanoate
			Pentanoic acid pentyl ester
			Amyl valerate
			Valeric acid amyl ester
$C_{10}H_{20}O_5$	33100-27-5	220.2658	1,4,7,10,13-Pentaoxacyclopentadecane
			15-Crown-5
$C_{10}H_{21}Br$	112-29-8	221.18074	1-Bromodecane
			Decyl bromide
$C_{10}H_{21}Cl$	1002-69-3	176.72944	1-Chlorodecane
			Decyl chloride
$C_{10}H_{21}NaO_4S$		260.330108	Decyl sodium sulfate
$C_{10}H_{22}$	124-18-5	142.28468	Decane
$C_{10}H_{22}$	4032-94-4	142.28468	2,4-Dimethyloctane
$C_{10}H_{22}$	2051-30-1	142.28468	2,6-Dimethyloctane
$C_{10}H_{22}$	1072-16-8	142.28468	2,7-Dimethyloctane
			Diisoamyl
			Diisopentyl
$C_{10}H_{22}$	871-83-0	142.28468	2-Methylnonane
$C_{10}H_{22}$	5911-04-6	142.28468	3-Methylnonane
$C_{10}H_{22}$	15869-85-9	142.28468	5-Methylnonane
$C_{10}H_{22}$	13475-81-5	142.28468	2,2,3,3-Tetramethylhexane
$C_{10}H_{22}$	1071-81-4	142.28468	2,2,5,5-Tetramethylhexane
$C_{10}H_{22}$	7154-80-5	142.28468	3,3,5-Trimethylheptane
$C_{10}H_{22}O$	544-01-4	158.28408	Bis(3-methylbutyl) ether
			Diisoamyl ether
			Diisopentyl ether
$C_{10}H_{22}O$	112-30-1	158.28408	1-Decanol
			Decyl alcohol
$C_{10}H_{22}O$	693-65-2	158.28408	Dipentyl ether
			Diamyl ether
$C_{10}H_{22}O$	55505-26-5	158.28408	8-Methyl-1-nonanol
			Isodecanol
$C_{10}H_{22}OS$		190.35008	Bis(3-methylbutyl) sulfoxide
			Diisopentyl sulfoxide
			Diisoamyl sulfoxide
$C_{10}H_{22}OS$		190.35008	Dipentyl sulfoxide
			Diamyl sulfoxide
$C_{10}H_{22}OS$		190.35008	Heptyl isopropyl sulfoxide
$C_{10}H_{22}OS$		190.35008	Heptyl propyl sulfoxide
$C_{10}H_{22}OS$		190.35008	Methyl nonyl sulfoxide
$C_{10}H_{22}O_2$	871-22-7	174.28348	6-Methyl-5,7-dioxaundecane
			Acetaldehyde dibutyl acetal
$C_{10}H_{22}O_3$	24083-03-2	190.28288	4,8-Dioxa-1-dodecanol
			Dipropylene glycol butyl ether
$C_{10}H_{22}O_4$	143-22-6	206.28228	3,6,9-Trioxa-1-tridecanol
			Triethylene glycol butyl ether
$C_{10}H_{22}O_5$	143-24-8	222.28168	2,5,8,11,14-Pentaoxapentadecane
			Tetraethylene glycol dimethyl ether
			Tetraglyme
			Bis(2-(2-methoxyethoxy)ethyl)ether
$C_{10}H_{22}O_5$	5650-20-4	222.28168	3,6,9,12-Tetraoxa-1-tetradecanol
			Tetraethylene glycol ethyl ether
$C_{10}H_{22}S$	544-02-5	174.35068	Bis(3-methylbutyl) sulfide
			Diisopentyl sulfide
			Diisoamyl sulfide
$C_{10}H_{22}S$	143-10-2	174.35068	1-Decanethiol
$C_{10}H_{22}S$	872-10-6	174.35068	6-Thiaundecane
			Dipentyl sulfide
			Diamyl sulfide
$C_{10}H_{23}N$	2016-57-1	157.29936	Decylamine
			1-Aminodecane
$C_{10}H_{23}NO_2$		189.29816	Tetraethylammonium ethanoate
			Tetraethylammonium acetate
$C_{10}Mn_2O_{10}$	10170-69-1	389.9801	Dimanganese decacarbonyl
			Manganese carbonyl
$C_{11}F_{20}$		512.089064	Icosafluoro(1-methylbicyclo-[4.4.0]decane)
			Perfluoro(1-methyldecahydronaphthalene)
			Perfluoro(1-methylbicyclo[4.4.0]-decane)

Formula	CAS RN	M_r	Name
$C_{11}H_{10}$	90-12-0	142.2004	1-Methylnaphthalene
$C_{11}H_{10}$	91-57-6	142.2004	2-Methylnaphthalene
$C_{11}H_{10}O$	93-04-9	158.1998	2-Methoxynaphthalene
$C_{11}H_{12}N_2O_2$	73-22-3	204.22856	(S)-(−)-Tryptophan
			L-Tryptophan
$C_{11}H_{14}O_2$	136-60-7	178.23096	Butyl benzoate
			Benzoic acid butyl ester
$C_{11}H_{16}$	98-51-1	148.24804	4-tert-Butyltoluene
$C_{11}H_{16}$	4132-72-3	148.24804	Isopropyl-1,4-dimethylbenzene
			Isopropyl-1,4-xylene
			2,5-Dimethylcumene
$C_{11}H_{16}$	538-68-1	148.24804	Pentylbenzene
			1-Phenylpentane
			Amylbenzene
$C_{11}H_{16}$	1196-58-3	148.24804	3-Phenylpentane
			(1-Ethylpropyl)benzene
$C_{11}H_{16}O$	98-27-1	164.24744	4-tert-Butyl-2-methylphenol
			4-tert-Butyl-2-cresol
$C_{11}H_{16}O$	2409-55-4	164.24744	2-tert-Butyl-4-methylphenol
			2-tert-Butyl-4-cresol
$C_{11}H_{16}O$	88-60-8	164.24744	2-tert-Butyl-5-methylphenol
			6-tert-Butyl-3-cresol
$C_{11}H_{16}O$	2219-82-1	164.24744	2-tert-Butyl-6-methylphenol
			6-tert-Butyl-2-cresol
$C_{11}H_{16}O_2S$		212.31284	Butyl tolyl sulfone
			<undefined positional isomer>
$C_{11}H_{18}O$	26533-38-0	166.26332	6-Methyl-3-isopropenyl-5-hepten-2-one
$C_{11}H_{18}O_2$		182.26272	2-Bornyl methanoate
			Isobornyl formate
			Formic acid isobornyl ester
			Methanoic acid 2-bornyl ester
$C_{11}H_{18}O_4$		214.26152	Propyl vinyl hexanedioate
			Propyl vinyl adipate
			Hexanedioic acid propyl vinyl ester
			Adipic acid propyl vinyl ester
$C_{11}H_{20}$	1606-08-2	152.2798	Cyclopentylcyclohexane
$C_{11}H_{20}$	2958-75-0	152.2798	1-Methylbicyclo[4.4.0]decane
			1-Methyldecahydronaphthalene
			<undefined stereoisomer>
$C_{11}H_{20}$	2958-76-1	152.2798	2-Methylbicyclo[4.4.0]decane
			2-Methyldecahydronaphthalene
			<undefined stereoisomer>
$C_{11}H_{20}O_2$	103-11-7	184.2786	2-Ethylhexyl acrylate
$C_{11}H_{20}O_2$	112-38-9	184.2786	10-Undecenoic acid
			Undecylenic acid
$C_{11}H_{20}O_4$	77-25-8	216.2774	Diethyl diethylpropanedioate
			Diethyl diethylmalonate
			Diethylmalonic acid diethyl ester
			Diethylpropanedioic acid diethyl ester
$C_{11}H_{20}O_4$	1732-10-1	216.2774	Dimethyl nonanedioate
			Dimethyl azelate
			Nonanedioic acid dimethyl ester
			Azelaic acid dimethyl ester
$C_{11}H_{20}O_4$		216.2774	Pentyl hexanedioate
$C_{11}H_{20}O_4$ (Cont.)			Amyl adipate
			Hexanedioic acid pentyl ester
			Adipic acid amyl ester
$C_{11}H_{21}N$	2244-07-7	167.29448	Undecanenitrile
			1-Cyanodecane
			1-Decanecarbonitrile
			Decyl cyanide
$C_{11}H_{22}$	4292-92-6	154.29568	Pentylcyclohexane
			Amylcyclohexane
$C_{11}H_{22}$	821-95-4	154.29568	1-Undecene
$C_{11}H_{22}N_2O_2$		214.30796	Bis(dimethylaminomethyl)-methyl 2-methylpropenoate
			Bis(dimethylaminomethyl)-methyl methacrylate
			Methacrylic acid bis(dimethylaminomethyl)methyl ester
			2-Methylpropenoic acid bis-(dimethylaminomethyl)methyl ester
$C_{11}H_{22}O$	112-44-7	170.29508	Undecanal
			1-Decanecarbaldehyde
$C_{11}H_{22}O$	112-12-9	170.29508	2-Undecanone
			Methyl nonyl ketone
$C_{11}H_{22}O$	927-49-1	170.29508	6-Undecanone
			Dipentyl ketone
			Diamyl ketone
$C_{11}H_{22}O_2$	123-29-5	186.29448	Ethyl nonanoate
$C_{11}H_{22}O_2$	110-42-9	186.29448	Methyl decanoate
			Methyl caprate
			Decanoic acid methyl ester
			Capric acid methyl ester
$C_{11}H_{22}O_2$	112-37-8	186.29448	Undecanoic acid
			1-Decanecarboxylic acid
$C_{11}H_{24}$	6975-98-0	156.31156	2-Methyldecane
$C_{11}H_{24}$	2847-72-5	156.31156	4-Methyldecane
$C_{11}H_{24}$	62016-38-0	156.31156	2,4,7-Trimethyloctane
$C_{11}H_{24}$	1120-21-4	156.31156	Undecane
$C_{11}H_{24}O$	32357-83-8	172.31096	Hexyl pentyl ether
			Amyl hexyl ether
$C_{11}H_{24}O$	112-42-5	172.31096	1-Undecanol
			Undecyl alcohol
$C_{11}H_{24}O$	1653-30-1	172.31096	2-Undecanol
$C_{11}H_{25}N$	7307-55-3	171.32624	Undecylamine
			1-Aminoundecane
$C_{11}H_{25}O_3P$	2452-70-2	236.291462	Bis(3-methylbutyl) methylphosphonate
			Diisopentyl methylphosphonate
			Diisoamyl methylphosphonate
			Methylphosphonic acid bis(3-methylbutyl) ester
			Methylphosphonic acid diisopentyl ester
			Methylphosphonic acid diisoamyl ester
$C_{12}Cl_2F_{24}$	103188-54-1	670.9990768	Tetracosafluoro(1,10-dichloro-2,9-dimethyldecane)
			Perfluoro(1,10-dichloro-2,9-dimethyldecane)
$C_{12}Co_4O_{12}$	17786-31-1	571.8576	Tetracobalt dodecacarbonyl
$C_{12}F_{26}$	103188-55-2	638.0904832	Hexacosafluoro(2,9-dimethyldecane)
			Perfluoro(2,9-dimethyldecane)

Formula	CAS RN	M_r	Name
$C_{12}F_{27}N$	311-89-7	671.0956264	Heptacosafluorotributylamine
			Perfluorotributylamine
$C_{12}Fe_3O_{12}$	17685-52-8	503.6658	Triiron dodecacarbonyl
$C_{12}H_8$	208-96-8	152.19552	Acenaphthylene
$C_{12}H_8N_2$	92-82-0	180.209	Phenazine
$C_{12}H_8O$	132-64-9	168.19492	Dibenzofuran
$C_{12}H_8S$	132-65-0	184.26152	Dibenzothiophene
$C_{12}H_9N$	86-74-8	167.2102	Carbazole
$C_{12}H_{10}$	83-32-9	154.2114	Acenaphthene
$C_{12}H_{10}$	92-52-4	154.2114	Biphenyl
			Diphenyl
$C_{12}H_{10}ClO_3P$	2524-64-3	268.636062	Diphenyl chlorophosphonate
			Diphenyl phosphorochloridate
			Diphenyl chlorophosphate
$C_{12}H_{10}N_2$	103-33-3	182.22488	Azobenzene
$C_{12}H_{10}N_2O$	495-48-7	198.22428	Azoxybenzene
$C_{12}H_{10}O$	101-84-8	170.2108	Diphenyl ether
			Phenyl ether
$C_{12}H_{10}O$	90-43-7	170.2108	2-Phenylphenol
$C_{12}H_{10}OS$	945-51-7	202.2768	Diphenyl sulfoxide
$C_{12}H_{11}N$	90-41-5	169.22608	2-Aminobiphenyl
			2-Biphenylamine
			(1,1'-Biphenyl)-2-amine
$C_{12}H_{11}N$	2116-65-6	169.22608	4-Benzylpyridine
$C_{12}H_{11}N$	122-39-4	169.22608	Diphenylamine
$C_{12}H_{12}$	575-43-9	156.22728	1,6-Dimethylnaphthalene
$C_{12}H_{12}$	581-40-8	156.22728	2,3-Dimethylnaphthalene
$C_{12}H_{12}$	581-42-0	156.22728	2,6-Dimethylnaphthalene
$C_{12}H_{12}$	1127-76-0	156.22728	1-Ethylnaphthalene
$C_{12}H_{12}N_2$	101-54-2	184.24076	4-Aminodiphenylamine
$C_{12}H_{12}N_2$	92-87-5	184.24076	4,4'-Biphenyldiamine
			Benzidine
$C_{12}H_{13}NO_2$		203.24076	Methyl 2-cyano-2-phenylbutanoate
			Methyl 2-cyano-2-phenylbutyrate
			2-Cyano-2-phenylbutanoic acid methyl ester
			2-Cyano-2-phenylbutyric acid methyl ester
$C_{12}H_{14}O_4$	84-66-2	222.24076	Diethyl 1,2-benzenedicarboxylate
			Diethyl phthalate
			1,2-Benzenedicarboxylic acid diethyl ester
			Phthalic acid diethyl ester
$C_{12}H_{16}$	827-52-1	160.25904	Cyclohexylbenzene
$C_{12}H_{16}O_2$	119-43-7	192.25784	Ethyl 2-phenylbutanoate
			Ethyl 2-phenylbutyrate
			2-Phenylbutanoic acid ethyl ester
			2-Phenylbutyric acid ethyl ester
$C_{12}H_{18}$	2765-29-9	162.27492	trans-trans-cis-1,5,9-Cyclododecatriene
$C_{12}H_{18}$	25321-09-9	162.27492	Diisopropylbenzene
			<undefined positional isomer>
$C_{12}H_{18}$	99-62-7	162.27492	1,3-Diisopropylbenzene
$C_{12}H_{18}$	100-18-5	162.27492	1,4-Diisopropylbenzene
$C_{12}H_{18}$	61827-89-2	162.27492	3,9-Dodecadiyne
$C_{12}H_{18}$	1120-29-2	162.27492	5,7-Dodecadiyne
$C_{12}H_{18}$	87-85-4	162.27492	Hexamethylbenzene
$C_{12}H_{18}$	1077-16-3	162.27492	Hexylbenzene
			1-Phenylhexane
$C_{12}H_{18}$	10222-95-4	162.27492	5-Isopropyl-1,2,4-trimethylbenzene
$C_{12}H_{18}O$	122-73-6	178.27432	Benzyl 3-methylbutyl ether
			Benzyl isopentyl ether
			Benzyl isoamyl ether
$C_{12}H_{18}O$	6382-14-5	178.27432	Benzyl pentyl ether
			Amyl benzyl ether
$C_{12}H_{18}O$	2078-54-8	178.27432	2,6-Diisopropylphenol
$C_{12}H_{18}O$	943-93-1	178.27432	1,2-Epoxy-5,9-cyclododecadiene
$C_{12}H_{19}N$		177.2896	2-Methyl-4,6-dipropylpyridine
$C_{12}H_{20}O_2$	26896-48-0	196.2896	4,8-Bis(hydroxymethyl)tricyclo[5.2.1.0|2.6]decane
			Tricyclo[5.2.1.0|2.6]decane-4,8-dimethanol
$C_{12}H_{20}O_2$		196.2896	Bis(hydroxymethyl)tricyclo-[5.2.1.0|2.6]decane
			Tricyclodecanedimethanol
			<undefined positional isomer>
$C_{12}H_{20}O_2$	125-12-2	196.2896	2-Bornyl ethanoate
			Isobornyl acetate
			Ethanoic acid 2-bornyl ester
			Acetic acid isobornyl ester
$C_{12}H_{20}O_2$	105-87-3	196.2896	trans-3,7-Dimethyl-2,6-octadienyl ethanoate
			Geranyl acetate
			Ethanoic acid trans-3,7-dimethyl-2,6-octadienyl ester
			Acetic acid geranyl ester
$C_{12}H_{20}O_4$		228.2884	Butyl vinyl hexanedioate
			Butyl vinyl adipate
			Hexanedioic acid butyl vinyl ester
			Adipic acid butyl vinyl ester
$C_{12}H_{20}O_4$		228.2884	Methyl vinyl nonanedioate
			Methyl vinyl azelate
			Nonanedioic acid methyl vinyl ester
			Azelaic acid methyl vinyl ester
$C_{12}H_{21}F_9O_3Si_3$		468.5390688	1,3,5-Tris(3,3,3-trifluoropropyl)-1,3,5-trimethylcyclotrisiloxane
$C_{12}H_{22}$	92-51-3	166.30668	Bicyclohexyl
			Dicyclohexyl
			Cyclohexylcyclohexane
$C_{12}H_{22}$	1750-51-2	166.30668	1,6-Dimethylbicyclo[4.4.0]-decane
			1,6-Dimethyldecahydronaphthalene
$C_{12}H_{22}$	1618-22-0	166.30668	2,6-Dimethylbicyclo[4.4.0]-decane
			2,6-Dimethyldecahydronaphthalene
$C_{12}H_{22}O$	830-13-7	182.30608	Cyclododecanone
$C_{12}H_{22}OS$		214.37208	Dicyclohexyl sulfoxide
$C_{12}H_{22}O_4$	117-47-5	230.30428	Diethyl (1-methylbutyl)-propanedioate
			Diethyl (1-methylbutyl)malonate

Formula	CAS RN	M_r	Name
$C_{12}H_{22}O_4$ (Cont.)			(1-Methylbutyl)malonic acid diethyl ester
			(1-Methylbutyl)propanedioic acid diethyl ester
$C_{12}H_{22}O_4$	106-79-6	230.30428	Dimethyl decanedioate
			Dimethyl sebacate
			Decanedioic acid dimethyl ester
			Sebacic acid dimethyl ester
$C_{12}H_{22}O_4$	106-19-4	230.30428	Dipropyl hexanedioate
			Dipropyl adipate
			Hexanedioic acid dipropyl ester
			Adipic acid dipropyl ester
$C_{12}H_{22}O_4$		230.30428	Hexyl hexanedioate
			Hexyl adipate
			Hexanedioic acid hexyl ester
			Adipic acid hexyl ester
$C_{12}H_{22}O_{11}$	69-79-4	342.30008	D-Maltose
			Maltose
$C_{12}H_{22}O_{11}$	57-50-1	342.30008	Sucrose
			Saccharose
$C_{12}H_{23}KO_2$		238.41172	Potassium dodecanoate
			Potassium laurate
			Dodecanoic acid potassium salt
			Lauric acid potassium salt
$C_{12}H_{23}N$	101-83-7	181.32136	Dicyclohexylamine
$C_{12}H_{23}N$	2437-25-4	181.32136	Dodecanenitrile
$C_{12}H_{23}NO$		197.32076	5-Octyl-2-pyrrolidone
			5-Octyl-2-pyrrolidinone
			4-Dodecanolactam
$C_{12}H_{24}$	294-62-2	168.32256	Cyclododecane
$C_{12}H_{24}$	112-41-4	168.32256	1-Dodecene
$C_{12}H_{24}O$	1724-39-6	184.32196	Cyclododecanol
$C_{12}H_{24}O$	2855-19-8	184.32196	Decyloxirane
			1,2-Epoxydodecane
$C_{12}H_{24}O$	112-54-9	184.32196	Dodecanal
$C_{12}H_{24}O_2$	589-75-3	200.32136	Butyl octanoate
			Octanoic acid butyl ester
$C_{12}H_{24}O_2$	143-07-7	200.32136	Dodecanoic acid
			Lauric acid
			1-Undecanecarboxylic acid
$C_{12}H_{24}O_5$		248.31956	3,6,9-Trioxatridecyl ethanoate
			Triethylene glycol butyl ether acetate
$C_{12}H_{24}O_6$	17455-13-9	264.31896	1,4,7,10,13,16-Hexaoxacyclooctadecane
			18-Crown-6
$C_{12}H_{25}Cl$	112-52-7	204.7832	1-Chlorododecane
			Dodecyl chloride
			Lauryl chloride
$C_{12}H_{25}I$	4292-19-7	296.23497	1-Iodododecane
			Dodecyl iodide
			Lauryl iodide
$C_{12}H_{25}NaO_4S$	151-21-3	288.383868	Dodecyl sodium sulfate
			Lauryl sodium sulfate
$C_{12}H_{26}$	112-40-3	170.33844	Dodecane
$C_{12}H_{26}$	13475-82-6	170.33844	2,2,4,6,6-Pentamethylheptane
$C_{12}H_{26}F_6O_4Si_4$		460.6684592	1,3-Bis(3,3,3-trifluoropropyl)-1,3,5,7-hexamethylcyclotetrasiloxane
$C_{12}H_{26}O$	112-58-3	186.33784	Dihexyl ether
$C_{12}H_{26}O$	112-53-8	186.33784	1-Dodecanol
			Dodecyl alcohol
			Lauryl alcohol
$C_{12}H_{26}O$	10203-32-4	186.33784	4-Dodecanol
$C_{12}H_{26}O$	6836-38-0	186.33784	6-Dodecanol
$C_{12}H_{26}OS$		218.40384	Dihexyl sulfoxide
$C_{12}H_{26}O_3$		218.33664	2,2,10,10-Tetramethyl-3,6,9-trioxaundecane
			3-Oxa-1,5-pentanediol di-tert-butyl ether
			Diethylene glycol di-tert-butyl ether
$C_{12}H_{26}O_3$	112-73-2	218.33664	5,8,11-Trioxapentadecane
			3-Oxa-1,5-pentanediol dibutyl ether
			Diethylene glycol dibutyl ether
$C_{12}H_{26}O_5$	4353-28-0	250.33544	3,6,9,12,15-Pentaoxaheptadecane
			Tetraethylene glycol diethyl ether
$C_{12}H_{26}S$	6294-31-1	202.40444	7-Thiatridecane
			Dihexyl sulfide
$C_{12}H_{26}S_2$	10496-15-8	234.47044	7,8-Dithiatetradecane
			Dihexyl disulfide
$C_{12}H_{27}Al$	100-99-2	198.327919	Tris(2-methylpropyl)aluminum
			Triisobutylaluminum
$C_{12}H_{27}BO_3$	688-74-4	230.15558	Tributoxyborane
			Tributyl borate
			Boric acid tributyl ester
$C_{12}H_{27}ClSn$	1461-22-9	325.50908	Tributylchlorostannane
			Tributyltin chloride
$C_{12}H_{27}N$	124-22-1	185.35312	Dodecylamine
			1-Aminododecane
			Laurylamine
$C_{12}H_{27}N$	102-82-9	185.35312	Tributylamine
$C_{12}H_{27}O_3P$	102-85-2	250.318342	Tributyl phosphite
			Tributoxyphosphine
			Phosphorous acid tributyl ester
$C_{12}H_{27}O_4P$	126-73-8	266.317742	Tributyl phosphate
			TBP
			Phosphoric acid tributyl ester
$C_{12}H_{27}O_4P$	126-71-6	266.317742	Tris(2-methylpropyl) phosphate
			Triisobutyl phosphate
$C_{12}H_{28}BrN$	1941-30-6	266.26506	Tetrapropylammonium bromide
$C_{12}H_{28}ClN$		221.81376	Dodecylamine hydrochloride
			Laurylamine hydrochloride
$C_{12}H_{28}ClN$	4499-86-9	221.81376	Tetrapropylammonium chloride
$C_{12}H_{28}IN$	631-40-3	313.26553	Tetrapropylammonium iodide
$C_{12}H_{28}O_4Si$	1992-48-9	264.43742	Tetraisopropoxysilane
			Tetraisopropyl orthosilicate
$C_{12}H_{28}O_4Si$	682-01-9	264.43742	Tetrapropoxysilane
			Tetrapropyl orthosilicate
$C_{12}H_{30}O_7Si_2$	2157-42-8	342.537	Hexaethoxydisiloxane
$C_{13}H_9BrO$	90-90-4	261.11786	4-Bromophenyl phenyl ketone
			4-Bromobenzophenone

Formula	CAS RN	M_r	Name
$C_{13}H_9ClO$	134-80-5	216.66656	4-Chlorophenyl phenyl ketone 4-Chlorobenzophenone
$C_{13}H_9N$	260-94-6	179.2212	Acridine
$C_{13}H_{10}$	86-73-7	166.2224	Fluorene
$C_{13}H_{10}O$	119-61-9	182.2218	Diphenyl ketone Benzophenone
$C_{13}H_{11}N$	1484-12-4	181.23708	9-Methylcarbazole
$C_{13}H_{11}NO$	607-00-1	197.23648	N,N-Diphenylmethanamide N,N-Diphenylformamide
$C_{13}H_{12}$	101-81-5	168.23828	Diphenylmethane
$C_{13}H_{12}O$	91-01-0	184.23768	Diphenylmethanol Benzhydrol Diphenylcarbinol
$C_{13}H_{13}O_4P$	115-89-9	264.217582	Methyl diphenyl phosphate Phosphoric acid methyl diphenyl ester
$C_{13}H_{14}$	2027-17-0	170.25416	2-Isopropylnaphthalene
$C_{13}H_{15}NO_2$		217.26764	Ethyl 2-cyano-2-phenylbutanoate Ethyl 2-cyano-2-phenylbutyrate 2-Cyano-2-phenylbutanoic acid ethyl ester 2-Cyano-2-phenylbutyric acid ethyl ester
$C_{13}H_{20}$	1078-71-3	176.3018	Heptylbenzene
$C_{13}H_{20}O$	141-10-6	192.3012	6,10-Dimethyl-3,5,9-undecatrien-2-one Pseudoionone
$C_{13}H_{20}O$		192.3012	Trimethyl-2-(3-oxo-1-butenyl)-cyclohexene Ionone <undefined positional isomer>
$C_{13}H_{20}O$	79-77-6	192.3012	1,3,3-Trimethyl-2-(3-oxo-1-butenyl)cyclohexene β-Ionone
$C_{13}H_{20}O$	127-41-3	192.3012	2,4,4-Trimethyl-3-(3-oxo-1-butenyl)cyclohexene α-Ionone
$C_{13}H_{20}OS$		224.3672	Benzyl hexyl sulfoxide
$C_{13}H_{22}Cl_2O_4$		313.22068	2,2-Bis(chloromethyl)propane-1,3-diol dibutanoate Pentaerythritedichlorohydrin dibutyrate
$C_{13}H_{22}O_4$		242.31528	Methyl vinyl decanedioate Methyl vinyl sebacate Decanedioic acid methyl vinyl ester Sebacic acid methyl vinyl ester
$C_{13}H_{22}O_4$		242.31528	Pentyl vinyl hexanedioate Amyl vinyl adipate Hexanedioic acid pentyl vinyl ester Adipic acid amyl vinyl ester
$C_{13}H_{24}$	3178-23-2	180.33356	Dicyclohexylmethane
$C_{13}H_{24}O_4$	72030-39-8	244.33116	Ethyl methyl ethyl(1-methylbutyl)propanedioate Ethyl methyl ethyl(1-methylbutyl)malonate Ethyl(1-methylbutyl)malonic acid ethyl methyl ester Ethyl(1-methylbutyl)propanedioic acid ethyl methyl ester
$C_{13}H_{24}O_4$		244.33116	Heptyl hexanedioate Heptyl adipate Hexanedioic acid heptyl ester Adipic acid heptyl ester
$C_{13}H_{26}N_2O_2$		242.36172	N,N,N',N'-Tetraethylpentanediamide Tetraethylglutaramide
$C_{13}H_{26}O$	462-18-0	198.34884	7-Tridecanone Dihexyl ketone
$C_{13}H_{26}O_2$	111-82-0	214.34824	Methyl dodecanoate Methyl laurate Dodecanoic acid methyl ester Lauric acid methyl ester
$C_{13}H_{28}$	629-50-5	184.36532	Tridecane
$C_{13}H_{28}O$	112-70-9	200.36472	1-Tridecanol Tridecyl alcohol
$C_{13}H_{30}BrN$	2082-84-0	280.29194	Decyltrimethylammonium bromide
$C_{14}H_2F_{26}$		664.1283632	1,2-Bis(tridecafluorohexyl)-ethene 1,2-Bis(perfluorohexyl)ethylene
$C_{14}H_{10}$	120-12-7	178.2334	Anthracene
$C_{14}H_{10}$	501-65-5	178.2334	Diphenylethyne Diphenylacetylene
$C_{14}H_{10}$	85-01-8	178.2334	Phenanthrene
$C_{14}H_{10}O_2$	134-81-6	210.2322	Benzil
$C_{14}H_{12}$	613-31-0	180.24928	9,10-Dihydroanthracene
$C_{14}H_{12}$	776-35-2	180.24928	9,10-Dihydrophenanthrene
$C_{14}H_{12}$	645-49-8	180.24928	cis-1,2-Diphenylethene cis-Stilbene cis-1,2-Diphenylethylene
$C_{14}H_{12}$	103-30-0	180.24928	trans-1,2-Diphenylethene trans-Stilbene trans-1,2-Diphenylethylene
$C_{14}H_{12}O_2$	120-51-4	212.24808	Benzyl benzoate Benzoic acid benzyl ester
$C_{14}H_{12}O_2$	117-34-0	212.24808	Diphenylethanoic acid Diphenylacetic acid
$C_{14}H_{13}N$	86-28-2	195.26396	9-Ethylcarbazole
$C_{14}H_{14}$	605-39-0	182.26516	2,2'-Dimethylbiphenyl 2,2'-Bitolyl 2,2'-Bitoluene
$C_{14}H_{14}$	612-00-0	182.26516	1,1-Diphenylethane
$C_{14}H_{14}$	103-29-7	182.26516	1,2-Diphenylethane Bibenzyl Dibenzyl
$C_{14}H_{14}O$	103-50-4	198.26456	Dibenzyl ether Benzyl ether
$C_{14}H_{14}OS$	621-08-9	230.33056	Dibenzyl sulfoxide
$C_{14}H_{18}O_6$	117-82-8	282.29332	Bis(3-oxabutyl) 1,2-benzenedicarboxylate Bis(ethylene glycol methyl ether)-phthalate 1,2-Benzenedicarboxylic acid bis-(3-oxabutyl) ester Phthalic acid bis(ethylene glycol methyl ether) ester
$C_{14}H_{20}O_2$		220.3116	Heptyl benzoate Benzoic acid heptyl ester
$C_{14}H_{21}BO_3$		248.12994	2-Octyloxy-1,3-dioxa-2-boraindan Octyl 1,2-phenylene borate

Formula	CAS RN	M_r	Name
$C_{14}H_{22}$	2189-60-8	190.32868	Octylbenzene
$C_{14}H_{22}O$	128-39-2	206.32808	2,6-Di-tert-butylphenol
$C_{14}H_{23}N$		205.34336	2,4-Dibutyl-6-methylpyridine
$C_{14}H_{24}O_4$		256.34216	Hexyl vinyl hexanedioate
			Hexyl vinyl adipate
			Hexanedioic acid hexyl vinyl ester
			Adipic acid hexyl vinyl ester
$C_{14}H_{26}O_3$	626-27-7	242.35864	Heptanoic anhydride
			Oenanthic anhydride
$C_{14}H_{26}O_4$	105-99-7	258.35804	Dibutyl hexanedioate
			Dibutyl adipate
			Adipic acid dibutyl ester
			Hexanedioic acid dibutyl ester
$C_{14}H_{26}O_4$	76-72-2	258.35804	Diethyl ethyl(1-methylbutyl)-propanedioate
			Diethyl ethyl(1-methylbutyl)-malonate
$C_{14}H_{26}O_4$		258.35804	Octyl hexanedioate
			Octyl adipate
			Hexanedioic acid octyl ester
			Adipic acid octyl ester
$C_{14}H_{27}KO_2$		266.46548	Potassium tetradecanoate
			Potassium myristate
			Tetradecanoic acid potassium salt
			Myristic acid potassium salt
$C_{14}H_{27}N$	629-63-0	209.37512	Tetradecanenitrile
			1-Cyanotridecane
			Myristonitrile
			1-Tridecanecarbonitrile
			Tridecyl cyanide
$C_{14}H_{28}$	1120-36-1	196.37632	1-Tetradecene
$C_{14}H_{28}$	10374-74-0	196.37632	7-Tetradecene
			<mixture of cis/trans isomers>
$C_{14}H_{28}O$	3234-28-4	212.37572	Dodecyloxirane
			1,2-Epoxytetradecane
$C_{14}H_{28}O_2$	106-33-2	228.37512	Ethyl dodecanoate
			Ethyl laurate
			Dodecanoic acid ethyl ester
			Lauric acid ethyl ester
$C_{14}H_{28}O_2$	544-63-8	228.37512	Tetradecanoic acid
			Myristic acid
			1-Tridecanecarboxylic acid
$C_{14}H_{30}$	629-59-4	198.3922	Tetradecane
$C_{14}H_{30}O$	629-64-1	214.3916	Diheptyl ether
			Heptyl ether
$C_{14}H_{30}O$	112-72-1	214.3916	1-Tetradecanol
			Tetradecyl alcohol
$C_{14}H_{30}OS$		246.4576	Dodecyl ethyl sulfoxide
			Ethyl lauryl sulfoxide
$C_{14}H_{30}O_4$		262.3898	3,6,9-Trioxa-1-heptadecanol
			Triethylene glycol octyl ether
$C_{14}H_{30}O_6$		294.3886	2-Methyl-3,6,9,12,15,18-hexaoxanonadecane
			Pentaethylene glycol methyl isopropyl ether
$C_{14}H_{30}S$	629-65-2	230.4582	8-Thiapentadecane
			Diheptyl sulfide
			Diheptyl thioether
$C_{14}H_{31}N$	112-18-5	213.40688	N,N-Dimethyldodecylamine
$C_{15}H_{10}N_2O_2$	101-68-8	250.25668	Bis(4-isocyanatophenyl)-methane
$C_{15}H_{10}N_2O_2$ (Cont.)			4,4'-Diphenylmethane diisocyanate
$C_{15}H_{12}$	779-02-2	192.26028	9-Methylanthracene
$C_{15}H_{16}O_2$	80-05-7	228.29084	2,2-Bis(4-hydroxyphenyl)-propane
			Bisphenol A
			4,4'-Isopropylidenediphenol
$C_{15}H_{18}$	93-22-1	198.30792	2-Pentylnaphthalene
			2-Amylnaphthalene
$C_{15}H_{24}$	1081-77-2	204.35556	Nonylbenzene
			1-Phenylnonane
$C_{15}H_{24}$		204.35556	Triisopropylbenzene
			<undefined positional isomer>
$C_{15}H_{24}$	717-74-8	204.35556	1,3,5-Triisopropylbenzene
$C_{15}H_{24}O$	128-37-0	220.35496	2,6-Di-tert-butyl-4-methylphenol
			2,6-Di-tert-butyl-4-cresol
$C_{15}H_{24}OS$		252.42096	Benzyl octyl sulfoxide
$C_{15}H_{26}O_6$	60-01-5	302.36784	1,2,3-Propanetriol tributanoate
			Glycerol tributyrate
			Tributyrin
$C_{15}H_{30}O_2$	124-10-7	242.402	Methyl tetradecanoate
			Methyl myristate
			Tetradecanoic acid methyl ester
			Myristic acid methyl ester
$C_{15}H_{30}O_2$	1002-84-2	242.402	Pentadecanoic acid
			1-Tetradecanecarboxylic acid
$C_{15}H_{31}Br$	629-72-1	291.31514	1-Bromopentadecane
			Pentadecyl bromide
$C_{15}H_{32}$	629-62-9	212.41908	Pentadecane
$C_{15}H_{32}O$		228.41848	Heptyl octyl ether
$C_{15}H_{32}O$	629-76-5	228.41848	1-Pentadecanol
			Pentadecyl alcohol
$C_{15}H_{33}BO_3$	621-78-3	272.23622	Tripentyloxyborane
			Tripentyl borate
			Boric acid tripentyl ester
$C_{15}H_{33}N$	2570-26-5	227.43376	Pentadecylamine
			1-Aminopentadecane
$C_{15}H_{33}N$	621-77-2	227.43376	Tripentylamine
			Triamylamine
$C_{15}H_{33}N$	645-41-0	227.43376	Tris(3-methylbutyl)amine
			Triisoamylamine
			Triisopentylamine
$C_{16}H_{10}$	206-44-0	202.2554	Fluoranthene
$C_{16}H_{10}$	129-00-0	202.2554	Pyrene
$C_{16}H_{12}$	605-02-7	204.27128	1-Phenylnaphthalene
$C_{16}H_{12}O_2$	84-51-5	236.27008	2-Ethylanthraquinone
$C_{16}H_{18}$	1520-44-1	210.31892	1,3-Diphenylbutane
$C_{16}H_{20}$	38640-62-9	212.3348	Diisopropylnaphthalene
			<undefined positional isomer>
$C_{16}H_{22}O_4$	84-69-5	278.34828	Bis(2-methylpropyl) 1,2-benzenedicarboxylate
			Diisobutyl phthalate
			Phthalic acid diisobutyl ester
			1,2-Benzenedicarboxylic acid bis-(2-methylpropyl) ester
$C_{16}H_{22}O_4$	84-74-2	278.34828	Dibutyl 1,2-benzenedicarboxylate
			Dibutyl phthalate
			1,2-Benzenedicarboxylic acid dibutyl ester

Formula	CAS RN	M_r	Name
$C_{16}H_{22}O_4$ (Cont.)			Phthalic acid dibutyl ester
$C_{16}H_{26}$	104-72-3	218.38244	Decylbenzene 1-Phenyldecane
$C_{16}H_{27}N$		233.39712	2-Methyl-4,6-dipentylpyridine 4,6-Diamyl-2-methylpyridine
$C_{16}H_{28}F_{12}O_4Si_4$		624.7187584	1,3,5,7-Tetrakis(3,3,3-trifluoropropyl)-1,3,5,7-tetramethylcyclotetrasiloxane
$C_{16}H_{30}O_2$	373-49-9	254.413	cis-9-Hexadecenoic acid Palmitoleic acid
$C_{16}H_{30}O_4$	14027-78-2	286.4118	Dipentyl hexanedioate Diamyl adipate Adipic acid diamyl ester Hexanedioic acid dipentyl ester
$C_{16}H_{32}$	629-73-2	224.43008	1-Hexadecene
$C_{16}H_{32}O_2$	57-10-3	256.42888	Hexadecanoic acid Palmitic acid 1-Pentadecanecarboxylic acid
$C_{16}H_{33}Cl$	4860-03-1	260.89072	1-Chlorohexadecane Cetyl chloride Hexadecyl chloride
$C_{16}H_{33}I$	544-77-4	352.34249	1-Iodohexadecane Cetyl iodide Hexadecyl iodide
$C_{16}H_{33}NO$	3352-87-2	255.44416	N,N-Diethyldodecanamide
$C_{16}H_{34}$	4390-04-9	226.44596	2,2,4,4,6,8,8-Heptamethylnonane
$C_{16}H_{34}$	544-76-3	226.44596	Hexadecane Cetane
$C_{16}H_{34}$	1560-93-6	226.44596	2-Methylpentadecane
$C_{16}H_{34}$	2801-87-8	226.44596	4-Methylpentadecane
$C_{16}H_{34}$	10105-38-1	226.44596	6-Methylpentadecane
$C_{16}H_{34}$	7249-32-3	226.44596	6-Pentylundecane 6-Amylundecane
$C_{16}{}^2H_{34}$		260.6554604	($^2H_{34}$)Hexadecane Hexadecane-d34
$C_{16}H_{34}O$	629-82-3	242.44536	Dioctyl ether Octyl ether
$C_{16}H_{34}O$	36653-82-4	242.44536	1-Hexadecanol Cetyl alcohol
$C_{16}H_{34}OS$		274.51136	Dioctyl sulfoxide
$C_{16}H_{34}S_2$	822-27-5	290.57796	9,10-Dithiaoctadecane Dioctyl disulfide
$C_{16}H_{35}N$	112-75-4	241.46064	N,N-Dimethyltetradecylamine
$C_{16}H_{35}O_4P$	298-07-7	322.425262	Bis(2-ethylhexyl) hydrogen phosphate Phosphoric acid bis(2-ethylhexyl)-ester
$C_{16}H_{36}BrN$	1643-19-2	322.37258	Tetrabutylammonium bromide
$C_{16}H_{36}ClN$	1112-67-0	277.92128	Tetrabutylammonium chloride
$C_{16}H_{36}ClNO_4$	1923-70-2	341.91888	Tetrabutylammonium perchlorate
$C_{16}H_{36}IN$	311-28-4	369.37305	Tetrabutylammonium iodide
$C_{16}H_{36}Sn$	1461-25-2	347.17184	Tetrabutylstannane Tetrabutyltin
$C_{17}H_{12}O_2$		248.28108	2-Benzoyl-1-naphthol
$C_{17}H_{14}$	611-45-0	218.29816	1-Benzylnaphthalene
$C_{17}H_{28}$	6742-54-7	232.40932	Undecylbenzene
$C_{17}H_{28}$ (Cont.)			1-Phenylundecane
$C_{17}H_{28}OS$		280.47472	Benzyl decyl sulfoxide
$C_{17}H_{30}ClN$	104-74-5	283.88464	1-Dodecylpyridinium chloride 1-Laurylpyridinium chloride
$C_{17}H_{32}O_4$	2917-73-9	300.43868	Dibutyl nonanedioate Dibutyl azelate Nonanedioic acid dibutyl ester Azelaic acid dibutyl ester
$C_{17}H_{34}O_2$	506-12-7	270.45576	Heptadecanoic acid Margaric acid 1-Hexadecanecarboxylic acid
$C_{17}H_{34}O_2$	112-39-0	270.45576	Methyl hexadecanoate Hexadecanoic acid methyl ester Methyl palmitate Palmitic acid methyl ester
$C_{17}H_{36}$	629-78-7	240.47284	Heptadecane
$C_{17}H_{36}O$	1454-85-9	256.47224	1-Heptadecanol Heptadecyl alcohol
$C_{17}H_{37}N$	4455-26-9	255.48752	N-Methyldioctylamine
$C_{17}H_{38}IN$		383.39993	Tributylpentylammonium iodide Tributylamylammonium iodide
$C_{18}H_{12}$	218-01-9	228.29328	Chrysene 1,2-Benzophenanthrene
$C_{18}H_{14}$	26314-60-3	230.30916	Diphenylbenzene Terphenyl <mixture of positional isomers>
$C_{18}H_{14}$	84-15-1	230.30916	1,2-Diphenylbenzene 1,2-Terphenyl
$C_{18}H_{14}$	92-06-8	230.30916	1,3-Diphenylbenzene 1,3-Terphenyl
$C_{18}H_{14}$	92-94-4	230.30916	1,4-Diphenylbenzene 1,4-Terphenyl
$C_{18}H_{15}O_3P$	101-02-0	310.289062	Triphenyl phosphite Phosphorous acid triphenyl ester
$C_{18}H_{15}P$	603-35-0	262.290862	Triphenylphosphine
$C_{18}H_{20}$	3910-35-8	236.3568	1,1,3-Trimethyl-3-phenylindan
$C_{18}H_{21}NO$	26227-73-6	267.37088	N-(4-Methoxybenzilidene)-4-butylaniline MBBA
$C_{18}H_{30}$	123-01-3	246.4362	1-Phenyldodecane Dodecylbenzene
$C_{18}H_{30}O_2$	3884-88-6	278.435	cis,cis,cis-9,11,13-Octadecatrienoic acid Eleostearic acid
$C_{18}H_{30}O_2$	463-40-1	278.435	9,12,15-Octadecatrienoic acid Linolenic acid
$C_{18}H_{30}O_3S$	27176-87-0	326.5004	Dodecylbenzenesulfonic acid <undefined positional isomer>
$C_{18}H_{32}O_2$	60-33-3	280.45088	9,12-Octadecadienoic acid Linoleic acid
$C_{18}H_{33}KO_2$	143-18-0	320.55712	Potassium cis-9-octadecenoate Potassium oleate cis-9-Octadecenoic acid potassium salt

Formula	CAS RN	M_r	Name
$C_{18}H_{33}KO_2$ (Cont.)			Oleic acid potassium salt
$C_{18}H_{33}NaO_2$	143-19-1	304.448588	Sodium cis-9-octadecenoate
			Sodium oleate
			cis-9-Octadecenoic acid sodium salt
			Oleic acid sodium salt
$C_{18}H_{34}O_2$	693-72-1	282.46676	trans-11-Octadecenoic acid
			trans-Vaccenic acid
$C_{18}H_{34}O_2$	112-80-1	282.46676	cis-9-Octadecenoic acid
			Oleic acid
$C_{18}H_{34}O_3$	141-22-0	298.46616	12-Hydroxy-cis-9-octadecenoic acid
			Ricinoleic acid
$C_{18}H_{34}O_4$	109-43-3	314.46556	Dibutyl decanedioate
			Dibutyl sebacate
			Decanedioic acid dibutyl ester
			Sebacic acid dibutyl ester
$C_{18}H_{34}O_4$	110-33-8	314.46556	Dihexyl hexanedioate
			Dihexyl adipate
			Hexanedioic acid dihexyl ester
			Adipic acid dihexyl ester
$C_{18}H_{35}KO_2$		322.573	Potassium octadecanoate
			Potassium stearate
			Octadecanoic acid potassium salt
			Stearic acid potassium salt
$C_{18}H_{36}$	112-88-9	252.48384	1-Octadecene
$C_{18}H_{36}O_2$	628-97-7	284.48264	Ethyl hexadecanoate
			Hexadecanoic acid ethyl ester
			Ethyl palmitate
			Palmitic acid ethyl ester
$C_{18}H_{36}O_2$	57-11-4	284.48264	Octadecanoic acid
			Stearic acid
			1-Heptadecanecarboxylic acid
$C_{18}H_{38}$	593-45-3	254.49972	Octadecane
$C_{18}H_{38}O$	112-92-5	270.49912	1-Octadecanol
			Stearyl alcohol
$C_{18}H_{38}O_2$		286.49852	7,14-Dioxaicosane
			7,14-Dioxaeicosane
$C_{18}H_{38}S$	2885-00-9	286.56572	1-Octadecanethiol
			Octadecyl mercaptan
$C_{18}H_{39}N$	102-86-3	269.5144	Trihexylamine
$C_{19}H_{16}$	519-73-3	244.33604	Triphenylmethane
$C_{19}H_{16}O$	76-84-6	260.33544	Triphenylmethanol
$C_{19}H_{20}O_4$	85-68-7	312.3654	Benzyl butyl 1,2-benzenedicarboxylate
			Benzyl butyl phthalate
			1,2-Benzenedicarboxylic acid benzyl butyl ester
			Phthalic acid benzyl butyl ester
$C_{19}H_{21}NO$		279.38188	4-Cyano-4'-hexyloxybiphenyl
$C_{19}H_{24}$		252.39956	Bis(4-isopropylphenyl)-methane
			Bis(4-cumyl)methane
$C_{19}H_{26}$	26438-26-6	254.41544	1-Nonylnaphthalene
$C_{19}H_{32}$	123-02-4	260.46308	1-Phenyltridecane
			Tridecylbenzene
$C_{19}H_{32}O_2$	301-00-8	292.46188	Methyl 9,12,15-octadecatrienoate
			Methyl linolenate
			Linolenic acid methyl ester
			9,12,15-Octadecatrienoic acid methyl ester
$C_{19}H_{34}O_2$	112-63-0	294.47776	Methyl 9,12-octadecadienoate
			Methyl linoleate
			9,12-Octadecadienoic acid methyl ester
			Linoleic acid methyl ester
$C_{19}H_{36}O_2$	112-62-9	296.49364	Methyl cis-9-octadecenoate
			Methyl oleate
			cis-9-Octadecenoic acid methyl ester
$C_{19}H_{37}N$	28623-46-3	279.50952	Nonadecanenitrile
			1-Cyanooctadecane
			1-Octadecanecarbonitrile
			Octadecyl cyanide
$C_{19}H_{38}O$	629-66-3	282.51012	2-Nonadecanone
			Heptadecyl methyl ketone
$C_{19}H_{38}O_2$	112-61-8	298.50952	Methyl octadecanoate
			Methyl stearate
			Octadecanoic acid methyl ester
			Stearic acid methyl ester
$C_{19}H_{38}O_2$	646-30-0	298.50952	Nonadecanoic acid
			Nonadecylic acid
			1-Octadecanecarboxylic acid
$C_{19}H_{40}$	629-92-5	268.5266	Nonadecane
$C_{19}H_{40}$	1921-70-6	268.5266	2,6,10,14-Tetramethylpentadecane
			Pristane
$C_{19}H_{42}BrN$	57-09-0	364.45322	Hexadecyltrimethylammonium bromide
			Cetyltrimetylammonium bromide
$C_{20}H_{12}$	198-55-0	252.31528	Perylene
$C_{20}H_{14}O_4$	77-09-8	318.32876	Phenolphtalein
$C_{20}H_{17}F_{25}$		732.31506	1,1,1,2,2,3,3,4,4,5,5,6,6,7,7,8,-8,9,9,10,10,11,11,12,12-Pentacosafluoroicosane
$C_{20}H_{21}F_{21}$		660.3532072	1,1,1,2,2,3,3,4,4,5,5,6,6,7,7,8,-8,9,9,10,10-Henicosafluoroicosane
$C_{20}H_{28}$	26438-27-7	268.44232	1-Decylnaphthalene
$C_{20}H_{30}O_2$	514-10-3	302.457	Abietic acid
			13-Isopropylpodocarpa-7,13-dien-15-oic acid
$C_{20}H_{30}O_6$	117-83-9	366.4546	Bis(3-oxaheptyl) 1,2-benzenedicarboxylate
			Bis(ethylene glycol butyl ether) phthalate
			1,2-Benzenedicarboxylic acid bis-(3-oxaheptyl) ester
			Phthalic acid bis(ethylene glycol butyl ether) ester
$C_{20}H_{32}O_2$	506-32-1	304.47288	5,8,11,14-Icosatetraenoic acid
			Arachidonic acid
			5,8,11,14-Eicosatetraenoic acid
$C_{20}H_{34}$	1459-10-5	274.48996	Tetradecylbenzene
			1-Phenyltetradecane
$C_{20}H_{36}O_2$	544-35-4	308.50464	Ethyl 9,12-octadecadienoate
			9,12-Octadecadienoic acid ethyl ester
			Ethyl linoleate
			Linoleic acid ethyl ester

Formula	CAS RN	M_r	Name
$C_{20}H_{38}O_2$	111-62-6	310.52052	Ethyl cis-9-octadecenoate Ethyl oleate cis-9-Octadecenoic acid ethyl ester Oleic acid ethyl ester
$C_{20}H_{38}O_4$	14697-48-4	342.51932	Diheptyl hexanedioate Diheptyl adipate Hexanedioic acid diheptyl ester Adipic acid diheptyl ester
$C_{20}H_{40}O_2$	111-61-5	312.5364	Ethyl octadecanoate Ethyl stearate Octadecanoic acid ethyl ester Stearic acid ethyl ester
$C_{20}H_{40}O_2$	506-30-9	312.5364	Icosanoic acid Eicosanoic acid Arachidic acid 1-Nonadecanecarboxylic acid
$C_{20}H_{40}O_2$	1731-94-8	312.5364	Methyl nonadecanoate Nonadecanoic acid methyl ester
$C_{20}H_{40}O_2$	14721-66-5	312.5364	3,7,11,15-Tetramethyl-2-hexadecenoic acid Phytanic acid
$C_{20}H_{42}$	112-95-8	282.55348	Icosane Didecyl Eicosane
$C_{20}H_{42}O$	629-96-9	298.55288	1-Icosanol Arachidic alcohol 1-Eicosanol Eicosyl alcohol
$C_{20}H_{42}OS$		330.61888	Didecyl sulfoxide
$C_{20}H_{44}BrN$	866-97-7	378.4801	Tetrapentylammonium bromide Tetraamylammonium bromide
$C_{21}H_{21}O_4P$	563-04-2	368.369102	Tris(3-tolyl) phosphate Phosphoric acid tris(3-tolyl) ester
$C_{21}H_{21}O_4P$	1330-78-5	368.369102	Tritolyl phosphate Phosphoric acid tritolyl ester <mixture of positional isomers>
$C_{21}H_{25}N$		291.43624	4-Cyano-4'-octylbiphenyl
$C_{21}H_{36}$	2131-18-2	288.51684	Pentadecylbenzene 1-Phenylpentadecane
$C_{21}H_{40}O_4$	3443-84-3	356.5462	2,3-Dihydroxypropyl cis-9-octadecenoate Glycerol oleate Monoolein cis-9-Octadecenoic acid 2,3-dihydroxypropyl ester
$C_{21}H_{44}$	629-94-7	296.58036	Henicosane Heneicosane
$C_{21}H_{44}N_2S$		356.65984	Tetrapentylammonium thiocyanate Tetraamylammonium thiocyanate
$C_{22}H_{26}N_4$		346.4754	Bis(N-ethyl-2-benzimidazolyl)butane <undefined positional isomer>
$C_{22}H_{34}O_2$		330.51076	Ethyl cis-5,8,11,14,17-icosapentaenoate cis-5,8,11,14,17-Icosapentaenoic acid ethyl ester Ethyl eicosapentaenoate Eicosapentaenoic acid ethyl ester
$C_{22}H_{42}O_2$	142-77-8	338.57428	Butyl cis-9-octadecenoate Butyl oleate cis-9-Octadecenoic acid butyl ester Oleic acid butyl ester
$C_{22}H_{42}O_2$	112-86-7	338.57428	cis-13-Docosenoic acid Erucic acid
$C_{22}H_{42}O_2$	506-33-2	338.57428	trans-13-Docosenoic acid Brassidic acid
$C_{22}H_{42}O_4$	103-23-1	370.57308	Bis(2-ethylhexyl) hexanedioate Bis(2-ethylhexyl) adipate Adipic acid bis(2-ethylhexyl) ester Hexanedioic acid bis(2-ethylhexyl) ester
$C_{22}H_{42}O_4$	123-79-5	370.57308	Dioctyl hexanedioate Dioctyl adipate Adipic acid dioctyl ester Hexanedioic acid dioctyl ester
$C_{22}H_{44}O_2$	123-95-5	340.59016	Butyl octadecanoate Butyl stearate Stearic acid butyl ester Octadecanoic acid butyl ester
$C_{22}H_{44}O_2$	112-85-6	340.59016	Docosanoic acid Behenic acid
$C_{22}H_{46}$	629-97-0	310.60724	Docosane
$C_{22}H_{46}O_6$	3055-95-6	406.60364	3,6,9,12,15-Pentaoxa-1-heptacosanol Pentaethylene glycol dodecyl ether
$C_{23}H_{48}$	638-67-5	324.63412	Tricosane
$C_{24}H_{12}$	191-07-1	300.35928	Coronene
$C_{24}H_{20}AsCl$		418.79709	Tetraphenylarsonium chloride
$C_{24}H_{20}BNa$	143-66-8	342.223568	Tetraphenylboron sodium Sodium tetraphenylborate
$C_{24}H_{20}ClP$	2001-45-8	374.849262	Tetraphenylphosphonium chloride
$C_{24}H_{25}F_{25}$		788.42258	1,1,1,2,2,3,3,4,4,5,5,6,6,7,7,8,-8,9,9,10,10,11,11,12,12-Pentacosafluorotetracosane
$C_{24}H_{26}$	17293-59-3	314.47044	1,3,5-Triphenylhexane
$C_{24}H_{36}O_2$		356.54864	Ethyl cis-4,7,10,13,16,19-docosahexaenoate cis-4,7,10,13,16,19-Docosahexaenoic acid ethyl ester Ethyl docosahexaenoate Docosahexaenoic acid ethyl ester
$C_{24}H_{38}O_4$	117-81-7	390.56332	Bis(2-ethylhexyl) 1,2-benzenedicarboxylate Bis(2-ethylhexyl) phthalate Dioctyl phthalate 1,2-Benzenedicarboxylic acid bis-(2-ethylhexyl) ester Phthalic acid bis(2-ethylhexyl) ester
$C_{24}H_{38}O_4$	117-84-0	390.56332	Dioctyl 1,2-benzenedicarboxylate

Formula	CAS RN	M_r	Name
$C_{24}H_{38}O_4$ (Cont.)			Dioctyl phthalate
			1,2-Benzenedicarboxylic acid dioctyl ester
			Phthalic acid dioctyl ester
$C_{24}H_{46}O_2$	506-37-6	366.62804	cis-15-Tetracosenoic acid
			Nervonic acid
$C_{24}H_{48}O_2$	557-59-5	368.64392	Tetracosanoic acid
			Lignoceric acid
$C_{24}H_{50}$	646-31-1	338.661	Tetracosane
$C_{24}H_{50}$		338.661	Tetracosane (c,I)
$C_{24}H_{50}$		338.661	Tetracosane (c,II)
$C_{24}H_{50}OP$	78-50-2	385.634162	Trioctylphosphine oxide
$C_{24}H_{50}OS$		386.7264	Didodecyl sulfoxide
			Dilauryl sulfoxide
$C_{24}H_{50}S$	2469-45-6	370.727	13-Thiapentacosane
			Didodecyl sulfide
			Didodecyl thioether
			Dodecyl sulfide
$C_{24}H_{51}N$	3007-31-6	353.67568	Didodecylamine
$C_{24}H_{51}N$	1116-76-3	353.67568	Trioctylamine
$C_{24}H_{51}O_4P$	78-42-2	434.640302	Tris(2-ethylhexyl) phosphate
			Phosphoric acid tris(2-ethylhexyl) ester
$C_{24}H_{51}P$	4731-53-7	370.642702	Trioctylphosphine
$C_{24}H_{52}BrN$	4328-13-6	434.58762	Tetrahexylammonium bromide
$C_{24}H_{52}O_4Si$	78-13-7	432.75998	Tetrakis(2-ethylbutoxy)silane
			Tetrakis(2-ethylbutyl) orthosilicate
$C_{24}H_{52}O_4Si$		432.75998	Tetrakis(4-methyl-2-pentoxy)-silane
			Tetrakis(4-methyl-2-pentyl) orthosilicate
$C_{24}H_{54}N_2O_{16}P_2U$		926.673064	Uranyl dinitrate-Tributyl phosphate(1/2)
			Uranyl nitrate tributyl phosphate addition compound
$C_{25}H_{44}O_6$		440.62076	3,6,9,12-Tetraoxa-14-(4-nonylphenyl)-1-tetradecanol
			Pentaethylene glycol (4-nonylphenyl) ether
$C_{25}H_{48}O_4$	103-24-2	412.65372	Bis(2-ethylhexyl) nonanedioate
			Bis(2-ethylhexyl) azelate
			Azelaic acid bis(2-ethylhexyl) ester
			Nonanedioic acid bis(2-ethylhexyl) ester
$C_{25}H_{52}$	629-99-2	352.68788	Pentacosane
$C_{26}H_{26}OSi_2$	807-28-3	410.66284	1,3-Dimethyl-1,1,3,3-tetraphenyldisiloxane
$C_{26}H_{36}O_2S$		412.63664	4-Pentylbenzenethiol 4--(octyloxy)benzoate
$C_{26}H_{42}O_4$	84-76-4	418.61708	Dinonyl 1,2-benzenedicarboxylate
			Dinonyl phthalate
			1,2-Benzenedicarboxylic acid dinonyl ester
			Phthalic acid dinonyl ester
$C_{26}H_{50}O_4$		426.6806	Bis(2-ethylhexyl) decanedioate
$C_{26}H_{50}O_4$ (Cont.)			Bis(2-ethylhexyl) sebacate
			Decanedioic acid bis(2-ethylhexyl) ester
			Sebacic acid bis(2-ethylhexyl) ester
			<undefined optical isomer>
$C_{26}H_{50}O_4$	122-62-3	426.6806	(±)-Bis(2-ethylhexyl) decanedioate
			(±)-Bis(2-ethylhexyl) sebacate
			(±)-Decanedioic acid bis(2-ethylhexyl) ester
			(±)-Sebacic acid bis(2-ethylhexyl) ester
$C_{26}H_{50}O_4$		426.6806	Dioctyl decanedioate
			Dioctyl sebacate
			Decanedioic acid dioctyl ester
			Sebacic acid dioctyl ester
$C_{26}H_{52}O_2$	506-46-7	396.69768	Hexacosanoic acid
			Cerotic acid
$C_{26}H_{54}$	630-01-3	366.71476	Hexacosane
			Cerane
$C_{27}H_{46}O$	57-88-5	386.66164	Cholesterol
$C_{27}H_{57}N$	2044-22-6	395.75632	Trinonylamine
$C_{28}H_{46}O_4$	89-16-7	446.67084	Bis(8-methylnonyl) 1,2-benzenedicarboxylate
			Diisodecyl phthalate
			Phthalic acid diisodecyl ester
			1,2-Benzenedicarboxylic acid bis-(8-methylnonyl) ester
$C_{28}H_{46}O_4$	84-77-5	446.67084	Didecyl 1,2-benzenedicarboxylate
			1,2-Benzenedicarboxylic acid didecyl ester
			Didecyl phthalate
			Phthalic acid didecyl ester
$C_{28}H_{56}O_2$	506-48-9	424.75144	Octacosanoic acid
			Montanic acid
$C_{28}H_{58}$	630-02-4	394.76852	Octacosane
$C_{28}H_{60}BrN$	4368-51-8	490.69514	Tetraheptylammonium bromide
$C_{29}H_{50}O_2$	10191-41-0	430.7148	Vitamin E
			DL-α-Tocopherol
			DL-2,5,7,8-Tetramethyl-2-(4',8',-12'-trimethyltridecyl)-6-chromanol
$C_{30}H_{50}$	111-02-4	410.727	2,6,10,15,19,23-Hexamethyl-2,6,10,14,18,22-tetracosahexaene
			Squalene
$C_{30}H_{54}Co_2O_6P_2$		690.569084	Di(tricarbonyltributylphosphinecobalt)
$C_{30}H_{62}$	111-01-3	422.82228	2,6,10,15,19,23-Hexamethyltetracosane
			Squalane
$C_{30}H_{62}$	638-68-6	422.82228	Triacontane
$C_{30}H_{63}BO_3$		482.63942	Tridecyl borate
$C_{30}H_{63}N$	1070-01-5	437.83696	Tris(decyl)amine
$C_{32}H_{41}F_{25}$		900.63762	1,1,1,2,2,3,3,4,4,5,5,6,6,7,7,8,-8,9,9,10,10,11,11,12,12-Pentacosafluorodotriacontane
$C_{32}H_{66}$	544-85-4	450.87604	Dotriacontane

INDEX

A

B

C

D

E

F

G

H